Environmental Footprints and Eco-design of Products and Processes

Series Editor

Subramanian Senthilkannan Muthu, Green Story Inc., Tung Chung, Hong Kong

Indexed by Scopus

This series aims to broadly cover all the aspects related to environmental assessment of products, development of environmental and ecological indicators and eco-design of various products and processes. Below are the areas fall under the aims and scope of this series, but not limited to: Environmental Life Cycle Assessment; Social Life Cycle Assessment; Organizational and Product Carbon Footprints; Ecological, Energy and Water Footprints; Life cycle costing; Environmental and sustainable indicators; Environmental impact assessment methods and tools; Eco-design (sustainable design) aspects and tools; Biodegradation studies; Recycling; Solid waste management; Environmental and social audits; Green Purchasing and tools; Product environmental footprints; Environmental management standards and regulations; Eco-labels; Green Claims and green washing; Assessment of sustainability aspects.

Swarup Roy · Mohammad Shahid
Editors

Biomaterials for Sustainable Food Packaging

Challenges and Solutions

Springer

Editors
Swarup Roy
Food Technology and Nutrition
Lovely Professional University
Phagwara, Punjab, India

Mohammad Shahid
Department of Chemistry
Koneru Lakshmaiah Education Foundation
Guntur, Andhra Pradesh, India

ISSN 2345-7651 ISSN 2345-766X (electronic)
Environmental Footprints and Eco-design of Products and Processes
ISBN 978-3-032-08026-4 ISBN 978-3-032-08027-1 (eBook)
https://doi.org/10.1007/978-3-032-08027-1

This Springer imprint is published by the registered company Springer Nature Switzerland AG
The registered company address is: Gewerbestrasse 11, 6330 Cham, Switzerland

Preface

The rising environmental burden of synthetic plastic waste, especially in food packaging, has necessitated a re-evaluation of the materials traditionally used for ensuring food safety and shelf life. Conventional plastics, while functionally efficient, pose serious challenges in terms of non-biodegradability, recycling limitations, and potential risks to human health through chemical leaching and in the form microplastics. The need for sustainable packaging materials that are biodegradable, renewable, and functionally competitive with existing solutions has become more pressing than ever. Against this backdrop, *Biomaterials for Sustainable Food Packaging—Challenges and Solutions* aims to present a comprehensive and up-to-date account of research advances, practical solutions, and emerging materials that contribute to the development of eco-friendly food packaging systems.

Chapter 1 provides a broad overview of bio-based polymers, focusing on their properties and applications in the context of food packaging. The discussion emphasizes the critical need to replace petrochemical-based plastics with sustainable alternatives sourced from biomass. The chapter also highlights recent developments in microbial polymers, polysaccharide-based films, protein-derived materials, and bio-polyesters—offering a foundation for understanding the breadth of available biopolymer resources. Chapters 2 and 3 focus specifically on starch-based materials, which are among the most studied biodegradable polymers for packaging. Chapter 2 outlines the production, structural characteristics, and functional properties of starch films, while also identifying key applications in food preservation. Chapter 3 extends this discussion by exploring chemical and physical modification strategies aimed at enhancing starch films' mechanical strength and barrier properties. It also examines the integration of nanomaterials and reinforcing agents to overcome inherent limitations of native starch. Chapter 4 addresses pectin, a polysaccharide with strong film-forming potential and excellent compatibility with food systems. The chapter discusses the incorporation of bioactive agents into pectin-based films to enhance their antimicrobial and antioxidant properties, thereby extending food shelf life. Chapter 5 turns to cellulose nanofibers, which offer a unique combination of biodegradability, high tensile strength, and large surface area. The chapter reviews production techniques, structural advantages, and the role of CNFs in the fabrication of functional packaging films, including those with active and intelligent properties. Chapters 6–9 examine marine-derived biopolymers—namely alginate, chitosan, carrageenan, and agar—recognized for their natural origin, biodegradability, and compatibility with food systems. Chapter 6 explores the extraction, structural properties, and packaging applications of alginate, while highlighting its water retention and film-forming abilities. Chapter 7 presents crustacean waste valorised chitosan as a viable matrix for bioactive packaging, with emphasis on the potential to develop smart systems. Chapter 8 presents carrageenan as a viable matrix for bioactive packaging, with emphasis on the potential to develop active and intelligent systems. Chapter 9 discusses agar-based packaging materials, emphasizing their physical integrity, natural decomposition, and safety for use in direct food contact. Chapter 10 shifts the focus to protein-based films derived from soy, whey, and other protein sources. These films exhibit favorable mechanical and barrier properties and are increasingly studied for their bioactive potential in food preservation. The chapter compares various protein sources, processing methods, and composite strategies that enhance the stability and performance of protein films. In a related context, Chap. 11 reviews the application of natural waxes—such as beeswax and carnauba wax—as hydrophobic additives to biopolymer films. Their role in improving moisture resistance and extending food shelf life is critically examined, alongside the challenges in processing and uniform blending. Chapter 12 introduces agro-waste as a promising source of raw material for sustainable packaging. Agricultural residues such as fruit peels and plant fibers are presented not only as renewable feedstocks but also as contributors to circular bioeconomy strategies. The chapter details processing techniques, practical applications, and the environmental implications of utilizing agro-waste in packaging systems. Chapter 13 continues this discussion by focusing on edible films and coatings derived from food and agricultural by-products. The chapter aligns these materials with zero-waste goals and circular economy principles, emphasizing their potential to minimize both packaging waste and food spoilage. Chapters 14 and 15 explore the role of nanotechnology in enhancing the functional properties of biopolymer-based packaging. Chapter 14 discusses the incorporation of nanoparticles into packaging films to improve mechanical strength, barrier function, and antimicrobial performance. It also highlights safety concerns related to nanoparticle migration and the need for comprehensive risk assessment. Chapter 15 offers a broader overview of nanotechnology-enhanced biomaterials, including both organic and inorganic nanoparticles, and their applications in smart packaging. The chapter gives particular attention to active systems that release bioactive agents and intelligent systems that monitor food freshness or contamination. The final Chap. 16, addresses the broader environmental and safety aspects of biodegradable packaging materials. It differentiates between bio-based and biodegradable plastics, reviews biodegradation mechanisms, and provides insights into life cycle assessments and global safety standards. The chapter also highlights consumer perception, market trends, and the need for regulatory alignment to support the safe and effective use of biodegradable packaging on a global scale.

This book is the outcome of collaborative contributions from researchers working across disciplines such as materials science, food technology, environmental science, and chemical engineering. By offering detailed insights into individual materials, processing techniques, application areas, and regulatory challenges, the volume serves as both a reference and a guide for scientists, students, and professionals interested in advancing sustainable packaging solutions. It is our hope that the material presented here will contribute meaningfully to ongoing discussions and developments toward environmentally sound, health-conscious, and technically viable alternatives to conventional food packaging.

Phagwara, India Swarup Roy
Guntur, India Mohammad Shahid

Contents

Editors and Contributors

About the Editors

Dr. Swarup Roy is currently working as an Assistant Professor at the Department of Food Technology and Nutrition, Lovely Professional University, India. Dr. Roy previously worked as an Assistant Professor at Shoolini University, India. He worked as a postdoctoral researcher at the BioNanocomposite Research Center, Kyung Hee University, Seoul, and Inha University, Incheon, South Korea. He also worked as a research associate at Indian Institute of Technology Indore. Dr. Roy completed his Ph.D. degree as a DST INSPIRE Fellow from University of Kalyani, India. His research is focused on the preparation of biopolymer, bioplastic-based nanocomposite materials for smart food packaging applications. Dr. Roy has already published over 200 research articles including research papers, reviews, editorials and book chapters with over 9700 citations and h-index 58 in Scopus. He has delivered 16 invited talks in conferences. He is serving as an editor/editorial board of several journals including Springer Nature, Wiley, Frontiers, etc. He has already edited two books and currently editing six books.

Dr. Mohammad Shahid Ph.D., is an Assistant Professor in the Department of Chemistry, Koneru Lakshmaiah Education Foundation, India. Before joining KLEF, he worked at various esteemed institutions, including Jamia Millia Islamia, New Delhi; Soochow University, China; University of Glasgow, UK; the Institute of Chemical Technology, Mumbai; Dr. K. N. Modi University, Tonk; and Mewar University, Chittorgarh. He earned his Master's degree in Chemistry from Shibli National College, Azamgarh, in 2006 and his Ph.D. in Chemistry from Jamia Millia Islamia, New Delhi, in 2014. Dr. Shahid has received several prestigious fellowships to support his research pursuits, including grants from UGC, India, China Postdoctoral Fellowship, EU Marie Curie Fellowship, and Dr. D. S. Kothari Fellowship. He has contributed to over 50 scientific publications with ~4300 citations and an h-index of 22. In addition, has also edited and authored several books with renowned international publishing houses.

Contributors

Krishna Aayush Department of Food Science and Technology, Graphic Era (Deemed to be University), Dehradun, Uttarakhand, India

Md. Adil Department of Chemical Engineering, Indian Institute of Technology Guwahati, Guwahati, Assam, India

Jameel M. Al-Khayri Department of Agricultural Biotechnology, College of Agriculture and Food Sciences, King Faisal University, Hofuf, Al-Ahsa, Saudi Arabia

Bader Alsubaie Department of Agricultural Biotechnology, College of Agriculture and Food Sciences, King Faisal University, Hofuf, Al-Ahsa, Saudi Arabia

Sadiya Anjum Department of Applied Sciences & Humanities, IMS Engineering College, Ghaziabad, Uttar Pradesh, India

Neha Naijo Areekal Department of Chemical Engineering, Indian Institute of Technology Tirupati, Tirupati, India

Behnam Bahramian Department of Food Science and Technology, Faculty of Nutrition and Food Sciences, Tabriz University of Medical Sciences, Tabriz, Iran

Vikas Bansal Department of Food Technology, School of Engineering and Technology, Jaipur National University, Jaipur, Rajasthan, India

Deblina Biswas Department of Instrumentation and Control Engineering, Dr B R Ambedkar National Institute of Technology Jalandhar, Jalandhar, Punjab, India

Ana Caroline Ferreira Carvalho Postgraduate Program in Food Engineering, Federal University of Santa Catarina, Technology Center, Trindade, Florianópolis, SC, Brazil

Garima Chetry Department of Life Sciences, School of Sciences, Christ University, Bengaluru, India

Dibyajyoti Das Pratiksha Institute of Pharmaceutical Sciences, Guwahati, Assam, India

Carolina Krebs de Souza Department of Chemical Engineering, University of Blumenau, Itoupava Seca, Blumenau, SC, Brazil

Manoj K. Dhal Department of Chemical Engineering, Indian Institute of Technology Guwahati, Guwahati, Assam, India

Saeid Fallahizadeh Department of Environmental Health Engineering, School of Public Health, Yasuj University of Medical Sciences, Yasuj, Iran

George Kwame Fobiri Department of Fashion Design and Textiles Studies, Kumasi Technical University, Kumasi, Ghana

Sajad Ghaderi Department of Nutrition, Faculty of Health and Nutrition Sciences, Yasuj University of Medical Sciences, Yasuj, Iran

Neetu Goel Department of Applied Sciences and Humanities, Inderprastha Engineering College, Ghaziabad, Uttar Pradesh, India

Priyanga Govindarasu Department of Food Science and Technology, Pondicherry University, Puducherry, India

Rakesh Kumar Gupta Agricultural and Food Engineering Department, Indian Institute of Technology Kharagpur, Kharagpur, India;
Agricultural and Food Engineering Department, Indian Institute of Technology Kharagpur, Kharagpur, India

Shouxiang Jiang School of Fashion and Textiles, The Hong Kong Polytechnic University, Kowloon, Hong Kong, China;
Research Centre of Textiles for Future Fashion, The Hong Kong Polytechnic University, Kowloon, Hong Kong, China

Geo Joseph Department of Botany, Carmel College (Autonomous), Mala, Thrissur, Kerala, India

Aswathi Raj K. V. Department of Food Technology and Nutrition, School of Agriculture, Lovely Professional University, Punjab, India

Pratap Kalita Pratiksha Institute of Pharmaceutical Sciences, Guwahati, Assam, India

Gurmeet Kaur Department of Food Technology and Nutrition, School of Agriculture, Lovely Professional University, Phagwara, Punjab, India

Behrooz Ebrahim Zadeh Koor Department of Nutrition, Faculty of Health and Nutrition Sciences, Yasuj University of Medical Sciences, Yasuj, Iran

Vasantha Veerappa Lakshmaiah Department of Life Sciences, School of Sciences, Christ University, Bengaluru, India

Łukasz Łopusiewicz School of Medical and Health Sciences, Vizja University, Warsaw, Poland;
Institute of Pharmacy, Department Pharmaceutical Biology, Greifswald University, Greifswald, Germany

Divya Mailk Department of Food Technology and Nutrition, School of Agriculture, Lovely Professional University, Phagwara, Punjab, India

Majid Majlesi Department of Nutrition, Faculty of Health and Nutrition Sciences, Yasuj University of Medical Sciences, Yasuj, Iran

Tintu Jose Manicketh Department of Botany, St. Teresa's College (Autonomous), Cochin, Kerala, India

Karine Marafon Postgraduate Program in Food Engineering, Federal University of Santa Catarina, Technology Center, Trindade, Florianópolis, SC, Brazil

P. Meenakshi Department of Food Technology and Nutrition, School of Agriculture, Lovely Professional University, Phagwara, Punjab, India

Caroline Meinert Department of Chemical Engineering, University of Blumenau, Itoupava Seca, Blumenau, SC, Brazil

Praveen Nagella Department of Life Sciences, School of Sciences, Christ University, Bengaluru, India

Debabrata Nath Pratiksha Institute of Pharmaceutical Sciences, Guwahati, Assam, India

Swathi Deepak Nayak Department of Life Sciences, School of Sciences, Christ University, Bengaluru, India

Thaís Costa Nihues Department of Chemical Engineering, University of Blumenau, Itoupava Seca, Blumenau, SC, Brazil

I. K. Nishitha Department of Botany, St. Teresa's College (Autonomous), Cochin, Kerala, India

Abhishek Parasar Pratiksha Institute of Pharmaceutical Sciences, Guwahati, Assam, India

M. Prasanna Kumar Department of Food Science and Technology, Pondicherry University, Puducherry, India

Júlia Prebianca Department of Chemical Engineering, University of Blumenau, Itoupava Seca, Blumenau, SC, Brazil

Amanda Alves Prestes Postgraduate Program in Food Engineering, Federal University of Santa Catarina, Technology Center, Trindade, Florianópolis, SC, Brazil

Elane Schwinden Prudêncio Postgraduate Program in Food Engineering, Federal University of Santa Catarina, Technology Center, Trindade, Florianópolis, SC, Brazil;
Department of Food Science and Technology, Federal University of Santa Catarina, Itacorubi, Florianópolis, SC, Brazil

Ramachandran Ramkumar Institute of Animal Nutrition, Tamil Nadu Veterinary and Animal Sciences University, Chengalpattu, Tamil Nadu, India

Jyoti Rani Parul Institute of Design, Parul University, Vadodara, Gujarat, India

K. T. Remisha Jahan Department of Food Technology and Nutrition, School of Agriculture, Lovely Professional University, Phagwara, Punjab, India

Swarup Roy Department of Food Technology and Nutrition, School of Agriculture, Lovely Professional University, Punjab, India;

Department of Food and Nutrition, Kyung Hee University, Seoul, Republic of Korea

Raphael Kanyire Seidu School of Fashion and Textiles, The Hong Kong Polytechnic University, Kowloon, Hong Kong, China

Gurvendra Pal Singh Department of Food Technology and Nutrition, Lovely Professional University, Jalandhar, India

Camila Vespúcio Bis Souza Department of Chemical Engineering, University of Blumenau, Itoupava Seca, Blumenau, SC, Brazil

Prem Prakash Srivastav Agricultural and Food Engineering Department, Indian Institute of Technology Kharagpur, Kharagpur, India

Milad Tavassoli Student Research Committee, Yasuj University of Medical Sciences, Yasuj, Iran;
Department of Nutrition, Faculty of Health and Nutrition Sciences, Yasuj University of Medical Sciences, Yasuj, Iran

Benjamin Tawiah Department of Industrial Art, Kwame Nkrumah University of Science and Technology, Kumasi, Ghana

Varee Tyagi Centre for Sustainable Polymers, Indian Institute of Technology, Guwahati, Assam, India

Sushil K. Verma Centre for Sustainable Polymers, Indian Institute of Technology, Guwahati, Assam, India

Wanli Zhang College of Food Science and Engineering, Hainan University, Haikou, PR China

Yiqin Zhang College of Food Science, Sichuan Agricultural University, Ya'an, PR China

Chapter 1
Bio-based Polymers in Food Packaging: Properties and Applications

Milad Tavassoli, Majid Majlesi, Sajad Ghaderi, Behrooz Ebrahim Zadeh Koor, and Saeid Fallahizadeh

Abstract The utility of plastic packaging in safeguarding food is undeniable. However, growing concerns have sparked significant attention regarding its non-biodegradability, recycling difficulties, and the potential for harmful chemicals to contaminate food and the environment. This has led to a rising global awareness and a call for the development of new packaging materials that prioritize environmental sustainability as well as food safety and quality. Consequently, there is an urgent need to explore alternative, biodegradable food packaging materials using bio-sourced polymers. These biopolymers should be cost-effective, easily obtainable, and derived from renewable biomass. By utilizing biopolymers, sustainable alternatives to petrochemical-based plastics can be developed. This chapter provides an overview of recent advancements in biopolymer-based coatings and films for active food packaging applications. Microbial polymers, polysaccharide-based polymers, and protein-based polymers have all been extensively used in the development of intelligent packaging films. Bio-based polyesters offer diverse alternatives, ranging from rigid to flexible materials, with some being partially or fully biodegradable. Despite the dominance of bio-based conventional plastic substitutes in the market due to their outstanding physical properties, cost-effectiveness, and durability, efforts are ongoing to develop biopolymer-based packaging for the future. This chapter focuses on the application of biopolymer-based packaging films and coatings in various food-related contexts. It also assesses the limitations, challenges, and safety considerations associated with the use of biopolymers within the packaging industry.

M. Tavassoli (✉)
Student Research Committee, Yasuj University of Medical Sciences, Yasuj, Iran
e-mail: mtavassoli2006@gmail.com

M. Tavassoli · M. Majlesi · S. Ghaderi · B. E. Z. Koor
Department of Nutrition, Faculty of Health and Nutrition Sciences, Yasuj University of Medical Sciences, Yasuj, Iran

S. Fallahizadeh
Department of Environmental Health Engineering, School of Public Health, Yasuj University of Medical Sciences, Yasuj, Iran

S. Roy and M. Shahid (eds.), *Biomaterials for Sustainable Food Packaging*,
Environmental Footprints and Eco-design of Products and Processes,
https://doi.org/10.1007/978-3-032-08027-1_1

Keywords Biopolymers · Bioplastics · Biopolyesters · Carbohydrate polymers · Food packaging

1.1 Introduction

Food packaging plays a crucial role in safeguarding food products from various types of contaminants, including chemical pollutants, biological agents, and physical damage. One of its primary objectives is to prolong the shelf life of food items while preserving their quality, taste, and nutritional value. However, a pressing concern surrounding food packaging is the substantial environmental impact associated with the use of synthetic polymers (Adeyeye et al. 2019; Agarwal et al. 2023). Recent statistics reveal a dramatic increase in global plastic consumption over the years, with packaging materials accounting for approximately one-third of total plastic usage. This significant reliance on plastics for packaging has rendered the food packaging sector a major contributor to the growing problem of plastic waste, which poses severe threats to ecosystems and wildlife (Aliabbasi et al. 2022; Alizadeh-Sani et al. 2020). Given these alarming trends, it has become imperative for the industry to explore sustainable alternatives to replace conventional synthetic polymers. Currently, many commonly used petroleum-based plastics in the packaging domain include materials such as polyethylene terephthalate (PET), high-density polyethylene (HDPE), polyvinyl chloride (PVC), low-density polyethylene (LDPE), polypropylene (PP), and polystyrene (PS). These plastics are favored for their various advantages, including cost-effectiveness, durability, lightweight characteristics, excellent resistance to chemicals and moisture, and suitable mechanical properties that make them ideal for food packaging applications (Andini et al. 2021; Bahramian et al. 2024a, b). Despite these benefits, a significant drawback of petroleum-based plastics is their environmental impact. These materials are not biodegradable or compostable, causing them to persist in the environment for hundreds of years. As a result, massive quantities of plastic waste continue accumulating globally, leading to detrimental effects on terrestrial and marine ecosystems, harming wildlife, and contributing to broader environmental pollution issues. Addressing these challenges by seeking out and implementing more sustainable packaging solutions is essential for mitigating the ecological footprint of the food packaging industry and fostering a more environmentally friendly future (Baishya et al. 2024; Balan and Kadeppagari 2021). It is anticipated that both natural and synthetic plastics will see an annual growth rate of approximately 13% from 2020 to 2023, highlighting a significant upward trend in plastic production. This increase raises environmental concerns, particularly regarding synthetic polymers, which pose a substantial threat due to their propensity to accumulate as waste. The pervasive presence of plastic particles throughout ecosystems, including their infiltration into living organisms, underscores the alarming reality of plastic pollution (Barra et al. 2020; Basavegowda and Baek 2021; Bayer 2021). Despite the challenges associated with synthetic polymers, recycling emerges as a viable and sustainable solution to mitigate these environmental impacts. However, it is important to note that current recycling methods are hindered by limitations in the range of usable materials and the properties of the recovered polymers. As a result, a considerable volume of these recyclable materials still ends up in landfills, defeating the purpose of recycling initiatives (Bayer et al. 2015; Bharathi et al. 2020). In response to these issues, the European Union (EU) Waste Management guidelines advocate for a reduction in the reliance on landfilling waste. This directive encourages researchers and industry stakeholders to investigate alternative solutions, particularly through the use of naturally occurring materials that do not pose similar waste disposal challenges (Bhaskar et al. 2023; Bhathiwal et al. 2025). This chapter examines advancements in biopolymer-based coatings and films for active food packaging. Microbial, polysaccharide, and protein-based polymers are explored for intelligent packaging films, while bio-based polyesters offer a range of biodegradable options. Despite the market's preference for bio-based conventional plastics due to their cost and durability, research continues into biopolymer alternatives. The chapter also assesses the applications, limitations, challenges, and safety considerations of using biopolymers in food packaging.

1.1.1 Overview of Bio-based Polymers in Food Packaging

Interestingly, it is estimated that around 45% of non-biodegradable plastics are utilized for food packaging, highlighting a critical area where sustainable practices can be implemented. This opens the door for biopolymer-based materials, which offer an opportunity to develop innovative, eco-friendly alternatives that can meet the demands of food packaging while minimizing environmental impact. By prioritizing biopolymers, we can move closer to creating sustainable solutions that align with both consumer needs and environmental responsibility (Bhattacharya 2015; Bindu et al. 2019). Biobased polymers are a class of polymers that are synthesized by living organisms, including plants and microbes, through processes that involve metabolic engineering techniques. These polymers are primarily derived from renewable materials, which can include a variety of natural resources such as wood, polysaccharides, proteins, lipids, and even organic waste products from food and agriculture. For instance, the use of agricultural residues or lignocellulosic materials, comprising cellulose, hemicellulose, and lignin, offers a sustainable pathway to produce these polymers (Nayak et al. 2025; Perumal et al. 2025). Sustainable food packaging innovation with biopolymer solutions is presented in Scheme 1.1.

The global adoption of biopolymers for packaging is on the rise, driven by their significant benefits compared to traditional plastics. These biopolymers stand out due to their biodegradability, environmentally friendly properties, non-toxicity, and biocompatibility, all contributing to more sustainable packaging solutions. Biopolymers, derived from renewable biological sources, form robust cohesive structures and can create thin yet effective protective film layers that enhance the preservation of packaged products (Carvalho and Conte-Junior 2021; Tavassoli et al. 2024). In food packaging, the main types of biopolymers utilized are polysaccharides, proteins, and

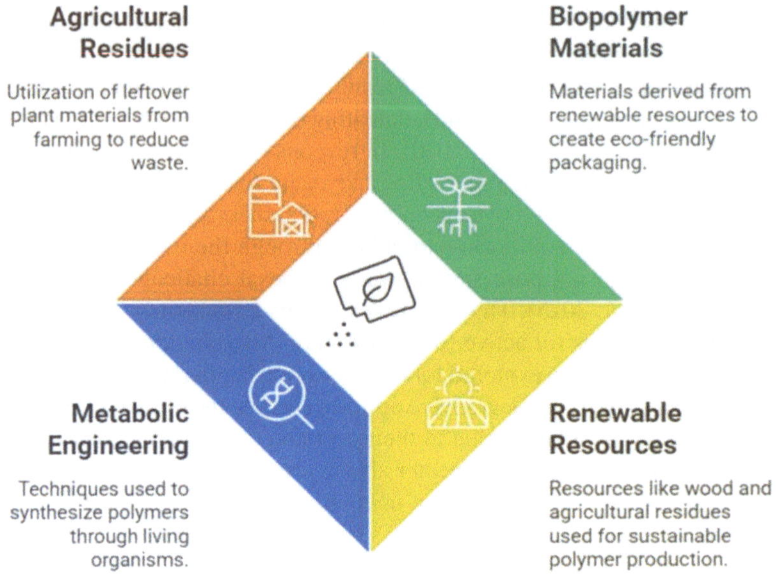

Scheme 1.1 Innovating sustainable food packaging with biopolymer solutions

aliphatic polyesters. These materials are essential for maintaining food quality and prolonging shelf life by offering critical barrier properties that manage the exchange of gases, moisture, aroma, and lipids between the food and its surroundings. This management is crucial for averting spoilage and maintaining the sensory qualities of food products. Moreover, numerous biopolymer-based packaging options display antimicrobial properties, which help safeguard food from external contaminants and pathogens, thereby further enhancing food safety. Sustainable food packaging solutions are presented in Scheme 1.2. They also help to minimize the loss of key characteristics such as flavor and texture, which are critical for consumer satisfaction (Bakhshizadeh et al. 2023a, b; Chandrababu et al. 2024).

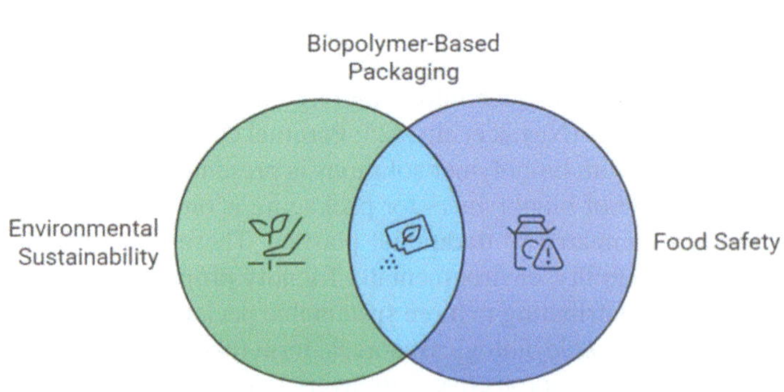

Scheme 1.2 Sustainable food packaging solutions

Currently, widely utilized biopolymer materials in food packaging include starch, cellulose, and polylactic acid (PLA), each offering unique benefits. Starch, for example, is derived from natural sources such as corn and potatoes and is known for its functional properties and biodegradability. Cellulose, another abundant biopolymer, is appreciated for its strong mechanical properties and renewability, while PLA is recognized for its strong performance in maintaining food freshness and its ability to be composted after use (Andini et al. 2021; Tavassoli et al. 2023). However, it is important to acknowledge the key limitations of using biopolymers for food packaging. One of the most significant challenges is their relatively low mechanical strength, which can hinder their ability to withstand physical stresses during handling and transportation. Biopolymers exhibit a high sensitivity to moisture, which can adversely affect their performance and undermine the protective barrier they are designed to establish. As research and development progress, it will be essential to address these limitations in order to optimize the applications of biopolymers in food packaging and improve their effectiveness for commercial use (Cherpinski et al. 2018; Chia et al. 2022; Chiu and Yang 2024). The advantages and disadvantages of biopolymers for food packaging can vary significantly depending on the specific type of biopolymer utilized. Common biopolymers include starch-based, protein-based, and polysaccharide-based materials, each with unique properties that can affect their performance in packaging applications. To overcome the inherent limitations of biopolymers, such as their mechanical strength, barrier properties, and biodegradation rates, extensive research has been conducted. This research involves the incorporation of reinforcing agents, which can include nanofillers like clay or silica, other biopolymers that enhance compatibility, plasticizers that improve flexibility and processability, and natural additives such as essential oils that can provide functional benefits (Bakhshizadeh et al. 2023a, b; Das et al. 2024). Biopolymer matrices can be designed to serve as carriers for a variety of active components, including antimicrobial agents that help extend the shelf life of food products, antioxidants that prevent spoilage, flavorings that enhance sensory appeal, and essential vitamins or nutrients that can fortify food items. This multifunctionality not only improves food quality and safety but also addresses consumer demands for healthier, more sustainable packaging solutions (de Matos Fonseca et al. 2021; Demchenko et al. 2024). Given the numerous advantages offered by biopolymers, particularly in terms of their potential to replace synthetic polymers like conventional plastics, there is a growing interest in reducing the environmental impacts typically associated with plastic waste. Biopolymers are celebrated for their ability to biodegrade and their reduced carbon footprint, which is crucial in the context of increasing global plastic pollution (Deng et al. 2025; Devi et al. 2024). With the rising application of biopolymers in the food packaging sector, it is imperative to scale up the biopolymer industry effectively. This scaling can be achieved through a combination of strategies. One promising approach involves leveraging agricultural waste, such as crop residues and other by-products, as feedstocks for biopolymer production, thus promoting a circular economy. Additionally, establishing efficient farming practices that focus on sustainable crop production can enhance the yield of raw materials needed for biopolymer synthesis. Finally, investing in advanced technologies

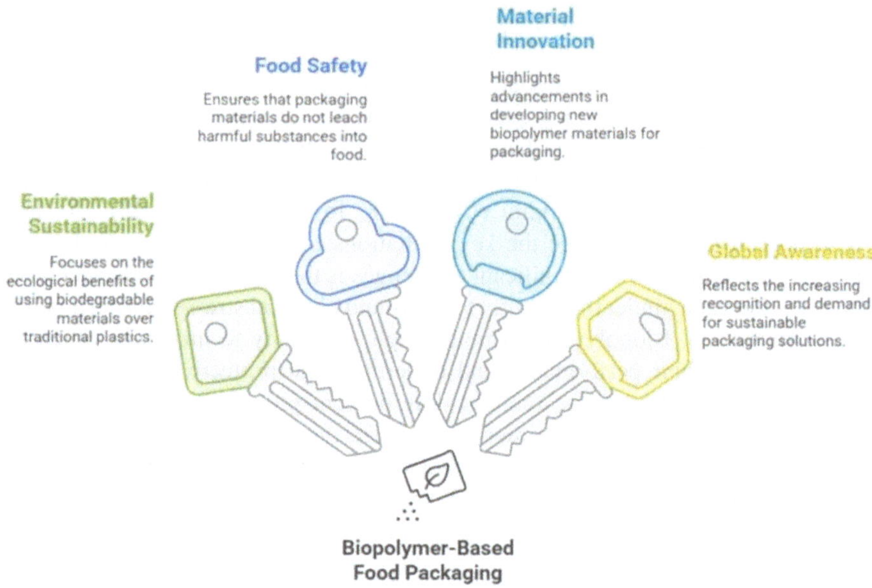

Scheme 1.3 Important advances in biopolymer-based food packaging

and innovative processes, such as bioreactors or enzymatic treatments, can further optimize biopolymer production, making it more cost-effective and environmentally friendly. By pursuing such strategies, the biopolymer industry can better meet the demands of the food packaging market while contributing positively to environmental sustainability (Khezerlou et al. 2023; Perumal et al. 2025). Some important advances in biopolymer-based food packaging are shown in Scheme 1.3.

This chapter provides a comprehensive overview of the recent advancements in biopolymer-based coatings and films utilized in active food packaging. It explores a variety of materials, including microbial polymers, polysaccharides, and protein-based polymers, highlighting their potential for developing intelligent packaging solutions that can enhance food preservation and safety. The discussion includes various bio-based polyesters, which offer flexible properties and often possess biodegradable characteristics that align with the current trends toward sustainability in packaging materials. While conventional bio-based plastics continue to dominate the market because of their favorable mechanical properties and cost efficiency, ongoing research is focused on improving the functionality and viability of biopolymer packaging alternatives. This chapter discusses the applications of biopolymer films and coatings in food packaging, exploring how they can interact with food products to extend shelf life, reduce spoilage, and indicate freshness. It also critically reviews the limitations and challenges faced in the widespread adoption of these materials, such as performance under real-world conditions, scalability of production, and potential security concerns. Moreover, safety considerations are addressed, emphasizing the need for rigorous testing and regulatory compliance to ensure that these biopolymer-based packaging solutions are safe for consumers and the environment. Finally, this chapter serves as a valuable resource for understanding the evolving landscape of biopolymer-based food packaging technologies.

1.2 Sources of Bio-based Polymers

Biobased polymers can be categorized into three primary groups based on their sources and methods of production, each with unique characteristics and applications (Fig. 1.1).

- *Polymers Extracted Directly from Biomass*: Polymers that are directly isolated from natural biomass.
- *Polymers Synthesized from Renewable Biobased Monomers*: Polymers that are prepared through traditional chemical synthesis using renewable monomers derived from biological sources.
- *Polymers Produced by Microorganisms or Genetically Modified Bacteria*: Biobased polymers that are synthesized through the metabolic processes of microorganisms.

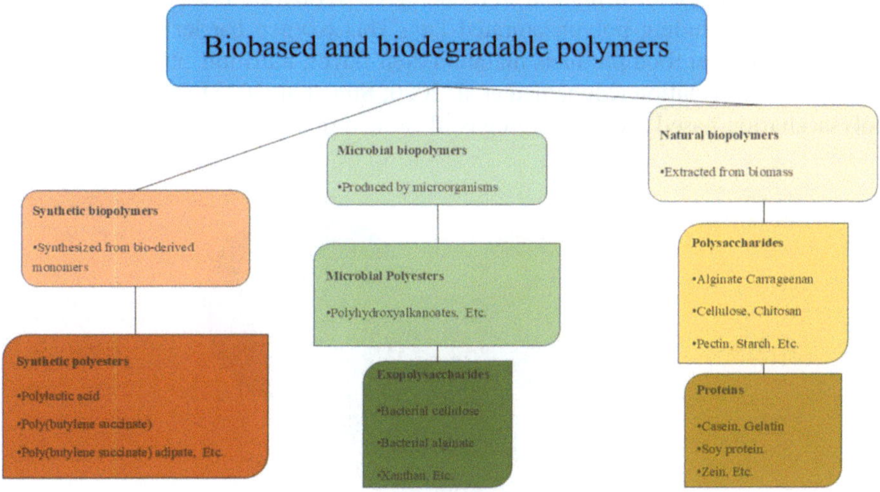

Fig. 1.1 Classification of bio-based polymers based on sources and characterizations

1.2.1 *Biomass-Based Polymers*

1.2.1.1 Polysaccharide Polymers

Polysaccharides, including cellulose derivatives, chitosan, starches, and various gums, have emerged as promising raw materials for developing edible films and coatings intended for food preservation. These materials serve as innovative packaging solutions that not only protect the integrity of food products but also enhance their shelf life and quality. When considering polysaccharide-based packaging materials, it is essential that they are not only economical and readily accessible but also exhibit specific functional characteristics tailored to the needs of the food industry. One of the primary roles of these films is to act as effective water barriers, which is critical for either reducing the dehydration of fresh produce or preserving the desired texture of dry goods. For instance, when packaging fruits and vegetables, it is important that the films slow down moisture loss to maintain freshness (Roy et al. 2024a, b; Zubair et al. 2022). Conversely, for items such as crackers or chips, the packaging must effectively prevent moisture from being absorbed from the environment to retain their crispness. Additionally, the permeation properties of these materials concerning other gases, particularly oxygen, cannot be overlooked. The control of oxygen levels within packaging is significant, as it plays a crucial role in managing the ripening process of fruits and in reducing the oxidation rates of sensitive food components, including polyunsaturated fats. This oxygen barrier characteristic is vital to ensure that food products maintain their flavor, nutrition, and quality during storage and distribution (Mohamed et al. 2020; Sun et al. 2025). The application of polysaccharide-based food packaging is illustrated in Fig. 1.2.

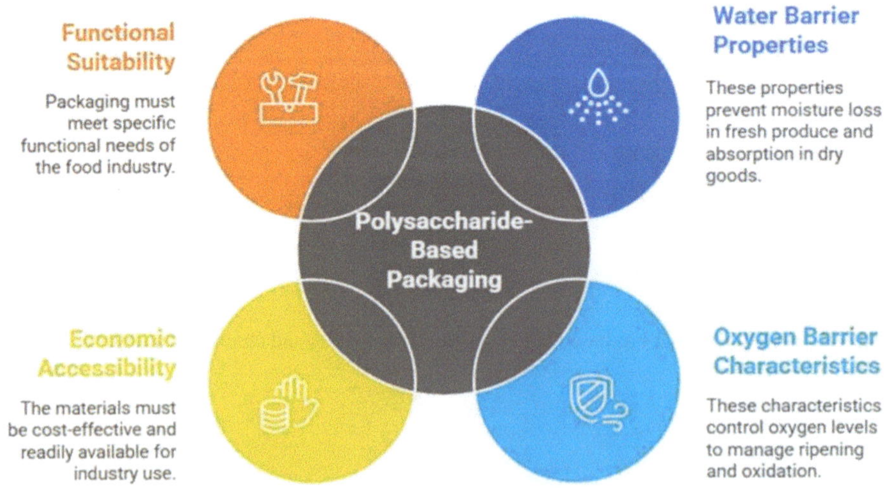

Fig. 1.2 Application of polysaccharide-based food packaging

Biodegradable films produced from polysaccharides are noteworthy for their proficient barrier properties regarding gas transfer, notably for gases such as O_2 and CO_2. However, one of the challenges associated with these materials is their high hydrophilicity, which often results in diminished water vapor barrier performance. For example, polysaccharides like alginate and carrageenan, although having limitations in water vapor barrier capabilities, exhibit a notable hygroscopic nature. This quality can be strategically utilized to create relatively thick coatings on food surfaces that absorb moisture from the environment, offering a temporary protective measure against further moisture loss. This dual function not only aids in prolonging the freshness of the food item but also ensures the product retains a desirable moisture content up until the point the coating itself becomes dehydrated (Chia et al. 2023; Tan et al. 2024). Moreover, the gas barrier properties of polysaccharide-based films are vital for minimizing the escape of volatile organic compounds, such as aroma compounds, during storage. This capability helps preserve the sensory attributes of food products and prevents potential solvent penetration, which could compromise food safety or degrade quality. The extensive variety of polysaccharides, combined with a wide range of additives that can be incorporated during the film production process, allows for the customization of biodegradable films. This flexibility enables manufacturers to create tailored packaging solutions that can significantly extend the shelf life of ready-to-eat products while maintaining their quality and safety over time (Liu et al. 2021; Mallakpour et al. 2021).

Cellulose is recognized as one of the most pervasive biopolymers in nature, forming a crucial component of the structure in plants and certain bacteria. This remarkable material is a biodegradable and sustainable resource derived primarily from cell walls, underscoring its vital role in the ecosystem. Structurally, cellulose is classified as a linear polysaccharide composed of numerous repeating units of cellobiose, each of which is formed from two anhydroglucose rings that are interconnected by β-1,4 glycosidic bonds (Khan et al. 2014; Stoica et al. 2018). The structural organization of cellulose is characterized by two distinct domains: a crystalline region and an amorphous region. The crystalline area is known for its orderly and tightly packed configuration, providing strength and stability, while the amorphous section has a less organized and more flexible structure. The degree of crystallinity in cellulose typically falls within a range of 40–60%, a value that is significantly influenced by the source of the cellulose and any preprocessing treatments it has undergone. The hydroxyl groups present in the amorphous regions are highly accessible and reactive, facilitating various interactions, while hydroxyl groups in the crystalline zones exhibit considerably lower accessibility due to their dense packing and high cohesive energy (Liu et al. 2021; Mallakpour et al. 2021). When considering interactions involving solid cellulose, the initial contact typically occurs at the amorphous regions or at the surfaces of the crystalline areas. The inherent stability of cellulose is attributed to a robust network of hydrogen bonds, which contributes to its fibrous structure, making it resistant to dissolution in common aqueous solvents. Importantly, cellulose lacks a definitive melting point, further emphasizing its unique physical characteristics (Mahmood Alizadeh-Sani et al. 2021; Stoica et al. 2018). In various industrial applications, materials and chemicals derived from cellulose are integral, particularly in the paper and textile industries. Within the paper production industry, cellulosic fibers serve as the foundational raw material for the manufacturing of paper and board. Cellulose and its synthesized derivatives play essential roles as coatings and films designed for paper applications, such as packaging, providing both functionality and durability. Despite its beneficial properties, cellulose in its native form presents several limitations, including its hydrophilic nature, insolubility in water, suboptimal film-forming capabilities, and its relatively high crystallinity. To overcome these challenges, a range of cellulose derivatives has been developed for commercial use. Key examples include carboxymethyl cellulose (CMC), methyl cellulose (MC), ethyl cellulose (EC), and hydroxypropyl and hydroxyethyl cellulose (HPC and HEC), along with hydroxypropyl methyl cellulose (HPMC). These derivatives are utilized effectively throughout various stages of paper production, from the wet-end to surface finishing, enhancing the barrier properties and functionality of the final product (Chia et al. 2023; Sun et al. 2025; Tavassoli et al. 2024). Cellulose and its derivatives exhibit notable advantages in providing mechanical reinforcement and improving the barrier characteristics of polymeric materials. Compared to native cellulose, these derivatives demonstrate increased resistance to microbial degradation and enzymatic breakdown, making them valuable not only in traditional applications but also in the development of innovative materials with enhanced longevity and durability (Khan et al. 2014; Stoica et al. 2018; Tan et al. 2024).

Chitin, following cellulose, ranks as the second most prevalent biopolymer in the natural world. This versatile substance is predominantly found in the exoskeletons of crustaceans, such as shrimp and crabs, as well as in the cell walls of various fungi, including mushrooms. It also appears in the structure of certain other biological materials, contributing to their strength and rigidity (Liao et al. 2023). Chitosan is derived from chitin through a process known as deacetylation, which involves the removal of acetyl groups from the chitin molecule. This transformation results in chitosan, a high molecular weight cationic polysaccharide that exhibits remarkable film-forming capabilities. Notably, chitosan possess documented antibacterial and antifungal properties, making it of particular interest for various applications (Patel et al. 2022). The versatility of chitosan allows it to be processed into various forms, including films, fibers, sponges, gels, nanoparticles, and beads. Among these, chitosan films have gained significant recognition as effective packaging materials. Their outstanding protective qualities make them suitable for preserving a variety of food items. When chitosan films are combined with other film-forming substances, their efficacy in food preservation is further enhanced, providing a reliable solution for maintaining food quality and extending shelf life (Kedir et al. 2024; Thiyagamoorthy et al. 2024). Alizadeh-Sani and colleagues created a pH-sensitive color indicator film by combining barberry anthocyanin with a nanofiber composite made of MC and chitosan nanofiber. This mixture enhanced the mechanical strength and water barrier capabilities of the film but led to a decrease in UV-vis light transmittance. The film exhibited a distinct color change in reaction to pH fluctuations and ammonia gas, shifting from reddish-pink to pale peach and yellow with various pH buffers, and from pink to pale green and yellow when exposed to ammonia vapor. It displayed notable antioxidant properties, making it an effective smart indicator for monitoring the freshness of meat and seafood in real-time (Mahmood Alizadeh-Sani et al. 2021).

Pectin is a complex array of polysaccharides primarily derived from various fruits and vegetables, including abundant sources like apple pomace and citrus peels. This naturally occurring polysaccharide acts as an anionic polymer, characterized by a structural backbone consisting mainly of (1 → 4)-linked α-d-galacturonic acid units. Due to its unique properties, pectin has garnered attention for its potential applications in the food packaging industry, especially in the creation of edible films made from pectin and its derivatives (Amiri Samani et al. 2025). These pectin-based edible films demonstrate notable mechanical properties and are recognized for their proficient barrier qualities against oils, aromas, and oxygen. They possess a high initial modulus, which contributes to their strength and durability; however, they also exhibit certain limitations. Notably, pectin films tend to lack moisture resistance, display low elongation capabilities, and can become quite brittle. To address these drawbacks, the incorporation of various plasticizers has been shown to significantly enhance their flexibility and performance (Mellinas et al. 2020). The functionality of pectin films can be enhanced through crosslinking with polyvalent cations, such as calcium ions, which imparts moderate mechanical properties to the resulting films and gels. These pectin-based formulations are particularly effective in preserving food products with low moisture content, making them valuable for extending the shelf life of such items (Syarifuddin et al. 2025). Bhatia and colleagues focused on creating composite films using potato starch and pectin enriched with juniper berry essential oil to enhance functional and structural properties. The study found that higher concentrations of juniper oil decreased moisture content, transparency, and mechanical strength, while increasing film thickness, water permeability, and elongation. Antioxidant assays showed a concentration-dependent increase in radical scavenging ability, and the films exhibited significant antibacterial and antifungal activities against pathogens like *Pseudomonas aeruginosa*, *Staphylococcus aureus*, and *Candida albicans*. This research highlights the potential of pectin-based films in sustainable food packaging and preservation (Bhatia et al. 2025).

Starch is a naturally occurring polysaccharide that plays a significant role in various applications, particularly in the production of biodegradable films. One of the key advantages of starch-based films is that they can be engineered to be translucent or even transparent, thereby providing aesthetic versatility for different packaging needs. These films are colorless, flavorless, and tasteless, making them suitable for a wide range of food packaging applications without altering the taste or appearance of the contents (Kedir et al. 2024; Matheus et al. 2023). Starch is composed of semi-crystalline hydrophilic granules, which means that these granules tend to attract water due to the presence of numerous hydroxyl groups on their surface. When these starch granules come into contact with water, the free hydroxyl groups form strong interactions with water molecules, initiating a process of swelling. This swelling occurs due to the absorption of water and can continue until a certain threshold, known as the critical concentration, is reached (Chia et al. 2023; Tarique et al. 2022). The critical concentration is crucial as it indicates the specific amount of starch necessary for the swollen granules to occupy the available volume when heated to 95 °C thoroughly. At this elevated temperature, the granules undergo a transition wherein they lose their crystalline structure and begin to gelatinize. Upon cooling, the mixture forms a gel, a process crucial for achieving the desired texture and consistency of the biodegradable film. The resulting gel exhibits unique properties that enhance the film's performance, including improved strength and flexibility, making it an excellent candidate for sustainable packaging options (Chia et al. 2022; Sone Aung et al. 2018). In this regard, Plaeyao and co-workers developed innovative packaging films using a blend of cassava starch (CV), chitosan, and CMC, while glycerol was incorporated to serve as a plasticizing agent. Among the various formulations tested, the Chitosan/CV/V/CMC/Gin-AgNPs1 variant stood out due to its remarkable mechanical properties, making it suitable for practical applications in the food packaging industry. The Chitosan/CV/V/CMC/Gin-AgNPs1 film showcased notable biodegradability. When buried in soil, it was able to lose more than 50% of its initial mass within just 21 days, indicating its potential to reduce environmental impact compared to conventional plastic packaging. In practical applications, this film also proved effective in maintaining the freshness of grapes stored at a temperature of 4 °C for a duration of 21 days, highlighting its viability as a sustainable packaging solution for perishable goods (Plaeyao et al. 2025).

Microbial polysaccharides play a vital role in food packaging. Pullulan, derived from the fungus Aureobasidium pullulans, consists of maltotriose units linked by α (1,6) glycosidic bonds and is highly water-soluble, making it ideal for creating transparent, edible films that are heat-sealable and minimally permeable to oxygen and oil (Ghosh et al. 2022; Kraśniewska et al. 2019). Xanthan gum is an exopolysaccharide from Xanthomonas campestris, known for its stability as a food thickener and emulsifier across various temperatures and pH levels. Its structure features β-d-glucose residues with a side chain of trisaccharides (Zambrano-Zaragoza et al. 2014). Edible coatings made from xanthan gum have been effective in prolonging the shelf life of fresh-cut fruits and delaying the ripening of acerola fruit (Freitas et al. 2013). Gellan, produced by the bacterium Sphingomonas elodea, is valued for its gelling and texturizing properties. It's increasingly used in food as a gelling agent and carrier for additives, enhancing the quality and shelf life of fresh-cut vegetables (Gomes et al. 2023). Yao and co-workers developed an innovative biofilm using a combination of pullulan and gellan gum, integrated with astaxanthin nanoemulsion, to significantly extend the shelf-life of strawberries. In their study, strawberries were stored for a period of 20 days under controlled conditions while being enveloped in this biofilm. The strawberries exhibited a weight loss of 12.48%, which indicates a moderate level of moisture retention attributable to the protective qualities of the biofilm (Fig. 1.3). The measurements taken revealed that the soluble solid content of the strawberries reached 7.4%, demonstrating the preservation of sugar and flavor components crucial for the fruit's palatability. Moreover, the ascorbic acid content, an important indicator of vitamin C retention, was recorded at 0.75%, reflecting the biofilm's effectiveness in maintaining the nutritional quality of the strawberries over the storage period. The titrable acidity, which provides insight into the fruit's taste profile, was measured at 36.45 mg of citric acid per 100 g. Significantly, the incorporation of biofilms infused

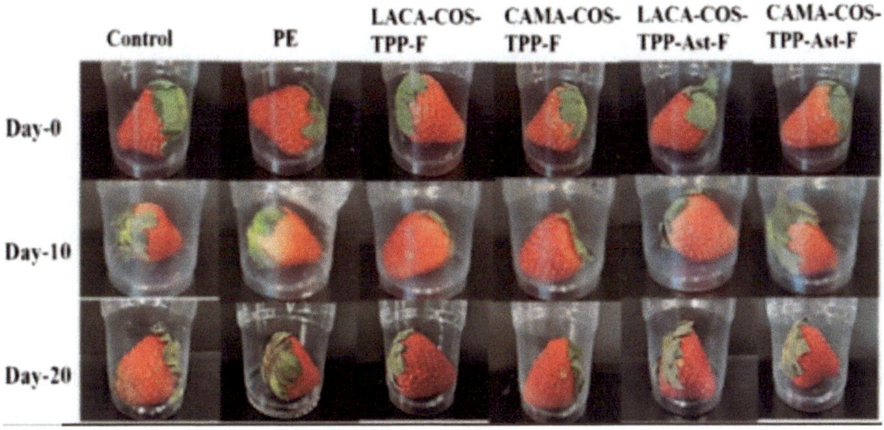

Fig. 1.3 Effect of various treatments of pullulan and gellan films on strawberry samples during storage at 4 °C for 20 days, Ref. Yao et al. (2025)

with astaxanthin nanoemulsion managed to prolong the shelf-life of strawberries to a minimum of 18 days, showcasing a remarkable improvement compared to the 12-day shelf-life observed when using the neat biofilm without the astaxanthin enhancement (Yao et al. 2025).

1.2.1.2 Protein Polymers

Protein-based polymers have emerged as highly effective materials for creating films that are not only nutritious but also environmentally friendly and biodegradable. These films offer significant advantages over those made from polysaccharides and lipids, particularly regarding gas barrier properties. This means they are more effective at preventing the ingress of oxygen and the escape of moisture, which is crucial for maintaining the freshness and shelf life of food products. Protein-based films not only have excellent barrier capabilities but also display superior mechanical properties, including higher tensile strength and flexibility compared to lipid and polysaccharide films. This makes them suitable for durable, lightweight packaging applications (Nayak et al. 2025; Perumal et al. 2025; Tammina et al. 2024). Some prominent features of protein-based food packaging films are shown in Fig. 1.4. Recent advancements in the food packaging industry have increasingly focused on enhancing the sustainability and safety of materials. There is a strong emphasis on improving the cost-effectiveness of protein-based films while also expanding their range of potential applications in food packaging. This development aligns with the growing demand for eco-friendly alternatives to traditional plastic packaging, which poses significant environmental challenges. Among the most commonly used proteins for the production of biodegradable films are gelatin, casein, zein, and whey protein. These proteins can be easily processed and transformed into films through

Fig. 1.4 Characteristics of protein-based food packaging films

various methods such as casting or extrusion, making them versatile for different packaging solutions. Derived from renewable resources, these proteins offer a more sustainable alternative to conventional synthetic plastics, which are typically made from fossil fuels. Ongoing research and development in protein-based films not only aim to improve their properties and functionalities but also play a crucial role in advancing sustainable food packaging solutions. As these materials gain popularity, they represent a significant step toward minimizing the environmental impact of food packaging while addressing consumer expectations for safety and quality (Ezati et al. 2023; Mohamed et al. 2020).

Protein-based biopolymers are emerging as some of the most promising materials for food packaging, largely due to their exceptional gas barrier properties and adequate mechanical characteristics. These films, primarily composed of natural proteins, exhibit favorable attributes that make them suitable for extending the shelf life of packaged food products. However, one of their significant limitations is their low resistance to water vapor, which can compromise product freshness. To enhance the performance of these films and tailor them for diverse applications, several modification techniques can be employed (Pushp et al. 2021). One effective approach is crosslinking, a process that improves the mechanical strength of protein films by covalently linking adjacent protein molecules through various chemical methods. Common reactants for crosslinking include glutaraldehyde, calcium chloride ($CaCl_2$), and the EDC/NHS (1-ethyl-3-(3-dimethylaminopropyl) carbodiimide/N-hydroxysuccinimide) system. This technique not only enhances the tensile strength and elasticity of the films but also significantly increases their resistance to solubility in water (Kumar et al. 2022a, b). Another strategy involves the incorporation of plasticizers or other polymers into the protein film matrix. This addition results in increased flexibility and greater elongation at break, which are crucial parameters for effective packaging. By blending different polymers, the physical properties and performance characteristics can be optimized, resulting in a more versatile packaging material (Oymaci and Altinkaya 2016; Shams et al. 2019). The integration of natural antioxidants, such as vitamin E, into protein films serves to improve their oxidative stability and prolong their shelf life. The antioxidant properties can help preserve the sensory qualities of the packaged food, preventing rancidity and degradation over time (Deng et al. 2025; Singh et al. 2024). To combat microbial spoilage, incorporating natural antimicrobial agents into protein films is a promising development. For instance, nisin, a natural peptide produced by the bacterium Lactococcus lactis, can inhibit the growth of certain bacteria, thereby enhancing the safety and longevity of the food products being packaged (Shams et al. 2019). Blending protein films with other biopolymers, such as chitosan, cellulose, and alginate, is another technique that can enhance both the mechanical and barrier properties of the films. This blending not only improves functionality but can also reduce production costs, making sustainable packaging solutions more economically viable. The surface modification of films using various molecules, such as lipids, polysaccharides, or proteins, can enhance their barrier properties and improve adhesion to other surfaces. These modifications are crucial for increasing the compatibility of protein films with different food products and packaging needs (Bakhshizadeh et al. 2023a, b). The introduction of functional groups into the side chains of amino acids in proteins allows for enhanced cross-linking possibilities, involving both covalent and non-covalent interactions. For example, whey proteins can effectively interact with Chitosan derivatives through hydrogen bonding as non-covalent crosslinkers. This results in films that exhibit improved insolubility, increased elongation, and enhanced hydrophobicity, which are vital for moisture-sensitive applications (Garavand et al. 2017; Zhang et al. 2023). Natural substances such as tannins, known for their antibacterial properties, can be incorporated to boost the antimicrobial activity of protein films. The efficacy of these antimicrobial agents can further strengthen the protective capabilities of the packaging (Garavand et al. 2017; Zhang et al. 2023).

Films that incorporate antioxidants play a crucial role in extending the shelf life of packaged foods by significantly mitigating oxidative reactions. These reactions often lead to various forms of spoilage, including the deterioration of flavor, loss of visual appeal such as color changes, and degradation of nutritional content. Like any specialized functionalized packaging, the efficacy of these antioxidant-infused films is heavily influenced by the selection of antioxidants used and their concentrations within the film matrix (Bharathi et al. 2020; Deng et al. 2025). Antioxidants can be effectively integrated into protein-based films, which not only enhances their mechanical properties but also improves their barrier functionalities. These naturally derived antioxidants can be obtained from a diverse range of sources, including vitamins, minerals, and plant extracts. Notable examples include ascorbic acid (vitamin C), alpha-tocopherol (vitamin E), coumarin, and ferulic acid, each contributing unique antioxidant activities (Guerrero et al. 2011; Han et al. 2018). Many studies have specifically evaluated the antioxidant efficacy of extracts from rosemary and thyme when incorporated into an active component derived from whey protein, highlighting their potential in food preservation. In practical observations, gelatin films infused with these extracts exhibited enhanced thickness, increased tensile strength, and improved clarity, which are desirable traits in packaging materials. In contrast, films made from soy protein isolate (SPI) showcased superior antioxidant properties and water resistance compared to alternatives (Guerrero et al. 2011). Interestingly, when placed side by side with traditional packaging materials such as methylcellulose, polyethylene, starch, and low-density pectin, the oxygen permeability of soy protein-based films was markedly reduced. This reduction, however, came at the cost of somewhat poorer mechanical properties; hence, finding a balance is crucial. To address these mechanical challenges, natural and/or biodegradable plasticizers were employed, which notably enhanced the films' viscoelasticity and extensibility. These plasticizers work through physical and chemical interactions: they infiltrate the molecular structure of the polymers and influence the bonding characteristics, subsequently improving critical properties like ductility, stiffness, and elasticity (Maurizzi et al. 2022; Shit and Shah 2014). When developing natural biofilm technology, employing water as the primary solvent can alter the free mass of the resultant biofilms, which may influence their performance. In addition to their antioxidant properties, protein-based antimicrobial films represent a significant advancement in food packaging technology. These films are engineered to incorporate natural antimicrobial agents within the protein film matrix, thereby extending the shelf life of food products while actively protecting against bacterial spoilage (Deng et al. 2025; Mishra et al. 2025a, b).

One of the prevalent methods for creating protein-based antimicrobial films involves embedding natural antimicrobial agents directly into the protein matrix. Several essential oils derived from herbs and spices, such as oregano, cinnamon, and cloves, have been documented to exhibit potent antibacterial properties (Kandasamy et al. 2021; Shams et al. 2019). Plant extracts such as thyme, rosemary, and grape seed are well-known for their antimicrobial properties (Bahramian et al. 2024a, b; Kandasamy et al. 2021). Fermentation byproducts like nisin, produced by the bacterium Lactococcus lactis, and pediocin from Pediococcus acidilactici, have demonstrated significant antibacterial effectiveness when incorporated into protein films. These agents employ various antibacterial mechanisms that disrupt essential microbial processes by targeting enzymatic functions, compromising the integrity of phospholipid membranes, or degrading bacterial genetic material (Jayakodi et al. 2022). Nanoparticles are another valuable component being utilized as nanofillers in antimicrobial films. Their broad spectrum of antibacterial activity, consistency, and low volatility make them highly effective. When applied in minimal quantities, nanoparticles such as silver and gold effectively inhibit microbial growth in food products. These nanoparticles are known to obstruct the production of ethylene, which is crucial in fruit ripening, thereby facilitating its absorption and breakdown, further extending the product's shelf life (Shams et al. 2019). For instance, the incorporation of zinc oxide (ZnO) nanoparticles into food packaging films significantly enhances their antimicrobial properties, effectively prolonging the shelf life of packaged goods by introducing active filler agents directly into the film matrix (Abedi-Firoozjah et al. 2025; Fahmy et al. 2020). A well-designed controlled release system allows for the sustained release of antimicrobial agents, ensuring a continuous protective effect throughout the product's shelf life. Such advancements in food packaging highlight an innovative convergence of material science and food technology aimed at improving food safety and quality (Bhaskar et al. 2023).

Gelatin (GEL) is an affordable, water-soluble protein derived from animal collagen, known for its film-forming capabilities, biodegradability, and biocompatibility. However, pure GEL films often have poor mechanical strength, high moisture sensitivity, and lack antibacterial properties, limiting their use in food packaging.

Blending polysaccharides with GEL can enhance these properties. Sun and collaborators utilized dialdehyde cellulose nanocrystal (DCNC) from coffee leaves and a coffee leaf extract-Fe^{3+} complex (CFe) as supplementing agents in GEL films for food packaging (Fig. 1.5). Adding 1 wt% CFe improved the film's microstructure, leading to better UV shielding, mechanical strength, and barrier properties. The composite films also exhibited strong antibacterial effectiveness against *S. aureus* and *E. coli*, with over 90% efficiency, and antioxidant propertics. When tested for packaging, these functional films extended the shelf life of fresh pork to 8 days at 4 °C when exposed to 808 nm near-infrared laser irradiation (Sun et al. 2025).

Zein is a unique protein derived primarily from corn, characterized by a high concentration of hydrophobic amino acids and a notable scarcity of polar amino acids. This composition makes zein insoluble in water, which positions it as an excellent candidate for various applications, particularly in the food packaging industry. Films produced from zein exhibit remarkable film-forming capabilities, creating a barrier that is effective against oxygen, which is crucial for preserving the quality and shelf life of food products. However, despite these advantageous properties, zein films present challenges, such as brittleness and rigidity, which can hinder their direct application in food packaging. To address these limitations, researchers are exploring strategies to enhance the mechanical properties and functionality of zein films (Nanda et al. 2024; Zhang et al. 2024). One such approach involves the reinforcement of zein with inorganic nanoparticles and the incorporation of functional additives, which can significantly improve the structural integrity and performance of the films. Cheng

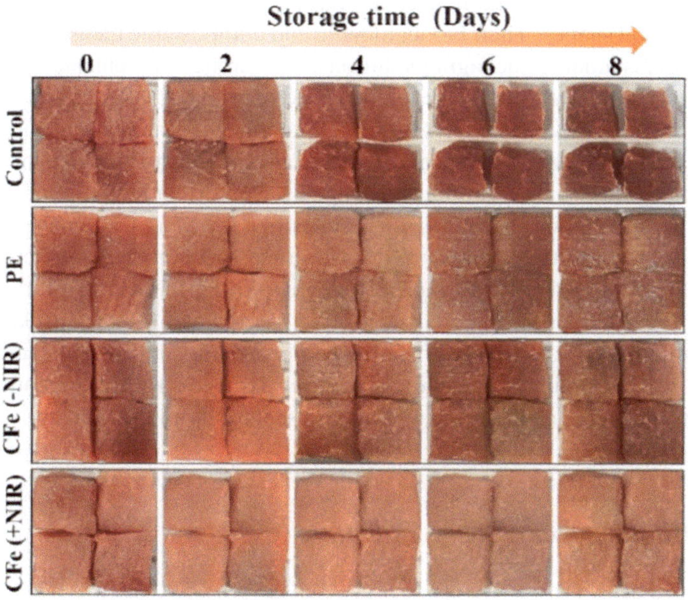

Fig. 1.5 Appearance changes of pork samples with different treatments during storage for 8 days at 4 °C, Ref. Sun et al. (2025)

and co-workers developed a flexible zein-based film integrated with titanium dioxide (TiO_2) nanotube arrays (TNTAs). Their research showed that incorporating TNTAs of varying lengths (700 nm–5.2 μm) led to a uniform dispersion in the zein matrix, which improved the mechanical strength and water resistance of the films. Analyses using XRD, FT-IR, and XPS confirmed intermolecular hydrogen bonding between zein and TNTAs, enhancing material properties. The films also demonstrated significant antimicrobial activity against E. coli under low-power UV irradiation (1600 mW/cm^2), suggesting their potential for improving food safety in packaging solutions (Cheng et al. 2024).

The combination of pectin from grape pomace (GPP) and kidney bean protein isolate (KPI) can create biodegradable films with notable properties. Adding grape pomace bioactive extract (GPE) may enhance these films for food packaging. Amiri Samani and co-workers found that a 50/50 blend of GPP and KPI improved hydration, barrier, and mechanical attributes. The films exhibited a water solubility of 14.96%, water vapor permeability of $1.86 \times 10 - 10$ g/m s Pa, tensile strength of 36.20 MPa, and elongation at break of 17.90%. Further analysis showed strong compatibility between KPI and GPP, along with enhanced structural characteristics. Thermogravimetric tests indicated improved thermal properties, and biodegradability studies found the films began to decompose within eight days in vegetal compost (Amiri Samani et al. 2025).

Protein-based polymeric films are gaining attention in food packaging due to their unique properties derived from natural proteins. These biopolymers offer exceptional barrier capabilities against oxygen and moisture, which helps maintain food freshness and shelf life. In addition, they are strong yet flexible, making them suitable for various food shapes. Their biodegradable and compostable nature aligns with the demand for sustainable packaging solutions. Researchers are enhancing the performance of these films through physical and chemical modifications, such as adding plasticizers to improve elasticity. Proteins like casein, whey, and gelatin serve as effective and environmentally friendly materials for film formation. Incorporating active components, such as antimicrobial agents, can further increase food safety and shelf life. Despite their advantages, challenges like lower transparency compared to synthetic films, sensitivity to external conditions, and higher manufacturing costs remain. The field is ripe for further research exploring innovative protein sources, such as insects and algae, to create cost-effective and eco-friendly packaging solutions. In summary, protein-based films offer a promising and sustainable alternative to traditional packaging, but ongoing research is necessary to overcome existing limitations and enhance their practicality and scalability in the industry.

1.2.2 Biopolyesters-Based Polymers

A variety of essential biomass-derived polyesters have found extensive applications in the development of industrial biological materials. Among these, polyhydroxyalkanoate (PHA) and polylactic acid (PLA) are particularly noteworthy for their versatility and sustainability. These biodegradable polymers are derived from renewable resources and are gaining popularity due to their environmentally friendly characteristics. In the sector of food packaging, several biodegradable polymeric biomaterials, including poly(lactic acid) (PLA), poly(e-caprolactone) (PCL), and poly(glycolic acid) (PGA) are frequently utilized (Stoica et al. 2018; Zia et al. 2016). These materials, along with their copolymers, are favored for their ability to decompose naturally, reducing the environmental impact associated with traditional petroleum-based plastics. They are commonly employed in the production of films and coatings, which serve as protective barriers for food products, enhancing shelf life while also aligning with eco-friendly packaging initiatives. The incorporation of these materials in food packaging not only demonstrates a commitment to sustainability but also offers opportunities for innovation in material science and product development (Dedieu et al. 2022).

1.2.2.1 Polylactic Acids and Poly(Butylene Adipate-Co-terephthalate)

PLA is an innovative and renewable aliphatic polyester that is synthesized from lactic acid monomers. The production of PLA begins with lactide, a high-purity monomer, which plays a crucial role in its polymerization process. Lactic acid, the building block of lactide, can be generated through two primary methods: chemical synthesis and microbial fermentation. These processes utilize renewable resources, such as cornstarch and sugarcane, making PLA an ecologically friendly alternative to traditional petroleum-based plastics (Swetha et al. 2023). Lactic acid is known for its biocompatibility, making it safe for various applications in medical and consumer products. It is also highly soluble in water, which contributes to its versatility in different environments. PLA belongs to the group of rigid thermoplastic polymers categorized as aliphatic polyesters. The specific types of PLA include poly(DL-lactide) (PDLLA), poly(D-lactide) (PDLA), poly(L-lactide) (PLLA), and poly(meso-lactide), along with a series of copolymers. Each variant possesses unique characteristics that influence the physical and chemical properties of the final product (Kirac et al. 2022). The polymerization of lactic acid involves the conversion of D-lactide and L-lactide into their respective polymers. This process enables the formulation of materials with varying mechanical strengths and thermal properties, which can be tailored for specific applications. Given its low-cost production and availability, PLA is regarded as one of the most promising bio-based polyesters, particularly in the fields of packaging and medicine (Bindu et al. 2019; Patel et al. 2022). PLA's adaptability and contributions to sustainability have positioned it as a viable alternative to traditional plastics. It is commonly utilized for packaging items, including food trays, beverage bottles, and disposable cups, due to its compostability and reduced environmental impact. Beyond these applications, PLA is also valuable in the electrical industry, where it frequently replaces polycarbonate (PC) in the production of enclosures and components for electronic devices. The combination of performance, safety, and environmental advantages makes PLA a leading choice for eco-conscious manufacturers seeking to innovate with sustainable materials (Kirac et al. 2022; Sone Aung et al. 2018).

The use of biomaterials in food packaging has garnered considerable attention, particularly PLA, due to its remarkable biocompatibility and a variety of desirable physical properties. PLA is known for its high tensile strength, which makes it suitable for maintaining the structural integrity of various packaging forms. Additionally, its thermoplastic nature allows for easy processing into a wide range of shapes and sizes, catering to different packaging needs. PLA is also non-toxic, making it a safer option for food contact applications. When enhanced with eco-friendly additives, PLA exhibits antimicrobial properties that can further prolong the shelf life of packaged foods by inhibiting the growth of harmful bacteria and fungi (Bindu et al. 2019).

Other types of polyesters comprise poly(butylene adipate-co-terephthalate) (PBAT), frequently utilized as a bioplastic. PBAT is recognized for creating high-quality packaging films and can be a favorable option because of its excellent flexibility, biocompatibility, and effective processability. PBAT is a man-made polymer that effectively merges the characteristics of both aromatic and aliphatic polyesters. It is produced through the polycondensation reaction involving 1,4-butanediol, adipic acid, and terephthalic acid. This distinctive structure grants PBAT significant flexibility and adaptability, making it especially appropriate for active food packaging applications, where sustaining product quality and prolonging shelf life are essential (Roy et al. 2024a, b). Regarding performance, packaging films made from PBAT show features that are on par with commonly used plastic films, such as low-density polyethylene (LDPE). As concerns about environmental issues rise, the demand for PBAT-based films has increased, largely due to a growing need for biodegradable packaging options that maintain excellent physical properties. Nevertheless, it is crucial to acknowledge that the mechanical strength of PBAT films is usually inferior when compared to other bioplastics, and their barrier characteristics, such as resistance to moisture and gases, do not reach the levels of traditional petroleum-based plastics. In addition, PBAT films often fall short in the functional attributes needed to enhance food shelf life during packaging, which may restrict their use in the food sector. To improve the usability of PBAT films in packaging, enhancing their physical properties through various methods is vital. One effective strategy involves the addition of fillers or additives that can strengthen the structure and functionality of the film. By blending PBAT with other compatible polymers, including thermoplastic starch, cellulose nanofibers, or polylactic acid (PLA), researchers have demonstrated that the performance of the films can be considerably improved. Combining thermoplastic starch (TPS) with PLA and PBAT offers a promising avenue for producing fully biodegradable and renewable biopolymers. Nevertheless, these composites frequently face challenges such as inadequate mechanical performance, low water vapor barrier properties, and limited antibacterial effectiveness due to the lack of compatibility between the hydrophilic TPS and the hydrophobic PLA/PBAT. Liu and co-workers developed silver nanoparticles incorporated into cellulose nanocrystals (Ag-CNC), which greatly improved the antibacterial functions and UV-shielding properties (a 283% enhancement) of TPS/PLA/PBAT films. Furthermore, the addition of OLA-g-ST and Ag-CNC resulted in better mechanical characteristics (with tensile strength and elongation at break rising by 77.95% and 63.23%, respectively) and enhanced water vapor barrier performance (a 40.8% improvement) of the films (Liu et al. 2025). Moreover, the incorporation of bioactive components can further refine these films. A variety of functional materials, including nanofillers, natural plant extracts, clay, and colorants, have been explored for their potential to enhance the physical and functional attributes of PBAT-based films. According to previous research, adding these polymeric and functional additives has shown significant promise in boosting the properties of PBAT films, thereby making them more suitable for active food packaging applications. This ongoing research aims not only to enhance the performance of PBAT films but also to support sustainable packaging solutions that align with eco-friendly practices. Olonisakin and colleagues developed an eco-friendly technique for producing a PBAT composite film utilizing epoxidized soybean oil (ESO) and encapsulated lignin (Lig)/tannic acid (Ta) (ESO-Lig-Ta) as a filler. The results revealed that the ESO-Lig-Ta increased tensile strength by more than 100%, and SEM indicated a consistent distribution of the encapsulated filler within the matrix. The transmission of UV light diminished with an increase in the quantity of ESO-Lig-Ta. In comparison to the unadulterated PBAT film, the incorporation of ESO-Lig-Ta resulted in a diminished rate of water vapor and oxygen transmission. Food products, including fresh potatoes and onions, preserved using the composite film remained fresh for a duration of up to 12 days at ambient temperature (Olonisakin et al. 2023). In a similar study, Xiao and co-workers developed hybrid particles called lignin-zinc oxide (LZn), consisting of lignin nanoparticles and zinc oxide, synthesized via a hydrothermal method at different pH levels. They grafted glyceryl methacrylate (GMA) onto poly(butylene adipate-co-terephthalate) (PBAT) and incorporated LZn hybrids to create master batches. These were catalyzed with trimethylamine to produce PBAT-G-xLZn composite films. The PBAT-G-3LZn composite showed a 35.2% increase in tensile modulus and a 28.6% increase in yield strength compared to pure PBAT. The composites provided UV protection, antioxidant properties, and antibacterial effects, with bacterial adhesion rates significantly lower than pure PBAT. The transmittance at $\lambda = 320$ nm was just 0.16%, and the radical scavenging activity (RSA) was 11.6%. Zn^{2+} migration in PBAT-G-3LZn was only 3.3 μg/L (Xiao et al. 2022).

Active packaging films derived from PBAT were developed using diverse active agents, including essential oils, encapsulated essential oils, nitrite, lignin-zinc oxide, and curcumin. The biodegradable polymer PBAT is seen as a promising option for producing food packaging to address the challenges associated with fossil fuel-based packaging materials. Packaging materials derived from fossil fuels contribute to several environmental issues, including solid waste generation, climate change, and greenhouse gas emissions. In this context, selecting PBAT as an alternative to fossil-derived packaging materials is a prudent choice, considering its superior biodegradability and exceptional packaging properties. A variety of PBAT-based goods is offered, including packing materials, courier bags, mulch films, and flatware. PBAT-based packaging significantly prolongs the shelf life of perishable food products (Pietrosanto et al. 2020).

Venkatesan and Rajeswari investigated the impact of zinc oxide (ZnO) on PBAT, emphasizing its packaging characteristics, such as barrier, thermal, and mechanical properties. They synthesized antimicrobial coatings with PBAT and ZnO nanoparticles, exhibiting enhanced mechanical and thermal stability, along with significant antibacterial efficacy against Escherichia coli and Staphylococcus aureus (Venkatesan and Rajeswari 2017). In another study, Pietrosanto and co-workers created a mixture of two bioplastics, PBAT and PLA, in a ratio of 60/40 to improve performance. They examined two types of PLA with different viscosity levels and stereoregularity and employed a chain extender (Joncryl®). By selecting the PLA that had a viscosity more similar to PBAT and exhibited lower stereoregularity, they achieved better morphology and interfacial adhesion. Including the chain extender also contributed to smaller sizes of the dispersed phase, which improved the mechanical properties of the films. The mixture of PBAT, amorphous PLA, and the compatibilizer demonstrated enhanced stiffness, ductility, transparency, and sealability suitable for flexible packaging uses (Pietrosanto et al. 2020). The PLA/PBAT-based functional film has emerged as a viable method for packaging soy protein-derived meat analogs. Studies indicate that including chemicals such as cinnamaldehyde and tea polyphenols into the PLA/PBAT matrix significantly enhances its antibacterial efficacy, successfully inhibiting bacterial proliferation on meat analogs. This novel packaging technique preserves the quality of vegan meat by prolonging its shelf life while ensuring consumer safety, in accordance with the rising demand for plant-based products (Wang et al. 2022). Yang et al. (2023) enhanced PBAT-based functional packaging by integrating modified clay with quercetin to improve the preservation of perishable fruits, including bananas and blueberries. The results indicated that this enhanced functional film markedly retarded the deterioration of these fruits for at least one week. This exceptional outcome can be attributed to the film's superior gas and water vapor barrier qualities, which diminish moisture loss and oxygen exposure, along with the strong antioxidant activity conferred by the supplementary chemicals (Yang et al. 2023). These talks indicate that PBAT, utilized independently or in conjunction with other polymers, significantly enhances the shelf life of packaged food products. PBAT acts as an efficient barrier material while also serving as a carrier for various bioactive compounds, thereby augmenting the package with additional health advantages. Consequently, the functional ingredients incorporated in PBAT-based packaging films are crucial for prolonging the shelf life of packaged goods, thereby maintaining their freshness and safety for extended durations.

Active food packaging is an innovative approach that integrates these biomaterials along with additives that provide antimicrobial and antioxidant benefits. This type of packaging not only serves to protect the food within but also actively works to maintain its quality, safety, and freshness. By using formulations that combine PLA with other natural compounds, packaging can be tailored for various food categories, improving consumer safety by reducing the risk of spoilage and foodborne illnesses (Sahraeian et al. 2024). Extensive research and development efforts have been focused on creating food-active packaging solutions derived from PLA tailored for an array of food products. These initiatives aim to leverage the unique properties of PLA as a sustainable alternative to conventional packaging materials, which often contribute to environmental pollution and are not biodegradable. Given the advantages offered by PLA, including its origin from renewable resources, utilizing this biopolymer in food packaging represents a promising advancement in sustainable practices within the food industry. This shift not only meets consumer demands for eco-friendly products but also aligns with global efforts to reduce plastic waste (Naskar et al. 2023).

PLA has inherent limitations regarding its mechanical strength, barrier properties, and thermal stability. To address these shortcomings, enhancing the crystallinity of PLA is a promising approach. One potential method to improve the crystallinity involves melt mixing PLA with a highly crystalline biopolymer, such as polyhydroxybutyrate (PHB). However, the broader industrial adoption of PHB faces significant barriers, primarily due to its high production costs, as well as complexities associated with its processing and forming techniques. In terms of solubility parameters, PLA has a range of 19.5–20.5 MPa1/2, while PHB's solubility parameters range from 18.5–20.1 MPa1/2 (Arrieta et al. 2014; Calvao et al. 2012). The minimal difference in these solubility parameters suggests that PLA and PHB possess good miscibility, making the blending of these two polymers a feasible strategy to enhance the properties of PLA while simultaneously increasing the market appeal of PHB. Research into the miscibility characteristics of PLA and PHB has involved examining blends formed from PLA of varying molecular weights, including low molecular weight PLA (Mn = 1579) and high molecular weight PLA (Mn = 159,400), combined with high molecular weight PHB (Mn = 222,000). Results indicated that low molecular weight PLA demonstrated favorable miscibility with PHB across the entire compositional spectrum. Conversely, blends created with high molecular weight PLA and high molecular weight PHB exhibited a biphasic system, indicating limited compatibility (Defieuw et al. 1989). In another study focusing on the crystallization behavior and miscibility of a blend comprising PLA (Mw = 20,000 g/mol) and low molecular weight PHB (Mw = 5000 g/mol), findings revealed that PLA maintains limited miscibility with PHB when the concentration of PHB reaches 25 wt%. Notably, no miscibility was observed at concentrations exceeding this threshold (Hu et al. 2008). During non-isothermal crystallization investigations, the incorporation of PHB at 30 wt% led to a noteworthy reduction in PLA's cold-crystallization temperature, suggesting alterations in the crystallization process due to the presence of PHB (Hu et al. 2008). The resultant materials derived from PLA/PHB blends exhibited several mechanical properties, including significant tensile strength and elastic modulus; however, they also demonstrated marked brittleness, which could limit their practical applications. Furthermore, the processing characteristics of PLA/PHB blends have been identified as suboptimal, presenting challenges particularly during film extrusion processes. Given these issues, ongoing research is dedicated to exploring various strategies to enhance the performance and processing efficacy of PLA/PHB blends. This includes the incorporation of plasticizers to improve flexibility and flow properties, the addition of functional additives to tailor specific characteristics, the use of nanofillers to strengthen the composite, and the development of copolymers derived from hydroxyalkanoic acid to further optimize the compatibility and properties of the blends (Armentano et al. 2015; Jandas et al. 2014).

The incorporation of PLA and PCL (polycaprolactone) blends in the creation of electrospun nanofibers has garnered significant attention due to their promising applications in various fields, including biomedical and packaging solutions (Dong et al. 2019). Research conducted by Dorati and colleagues demonstrated that the integration of chitosan into the PLA-PCL copolymer not only improved the biological properties of the resulting nanofibers but also enhanced their wettability characteristics (Dorati et al. 2018). Importantly, this modification did not compromise the structural integrity or stability of the electrospun fibers, suggesting potential for enhanced performance in practical applications. Moreover, the utilization of essential oils (EO) within electrospun nanofiber materials has been investigated to leverage their antimicrobial properties. However, challenges persist regarding the efficiency of encapsulating these oils using conventional blending techniques in electrospinning, which may limit their effectiveness. In this regard, Shi and colleagues developed a novel approach to active food packaging by creating electrospun nanofibers composed of PLA and PCL. These nanofibers incorporated oregano essential oil (OEO) that was effectively encapsulated in β-cyclodextrin (β-CD). The integration of OEO within the nanofibers aimed to enhance the antimicrobial properties of the packaging material. The resultant OEO@β-CD/PLA/PCL nanofibers showed marked improvements in thermal stability and resistance to deformation, which are critical characteristics for maintaining the integrity of food packaging under various storage conditions. However, it is noteworthy that the tensile strength of the fibers experienced a decline, which could impact their mechanical performance during handling and application. Moreover, the addition of OEO@β-CD altered the surface properties of the nanofibers, leading to a notable decrease in hydrophobicity. This change is attributed to the formation of hydrogen bonds between the components, which can affect moisture transmission through the packaging material. Despite this shift towards a more hydrophilic surface, the water vapor barrier properties of the nanofibers remained consistent, which is beneficial for controlling moisture levels in packaged food products. In terms of functional benefits, the OEO@β-CD/PLA/PCL nanofibers exhibited significant antibacterial and antifungal properties, making them particularly suitable for food preservation. This was further validated through interaction studies with zebrafish, which confirmed the biosafety of the nanofibers for potential application in food contact materials. In practical application trials, these active nanofibers demonstrated their efficacy by significantly slowing down postharvest decay and deterioration in blackberries (Fig. 1.6). This preservation led to an improvement in the storage quality of the fruits, highlighting the promising potential of OEO@β-CD/PLA/PCL nanofibers as a viable solution for enhancing the shelf life and safety of food products. This research indicates a significant advancement in the development of smart packaging solutions that not only protect food items but also contribute to food safety and sustainability (Shi et al. 2022).

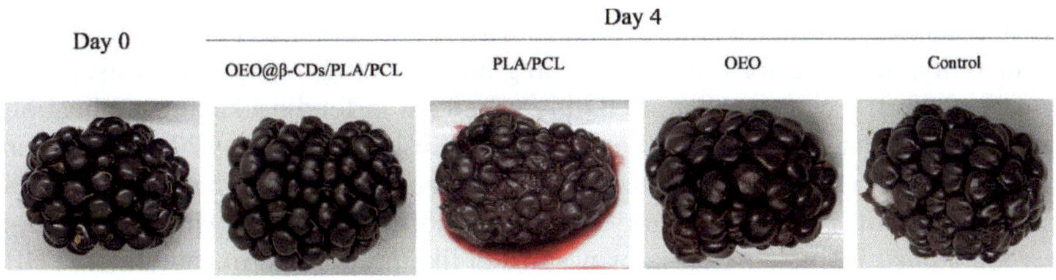

Fig. 1.6 Effect of PLA/PCL/OEO@β-CD nanofiber films on the appearance changes of blackberry fruit stored at 25 °C for 4 days, Ref. Shi et al. (2022)

1.2.2.2 Polyhydroxyalkanoates (PHA)

PHAs are increasingly recognized as highly advantageous materials for food packaging applications. These bioplastics offer a variety of beneficial characteristics that make them an appealing alternative to traditional petroleum-based plastics. One of the most significant advantages of PHAs is their inherent biodegradability, which means they can break down naturally in the environment, reducing plastic pollution and minimizing landfill waste. Additionally, PHAs possess excellent water-repellent properties, making them an appropriate material for packaging perishable food items, as they help maintain freshness by providing a barrier against moisture (Masood 2017). Furthermore, PHAs are non-toxic, ensuring that they pose no harm to human health or the environment, which is especially important in applications involving food contact. They also exhibit unique nonlinear optical characteristics, allowing for potential uses in advanced packaging technologies that can enhance product visibility or freshness indicators. The thermal plasticity of PHAs means they can be easily molded into various shapes and forms, enabling versatile design options for packaging solutions (Dejene and Gudayu 2024; Tripathi et al. 2022). These bioplastics show piezoelectric properties, which could lead to innovative applications in smart packaging systems that respond to changes in pressure or motion. Lastly, PHAs demonstrate resistance to gases and moisture, further enhancing their suitability for food preservation by reducing the risk of spoilage caused by external elements. The combination of these properties positions PHAs as a compelling choice for sustainable food packaging solutions (Bhardwaj et al. 2014). PHAs are a class of biopolyesters that are generated through the fermentation process of sugars and lipids by specific strains of bacteria. These natural polyesters possess several functional properties that make them suitable for various applications, particularly in the food packaging industry. PHAs are regarded as safe for ingestion, non-tacky in texture, and insoluble in water. Their biodegradability, biocompatibility, and renewability significantly contribute to their appeal as environmentally friendly alternatives to traditional petroleum-based plastics (Plackett and Siró 2011). One of the remarkable characteristics of PHAs is their excellent resistance to UV radiation, which enhances their stability when exposed to sunlight. Additionally, they exhibit high tensile strength, which is a crucial attribute for materials used in packaging, as it ensures durability and structural integrity. Furthermore, PHAs have notable printability, allowing for effective labeling and branding, along with resistance to grease and oil, properties that are highly beneficial in food packaging applications where contamination and leakage are concerns (Keshavarz and Roy 2010). PHAs are categorized into three primary groups based on the number of carbon atoms in their monomeric units. The first category comprises short-chain PHAs, which contain 3–5 carbon atoms, including well-known variants like poly(3-hydroxybutyrate) (PHB) and poly(4-hydroxybutyrate). The second category consists of medium-chain PHAs, which encompass those with 6–14 carbon atoms, such as poly(3-hydroxyhexanoate). The third category includes long-chain PHAs, characterized by having more than 14 carbon atoms (Abdelwahab et al. 2012). Among the various types of PHAs, PHB has been the subject of extensive research and analysis. It is a linear and isotactic semi-crystalline polymer that displays impressive tensile strength as well as water vapor permeability, allowing it to serve as an effective barrier against moisture. These properties make PHB comparable to polyvinyl chloride (PVC), a commonly used thermoplastic, thus contributing to its recognition and popularity in the food packaging industry. Additionally, PHB has a thermal decomposition temperature of approximately 270 °C, while its melting temperature ranges between 170 and 180 °C. This narrow processing temperature range poses considerations for its application, particularly during the melting process. However, one limitation of PHB is its low elongation at break, which ranges from 3 to 6%, indicating that it may not stretch significantly before breaking (Abdelwahab et al. 2012; Armentano et al. 2015; Zare et al. 2019). Through the strategic modification of monomers and the use of constitutional isomerism techniques, a diverse array of PHAs exhibiting unique properties can be synthesized. Within the PHA family, various homopolymers can be produced, including poly(3-hydroxypropanoate), poly(3-hydroxyvalerate), poly(3-hydroxyhexanoate), and poly(3-hydroxydecanoate). A variety of copolymers are part of the PHA family, including poly(3-hydroxybutyrate-co-3-hydroxyvalerate) (PHBHV), poly(3-hydroxybutyrate-co-3-hydroxyhexanoate) (PHBHHx), poly(3-hydroxybutyrate-co-3-hydroxyoctanoate), and poly(3-hydroxybutyrate-co-3-hydroxydecanoate). Notably, poly(4-hydroxybutyrate) stands out for its exceptional flexibility, exhibiting an impressive elongation at break of approximately 1000%. This remarkable characteristic greatly exceeds that of many comparable materials, making it ideal for applications that require enhanced stretchability and durability (Pryadko et al. 2021; Vandewijngaarden et al. 2016). Manikandan and colleagues conducted an in-depth investigation into the solution casting technique used for producing PHB nanocomposites by incorporating various concentrations of graphene nanoplatelets (Gr-NPs), ranging from 0 to 1.3 weight percent (wt%). The research revealed that the incorporation of Gr-NPs into the PHB matrix resulted in significant enhancements in several critical material properties. Specifically, the melting point of the PHB/Gr-NPs nanocomposite increased by 10 °C compared to that of pristine PHB, indicating improved thermal characteristics. Additionally, the thermal stability of the nanocomposite showed a similar increase of 10 °C, showcasing its ability to maintain integrity under elevated temperature conditions. Impressively, the tensile strength of the material doubled, demonstrating a substantial improvement in mechanical performance. The study also quantified the barrier properties of the nanocomposite, revealing that the addition of Gr-NPs led to a threefold reduction in oxygen permeability and a twofold decrease in water vapor permeability. These enhancements indicate a significant improvement in the material's ability to protect contents from both oxygen and moisture, which are critical factors in packaging applications. Moreover, the inclusion of Gr-NPs was found to considerably reduce the transmission of UV and visible light through the nanocomposite, further enhancing its suitability for protecting sensitive food products from light-induced degradation. In terms of safety, the cytotoxicity tests performed on the synthesized nanocomposite indicated that its toxicity levels were statistically negligible when compared to those of pristine PHB, suggesting that the addition of Gr-NPs does not pose significant health risks. To assess the practical implications of these findings, a simulation study was carried out focusing on food items sensitive to moisture and oxygen, specifically potato chips and dairy products. The results of this study were promising, demonstrating a remarkable fourfold increase in shelf life for these products when packaged using the PHB/Gr-NPs nanocomposite. This outcome highlights the potential of this innovative material in extending the freshness and safety of perishable food items in commercial applications (Manikandan et al. 2020).

1.3 Applications of Bio-based Polymers in Food Packaging

The application of biodegradable and bio-based packaging in different types of food is summarized in Table 1.1. A biodegradable film can be characterized as a sophisticated primary packaging material crafted from biodegradable polymers combined with food-grade additives. This innovative thin layer of material is engineered to form a flexible film that can be utilized effectively as a food wrap, ensuring that the original ingredients and processing methods of the food remain unaltered (Mishra et al. 2025a, b; Syarifuddin et al. 2025). The primary purpose of biodegradable films is to serve as a protective barrier for food products, helping to extend their shelf life while minimizing environmental impact. The production of biodegradable films begins with the careful selection and preparation of raw materials, which need to be dissolved or dispersed in a suitable solvent. Common choices for these solvents include water, alcohol, or a tailored mixture of both along with other solvents to achieve the desired consistency. This phase might also involve the incorporation of various additives such as plasticizers, which enhance the flexibility and workability of the film; antimicrobial agents, which help inhibit the growth of spoilage microorganisms; and coloring or flavoring agents to improve the aesthetic appeal of the final product. To optimize the solubility of certain biopolymers effectively, it may be necessary to adjust the pH of the solution or apply heat during the preparation phase. Following the formulation of the film-forming solution, the next step involves casting the solution onto a suitable surface and allowing it to dry under meticulously controlled temperature and humidity conditions. This careful drying process results in the formation of free-standing films, which possess the desired mechanical and barrier properties for packaging applications (Deng et al. 2025; Koirala et al. 2025; Marcoaldi et al. 2025). In addition to being used as standalone films, these biodegradable film-forming solutions can also be applied directly to food items as a protective coating. This application can be executed through various techniques, including dipping, spraying, brushing, or panning, each followed by a drying process to ensure optimal adhesion and performance. Biodegradable films represent a significant advancement in food packaging technology, merging functionality with sustainability to address the growing environmental concerns associated with conventional plastic use (Bhatia et al. 2025; Thiyagamoorthy et al. 2024; Yang et al. 2024). Sustainable development has emerged as a focal point of research and discussion among global scholars and policymakers, particularly in the context of environmental preservation and resource management. It is often articulated as "development that meets the needs of the present without compromising the ability of future generations to meet their own needs." This concept is fundamental in the food packaging industry, where achieving sustainability goals can be addressed at different stages of the packaging lifecycle (Nayak et al. 2025; Perumal et al. 2025). At the initial stage, raw materials acquisition, the use of renewable or recycled materials is critical. This approach not only reduces carbon dioxide emissions but also lessens reliance on fossil fuels, which are traditionally used in plastic production. Among the sustainable materials being explored, biopolymers such as starch, cellulose, chitosan, PLA, and polyhydroxyalkanoates (PHAs) stand out as promising alternatives to conventional plastics. These materials are derived from renewable biological resources, including agricultural crops and microorganisms, and are especially notable for their neutral carbon life cycles, which can significantly mitigate CO_2 emissions in the atmosphere. The Intergovernmental Panel on Climate Change (IPCC) emphasizes the potential of biopolymers to reduce global warming impacts by as much as 1.5 °C. Researchers are also actively investigating innovative renewable and cost-effective feedstocks, such as agricultural waste, food industry byproducts, and textile mill discards, to minimize both the economic and environmental footprints of biopolymer production (Marcoaldi et al. 2025; Mishra et al. 2025a, b). The production stage is where the choice of manufacturing methods becomes crucial. Opting for energy-efficient processes can lead to lower resource consumption and reduced emissions, thereby enhancing the sustainability of the packaging lifecycle. However, advancing to the waste management stage reveals the stark contrast between conventional plastics and biopolymers. Conventional plastics present significant environmental challenges, particularly because they are not biodegradable. When it comes to recycling, food packaging often becomes contaminated with residual food, rendering the recycling process economically and logistically burdensome. The biodegradability of polymers can be assessed through various methods, including monitoring their physical and chemical property changes, as well as measuring the rates of CO_2 evolution and O_2 consumption during decomposition. The rate at which a polymer biodegrades is influenced by multiple environmental factors such as soil composition, compost conditions, humidity levels, temperature variations, and the diversity and quantity of microorganisms present in the environment (dos Santos Lima et al. 2025; Iqbal et al.

2025; Yang et al. 2024). In addition to these external factors, the intrinsic characteristics of the biopolymers play an essential role in their degradation rates. Elements such as the types of functional groups, the chemical structures, crystallinity, the length of polymer chains, and the nature and concentration of plasticizers or additives incorporated into the formulation all contribute to how quickly and efficiently a biopolymer breaks down (Tammina et al. 2024; Tan et al. 2024). The specific requirements of food packaging mean that the necessary durability of biopolymers must align with their intended use. For example, highly biodegradable biopolymers like polysaccharides, proteins, and biopolyesters are ideal for packaging perishable food items with a short shelf life. Conversely, biopolymers that exhibit slower biodegradation, such as PLA, may be more suitable for products with a longer shelf life and provide the added benefit of recyclability (Sahraeian et al. 2024; Sen and Singh 2024). Ultimately, effective packaging solutions must meet environmental standards and enhance social conditions for individuals handling and using the packaging, encompassing both workers and consumers. Furthermore, for sustainable packaging to make a significant impact in the market, it must also be economically viable and designed to thrive alongside traditional packaging alternatives, thus contributing positively to broader development goals.

Biodegradable packaging from food waste and by-products demonstrates significant potential for meat and seafood products. For instance, gelatin-based films with grapefruit seed extract showed strong antimicrobial effects against common spoilage bacteria, extending the shelf life of beef patties by 6 days. Chitosan films infused with clove bud essential oil enhanced the shelf life of sliced ham by over 21 days (Peng and Li 2014), while peanut skin extract in chitosan maintained precooked tuna fish balls for 12 days (Kanatt et al. 2010). Whey protein isolate films with mango peel powder and green tea extract prevented lipid oxidation in raw chicken for 12 days (Muppalla et al. 2014), and cassava starch-based films with red grape pomace extended shelf life in salmon patties by 21 days (Piñeros-Hernandez et al. 2017). Gelatin films containing oregano essential oil and chitosan films with clove oil also inhibited bacteria growth, prolonging the shelf life of shrimp by 6 and 5 days, respectively. While these findings are promising, further research is needed to address film properties and the feasibility of large-scale production for effective commercialization (Chi et al. 2006).

Effective packaging solutions need to comply with environmental standards and enhance the social conditions for all parties involved, including both workers and consumers. For sustainable packaging to really make a difference in the market, it must also be financially viable and designed to evolve alongside traditional packaging methods, contributing to broader development objectives. As a result, bio-based packaging is being created to substitute conventional plastics in various food items while ensuring that quality, safety, integrity, and shelf life are preserved. The following discusses the use of bio-based polymers in specific food products.

Fresh fruits and vegetables face high respiration rates, leading to weight loss, quality deterioration, and spoilage. Biodegradable films can modify the in-package atmosphere, reducing condensation. Chitosan films infused with EOs and AgNPs show significant antimicrobial effectiveness against pathogens like *E. coli* and *Salmonella*, enhancing the shelf life of strawberries (Shankar et al. 2021). When

Table 1.1 Application of biodegradable and bio-based packaging in various types of food

Matrix type	Packaging method	Food sample	Storage conditions	Key results	References
Alginate/EOs/citrus extract and Gamma irradiation	Film wrapping	Merluccius sp. fillets	4 °C, 21 days	Extended shelf life from 7 to 28 days	Shankar et al. (2019)
CMC/gelatin	Film wrapping	Cherry, tomatoes and grapes	Room temperature, 14 days	Reduced browning and minimized weight reduction by incorporating CMC at levels up to 75%	Samsi et al. (2019)
CMC/okra mucilage/ZnO nanoparticles	Film wrapping	Chicken breast fillet	4 °C for 12 days	ZnO and okra extended the shelf life from 6 days in CMC to 12 days. Additionally, total volatile nitrogen and lipid oxidation were decreased by as much as 40%	Mohammadi et al. (2019)
Chitosan/cellulose acetate phthalate/ZnO nanoparticles	Film wrapping	Black grapefruits	Room temperature, 9 days	Cellulose/chitosan/ZnO increased the shelf life from 3 days in polyethylene to 6 days	Indumathi et al. (2019)
Potato starch/butylated Hydroxytoluene or green tea extract	Film wrapping	Beef	4 °C and room temperature, 10 days	Decreased lipid oxidation (thiobarbituric acid-reactive substances [TBARS]) and reduced protein oxidation (formation of metmyoglobin)	u Nisa et al. (2015)
Alginate/β CD/carvacrol	Film wrapping	Mushroom	4 °C, 12 days	Decreased weight loss, prolonged microbial and sensory shelf life	Cheng et al. (2019)
Tapioca starch film/chitosan nanoparticles	Film wrapping	Cherry tomatoes	10 °C for 10 days	The inclusion of chitosan suppressed the growth of gram-positive bacteria	Shapi'i et al. (2020)
Cassava starch/EVOH/alginate	Film wrapping	Lettuce	6 °C for 8 days	Equal functionality compared to PVC	Brandelero et al. (2016)
Alginate/EOs/citrus extract and Ozonation	Film wrapping	Merluccius sp. fillets	4 °C, 14 days	The shelf life has been extended from 7 to 21 days	Shankar et al. (2019)

(continued)

Table 1.1 (continued)

Matrix type	Packaging method	Food sample	Storage conditions	Key results	References
Chitosan/apricot kernel oil	Pouches	Bread slices	Room temperature, 10 days	Reduced fungal development compared to packaging in LDPE. Increased shelf life	Priyadarshi et al. (2018a)
Chitosan	Pouches	Green chili	27 °C, 7 days	Prolonged shelf life and reduced moisture loss in comparison to products that are not packaged	Priyadarshi et al. (2018b)
Chitosan with kombucha tea	Film wrapping	Minced beef	4 °C, 6 days	Shelf life has been increased by three days	Ashrafi et al. (2018)
Zein/pomegranate peel extract (PPE)	Film wrapping	Fresh Himalayan cheese	Refrigerated storage, 30 days	PPE slowed down the spoilage of cheese as well as the oxidation of fats and proteins in the cheese	Mushtaq et al. (2018)
Gelatin/catechin lysozyme	Film wrapping	Minced pork	4 °C, 7 days	Decreased weight loss, slowed formation of TBARS; delayed growth of vital plate count, yeast, and mold in comparison to the control in PVC wrap. Preservation of color	Kaewprachu et al. (2015)
Tilapia and squid skin Gelatin film/nanoclay/ethanolic extract from coconut husk	Film wrapping	Mackerel meat powder	28–30 °C, 30% RH, 30 days	Slowed lipid oxidation; decreased degradation of sensory quality	Nagarajan et al. (2015)
Whey protein isolate/ cellulose nanofiber/TiO_2 nanoparticle incorporated with rosemary EO	Film wrapping	Lamb meat	4 °C, 15 days	Shelf life is increased twofold in comparison to plastic, along with enhanced quality (slowed development of spoilage bacteria and better organoleptic characteristics)	Alizadeh Sani et al. (2017)

(continued)

Table 1.1 (continued)

Matrix type	Packaging method	Food sample	Storage conditions	Key results	References
PLA/poly butylene adipate-coterephthalate (PBAT) blend (compared with PE and unpacked)	Film wrapping	Passion fruit	20 °C, 21 days	Enhanced retention of sensory attributes, taste, and consumable quality during the storage period	Zhang et al. (2019)
PLA/WPI/PLA (Control: PLA)	Pouch	Baby formula (dry product)	4, 25, and 35 °C, 50–59% RH, 19 days	Lipid oxidation occurs at a slower rate when compared to pure PLA	Phupoksakul et al. (2017)
PLA films containing oregano essential oil (Origanum vulgare L virens) (compared with PP)	Extruded films placed inside PP bags with MAP (10% O_2, 10% CO_2, 80% N_2)	Iceberg salad	4 °C, 7 days	The antifungal properties of EOs against fungi and yeast	Llana Ruiz-Cabello et al. (2016)
PLA/PBS with oregano EO (compared with PLA/PBS)	Packaging bags with MAP (10% O_2, 10% CO_2, 80% N_2)	Lettuce	4 °C, 8 days	Extended shelf life was achieved with the incorporation of essential oils, lasting up to 8 days; essential oils enabled the preservation of lettuce for as long as 8 days	Llana-Ruíz-Cabello et al. (2019)
PBSA/PLA/carvacrol or thymol	Film wrapping	Salmon slice	4 °C, 3–4 days	Extended the shelf life by an additional 3–4 days while in cold storage	Yang et al. (2019)
PBS/PBSA	Film wrapping	Raw chicken and turkey meat	4 °C, 5 days for raw poultry and 15 days for smoked poultry	Prolonged shelf life	Vytejčková et al. (2017)

combined with gamma irradiation, these films help maintain weight and preserve phenols and anthocyanins over 12 days. Starch coatings retain firmness in bell peppers, PLA films manage moisture in fresh-cut pineapples, and cellulose casings extend cucumber shelf life (Ali et al. 2010; Sothornvit and Krochta 2001). Biopolymers like polysaccharides and proteins can sustain the quality of fresh produce and decompose after disposal.

Dos Santos Pires et al. developed antimicrobial films with nisin and natamycin in cellulose polymer, which extended the shelf life of sliced mozzarella by 6 days compared to controls (dos Santos Pires et al. 2008). Gammariello et al. created bio-based polyethylene packaging with ginger essential oil coatings, successfully reducing microbial counts in cheese after 60 days of refrigerated storage. Fajardo et al. studied edible films with natamycin for Queso Blanco and found them effective for 28 days against yeasts and molds (Gammariello et al. 2008). Scannell et al. evaluated cellulose-based paper packaging with nisin between cheddar cheese slices under modified atmosphere packaging. Their findings showed that nisin inserts reduced Listeria innocua by $\geq$2 log units and Staphylococcus aureus by $\sim$1.5 log units, significantly extending shelf life compared to control packaging without nisin (Scannell et al. 2000). Holm et al. studied PLA packaging for Danbo cheese and found that moisture loss was 10 times greater in PLA compared to polyethylene, although dry spots didn't form until 56 days. They noted that lipid oxidation was limited in PLA for 56 days in light and negligible after 84 days in the dark. Additionally, they discovered that while oxygen scavengers reduced oxidation, more occurred in PLA due to its higher oxygen transmission rate, emphasizing the need for moisture loss protection and dark storage (Holm et al. 2006). Research indicates that biodegradable packaging like PLA and bio-based polyethylene could serve as alternatives to conventional dairy packaging. Incorporating antimicrobials can enhance their properties, though challenges such as moisture loss and lipid oxidation remain. Further research is needed to optimize these materials for shelf-life extension while ensuring product durability before biodegrading.

1.4 Challenges, Limitations, and Safety Aspects

Biopolymer materials present a promising alternative for food packaging solutions, offering a sustainable path away from traditional petroleum-based products. However, several critical factors must be comprehensively addressed to attract investment and meet the evolving demands of the marketplace. Firstly, a significant barrier to the widespread adoption of biobased materials is the regulatory approval process. Many such materials have yet to receive clearance from the Food and Drug Administration (FDA) for commercial use within food packaging. This lack of approval creates uncertainty for manufacturers and investors who are hesitant to engage with products that may not meet federal safety standards. Secondly, the production of biobased materials often involves innovative methods, such as the genetic engineering of source materials, which can lead to public apprehension regarding their safety and ethical implications. These concerns underscore the importance of transparent communication and education about the safety and benefits of biobased products. In addition to regulatory and public acceptance challenges, extensive safety evaluations are vital (Mishra et al. 2025a, b; Nayak et al. 2025; Perumal et al. 2025). This includes conducting in vivo degradation studies to assess how these materials break down in real-world conditions, as well as evaluating the safety of any metabolites that may be produced during degradation. Thorough toxicity assessments are equally necessary to ensure that these materials do not pose health risks to consumers. To facilitate the immediate commercial viability of biobased food packaging, a multifaceted approach is required. This approach should encompass process evaluations, comprehensive safety assessments, toxicological evaluations, and adherence to regulatory frameworks. Collecting consumer feedback on packaged food products is essential for understanding acceptance and preferences. This will help create a more tailored approach to product development. Additionally, cost is a significant challenge; the high production costs of certain biobased materials require dedicated efforts to optimize processes (Bhathiwal et al. 2025; Deng et al. 2025; dos Santos Lima et al. 2025; Yang et al. 2024). Exploring alternative raw materials, which may be more cost-effective while still providing the desired properties, can enhance the economic feasibility of these innovative solutions. In terms of functionality, biobased materials demonstrate considerable potential to replace traditional packaging options. Specific additives have shown beneficial properties, enabling them to act as effective packaging and coating agents. These materials not only protect against microbial contamination but do so in a manner that remains non-toxic and safe for consumers. As the food industry increasingly seeks scalable solutions, the development of packaging formats specifically designed for large-scale production becomes essential, particularly for coating applications. Moreover, the integration of natural materials with other additives, such as plant oils or yeast, expands the horizons of biobased materials beyond food packaging into diverse sectors like cosmetics and pharmaceuticals. The potential for creating functional materials is particularly exciting; for example, the development of biosensors utilizing various bionanomaterials can greatly enhance the food industry's capacity for early detection of pathogens and contaminants, thereby improving food safety (Tan et al. 2024; Thiyagamoorthy et al. 2024; Wardejn et al. 2024).

Another innovative area lies in the formulation of composites created from biopolymers combined with both natural and synthetic additives. These composites offer diverse functionalities and applications beyond traditional membranes. Advancements in synthesis techniques, including 3D bioprinting and regeneration technologies, hold the promise of creating packaging materials that boast superior properties and capabilities. In situ and ex situ processing methods, along with the development of cell-free enzyme systems, can further elevate the functionality of these biobased materials. While the potential of biobased materials for food packaging is immense, a thorough, multi-disciplinary approach encompassing safety, regulatory concerns, public perception, and cost optimization is essential for their successful commercialization. By addressing these critical factors, the food industry can embrace a more sustainable future through innovative packaging solutions (Sahraeian et al. 2024; Tan et al. 2024).

From a regulatory perspective, the current landscape regarding bio-based materials and intelligent packaging in the food sector is characterized by a significant lack of harmonized standards and guidelines, which creates considerable uncertainty for both manufacturers and consumers alike. One of the primary challenges in this sector is the need for transparent and reliable compostability certification processes. Certifications are essential in ensuring that these materials can indeed break down safely in composting environments. There are pressing food safety issues that arise with the introduction of innovative materials, which may not have been thoroughly tested or classified under existing regulations. This can lead to concerns about how these new packaging technologies may interact with food products, ultimately jeopardizing public health (Iqbal et al. 2025; Yang et al. 2024).

Challenges are further compounded by the complex waste management strategies required for packaging solutions that combine biodegradable components with electronic features—such as sensors or indicators that monitor food quality. The integration of these technologies necessitates consistent regulatory frameworks to ensure that they not only meet compostability criteria but also do not disrupt existing waste management systems. Furthermore, manufacturers face the dual burden of adhering to stringent environmental standards while simultaneously guaranteeing the safety, efficacy, and performance of their packaging solutions. The role of consumer acceptance cannot be underestimated, as it is pivotal for the successful implementation and widespread adoption of smart packaging technologies (Leya et al. 2024; Ma et al. 2024; Sukumaran and Jano 2024). While these advanced systems promise substantial benefits by enhancing food quality tracking, ensuring safety through monitoring, and potentially reducing waste, significant challenges related to consumer trust and understanding persist. For instance, consumers often express concerns about the financial implications of adopting such technologies, particularly if they lead to increased product prices. There is a critical need for packaging designs that effectively communicate important information to consumers in a manner that is both intuitive and accessible. To foster broader acceptance and use, upcoming research should focus on exploring innovative strategies aimed at building consumer confidence in smart packaging technologies, effectively navigating complex regulatory frameworks, and maximizing the multifunctionality and usability of smart packaging solutions to secure their success in the market. Market dynamics play a crucial role in shaping the future of bio-based materials in the food packaging sector. One of the most significant obstacles to widespread adoption is the relatively high cost associated with bio-based materials compared to traditional plastics. This price disparity presents a formidable challenge, especially in price-sensitive sectors such as food production, where cost-effectiveness is critical. Despite an increasing consumer appetite for sustainable alternatives, conventional plastics continue to dominate due to their affordability, well-established supply chains, and widespread availability (Leya et al. 2024; Mishra et al. 2025a, b; Przybyszewska and Galus 2024; Rahman et al. 2024). Moreover, the existing infrastructure for recycling and composting bio-based materials remains limited, particularly in regions where waste management systems are underdeveloped or lacking altogether. Looking ahead, the future of packaging within the food industry is poised for transformative advancements. As technological innovation continues to progress, it is anticipated that the costs associated with bio-based materials will decrease, enhancing their competitiveness with traditional plastics. Developments in nanotechnology and material sciences are likely to lead to substantial improvements in the mechanical strength, gas barrier properties, and performance of bio-based materials. This progress is expected to broaden their application across high-demand sectors, particularly in the packaging of perishable goods, where the combination of functionality and sustainability will be increasingly sought after by manufacturers and consumers alike.

1.5 Conclusion and Future Prospects

Integrating bio-based polymer films into food packaging technology presents a promising approach, offering biodegradable and environmentally friendly alternatives to conventional petrochemical plastics. This strategy directly tackles the pressing issue of waste accumulation that results from the widespread use of non-biodegradable plastics, which contribute significantly to environmental pollution. This chapter reviews the latest developments in biomaterials specifically designed for food packaging, emphasizing their key properties, performance criteria, and potential applications. Various bio-based materials—including biobased polymers, biobased nanomaterials, biobased fibers, and their composites—have demonstrated efficacy in preserving food quality. These materials are used in various forms within food packaging, such as coatings, films, hydrogels, aerogels, and emulsions. The chapter thoroughly examines several critical properties of these bio-based materials, including mechanical strength, gas barrier effectiveness, moisture permeability, and antimicrobial properties, assessing their suitability for various food preservation applications. Additionally, the discussion covers the functional attributes of these materials, particularly the efficiency and potential of active and smart packaging solutions aimed at extending shelf life and enhancing food safety.

Continued research and development in the biomaterials field for food packaging should prioritize enhancing the effectiveness and practicality of these materials. This need arises from globalization and the growing demand for efficient solutions in food preservation and transportation. Various key factors must be evaluated when considering the implementation of biobased materials in real-world scenarios. For optimal results, the selection of the appropriate biobased materials is vital, and their properties must be finely tuned through physical or chemical modification techniques, tailored specifically to meet the unique storage conditions required for diverse food products. Despite the substantial progress in the development and application of biobased materials within the food packaging sector, several challenges continue to impede their wide-scale commercialization. A primary hurdle lies in identifying methods to lower the costs and energy consumption linked to the production and processing of these materials. This encompasses the extraction of biopolymers, isolation of biobased nanomaterials, and the comprehensive preparation of bio-based materials suited for food packaging applications. Consequently, there is an urgent need for the development and dissemination of multifunctional commercial technologies and equipment capable of facilitating large-scale production processes of biobased materials. Another critical area for future investigation involves enhancing our understanding of the biodegradation processes and sustainability profiles of these materials. This knowledge is essential not only for minimizing their environmental impact once they are discarded but also for adhering to the principles of the circular economy. More intensive efforts are required to innovate greener methodologies that mitigate or eliminate the reliance on toxic organic solvents commonly used in the production processes. Food safety remains a paramount concern in the development of certain biobased materials for food packaging. One significant challenge relates to the potential use of harmful solvents or hazardous production processes that could compromise the safety of the packaging materials. Therefore, particular attention must be devoted to active and intelligent packaging solutions, especially those that incorporate nano-technology. It is vital to quantitatively assess the migration potential of substances from packaging into food products and ensure compliance with specific food safety regulations designed to protect consumer health. The deployment of advanced detection technologies will be crucial in identifying the presence of any migrating substances, thereby ensuring the safety of packaged foods. Research efforts are increasingly focused on optimizing the composite structures of mixed materials, as the effectiveness of biopolymeric films is heavily dependent on the specific polymers and additive materials used to enhance their functional properties. Further investigations are necessary to enhance the physical and chemical characteristics of bioplastics, enabling them to align with the performance metrics of traditional petroleum-based plastics more closely. The ambition of replacing synthetic plastics entirely with bio-based alternatives remains a long-term goal, necessitating continued exploration, increased awareness, and thoughtful consideration within future research initiatives. By addressing these complexities, the potential for widespread adoption of sustainable, biodegradable, and high-performance biopolymer films in food packaging can be realized.

Acknowledgements This work was supported by the Student Research Committee, Yasuj University of Medical Sciences, Yasuj, Iran.

Declaration The authors declare that AI tools were used in the preparation of this manuscript for the purpose of improving the readability and clarity. All content was critically reviewed and verified by the authors to ensure accuracy and originality. The authors take full responsibility for the final content of this work.

References

Abdelwahab MA, Flynn A, Chiou B-S, Imam S, Orts W, Chiellini E (2012) Thermal, mechanical and morphological characterization of plasticized PLA–PHB blends. Polym Degrad Stability 97(9):1822–1828

Abedi-Firoozjah R, Bahramian B, Tavassoli M, Ahmadi N, Noori SMA, Hashemi M et al (2025) A comprehensive review of gum-based electrospun nanofibers for food packaging: preparation, developments, and potential applications. Int J Biol Macromol 288:138717. https://doi.org/10.1016/j.ijbiomac.2024.138717

Adeyeye OA, Sadiku ER, Babu Reddy A, Ndamase AS, Makgatho G, Sellamuthu PS et al (2019) The use of biopolymers in food packaging. In: Materials horizons: from nature to nanomaterials, pp 137–158

Agarwal A, Shaida B, Rastogi M, Singh NB (2023) Food packaging materials with special reference to biopolymers-properties and applications. Chem Africa 6(1):117–144. https://doi.org/10.1007/s42250-022-00446-w

Ali A, Maqbool M, Ramachandran S, Alderson PG (2010) Gum Arabic as a novel edible coating for enhancing shelf-life and improving postharvest quality of tomato (Solanum lycopersicum L.) fruit. Postharvest Biol Technol 58(1):42–47. https://doi.org/10.1016/j.postharvbio.2010.05.005

Aliabbasi N, Faraji B, Emam-Djomeh Z, Salami M, Askari G (2022) Biopolymer-based active and intelligent packaging for food applications. In: Biodegradable polymer-based food packaging, pp 245–275

Alizadeh Sani M, Ehsani A, Hashemi M (2017) Whey protein isolate/cellulose nanofibre/TiO_2 nanoparticle/rosemary essential oil nanocomposite film: its effect on microbial and sensory quality of lamb meat and growth of common foodborne pathogenic bacteria during refrigeration. Int J Food Microbiol 251:8–14. https://doi.org/10.1016/j.ijfoodmicro.2017.03.018

Alizadeh-Sani M, Mohammadian E, Rhim JW, Jafari SM (2020) PH-sensitive (halochromic) smart packaging films based on natural food colorants for the monitoring of food quality and safety. Trends Food Sci Technol 105:93–144. https://doi.org/10.1016/j.tifs.2020.08.014

Alizadeh-Sani M, Tavassoli M, Mohammadian E, Ehsani A, Khaniki GJ, Priyadarshi R, Rhim J-W (2021) PH-responsive color indicator films based on methylcellulose/chitosan nanofiber and barberry anthocyanins for real-time monitoring of meat freshness. Int J Biol Macromol 166:741–750. https://doi.org/10.1016/j.ijbiomac.2020.10.231

Amiri Samani S, PourvatanDoust S, Savarolyia M, Aboutalebzadeh S, Khezri M, Kazemi M et al (2025) Valorization of red grape pomace for sustainable food packaging: development of pectin/kidney bean protein based biocomposite films enriched with grape pomace polyphenols. Food Hydrocolloids 160:110806. https://doi.org/10.1016/j.foodhyd.2024.110806

Andini R, Sulaiman MI, Umam AH, Olivia M, Endres HJ (2021) Biopolymer nanocomposites: their mechanical, thermal, and gas barrier properties for food packaging. IOP Conf Ser: Earth Environ Sci

Armentano I, Fortunati E, Burgos N, Dominici F, Luzi F, Fiori S et al (2015) Bio-based PLA_PHB plasticized blend films: processing and structural characterization. LWT-Food Sci Technol 64(2):980–988

Arrieta MP, Samper MD, López J, Jiménez A (2014) Combined effect of poly (hydroxybutyrate) and plasticizers on polylactic acid properties for film intended for food packaging. J Polym Environ 22:460–470

Ashrafi A, Jokar M, Mohammadi Nafchi A (2018) Preparation and characterization of biocomposite film based on chitosan and kombucha tea as active food packaging. Int J Biol Macromol 108:444–454. https://doi.org/10.1016/j.ijbiomac.2017.12.028

Bahramian B, Abedi-Firoozjah R, Ebrahimi A, Tavassoli M, Ehsani A, Naebe M (2024a) Resveratrol-biopolymer materials: a sustainable approach to food packaging. Trends Food Sci Technol 154. https://doi.org/10.1016/j.tifs.2024.104761

Bahramian B, Abedi-Firoozjah R, Kiani-Salmi N, Azizian A, Dorud N, Mohammad Ali Noori S et al (2024b) Multifunctional performance of gallic acid in biodegradable food packaging films

and coatings: Mechanisms, developments, applications, and horizons. Eur Polym J 221:113559. https://doi.org/10.1016/j.eurpolymj.2024.113559
Baishya H, Kalita S, Mukherjee A, Kumar S (2024) Biopolymer-based nanocomposites for intelligent food packaging. In: Smart food packaging systems: smart food packaging systems, pp 387–414
Bakhshizadeh M, Ayaseh A, Hamishehkar H, Samadi Kafil H, Niknazar Moghaddam T, Baghban Haghi P et al (2023a) Multifunctional performance of packaging system based on gelatin/alove vera gel film containing of rosemary essential oil and common poppy anthocyanins. Food Control 154:110017. https://doi.org/10.1016/j.foodcont.2023.110017
Bakhshizadeh M, Moghaddam TN, Tavassoli M, Ayaseh A, Bangar SP (2023b) Gelatin/chitosan nanofibres containing β-cyclodextrin complex and corn poppy (Papaver rhoeas L.) for intelligent packaging. Int J Food Sci Technol 58(5):2360–2368. https://doi.org/10.1111/ijfs.16369
Balan A, Kadeppagari RK (2021) Biopolymers for Nano-enabled packaging of foods. In: Handbook of polymer and eramic nanotechnology, vol 2, pp 839–854
Barra A, Santos JDC, Silva MRF, Nunes C, Ruiz-Hitzky E, Gonçalves I et al (2020) Graphene derivatives in biopolymer-based composites for food packaging applications. Nanomaterials 10(10):1–32. https://doi.org/10.3390/nano10102077
Basavegowda N, Baek KH (2021) Advances in functional biopolymer-based nanocomposites for active food packaging applications. Polymers 13(23). https://doi.org/10.3390/polym13234198
Bayer IS (2021) Biopolymers in multilayer films for long-lasting protective food packaging: a review. In: Sustainable food packaging technology, pp 397–426
Bayer IS, Athanassiou A, Cingolani R (2015) Biopolymers for food packaging. Food Sci Technol (London) 29(3):16–19
Bharathi SKV, Leena MM, Moses JA, Anandharamakrishnan C (2020) Nanofibre-based bilayer biopolymer films: enhancement of antioxidant activity and potential for food packaging application. Int J Food Sci Technol 55(4):1477–1484. https://doi.org/10.1111/ijfs.14492
Bhardwaj U, Dhar P, Kumar A, Katiyar V (2014) Polyhydroxyalkanoates (PHA)-cellulose based nanobiocomposites for food packaging applications. In Food additives and packaging. ACS Publications, pp 275–314
Bhaskar R, Zo SM, Narayanan KB, Purohit SD, Gupta MK, Han SS (2023) Recent development of protein-based biopolymers in food packaging applications: a review. Polym Testing 124. https://doi.org/10.1016/j.polymertesting.2023.108097
Bhathiwal AS, Rao GBD, Bendi A, Tiwari A (2025) protein-based biopolymers of plant origin for food packaging. In: Green biopolymers for packaging applications, pp 272–302
Bhatia S, Jawad M, Chinnam S, Al-Harrasi A, Shah YA, Khan TS et al (2025) Development and characterization of potato starch–pectin-based active films enriched with juniper berry essential oil for food packaging applications. Food Sci Nutr 13(1):e4688. https://doi.org/10.1002/fsn3.4688
Bhattacharya A (2015) Biopolymer based paper coatings towards sustainable packaging. IPPTA: Q J Indian Pulp Pap Tech Assoc 27(2):63–70
Bindu J, Sathish Kumar K, Panda SK, Katiyar V (2019) Biopolymer dispersed poly lactic acid composites and blends for food packaging applications. In: Materials horizons: from nature to nanomaterials, pp 209–235
Brandelero RPH, Brandelero EM, Almeida FMD (2016) Biodegradable films of starch/PVOH/alginate in packaging systems for minimally processed lettuce (Lactuca sativa L.). Ciência e Agrotecnologia 40(5):510–521
Calvao PS, Chenal JM, Gauthier C, Demarquette NR, Bogner A, Cavaille JY (2012) Understanding the mechanical and biodegradation behaviour of poly (hydroxybutyrate)/rubber blends in relation to their morphology. Polym Int 61(3):434–441
Carvalho APAD, Conte-Junior CA (2021) Food-derived biopolymer Kefiran composites, nanocomposites and nanofibers: emerging alternatives to food packaging and potentials in nanomedicine. Trends Food Sci Technol 116:370–386. https://doi.org/10.1016/j.tifs.2021.07.038

Chandrababu V, Parameswaranpillai J, Gopi JA, Pathak C, Midhun Dominic CD, Feng NL et al (2024) Progress in food packaging applications of biopolymer-nanometal composites—A comprehensive review. Biomater Adv 162. https://doi.org/10.1016/j.bioadv.2024.213921

Cheng K, Du J, Xu F, Wang Z, Zhang L, Bai M et al (2024) Preparation and characterization of zein-based active packaging films integrated with TiO_2 nanotube arrays. Food Packag Shelf Life 45:101348. https://doi.org/10.1016/j.fpsl.2024.101348

Cheng M, Wang J, Zhang R, Kong R, Lu W, Wang X (2019) Characterization and application of the microencapsulated carvacrol/sodium alginate films as food packaging materials. Int J Biol Macromol 141:259–267. https://doi.org/10.1016/j.ijbiomac.2019.08.215

Cherpinski A, Torres-Giner, S., Cabedo, L., Méndez, J. A., & Lagaron, J. M. (2018). Multilayer structures based on annealed electrospun biopolymer coatings of interest in water and aroma barrier fiber-based food packaging applications. J Appl Polym Sci 135(24). https://doi.org/10.1002/app.45501

Chi S, Zivanovic S, Penfield MP (2006) Application of chitosan films enriched with oregano essential oil on bologna—Active compounds and sensory attributes. Food Sci Technol Int 12(2):111–117. https://doi.org/10.1177/1082013206063845

Chia MR, Ahmad I, Phang SW (2022) Starch/polyaniline biopolymer film as potential intelligent food packaging with colourimetric ammonia sensor. Polymers 14(6). https://doi.org/10.3390/polym14061122

Chia MR, Phang SW, Ahmad I (2023) Influence of polyaniline and cellulose nanocrystals on starch biopolymer film for intelligent food packaging. Food Biosci 56. https://doi.org/10.1016/j.fbio.2023.103212

Chiu I, Yang T (2024) Biopolymer-based intelligent packaging integrated with natural colourimetric sensors for food safety and sustainability. Anal Sci Ad 5(5–6). https://doi.org/10.1002/ansa.202300065

Das P, Nayak PK, Kesavan RK (2024) Biopolymers for food packaging. In: ACS symposium series, vol 1486, pp 133–154

de Matos Fonseca J, Koop BL, Trevisol TC, Capello C, Monteiro AR, Valencia GA (2021) An overview of biopolymers in food packaging systems. In: Nanotechnology-enhanced food packaging, pp 19–53

Dedieu I, Peyron S, Gontard N, Aouf C (2022) The thermo-mechanical recyclability potential of biodegradable biopolyesters: perspectives and limits for food packaging application. Polym Test 111:107620. https://doi.org/10.1016/j.polymertesting.2022.107620

Defieuw G, Groeninckx G, Reynaers H (1989) Miscibility, crystallization and melting behaviour, and semicrystalline morphology of binary blends of polycaprolactone with poly (hydroxy ether of bisphenol A). Polymer 30(12):2164–2169

Dejene BK, Gudayu AD (2024) Eco-friendly packaging innovations: integrating natural fibers and ZnO nanofillers in polylactic acid (PLA) based green composites—A review. Polym-Plastics Technol Mater 63(12):1645–1681. https://doi.org/10.1080/25740881.2024.2350686

Demchenko V, Iurzhenko M, Rybalchenko N, Marynin A, Demchenko O, Sytnyk I, Shtepa D (2024) Formation of antimicrobial packaging biopolymer nanocomposite materials for food storage. In: Proceedings of the 2024 IEEE 14th international conference "Nanomaterials: Applications and Properties", NAP 2024

Deng H, Zhang W, Ramezan Y, Riahi Z, Khan A, Huang Z (2025) Antibacterial and antioxidant plant-derived aldehydes: a new role as cross-linking agents in biopolymer-based food packaging films. Compr Rev Food Sci Food Saf 24(1). https://doi.org/10.1111/1541-4337.70089

Devi LS, Das B, Dutta D, Kumar S (2024) Essential oils as functional agents in biopolymer-based sustainable food packaging system: a review. Sustain Chem Pharm 39. https://doi.org/10.1016/j.scp.2024.101563

Dong X, Zhang J, Pang L, Chen J, Qi M, You S, Ren N (2019) An anisotropic three-dimensional electrospun micro/nanofibrous hybrid PLA/PCL scaffold. RSC Adv 9(17):9838–9844

Dorati R, Pisani S, Maffeis G, Conti B, Modena T, Chiesa E et al (2018) Study on hydrophilicity and degradability of chitosan/polylactide-co-polycaprolactone nanofibre blend electrospun membrane. Carbohydr Polym 199:150–160

dos Santos Lima K, de Matos Fonseca J, Monteiro A, Valencia G (2025) Progress in biopolymer coatings for achieving sustainable paper and paperboard packaging for food applications: a brief review. Packag Technol Sci 38(2):109–116. https://doi.org/10.1002/pts.2861

dos Santos Pires AC, de Fátima Ferreira Soares N, de Andrade NJ, da Silva LHM, Camilloto GP, Bernardes PC (2008) Development and evaluation of active packaging for sliced mozzarella preservation. Pack Technol Sci 21(7):375–383. https://doi.org/10.1002/pts.815

Ezati P, Khan A, Priyadarshi R, Bhattacharya T, Tammina SK, Rhim JW (2023) Biopolymer-based UV protection functional films for food packaging. Food Hydrocoll 142. https://doi.org/10.1016/j.foodhyd.2023.108771

Fahmy HM, Salah Eldin RE, Abu Serea ES, Gomaa NM, AboElmagd GM, Salem SA et al (2020) Advances in nanotechnology and antibacterial properties of biodegradable food packaging materials. RSC Adv 10(35):20467–20484. https://doi.org/10.1039/D0RA02922J

Freitas IR, Cortez-Vega WR, Pizato S, Prentice-Hernández C, Borges CD (2013) Xanthan gum as a carrier of preservative agents and calcium chloride applied on fresh-cut apple. J Food Saf 33(3):229–238. https://doi.org/10.1111/jfs.12044

Gammariello D, Di Giulio S, Conte A, Del Nobile MA (2008) Effects of natural compounds on microbial safety and sensory quality of Fior di Latte Cheese, a Typical Italian Cheese. J Dairy Sci 91(11):4138–4146. https://doi.org/10.3168/jds.2008-1146

Garavand F, Rouhi M, Razavi SH, Cacciotti I, Mohammadi R (2017) Improving the integrity of natural biopolymer films used in food packaging by crosslinking approach: a review. Int J Biol Macromol 104:687–707. https://doi.org/10.1016/j.ijbiomac.2017.06.093

Ghosh T, Priyadarshi R, Krebs de Souza C, Angioletti BL, Rhim JW (2022) Advances in pullulan utilization for sustainable applications in food packaging and preservation: a mini-review. Trends Food Sci Technol 125:43–53. https://doi.org/10.1016/j.tifs.2022.05.001

Gomes D, Batista-Silva JP, Sousa A, Passarinha LA (2023) Progress and opportunities in Gellan gum-based materials: a review of preparation, characterization and emerging applications. Carbohyd Polym 311:120782. https://doi.org/10.1016/j.carbpol.2023.120782

Guerrero P, Nur Hanani ZA, Kerry JP, de la Caba K (2011) Characterization of soy protein-based films prepared with acids and oils by compression. J Food Eng 107(1):41–49. https://doi.org/10.1016/j.jfoodeng.2011.06.003

Han Y, Yu M, Wang L (2018) Preparation and characterization of antioxidant soy protein isolate films incorporating licorice residue extract. Food Hydrocoll 75:13–21. https://doi.org/10.1016/j.foodhyd.2017.09.020

Holm VK, Ndoni S, Risbo J (2006) The stability of poly(lactic acid) packaging films as influenced by humidity and temperature. J Food Sci 71(2):E40–E44. https://doi.org/10.1111/j.1365-2621.2006.tb08895.x

Hu Y, Sato H, Zhang J, Noda I, Ozaki Y (2008) Crystallization behavior of poly (l-lactic acid) affected by the addition of a small amount of poly (3-hydroxybutyrate). Polymer 49(19):4204–4210

Indumathi MP, Saral Sarojini K, Rajarajeswari GR (2019) Antimicrobial and biodegradable chitosan/cellulose acetate phthalate/ZnO nano composite films with optimal oxygen permeability and hydrophobicity for extending the shelf life of black grape fruits. Int J Biol Macromol 132:1112–1120. https://doi.org/10.1016/j.ijbiomac.2019.03.171

Iqbal Y, Ameer ME, Ahmed I, Ullah A, Irfan MF, Shahid Chatha SA (2025) Green biopolymers in food packaging. In: Green biopolymers for packaging applications, pp 93–148

Jandas P, Mohanty S, Nayak SK (2014) Morphology and thermal properties of renewable resource-based polymer blend nanocomposites influenced by a reactive compatibilizer. ACS Sustain Chem Eng 2(3):377–386

Jayakodi S, Kim H, Menon S, Shanmugam VK, Choi I, Sekhar MR et al (2022) Preparation of novel nanoformulation to enhance efficacy in the treatment of cardiovascular disease. Biomimetics 7(4). https://doi.org/10.3390/biomimetics7040189

Kaewprachu P, Osako K, Benjakul S, Rawdkuen S (2015) Quality attributes of minced pork wrapped with catechin–lysozyme incorporated gelatin film. Food Packag Shelf Life 3:88–96. https://doi.org/10.1016/j.fpsl.2014.11.002

Kandasamy S, Yoo J, Yun J, Kang H-B, Seol K-H, Kim H-W, Ham J-S (2021) Application of whey protein-based edible films and coatings in food industries: an updated overview. Coatings 11(9). https://doi.org/10.3390/coatings11091056

Kanatt SR, Chander R, Sharma A (2010) Antioxidant and antimicrobial activity of pomegranate peel extract improves the shelf life of chicken products. Int J Food Sci Technol 45(2):216–222. https://doi.org/10.1111/j.1365-2621.2009.02124.x

Kedir WM, Geletu AK, Weldegirum GS (2024) Spider web-reinforced chitosan/starch biopolymer for active biodegradable food packaging. Appl Food Res 4(2). https://doi.org/10.1016/j.afres.2024.100526

Keshavarz T, Roy I (2010) Polyhydroxyalkanoates: bioplastics with a green agenda. Curr Opin Microbiol 13(3):321–326

Khan A, Huq T, Khan RA, Riedl B, Lacroix M (2014) Nanocellulose-based composites and bioactive agents for food packaging. Crit Rev Food Sci Nutr 54(2):163–174. https://doi.org/10.1080/10408398.2011.578765

Khezerlou A, Alizadeh Sani M, Tavassoli M, Abedi-Firoozjah R, Ehsani A, McClements DJ (2023) Halochromic (pH-Responsive) indicators based on natural anthocyanins for monitoring fish freshness/spoilage. J Compos Sci 7(4). https://doi.org/10.3390/jcs7040143

Kirac FT, Dagdelen AF, Saricaoglu FT (2022) Recent advances in polylactic acid biopolymer films used in food packaging systems. J Food Nutr Res 61(1):1–15

Koirala P, Sagar NA, Thuanthong A, Al-Asmari F, Jagtap S, Nirmal N (2025) Revolutionizing seafood packaging: advancements in biopolymer smart nano-packaging for extended shelf-life and quality assurance. Food Res Int 203. https://doi.org/10.1016/j.foodres.2025.115826

Kraśniewska K, Pobiega K, Gniewosz M (2019) Pullulan-biopolymer with potential for use as food packaging. Int J Food Eng 15(9). https://doi.org/10.1515/ijfe-2019-0030

Kumar S, Mukherjee A, Dutta J (2022a) Biopolymer-based food packaging: innovations and technology applications

Kumar S, Mukherjee A, Kalita S, Singh N, Katiyar V, Mitra A, Halder D (2022b) Nanomaterials in food packaging. In: Biopolymer-based food packaging: innovations and technology applications, pp 336–367

Leya B, Shan Franklin R, Pragalyaashree MM, Asha Monicka A, Tiroutchelvame D, Blessy C, Freeda Blessie R (2024) Biopolymer-based edible packaging: a critical review on the biomaterials, formation, and applications on food products. J Appl Biol Biotechnol 12(6):42–57. https://doi.org/10.7324/JABB.2024.145531

Liao J, Zhou Y, Hou B, Zhang J, Huang H (2023) Nano-chitin: Preparation strategies and food biopolymer film reinforcement and applications. Carbohyd Polym 305. https://doi.org/10.1016/j.carbpol.2023.120553

Llana-Ruiz-Cabello M, Pichardo S, Bermúdez JM, Baños A, Núñez C, Guillamón E et al (2016) Development of PLA films containing oregano essential oil (Origanum vulgare L. virens) intended for use in food packaging. Food Addit Contamin Part A 33(8):1374–1386. https://doi.org/10.1080/19440049.2016.1204666

Llana-Ruíz-Cabello M, Puerto M, Pichardo S, Jiménez-Morillo NT, Bermúdez JM, Aucejo S et al (2019) Preservation of phytosterol and PUFA during ready-to-eat lettuce shelf-life in active bio-package. Food Packag Shelf Life 22:100410. https://doi.org/10.1016/j.fpsl.2019.100410

Liu Y, Ahmed S, Sameen DE, Wang Y, Lu R, Dai J et al (2021) A review of cellulose and its derivatives in biopolymer-based for food packaging application. Trends Food Sci Technol 112:532–546. https://doi.org/10.1016/j.tifs.2021.04.016

Liu Y, Yang M, Wang J, Yue H, Du Z, Cheng X et al (2025) Ag-CNC nanoparticles decorated thermoplastic starch/PLA/PBAT composite film with superior antibacterial property, enhanced barrier capacity and high UV-shielding effect. Ind Crops Prod 225:120510. https://doi.org/10.1016/j.indcrop.2025.120510

Ma K, Li F, Zhe T, Sun X, Zhang X, Wan P et al (2024) Biopolymer films incorporated with chlorogenic acid nanoparticles for active food packaging application. Food Chem 435. https://doi.org/10.1016/j.foodchem.2023.137552

Mallakpour S, Sirous F, Hussain CM (2021) A journey to the world of fascinating ZnO nanocomposites made of chitosan, starch, cellulose, and other biopolymers: Progress in recent achievements in eco-friendly food packaging, biomedical, and water remediation technologies. Int J Biol Macromol 170:701–716. https://doi.org/10.1016/j.ijbiomac.2020.12.163

Manikandan NA, Pakshirajan K, Pugazhenthi G (2020) Preparation and characterization of environmentally safe and highly biodegradable microbial polyhydroxybutyrate (PHB) based graphene nanocomposites for potential food packaging applications. Int J Biol Macromol 154:866–877. https://doi.org/10.1016/j.ijbiomac.2020.03.084

Marcoaldi C, Acampora V, Venezia V, Prieto C, Grappa R, Silvestri B et al (2025) Challenges in lignin integration within biopolymer matrices: toward stable and effective lignin nanoparticles as additives for sustainable food packaging. Ind Crops Prod 224. https://doi.org/10.1016/j.indcrop.2024.120336

Masood F (2017) Chapter 8—Polyhydroxyalkanoates in the food packaging industry. In: Oprea AE, Grumezescu AM (eds) Nanotechnology applications in food. Academic Press, pp 153–177

Matheus JRV, Dalsasso RR, Rebelatto EA, Andrade KS, Andrade LMD, Andrade CJD et al (2023) Biopolymers as green-based food packaging materials: a focus on modified and unmodified starch-based films. Compr Rev Food Sci Food Safety 22(2):1148–1183. https://doi.org/10.1111/1541-4337.13107

Maurizzi E, Bigi F, Quartieri A, De Leo R, Volpelli LA, Pulvirenti A (2022) The green era of food packaging: general considerations and new trends. Polymers 14(20). https://doi.org/10.3390/polym14204257

Mellinas C, Ramos M, Jiménez A, Garrigós MC (2020) Recent trends in the use of pectin from agro-waste residues as a natural-based biopolymer for food packaging applications. Materials 13(3). https://doi.org/10.3390/ma13030673

Mishra N, Singh RK, Varshney B (2025a) Biopolymers as antimicrobial food packaging. In: Green biopolymers for packaging applications, pp 365–377

Mishra P, Sharma N, Mishra A, Gupta PC (2025b) Green biopolymers in food packaging: general consideration and novel trends. In: Green Biopolymers for packaging applications, pp 193–211

Muppalla SR, Kanatt SR, Chawla SP, Sharma A (2014) Carboxymethyl cellulose–polyvinyl alcohol films with clove oil for active packaging of ground chicken meat. Food Packag Shelf Life 2(2):51–58. https://doi.org/10.1016/j.fpsl.2014.07.002

Mushtaq M, Gani A, Gani A, Punoo HA, Masoodi FA (2018) Use of pomegranate peel extract incorporated zein film with improved properties for prolonged shelf life of fresh Himalayan cheese (Kalari/kradi). Innov Food Sci Emerg Technol 48:25–32. https://doi.org/10.1016/j.ifset.2018.04.020

Mohamed SAA, El-Sakhawy M, El-Sakhawy MA-M (2020) Polysaccharides, protein and lipid-based natural edible films in food packaging: a review. Carbohyd Polym 238:116178. https://doi.org/10.1016/j.carbpol.2020.116178

Mohammadi H, Kamkar A, Misaghi A, Zunabovic-Pichler M, Fatehi S (2019) Nanocomposite films with CMC, okra mucilage, and ZnO nanoparticles: extending the shelf-life of chicken breast meat. Food Packag Shelf Life 21:100330. https://doi.org/10.1016/j.fpsl.2019.100330

Nagarajan M, Benjakul S, Prodpran T, Songtipya P (2015) Effects of bio-nanocomposite films from tilapia and squid skin gelatins incorporated with ethanolic extract from coconut husk on storage stability of mackerel meat powder. Food Packag Shelf Life 6:42–52. https://doi.org/10.1016/j.fpsl.2015.09.001

Nanda A, Pandey P, Rajinikanth PS, Singh N (2024) Revolution of nanotechnology in food packaging: harnessing electrospun zein nanofibers for improved preservation—A review. Int J Biol Macromol 260:129416. https://doi.org/10.1016/j.ijbiomac.2024.129416

Naskar A, Sanyal I, Nahar N, Ghosh DD, Chakraborty S (2023) Bionanocomposites films applied as active and smart food packaging: a review. Polym Eng Sci 63(9):2675–2699. https://doi.org/10.1002/pen.26415

Nayak P, Adhikary M, Aparnna VP, Parmar M (2025) Green biopolymers in food and dairy packaging. In: Green biopolymers for packaging applications, pp 344–364

Olonisakin K, Wen A, He S, Lin H, Tao W, Chen S et al (2023) The development of biodegradable PBAT-lignin-tannic acid composite film: properties, biodegradability, and potential barrier application in food packaging. Food Bioprocess Technol 16(7):1525–1540. https://doi.org/10.1007/s11947-023-02997-3

Oymaci P, Altinkaya SA (2016) Improvement of barrier and mechanical properties of whey protein isolate based food packaging films by incorporation of zein nanoparticles as a novel bionanocomposite. Food Hydrocoll 54:1–9. https://doi.org/10.1016/j.foodhyd.2015.08.030

Patel MK, Hansson F, Pitkänen O, Geng S, Oksman K (2022) Biopolymer blends of poly(lactic acid) and poly(hydroxybutyrate) and their functionalization with glycerol triacetate and chitin nanocrystals for food packaging applications. ACS Appl Polym Mater 4(9):6592–6601. https://doi.org/10.1021/acsapm.2c00967

Peng Y, Li Y (2014) Combined effects of two kinds of essential oils on physical, mechanical and structural properties of chitosan films. Food Hydrocoll 36:287–293. https://doi.org/10.1016/j.foodhyd.2013.10.013

Perumal T, Krebs de Souza C, Nihues TC, Jain P, Gaikwad KK, Roy S (2025) A review on biopolymer-based oil and water-resistant functional paper coating for food packaging. Food Biosci 63. https://doi.org/10.1016/j.fbio.2024.105656

Pietrosanto A, Scarfato P, Di Maio L, Incarnato L (2020) Development of eco-sustainable PBAT-based blown films and performance analysis for food packaging applications. Materials 13(23). https://doi.org/10.3390/ma13235395

Piñeros-Hernandez D, Medina-Jaramillo C, López-Córdoba A, Goyanes S (2017) Edible cassava starch films carrying rosemary antioxidant extracts for potential use as active food packaging. Food Hydrocoll 63:488–495. https://doi.org/10.1016/j.foodhyd.2016.09.034

Phupoksakul T, Leuangsukrerk M, Somwangthanaroj A, Tananuwong K, Janjarasskul T (2017) Storage stability of packaged baby formula in poly(lactide)-whey protein isolate laminated pouch. J Sci Food Agric 97(10):3365–3373. https://doi.org/10.1002/jsfa.8187

Plackett D, Siró I (2011) Polyhydroxyalkanoates (PHAs) for food packaging. In: Lagarón J-M (ed) Multifunctional and nanoreinforced polymers for food packaging. Woodhead Publishing, pp 498–526

Plaeyao K, Talodthaisong C, Yingyuen W, Kaewbundit R, Tun WST, Saenchoopa A et al (2025) Biodegradable antibacterial food packaging based on carboxymethyl cellulose from sugarcane bagasse/cassava starch/chitosan/gingerol extract stabilized silver nanoparticles (Gin-AgNPs) and vanillin as cross-linking agent. Food Chem 466:142102. https://doi.org/10.1016/j.foodchem.2024.142102

Priyadarshi R, Sauraj, Kumar B, Deeba F, Kulshreshtha A, Negi YS (2018a) Chitosan films incorporated with Apricot (Prunus armeniaca) kernel essential oil as active food packaging material. Food Hydrocoll 85:158–166. https://doi.org/10.1016/j.foodhyd.2018.07.003

Priyadarshi R, Sauraj, Kumar B, Negi YS (2018b) Chitosan film incorporated with citric acid and glycerol as an active packaging material for extension of green chilli shelf life. Carbohydr Polym 195:329–338. https://doi.org/10.1016/j.carbpol.2018.04.089

Pryadko A, Surmeneva MA, Surmenev RA (2021) Review of hybrid materials based on polyhydroxyalkanoates for tissue engineering applications. Polymers 13(11):1738

Przybyszewska A, Galus S (2024) Green approach for biopolymer-based food packaging films enhanced by zinc oxide nanoparticles. In: Nanotechnology in the life sciences, vol Part F3579, pp 319–342

Pushp P, Bhaskar R, Kelkar S, Sharma N, Pathak D, Gupta MK (2021) Plasticized poly(vinylalcohol) and poly(vinylpyrrolidone) based patches with tunable mechanical properties for cardiac tissue engineering applications. Biotechnol Bioeng 118(6):2312–2325. https://doi.org/10.1002/bit.27743

Rahman S, Gogoi J, Dubey S, Chowdhury D (2024) Animal derived biopolymers for food packaging applications: a review. Int J Biol Macromol 255. https://doi.org/10.1016/j.ijbiomac.2023.128197

Roy S, Ghosh T, Zhang W, Rhim J-W (2024a) Recent progress in PBAT-based films and food packaging applications: a mini-review. Food Chem 437:137822. https://doi.org/10.1016/j.foodchem.2023.137822

Roy S, Malik B, Chawla R, Bora S, Ghosh T, Santhosh R et al (2024b) Biocompatible film based on protein/polysaccharides combination for food packaging applications: a comprehensive review. Int J Biol Macromol 278. https://doi.org/10.1016/j.ijbiomac.2024.134658

Sahraeian S, Abdollahi B, Rashidinejad A (2024) Biopolymer-polyphenol conjugates: novel multifunctional materials for active packaging. Int J Biol Macromol 280. https://doi.org/10.1016/j.ijbiomac.2024.135714

Samsi MS, Kamari A, Din SM, Lazar G (2019) Synthesis, characterization and application of gelatin–carboxymethyl cellulose blend films for preservation of cherry tomatoes and grapes. J Food Sci Technol 56(6):3099–3108. https://doi.org/10.1007/s13197-019-03809-3

Sen M, Singh HP (2024) Edible food coatings are biopolymers: packaging and preservation. In: Food coatings and preservation technologies, pp 1–25

Scannell AGM, Hill C, Ross RP, Marx S, Hartmeier W, Arendt EK (2000) Development of bioactive food packaging materials using immobilised bacteriocins Lacticin 3147 and Nisaplin®. Int J Food Microbiol 60(2):241–249. https://doi.org/10.1016/S0168-1605(00)00314-7

Shams B, Ebrahimi NG, Khodaiyan F (2019) Development of antibacterial nanocomposite: whey protein-gelatin-nanoclay films with orange peel extract and tripolyphosphate as potential food packaging. Adv Polym Technol 2019(1):1973184. https://doi.org/10.1155/2019/1973184

Shankar S, Danneels F, Lacroix M (2019) Coating with alginate containing a mixture of essential oils and citrus extract in combination with ozonation or gamma irradiation increased the shelf life of Merluccius sp. fillets. Food Pack Shelf Life 22:100434. https://doi.org/10.1016/j.fpsl.2019.100434

Shankar S, Khodaei D, Lacroix M (2021) Effect of chitosan/essential oils/silver nanoparticles composite films packaging and gamma irradiation on shelf life of strawberries. Food Hydrocoll 117:106750. https://doi.org/10.1016/j.foodhyd.2021.106750

Shapi'i RA, Othman SH, Nordin N, Kadir Basha R, Nazli Naim M (2020) Antimicrobial properties of starch films incorporated with chitosan nanoparticles: in vitro and in vivo evaluation. Carbohydr Polym 230:115602. https://doi.org/10.1016/j.carbpol.2019.115602

Shi C, Zhou A, Fang D, Lu T, Wang J, Song Y et al (2022) Oregano essential oil/β-cyclodextrin inclusion compound polylactic acid/polycaprolactone electrospun nanofibers for active food packaging. Chem Eng J 445:136746. https://doi.org/10.1016/j.cej.2022.136746

Shit SC, Shah PM (2014) Edible polymers: challenges and opportunities. J Polym 2014(1):427259. https://doi.org/10.1155/2014/427259

Singh AK, Itkor P, Lee M, Saenjaiban A, Lee YS (2024) Synergistic integration of carbon quantum dots in biopolymer matrices: an overview of current advancements in antioxidant and antimicrobial active packaging. Molecules 29(21). https://doi.org/10.3390/molecules29215138

Sone Aung SP, Htet Shein HH, Aye KN, Nwe N (2018) Environment-friendly biopolymers for food packaging: Starch, protein, and poly-lactic acid (PLA). In: Bio-based materials for food packaging: green and sustainable advanced packaging materials, pp 173–195

Sothornvit R, Krochta JM (2001) Plasticizer effect on mechanical properties of β-lactoglobulin films. J Food Eng 50(3):149–155. https://doi.org/10.1016/S0260-8774(00)00237-5

Stoica M, Dima CV, Alexe P (2018) Eco-friendly nanocomposites from bacterial cellulose and biopolyesters as a sustainable alternative for plastic food packaging. In: Food packaging and preservation: techniques, applications and technology, pp 113–127

Sukumaran L, Jano NA (2024) Edible biopolymer coatings/films used for packaging and preserving foods. In: Food coatings and preservation technologies, pp 301–338

Sun F, Zhao J, Shan P, Wang K, Li H, Peng L (2025) Gelatin-based composite film integrated with nanocellulose and extract-metal complex derived from coffee leaf for sustainable and active food packaging. Food Hydrocoll 159:110610. https://doi.org/10.1016/j.foodhyd.2024.110610

Swetha TA, Bora A, Mohanrasu K, Balaji P, Raja R, Ponnuchamy K et al (2023) A comprehensive review on polylactic acid (PLA)—Synthesis, processing and application in food packaging. Int J Biol Macromol 234:123715. https://doi.org/10.1016/j.ijbiomac.2023.123715

Syarifuddin A, Muflih MH, Izzah N, Fadillah U, Ainani AF, Dirpan A (2025) Pectin-based edible films and coatings: From extraction to application on food packaging towards circular economy-a review. Carbohydrate Polym Technol Appl 9:100680. https://doi.org/10.1016/j.carpta.2025.100680

Tammina SK, Sahithya K, Khan A (2024) Protein as a sustainable biopolymer for food packaging and preservation. In: Sustainable materials for food packaging and preservation: food security and sustainability, pp 105–125

Tan W, McClements DJ, Chen J, Ma D (2024) Novel biopolymer-based active packaging material for food applications: cinnamaldehyde-loaded calcium nanoparticles incorporated into alginate-carboxymethyl cellulose films. Food Packag Shelf Life 45. https://doi.org/10.1016/j.fpsl.2024.101351

Tarique J, Sapuan SM, Khalina A, Ilyas RA, Zainudin ES (2022) Thermal, flammability, and antimicrobial properties of arrowroot (Maranta arundinacea) fiber reinforced arrowroot starch biopolymer composites for food packaging applications. Int J Biol Macromol 213:1–10. https://doi.org/10.1016/j.ijbiomac.2022.05.104

Tavassoli M, Khezerlou A, Firoozy S, Ehsani A, Punia Bangar S (2023) Chitosan-based film incorporated with anthocyanins of red poppy (Papaver rhoeas L.) as a colorimetric sensor for the detection of shrimp freshness. Int J Food Sci Technol 58(6):3050–3057. https://doi.org/10.1111/ijfs.16432

Tavassoli M, Khezerlou A, Sani MA, Hashemi M, Firoozy S, Ehsani A et al (2024) Methylcellulose/chitosan nanofiber-based composites doped with lactoferrin-loaded Ag-MOF nanoparticles for the preservation of fresh apple. Int J Biol Macromol 259:129182. https://doi.org/10.1016/j.ijbiomac.2023.129182

Thiyagamoorthy UM, Sadayandi G, Jeyaraj S, Thangarasu S, Wadaan MA, Baabbad A et al (2024) Exploring the efficacy of various essential oils in chitosan-based composite biopolymer films for food packaging. Polym Adv Technol 35(4). https://doi.org/10.1002/pat.6371

Tripathi AD, Mishra PK, Darani KK, Agarwal A, Paul V (2022) Hydrothermal treatment of lignocellulose waste for the production of polyhydroxyalkanoates copolymer with potential application in food packaging. Trends Food Sci Technol 123:233–250

u Nisa I, Ashwar BA, Shah A, Gani A, Gani A, Masoodi FA (2015) Development of potato starch based active packaging films loaded with antioxidants and its effect on shelf life of beef. J Food Sci Technol 52(11):7245–7253. https://doi.org/10.1007/s13197-015-1859-3

Xiao L, Yao Z, He Y, Han Z, Zhang X, Li C et al (2022) Antioxidant and antibacterial PBAT/lignin-ZnO nanocomposite films for active food packaging. Ind Crops Prod 187:115515. https://doi.org/10.1016/j.indcrop.2022.115515

Vandewijngaarden J, Wauters R, Murariu M, Dubois P, Carleer R, Yperman J et al (2016) Poly (3-hydroxybutyrate-co-3-hydroxyhexanoate)/organomodified montmorillonite nanocomposites for potential food packaging applications. J Polym Environ 24:104–118

Venkatesan R, Rajeswari N (2017) ZnO/PBAT nanocomposite films: investigation on the mechanical and biological activity for food packaging. Polym Adv Technol 28(1):20–27. https://doi.org/10.1002/pat.3847

Vytejčková S, Vápenka L, Hradecký J, Dobiáš J, Hajšlová J, Loriot C et al (2017) Testing of polybutylene succinate based films for poultry meat packaging. Polym Test 60:357–364. https://doi.org/10.1016/j.polymertesting.2017.04.018

Wang L, Xu J, Zhang M, Zheng H, Li L (2022) Preservation of soy protein-based meat analogues by using PLA/PBAT antimicrobial packaging film. Food Chem 380:132022. https://doi.org/10.1016/j.foodchem.2021.132022

Wardejn S, Wacławek S, Dudek G (2024) Improving antimicrobial properties of biopolymer-based films in food packaging: key factors and their impact. Int J Mol Sci 25(23). https://doi.org/10.3390/ijms252312580

Yang C, Tang H, Wang Y, Liu Y, Wang J, Shi W, Li L (2019) Development of PLA-PBSA based biodegradable active film and its application to salmon slices. Food Packag Shelf Life 22:100393. https://doi.org/10.1016/j.fpsl.2019.100393

Yang F, Chen G, Li J, Zhang C, Ma Z, Zhao M et al (2023) Effects of quercetin and organically modified montmorillonite on the properties of poly(butylene adipate-co-terephthalate)/thermoplastic starch active packaging films. ACS Omega 8(1):663–672. https://doi.org/10.1021/acsomega.2c05836

Yang J, Punia Bangar S, Rizwan Khan M, Hammouda GA, Alam P, Zhang W (2024) Biopolymer-based packaging films/edible coatings functionalized with ε-polylysine: new options for food preservation. Food Res Int 187. https://doi.org/10.1016/j.foodres.2024.114390

Yao L, Tao Q, Xian F, Chen Z, Huang L, Zhong N, Gao J (2025) Development of pullulan/gellan gum films loaded with astaxanthin nanoemulsion for enhanced strawberry preservation. Food Res Int 201:115644. https://doi.org/10.1016/j.foodres.2024.115644

Zambrano-Zaragoza ML, Mercado-Silva E, Del Real LA, Gutiérrez-Cortez E, Cornejo-Villegas MA, Quintanar-Guerrero D (2014) The effect of nano-coatings with α-tocopherol and xanthan gum on shelf-life and browning index of fresh-cut "Red Delicious" apples. Innov Food Sci Emerg Technol 22:188–196. https://doi.org/10.1016/j.ifset.2013.09.008

Zare M, Namratha K, Ilyas S, Hezam A, Mathur S, Byrappa K (2019) Smart fortified PHBV-CS biopolymer with ZnO–Ag nanocomposites for enhanced shelf life of food packaging. ACS Appl Mater Interfaces 11(51):48309–48320. https://doi.org/10.1021/acsami.9b15724

Zhang R, Lan W, Ding J, Ahmed S, Qin W, He L, Liu Y (2019) Effect of PLA/PBAT antibacterial film on storage quality of passion fruit during the shelf-life. Molecules 24:3378. https://doi.org/10.3390/molecules24183378

Zhang W, Roy S, Ezati P, Yang DP, Rhim JW (2023) Tannic acid: a green crosslinker for biopolymer-based food packaging films. Trends Food Sci Technol 136:11–23. https://doi.org/10.1016/j.tifs.2023.04.004

Zhang M, Xu L, Liu F (2024) Batch preparation and characterization of zein-based beaded nanofiber membranes for active food packaging. Int J Biol Macromol 276:133966. https://doi.org/10.1016/j.ijbiomac.2024.133966

Zia KM, Noreen A, Zuber M, Tabasum S, Mujahid M (2016) Recent developments and future prospects on bio-based polyesters derived from renewable resources: a review. Int J Biol Macromol 82:1028–1040. https://doi.org/10.1016/j.ijbiomac.2015.10.040

Zubair M, Shahzad S, Hussain A, Pradhan RA, Arshad M, Ullah A (2022) Current trends in the utilization of essential oils for polysaccharide-and protein-derived food packaging materials. Polymers 14(6). https://doi.org/10.3390/polym14061146

Chapter 2
Starch-Based Polymers for Sustainable Food Packaging: Advances and Applications

Neha Naijo Areekal, Ramachandran Ramkumar, Priyanga Govindarasu, M. Prasanna Kumar, Rakesh Kumar Gupta, and Prem Prakash Srivastav

Abstract The growing environmental concern over synthetic plastic pollution has witnessed an increasing trend in demand for eco-friendly alternatives in food packaging. Starch-based films are a promising solution due to their biodegradable, renewable, and sustainable nature. Derived from abundant natural resources such as corn, potato, and cassava, starch-based films offer an excellent potential to replace conventional plastics in food packaging, significantly reducing synthetic waste and its adverse effects on biodiversity. This review article comprehensively discusses the recent advances in the development of starch-based films. Various production methods, such as casting, forming, and extrusion, are elaborated, highlighting their influence on film characteristics. Furthermore, the article explores the applications of starch-based films in preserving food quality and extending shelf life. Challenges such as film brittleness, hydrophilicity, and scalability are addressed, along with the prospects for overcoming these limitations through innovations in biopolymer technology. By replacing synthetic plastics with starch-based films, the food industry can play a pivotal role in mitigating pollution, protecting ecosystems, and promoting sustainability in packaging solutions.

Keywords Starch-based films · Biopolymer · Packaging · Sustainability

N. N. Areekal
Department of Chemical Engineering, Indian Institute of Technology Tirupati, Tirupati, India

R. Ramkumar
Institute of Animal Nutrition, Tamil Nadu Veterinary and Animal Sciences University, Chengalpattu, Tamil Nadu, India

P. Govindarasu · M. Prasanna Kumar
Department of Food Science and Technology, Pondicherry University, Puducherry, India

R. K. Gupta (✉) · P. P. Srivastav
Agricultural and Food Engineering Department, Indian Institute of Technology Kharagpur, Kharagpur, India
e-mail: rakeshgupta.iitkgp@gmail.com

S. Roy and M. Shahid (eds.), *Biomaterials for Sustainable Food Packaging*, Environmental Footprints and Eco-design of Products and Processes, https://doi.org/10.1007/978-3-032-08027-1_2

2.1 Introduction

In the present era, environmental pollution is a serious concern for the health and sustainability of global biodiversity. Among the various contributors to this alarming issue, improper disposal of pollutants, particularly plastics, is one of the most critical concerns. The production and use of plastic has seen an exponential growth over the decades, with studies reporting an increase from 464 Mt in 2020 up to 884 Mt in 2050 (Dokl et al. 2024). Excessive plastic usage and subsequent disposal have long-lasting consequences, especially for ecosystems. Plastics accumulate in terrestrial and aquatic environments, severely impacting soil health and water bodies (Ahmadi et al. 2018). Plastic pollution is an alarming threat to the aquatic ecosystem. Microplastics and other plastic debris are transported to the body through the food and infiltrate the water consumed by marine organisms. These plastics are passed on by smaller marine organisms that are eaten by larger ones, they progressively enter the food chain (Waring et al. 2018). This bioaccumulation poses serious threats to marine biodiversity and raises concerns about human food safety. On land, plastic particles deposited in soil can affect soil health, reducing its nutrient profile. It also poses a serious risk to agricultural land and arouses the need to adopt safety measures for human consumption (Yu et al. 2021). In this regard, to address the environmental concerns, researchers and industries have been exploring alternatives for synthetic plastics, and a promising strategy adopted is the development of biodegradable polymers. Among these, biodegradable polymers based on starch have drawn more attention due to their extensive use in the food packaging industry and their environmentally beneficial qualities (Imam et al. 2012).

Starch, a renewable and biodegradable natural polymer, is an attractive alternative to conventional plastics because of its abundance, affordability, and superior film-forming properties (Chen et al. 2020; Gupta et al. 2022). Starch-based biodegradable polymers are derived from various sources such as corn, potato, rice, cassava, and taro (Gupta et al. 2023). These polymers naturally break down in the environment, and the long-term environmental impact of synthetic plastics is lessened. (Kringel et al. 2020). Beyond starch, other natural polymers like proteins, polysaccharides, and lipids have also been explored for their potential in replacing synthetic polymers. The production of starch-based biodegradable films involves various methods, including film solution casting, foam processing, and extrusion processing (Nayak 2010). These traditional methods are improved by advanced techniques such as electrospinning, reactive extrusion, nanotechnology, and 3-D printing (Cui et al. 2022). Each method offers unique advantages in enhancing the properties and performance of the films, enabling an aid for diverse applications.

This chapter primarily focuses on the critical role of starch in advancing sustainable food packaging solutions. It discusses the various processing and production methods for starch-based biodegradable films and highlights recent advancements in this field. Additionally, the review explores the wide-ranging applications of starch-based films, emphasizing the plethora of food applications. By adopting starch-based biodegradable polymers, industries can take significant strides in achieving sustainability and combating environmental pollution. It is considered the best alternative as it aligns with global efforts to protect biodiversity and ensure food safety. Furthermore, it holds a promise for revolutionizing the packaging industry with a cleaner and sustainable future.

2.2 Relevance of Starch Incorporation for Sustainable Food Packaging

Starch is a well-known polymer produced naturally. It is known for its abundance, cost-effectiveness, nontoxicity, and biocompatibility (Gupta et al. 2024d). These polymers are extensively employed for the development of films. Different methods for the development of films include modulating their properties using chemical or enzymatic methods (Chen et al. 2020). Furthermore, these films are used to enhance antioxidant and oxygen barrier properties. Recently, researchers have made efforts to develop biodegradable films from natural sources such as corn, potatoes, and tapioca. These films are biodegradable and compostable, making them an environmentally friendly alternative to the plastics widely consumed. Different characteristic features of these films include sufficient barrier properties against moisture, oxygen, and light, thereby preserving the quality of the packaged goods. Starch-based films also serve as eco-friendly products, aligning with global sustainability goals (Guillard et al. 2018). Starch-based films are extensively used in the food industry for enhancing the film properties for storage and shelf-life enhancement. It also promotes a circular economy where materials can be reused and recycled (Akinsemolu et al. 2024).

2.3 Production and Processing of Starch-Based Packaging

There are many different sources of starches, such as roots, tubers, and grains. The glucose molecules that glycosidic bonds join make starch to form a polymer. The molecular components of starch comprise amylose and amylopectin. According to Liu et al. (2020), starch particles are impermeable in cold water; they absorb water, expand, and release some amylose when heated to a suitable temperature above water. Starch gelatinization requires a high temperature and additional water, which does not allow all nutritional starch applications. At minimal temperatures, the crystalline part of the starch granules maintains its reliability. On the other hand, the starch particles are about to absorb water, and the crystalline regions of starch will enlarge if the starch solution is heated (Gupta et al. 2025). The starch granules will eventually disintegrate, forming a hydrophilic colloidal solution. This process is known as gelatinization of starch. Cooling a starch suspension to a very low temperature after the gelatinization process causes the starch molecules to reorganize, the presence of hydrogen bonds

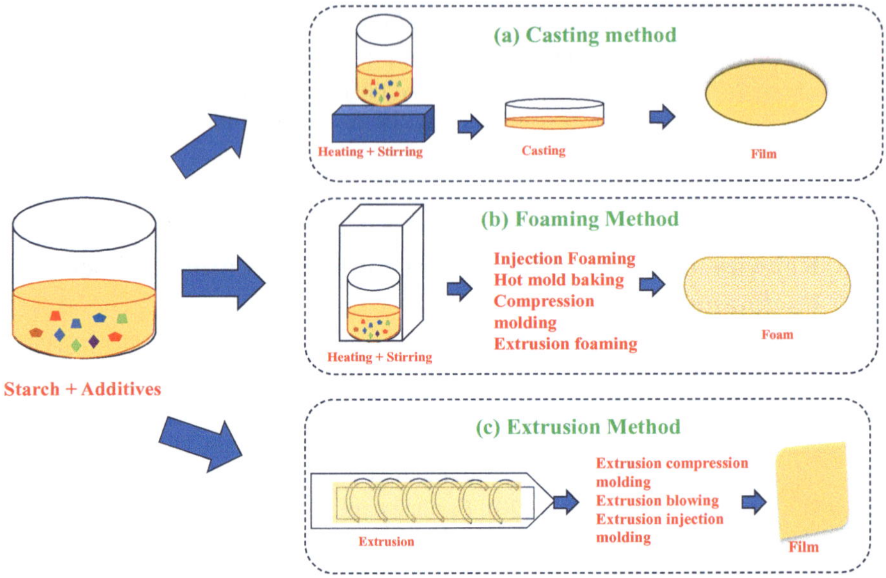

Fig. 2.1 Production and processing of starch

to form, and arranged formations to return. At the same time, the gel forms, and the viscosity of the starch solution increases (Li et al. 2021). Starch retrogradation is the term for this process. The properties and structure of biodegradable polymers made from starch are closely related to those of starch. Three traditional types of processing the starch-based biopolymers include the Casting method, the Foaming method, and the Extrusion method, as shown in Fig. 2.1, and the various methods employed for starch polymer processing are summarized in Table 2.1.

2.3.1 Casting Method

The casting method is commonly used to develop starch-based biopolymers. This method is well known for its simple preparation method, cost-effectiveness, and suitability for laboratory-scale research. The processing method involves the dissolution of starch into the solvent (starch), plasticization, solution casting, and drying. Briefly, the starch from renewable sources such as corn, potato, cassava, and other tubers serves as a primary matrix. To induce gelatinization, the starch will be dissolved in water at elevated temperatures, between 60 and 90. During this process, the starch absorbs water, swells, and ruptures. This process releases the amylose and amylopectin chains, forming a homogeneous polymer solution. Plasticizers such as glycerol, sorbitol, or polyethylene glycol are added to focus on the characteristic brittleness of starch films. Plasticizers reduce intermolecular hydrogen bonding,

Table 2.1 Various methods of starch polymer processing

Raw materials	Plasticizers	Processing method	Inference	References
Potato starch Corn starch Tapioca starch Rice starch Wheat starch	Glycerol, 30% (v/w of starch)	Casting	Tapioca, rice, and potato starches made film has been better film properties than corn and wheat starch film	Brain Wilfer et al. (2021)
Anchote starch	Glycerol, 1-ethyl-3 methylimidazolium acetate, sorbitol or triethylene glycol	Casting	Sorbitol as a plasticizer revealed its high tensile strength of the film	Abera et al. (2020)
Corn starch and 3% chitosan-starch Nanoparticles (SC-NP)	–	Casting	3% of SC-NP lowers the water vapor permeability, water solubility, and swelling ratio	Liu et al. (2023)
Potato starch	Kaolin	Casting	Increasing the concentration of kaolin increases its roughness and wettability of the film surfaces	Waśniewska et al. (2020)
Corn starch, tapioca starch, and gelatin	–	Foaming	The foams with densities of 45–50 kg/m^3 were prepared with denser foams and greater compression	Martin Torrejon ct al. (2023)
Sweet potato starch	Glycerol, no foaming agent	Foaming	Low water solubility, and low mechanical resistance	Cruz-Tirado et al. (2020)
Cassava starch	–	Foaming	Lightweight, density: 124–245 kg/m^3	Amaraweera et al. (2022)
Cassava starch	Citric acid	Extrusion	Lower strength than control hydrogels; storage modulus increased	Simões et al. (2020)

(continued)

Table 2.1 (continued)

Raw materials	Plasticizers	Processing method	Inference	References
Cassava starch and grape stalks	Glycerol	Compression foaming	No residual agglomeration and minimal cell structural disintegration	Engel et al. (2019)
Carboxymethyl chitosan	Poly vinyl alcohol	Injection foaming	Improved hydrophobic and antimicrobial properties	Liu et al. (2025)
Cellulose & starch	Poly vinyl alcohol & glycerol	Extrusion foaming	Improved Viscoelastic properties and higher foaming properties	Duan et al. (2022)
Cassava starch & chitosan	Glycerol & poly ethylene glycol	Compression moulding	Reduction in coliforms and total aerobic counts of pork meats	Wu et al. (2023)
Pineapple stem starch	Glycerol	Compression moulding	Biodegradable and enhanced water resistance	Tessanan et al. (2023)
Cassava starch	Glycerol & Agar from *Gracilaria fisheri*	Hot mold baking	The enhanced physical, mechanical and thermal properties for food packaging	Trongnit et al. (2024)
Cassava starch	Fique bagasse microparticles		Addition of microparticles, increasing the thermal properties of starch-based foams	Parra-Campos et al. (2022)

increasing film flexibility and elongation at break. A smooth surface, either a glass plate or a sheet of Polytetrafluoroethylene (PTFE), is cast with the prepared starch solution for a smooth, uniform distribution. Its uniform distribution obtains the starch-based polymer's consistent biomechanical and optical characteristics. The optimum drying temperature range for the cast film was between 40 and 60 °C. The solvent's volatility, humidity, and film thickness all affect the amount of time it takes to dry. Corn starch polymers have been prepared using the film solution casting process by Patil et al. (2021), Nordin et al. (2020) starch-agar based films by Ezati and Rhim (2021), Dioscorea hispida (Indian three-leaved yam) based films by Hazrati et al. (2021), the cationic potato loaded with curcumin films by Liu et al. (2022a, b). The film solution casting method does not require specialized equipment, is cost-effective, has flexibility in formulation, and is uniform. The limitation of this method is that industrial scale-up is challenging, since starch films are hydrophilic, their performance in humid conditions is limited. Pure starch films exhibit low tensile strength and poor elasticity.

2.3.2 Foaming Method

The foaming method of processing is a conventional technique for producing starch-based foams. The foam is converted into thermoplastic material, which retains its biodegradability. Moreover, starch softens and expands into foam, similar to expanded polystyrene (EPS). These foams have significant attention as an alternative to petroleum-based synthetic foams, such as EPS. The foam formation was divided into two steps: gelatinization of starch and evaporation of water from the paste, batter, or gel. The standard methods for producing foams are injection, compression molding, extrusion, hot mold baking, and reactive extrusion. The foams are classified into open or closed cell structures based on the gas bubble entrapment during the foaming process. The factors that affect the porosity are starch concentration, blowing agent, and drying conditions (Amaraweera et al. 2022).

2.3.2.1 Injection Foaming

Injection Foaming is the most widely used technique in the large production of starch-based foams due to its accuracy, uniformity, and to create intricate forms. When molten starch is injected into the mold's cavity, it cools and takes on the required shape. The starch is exhibited above the glass transition temperature; thus, it is converted into thermoplastics using plasticizers. The gas bubbles are formed and dispersed in molten starch during injection molding. This results in foam formation upon cooling and solidifying molten starch (Soykeabkaew et al. 2015).

2.3.2.2 Hot Mold Baking

The baking technique includes the addition of starch and a foaming agent in a mold, followed by heating in the oven. During baking, the vaporization of water leads to foam formation from starch-based polymer, and the viscous property of starch plays a prominent part in the character of gas bubbles (Soykeabkaew et al. 2015).

2.3.2.3 Compression Molding

The molding technique is similar to the baking method of foaming. The molten starch is filled into the hydraulic pressure container, which has a temperature controller, and the foam is developed by temperature and pressure changes or thermal expansion. The cassava starch-based foams, which are biodegradable and incorporated with grape stalks, were obtained for the packaging material (Engel et al. 2019).

2.3.2.4 Extrusion Foaming

By heating and mixing the raw materials under pressure and then pushing them through a screw extruder, extrusion foaming produces foam. According to Zeng et al. (2019), bubbles are produced by dissolved gases stuck in the starch matrix when the pressure is released. According to Combrzyński et al. (2019), the technological features of starch-based foams in the extrusion method of foaming are affected by the operating factors, including screw speed and temperature, as well as the amylose to amylopectin ratio of the starch used. Extrusion compression molding, extrusion blowing, and extrusion injection molding are the three main processes used in extrusion processing, a popular dry method for creating starch-based films that depend on the thermoplastic properties of biopolymers. Heat and shear pressures are used to melt the combination of film-forming ingredients that is first introduced into a screw extruder. The combination is frequently pelletized and pre-extruded in industrial settings to create master batches, which are then reintroduced for use in filmmaking. Using different die geometries, the molten material is forced to the die head, which is further processed into films. While extrusion compression molding forms films by hot pressing and chilling, extrusion blowing uses pressure to produce a thin film.

2.4 Raw Material Preparation

Biopolymers like starch offer a wide opening for biodegradable food packaging material manufacturing that reduces environmental pollution. Starch biopolymeric materials can be enhanced by incorporating additional biopolymers, additives, and innovative production procedures. However, producing high-performance starch-based biodegradable polymers on a wide scale remains challenging, necessitating additional research (Cheng et al. 2021). Starch-based films are biodegradable, reusable, economical, and suitable for food packaging. Researchers have recently focused heavily on functional starch-based films. Biologically active starch used in biofilms with inclined bacterial resistance, oxidation barrier, ultra-violet, and water vapor permeation prevention properties ensures protection of packed goods from microbes and extends their shelf life. Films manufactured from biomolecules like starch, containing sensors, potentially ensure food quality and deliver functional ingredients and medications. Therefore, developing renewable, biologically active, and advanced green films from starch is welcomed. Adding extract matrices, oils from green sources, and nano-sized particles to these biofilms enhances their resistance against bacteria, oxygen, and other barriers. Adding anthocyanins to starch films can detect deterioration in fish items and alert consumers. Future research should focus on various films manufactured from starches, along with certain properties pertaining to enzyme and salt-responsive properties (Cui et al. 2021). Raw material preparation commences from the starch extraction which is achieved by selection of suitable raw materials like millets and cereals tubers and other commodities as they are the essential component for film formation (Cheng et al. 2021; Wang et al. 2022a, b; Gupta et al. 2024a). Mostly these commodities are initially cleansed with running water followed by soaking it in suitable alkalis for elongated time period and grinding it with suitable blenders along with decantation process or else mostly by freeze drying followed by sieving it with suitable mesh and drying to get isolated starch (Gautam et al. 2021; Oliveira et al. 2020). Additionally non-thermal treatments can also be used for enhancing the yield of starch for instance, the material is added to water to get a starch slurry followed by ultrasound treatment where the ultrasound probe is immersed in such a way that it is in contact with the slurry in the beaker with the accurate optimization of sonication amplitude, frequency, on-duty cycles, pulses mode followed by the filtration and drying with optimum time and temperature the starch is obtained which infers that the integration of ultrasonic extraction and low-cost ingredients rendered this technique economically viable and possibly enhance sustainable manufacturing of bioplastic films (Tan et al. 2021). Alternatively, an electric field generated in a pulsed form can enhance the extractability of starch biopolymers, along with effective modification of its structural integrity and characteristics by reducing gelatinization temperatures, crystallinity, enthalpies, pasting temperature, and flow properties. Electrical field application can change the easily digestible starch content while preserving the digestion-escaping starch (Castro et al. 2023). On the other hand, exposure to a dielectric barrier discharge-based plasma system with aluminum electrodes and acrylic barriers, exposing starch and film-forming solution to three stimulation frequencies at 100, 200, and 300 Hz for a time. Plasma treatment produces films with inferior solubility and superior amylose-amylopectin ratio, and reduces cost. Film formation rate is higher in a mediated process than casting, which relies on water evaporation and can be done concurrently in industrial facilities. Plasma systems can be installed at either end (Goiana et al. 2024). The conjugating starch and carrot nano-fibres (CNF) film process involves solution casting and evaporation of the liquid phase. The mixture of thermoplastic starch with distilled water is stirred, treated with varying CNF percentages, and gelatinized. The films are then placed on glass petri plates and evaporated, which offers potential in food packaging as a barrier against chemical and biological stress factors (Ghoshal and Singh 2020). Conversely Starch hydrolysis mediated by acids can greatly impart changes on films for instance, Cassava Starch was hydrolysed with three different molar hydrochloric acid solutions, neutralized with NaOH, rinsed with distilled water, and freeze-dried until powdered to get acid hydrolysed cassava starch and films formulated exhibited reduced the hydrophilicity and water absorption rate of the thin films. Acid hydrolysed cassava starch films have potential for use in food packaging due to their inexpensive cost and easy preparation, as well as their improved structural, mechanical, and thermal qualities (Amaraweera et al. 2021). A study which proclaims that the bioplastics can be synthesised from basic raw materials like banana peel starch where the peel was subjected to extraction with water containing alkali like sodium bisulphate followed by conventional drying process which was to be conjugated with cellulose fibres from banana pseudostem scraping away plant tissue with an iron ribbon until totally separated followed by the elimination of non-cellulosic components, the sample was treated with sodium hydroxide at a slightly higher temperature for 1 h. Cleansing it with deionized water followed by drying upto 48 h, which paved the way of preparing bioplastic films from waste biopolymers (Abera et al. 2024). Glycerol, a plasticizer for starch-based polymers, can migrate rapidly during retrogradation, affecting the film's features. Co-plasticizers reduce melting fluidity, induce hydrogen bonds, and produce a hard film structure. Sucrose aids in B-type starch crystal development, while glucose and sucrose convert bound water to entrapped water. Sorbitol has a similar plasticizing capacity to glycerol. The structure, hydroxyl group percentage, and melting point of sugar/sugar alcohol influence film characteristics (Wu et al. 2025). Type and nature of starch along with raw material, film formation temperature and time, plasticizers, co- biopolymers, and storage conditions can all have an impact pertaining to the attributes of films manufactured from the starches of different sources as shown in Fig. 2.2.

The optimization of these elements can result in superior quality films that meet the criteria for their application in each type of fresh food. Future research should address the best conditions for gelatinization, which allows successful binding with co-biopolymers or functional components to get superior film characteristics (Thakur et al. 2019). Additionally, the molecular investigation of the interactions of different starches from diverse raw materials of biological origin will help to increase our comprehension of this sustainable biopolymer in addition to the mono or poly-modification techniques employed are used to improve technological, structural as well as film formation capability (Kumari and Sit 2023; Singh et al. 2024).

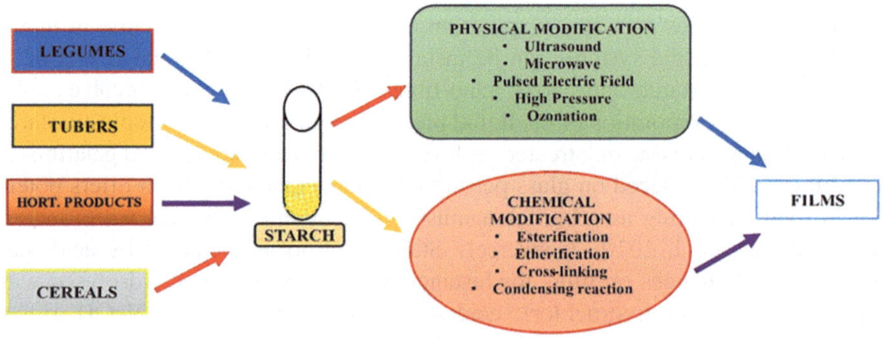

Fig. 2.2 Raw materials employed for film preparation and modifications techniques (Schmiele et al. 2019; Wang et al. 2020)

2.5 Advances in Starch-Based Polymer

Starch from native plant sources is utilized in the packaging of consumables due to its widespread accessibility, affordability, and film-forming capabilities. Starch's hydrophilic properties and high moisture sensitivity limit its use. Modifying NS (native starch) into Resistant Starch (RS) enhances film formation and expands the use of starch in packaging formulations. This modified biopolymer is combined with other polymeric components, nano-sized particles, phytoextracts, and oils in creating novel replenishable packing materials with poorer vapor penetrability, gas permeability, enhanced microbial resistance, and physical attributes. This modified starch used in fabricating intelligent packing materials is noted to possess superior film-fabrication capacity, increased physico-mechanical strength, and water resistance (Fatima et al. 2024). Starch supply, foamability factors, the quantity of plasticizing components, concentration and nature of the fibres incorporated, and the inclusion of additional polymeric components in making blended mixtures potentially impact the characteristics and shape of fabricated bio-foams. Hence, these variables should be investigated and optimized further. Agro-industrial wastes have emerged as a promising alternative for lowering the cost of foams. Studies have been conducted on biologically fabricated active starch-mediated foams for antiviral activity against the virus, and detailed research is also required (Tapia-Blácido et al. 2022).

2.6 Starch Blending and Conjugates for Advanced Food Packaging

The spread of single-use plastic packaging has culminated in rising waste and environmental degradation. Biobased products, such as starch, can serve as a renewable and biodegradable alternative (Gupta et al. 2024b). However, the low processability of starches makes it difficult to employ. Plasticized starch (PS) is a solution that can be combined with more durable polymers to increase its flexibility (Surendren et al. 2022). Alternatively, blending starch with other organic compounds like ethyl maltol, which imparts a caramel-like flavour and antibacterial characteristics, works perfectly with baked products as presented in Table 2.2. Films fabricated from extracted starch conjugated with ethyl-maltitol can be potentially regarded as intelligent consumable-packing materials that have antibacterial capabilities and significantly extend the shelf life of the packed consumables efficiently with antifungal properties. These benefits make them be utilized in baked goods like butter cakes, increasing their shelf life up to 3 times (Promhuad and Harnkarnsujarit 2025). Adding chitosan (CH) improves the stability of film-forming dispersions (FFDs) in refrigerated circumstances, and promotes ultraviolet radiation barrier characteristics. Yet, the price of materials are higher. Films fabricated have exhibited greater tensile strength along with elastic modulus levels with increased proportion of starch levels. The percentage of starch obtained from corn applied in the film fabrication enhances the vapor penetration and thermo-sealing capability. Composite films containing 75% corn starch and 25% chitosan (medium molecular weight) are a viable solution for consumable packaging for their renowned properties and economic feasibility (Bof et al. 2021). Furthermore, employing biopolymers as ecologically acceptable packing material remains one of the top focuses for researchers. PVA/starch films exhibiting significant antioxidant and antibacterial activity could increase consumable security and visual properties. The film-forming capability of Polyvinyl Alcohol and starch composites depends on various aspects, including the origin, variety, proportion of essential components, as well as their interaction with polymers within the fabricated films (Abedi-Firoozjah et al. 2023). Advanced research on starch-based packaging systems involves the incorporation of inorganic carbon materials due to the superior qualities they possess that can be used to improve the physicochemical properties and functional performance of starch-based films, thereby allowing for the fabrication of modern packaging films (Xu et al. 2022). Active films manufactured from high amylose corn starch-cinnamaldehyde complex demonstrated superior mechanical properties together with greater tensile strength and elongation attributes in addition to enhanced Ultra-Violet resistance, microbial resistance, especially against *Staphylococcus aureus* and *Escherichia coli* among other microbes which rendered them in experimenting its effectiveness on shelf-life extension of strawberries resulting in effective strawberry preservation which proves that complexing starch with other polymers can potentially be used in manufacturing films for food shelf life prolongation (Wan et al. 2023).

2.7 Applications of Hydrogels, Cryogels, and Aerogels from Starch for Sustainable Food Packaging

On the other hand, hydrogels, along with aerogels, are colloids manufactured from biopolymers, mostly polysaccharides, that form 3-dimensional networking systems. They exhibit outstanding mechanical qualities and biodegradability, with potential uses in diverse industries. Optimizing starch hydrogels and aerogels requires careful consideration of starch type, concentration, modification, gelling improvers, and preparation procedures. Drying procedures are important to consider when preparing aerogel. Packaging materials incorporated with hydrogels and aerogels extend the keeping quality of consumable matrices by imbibing water, inhibiting O_2, and ultra-violet radiation. Intelligent, modern packaging systems fabricated with incorporated gel phase systems can be used as a suitable vector containing oxidation-preventing compounds and microbial-resistant substrates, which not only serve as a mechanical preventor against surrounding spoilage factors but also aid in food preservation. When it comes to encapsulating active chemicals, starch-based aerogels outperform hydrogels. Starch-based hydrogels have a unique trait called the stimulus-response system. Hydrogels' ability to respond to specific stimuli makes them ideal for smart

Table 2.2 Starch-polymer conjugates and their efficiency and applications

Starch-polymer conjugates	Efficiency and application	References
Starch extracted from corn and chitosan based nano sized particles with the incorporation of thymol fabricating bionanocomposite films	• Films were exposed to the decrement in their mechanical properties with the addition of thymol but its increment hasn't altered the physico-chemical characteristics in the fabricated film • Increased levels of Water Vapor Permeability of the films but its increment undesirably affected WVP of films • The keeping quality of cherry tomatoes packed with this conjugated film prolonged, with barely any variations in firmness. Furthermore, no development of mold on the sliced cherry tomatoes was observed during storage of conjugated films up to a week	Othman et al. (2021)
Thermoplastic starch poly-lactic acid blend	• The incorporation of Starch-based Polylactic acid compatibilizer substantially decreases the size of poly-lactic acid crystal spherulite, enhances the microstructure and morphology, and results in greater oxygen barrier and thermomechanical properties, rendering it ideal for single layer packaging that is flexible and recyclable	Trinh et al. (2022)
Starch from rice conjugated with proteins from fish	• These biopolymer conjugates were used to create environmentally desirable and economically-effective materials • Equal proportions of these bio-polymer blending conjugates displayed intriguing characteristics • Oregano oil has been successfully added to these biopolymer conjugates demonstrated effective resistance against browning reaction by resisting the enzyme peroxidase activity which potentially renders them in developing intelligent packing systems for horticultural produces, making it a viable option for anti-browning packaging	Romani et al. (2017)
Microcrystalline cellulose, corn starch-based active packaging improved with resveratrol, β-cyclodextrin complex	• Thermal durability and moisture-vapor resistance properties improved in the complex-incorporated composite films • The combination of resveratrol and β-cyclodextrin can enhance the antibacterial and antioxidant properties of manufactured films • Microcrystalline cellulose-cornstarch films containing resveratrol and β- cyclodextrin complex were found effective as modern packaging films	Yang et al. (2025)

(continued)

Table 2.2 (continued)

Starch-polymer conjugates	Efficiency and application	References
Grape seed oil nanoliposomes incorporated in starch extracted from tubers like tapioca conjugated with konjac gums fabricated into active composite films	• Used in assessing the shelf life of refrigerated mutton • Different components exhibited the better affinity, with inclined physical and resistance • characteristics by the bonds within ionic matrices, which even improved the crystalline nature. The packing films possessed optimal oxidation resistance features along with • antibacterial activity, in addition to tardy grape seed oil release • This conjugated film preserves the refrigerated meat carcasses and extend its keeping quality up to 2 weeks which can be used in designing ecologically friendly intelligent meat packing films	Song et al. (2025)
Chitosan and Starch based Films incorporated with nitrogen and phosphorus doped carbon dots from tea	• Utilized to enhance the keeping quality of meat • Anti-microbial and antioxidant properties of the films fabricated were enhanced with the incorporation of phosphorus, nitrogen added carbon dots from tea • Effective against the strains of common microbes inducing meat spoilage like *E. coli* and others	Khan et al. (2023)
Eugenol incorporated Yam starch based active packaging films	• The mechanical and other physical properties of the fabricated packaging materials were dependent upon the plasticizer used • These anti-bacterial films showed resistance against L. Monocytogenes and others, followed by increasing the shelf life of pork meat beyond 50%	Cheng et al. (2019)
Microporous starch-polyvinyl alcohol loaded oregano essential oil-based films	• Inhibits lipid oxidation, microbial growth and protein degradation in sea bass • Oxidation resistance properties were enhanced by oregana-essential oil and microporous starch addition	Wang et al. (2022a, b)

packaging applications (Zhao et al. 2024). Starch aerogels are an uprising clean-label solution in active packaging employed for preserving foodstuffs. Starch aerogels from potatoes and beans can be reused as water absorbents up to 12 times, demonstrating a consistent H_2O-retention capability after the first immersion. Aerogels fabricated from starch extracted from maize is applied in breast exudates from chicken, with a thirty-nine percentage of absorption rate, demonstrating its suitability in fabricating novel innovative food packaging materials for fresh consumable preservation (Costa et al. 2024). Two starch sources with varying levels of amylose/amylopectin were used to create antimicrobial cryogels and hydrogels. The microstructural stability was found to be greater at pH 4.6 than at pH 6.9. The cryogels were designed for loading and releasing volatile antibacterial compounds, with potato starch samples absorbing more diacetyl than pea starch samples. Tests showed that cryogels filled with mint oil were more active than hydrogels. These materials offer potential socioeconomic advantages, including reducing food waste and preventing foodborne diseases. Through the use of ultrasonic treatment and the introduction of TiO_2 into the SC, a starch-based ethylene scavenger was effectively produced. The SC's polyhydroxy group, greater porosity with enhanced specific surface area made it an excellent vehicle for distributing TiO_2, which helped to raise the ethylene removal effectiveness from 43.55 to 89.50%. By breaking down and reordering the starch molecular chains, ultrasonication restored the porous structure and enhanced TSC's ethylene elimination rate by 63.23%. The right away synthesized a new biomolecule-based scavenger. It also contributes to ethylene photodegradation; its effective removal in packaging and extension of their shelf life offers a useful tactic. The newly prepared carbohydrate-based scavenger serves as a non-food contact inner filler to buffer collisions during the transportation of fruits and vegetables. It also contributes to ethylene photodegradation; its effective removal of ethylene in bananas within the packaging and extension of their keeping quality offers a useful fruit and vegetable packaging technique. The effectiveness of antimicrobial aerogels made with inexpensive ingredients (cellulose from sunflower stalks and acetylated potato starch) and the addition of essential oil from Thymus daenensis as an antibacterial vehicle in preserving koopeh cheese was examined. According to this study, aerogels formulated from cellulose with a starch incorporation rate of three percent or less showed more holes, which caused the essential oil to escape more quickly than anticipated. Psychrophilic microbes and yeast-mold colonies in cheese were shown to be gradually inhibited by the aerogels, with psychrophilic microorganisms being significantly vulnerable to TDEO than yeast-mold (Mirmoeini et al. 2023).

2.8 Starch and Pigments in Developing Smart-Intelligent Packaging Material

Conversely, as modern consumable packaging progresses, consumer choices dynamically change from conventional petrochemicals to biodegradable food packaging systems with infused pigments, rendering them in developing modern packaging systems. Upcoming packaging media consist of biologically active substances that exhibit dynamic alteration in a visually interactive manner that end users may perceive, which allows for visual checking on the food's quality while also increasing clarity, for instance, films fabricated from compounds that change colour according to the fluctuations in pH. Natural color pigments have been extensively studied for use in the development of biodegradable pH-sensitive starch films (Kumar et al. 2024). The physical characteristics of anthocyanin-based films are primarily determined by the supply and concentration of anthocyanins and their interactions with biopolymers. In addition, appropriate preparation processes (3D printing, electrostatic spinning, dip coating, casting, nanoparticle embedding and hot-pressing extrusion) have a substantial impact on the formulated properties of films. Anthocyanin-based films paired with suitable biopolymers like starch mediated by creative technologies can check the freshness of food goods while also extending their shelf life (Xu et al. 2024). A 3D smart film (CF) has been developed using blueberry anthocyanin and phycocyanin as indicators and chondroitin sulphate in a potato starch matrix. The film responds well to salmon and beef freshness, reducing visual detection disparities and expanding the use of 3D printing technology in intelligent food packaging (Bao et al. 2024). Using curcumin and the starch of Artocarpus L., compostable and pH-sensitive films were fabricated. The films' sensitivity was demonstrated by their pH-dependent color shifts, which exhibited red coloration at pH value ranging up to 2 and yellow-green coloration at pH value ranging up to 11. The Physico-mechanical characteristics started to diminish with the curcumin incorporation, but vapor transfer rate and crystalline property was enhanced considerably. Additionally, these composites has demonstrated its potential antibacterial and antioxidant properties, making them appropriate for clever meat packaging. Enhancing mechanical qualities is still a priority for the future (Ranade et al. 2025). Red pitaya peel extract, which is high in betacyanins, was added in varying amounts to starch and polyvinyl alcohol to create innovative food packaging films. The extract's primary constituents were betacyanins, which in alkaline circumstances showed notable color changes. The inclusion of the extract enhanced the films' compactness and the extract's betacyanins could form hydrogen bonds with glycerol, starch, and polyvinyl alcohol. The extract improved the films' mechanical, antioxidant, and antibacterial properties as well as their capacity to restrict water vapor in addition to UV light. Compared to other films, the one that included the extract was more ammonia-sensitive. Because of the collected volatile nitrogen compounds throughout the shrimp's rotting process, the film containing the extract showed noticeable color changes when used to monitor the freshness of the shrimp. The findings imply that the film with the extract can be utilized in the food business as intelligent and active packaging. Both betalain and oat starch were effective in using the smart food packaging material. The subcritical water extractor system's modification of oat starch enhanced the manufactured smart packaging film's practical qualities. The subcritical water extractor system effectively changes the starch in a short amount of time without the application of toxicological chemicals, in contrast to the chemical and biological approaches. The subcritical water extractor system's elevated reaction temperature effectively damaged the native oat starch's structure, lowering its molecular weight, viscosity, and crystallinity while enhancing its water solubility and film-forming ability. The cleverness of the modified oat starch/betalain sensing property was demonstrated when it changed the color of the chicken from red to yellow, indicating that betalain derived from vegetable waste could take the place of artificial dyes to visually check the freshness of food while it is being stored. The mechanical and physical characteristics of modified oat starch-based smart films were greatly boosted by the addition of CNF, which also increased betalain's heat sensitivity. Therefore, the current study sheds light on how to effectively modify starch using SCW and use modified oat starch with natural pigment for clever food packaging at the commercial level (Surendhiran et al. 2023).

2.9 Nano-Technological Interventions

Nanotechnology in food operations improves the functionality and quality of food packaging. Recent advances in food nano packaging led to the development of smart biopolymer-based materials, offering an alternative to non-degradable polymer compounds (Prakash et al. 2024; Gupta et al. 2024c). Starch nanoparticles are nano-sized starch particles. They can be created using different methodologies, notably physico-chemical methodologies. These innovative fabricated materials have been demonstrated to exhibit distinguishing features like the greater area to volume ratio, making them suitable for use in a variety of applications, including medicine implementation and food packaging (Gupta et al. 2024e). In consumable packaging sectors, starch nanoparticles can be utilized to improve the barrier qualities in the packaging material, extending the shelf life of the consumables. Starch nanoparticles are quite a promising tool to be utilized as flexible and renewable materials with numerous potential applications (Mittu et al. 2024; Saeed et al. 2024). Nanotechnological interventions elevated the mechanical robustness and microbial resistance of films fabricated from starch of nano size and nano-starches coated with tannic acid. In just ten days, these nano-based biofilms demonstrate increased tensile strength and biodegradability (Karmakar et al. 2024). Catfish bone side streams used in synthesizing hydroxyapatite of nano-size (n-HAP), resulting in superior mechanical and barrier properties in composite films fabricated from cassava starch nano-hydroxyapatite (Thool et al. 2025). Biologically active nanocomplexes from curcumin have been extensively fabricated in bio-composite films from corn starch (CS) that exhibited uniform dispersion in the matrix of polymers through physical interaction. Furthermore, the physical, air resistance, hydrophobic attributes, visual, and thermal attributes, as well as the oxidation resistance property of the formulated films has exponentially improvised by the supplementation of curcumin loaded bioactive nano complexes, that revealed well to preserve the after-harvest characteristics of blueberry fruits through enhancing antioxidant enzyme activity and suppressing excessive reactive oxygen species accumulation, which contributes to the progression of bioactive packaging along with offering new perspectives. The creation of active packaging promises to absorb reactive radicals from food and protect it from intrinsic and external causes, hence improving holding time and safety of the consumables (Li et al. 2024). Bioactive edible films made from encapsulated quercetin along with starch nano-sized particles from lotus stem, in addition to gellan-gums and other polymers. The films formulated were observed with enhanced durability, thermal endurance, barrier characteristics, and antioxidant activity. It also demonstrated effective grape preservation during 21 days of storage. The film also shows an increase in oxygen and fungi-resistant properties, prolonging the keeping quality of horticultural commodities like grapes (Zeng et al. 2024).

Carbon based dots (CDs) synthesized and integrating it in a film composite composed of other polymers polyvinyl alcohol and corn starch to generate a packaging material to prolong the keeping quality of mushroom demonstrated effective mechanical property, ultraviolet radiation resistibility, and antimicrobial and antioxidant properties that potentially retarded mushroom browning reactions asserting which carbon dots could be an progressive tool in prolonging consumable shelf-life (Liu et al. 2024). High amylose starch (HAMS) was used as an efficient vehicle for integrating thymol into pullulan nanofibers. Additional studies into the interaction of high amylose starch, thymol, and pullulan in nanofibers were conducted. HAMS demonstrated efficacy in increasing thymol bioavailability and offers an exciting option to create palatable, edible polysaccharide anti-microbial nanofibrous films in consumable processing sectors (Liang et al. 2025). The fabrication of nano-bio composites with intelligent amenities and reducing environmental impacts is currently vital to humanity's future growth. Furthermore, consumer awareness of environmental issues drives industries to seek alternative "green" solutions, and the proper application of appropriate nanomaterials plays an essential function in enhancing consumable keeping quality and protection, providing a permanent solution to petro-based conventional food packaging materials by fabrication of packing films with enhanced physio-chemical characteristics (Kaur et al. 2021; Pattnaik et al. 2024).

2.10 Applications of Starch-Based Polymers in Food Packaging

The study of sustainable alternatives for various industrial uses has become essential due to the increasing worry about the environmental damage that conventional polymer compounds from petroleum cause. Among those alternatives, biopolymers from renewable resources generate considerable interest, particularly in the food packaging industry. Starch-based polymers have become an appropriate choice because of their intrinsic functional qualities, availability, affordability, and biodegradability (Liu et al. 2022a, b).

Starch is a natural polymer derived from various plant sources, including maize, potatoes, wheat, and cassava. Starch provides a flexible substrate for synthesizing polymers. It is a desirable alternative for the packaging sector because of its film-forming capabilities, excellent oxygen barrier characteristics, and compatibility with other biopolymers. Starch-based polymers offer advantages, though they have several limitations for food packaging. Several investigations on starch modification and composite material creation have been prompted by difficulties, including low mechanical qualities, high hydrophilicity, and restricted heat stability. Innovations like chemically altering its structure, combining starch with other biopolymers, or reinforcing it with nanoparticles have significantly enhanced its performance and extended its uses.

This section analyses improvements in starch-based polymer technology utilized in food packaging. It investigates the fundamental characteristics of starch, the various methods used to enhance them, and the particular functions they serve in

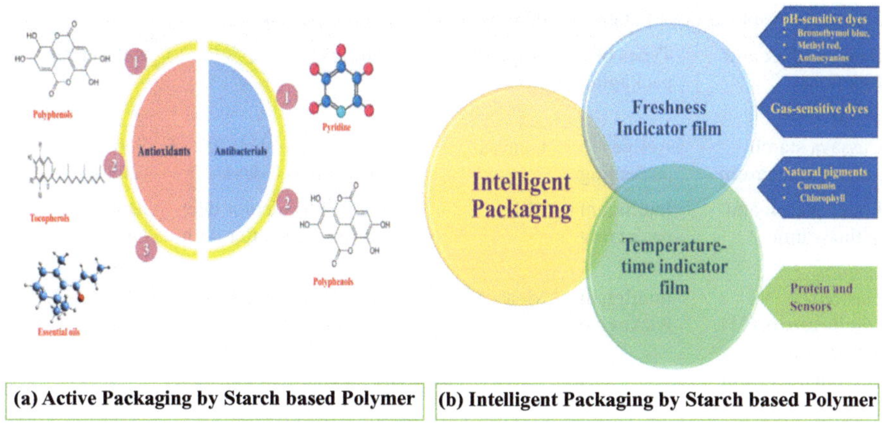

Fig. 2.3 Applications of Starch-based polymers based on their functionality, (a) Active packaging by starch-based polymer (b) Intelligent packaging by starch-based films and different types of films

food packaging systems. Figure 2.3 illustrates the applications of starch-based polymers based on the functionality, which demonstrates its application towards food products mentioned in Table 2.3.

2.10.1 Active Packaging

Deterioration of food is mainly caused by oxidation and bacterial contamination. Similar to standard packaging, starch packaging focuses on the issue of extending the shelf life of goods. The primary method of preserving food is by releasing active ingredients from starch substrates into the packaged food and adding antioxidants or antimicrobials. Thus, it is possible to increase the quality and shelf life of food. In brief, the potential release of active constituents from the film environment mainly determines a film's antioxidant activity. This is thoroughly associated with the cause of the active substances, the environments under which they are obtained, the additional quantity, the interrelation between the functional compounds and the film material, and the microstructures of the film.

2.10.2 Active Packaging by Antioxidants

Active packaging uses elements or techniques that may absorb or release materials (such as moisture, oxygen, or antibacterial chemicals) to preserve food quality (Fig. 7.1). The common types of active packaging are oxygen scavengers, moisture regulators, antimicrobial agents, antioxidants, and ethylene absorbers. By incorporating

Table 2.3 Applications of starch-based polymers in the food packaging

Base product and functional compounds	Types of packaging	Applications	Findings	References
Cassava starch; gelatin and casein	Active packaging	Guavas	Increased the shelf life of guavas up to 2 days	Pellá et al. (2020)
Potato starch and anthocyanin	Intelligent packaging	Milk	Changed film color deep blue to purple after 48 h	Moazami Goodarzi et al. (2020)
Dioscorea zingiberensis C.H. Wright starch; polyphenols from butterfly pea extract	Intelligent Packaging	Meat	The color change from blue-violet to green-yellow monitors the freshness of chicken, when pH changes from 2 to 13	Wang et al. (2023)
Corn starch; tea polyphenols	Active packaging	Food matrix	Excellent antioxidant properties, thermal stability	Miao et al. (2021)
Corn starch; red cabbage extract and sweet whey	Active packaging	Meat	Excellent antioxidant property, good structural integrity	Ribeiro Sanches et al. (2021)
Starch and agar; silver nanoparticles from mushroom extract	Active packaging	–	Antibacterial activity against *E. coli* and *L. monocytogenes*	Roy and Rhim (2022)
Tapioca Starch; lemon juice	Active packaging	Chicken	Delays the oxidation changes in chicken	Kanatt (2020)
Corn starch; curcumin	Intelligent packaging	Fish	The color change from yellow to red indicates the spoilage of fish	Liu et al. (2021)
Starch/polyvinyl alcohol; Cu-based nanocrystals	Active and intelligent packaging	Shrimp	Antibacterial, discoloration capability	Huang et al. (2024)

antioxidants into the packaging materials, the antioxidant active film can prevent or postpone food oxidation and produce an antioxidant effect. Antioxidants are released into the food, which is packaged by the antioxidant-enriched polymer material, extending its shelf life and, to some extent, maintaining its quality. There are two types of antioxidants used in food packaging: synthetic and naturally derived. Synthetic antioxidants, such as butyl-hydroxyanisole (BHA) and butyl-hydroxytoluene (BHT), starch structures to inhibit lipid oxidation have been frequently utilized in food packaging. Chemically synthesized antioxidants are typical components for toxicity and carcinogenicity. As a result, the focus of current research has shifted to naturally derived antioxidants, such as plant extracts, tocopherols, polyphenols, and vital oils, which are ideal for active packaging materials. These include packaging that incorporates antioxidants to prevent the oxidation of food and staleness and packaging that uses antioxidant dissemination to the food surface. The starch films incorporated with tea polyphenols microcapsules with a controlled release were developed by Miao et al. (2021). The biodegradable carboxymethyl cellulose and cassava starch films (3:7 w/w films) loaded with quercetin and tertiary butylhydroquinone as antioxidants show their improved tensile strength, retards its lard oxidation and delaying the pork discoloration (Tongdeesoontorn et al. 2020). The development of poly (butylene adipate-co-terephthalate) (PBAT)/thermoplastic starch-based films incorporated with quercetin for fruit storage has excellent antioxidant properties towards its shelf life (Yang et al. 2023).

2.10.3 Active Packaging by Antibacterials

Compared to conventional packaging, the starch-based polymer has weak anti-bacterial activity. To overcome these difficulties, active packaging loaded with anti-bacterial material plays a vital role in food industry packaging applications. Due to electrostatic absorption, anti-bacterial agents such as ammonium salts, pyridine, polyphenols, etc., act as anti-bacterial properties on the food packaging. According to the study Fonseca et al. (2019), the nanofibers incorporated with carvacrol encapsulates perform their antimicrobial properties towards *S. aureus.* in the potato starch films. In this study (Yusof et al. 2021), polybutylene succinate/tapioca starch (PBS/TPS) and biopolymer polybutylene succinate (PBS) films incorporating Bio master-silver and SANOFER for the chicken breasts have been delayed the hardening of the breasts due to its antimicrobial property of SANOFER. Chitosan nanoparticles (CNPs) were added in tapioca starch films at 5, 10, 15, and 20 weight percentages (Shapi'i et al. 2020). The disc diffusion experiment showed that more potent antibacterial activity was correlated with higher CNP concentration. Gram-positive bacteria (B. cereus and B. aureus) were suppressed at 15 and 20% CNP. Still, no inhibition of gram-negative bacteria (E. coli and S. typhi) was shown for any of the films made. The ability of the 15-weight percent CNP film to inhibit bacteria in cherry tomatoes was also examined. The bacterial growth in a clean starch film was 5.3 times lower than that of exposed tomatoes after 10 days of cold storage. In contrast, the bacterial growth in the film containing CNP was 16 times lower than in uncovered tomatoes.

2.10.4 Controlled Release Starch-Based Active Films

Commonly used antibacterial and antioxidant materials have strong antibacterial and antioxidant properties (Chen et al. 2018). However, if the rate of release is accelerated, the active substance's concentration is too high, which can cause deterioration or interaction with the food substituents; if the rate of the discharge is slow, the absorption of the active components is insufficient, making the food susceptible to spoiling. As a result, controlling the solvation of active components from polymers is the primary goal of both present and future research. Controlled release packaging (CRP) regulates the solvation of active ingredients during food packing, extending the shelf life of products (Almasi et al. 2021). Several approaches to designing CRP include cross-linkers, multilayer preparation, and chemical modification. The most current approaches for creating CRP systems have recently involved using a metal–organic framework (MOF) in an active starch film matrix and microencapsulating active chemicals (Vasile and Baican 2021). The regulated release system may be separated towards the inside of an absorption complex and a release system based on the mechanism of action. Antimicrobial agents are applied to the food's exterior as active ingredients in the release operations to stop food from spoiling and losing quality. Composite films made of surfactants and starch matrix are made to release active ingredients by disintegration, expansion, and diffusion. The film made of starch molecules is a sort of bottom-induced release. Since the starch-based film system has a modest diffusion coefficient, the additional active substance cannot diffuse within the starch matrix. When the material from starch compounds is submerged in an aqueous medium, the starch swells into the matrix through the water, since most meals are humid and contain a lot of water. When the active agent is in the expansion stage, its diffusion coefficient rises and spreads outwards (Kumar et al. 2015). Considering this further, as food's moisture content increases, so does the rate at which it destroys and releases active ingredients into starch-based films, demonstrating that films can extend the shelf life of goods. Prepared maize starch (CS) and Zein-Rutin composite nanoparticles (RNs) are essential to creating active films. According to the experimental results (Zhang and Zhao 2017), the mild binding between CS and RNs caused the initial release rate to be rapid. After two hours, the release amount reached 19.8 ~ 27.1%. In the stage of sustained release, Rutin is released from CS. Due to Rutin migrating from nanoparticles to solution, only 27–37.1% of the total Rutin was released after 12 h. RNs distributed in CS film may be managed and released in liquid food packaging. Maize starch and peas are essential for creating a starch film with doughnut-like quercetin-starch particles. Quercetin films' in vitro release in aqueous ethanol was studied. The release of quercetin to the grain starch film takes about 1–4 days, whereas a week is required to release the legume starch film (Farrag et al. 2018).

2.10.5 Intelligent Packaging

Packaging technologies have advanced significantly due to the growing complexity of global food supply networks and consumer demand for safer, higher-quality goods. Intelligent packaging is one of these developments that has proven to be a game-changer, allowing for real-time communication and monitoring of the condition and quality of packaged food. Intelligent packaging concentrates on obtaining, processing, and communicating vital information about the product and its surroundings, in contrast to traditional and active packaging, which mainly seek to safeguard and preserve food. Components, including sensors, indicators, and communication technologies, are used in intelligent packaging systems to monitor key factors like freshness, temperature, humidity, and gas composition. These systems give producers, retailers, and consumers helpful information on food items' quality, safety, and use. For example, freshness indicators can identify microbial activity or spoilage, lowering the risk of foodborne infections. In contrast, time-temperature indicators can indicate if a product has been subjected to unfavorable storage conditions. There are several benefits to incorporating intelligent packaging into food systems. It reduces food waste by optimizing consumption choices, increases supply chain efficiency through better traceability, and boosts customer confidence through transparency. Furthermore, the application of cutting-edge technology like biosensors, QR codes, and RFID tags is consistent with the food industry's increasing focus on digital transformation and sustainable practices.

2.10.5.1 Freshness Indicator

To determine the quality of the food product within the package, freshness indicators monitor or react with metabolites produced in the fresh food (such as CO_2, organic acids, volatile elemental nitrogen, or sulfur derivatives) due to microbial growth. This results in a noticeable color change. A hydrogel loaded with baggage nanocellulose acts as a freshness indicator for the spoilage of chicken to change its color from green to red (Lu et al. 2020). The pH-sensitive dye-based indicator films were prepared for kimchi storage, where the color changes from blue to yellow (Baek et al. 2020). The freshness indicators such as methyl red, bromothymol 368 blue, and a mixture of bromothymol blue and bromocresol purple were added to the packaging material, indicating the storage life of lean pork from the changed its color from red to green (Chen et al. 2019). The halochromic freshness indicator containing poly-lactic acid and anthocyanin indicates the freshness of shrimp, minced beef, chicken fillets, and fish roe (Ghorbani et al. 2021)Anthocyanin as a freshness indicator used to develop the pH-sensitive intelligent packaging for pork, beef, cheese, and yogurt (Zhao et al. 2022).

In pH sensors, chemical data is transformed into electrical signals for analytical tests using electrochemical processes. The sensor transforms chemical data into useful energy, which is then transformed into electrical impulses. Chemical alterations are produced when food spoilage chemical byproducts come into contact with the electrodes in food packaging. Colorimetric sensors and pH sensors have a stronger correlation in terms of sensitivity than electrochemical sensors do. For colorimetric sensors, the primary sources of visual experiences are the pH range, sensitivity, or color intensity. This increase in color intensity can be facilitated by reducing the number of binding substrates and increasing the percentage of pH-sensitive compounds (Benjamin et al. 2023). The sensitivity increases with the amount of porous starch; the mechanical strength of the sensor increases with the amount of cellulose binder; the color intensity increases with the amount of anthocyanin.

2.10.5.2 *Temperature-Time Indicator*

Since perishable food is temperature-sensitive, storing it at a low temperature will extend its shelf life. High temperatures cause food to deteriorate more quickly and reach the end point of its storage life earlier. Temperature and time have a significant impact on the quality of most meals. The temperature-time indicator (TTI) can record and show the temperature change of a food's remaining shelf life in circulation. The shift in product temperature in each connection may be tracked using product time and temperature data, ensuring food safety and quality. According to temperature variations brought on by chemical, microbial, enzyme, or physical changes, this offers irreversible, observable color changes. The shift in color correlating to the accompanying shelf life of food often represents the range from activation to termination.

The study of temperature and time indicators started later than other fields, and there doesn't appear to be a significant relevant study in these areas. Santos et al. (2020) included nitrite and myoglobin extracts into the cassava starch thermoplastic sensor sheet to replace conventional electrical time-temperature sensors. They developed a thermochromic protein sensor that is natural, non-toxic, and biodegradable to investigate how myoglobin and nitrite proteins change color in response to temperature. According to currently known research, the color principle of CR and ATH is mainly utilized in creating intelligent packaging materials because it has a good indicator influence on food with a significant shift in pH value during food rotting. However, further research is needed to determine the exact association between changes in pH values and quality. Furthermore, intelligent packaging has been used to manufacture film material with antibacterial properties. Thus, future studies will focus on creating intelligent, active packaging materials.

2.11 Limitations and Future Prospects

Starch-based films are widely employed for food applications; however, retrogradation of starch is a major limitation that needs to be addressed (Li 2024). Studies relying on characteristic features of these films should be broadened for a better understanding of the properties exhibited by starch-based films. A few other bottlenecks associated with these starch-based films include poor mechanical properties. Inadequate techniques or lack of production facilities are a few limitations to thin film production. The strength of these developed films is an integral and significant factor for myriad food applications. Different ways of increasing the durability of these films include blending other polymers and incorporating starch-based films (Fredi and Dorigato 2024). Properties of film such as water resistance and mechanical properties can be enhanced with the help of modifications adopted with the help of physical and chemical methods. The usage of improved plasticizers is another way by which functional properties can be improved. Though high-strength starch-based films are developed, scale-up, sustainable raw material sourcing, and integration of circular models are an integral part of future takeaways for any developed product to be utilized to its full potential (Dziuba et al. 2021). Employing appropriate tools such as Artificial Intelligence (AI) and machine learning tools to optimize the composition of different plasticizers and polymers will also aid in choosing the appropriate material while reducing the wastage of raw materials with fewer experiments undertaken.

2.12 Conclusion

Starch-based films are an eco-friendly alternative polymer in the packaging industry. Starch, being abundant, cost-effective, and versatile, can be used to form these biodegradable films. These films are obtained from bio-sources, which are renewable and biodegradable sources. These films address the growing concerns about plastic pollution. The applications of these biodegradable films in the food sector are diverse and have been used increasingly for packaging fresh produce, bakery products, dairy items, and dry foods, where biodegradability and environmental sustainability are crucial. Moreover, their potential for reducing food wastage and plastic pollution aligns with global goals for sustainability. Functional properties of starch-based films are yet another significant area of studies, where the properties are enhanced with the help of nanoparticles and other polymers, etc., overcoming bottlenecks which include brittleness, high hydrophilicity, and low mechanical strength. Techniques like incorporating starch with other biopolymers (e.g., chitosan, gelatin, or polyvinyl alcohol) and incorporating nanomaterials (e.g., cellulose nanocrystals or clay nanoparticles) have improved their flexibility, mechanical robustness, and barrier properties. Additionally, functionalization was carried out using essential oils, antimicrobial agents, or antioxidants has expanded their potential beyond passive packaging. All these innovations have enabled the formation of smart and intelligent packaging that can monitor, preserve, and enhance the shelf life of food products. Despite these advancements, challenges such as moisture sensitivity, thermal instability, and the need for scalable production methods persist. However, researchers have made significant efforts to address these issues. Collaborations between academia, industry, and policymakers hold a prominence in driving innovation, reducing cost, and fostering the regulatory frameworks necessary to support the widespread use. Therefore, starch-based films offer a potential remedy for the problems of environmental sustainability and food preservation. Furthermore, these materials can be investigated to influence the direction of environmentally friendly food packaging in the future and thereby build a more robust and sustainable global food system with continuous advancements.

Declaration Regarding AI Assistance The authors acknowledge that AI-based language tools were employed solely for the purpose of refining grammar, syntax, readability, and overall linguistic clarity. The authors have reviewed and verified the final content after using these tools and take full responsibility for its accuracy, interpretation, and overall presentation in this chapter.

References

Abedi-Firoozjah R, Chabook N, Rostami O, Heydari M, Kolahdouz-Nasiri A, Javanmardi F, Abdolmaleki K, Mousavi Khaneghah A (2023) PVA/starch films: an updated review of their preparation, characterization, and diverse applications in the food industry. Polym Test 118:107903. https://doi.org/10.1016/j.polymertesting.2022.107903

Abera G, Woldeyes B, Demash HD, Miyake G (2020) The effect of plasticizers on thermoplastic starch films developed from the indigenous Ethiopian tuber crop Anchote (Coccinia abyssinica) starch. Int J Biol Macromol 155:581–587. https://doi.org/10.1016/j.ijbiomac.2020.03.218

Abera WG, Kasirajan R, Majamo SL (2024) Synthesis and characterization of bioplastic film from banana (Musa Cavendish species) peel starch blending with banana pseudo-stem cellulosic fiber. Biomass Convers Biorefinery 14(17):20419–20440. https://doi.org/10.1007/s13399-023-04207-8

Ahmadi F, Pirdashti M, Rostami AA (2018) Density, refractive index and liquid – liquid equilibrium data of polyethylene glycol 3000 + potassium formate + water at different pH values. Chin J Chem Eng 26:168–174. https://doi.org/10.1016/j.cjche.2017.07.003

Akinsemolu AA, Idowu AM, Onyeaka HN (2024) Recycling technologies for biopolymers: current challenges and future directions. Polymers 16(19):1–18. https://doi.org/10.3390/polym16192770

Almasi H, Jahanbakhsh Oskouie M, Saleh A (2021) A review on techniques utilized for design of controlled release food active packaging. Crit Rev Food Sci Nutr 61(15):2601–2621. https://doi.org/10.1080/10408398.2020.1783199

Amaraweera SM, Gunathilake C, Gunawardene OH, Fernando NM, Wanninayaka DB, Manamperi A, Dassanayake RS, Rajapaksha SM, Gangoda M, Fernando CAN, Kulatunga AK, Manipura A (2021) Preparation and characterization of biodegradable cassava starch thin films for potential food packaging applications. Cellulose 28(16):10531–10548

Amaraweera SM, Gunathilake C, Gunawardene OHP, Dassanayake RS, Fernando NML, Wanninayaka DB, Rajapaksha SM, Manamperi A, Gangoda M, Manchanda A, Fernando CAN, Kulatunga AK, Manipura A (2022) Preparation and characterization of dual-modified cassava starch-based biodegradable foams for sustainable packaging applications. ACS Omega 7(23):19579–19590. https://doi.org/10.1021/acsomega.2c01292

Baek S, Maruthupandy M, Lee K, Kim D, Seo J (2020) Freshness indicator for monitoring changes in quality of packaged kimchi during storage. Food Packag Shelf Life 25:100528. https://doi.org/10.1016/j.fpsl.2020.100528

Bao Y, Wang M, Li J, Yang X, Wen B, Wang L, He Y, Si X, Li B (2024) A starch-based 3D printed intelligent colorimetric film co-loaded natural pigments for visualizing food freshness: effect of nozzle size on gel structure formation. Food Hydrocoll 155:110218. https://doi.org/10.1016/j.foodhyd.2024.110218

Benjamin S, de Lima F, Nascimento V, de Andrade G, Oriá R (2023) Advancement in paper-based electrochemical biosensing and emerging diagnostic methods. Biosensors 13(7):689. https://doi.org/10.3390/bios13070689

Bof MJ, Locaso DE, García MA (2021) Corn starch chitosan proportion affects biodegradable film performance for food packaging purposes. Starch-Stärke 73(5–6). https://doi.org/10.1002/star.202000104

Brain Wilfer P, Giridaran G, Jeya Jeevahan J, Britto Joseph G, Senthil Kumar G, Thykattuserry NJ (2021) Effect of starch type on the film properties of native starch based edible films. Mater Today: Proc 44:3903–3907. https://doi.org/10.1016/j.matpr.2020.12.1118

Castro LMG, Alexandre EMC, Saraiva JA, Pintado M (2023) Starch extraction and modification by pulsed electric fields. Food Rev Int 39 (4):2161–2182. https://doi.org/10.1080/87559129.2021.1945620

Chen C, Xu Z, Ma Y, Liu J, Zhang Q, Tang Z, Fu K, Yang F, Xie J (2018) Properties, vapour-phase antimicrobial and antioxidant activities of active poly(vinyl alcohol) packaging films incorporated with clove oil. Food Control 88:105–112. https://doi.org/10.1016/j.foodcont.2017.12.039

Chen H, Zhang M, Bhandari B, Yang C (2019) Development of a novel colorimetric food package label for monitoring lean pork freshness. LWT 99:43–49. https://doi.org/10.1016/j.lwt.2018.09.048

Chen J, Wang Y, Liu J, Xu X (2020) Preparation, characterization, physicochemical property and potential application of porous starch: a review. Int J Biol Macromol 148:1169–1181. https://doi.org/10.1016/j.ijbiomac.2020.02.055

Cheng J, Wang H, Kang S, Xia LI, Jiang S, Chen M, Jiang S (2019) An active packaging film based on yam starch with eugenol and its application for pork preservation. Food Hydrocoll 96:546–554. https://doi.org/10.1016/j.foodhyd.2019.06.007

Cheng H, Chen L, McClements DJ, Yang T, Zhang Z, Ren F, Miao M, Tian Y, Jin Z (2021) Starch-based biodegradable packaging materials: a review of their preparation, characterization and diverse applications in the food industry. Trends Food Sci Technol 114:70–82. https://doi.org/10.1016/j.tifs.2021.05.017

Combrzyński M, Wójtowicz A, Mitrus M, Oniszczuk T, Matwijczuk A, Pawelczyk P, Mościcki L (2019) Effect of starch type and screw speed on mechanical properties of extrusion-cooked starch-based foams. Int Agrophys 33(2):233–240. https://doi.org/10.31545/intagr/109517

Costa DA, da Silva FT, da Cruz EP, da Fonseca LM, Jansen ET, Oliveira RM, Souza AS, da Zavareze ER, Zambiazi RC (2024) Water absorbent biodegradable aerogels based on potato, cassava, bean, and maize starches applied in the absorption of chicken exudate. Starch-Stärke 76(11–12). https://doi.org/10.1002/star.202300070

Cruz-Tirado JP, Barros Ferreira RS, Lizárraga E, Tapia-Blácido DR, Silva NCC, Angelats-Silva L, Siche R (2020) Bioactive Andean sweet potato starch-based foam incorporated with oregano or thyme essential oil. Food Packag Shelf Life 23:100457. https://doi.org/10.1016/j.fpsl.2019.100457

Cui C, Ji N, Wang Y, Xiong L, Sun Q (2021) Bioactive and intelligent starch-based films: a review. Trends Food Sci Technol 116:854–869. https://doi.org/10.1016/j.tifs.2021.08.024

Cui Y, Li C, Guo Y, Liu X, Zhu F, Liu Z, Liu X, Yang F (2022) Rheological & 3D printing properties of potato starch composite gels. J Food Eng 313(May 2021), 110756. https://doi.org/10.1016/j.jfoodeng.2021.110756

Dokl M, Copot A, Krajnc D, Van Fan Y, Vujanović A, Aviso KB, Tan RR, Kravanja Z, Čuček L (2024) Global projections of plastic use, end-of-life fate and potential changes in consumption, reduction, recycling and replacement with bioplastics to 2050. Sustain Prod Consum 51:498–518

Duan Q, Zhu Z, Chen Y, Liu H, Yang M, Chen L, Yu L (2022) Starch-based foams nucleated and reinforced by polysaccharide-based crystals. ACS Sustain Chem Eng 10(6):2169–2179. https://doi.org/10.1021/acssuschemeng.1c07738

Dziuba R, Kucharska M, Madej-Kiełbik L, Sulak K, Wiśniewska-Wrona M (2021) Biopolymers and biomaterials for special applications within the context of the circular economy. Materials 14(24):1–15. https://doi.org/10.3390/ma14247704

Engel JB, Ambrosi A, Tessaro IC (2019) Development of biodegradable starch-based foams incorporated with grape stalks for food packaging. Carbohydr Polym 225. https://doi.org/10.1016/j.carbpol.2019.115234

Ezati P, Rhim J-W (2021) Starch and agar-based color-indicator films integrated with shikonin for smart packaging application of shrimp. ACS Food Sci Technol 1(10):1963–1969. https://doi.org/10.1021/acsfoodscitech.1c00292

Farrag Y, Ide W, Montero B, Rico M, Rodríguez-Llamazares S, Barral L, Bouza R (2018) Starch films loaded with donut-shaped starch-quercetin microparticles: characterization and release kinetics. Int J Biol Macromol 118:2201–2207. https://doi.org/10.1016/j.ijbiomac.2018.07.087

Fatima S, Khan MR, Ahmad I, Sadiq MB (2024) Recent advances in modified starch based biodegradable food packaging: a review. Heliyon 10(6):e27453. https://doi.org/10.1016/j.heliyon.2024.e27453

Fonseca LM, dos Santos Cruxen CE, Bruni GP, Fiorentini ÂM, da Rosa Zavareze E, Lim LT, Dias AR (2019) Development of antimicrobial and antioxidant electrospun soluble potato starch nanofibers loaded with carvacrol. Int J Biol Macromol 139:1182–1190. https://doi.org/10.1016/j.ijbiomac.2019.08.096

Fredi G, Dorigato A (2024) Compatibilization of biopolymer blends: a review. Adv Ind Eng Polym Res 7(4):373–404. https://doi.org/10.1016/j.aiepr.2023.11.002

Gautam N, Garg S, Yadav S (2021) Underutilized finger millet crop for starch extraction, characterization, and utilization in the development of flexible thin film. J Food Sci Technol 58(11):4411–4419. https://doi.org/10.1007/s13197-020-04926-0

Ghorbani M, Divsalar E, Molaei R, Ezati P, Moradi M, Tajik H, Abbaszadeh M (2021) A halochromic indicator based on polylactic acid and anthocyanins for visual freshness monitoring of minced meat, chicken fillet, shrimp, and fish roe. Innov Food Sci Emerg Technol 74:102864. https://doi.org/10.1016/j.ifset.2021.102864

Ghoshal G, Singh D (2020) Synthesis and characterization of starch nanocellulosic films incorporated with eucalyptus globulus leaf extract. Int J Food Microbiol 332:108765. https://doi.org/10.1016/j.ijfoodmicro.2020.108765

Goiana ML, Mattos ALA, de Rosa MF, Fernandes FAN (2024) Use of cold plasma as an alternative to improve corn starch-based films: effect of the plasma application strategy. Processes 12(7):1429. https://doi.org/10.3390/pr12071429

Guillard V, Gaucel S, Fornaciari C, Angellier-Coussy H, Buche P, Gontard N (2018) The next generation of sustainable food packaging to preserve our environment in a circular economy context. Front Nutr 5(December):1–13. https://doi.org/10.3389/fnut.2018.00121

Gupta RK, Guha P, Srivastav PP (2022) Natural polymers in bio-degradable/edible film: a review on environmental concerns, cold plasma technology and nanotechnology application on food packaging-a recent trends. Food Chem Adv 1:100135

Gupta RK, Guha P, Srivastav PP (2023) Effect of high voltage dielectric barrier discharge (DBD) atmospheric cold plasma treatment on physicochemical and functional properties of taro (Colocasia esculenta) starch. Int J Biol Macromol 253:126772

Gupta RK, Guha P, Srivastav PP (2024a) Dielectric barrier discharge plasma: a green and novel method to change the structure of taro peel starch and improve the physicochemical properties of taro peel starch films. Plasma Processes Polym 21(9):2400047

Gupta RK, Guha P, Srivastav PP (2024b) Exploring the potential of taro (Colocasia esculenta) starch: recent developments in modification, health benefits, and food industry applications. Food Bioeng 3(3):365–379

Gupta RK, Abd El Gawad F, Ali EA, Karunanithi S, Yugiani P, Srivastav PP (2024c) Nanotechnology: current applications and future scope in food packaging systems. Meas Food 13:100131

Gupta RK, Guha P, Srivastav PP (2024d) Insights into the physicochemical, crystalline, multi-structural, thermal and digestibility behavior of taro starch (Colocasia esculenta) modified by cold plasma treatment with high voltage dielectric barrier discharge (HVDBD): an alternative technique for chemical modification of starch and environmental sustainability. J Clean Prod 461:142710

Gupta RK, Guha P, Srivastav PP (2024e) Investigating the toxicological effects of nanomaterials in food packaging associated with human health and the environment. J Hazard Mater Lett 5:100125

Gupta RK, Gnana Moorthy Eswaran U, Kumar S, Pipliya S, Kaur B, Srivastav PP (2025) Synergistic modification of starch polysaccharide via sequential treatments with ultrasonic and cold plasma technology. J Food Process Preserv 2025(1):9063007

Hazrati KZ, Sapuan SM, Zuhri MYM, Jumaidin R (2021) Effect of plasticizers on physical, thermal, and tensile properties of thermoplastic films based on Dioscorea hispida starch. Int J Biol Macromol 185:219–228. https://doi.org/10.1016/j.ijbiomac.2021.06.099

Huang K, Wang Y, Xu Z, Zou Z, Tang Q, Li H, Peng D (2024) Novel intelligent packaging films based on starch/PVA with Cu-ICA nanocrystal as functional compatibilizer for monitoring food freshness. Int J Biol Macromol 271:132373. https://doi.org/10.1016/j.ijbiomac.2024.132373

Imam SH, Glenn GM, Chiellini E (2012) Utilization of biobased polymers in food packaging: assessment of materials, production and commercialization. In: Emerging food packaging technologies. Woodhead Publishing Limited. https://doi.org/10.1533/9780857095664.4.435

Kanatt SR (2020) Irradiation as a tool for modifying tapioca starch and development of an Active food packaging film with irradiated starch. Radiat Phys Chem 173:108873. https://doi.org/10.1016/j.radphyschem.2020.108873

Karmakar B, Sarkar S, Chakraborty R, Saha SP, Thirugnanam A, Roy PK, Roy S (2024) Starch-based biodegradable films amended with nano-starch and tannic acid-coated nano-starch exhibit enhanced mechanical and functional attributes with antimicrobial activity. Carbohydr Polym 341:122321. https://doi.org/10.1016/j.carbpol.2024.122321

Kaur G, Sharma S, Mir SA, Dar BN (2021) Nanobiocomposite films: a "Greener Alternate" for food packaging. Food Bioprocess Technol 14(6):1013–1027. https://doi.org/10.1007/s11947-021-02634-x

Khan A, Ezati P, Rhim JW (2023) Chitosan/starch-based active packaging film with N, P-doped carbon dots for meat packaging. ACS Appl Bio Mater 6(3):1294–1305. https://doi.org/10.1021/acsabm.3c00039

Kringel DH, Dias ARG, da Zavareze ER, Gandra EA (2020) Fruit wastes as promising sources of starch: extraction, properties, and applications. Starch/Staerke 72(3–4):1–9. https://doi.org/10.1002/star.201900200

Kumar Y, Yadav DN, Ahmad T, Narsaiah K (2015) Recent trends in the use of natural antioxidants for meat and meat products. Compr Rev Food Sci Food Saf 14(6):796–812. https://doi.org/10.1111/1541-4337.12156

Kumar Y, Bist Y, Thakur D, Nagar M, Saxena DC (2024) A review on the role of pH-sensitive natural pigments in biopolymers based intelligent food packaging films. Int J Biol Macromol 276:133869. https://doi.org/10.1016/j.ijbiomac.2024.133869

Kumari B, Sit N (2023) Comprehensive review on single and dual modification of starch: methods, properties and applications. Int J Biol Macromol 253:126952. https://doi.org/10.1016/j.ijbiomac.2023.126952

Li C, Hu Y, Li E (2021) Effects of amylose and amylopectin chain-length distribution on the kinetics of long-term rice starch retrogradation. Food Hydrocoll 111:106239. https://doi.org/10.1016/j.foodhyd.2020.106239

Li C (2024) Unraveling the complexities of starch retrogradation: insights from kinetics, molecular interactions, and influences of food ingredients. Food Rev Int 40(9):3159–3182. https://doi.org/10.1080/87559129.2024.2347467

Li X, Zhang X, Lv J, Zhang X, Li Y, Han X, Zhang W (2024) Development of starch-based films reinforced with curcumin-loaded nanocomplexes: characterization and application in the preservation of blueberries. Int J Biol Macromol 264:130464. https://doi.org/10.1016/j.ijbiomac.2024.130464

Liang Q, Kilmartin PA, Gao Q (2025) Effect of high amylose starch on the incorporation of thymol into electrospun pullulan nanofibers. Food Chem 465:142070. https://doi.org/10.1016/j.foodchem.2024.142070

Liu Y, Chao C, Yu J, Wang S, Wang S, Copeland L (2020) New insights into starch gelatinization by high pressure: comparison with heat-gelatinization. Food Chem 318:126493. https://doi.org/10.1016/j.foodchem.2020.126493

Liu D, Dang S, Zhang L, Munsop K, Li X (2021) Corn starch/polyvinyl alcohol based films incorporated with curcumin-loaded Pickering emulsion for application in intelligent packaging. Int J Biol Macromol 188:974–982. https://doi.org/10.1016/j.ijbiomac.2021.08.080

Liu D, Zhao P, Chen J, Yan Y, Wu Z (2022a) Recent advances and applications in starch for intelligent active food packaging: a review. Foods 11(18):2879. https://doi.org/10.3390/foods11182879

Liu Y, Liu M, Zhang L, Cao W, Wang H, Chen G, Wang S (2022b) Preparation and properties of biodegradable films made of cationic potato-peel starch and loaded with curcumin. Food Hydrocoll 130:107690. https://doi.org/10.1016/j.foodhyd.2022.107690

Liu Q, Gao L, Qin Y, Ji N, Dai L, Xiong L, Sun Q (2023) Incorporation of oxidized debranched starch/chitosan nanoparticles for enhanced hydrophobicity of corn starch films. Food Packag Shelf Life 35:101032. https://doi.org/10.1016/j.fpsl.2023.101032

Liu Z, Cui M, Weng R, Li H, Hati S, Hu L, Mo H (2024) Incorporation of carbon dots into polyvinyl alcohol/corn starch based film and its application on shiitake mushroom preservation. Int J Biol Macromol 280:135998. https://doi.org/10.1016/j.ijbiomac.2024.135998

Liu F, Cao Y, Ren J, Xie Y, Xiao X, Zou Y, Bai H, Zhang X, Chen Y (2025) Optimization of starch foam extrusion through PVA polymerization, moisture content control, and CMCS incorporation for enhanced antibacterial cushioning packaging. Carbohyd Polym 347:122763. https://doi.org/10.1016/j.carbpol.2024.122763

Lu P, Yang Y, Liu R, Liu X, Ma J, Wu M, Wang S (2020) Preparation of sugarcane bagasse nanocellulose hydrogel as a colourimetric freshness indicator for intelligent food packaging. Carbohydr Polym 249:116831. https://doi.org/10.1016/j.carbpol.2020.116831

Martin Torrejon V, Song H, Wu B, Luo G, Song J (2023) Effect of starch type and pre-treatment on the properties of gelatin-starch foams produced by mechanical foaming. Polymers 15(7):1775. https://doi.org/10.3390/polym15071775

Miao Z, Zhang Y, Lu P (2021) Novel active starch films incorporating tea polyphenols-loaded porous starch as food packaging materials. Int J Biol Macromol 192:1123–1133. https://doi.org/10.1016/j.ijbiomac.2021.09.214

Mirmoeini SS, Hosseini SH, Javid AL, Koutamehr ME, Sharafi H, Molaei R, Moradi M (2023) Essential oil-loaded starch/cellulose aerogel: preparation, characterization and application in cheese packaging. Int J Biol Macromol 244:125356

Mittu B, Lily MK, Begum Z, Bala R, Singh M, Kaur M (2024) Starch-based nanomaterials, their properties and sources. In: Starch based nanomaterials for food packaging. Elsevier, pp 1–22. https://doi.org/10.1016/B978-0-443-18967-8.00008-6

Moazami Goodarzi M, Moradi M, Tajik H, Forough M, Ezati P, Kuswandi B (2020) Development of an easy-to-use colorimetric pH label with starch and carrot anthocyanins for milk shelf life assessment. Int J Biol Macromol 153:240–247. https://doi.org/10.1016/j.ijbiomac.2020.03.014

Nayak SK (2010) Biodegradable PBAT/Starch nanocomposites. Polym Plast Technol Eng 49(14):1406–1418. https://doi.org/10.1080/03602559.2010.496397

Nordin N, Othman SH, Rashid SA, Basha RK (2020) Effects of glycerol and thymol on physical, mechanical, and thermal properties of corn starch films. Food Hydrocoll 106:105884. https://doi.org/10.1016/j.foodhyd.2020.105884

Oliveira G, Gonçalves I, Barra A, Nunes C, Ferreira P, Coimbra MA (2020) Coffee silverskin and starch-rich potato washing slurries as raw materials for elastic, antioxidant, and UV-protective biobased films. Food Res Int 138:109733. https://doi.org/10.1016/j.foodres.2020.109733

Othman SH, Othman NFL, Shapi'i RA, Ariffin SH, Md. Yunos KF (2021) Corn starch/chitosan nanoparticles/thymol bio-nanocomposite films for potential food packaging applications. Polymers 13(3):390. https://doi.org/10.3390/polym13030390

Parra-Campos A, Serna-Cock L, Solanilla-Duque JF (2022) Effect of the addition of fique bagasse microparticles in obtaining a biobased material based on cassava starch. Int J Biol Macromol 207:289–298. https://doi.org/10.1016/j.ijbiomac.2022.03.016

Patil S, Bharimalla AK, Mahapatra A, Dhakane-Lad J, Arputharaj A, Kumar M, Raja ASM, Kambli N (2021) Effect of polymer blending on mechanical and barrier properties of starch-polyvinyl alcohol based biodegradable composite films. Food Biosci 44:101352. https://doi.org/10.1016/j.fbio.2021.101352

Pattnaik R, Panda SK, Biswas S, De S, Satahrada S, Kumar S (2024) Prospects and challenges of nanomaterials in sustainable food preservation and packaging: a review. Discover Nano 19(1):178. https://doi.org/10.1186/s11671-024-04142-1

Pellá MCG, Silva OA, Pellá MG, Beneton AG, Caetano J, Simões MR, Dragunski DC (2020) Effect of gelatin and casein additions on starch edible biodegradable films for fruit surface coating. Food Chem 309:125764. https://doi.org/10.1016/j.foodchem.2019.125764

Prakash S, Kumari M, Chauhan AK (2024) The intervention of nanotechnology in food packaging: a review. J Mater Sci 59(7):2585–2601. https://doi.org/10.1007/s10853-024-09360-7

Promhuad K, Harnkarnsujarit N (2025) Effects of ethyl maltol on properties of starch-based active packaging: morphology, volatile release and mold inhibition. Food Control 167:110840. https://doi.org/10.1016/j.foodcont.2024.110840

Ranade T, Sati A, Pratap A, Mali SN (2025) Curcumin-integrated biopolymer films for active packaging: current trends and future directions. Chem Pap 79(3):1303–1334

Ribeiro Sanches MA, Camelo-Silva C, da Silva Carvalho C, Rafael de Mello J, Barroso NG, da Silva L, Barros E, Silva PP, Pertuzatti PB (2021) Active packaging with starch, red cabbage extract and sweet whey: characterization and application in meat. LWT 135:110275. https://doi.org/10.1016/j.lwt.2020.110275

Romani VP, Prentice-Hernández C, Martins VG (2017) Active and sustainable materials from rice starch, fish protein and oregano essential oil for food packaging. Ind Crops Prod 97:268–274. https://doi.org/10.1016/j.indcrop.2016.12.026

Roy S, Rhim J-W (2022) Starch/agar-based functional films integrated with Enoki mushroom-mediated silver nanoparticles for active packaging applications. Food Biosci 49:101867. https://doi.org/10.1016/j.fbio.2022.101867

Saeed F, Afzaal M, Khan MA, Niaz B, Rasheed A, Raza MA, Islam M, Arooj H (2024) Cereal starch-based nanoparticles in food packaging applications. In: Starch based nanomaterials for food packaging. Elsevier, pp 227–246. https://doi.org/10.1016/B978-0-443-18967-8.00003-7

Santos CT, Veiga Santos P, Sestari P, Sorrini NC, Oliveira Roça R (2020) Protein time–temperature sensor for intelligent starch polymers. J Food Process Preserv 44(5). https://doi.org/10.1111/jfpp.14428

Schmiele M, Sampaio UM, Gomes PTG, Clerici MTPS (2019) Physical modifications of starch. In: Starches for food application. Academic Press, pp 223–269

Shapi'i RA, Othman SH, Nordin N, Kadir Basha R, Nazli Naim M (2020) Antimicrobial properties of starch films incorporated with chitosan nanoparticles: in vitro and in vivo evaluation. Carbohydr Polym 230:115602. https://doi.org/10.1016/j.carbpol.2019.115602

Simões BM, Cagnin C, Yamashita F, Olivato JB, Garcia PS, de Oliveira SM, Eiras Grossmann MV (2020) Citric acid as crosslinking agent in starch/xanthan gum hydrogels produced by extrusion and thermopressing. LWT 125:108950. https://doi.org/10.1016/j.lwt.2019.108950

Singh S, Gaur S, Sharma N (2024) Starchy films as a sustainable alternative in food industry: current research and applications. Starch-Stärke 76(11–12). https://doi.org/10.1002/star.202300078

Song Z, Zang Z, Cao Y, Ma Y, Li B, Han L, Yu Q (2025) Tapioca starch/konjac gum-based composite film incorporated with nanoliposomes encapsulated grape seed oil: structure, functionality, controlled release and its preservation role for chilled mutton. Food Chem 463:141081. https://doi.org/10.1016/j.foodchem.2024.141081

Soykeabkaew N, Thanomsilp C, Suwantong O (2015) A review: starch-based composite foams. Compos A Appl Sci Manuf 78:246–263. https://doi.org/10.1016/j.compositesa.2015.08.014

Surendren A, Mohanty AK, Liu Q, Misra M (2022) A review of biodegradable thermoplastic starches, their blends and composites: recent developments and opportunities for single-use plastic packaging alternatives. Green Chem 24(22):8606–8636. https://doi.org/10.1039/D2GC02169B

Surendhiran D, Roy VC, Ho TC, Park JS, Lee SC, Chun BS (2023) Smart packaging film prepared from subcritical water-modified oat starch and betalain of beetroot extract reinforced with cellulose nanofibrils. Sustain Chem Pharm 36:101349. https://doi.org/10.1016/j.scp.2023.101349

Tan SX, Andriyana A, Lim S, Ong HC, Pang YL, Ngoh GC (2021) Rapid ultrasound-assisted starch extraction from sago pith waste (SPW) for the fabrication of sustainable bioplastic film. Polymers 13(24):4398. https://doi.org/10.3390/polym13244398

Tapia-Blácido DR, Aguilar GJ, de Andrade MT, Rodrigues-Júnior MF, Guareschi-Martins FC (2022) Trends and challenges of starch-based foams for use as food packaging and food container. Trends Food Sci Technol 119:257–271. https://doi.org/10.1016/j.tifs.2021.12.005

Tessanan W, Phinyocheep P, Amornsakchai T (2023) Development of biodegradable thermosetting plastic using dialdehyde pineapple stem starch. Polymers 15(18):3832. https://doi.org/10.3390/polym15183832

Thakur R, Pristijono P, Scarlett CJ, Bowyer M, Singh SP, Vuong QV (2019) Starch-based films: major factors affecting their properties. Int J Biol Macromol 132:1079–1089. https://doi.org/10.1016/j.ijbiomac.2019.03.190

Thool OK, Sasidharan A, Krishna BM, Sabu S, Navaf M, Sunooj KV (2025) Nano-hydroxyapatite (n-HAP) from Pangasius bone side streams and its application as a reinforcing agent in biodegradable food packaging films. Sustain Food Technol. https://doi.org/10.1039/D4FB00264D

Tongdeesoontorn W, Mauer LJ, Wongruong S, Sriburi P, Rachtanapun P (2020) Physical and antioxidant properties of cassava starch-carboxymethyl cellulose incorporated with quercetin and TBHQ as active food packaging. Polymers 12(2):366. https://doi.org/10.3390/polym12020366

Trinh BM, Tadele DT, Mekonnen TH (2022) Robust and high barrier thermoplastic starch—PLA blend films using starch-graft-poly(lactic acid) as a compatibilizer. Mater Adv 3(15):6208–6221. https://doi.org/10.1039/D2MA00501H

Trongnit J, Mayakun J, Kaewtatip K (2024) The effect of agar from the seaweed Gracilaria fisheri on properties of biodegradable starch foam. Int J Biol Macromol 273:132952. https://doi.org/10.1016/j.ijbiomac.2024.132952

Vasile C, Baican M (2021) Progresses in food packaging, food quality, and safety—Controlled-release antioxidant and/or antimicrobial packaging. Molecules 26(5):1263. https://doi.org/10.3390/molecules26051263

Wan S, Liu Q, Yang D, Guo P, Gao Y, Mo R, Zhang Y (2023) Characterization of high amylose corn starch-cinnamaldehyde inclusion films for food packaging. Food Chem 403:134219. https://doi.org/10.1016/j.foodchem.2022.134219

Wang X, Huang L, Zhang C, Deng Y, Xie P, Liu L, Cheng J (2020) Research advances in chemical modifications of starch for hydrophobicity and its applications: a review. Carbohydr Polym 240:11629. https://doi.org/10.1016/j.carbpol.2020.116292

Wang B, Yu B, Yuan C, Guo L, Liu P, Gao W, Li D, Cui B, Abd El-Aty AM (2022a) An overview on plasticized biodegradable corn starch-based films: the physicochemical properties and gelatinization process. Critical Rev Food Sci Nutr 62(10):2569–2579. https://doi.org/10.1080/10408398.2020.1868971

Wang J, Chen C, Xie J (2022b) Loading oregano essential oil into microporous starch to develop starch/polyvinyl alcohol slow-release film towards sustainable active packaging for sea bass (Lateolabrax japonicus). Ind Crops Prod 188:115679. https://doi.org/10.1016/j.indcrop.2022.115679

Wang L, Yang C, Deng X, Peng J, Zhou J, Xia G, Zhou C, Shen Y, Yang H (2023) A pH-sensitive intelligent packaging film harnessing Dioscorea zingiberensis starch and anthocyanin for meat freshness monitoring. Int J Biol Macromol 245:125485. https://doi.org/10.1016/j.ijbiomac.2023.125485

Waring RH, Harris RM, Mitchell SC (2018) Plastic contamination of the food chain: a threat to human health? Maturitas 115(May):64–68. https://doi.org/10.1016/j.maturitas.2018.06.010

Waśniewska A, Chocyk D, Gładyszewski G, Borc J, Świetlicki M, Gładyszewska B (2020) The influence of kaolin clay on the mechanical properties and structure of thermoplastic starch films. Polymers 12(1):73. https://doi.org/10.3390/polym12010073

Wu H, Li T, Peng L, Wang J, Lei Y, Li S, Li Q, Yuan X, Zhou M, Zhang Z (2023) Development and characterization of antioxidant composite films based on starch and gelatin incorporating resveratrol fabricated by extrusion compression moulding. Food Hydrocoll 139:108509. https://doi.org/10.1016/j.foodhyd.2023.108509

Wu Y, Gao S, Zhao J, Kong S, Wang H, Wang W, Hou H (2025) Sugar/sugar alcohol with glycerol as co plasticizers for high content starch/PBAT blown films: from fine structure to physicochemical properties. J Sci Food Agric 105(2):1105–1115. https://doi.org/10.1002/jsfa.13901

Xu H, Cheng H, McClements DJ, Chen L, Long J, Jin Z (2022) Enhancing the physicochemical properties and functional performance of starch-based films using inorganic carbon materials: a review. Carbohyd Polym 295:119743. https://doi.org/10.1016/j.carbpol.2022.119743

Xu M, Fang D, Kimatu BM, Lyu L, Wu W, Cao F, Li W (2024) Recent advances in anthocyanin-based films and its application in sustainable intelligent food packaging: a review. Food Control 162:110431. https://doi.org/10.1016/j.foodcont.2024.110431

Yang F, Chen G, Li J, Zhang C, Ma Z, Zhao M, Yang Y, Han Y, Huang Z, Weng Y (2023) Effects of quercetin and organically modified montmorillonite on the properties of poly(butylene adipate-co-terephthalate)/thermoplastic starch active packaging films. ACS Omega 8(1):663–672. https://doi.org/10.1021/acsomega.2c05836

Yang N, Sha H, Bi W, Li S, Wu S, Su D (2025) Preparation and characterization of microcrystalline cellulose/corn starch-based active packaging enhanced by resveratrol/β-cyclodextrin complex. Food Packag Shelf Life 47:101427. https://doi.org/10.1016/j.fpsl.2024.101427

Yu H, Zhang Z, Zhang Y, Song Q, Fan P, Xi B, Tan W (2021) Effects of microplastics on soil organic carbon and greenhouse gas emissions in the context of straw incorporation: a comparison with different types of soil. Environ Pollut 288(May):117733. https://doi.org/10.1016/j.envpol.2021.117733

Yusof NL, Mutalib N-AA, Nazatul UK, Nadrah AH, Aziman N, Fouad H, Jawaid M, Ali A, Kian LK, Sain M (2021) Efficacy of biopolymer/starch based antimicrobial packaging for chicken breast fillets. Foods 10(10):2379. https://doi.org/10.3390/foods10102379

Zeng GS, Hu C, Zou S, Zhang L, Sun G (2019) BP neural network model for predicting the mechanical performance of a foamed wood fiber reinforced thermoplastic starch composite. Polym Compos 40(10):3923–3928. https://doi.org/10.1002/pc.25252

Zeng Y-F, Chen Y-Y, Deng Y-Y, Zheng C, Hong C-Z, Li Q-M, Yang X-F, Pan L-H, Luo J-P, Li X-Y, Zha X-Q (2024) Preparation and characterization of lotus root starch based bioactive edible film containing quercetin-encapsulated nanoparticle and its effect on grape preservation. Carbohyd Polym 323:121389. https://doi.org/10.1016/j.carbpol.2023.121389

Zhang S, Zhao H (2017) Preparation and properties of zein–rutin composite nanoparticle/corn starch films. Carbohyd Polym 169:385–392. https://doi.org/10.1016/j.carbpol.2017.04.044

Zhao L, Liu Y, Zhao L, Wang Y (2022) Anthocyanin-based pH-sensitive smart packaging films for monitoring food freshness. J Agric Food Res 9:100340. https://doi.org/10.1016/j.jafr.2022.100

Zhao D, Zhang X, Zhang Y, Xu E, Yan S, Xu H, Li M (2024) Recent advances in the fabrication, characterization and application of starch-based materials for active food packaging: hydrogels and aerogels. Sustain Food Technol 2(3):615–634. https://doi.org/10.1039/D4FB00030G

Chapter 3
Starch-Based Polymers in Food Packaging: Recent Advances

Milad Tavassoli, Behnam Bahramian, Majid Majlesi, and Sajad Ghaderi

Abstract The field of research and innovation in packaging materials is currently undergoing significant transformation, driven by an increasing awareness of sustainability and environmental conservation. Among the various materials being explored, starch stands out as a promising candidate for eco-friendly packaging solutions. As a naturally occurring polymer, starch is sustainable and abundant and offers cost-effective alternatives to conventional petroleum-based packaging materials. Its biodegradability and renewability are crucial in addressing pressing environmental issues, thus reducing our reliance on fossil fuels. However, despite these advantageous properties, the application of starch in the packaging industry faces certain limitations. One of the primary challenges associated with starch-based materials is their relatively low mechanical strength, which can compromise the durability and reliability of the packaging. Starch materials often exhibit inadequate barrier properties, making them less effective in preserving food quality and extending shelf life. Several strategies have emerged to tackle these drawbacks to improve starch-based films' performance. One approach is the modification of starch itself, which can enhance its physical and chemical properties. Incorporating various reinforcing agents, such as natural fibers, nanomaterials, or synthetic polymers, can bolster the mechanical strength of starch films. Another effective strategy is blending starch with other polymers to create composite materials that maintain the environmental benefits of starch while improving performance. Exploring starch as a potentially active and intelligent packaging material is gaining traction. This chapter reviews recent advancements in starch-based films, focusing on methodologies to improve their mechanical and barrier properties. It highlights innovations in active

M. Tavassoli (✉)
Student Research Committee, Yasuj University of Medical Sciences, Yasuj, Iran
e-mail: mtavassoli2006@gmail.com

M. Tavassoli · M. Majlesi · S. Ghaderi
Department of Nutrition, Faculty of Health and Nutrition Sciences, Yasuj University of Medical Sciences, Yasuj, Iran

B. Bahramian
Department of Food Science and Technology, Faculty of Nutrition and Food Sciences, Tabriz University of Medical Sciences, Tabriz, Iran

S. Roy and M. Shahid (eds.), *Biomaterials for Sustainable Food Packaging*,
Environmental Footprints and Eco-design of Products and Processes,
https://doi.org/10.1007/978-3-032-08027-1_3

and intelligent packaging for food applications, noting that using different fillers and starch sources can significantly improve thermoplastic starch properties. A multidisciplinary approach involving chemistry, materials science, and food technology is crucial for developing cost-effective and eco-friendly materials. The chapter also addresses potential challenges in commercializing starch-based films in the food industry and suggests future research directions in this promising field.

Keywords Starch · Biopolymer · Smart food packaging · Films · Biodegradable · Food

3.1 Introduction

In recent years, there has been a remarkable surge in the development of degradable food packaging, driven by growing environmental concerns and the need for sustainable alternatives to traditional plastic wrappers. Various innovative degradation technologies have emerged, offering both photodegradable and biodegradable options. Photodegradable materials are designed to break down under exposure to sunlight, while biodegradable materials decompose through natural processes involving microorganisms (Khezerlou et al. 2023; Tavassoli et al. 2022). Despite these advancements, there are important considerations regarding the safety of certain degradable materials. Some can be toxic and pose significant risks to human health, which raises concerns about their suitability for food packaging. This highlights the necessity for careful selection of materials.

Biodegradable materials are particularly well-suited for food packaging, as they can decompose without leaving harmful residues, offering a safer alternative. The primary functions of biodegradable packaging films mirror those of traditional packaging: they safeguard food quality, facilitate efficient food distribution, and enhance the value of the product. These films serve to protect food from external contaminants, extend shelf life, and minimize food spoilage all while being environmentally friendly (Bakhshizadeh et al. 2023; Roy et al. 2023a, b). The formulation of completely biodegradable films utilizes a range of natural substances, including polysaccharides such as starch, chitosan, and cellulose; proteins sourced from whey, soy, and silk; and various lipids, including beeswax and lauric acid. Each of these components contributes different beneficial properties, such as moisture barrier capabilities and mechanical strength, which are essential for effective food packaging. However, the widespread adoption of biodegradable materials for food packaging faces hurdles, notably the high production and utilization costs which can deter manufacturers (Roy et al. 2023a, b).

Biopolymers are increasingly recognized as one of the most promising alternatives to synthetic plastics, primarily due to their ability to decompose more rapidly and their superior biocompatibility with a wide range of materials. Unlike conventional plastics, which can persist in the environment for hundreds of years, biopolymers typically break down more quickly, making them an appealing option for sustainable practices. These materials are highly accessible, derived from renewable resources, and exhibit properties that allow them to mimic traditional plastics, contributing to their viability in various applications (Suryavanshi et al. 2025; Tavassoli et al. 2024a, b). The production of biodegradable polymers employs a diverse array of natural substances. Among the most predominant sources are starch, cellulose, chitosan, and various protein-based materials. Starch and cellulose are particularly favored in the food packaging industry. Their biodegradability combined with non-toxic characteristics renders them suitable for use in products that come into direct contact with food. These biopolymers are cost-effective due to their natural abundance and renewability, aligning with the growing demand for sustainable materials in packaging solutions (Zhang et al. 2025; Zhu 2025). Despite starch's inherent limitations as a thermoplastic material, it can be modified to display thermoplastic properties through the addition of plasticizers such as water and polyols. When subjected to heat and shear stresses, starch can be transformed into a pliable substance suitable for molding and film formation. This plasticization process effectively disrupts the granular structure of starch, promoting the diffusion of plasticizers within the starch matrix. The resulting material can be formulated as either intact granules or in a gelatinized state, which is crucial in creating films utilizing thermoplastic methods (Frangopoulos et al. 2025; Li et al. 2025; Zubair et al. 2024). However, one significant challenge with starch-based films is their sensitivity to moisture, which can compromise their mechanical integrity and shelf life. To address this issue, researchers have explored various approaches, including blending starch with other synthetic polymers, such as polyvinyl alcohol (PVA), polycaprolactone (PCL) polylactic acid (PLA), or utilizing other biodegradable polymers. Chemical modifications are also employed to enhance moisture resistance and performance (Singh et al. 2024; Wang et al. 2024a, b; Xia et al. 2024). The properties and applicability of starch-based materials are influenced by a multitude of factors, such as the botanical source, different plants yield varying starch characteristics, along with growing conditions, harvest time, and the genetic diversity of the starch source. These variables can impact the material's strength, flexibility, and biodegradation rates, emphasizing the importance of selecting the right type of starch for specific applications (Mhaske et al. 2024; Niranjana Prabhu and Prashantha 2018).

The aim of this chapter is to review the latest advances in preparation, modification, and processing techniques related to starch-based materials, with a particular emphasis on their customized applications in food packaging. By reviewing these advances, this chapter demonstrates how innovations in biopolymer technology can lead to more sustainable packaging solutions that are in line with environmental conservation efforts.

3.2 Overview of Starch

Starch is a naturally occurring biodegradable polysaccharide that plays a pivotal role in the plant kingdom, primarily as a stored energy source produced during the process of photosynthesis. It is widely regarded as the second most abundant organic compound in nature, following cellulose. Within the plant ecosystem, starch serves as the main conserved carbohydrate, essential for energy storage and supply. Its production is predominantly attributed to a variety of plant sources, with cereal grains being recognized as the primary source. Starch can also be extracted from certain roots, tubers, seeds of fruits, and rhizomes. The extraction process is relatively straightforward and economically feasible, often employing methods that require minimal technological intervention (Ma et al. 2025; Nazrin et al. 2025; Zubair et al. 2024). The wet extraction of starch granules involves several key stages that are crucial for yielding a high-quality product. These stages include washing the raw materials to remove impurities, milling to break down the cell walls and liberate starch, extraction, where starch is solubilized in water, decantation to separate the liquid starch from the solid residues, and finally, drying to achieve the desired moisture content in the final starch product (Deng et al. 2024; Diah et al. 2024; Fatima et al. 2024). The progress in starch-based polymers for food packaging is outlined in Scheme 3.1.

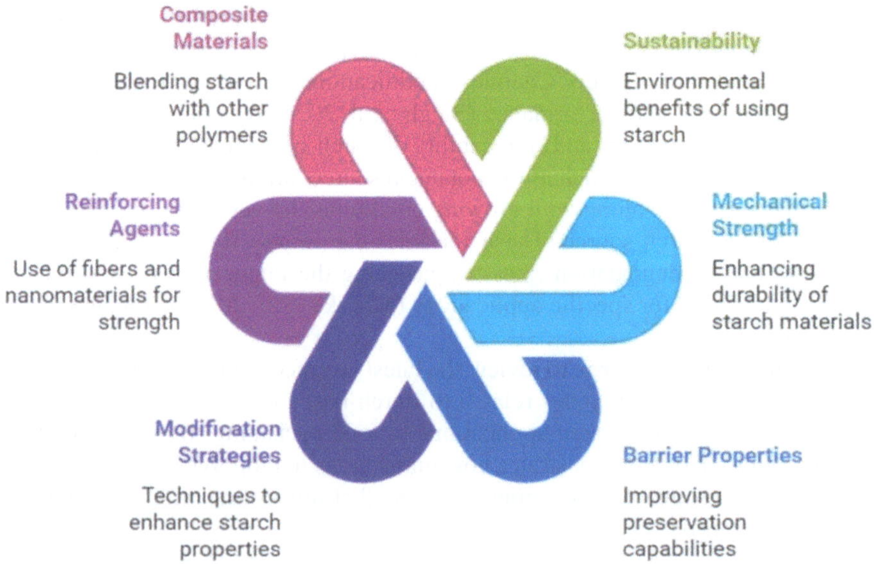

Scheme 3.1 Advancements in starch-based polymers for food packaging

3.2.1 Starch-Based Biodegradable Food Packaging

Biodegradable materials derived from starch undergo a systematic development through three distinct phases: the starch-filled plastics phase, the blended starch plastics phase, and finally, the all starch plastics phase. As the concentration of starch within these materials increases, significant advancements in starch-based plastics have emerged, with all starch plastics standing out due to their ability to achieve complete biodegradation. These plastics are not only environmentally friendly but also utilize a broad array of raw materials that are typically low-cost, positioning them at the forefront of innovation in the biodegradable materials sector (Picolotto et al. 2024; Plaeyao et al. 2025). The first phase, starch-filled plastics, involves the incorporation of either original or modified starch into a polymer matrix—usually polyethylene or other types of thermoplastics. This blend may also include various additives tailored to enhance biodegradability. The primary objective of this phase is to improve the environmental profile of conventional petroleum-based plastics by integrating natural components. However, it is important to note that while these starch-filled plastics demonstrate improved biodegradability compared to their petroleum-derived counterparts, they still necessitate several years for decomposition and do not achieve full breakdown (Fatima et al. 2024; Weligama Thuppahige et al. 2023). In the second phase, known as blended starch plastics, a combination of starch is mixed with synthetic resins or other natural polymers. Typically, these blends comprise 30–60% starch or modified starch integrated with synthetic biodegradable materials. This blending results in products that are capable of complete biodegradation without inflicting harm on the environment. Compared to entirely synthetic polymers, these blended materials display enhanced degradation rates and superior mechanical properties, making them a more viable alternative. However, one must consider the challenges that arise from using synthetic resins or natural polymers, which are predominantly polar compounds. Their affinity for moisture can lead to a significant degradation of the plastic's integrity when exposed to humid conditions over extended periods. Compatibility issues often arise between starch and the additives, which can complicate the formulation process (Shanbhag et al. 2023; Wang et al. 2023a, b). The final phase involves all starch plastics, also known as thermoplastic starch plastics. This category is characterized by a high starch content of over 90%, only supplemented with a minimal amount of non-toxic additives that can be fully decomposed. These materials are engineered by integrating degradable plasticizers along with other additives using various processing techniques such as extrusion, injection molding, blow molding, and calendering. This comprehensive processing results in a "disordered" arrangement of starch molecules, allowing for enhanced material properties and processing flexibility. One of the notable advantages of all starch plastics is their genuine and complete biodegradability, as they can break down entirely into natural components under suitable conditions. These plastics can be processed using nearly all established plastic processing methods, making them a versatile and promising category in the world of sustainable materials (Promhuad and Harnkarnsujarit 2022; Todhanakasem et al. 2022). The role

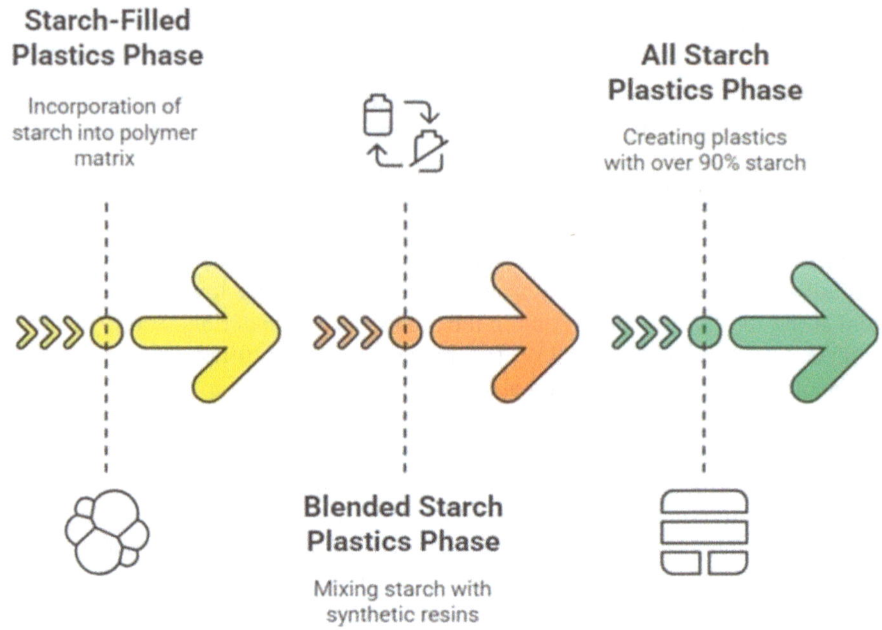

Scheme 3.2 Starch's role in producing biopolymers with biodegradability

of starch in the production of biopolymers with biodegradability is illustrated in Scheme 3.2.

3.3 Physical and Chemical Properties

At the molecular level, starch is characterized by a complex multi-scale structure that consists of two main macromolecular components: amylose and amylopectin. Amylose is a linear polymer in which glucose units are interconnected through α-(1,4) glycosidic bonds, resulting in a relatively straight chain structure. Amylopectin, the larger and more predominant component of starch, has a branched architecture. It features a backbone of linear glucose units linked by α-(1,4) bonds, along with branched chains that are formed via α-(1,6) glycosidic bonds. The structural intricacies of amylopectin give rise to short chains with a degree of polymerization (DP) typically greater than 10, which form double helices that contribute to the crystalline regions of starch granules. It is hypothesized that these crystalline regions are a result of the crosslinking of amylopectin side chains, while amylose predominantly resides in the amorphous regions of the granules, distributed somewhat randomly among the more structured amylopectin (Rahardiyan et al. 2023; Su et al. 2023).

Starches can be classified based on the amylose-to-amylopectin ratio, which influences their functionality in various applications. For instance, "waxy" starches contain less than 5% amylose and are known for their unique gelatinization properties. "Normal" starches possess an amylose content ranging from 20 to 30%, whereas "high amylose" starches exhibit amylose levels of 40% or greater. The amylose content varies widely among different source materials: starch obtained from cereals typically features amylose percentages between 15 and 33%, roots provide between 15 and 30%, and pulses like lentils and chickpeas can contain amylose levels ranging from 25 to 50%. Notably, in cereals, amylose often exists in association with lipids, while roots and tubers generally have little to no lipid content (Muñoz-Gimena et al. 2023; Othman et al. 2023). The characteristics of starch granules, including their shape, size, composition, and molecular structure, are significantly influenced by their botanical origin, the period of formation, and the conditions under which the plants grow. These granules exhibit a variety of shapes, including spherical, oval, polygonal, and lenticular forms, with sizes that can range considerably from 1 to 110 μm in diameter. The shape and size of starch granules are crucial parameters as they affect the physical properties of the starch, including its gelatinization behavior and digestibility. For instance, cereal starches typically have distinct grain shapes and display an amylose content ranging from 15 to 27%, exhibiting a type A (monoclinic) crystal structure, which influences their functionality in food and industrial applications (Liu et al. 2022; Niranjana Prabhu and Prashantha 2018; Vianna et al. 2021). While much research has historically focused on starches derived from wheat, corn, potato, and rice due to their economic importance, there is an increasing interest in exploring alternative sources of starch. This shift is motivated by the desire to develop bio-based and compostable materials. Promising sources under investigation include pehuen (also known as araucaria), chestnuts, jackfruit seeds, avocado seeds, as well as legumes such as peas, lentils, faba beans, and chickpeas. Each of these alternative starch sources presents unique properties that may offer innovative solutions in various applications ranging from food products to biodegradable materials, thereby meeting the growing demand for sustainable and eco-friendly options in both household and industrial uses (Deng et al. 2024; Dutta and Sit 2024; Fatima et al. 2024; Fattahi and Bahrami 2018).

There are several effective strategies that have been identified to enhance the performance of starch-based materials, particularly in humid environments where water resistance is crucial. One notable method involves the incorporation of hydrophobic polymers such as PLA and polyhydroxybutyrate (PHB) into starch matrices. This blending can significantly improve the water resistance of starch by creating a barrier to moisture. However, it is important to note that while these polymer-starch mixtures enhance water resistance, they often exhibit subpar mechanical properties, especially when the starch content remains high (Lauer and Smith 2020). Another promising approach to improve the performance of starch-based composites is the addition of natural fibers. These fibers, which can include materials such as cellulose, jute, or kenaf, play a vital role in reducing the water absorption capacity of the composites. However, achieving a substantial reduction in hydrophilicity typically requires a considerable proportion of natural fibers, which

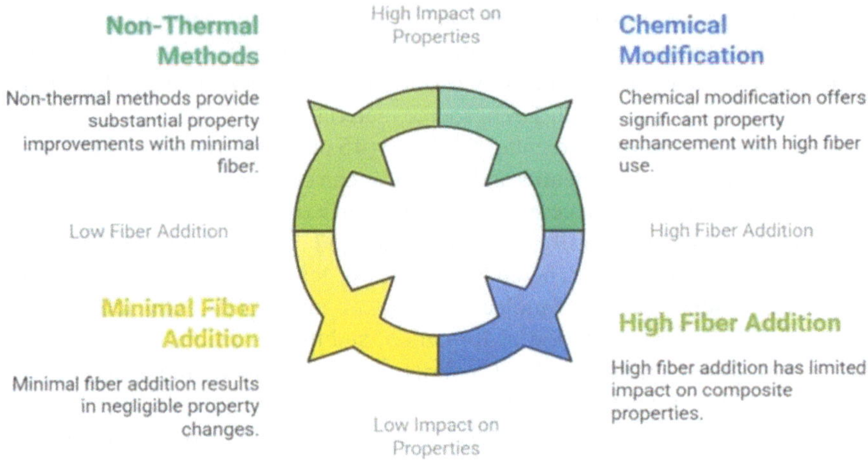

Scheme 3.3 Strategies for improving starch-based polymers

may pose challenges regarding the processing and structural integrity of the material (Díaz-Galindo et al. 2020). Chemical and physical modifications offer alternative ways to enhance the characteristics of starch-based materials. One method of chemical modification involves the esterification of the hydroxyl groups present in α-D-glucan with methylene groups. This chemical reaction can effectively reduce the material's tendency to absorb water and dissolve, thereby improving its stability and durability in moist environments. The combination of these strategies can lead to the development of advanced starch-based materials that possess both improved mechanical properties and enhanced water resistance, making them suitable for a wider range of applications (Arooj et al. 2024; Jayakumar et al. 2019). Strategies for improving starch-based polymers are summarized in Scheme 3.3

Chemical modifications of starches are crucial processes that involve the introduction of various functional groups to the starch macromolecules. This modification can be achieved through the breaking of glycosidic bonds, leading to either partial or total alterations in the physicochemical properties of starch. These properties include but are not limited to the structure, composition, gelatinization, retrogradation, and pasting characteristics of starch. The main categories of chemical modifications that are commonly applied to starch include reduction (which can entail oxidation, hydrolytic, or enzymatic reactions), substitution (such as etherification and esterification), and crosslinking (utilizing di-ethers and di-esters) (Pei et al. 2024; Zubair et al. 2024). In the substitution modification technique, certain hydroxyl groups present in the anhydro-glucose units constituting the starch chain are replaced with alternative groups, for example, acetyl groups. This substitution process leads to significant changes in the molecular structure of the starch. Research has indicated that utilizing organic acids and acid anhydrides for starch esterification effectively reduces retrogradation while simultaneously enhancing the physicochemical properties of thermoplastic starch (TPS). This modified starch is increasingly important in the plastics industry, especially in applications related to packaging materials, where improved performance characteristics are vital (Diah et al. 2024; Dutta and Sit 2024; Fatima et al. 2024). The method of starch reduction involves the introduction of additional functional groups to the starch molecule. Through oxidation processes, for instance, carbonyl and carboxyl functional groups are introduced, which serve to increase the functionality and reactivity of the modified starch. This enhanced functionality can be advantageous in various applications, as it allows for improved interactions with other materials and modifications (Deng et al. 2024; Rahardiyan et al. 2023). Crosslinking is yet another important modification approach that entails the covalent linkage of linear or branched starch chains. This process is significant because it enhances the hydrophobicity and structural integrity of the starch granules, which in turn bolsters their hydration properties. Various chemical agents have been deployed for the crosslinking modification of starch, including sodium trimetaphosphate (STMP), sodium tripolyphosphate (STPP), epichlorohydrin (ECH), phosphorus chloride ($POCl_3$), and various acids. However, it's important to note that certain agents like glutaraldehyde and epichlorohydrin are considered hazardous due to their toxic and irritant properties, posing potential health risks (Muñoz-Gimena et al. 2023; Othman et al. 2023). Citric acid stands out as one of the most widely utilized and favorable crosslinking agents for starch granule modifications. Its ability to form intermolecular diester covalent bonds is facilitated by the reaction of its carboxyl groups with the hydroxyl groups present in starch molecules. This makes citric acid not only an inexpensive option but also a non-toxic one, making it suitable for applications requiring high safety standards. It has been demonstrated that citric acid-treated starch exhibits enhanced film stability during storage and reduced water vapor transmittance, making it a valuable component in the development of starch-based materials for various practical applications (Cheng et al. 2021; Liu et al. 2022; Rahardiyan et al. 2023).

Interest in non-thermal methods for material treatment has increased due to their low energy use and positive environmental impact. These technologies can significantly modify material properties, especially starches. For example, high hydrostatic pressure treatments can change starch gelatinization patterns and crystallinity, affecting functional characteristics based on processing conditions (Fattahi and Bahrami 2018; Niranjana Prabhu and Prashantha 2018). Similarly, ozonation is another non-thermal technique that disrupts the structure of starch granules. This process not only affects the physical integrity of the starch but also influences critical parameters such as particle size and weight. Interestingly, a category of starch that remains chemically unmodified yet exhibits behaviors analogous to chemically modified varieties has garnered considerable attention in research and industry due to its potential applications (Bangar et al. 2021; Cheng et al. 2021; Rahardiyan et al. 2023). The impact of non-thermal methods extends to the chemical properties of starch as well. Research indicates that plasma treatments can instigate a variety of chemical transformations within starch molecules, such as crosslinking. These transformations can lead to important changes in thermal transitions, including alterations in melting temperature and enthalpy, which are crucial for understanding starch behavior under different processing and storage conditions (Lee et al. 2021;

Xu et al. 2022). Ultrasound treatment introduces additional complexities to starch modification. This technique subjects starch to intense shear forces and elevated temperatures produced by ultrasound waves. These factors play a pivotal role in determining the alterations to the starch properties, which can depend significantly on variables such as ultrasonic frequency, intensity, and duration, along with the specific type of starch being treated. Generally, ultrasound applications tend to cause structural damage to the starch granules, manifested as pores, fissures, and cracks. At the molecular level, the experimental conditions can lead to variations in the composition and arrangement of amylose and amylopectin, the two main components of starch (Díaz-Galindo et al. 2020; Tongdeesoontorn et al. 2020). Research has yielded varied results regarding the effects of ultrasound on the physicochemical characteristics of treated starch. Notably, there is no consistent trend observed across studies, with some reporting increases while others indicate decreases in properties such as gelatinization, retrogradation, swelling, pasting behavior, and viscosity. This variability underscores the importance of considering factors including the starch source, type, and inherent structure, as well as the specific conditions of the ultrasonic treatment, including temperature and moisture content (Sanyang et al. 2016). In terms of Pulsed Electric Fields (PEF), diverse outcomes have been observed concerning the duration of treatments, pulse characteristics, and intensity related to starch modifications. Despite the variation in results, the majority of research surrounding PEF has focused on the damage inflicted on starch granules. Noteworthy findings include the increased agglomeration of granules, reductions in relative crystallinity, and viscosity, as well as significant decreases in gelatinization temperatures. Changes in birefringence have been reported, reflecting alterations in the molecular structure and organization of the starch following PEF treatment. These insights are vital for optimizing starch properties for specific applications in food technology and other industries (García-Guzmán et al. 2022).

3.4 Preparation and Processing Techniques of Starch for Making Food Packaging Films

Various processing techniques have been employed to develop starch-based films, with casting and extrusion being two of the most prominent methods. The casting method is the predominant technique employed in the production of biodegradable films. This process begins with the preparation of a homogeneous solution, which is achieved by heating and stirring a mixture of film-forming polymers in an appropriate solvent. This initial step is crucial, as it allows the polymers to dissolve adequately and attain a uniform consistency (Othman et al. 2023). Once the solution is prepared, it undergoes a degassing procedure to eliminate any trapped air bubbles, which can negatively impact the film's quality. Following this, the degassed solution is carefully poured onto a flat, level surface, where it spreads out to form a thin layer. The drying phase can occur either at room temperature or in an oven where the temperature and humidity are meticulously controlled to ensure optimal evaporation of the solvent (do Val Siqueira et al. 2021). There are several advantages to using the casting method for film production. It is generally considered user-friendly and cost-effective, making it accessible for smaller operations or research applications. However, it also presents notable challenges. One significant drawback is managing the film's thickness and uniformity; variations in these parameters can lead to films with inconsistent qualities that may not meet the desired specifications for certain applications. Because of these limitations, the casting method is often deemed unsuitable for large-scale production, leading to its infrequent use in industrial contexts (Gulati et al. 2022). To enhance the performance characteristics of the produced films, researchers and manufacturers have employed various strategies. These include experimenting with different polymer blends, implementing acid modifications, and exploring both simple and complex modifications of native starches. Such approaches aim to tailor the properties of the films to achieve specific requirements, such as improved mechanical strength, barrier properties, or biodegradability, thereby expanding the potential applications of biodegradable films in various industries. YousefniaPasha and colleagues developed a method to improve the brittleness and rigidity of polylactic acid (PLA) for food packaging by incorporating plasticizers triacetin (TA) and polyethylene glycol (PEG) at levels of 10, 20, and 30 wt% using the solvent-casting technique. Their results showed that both TA and PEG increased elongation at break (Eab) while decreasing tensile strength (TS). Notably, the films with 10 and 30 wt% of TA exhibited the highest TS (26.6 MPa) and Eab (81.1%), respectively. Incorporating plasticizers reduced light transmission and increased water vapor permeability (WVP), moisture content (MC), and opacity compared to neat PLA. The films with 10 wt% TA and PEG had the lowest WVP (4.9×10^{-10} g/m s^{-1} Pa^{-1}) and MC (5.5%). Thermogravimetric analysis (TGA) indicated improved thermal stability in plasticized films, with PLA-TA10 showing the best properties for food packaging (YousefniaPasha et al. 2021).

The casting technique is particularly noted for its straightforward nature and user-friendly operation. However, it is not without its drawbacks; specifically, the method encounters significant obstacles when it comes to large-scale production of continuous films. One of the primary limitations is the extended drying times required for effective film formation, which in turn restricts productivity and efficiency in a manufacturing context (Pacheco et al. 2023). On the other hand, the extrusion technique presents a more dynamic approach. It necessitates the integration of various plasticizers or biopolymers, such as polylactic acid (PLA), polyvinyl alcohol (PVA), polycaprolactone (PCL), polybutylene adipate terephthalate (PBAT), and polybutylene succinate adipate (PBSA), into the starch formulation. This incorporation is critical to generating starch-based films that meet the necessary material property specifications for various applications, particularly in packaging and sustainable materials (Othman et al. 2023; YousefniaPasha et al. 2021). An emerging and innovative method in this field is electrospinning, which is recognized as a highly efficient and versatile nanotechnology for producing non-woven nanofibrous films. One of electrospinning's key advantages lies in its ability to create films with a high surface-to-volume ratio, enhanced porosity, and customizable morphology. These attributes make electrospinning a widely applicable technique for both natural and synthetic polymers (Dejene et al. 2024; Deng et al. 2024).

In comparison with traditional film-forming methods, electrospinning, especially when utilizing multi-needle or non-needle systems, offers significant advantages in terms of scalability and productivity. This capability is particularly advantageous for large-scale production, facilitating the creation of films in higher volumes without compromising the structural integrity of the materials used. Moreover, electrospinning is a non-thermal process, which is crucial for maintaining the stability of additives that are sensitive to heat. For instance, many plastics and biopolymers can degrade or change properties at elevated temperatures, but electrospinning preserves their structural characteristics (Roveda et al. 2024). The electrospinning process can be systematically broken down into three distinct phases: First, in the presence of an electric field, the droplet of polymer solution at the needle tip adopts a conical shape, referred to as a "Taylor cone." Second, when the strength of the electric field exceeds the viscoelastic forces acting on the solution, a thin jet of polymer emerges from the apex of the cone. Finally, as the solvent evaporates rapidly during the process, a solid polymer jet is deposited onto the surface of a collector, resulting in the formation of numerous fine fibers (Lv et al. 2024). In this regard, Zhu and colleagues developed a streamlined method for producing crosslinked starch nanofibrous films (C-SNFs) by initially electrospinning a starch solution and subsequently crosslinking it using glutaraldehyde (GTA) vapor. Their study rigorously examined how various electrospinning parameters, such as starch concentration (typically ranging from 20 to 30% by weight), applied voltage (between 15 and 25 kV), and the distance between the needle and collector (10–20 cm), influenced the resulting fiber morphology. When the electrospun starch nanofibrous films (SNFs) were compared to solution-cast starch films, notable differences emerged: while the SNFs displayed greater flexibility, they also exhibited lower mechanical strength and heightened sensitivity to water. Following the crosslinking process with GTA, significant improvements were observed in both the tensile strength and the water contact angle (WCA) of the SNFs, enhancing these properties to approximately 9.65 MPa and 94.3°, respectively. The results of this study imply a straightforward and effective pathway for fabricating starch-based film materials that possess promising applications in various sectors, particularly in food packaging, where flexibility, strength, and moisture resistance are crucial factors. These advancements could potentially revolutionize the use of biopolymer films, contributing to more sustainable packaging solutions in the industry (Zhu et al. 2022). The preparation methods for starch-based biodegradable materials have become a focal point of research and development due to their potential environmental benefits. Various approaches to creating these materials are influenced by several key factors, including the time required for production, costs, and the complexity of the operational processes involved (Lv et al. 2024; Şen et al. 2017).

One widely used technique is the film solution casting method. This method, while straightforward and easy to execute, involves a time-consuming process that often results in limited output. Consequently, it is primarily used for small-scale laboratory research rather than industrial applications. Researchers appreciate its simplicity, but the long processing times can hinder its scalability. The extrusion processing method presents a viable solution for large-scale production of starch-based biodegradable materials. This method is characterized by its high processing speed, allowing for higher throughputs. However, it often comes with a trade-off, as increased energy consumption becomes a significant consideration in the sustainability of the production process (Baysal and Doğan 2020; Mallick et al. 2020). Another innovative technique is electrospinning, which is notable for its ability to produce multi-layered thin films. This process is particularly effective in promoting adhesion between hydrophilic natural polymers and hydrophobic layers, thereby enhancing the functionality and performance of the resulting materials. The versatility of electrospinning is beneficial in various applications, especially in packaging and protective coatings (Kanatt 2020; Kumar et al. 2019; Sariningsih et al. 2019). 3D printing technology also provides significant advantages by allowing for the customization and tailoring of starch-based biodegradable materials to fit specific design requirements or consumer preferences. This capability not only facilitates the production of unique shapes and structures but also enhances the potential for innovative applications in diverse fields (Huang et al. 2024; Singh et al. 2024). Reactive extrusion further enhances the production efficiency of starch-based biodegradable materials. This method leverages the combined effects of enzymatic and thermo-mechanical actions during the extrusion process. By optimizing these conditions, producers can achieve high yields and improved material properties, making this approach an attractive option for large-scale manufacturing (Bhat et al. 2013; Mangaraj et al. 2022). Moreover, the incorporation of nanotechnology into food packaging presents exciting possibilities for enhancing functionality. By utilizing nano carriers, bioactive substances can be integrated into packaging materials, offering features such as antimicrobial properties, improved barrier performance, and extended shelf life for packaged goods (Huang et al. 2024; Shanbhag et al. 2023). Looking ahead, it will be crucial to strike a balance between the various processing technologies, production efficiencies, and associated costs. Manufacturers must carefully evaluate their options and select the most appropriate preparation methods for starch-based biodegradable materials, considering their unique requirements and operational circumstances. This strategic approach will be key to advancing the development and adoption of sustainable packaging solutions in the industry.

3.5 Advanced Strategies for Improvement of Starch-Based Food Packaging

3.5.1 *Blends with Other Biopolymers*

3.5.1.1 Blends with Polysaccharides

A polymer/matrix blend is a group of dissimilar matrices that chemically interact with each other to achieve desired properties that would not be available on their own. Studies on the combination of two polymers together in biodegradable packaging films have increased in recent years because (1) it improves the processability of a polymer (such as starch) based on the industrial application, (2) and increases the chemical, physical, mechanical, and functional properties of starch-based polymers. Polymer blending is less expensive than other processes such as copolymerization and does not require complex chemical techniques, and the desired properties can be achieved with simple physical techniques (Abedi-Firoozjah et al. 2025; Tavassoli et al. 2024a, b). In this regard, starch-based films have poor processability due to their poor thermomechanical properties. Therefore, combining starch with other biopolymers has become an attractive technique to compensate for the major disadvantages of starch while achieving specific properties for an application. Starch blending with biodegradable polysaccharide polymers is a strategic approach in food packaging. This process involves blending starch with various natural polymers such as chitosan, cellulose, agar, gums, etc. to reduce the disadvantages of using starch alone for food packaging. For example, natural fibers as fillers in starch films have been considered as a more environmentally friendly way to develop materials. Natural fibers are essentially composite materials, strengthened by tough, crystalline cellulose microfibrils held together within a more flexible hemicellulose/lignin structure (Dutta and Sit 2024). A common example of a natural polymer composite is reinforcing starch with cellulose (Kumari et al. 2022). Starch-based materials tend to be susceptible to moisture. Incorporating natural fibers can enhance thermoplastic materials, improving the properties of starch films. Cellulose fibers are mechanically mixed together to reinforce starch films and to reduce deformation and increase the tensile strength of the starch-based film (Pérez-Pacheco et al. 2016). The addition of 15% carboxymethyl cellulose resulted in a reduction in moisture absorption, vapor permeability, and film solubility (Onyeaka et al. 2022). Combining starch polymer with other biodegradable polymers not only results in a cost-effective and biodegradable film product, but also increases the processability and mechanical properties of the starch film. In this context, the researchers reported that the strength of the starch film increased by 37.5% with the addition of cassava fibers, which was due to the intermolecular interaction between the reinforcing component and corn starch (Travalini et al. 2019). Liu et al. (2022) showed that the addition of lignin, cellulose, and keratin to starch-based films improved the physical and chemical properties of the film (Liu et al. 2022).

3.5.1.2 Blends with Proteins

The blending of starch with protein-based polymers is a suitable strategy to enhance the physicochemical and functional properties of non-woven films. Proteins, due to their ability to form three-dimensional macromolecular networks stabilized by hydrophobic interactions, hydrogen bonds, and disulfide bonds, can be considered as one of the strategies for combining with starch polymers to overcome the disadvantages of the physicochemical properties of starch films alone (Romani et al. 2017). Studies have shown that the combination of polysaccharide-based films and proteins leads to enhanced intrinsic polyelectrolyte functional chemical groups, ultimately enhancing the physicochemical and mechanical properties of the film. Researchers reported that casein in the starch film reduced the crystallization of the starch layer and prevented starch recrystallization during storage (Lara-Castellanos et al. 2021). In a 2017 study, Ferreira et al. found that adding soy protein isolate to starch film increased the crystallinity of the film due to crosslinking between the protein and starch chains. Tensile strength (1.66 → 1.52 MPa) and elongation (84.98 → 8.27%) decreased after adding soy protein isolate, while elastic modulus increased from 2 to 4.4 MPa (Ferreira et al. 2021). In another study, Huntrakul et al. (2020) showed that the addition of chickpea protein isolates to acetylated cassava starch films increased tensile strength, water vapor and oxygen barrier properties, thermal stability, and hydrophobicity, while light transmission and solubility of the film were also reduced (Huntrakul et al. 2020). Similarly, combining starch with gelatin improves water solubility, water vapor permeability, transparency, and mechanical strength due to the formation of hydrogen bonds between starch and gelatin (Zhang et al. 2023).

3.5.2 *Incorporating Bioactive Materials*

3.5.2.1 Bioactive Compounds

The addition of bioactive ingredients to starch-based films serves a variety of purposes, including quality, shelf life, and safety considerations as well as consumer health concerns. The purposes for incorporating bioactive ingredients into starch films include the following.

Improved shelf life: The addition of bioactive compounds such as chlorogenic acid, quercetin, carvacrol, butylated, gallic acid, ascorbic acid, citric acid, tea polyphenols, etc., due to their antioxidant and antimicrobial potential, can prevent the growth of microorganisms and oxidative degradation and increase the shelf life of products packaged with starch films. Starch films containing bioactive compounds help maintain the taste, freshness, and quality of food during distribution and storage by delaying the deterioration process (Pei et al. 2024). For example, Almeida et al. (2023) reported that gallic acid starch film and bacterial nanocellulose delayed browning and weight loss of fresh apples at 4 °C for 7 days (Almeida et al. 2023). In another study, Tongdeesoontorn et al. (2020) found that adding quercetin and

TBHQ to starch/carboxymethylcellulose-based films delayed the oxidation of pork for 35–70 days (Tongdeesoontorn et al. 2020).

Nutritional Fortification: Starch-based films can contain bioactive compounds such as probiotics, vitamins, prebiotics, omega-3 fatty acids, plant sterols, dietary fibers, etc. These bioactive compounds with therapeutic or medicinal properties, in addition to nutritional properties, can also ensure consumers' health and maintain functional properties and nutritional value. Starch-based biodegradable films also protect sensitive materials from environmental conditions during transportation and storage. Starch-based biodegradable films, which offer nutritional enhancement properties, are suitable for packaging dietary supplements, functional foods, and fortified products. Starch-based films containing bioactive compounds enhance bioavailability and absorption in the body by controlled release and targeted delivery of bioactive compounds (Chen et al. 2022; Li et al. 2023).

Increasing the safety of food products: Bioactive compounds, due to their antimicrobial potential, can be incorporated into starch-based films by creating a protective barrier to prevent the growth of microorganisms and increase the safety of packaged food products, especially for minimally processed and perishable products (Adame et al. 2024; Pei et al. 2024). The antimicrobial mechanism of starch films containing antimicrobial compounds is that the slow release of antimicrobial compounds disrupts the microorganism's cell membrane, microbial enzyme systems, and metabolic pathways, ultimately leading to the death of the microorganism. Biodegradable starch films that contain antimicrobial compounds are suitable for packaging food products that are susceptible to microbial spoilage, including seafood, meat, and fresh produce (Al-Tayyar et al. 2020). For example, researchers reported that starch films containing carvacrol and chitosan inhibited the growth of *Listeria monocytogenes* bacteria for 4 weeks (Zhao et al. 2019). In another study, Chen et al. (2023) found that the inhibition rates of *Staphylococcus aureus* and *Escherichia coli* by starch/polyvinyl alcohol nanofibers containing 15% tea polyphenols were 42.29 and 29.97%, respectively (Chen et al. 2023).

Antioxidant potential: By adding bioactive compounds such as vitamin E, flavonoids, vitamin C, carotenoids, phenolic acids, etc. to biodegradable starch-based films, they prevent oxidation of packaged food products due to their antioxidant potential by neutralizing free radicals and reactive oxygen species and inhibiting oxidative reactions. The antioxidant potential of starch-based films can be enhanced through formulation and additive optimization. By adding bioactive compounds with antioxidant potential to the starch film matrix, a packaging can be produced that maintains the quality of packaged food products and ensures the health of consumers. Starch-based biodegradable films containing compounds with antioxidant potential are suitable for packaging food products such as oils, snacks and nuts, and processed foods (Cui et al. 2021; Ribeiro et al. 2021). For example, researchers reported that the antioxidant potential of a starch film containing rosemary extract was 81.9% and a polyvinyl alcohol/starch film containing pineapple peel extract was 40% (Kumar et al. 2021; Piñeros-Hernandez et al. 2017).

Clean Labeling: The addition of bioactive compounds to starch-based films aligns with consumer demands for natural additives and clean label ingredients. Starch-based films containing bioactive compounds display a clear and clean labeling, providing consumers with assurance of the quality, safety, and wholesomeness of the packaged products (Khan et al. 2022).

The findings show that the addition of bioactive compounds such as chlorogenic acid, quercetin, carvacrol, butylated, gallic acid, ascorbic acid, citric acid, tea polyphenols, etc. to biodegradable starch films leads to improved barrier, mechanical, antioxidant and antimicrobial properties of the film (Pei et al. 2024). Bioactive compounds such as plasticizers improve the tensile strength and flexibility of starch-based films, making them stronger for packaging and transportation. Bioactive compounds improve the barrier properties and water vapor permeability of starch-based films, and extend the shelf life of packaged food products by reducing oxidation and water vapor permeability (Cui et al. 2021; García-Guzmán et al. 2022). For example, Shi et al. (2024) showed that the addition of Litsea cubeba essential oil to cassava starch (CS)/sodium carboxymethyl cellulose film reduced the moisture content and solubility of the film by 24.37 and 29.68%, respectively. The water vapor permeability ($1.23 \rightarrow 1.11 \pm 0.01$ g mm m^2 d^{-1} KPa^{-1}) and oxygen permeability ($1.47 \rightarrow 1.35 \times 10^{-3}$ cm^{-1} d^{-1} atm^{-1}) of the film also decreased (Shi et al. 2024). In another study, Chen et al. (2023) found that the tensile strength, water vapor permeability, and thermal stability of starch/polyvinyl alcohol nanofibers containing 15% tea polyphenols were improved compared to the same nanofibers (Chen et al. 2023).

3.5.2.2 Nanoparticles

The use of nanotechnology is essential for inhibiting the growth of microorganisms and for long-term storage. The addition of nanoparticle in starch-based biodegradable packaging can increase the safety and quality of packaged products (Bahramian et al. 2024a). Nanoparticles such as montmorillonite, cellulose, chitosan, silver, copper, zinc oxide, titanium oxide, magnesium oxide have been used in starch-based polymers, which leads to improved functional and physicochemical properties of starch-based polymers. It should be noted that good compatibility between nanoparticles and starch polymer is essential to create a film with desirable properties (Bahramian et al. 2024b). The uniform distribution of nanoparticles in starch-based polymers increases the interfacial area between the starch matrix and nanoparticles, which leads to changes in the relaxation behavior of molecules and their mobility and ultimately improves the thermal and mechanical properties of the starch film. Nanoparticles lead to an increase in the tortuous path of gas molecules and lead to a decrease in the gas permeability in the starch film (Kumari et al. 2022). Researchers reported that the use of nanomaterials in starch packaging films is due to the main characteristics of nanomaterials, including physicochemical properties (optical, solubility, magnetic, etc.), high surface-to-volume ratio, thermodynamic properties, and stability at high pressure and temperature, which increase the shelf life and safety of packaged products (Onyeaka et al. 2022). Several studies show that the use of nanoparticles in starch-based films reduces the permeability of gases, including oxygen and carbon dioxide, which reduces the shelf life of packaged foods and acts as a barrier to the growth of microorganisms (Pandian et al. 2023; Peighambardoust et al. 2019; Roy and Rhim 2022). Sun et al. (2023) reported that by increasing the concentration of lignin nanoparticles from 0 to 2%, the tensile strength of the starch film increased from 12.1 to 48.9 MPa. The authors attributed the improvement in mechanical properties after the addition of nanoparticles to the formation of hydrogen bonds. The starch film containing 2% nanoparticles had complete protection against the transmission of ultraviolet radiation. The authors attributed the blocking of ultraviolet radiation by lignin nanoparticles to the conjugated double bond of the benzene ring, the carbonyl group of the phenylpropane side chain, and the phenolic hydroxyl group in the structure of lignin nanoparticles (Sun et al. 2023). In another study, Roy and Rhim (2022) prepared a starch/agar-based film containing green silver nanoparticles. The results showed that the addition of silver nanoparticles increased hydrophobicity and water vapor barrier. They showed excellent antimicrobial activity against *Listeria monocytogenes* and *Escherichia coli* (Roy and Rhim 2022). Similarly, the addition of zinc oxide nanoparticles and nanoclay decreased the water vapor permeability, oxygen permeability and solubility of the starch film and increased the thermal stability and tensile strength (Arifin et al. 2022).

3.5.2.3 Essential Oils

Essential oils (EOs) are an excellent option for optimizing and enhancing the performance of biodegradable films due to their bioactive compounds. In this regard, many studies have been conducted to enhance the performance of starch-based films using essential oils. A mixture of starch with essential oils, polymers, and nanoparticles was developed by researchers to produce starch-based biofilms that are comparable to commercial packaging (Dutta and Sit 2024). Studies have shown that the incorporation of essential oils into starch films significantly increases the chemical, physical, antioxidant and antimicrobial properties (Kalateh-Seifari et al. 2021; Romani et al. 2017; Yang et al. 2024a, b, c). For example, adding cinnamon bark, clove and oregano essential oils to gelatin and cassava starch films improved functional properties including water vapor, transparency and oxygen barrier properties (Sharma et al. 2021). Similar findings were obtained by adding emulsified thyme essential oil to starch/gellan films, which improved the oxygen barrier of the film. The use of antimicrobial compounds such as essential oils in starch-based films is a novel and safe technique to reduce the use of synthetic additives and preservatives (Júnior et al. 2021). Starch-based films containing essential oils, due to their bioactive compounds such as thymol, carvacrol, and eugenol, delay oxidation by preventing the formation of free radicals. The researchers found that increasing the essential oil content increased the antioxidant and antimicrobial potential of starch-based films. However, increasing the concentration of essential oils may affect the mechanical properties of starch-based films. For example, researchers found that films become brittle with increasing essential oil content and it is necessary to change the softener concentration (Vianna et al. 2021). In general, gram-positive bacteria are more sensitive to essential oils than gram-negative bacteria because the outer membrane around the cell wall is absent in gram-positive bacteria, and hydrophobic compounds such as essential oils easily penetrate into the bacterial cell. However, the antimicrobial mechanism of essential oils against gram-negative bacteria is through the phenolic compounds present in the essential oil, which damage the phospholipid cell membrane, leading to permeability into the bacterial cell and leakage of cytoplasmic contents (Syafiq et al. 2020). In this regard, Campos-Requena et al. (2017) reported that the addition of carvacrol and thymol essential oils to starch/montmorillonite films resulted in a 2- and 9-fold increase in thermal and mechanical properties, respectively (Campos-Requena et al. 2017).

3.5.2.4 Pigments and Dyes

In recent years, the use of pigments and dyes such as anthocyanins, betalains, etc. in starch-based smart films that are pH-responsive and can detect the spoilage status of packaged products has been promoted, allowing consumers to avoid purchasing spoiled products. Because the pH of packaged food products changes, it displays different colors to indicate the spoilage status of the products. Currently, studies on smart pH-responsive starch-based smart films are still in their early stages, and typically use anthocyanins and betalains and the addition of acid/base indicators to achieve pH response (Liu et al. 2022). In this regard, anthocyanins are natural non-toxic, safe and water-soluble pigments that are sensitive to pH changes. These natural pigments are extracted from anthocyanin-rich wastes such as grapes, red cabbage, blueberries, beets, etc (Wang et al. 2023a, b). For example, researchers prepared a Guinea fowl starch-based film containing grape waste powder with pH-responsive properties. They reported that at pH = 1, it showed a pink color due to the anthocyanin flavonoid cations, while at pH = 7, it showed a blue color due to the anthocyanin quinoline structure (Gutiérrez et al. 2018). In another study, Choi et al. (2017) added purple sweet potato anthocyanin to agar/potato starch film to monitor pork spoilage. The authors reported that changing the pH of pork resulted in a change in the film color from red to green (Choi et al. 2017). In general, shrimp and fish are perishable and enzymatic hydrolysis of trimethylamine oxide results in the production of volatile nitrogen compounds in these products, the release of which increases the pH of the fish sample (Bahramian et al. 2024a, b). For example, Jiang et al. (2020) developed a carboxymethyl cellulose/starch-based smart film containing purple sweet potato anthocyanin to monitor fish freshness. The researchers reported that the film's color changed from red to blue when the fish spoiled (Jiang et al. 2020). In another study, Luchese et al., by adding blueberry waste to starch film, concluded that the film is blue and red at alkaline and acidic pH, respectively, which can be used for starch-based smart packaging (Luchese et al. 2017).

3.5.2.5 Other Materials

In addition to the additives mentioned in the above discussions, other materials have also been used to improve the performance and properties of starch-based films. In this regard, several studies have been conducted on the preparation of emulsion-based starch films or the addition of wax compounds. For example, in one study, candelilla wax, beeswax, and carnauba wax were used to prepare a starch/gelatin-based film. They reported that the tensile strength of the films containing beeswax decreased due to the weakening of intermolecular hydrogen bonds. The elongation at break of the films containing the waxes was lower than that of the control film. The films containing candelilla wax showed the lowest solubility compared to the control film, while the films containing beeswax showed higher water vapor barrier, hydrophobicity, and thermal stability than the other films (Cheng et al. 2023). Another technique that can be used to improve the physicochemical and functional properties of starch polymers is the emulsion system. For example, researchers reported that emulsification of essential oils increased the hydrophobicity of starch/gelatin films (Acosta et al. 2016). Recently, metal-organic frameworks, which are one of the classes of porous hybridization materials, have attracted the attention of researchers for use in food packaging due to their affordability, easy preparation, stable structure, and antimicrobial properties (Bahramian et al. 2025). For example, Wang et al. (2024a, b) showed that after adding zinc-copper bimetallic organic framework nanorods to potato starch/polyvinyl alcohol film, the tensile strength increased and the UV-blocking and antimicrobial efficiency of the films reached 95% and 99%, respectively. The water vapor permeability ($12.37 \rightarrow 8.81 \times 10^{-8}$ g/m h Pa) and oxygen permeability ($3.32 \rightarrow 2.05 \times 10^{-3}$ g/m^2 s) decreased, which was due to the uniform distribution of zinc-copper bimetallic organic framework nanorods (Wang et al. 2024a, b). It is worth noting that another group of fillers such as carbon dots have been used to meet the packaging requirements in biodegradable films. Carbon dots are zero-dimensional, biocompatible, cost-effective, and have functions such as antimicrobial, antioxidant, and UV-blocking, which have been used in food packaging in recent years. Studies have shown that the addition of carbon dots to the gas barrier film improved the mechanical strength, UV blocking, and UV light blocking properties (Hadavifar et al. 2025). For example, in a study by Wu et al. (2024), it was found that adding carbon dots to corn starch film increased its thermal stability and elasticity and reduced its water vapor permeability. The film's antimicrobial potential against Staphylococcus aureus and Escherichia coli reached 99.9% (Wu et al. 2024).

3.6 Application of Starch-Based Polymers as Food Packaging Films

Starch-based films and coatings can be engineered to incorporate specific antibacterial and antioxidant agents, significantly enhancing their protective features. This innovative packaging approach not only improves the preservation of various food items, such as meat and meat products, dairy products, fruits, and vegetables but also plays a crucial role in extending their shelf life. These starch-based solutions help maintain optimal internal conditions within the packaging by effectively halting or detecting harmful chemical and biological agents. As a result, they provide a safeguard against spoilage and degradation, ensuring that food remains fresher for longer periods. A summary of the applications of starch-based polymers in food packaging is presented in Table 3.1.

The advancement of bioactive food packaging is rising due to the escalating demand for sustainable products and the necessity to prolong food shelf life. Zhao and his colleagues created novel antibacterial and antioxidant bioactive films utilizing subcritical water (SCW) technology, integrating chitosan and gallic acid with cassava starch. The antimicrobial efficacy of these films against spoiling microorganisms in pre-inoculated ham extended the ham's shelf life to 25 days, in contrast to only 7 days for the control group (Zhao et al. 2018). Cassava starch has gained recognition among various starches for its unique properties that make it particularly suitable for film development. One of its standout attributes is its remarkable clarity when pasteurized, which allows for an aesthetically pleasing appearance. Cassava starch exhibits a relatively low gelatinization temperature, making it easier to process, and it maintains good stability when formed into a gel (Plaeyao et al. 2025; Prithiviraj et al. 2024). Standard films made from cassava starch have several challenges, including low mechanical strength, inadequate barrier properties against water vapor, and limited antioxidant capabilities. To improve these issues, researchers are incorporating various natural additives, particularly plant extracts. These enhancements aim to boost physical properties like tensile strength and elasticity, as well as functional characteristics, such as water resistance and antioxidant potential, ultimately leading to more effective applications in food packaging (Mueller et al. 2024). Qin and co-workers developed smart packaging from cassava starch and anthocyanins derived from *L. ruthenicum*. Their research revealed that these starch-LRA films exhibit a remarkable sensitivity to changes in pH levels, demonstrating the ability to shift colors across a broad spectrum of buffer solutions with pH levels ranging from 2 to 13. This unique property allows the films to be a real-time food freshness indicator. Notably, when monitoring pork quality, the starch-LRA films underwent significant and discernible color transformations that closely aligned with the variations in the pork's freshness (Qin et al. 2019). Jayakumar and co-workers created an innovative bio nanocomposite utilizing a blend of PVA, starch, and titanium dioxide nanoparticles combined with elderberry extract. This composite was meticulously fabricated through the solvent casting technique, which allowed for a homogeneous distribution

Table 3.1 Summary of applications of starch-based polymers in food packaging

Starch type	Active ingredient	Fabrication method	Function	Food model	Key results	References
Cassava starch	Carboxymethyl cellulose lactate	Casting method	Antioxidants and antimicrobial	Pork	Adding antioxidants enhanced the tensile strength of cassava starch–CMC films while reducing elongation at break. Films with quercetin had higher tensile strength and less elongation than those with TBHQ. Increasing quercetin and TBHQ concentrations decreased water solubility. Over 30 days, total phenolic content and antioxidative activity (DPPH assay) remained stable. Quercetin and TBHQ-infused films effectively slowed lard oxidation (35–70 days) and delayed pork discoloration	Tongdeesoontorn et al. (2020)
Cassava starch	Cinnamon bark	Solution casting	Antioxidant and antimicrobial	Fuji apple slices	Burial tests indicated that starch and glycerol components degraded preferentially, as shown by the absence of their FTIR signatures, and the amorphous phase of PVA degraded, reflected in the increased crystallinity index of the samples. The blends' rheological properties influence their morphology, significantly affecting their mechanical properties	Boonsuk et al. (2020)

(continued)

Table 3.1 (continued)

Starch type	Active ingredient	Fabrication method	Function	Food model	Key results	References
Cowpea starch	Maqui berry extract	Casting method	Antioxidant	Salmon	The lipid oxidation of salmon samples stored at 4 °C was slowed down, and the antioxidant activity rose with a higher concentration of maqui berry extract	Baek et al. (2019)
Rye starch	Rosehip extract	Casting method	Antioxidant properties, Flexibility, optical	Chicken breast	The inclusion of rosehip extract enhanced the flexibility and light barrier characteristics of the films while also preventing lipid oxidation in chicken breast	Hassan et al. (2019)
Cassava starch	Mango pulp and Yerba mate extract	Casting method	Antioxidant and antimicrobial	Palm oil	Bio-based films were employed to package palm oil for a duration of 90 days under conditions that accelerated oxidation (63% RH/30 °C). The films diminished the rate of the oxidative process, with yerba mate and mango pulp contributing to enhanced stability in a concentration-dependent manner. Greater concentrations of additives led to improved stability of the palm oil, while the incorporation of these bioactive compounds notably altered the mechanical, physical, and barrier properties of the films ($p < 0.05$)	Reis et al. (2015)

(continued)

Table 3.1 (continued)

Starch type	Active ingredient	Fabrication method	Function	Food model	Key results	References
Corn starch/ Xanthan gum	Clove	Casting method	Antioxidant and antimicrobial	Red meat	A reduction in microbial populations in meat samples enhanced the color stability of the meat after being stored for 15 days	Radha Krishnan et al. (2015)
Yam starch	Eugenol	Casting method	Antimicrobial	Pork	Incorporating 3% eugenol into the films can extend the shelf life of pork by over 50%, and the release of eugenol resulted in antibacterial properties	Cheng et al. (2019)
Acetylated cassava thermoplastic starch/linear low-density polyethylene	Green tea	Blown extrusion	Antioxidant activity	Meat and oil	Microstructures controlled the release of phenolic and antioxidant substances Blown films containing green tea maintained the color and quality of packaged meat and oil	Panrong et al. (2019)
Longan seed starch	Banana flower bract anthocyanin extracts	Solution casting	Antioxidant and antimicrobial	Shrimp and beef	Improved UV light barrier capacities Increased water vapor permeability, and lowered thermal stability Reduced tensile strength and increased elongation at break Great antioxidant activities and pH-sensitive and ammonia-sensitive discoloration capacities	Jiang et al. (2023)

(continued)

Table 3.1 (continued)

Starch type	Active ingredient	Fabrication method	Function	Food model	Key results	References
Cassava starch	Zinc-nanoparticles	Casting method	Antimicrobial and oxygen barrier properties	Tomatoes	For a duration of 9 days at a temperature of 27 °C, the incorporation of zinc nanoparticles enhanced both the oxygen barrier characteristics and the antimicrobial properties of the films	Fadeyibi et al. (2017)
Cassava starch	Grape stalks	Thermal expansion	Flexural mechanical properties	English cake	Storage for 9 days at an average temperature of 23 °C and 55% relative humidity resulted in flexural mechanical properties that were comparable to those of foams produced from petroleum-based polymers	Engel et al. (2019)
Pine seeds (pinhão) starch	Citric pectin and feijoa peel flour	Casting method	Antioxidants and antimicrobial	Apple	After being stored for fifteen days, the inclusion of feijoa peel flour enhanced the antimicrobial activity and antioxidant properties of the films	Sganzerla et al. (2020)

(continued)

Table 3.1 (continued)

Starch type	Active ingredient	Fabrication method	Function	Food model	Key results	References
Pea starch/ polylactic acid	Maqui berry extract	Casting method	Tensile properties and barrier properties to water vapor and oxygen	Cherry tomatoes	Improved durability, thermal stability, and barrier properties Enhanced water resistance and tensile strength The bilayer films decreased the weight loss percentage of cherry tomatoes and prolonged the retention of organic acids and vitamin C	Zhou et al. (2019)
Corn starch/ gelatin	N-α-lauroyl-l-arginine ethyl ester monohydrochloride	Thermo-pressing and casting	Antimicrobial	Chicken breast fillets	The incorporation of N-α-lauroyl-l-arginine ethyl ester monohydrochloride enhanced the antimicrobial properties of films and prolonged the shelf life of chicken breast fillets	Moreno et al. (2018)

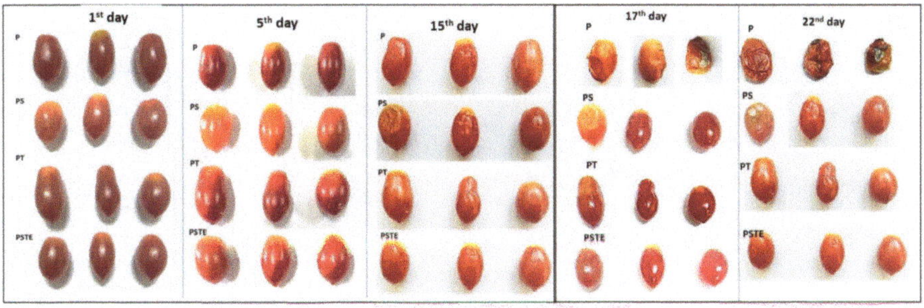

Fig. 3.1 Changes in cherry tomatoes during storage for 22 days. P: PVA, PS: PVA/starch, PT: PVA/TiO_2 nanoparticles and PSTE: PVA/starch/ TiO_2 nanoparticles/elderberry extract, Ref. Aswathy Jayakumar et al. (2022)

of its components, enhancing its properties for practical applications. Their extensive studies discovered that cherry tomatoes coated with the resulting polyvinyl alcohol-starch-titanium dioxide-elderberry solution exhibited remarkable benefits (Fig. 3.1). Specifically, these tomatoes displayed significantly reduced weight loss and remained free from microbial infections for an impressive duration of 22 days (Aswathy Jayakumar et al. 2022).

Starch, recognized for its biodegradable characteristics, functions as a superior thermoplastic polymer and has been successfully employed in the manufacture of active packaging films via the electrospinning method. For instance, Yang and colleagues investigated a novel approach to fabricate starch composite nanofibrous films (SNFs/TA/Fe^{3+}) utilizing starch, tannic acid, and Fe^{3+} via the temperature-assisted electrospinning technique. The incorporation of TA enhances the antioxidant capabilities of the SNFs, and the robust SNFs/TA/Fe^{3+} exhibited antioxidant activity comparable to that of SNFs/TA. Nonetheless, the SNFs/TA/Fe^{3+} exhibited a reduction in antibacterial efficacy, presumably attributable to the elevated valence state of the metal ions. In conclusion, the results highlighted that a simple electrospinning technique was employed to produce SNFs/TA/Fe^{3+}, resulting in improved mechanical properties and antioxidant efficacy, hence offering an innovative method for the fabrication of active food packaging utilizing SNFs (Yang et al. 2024a, b, c). Huanga and colleagues developed a novel copper-based metal–organic framework (Cu-ICA) nanocrystal exhibiting ammonia-sensitive characteristics, which was utilized as a functional compatibilizer in starch/PVA blends to fabricate high-performance intelligent packaging films designed for monitoring food freshness. Their findings demonstrated that the starch/PVA/Cu-ICA films had enduring color stability, remarkable antibacterial efficacy (surpassing 99.5%) against both Escherichia coli and Staphylococcus aureus, along with notable ammonia-sensitive discoloration capability. The starch/PVA/Cu-ICA films exhibited a discernible color shift throughout the monitoring of shrimp deterioration (Fig. 3.2) (Huang et al. 2024).

Cui and colleagues developed MCM-41 as a delivery mechanism for thymol (THY), which was integrated into a potato starch (PS) matrix to create an intelligent active packaging film tailored for the storage microenvironment. The THY-MCM-41/

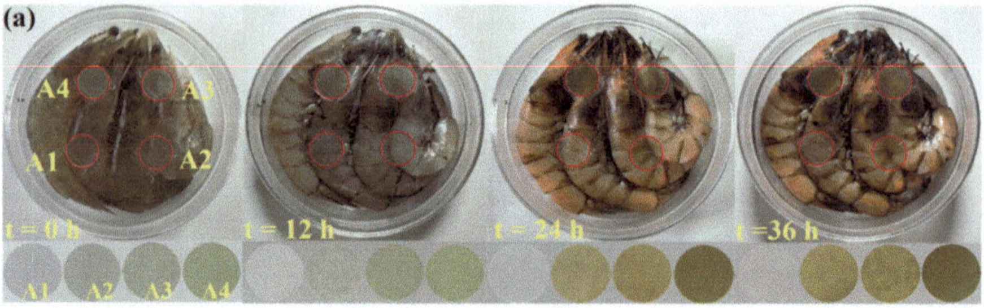

Fig. 3.2 Colour changes of the starch/PVA/Cu-ICA films to shrimp samples at different time (0–36 h): **a** actual view (A1: virgin starch/PVA, A2: starch/PVA/Cu-ICA0.5, A3: starch/PVA/Cu-ICA1.0, A4: starch/PVA/Cu-ICA2.0), Ref. Huang et al. (2024)

PS film demonstrated prolonged antibacterial efficacy during a 10-day storage period attributable to the continuous release of THY. Moreover, strawberries encased in the THY–MCM-41/PS film exhibited improved sensory attributes following a 5-day storage period at 25 °C and 50% relative humidity (Fig. 3.3) (Cui et al. 2024).

Nano-cellulose is commonly added as a reinforcing component to enhance the mechanical characteristics of starch film. This improvement boosts the mechanical strength and thermal stability of the film, making the composite film better suited for food packaging uses. In this regard, Yang and colleagues created a multifunctional smart biomass film by integrating chitosan-adsorbed laurate esterified starch curcumin Pickering emulsion into the starch film matrix, utilizing nano-cellulose as reinforcing agents. Their findings indicated that the free radical scavenging rate attained 56.01%, maintaining a high antioxidant capability even after 8 days. The inclusion of curcumin endowed the film with pH indicator properties and postponed the deterioration of pork (Yang et al. 2024a, b, c).

3.7 Biodegradability Properties of Starch-Based Polymers

Films created using starch must be authentically biodegradable to attain this classification. Starch-based materials are generally regarded as biodegradable, leading to comprehensive scientific inquiries into the effects of diverse additives and formulations on processing attributes and the efficacy of the final products. Biodegradability, as delineated by ASTM standards, denotes the ability of a substance to break down into carbon dioxide, methane, water, inorganic compounds, or biomass, chiefly via the enzymatic activity of microorganisms. This degradation can be assessed through standardized tests performed over a predetermined duration, mirroring the existing disposal conditions (Dutta and Sit 2024; Trotta et al. 2024; Zhang et al. 2025).

The soil-burial test is a widely utilized method for evaluating biodegradability. This method meticulously observes the progressive breakdown of starch and its composites in a soil matrix over several days, facilitating direct examination of

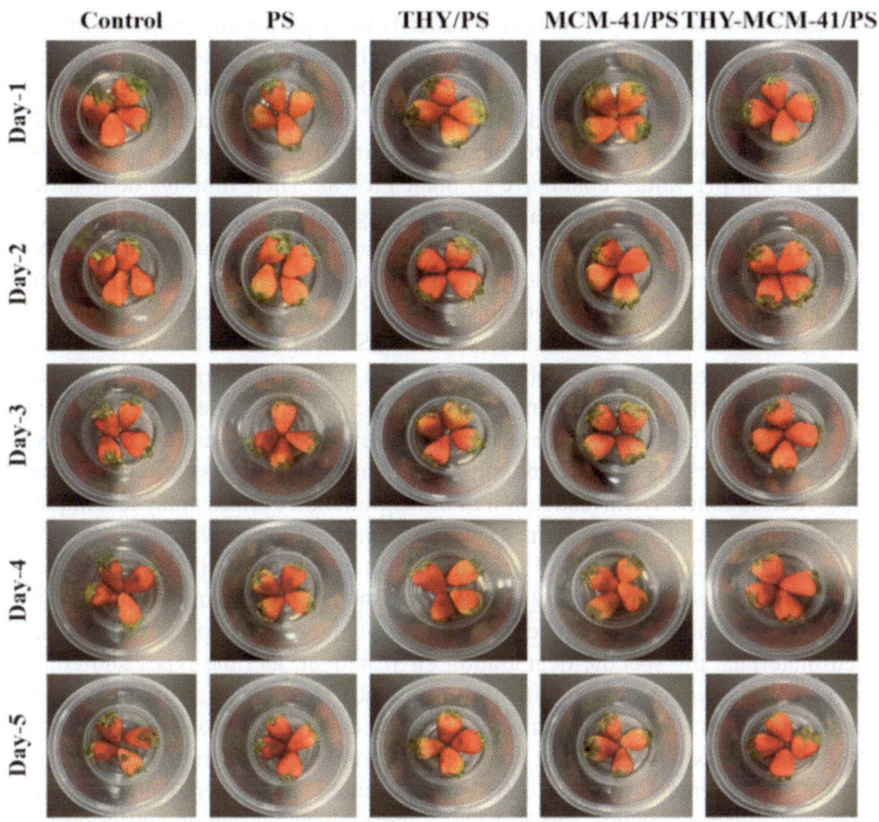

Fig. 3.3 Investigating quality change indices of stored strawberries during a 5-day storage period, Ref. Cui et al. (2024)

the degradation process. Although it is a laborious method, it reflects the natural environmental conditions that materials face during biodegradation, offering essential insights into the intricacies of the biodegradation process. Besides soil burial, the biodegradability of starch can be evaluated using different approaches including composting, aquatic settings, and enzymatic degradation assessments (Frangopoulos et al. 2025; Plaeyao et al. 2025). The degradation of composite or blended films in soil occurs in two phases: the first phase entails water absorption by the film, leading to swelling and promoting microbial colonization; the second phase involves enzymes and other biological secretions that diminish the film's weight and structural integrity. The extent and speed of biodegradation are determined by both the chemical characteristics of the polymer and the active biological processes in the degradation environment. Moreover, modifying starch or augmenting it with additional elements to form a nanocomposite can substantially influence its biodegradability characteristics. This element addresses the practical issues of the disposal locations of biodegradable products, such as home items, after usage (Xia et al. 2024; Yang et al. 2024a, b, c). A thorough investigation by Julinová et al. (2020) examined the impact of several fillers on the biodegradation rates of thermoplastic starch in aqueous and soil environments (Julinová et al. 2020). In a similar vein, Sen and Das (2018) conducted respirometric experiments to evaluate the biodegradability of starch-based antimicrobial films, quantifying biological oxygen demand to confirm that biodegradation processes preserved soil pH values conducive to plant growth. Their findings demonstrated enhancements in essential soil qualities such as organic carbon content, total nitrogen levels, accessible nitrogen, and water retention capacity, consequently underscoring the beneficial effects of biodegradation on soil health (Sen and Das 2018). Generally, starch blends disintegrate into tiny particles that can be readily broken within one week of environmental contact. Research by Su et al. (2023) indicates an inverse correlation between the hydrophobicity of the film and its biodegradability; increased hydrophobicity is frequently essential for food packaging materials while preserving biodegradability. Therefore, it is essential to achieve an appropriate balance between the hydrophobic characteristics and the degradability of starch-based films (Su et al. 2023). A comprehensive investigation of biodegradation utilizes many techniques in conjunction with visual assessments, such as Scanning Electron Microscopy (SEM), Atomic Force Microscopy (AFM), thermogravimetric analysis (TGA), and Fourier Transform Infrared Spectroscopy (FTIR). The interaction dynamics of starch with its additives, the film's structural compactness, the hygroscopic properties of starch-based materials, the specific degradation environment, and the types and concentrations of microorganisms significantly influence the biodegradability of starch-based composites. Comprehending these variables is crucial for precisely categorizing materials as biodegradable components, guaranteeing compliance with environmental requirements and consumer expectations. This chapter explores the novel methods utilized to enhance the properties and performance of starch-based films, highlighting the significant potential and many options offered by starch as a material. The global market for starch-derived products is projected to achieve a compound annual growth rate (CAGR) of 5% from 2020 to 2025, surpassing $300 million by 2025. Starch has become a valued substance in this field, principally owing to its affordability, superior film-forming characteristics, biocompatibility, abundant availability, renewability, and eco-friendly biodegradability (Alves et al. 2024; Zannou et al. 2023).

A notable advancement in this domain is the development of starch-based films tailored for food packaging purposes. These films are a sustainable and environmentally beneficial substitute for conventional plastic packaging, tackling the escalating worldwide issues related to environmental pollution and dependence on petrochemical materials. To efficiently function in food packaging, these biodegradable starch-based polymers must satisfy a series of rigorous performance standards. These criteria include various mechanical qualities, effective barrier functions against gases and moisture, and the integration of antibacterial and antioxidant capabilities to guarantee food safety and longevity (Cai et al. 2024; Wang et al. 2023a, b). Packaging plays a crucial role in improving food quality, safety, and total shelf life within the food business. Current research endeavors with starch-based materials predominantly focus on two principal ways to enhance the shelf life of food products.

The initial method entails augmenting the oxygen and water vapor barrier characteristics of packing materials by amalgamating starch with diverse components to manufacture a composite material that demonstrates superior protective attributes. The second technique emphasizes the incorporation of antioxidants or antibacterial agents directly into the film-forming matrix, thus imparting the films with the capacity to actively maintain food quality and safety. Through the examination of these approaches, researchers seek to devise sustainable solutions that both safeguard food and promote an environmentally friendly future. The progressive addition of preservatives to food during storage can markedly prolong its shelf life and durability. Starch-based biodegradable materials have recently emerged as viable alternatives for packaging a diverse range of food products, including fresh vegetables, fruits, various meats, and baked goods such as bread. This signifies a robust potential for commercial adoption, propelled by the rising need for sustainable packaging solutions. Ongoing research endeavors seek to improve the moisture resistance of starch films, essential for preserving the quality and safety of packed products. Strategies encompass the incorporation of hydrophobic chemicals that deter water, alongside advanced surface modification techniques to establish a more impermeable barrier. Moreover, researchers are concentrating on enhancing the processing methods for these starch films. Innovative techniques such as extrusion, casting, and blowing are being investigated to enhance the physical qualities and performance attributes of these materials (Su et al. 2023; Zannou et al. 2023). Furthermore, breakthroughs in nanotechnology are crucial for the creation of starch-based nanocomposites that exhibit greater mechanical strength and improved barrier characteristics. These improvements seek to overcome the shortcomings of conventional starch films, enhancing their ability to safeguard food from environmental influences. A vital research domain focuses on enhancing the stability of active compounds integrated into starch films. Researchers are examining sophisticated encapsulation methods to protect these compounds and enhance their release properties over time. Utilizing nanoparticles, they intend to develop a supporting matrix that guarantees the efficient transport of these active components, thus improving the preservation characteristics of the packaging (Arooj et al. 2024; Chavan et al. 2024). Starch-based biodegradable polymers offer numerous benefits, including sustainability, biodegradability, superior food quality preservation, controlled active agent release, greater safety, and cost efficiency. Nonetheless, considerable obstacles persist. Concerns related to mechanical strength, moisture sensitivity, film production intricacies, active agent stability, and the inadequate barrier qualities of current films require additional investigation. To address these difficulties, research and development initiatives must progress, perhaps resulting in the development of novel active and intelligent packaging solutions for the food sector. The design of such films requires eco-friendly procedures, including the alteration of graphene surfaces to improve their dispersibility in starch matrices (Diah et al. 2024; Fatima et al. 2024). This advanced method enhances the mechanical properties of the films and broadens their potential applications in several areas. Ongoing advancements in starch-based biodegradable active packaging involve both the materials and the incorporation of bio-based innovations alongside improved active and smart packaging technologies. These innovations enhance the preservation and protection of food products. The integration of active packaging technologies, including antibacterial agents and oxygen scavengers, shows potential for prolonging the shelf life of perishable goods, hence improving food safety for customers. Notwithstanding these encouraging advancements, researchers encounter significant challenges. Achieving an optimal equilibrium among critical mechanical properties, barrier functions, and biodegradability of the films constitutes a significant challenge (Liu et al. 2022; Rahardiyan et al. 2023). Researchers are continually challenged to develop films that exhibit sufficient tensile strength and flexibility while being environmentally sustainable. Guaranteeing the scalable production of these unique films for commercial application necessitates extensive research and development. Current endeavors to enhance these films involve investigating various starch sources, additional changes of graphene surfaces, and the integration of active packaging technologies. Despite ongoing hurdles, progress in this subject fosters a more sustainable and ecologically responsible packaging business, facilitating a future where packaging solutions are closely aligned with ecological principles. A thorough assessment of the life cycle of starch-based products relative to synthetic polymers is essential for comprehending their environmental and ecological impacts. This analysis includes several essential stages: the cultivation and harvesting of raw materials, the manufacturing processes of starch-based films, their distribution, usage in diverse contexts, and their disposal techniques (Cheng et al. 2021). By methodically evaluating the environmental impact at each stage, we may more effectively pinpoint critical areas for enhancement, potentially resulting in a substantial decrease in the ecological footprint of starch-based films. Analyzing the agricultural methods employed in the cultivation of starch sources such as maize or potatoes can uncover prospects for sustainable farming practices that reduce pesticide application and improve soil health. Examining the manufacturing processes may reveal more energy-efficient techniques or alternative biodegradable chemicals that improve film performance while reducing waste (Niranjana Prabhu and Prashantha 2018; Sadeghizadeh-Yazdi et al. 2019). Moreover, the research must not solely concentrate on environmental repercussions but also take into account economic aspects, particularly the cost structures related to the production of starch-based films. Comprehending these financial ramifications will be essential for augmenting output and rendering these materials competitive in the market. A comprehensive life cycle assessment of starch-based products is essential, as it facilitates informed decisions that enhance sustainability and innovation in future production methods.

3.8 Limitations and Challenges

Starch-based biodegradable materials have the potential to significantly enhance the sustainability of food packaging solutions moving forward. These materials are derived from renewable resources, making them an eco-friendlier alternative to conventional petroleum-based plastics. However, one of the primary challenges associated with starch is its retrogradation, essentially the process of starch molecules re-associating after they have been gelatinized, which can lead to mechanical weaknesses and a decline in performance over time. This issue has sparked considerable research aimed at improving the mechanical properties of starch-based films (Ma et al. 2025; Nazrin et al. 2025). To combat the challenges presented by retrogradation and inadequate mechanical properties, researchers have explored a variety of modification techniques. These methods include chemical modifications such as esterification and etherification, enzymatic modifications employing specific enzymes to alter starch properties, and physical modifications involving heat and shear processing. Blending starch with other biopolymers, such as polylactic acid (PLA) or chitosan, as well as the incorporation of functional additives, can enhance the performance of the resulting films (Tavassoli et al. 2024a, b; Zubair et al. 2024). One of the most notable limitations of starch-based packaging is their mechanical strength and barrier properties, which directly impact their efficacy in protecting food. However, by utilizing various modification techniques and blending strategies, scientists are working to create packaging materials that do not compromise on these properties. Moreover, the development of active packaging, a system that incorporates components that actively respond to environmental factors, can be achieved by integrating antioxidants or antibacterial agents into starch-based films. Such advancements could not only extend the shelf life of perishable food items but also significantly reduce food waste (Fatima et al. 2024; Pei et al. 2024). Innovative developments in this field have led to the creation of films that can indicate freshness by responding to time and temperature changes. These smart films can visually signal when food is no longer fresh through a change in color, providing consumers with real-time information about product quality. Currently, starch-based films are commonly employed in food packaging and have proven effective at safeguarding fresh produce against spoilage (Diah et al. 2024; Dutta and Sit 2024). Several factors influence the functional performance of starch-based biodegradable materials, including the choice of starch source (such as corn, tapioca, or potato), as well as the specific conditions during film formation, such as temperature, humidity, and duration. The use of plasticizers, such as glycerol or sorbitol, which enhance flexibility and workability, also has a significant impact. To enhance the properties and functionalities of starch-based materials, researchers are increasingly integrating functional components. These include antibacterial agents that can prevent microbial growth and antioxidants that counteract oxidative processes, ultimately improving food safety and quality (Dejene et al. 2024; Deng et al. 2024). A pressing challenge that the industry must address is developing efficient and scalable production methods for starch-based materials. Though much progress has been made in laboratory settings, the transition to large-scale, economically viable manufacturing remains a hurdle. Research efforts are needed to create processing techniques that are adaptable to industrial settings while also ensuring that the materials produced meet the functional standards required for commercial applications (Rahardiyan et al. 2023; Su et al. 2023). Despite advancements in certain techniques to improve the properties of starch-based films, additional research is vital to achieve mechanical and barrier properties that are comparable, if not superior, to those of traditional plastic packaging. The performance of these biodegradable materials can be influenced by various factors, such as starch type, specific preparation methods, and environmental storage conditions, which can lead to inconsistencies in product quality and stability (Muñoz-Gimena et al. 2023; Othman et al. 2023).

3.9 Safety Aspects

Currently, most research regarding the application of starch in food packaging occurs at the laboratory level, and several factors, including limitations in industrial scalability, regulatory safety standards, environmental concerns, and the need for consumer acceptance, pose challenges to the commercialization of starch-based films. Exploring the synergistic properties of starch when combined with other materials requires further investigation to meet practical packaging requirements (Niranjana Prabhu and Prashantha 2018; Zubair et al. 2024). Many additives present exciting opportunities in the development of films with antibacterial and antioxidant properties. There remains considerable room for exploration in identifying and formulating effective antimicrobial agents compatible with starch matrices. Although the field of starch-based smart films looks promising, functionality to date has been limited. Currently, pH indicators are the primary means of achieving intelligent packaging. However, care must be taken, as active substances that modify the package's environment may also alter the pH, potentially leading to inconsistencies in indicator function, which could mislead consumers regarding product freshness (Carneiro da Silva et al. 2023; Muñoz-Gimena et al. 2023). For the advancement of intelligent packaging, integrating functional food components with multi-reactive starch membranes will be crucial. By incorporating heat-sensitive, moisture-sensitive, gas-sensitive, and other multifunctional properties, new forms of intelligent packaging could emerge, providing diverse functionalities and enabling a controlled release of active substances incorporated in the packaging film (Liu et al. 2022; Vianna et al. 2021). In conclusion, while the development of active and intelligent packaging made from starch-based materials faces various obstacles, including material selection, method of preparation, and functionality, it is evident that starch presents the most promising substitute for traditional petroleum-based packaging due to its inherent structural advantages and additional benefits. The future evolution of intelligent food packaging is likely to be spearheaded by starch-based innovations, paving the way for safer and more sustainable food storage solutions.

3.10 Conclusion and Future Prospective

Starch-based polymer materials have emerged as a potentially transformative solution for the future of sustainable food packaging. These materials can significantly reduce the reliance on petroleum-based packaging, thereby minimizing the energy consumption and environmental footprint often associated with conventional plastic options. However, the primary hurdle in the advancement of starch-based packaging lies in the insufficient mechanical strength and barrier properties inherent to starch itself. This limitation poses challenges in protecting food products adequately against external factors such as moisture, gases, and microbial contamination. Researchers are actively exploring various strategies to overcome these obstacles. One promising approach involves physical or chemical modifications of starch to enhance its properties. For instance, crosslinking techniques or plasticizers can be employed to improve flexibility and tensile strength. Blending starch with other biopolymers, such as polyvinyl alcohol or polylactic acid, or incorporating functional additives can significantly augment its performance. This multifaceted approach aims to meet the essential requirements for effective food packaging, which can eventually lead to the development of active packaging systems. Active packaging made from starch can integrate antioxidant or antibacterial agents, contributing to a prolonged shelf life for perishable goods and ultimately reducing food waste. Films manufactured from starch can serve as time and temperature indicators. These smart films have the potential to provide real-time assessments of food freshness through colorimetric changes, alerting consumers to the state of the product inside. Despite some advancements in enhancing the properties of starch-based films, further research and development are necessary to achieve mechanical and barrier properties that can rival those of traditional plastic packaging. The performance metrics of starch-based biodegradable materials can vary significantly based on factors such as the type of starch used, the specific preparation methods, the storage conditions, and a variety of other influencing variables. Such inconsistencies can lead to instability in the final product, which poses challenges for large-scale manufacturing and commercial viability.

Currently, much experimentation is conducted in controlled laboratory settings, which limits the translation of innovative findings into practical applications. This transition is obstructed by several factors, including industrial production constraints, stringent safety regulations, ecological considerations, and consumer preferences, which collectively hinder starch-based films' broader acceptance and marketability. Integrating starch with various additives and other materials necessitates further investigation to ensure that practical requirements are met while enhancing film performance. Various additives such as natural extracts, essential oils, and nanoparticles show great potential for fabricating films with enhanced antibacterial and antioxidant characteristics. There remains a vast opportunity for innovative developments, especially in formulating antimicrobial agents suitable for incorporation into starch matrices. The field of starch-based smart films presents significant promise; however, its functionalities have yet to reach their full potential. Currently, pH indicators are primarily utilized in intelligent packaging solutions. Active agents within the packaging can modify the environment surrounding the food, potentially influencing the pH of the system. This interaction could lead to irreversible changes or time-sensitive transformations in the indicators embedded within the film, which might mislead consumers regarding the product's quality. To advance this field, further research is warranted to explore the integration of functional food components with multireactive starch membranes. By interlinking heat-sensitive, moisture-sensitive, gas-sensitive, and other multifunctional materials, it is feasible to create innovative intelligent packaging solutions that offer diverse functionalities. Such advancements would enable the development of stimulus-responsive packaging tailored to specific food requirements, as well as controlling the gradual release of active substances embedded within the film. In conclusion, while starch-based active and intelligent packaging confronts several challenges related to material selection, production methodologies, and efficacy, it is evident that starch possesses substantial structural advantages and beneficial properties. Thus, starch stands out as a leading candidate to replace traditional petroleum-based packaging. With continued research and innovation, starch-based materials have the potential to become frontrunners in the evolution of smart food packaging, paving the way for a more sustainable future in food preservation and distribution.

Acknowledgements This chapter was supported by the Student Research Committee, Yasuj University of Medical Sciences, Yasuj, Iran.

Declaration The authors declare that AI tools were used in the preparation of this manuscript for the purpose of improving the readability and clarity. All content was critically reviewed and verified by the authors to ensure accuracy and originality. The authors take full responsibility for the final content of this work.

References

Abedi-Firoozjah R, Ahmadi N, Valaee S, Zamanpour S, Moradiyan MH, Bahramian B et al (2025) Unlocking the potential of cyclodextrin/metal–organic frameworks in food packaging systems: an exclusive review on the latest trends. Food Bioprocess Technol 18:5010–5037. https://doi.org/10.1007/s11947-025-03769-x

Acosta S, Chiralt A, Santamarina P, Rosello J, González-Martínez C, Cháfer M (2016) Antifungal films based on starch-gelatin blend, containing essential oils. Food Hydrocoll 61:233–240. https://doi.org/10.1016/j.foodhyd.2016.05.008

Adame MY, Wang Y, Shi C, Aziz T, Al-Asmari F, Sameeh MY et al (2024) Fortification of pullulan/cassava starch-based edible films incorporated with LC-EO nanoparticles and the application for beef meat preservation. Int J Biol Macromol 279:135629. https://doi.org/10.1016/j.ijbiomac.2024.135629

Al-Tayyar NA, Youssef AM, Al-Hindi RR (2020) Edible coatings and antimicrobial nanoemulsions for enhancing shelf life and reducing foodborne pathogens of fruits and vegetables: a review. Sustain Mater Technol 26:e00215. https://doi.org/10.1016/j.susmat.2020.e00215

Almeida T, Karamysheva A, Valente BF, Silva JM, Braz M, Almeida A et al (2023) Biobased ternary films of thermoplastic starch, bacterial nanocellulose and gallic acid for active food packaging. Food Hydrocoll 144:108934. https://doi.org/10.1016/j.foodhyd.2023.108934

Alves Z, Brites P, Ferreira NM, Figueiredo G, Otero-Irurueta G, Gonçalves I et al (2024) Thermoplastic starch-based films loaded with biochar-ZnO particles for active food packaging. J Food Eng 361. https://doi.org/10.1016/j.jfoodeng.2023.111741

Arifin HR, Djali M, Nurhadi B, Azlin-Hasim S, Masruchin N, Vania PA, Hilmi A (2022) Corn starch-based bionanocomposite film reinforced with ZnO nanoparticles and different types of plasticizers. Front Sustain Food Syst 6:886219. https://doi.org/10.3389/fsufs.2022.886219

Arooj A, Khan M, Munawar KS (2024) Preparation and physicochemical characterization of starch/pectin and chitosan blend bioplastic films as future food packaging materials. J Environ Chem Eng 12(1). https://doi.org/10.1016/j.jece.2023.111825

Baek S-K, Kim S, Song KB (2019) Cowpea starch films containing maqui berry extract and their application in salmon packaging. Food Packag Shelf Life 22:100394. https://doi.org/10.1016/j.fpsl.2019.100394

Bahramian B, Abedi-Firoozjah R, Ebrahimi A, Tavassoli M, Ehsani A, Naebe M (2024a) Resveratrol-biopolymer materials: a sustainable approach to food packaging. Trends Food Sci Technol 154:104761. https://doi.org/10.1016/j.tifs.2024.104761

Bahramian B, Abedi-Firoozjah R, Kiani-Salmi N, Azizian A, Dorud N, Noori SMA et al (2024b) Multifunctional performance of gallic acid in biodegradable food packaging films and coatings: mechanisms, developments, applications, and horizons. Eur Polym J 221:113559. https://doi.org/10.1016/j.eurpolymj.2024.113559

Bahramian B, Abedi-Firoozjah R, Kiani-Salmi N, Ebrahimi A, Oladzadabbasabadi N, Moradiyan MH et al (2025) Development of polycaprolactone-based electrospun nanofiber incorporated lemon beebrush essential oil-loaded metal-organic frameworks as a novel active food packaging for meat preservation. Food Control 168:110981. https://doi.org/10.1016/j.foodcont.2024.110981

Bakhshizadeh M, Ayaseh A, Hamishehkar H, Samadi Kafil H, Niknazar Moghaddam T, Baghban Haghi P et al (2023) Multifunctional performance of packaging system based on gelatin/alove vera gel film containing of rosemary essential oil and common poppy anthocyanins. Food Control 154:110017. https://doi.org/10.1016/j.foodcont.2023.110017

Bangar SP, Whiteside WS, Ashogbon AO, Kumar M (2021) Recent advances in thermoplastic starches for food packaging: a review. Food Packag Shelf Life 30. https://doi.org/10.1016/j.fpsl.2021.100743

Baysal G, Doğan F (2020) Investigation and preparation of biodegradable starch-based nanofilms for potential use of curcumin and garlic in food packaging applications. J Biomater Sci Polym Ed 31(9):1127–1143. https://doi.org/10.1080/09205063.2020.1743947

Bhat R, Abdullah N, Din RH, Tay GS (2013) Producing novel sago starch based food packaging films by incorporating lignin isolated from oil palm black liquor waste. J Food Eng 119(4):707–713. https://doi.org/10.1016/j.jfoodeng.2013.06.043

Boonsuk P, Sukolrat A, Kaewtatip K, Chantarak S, Kelarakis A, Chaibundit C (2020) Modified cassava starch/poly(vinyl alcohol) blend films plasticized by glycerol: structure and properties. J Appl Polym Sci 137(26):48848. https://doi.org/10.1002/app.48848

Cai J, Feng X, Zhang D, Ma S, Xie F (2024) Research progress on starch-based fibrous films: electrospinning and food packaging. Shipin Kexue/Food Sci 45(7):272–280. https://doi.org/10.7506/spkx1002-6630-20230728-309

Campos-Requena VH, Rivas BL, Pérez MA, Figueroa CR, Figueroa NE, Sanfuentes EA (2017) Thermoplastic starch/clay nanocomposites loaded with essential oil constituents as packaging for strawberries—In vivo antimicrobial synergy over Botrytis cinerea. Postharvest Biol Technol 129:29–36. https://doi.org/10.1016/j.postharvbio.2017.03.005

Carneiro da Silva LR, Rios ADO, Campomanes Santana RM (2023) Polymer blends of poly(lactic acid) and starch for the production of films applied in food packaging: a brief review. Polym Renew Resour 14(2):108–153. https://doi.org/10.1177/20412479231154924

Chavan P, Singh GP, Aayush K, Sharma S, Chauhan PK, Thory R et al (2024) Development and characterization of pearl millet (AHB 1200) starch nanoparticle-based edible films: a paradigm shift in food packaging. J Food Process Eng 47(3). https://doi.org/10.1111/jfpe.14575

Chen W, Ma S, Wang Q, McClements DJ, Liu X, Ngai T, Liu F (2022) Fortification of edible films with bioactive agents: a review of their formation, properties, and application in food preservation. Crit Rev Food Sci Nutr 62(18):5029–5055. https://doi.org/10.1080/10408398.2021.1881435

Chen L, Wu F, Xiang M, Zhang W, Wu Q, Lu Y et al (2023) Encapsulation of tea polyphenols into high amylose corn starch composite nanofibrous film for active antimicrobial packaging. Int J Biol Macromol 245:125245. https://doi.org/10.1016/j.ijbiomac.2023.125245

Cheng J, Wang H, Kang S, Xia L, Jiang S, Chen M, Jiang S (2019) An active packaging film based on yam starch with eugenol and its application for pork preservation. Food Hydrocoll 96:546–554. https://doi.org/10.1016/j.foodhyd.2019.06.007

Cheng H, Chen L, McClements DJ, Yang T, Zhang Z, Ren F et al (2021) Starch-based biodegradable packaging materials: a review of their preparation, characterization and diverse applications in the food industry. Trends Food Sci Technol 114:70–82. https://doi.org/10.1016/j.tifs.2021.05.017

Cheng Y, Zhai X, Wu Y, Li C, Zhang R, Sun C et al (2023) Effects of natural wax types on the physicochemical properties of starch/gelatin edible films fabricated by extrusion blowing. Food Chem 401:134081

Choi I, Lee JY, Lacroix M, Han J (2017) Intelligent pH indicator film composed of agar/potato starch and anthocyanin extracts from purple sweet potato. Food Chem 218:122–128

Cui C, Ji N, Wang Y, Xiong L, Sun Q (2021) Bioactive and intelligent starch-based films: a review. Trends Food Sci Technol 116:854–869

Cui Y, Zhang R, Cheng M, Li M, Wang X (2024) Development and application of mathematical modeling of thymol release from environmental-responsive potato starch active packaging films. Int J Biol Macromol 271:132353. https://doi.org/10.1016/j.ijbiomac.2024.132353

Dejene BK, Birilie AA, Yizengaw MA, Getahun SA (2024) Thermoplastic starch-ZnO nanocomposites: a comprehensive review of their applications in functional food packaging. Int J Biol Macromol 282. https://doi.org/10.1016/j.ijbiomac.2024.137099

Deng H, Su J, Zhang W, Khan A, Sani MA, Goksen G et al (2024) A review of starch/polyvinyl alcohol (PVA) blend film: a potential replacement for traditional plastic-based food packaging film. Int J Biol Macromol 273. https://doi.org/10.1016/j.ijbiomac.2024.132926

Diah SBM, Abdullah SHYS, Seok YB, Fatah IYA, Rahman NIA, Hafizulhaq F, Alias NN (2024) Towards sustainable food packaging: a review of thermoplastic starch (TPS) as a promising bioplastic material, its limitations, and improvement strategies with bio-fillers and essential oils. J Adv Res Fluid Mech Thermal Sci 119(1):80–104. https://doi.org/10.37934/arfmts.119.1.80104

Díaz-Galindo EP, Nesic A, Cabrera-Barjas G, Mardones C, Von Baer D, Bautista-Baños S, Garcia OD (2020) Physical-chemical evaluation of active food packaging material based on thermoplastic starch loaded with grape cane extract. Molecules 25(6). https://doi.org/10.3390/molecules25061306

do Val Siqueira L, Arias CILF, Maniglia BC, Tadini CC (2021) Starch-based biodegradable plastics: methods of production, challenges and future perspectives. Curr Opin Food Sci 38:122–130. https://doi.org/10.1016/j.cofs.2020.10.020

Dutta D, Sit N (2024) Comprehensive review on developments in starch-based films along with active ingredients for sustainable food packaging. Sustain Chem Pharm 39:101534. https://doi.org/10.1016/j.scp.2024.101534

Engel JB, Ambrosi A, Tessaro IC (2019) Development of biodegradable starch-based foams incorporated with grape stalks for food packaging. Carbohyd Polym 225:115234. https://doi.org/10.1016/j.carbpol.2019.115234

Fadeyibi A, Osunde ZD, Egwim EC, Idah PA (2017) Performance evaluation of cassava starch-zinc nanocomposite film for tomatoes packaging. J Agric Eng 48(3):137–146

Fatima S, Khan MR, Ahmad I, Sadiq MB (2024) Recent advances in modified starch based biodegradable food packaging: a review. Heliyon 10(6). https://doi.org/10.1016/j.heliyon.2024.e27453

Fattahi R, Bahrami A (2018) Application of edible and biodegradable starch films in food packaging: a review. J Babol Univ Med Sci 20

Ferreira LF, de Oliveira ACS, de Oliveira Begali D, de Sena Neto AR, Martins MA, de Oliveira JE et al (2021) Characterization of cassava starch/soy protein isolate blends obtained by extrusion and thermocompression. Ind Crops Prod 160:113092

Frangopoulos T, Marinopoulou A, Petridis D, Rhoades J, Likotrafiti E, Goulas A et al (2025) Films from starch inclusion complexes with bioactive compounds as food packaging material. Food Bioprocess Technol. https://doi.org/10.1007/s11947-025-03757-1

García-Guzmán L, Cabrera-Barjas G, Soria-Hernández CG, Castaño J, Guadarrama-Lezama AY, Rodríguez Llamazares S (2022) Progress in starch-based materials for food packaging applications. Polysaccharides 3(1):136–177

Gulati S, Kumar S, Chandra P, Jain A, Ahuja L, Batra K, Sharma N (2022) Starch based bio-nanocomposites: modern and benign materials in food packaging industry. In: Handbook of consumer nanoproducts, pp 881–909

Gutiérrez TJ, Herniou-Julien C, Álvarez K, Alvarez VA (2018) Structural properties and in vitro digestibility of edible and pH-sensitive films made from guinea arrowroot starch and wastes from wine manufacture. Carbohyd Polym 184:135–143

Hadavifar S, Abedi-Firoozjah R, Bahramian B, Jafari N, Sadeghi SM, Majnouni S et al (2025) Multifunctional performance of chitosan/soy protein isolation-based films impregnated carbon dots/anthocyanin derived from purple hull pistachio for tracking and extending the shelf life of fish. Food Hydrocoll 159:110678

Hassan M, Dave K, Chandrawati R, Dehghani F, Gomes VG (2019) 3D printing of biopolymer nanocomposites for tissue engineering: nanomaterials, processing and structure-function relation. Eur Polymer J 121:109340. https://doi.org/10.1016/j.eurpolymj.2019.109340

Huang K, Wang Y, Xu Z, Zou Z, Tang Q, Li H, Peng D (2024) Novel intelligent packaging films based on starch/PVA with Cu-ICA nanocrystal as functional compatibilizer for monitoring food freshness. Int J Biol Macromol 271:132373. https://doi.org/10.1016/j.ijbiomac.2024.132373

Huntrakul K, Yoksan R, Sane A, Harnkarnsujarit N (2020) Effects of pea protein on properties of cassava starch edible films produced by blown-film extrusion for oil packaging. Food Packag Shelf Life 24:100480

Jayakumar A, Heera KV, Sumi TS, Joseph M, Mathew S, Praveen G et al (2019). Starch-PVA composite films with zinc-oxide nanoparticles and phytochemicals as intelligent pH sensing wraps for food packaging application. Int J Biol Macromol 136:395–403. https://doi.org/10.1016/j.ijbiomac.2019.06.018

Jayakumar A, Radoor S, Kim JT, Rhim JW, Parameswaranpillai J, Nandi D et al (2022) Titanium dioxide nanoparticles and elderberry extract incorporated starch based polyvinyl alcohol films as active and intelligent food packaging wraps. Food Packag Shelf Life 34:100967. https://doi.org/10.1016/j.fpsl.2022.100967

Jiang G, Hou X, Zeng X, Zhang C, Wu H, Shen G et al (2020) Preparation and characterization of indicator films from carboxymethyl-cellulose/starch and purple sweet potato (Ipomoea batatas (L.) lam) anthocyanins for monitoring fish freshness. Int J Biol Macromol 143:359–372

Jiang H, Zhang W, Cao J, Jiang W (2023) Development of biodegradable active films based on longan seed starch incorporated with banana flower bract anthocyanin extracts and applications in food freshness indication. Int J Biol Macromol 251:126372. https://doi.org/10.1016/j.ijbiomac.2023.126372

Julinová M, Vaňharová L, Jurča M, Minařík A, Duchek P, Kavečková J et al (2020) Effect of different fillers on the biodegradation rate of thermoplastic starch in water and soil environments. J Polym Environ 28:566–583

Júnior LM, Vieira RP, Jamróz E, Anjos CAR (2021) Furcellaran: an innovative biopolymer in the production of films and coatings. Carbohyd Polym 252:117221

Kalateh-Seifari F, Yousefi S, Ahari H, Hosseini SH (2021) Corn starch-chitosan nanocomposite film containing nettle essential oil nanoemulsions and starch nanocrystals: optimization and characterization. Polymers 13(13):2113

Kanatt SR (2020) Irradiation as a tool for modifying tapioca starch and development of an active food packaging film with irradiated starch. Radiat Phys Chem 173. https://doi.org/10.1016/j.radphyschem.2020.108873

Khan AW, Roobab U, Shehzadi K, Inam-Ur-Raheem M, Aadil RM (2022) Clean label interventions in active and intelligent food packaging. In: The age of clean label foods. Springer, pp 161–208

Khezerlou A, Alizadeh Sani M, Tavassoli M, Abedi-Firoozjah R, Ehsani A, McClements DJ (2023) Halochromic (pH-Responsive) indicators based on natural anthocyanins for monitoring fish freshness/spoilage. J Compos Sci 7(4). https://doi.org/10.3390/jcs7040143

Kumar P, Tanwar R, Gupta V, Upadhyay A, Kumar A, Gaikwad KK (2021) Pineapple peel extract incorporated poly (vinyl alcohol)-corn starch film for active food packaging: preparation, characterization and antioxidant activity. Int J Biol Macromol 187:223–231

Kumar R, Ghoshal G, Goyal M (2019) Synthesis and functional properties of gelatin/CA-starch composite film: excellent food packaging material. J Food Sci Technol 56(4):1954–1965. https://doi.org/10.1007/s13197-019-03662-4

Kumari SVG, Pakshirajan K, Pugazhenthi G (2022) Recent advances and future prospects of cellulose, starch, chitosan, polylactic acid and polyhydroxyalkanoates for sustainable food packaging applications. Int J Biol Macromol 221:163–182

Lara-Castellanos M, Azuara E, Jimenez-Fernandez V, Luna-Solano G, Jimenez M (2021) Effect of casein replacement by modified casein on physicochemical, textural, sensorial properties and microbiological stability of fresh cheese. Int Dairy J 112:104864

Lauer MK, Smith RC (2020) Recent advances in starch-based films toward food packaging applications: physicochemical, mechanical, and functional properties. Compr Rev Food Sci Food Saf 19(6):3031–3083. https://doi.org/10.1111/1541-4337.12627

Lee SW, Said NS, Sarbon NM (2021) The effects of zinc oxide nanoparticles on the physical, mechanical and antimicrobial properties of chicken skin gelatin/tapioca starch composite films in food packaging. J Food Sci Technol 58(11):4294–4302. https://doi.org/10.1007/s13197-020-04904-6

Li X-L, Shen Y, Hu F, Zhang X-X, Thakur K, Rengasamy KR et al (2023) Fortification of polysaccharide-based packaging films and coatings with essential oils: a review of their preparation and use in meat preservation. Int J Biol Macromol 242:124767

Li T, Guan L, Zeng M, Li T, Lei S, Tang X et al (2025) Fabrication of corn starch-based composite films functionalized by Rosa roxburghii Tratt fruit pomace via extrusion compression molding for active food packaging. Int J Biol Macromol 284. https://doi.org/10.1016/j.ijbiomac.2024.138091

Liu D, Zhao P, Chen J, Yan Y, Wu Z (2022) Recent advances and applications in starch for intelligent active food packaging: a review. Foods 11(18):2879. https://doi.org/10.3390/foods11182879

Luchese CL, Sperotto N, Spada JC, Tessaro IC (2017) Effect of blueberry agro-industrial waste addition to corn starch-based films for the production of a pH-indicator film. Int J Biol Macromol 104:11–18

Lv H, Xu H, Xu E, Jin Z, Zhao H, Yuan C et al (2024) Improving structural and functional properties of starch-catechin-based green nanofiber mats for active food packaging by electrospinning and crosslinking techniques. Int J Biol Macromol 267. https://doi.org/10.1016/j.ijbiomac.2024.131460

Ma S, He J, Chen Q, Zhou J, Xie F, Cai J (2025) Advancements in starch-based nanomaterials for functional delivery and food packaging: a comprehensive review and future directions. J Future Foods 5(5):443–454. https://doi.org/10.1016/j.jfutfo.2024.08.002

Mallick N, Pal D, Soni AB, Jhariya DC, Singh D (2020) Starch based antimicrobial food packaging film towards a sustainable environment. IOP Conf Ser: Earth Environ Sci

Mangaraj S, Thakur RR, Yadav A (2022) Development and characterization of PLA and Cassava starch-based novel biodegradable film used for food packaging application. J Food Process Preserv 46(9). https://doi.org/10.1111/jfpp.16314

Mhaske ST, Mohanty JD, Borse PY (2024) Starch from agro-waste for food packaging applications. In: Agro-waste derived biopolymers and biocomposites: innovations and sustainability in food packaging, pp 241–266
Moreno O, Atarés L, Chiralt A, Cruz-Romero MC, Kerry J (2018) Starch-gelatin antimicrobial packaging materials to extend the shelf life of chicken breast fillets. LWT 97:483–490. https://doi.org/10.1016/j.lwt.2018.07.005
Mueller E, Hoffmann TG, Schmitz FRW, Helm CV, Roy S, Bertoli SL, de Souza CK (2024) Development of ternary polymeric films based on cassava starch, pea flour and green banana flour for food packaging. Int J Biol Macromol 256. https://doi.org/10.1016/j.ijbiomac.2023.128436
Muñoz-Gimena PF, Oliver-Cuenca V, Peponi L, López D (2023) A Review on reinforcements and additives in starch-based composites for food packaging. Polymers 15(13). https://doi.org/10.3390/polym15132972
Nazrin A, Ilyas RA, Rajeshkumar L, Hazrati KZ, Jamal T, Mahardika M et al (2025) Lignocellulosic fiber-reinforced starch thermoplastic composites for food packaging application: a review of properties and food packaging abetted with safety aspects. Food Packag Shelf Life 47. https://doi.org/10.1016/j.fpsl.2025.101431
Niranjana Prabhu T, Prashantha K (2018) A review on present status and future challenges of starch based polymer films and their composites in food packaging applications. Polym Compos 39(7):2499–2522. https://doi.org/10.1002/pc.24236
Onyeaka H, Obileke K, Makaka G, Nwokolo N (2022) Current research and applications of starch-based biodegradable films for food packaging. Polymers 14(6):1126
Othman SH, Kahar NS, Nordin N, Ahmad Shapi'i R (2023) Properties and food packaging applications of solvent casting-made starch-based films incorporated with essential oils: a review. Starch/Staerke 75(5–6). https://doi.org/10.1002/star.202200152
Pacheco N, Martín-López H, Uribe-Calderón JA, Cervantes-Uc JM, Cuevas-Bernardino JC, Ayora-Talavera T, Pech-Cohuo SC (2023) Rheological characterization of Ramon starch (Brosimum alicastrum) suspensions with potential use for food-packaging film. MRS Adv 8(3):71–76. https://doi.org/10.1557/s43580-023-00491-8
Pandian H, Senthilkumar K, Naveenkumar M, Samraj S (2023) Azadirachta indica leaf extract mediated silver nanoparticles impregnated nano composite film (AgNP/MCC/starch/whey protein) for food packaging applications. Environ Res 216:114641
Panrong T, Karbowiak T, Harnkarnsujarit N (2019) Thermoplastic starch and green tea blends with LLDPE films for active packaging of meat and oil-based products. Food Packag Shelf Life 21:100331. https://doi.org/10.1016/j.fpsl.2019.100331
Pei J, Palanisamy CP, Srinivasan GP, Panagal M, Kumar SSD, Mironescu M (2024) A comprehensive review on starch-based sustainable edible films loaded with bioactive components for food packaging. Int J Biol Macromol 274:1332. https://doi.org/10.1016/j.ijbiomac.2024.133332
Peighambardoust SJ, Peighambardoust SH, Pournasir N, Pakdel PM (2019) Properties of active starch-based films incorporating a combination of Ag, ZnO and CuO nanoparticles for potential use in food packaging applications. Food Packag Shelf Life 22:100420
Pérez-Pacheco E, Canto-Pinto JC, Moo-Huchin VM, Estrada-Mota IA, Estrada-León RJ, Chel-Guerrero L (2016) Thermoplastic starch (TPS)-cellulosic fibers composites: mechanical properties and water vapor barrier: a review. Compos Renew Sustain Mater 85:89–107
Picolotto A, Rossa Beltrami LV, Dallé D, Ornaghi Júnior HL, Zattera AJ, Catto AL, Borsoi C (2024) Development and evaluation of biodegradable starch-based films containing cellulose nanocrystals/titanium dioxide nanoparticles as an alternative for food packaging. J Thermoplast Compos Mater 37(6):1991–2012. https://doi.org/10.1177/08927057231208138
Piñeros-Hernandez D, Medina-Jaramillo C, López-Córdoba A, Goyanes S (2017) Edible cassava starch films carrying rosemary antioxidant extracts for potential use as active food packaging. Food Hydrocoll 63:488–495
Plaeyao K, Talodthaisong C, Yingyuen W, Kaewbundit R, Tun WST, Saenchoopa A et al (2025) Biodegradable antibacterial food packaging based on carboxymethyl cellulose from sugarcane

bagasse/cassava starch/chitosan/gingerol extract stabilized silver nanoparticles (Gin-AgNPs) and vanillin as cross-linking agent. Food Chem 466. https://doi.org/10.1016/j.foodchem.2024.142102

Prithiviraj M, Sasidharan A, Krishna BM, Sabu S, Sunooj KV, Anoop K, George J (2024) Characterization and qualitative evaluation of cassava starch-chitosan edible food wrap enriched with culinary leaf powders for eco-friendly food packaging applications. Food Sci Technol Int 30(8):751–763. https://doi.org/10.1177/10820132231179492

Promhuad K, Harnkarnsujarit N (2022) Development and characterization of ethyl maltol incorporated starch-based films for active biodegradable food packaging. In: Conference proceedings—23rd IAPRI World conference on packaging, IAPRI Bangkok 2022

Qin Y, Liu Y, Yong H, Liu J, Zhang X, Liu J (2019) Preparation and characterization of active and intelligent packaging films based on cassava starch and anthocyanins from Lycium ruthenicum Murr. Int J Biol Macromol 134:80–90. https://doi.org/10.1016/j.ijbiomac.2019.05.029

Rahardiyan D, Moko EM, Tan JS, Lee CK (2023) Thermoplastic starch (TPS) bioplastic, the green solution for single-use petroleum plastic food packaging—A review. Enzyme Microb Technol 168. https://doi.org/10.1016/j.enzmictec.2023.110260

Radha Krishnan K, Babuskin S, Rakhavan KR, Tharavin R, Azhagu Saravana Babu P, Sivarajan M, Sukumar M (2015) Potential application of corn starch edible films with spice essential oils for the shelf life extension of red meat. J Appl Microbiol 119(6):1613–1623. https://doi.org/10.1111/jam.12932

Reis LCB, de Souza CO, da Silva JBA, Martins AC, Nunes IL, Druzian JI (2015) Active biocomposites of cassava starch: the effect of yerba mate extract and mango pulp as antioxidant additives on the properties and the stability of a packaged product. Food Bioprod Process 94:382–391. https://doi.org/10.1016/j.fbp.2014.05.004

Ribeiro AM, Estevinho BN, Rocha F (2021) Preparation and incorporation of functional ingredients in edible films and coatings. Food Bioprocess Technol 14:209–231. https://doi.org/10.1007/s11947-020-02528-4

Romani VP, Prentice-Hernandez C, Martins VG (2017) Active and sustainable materials from rice starch, fish protein and oregano essential oil for food packaging. Ind Crops Prod 97:268–274

Roveda AC, Dias BC, Passini LN, Manzani D, Petruci JFDS (2024) Transparent, flexible, and eco-friendly starch-based films for reversible optoelectronic noses for food spoilage monitoring in smart packaging. Microchimica Acta 191(6). https://doi.org/10.1007/s00604-024-06426-w

Roy S, Min S-J, Biswas D, Rhim J-W (2023a) Pullulan/chitosan-based functional film incorporated with curcumin-integrated chitosan nanoparticles. Colloids Surf A 660:130898. https://doi.org/10.1016/j.colsurfa.2022.130898

Roy S, Min S-J, Rhim J-W (2023b) Essential oil-added chitosan/gelatin-based active packaging film: a comparative study. J Compos Sci 7(3). https://doi.org/10.3390/jcs7030126

Roy S, Rhim J-W (2022) Starch/agar-based functional films integrated with enoki mushroom-mediated silver nanoparticles for active packaging applications. Food Biosci 49:101867

Sadeghizadeh-Yazdi J, Habibi M, Kamali AA, Banaei M (2019) Application of edible and biodegradable starch-based films in food packaging: a systematic review and meta-analysis. Curr Res Nutr Food Sci 7(3):624–637. https://doi.org/10.12944/CRNFSJ.7.3.03

Sanyang ML, Sapuan SM, Jawaid M, Ishak MR, Sahari J (2016) Effect of plasticizer type and concentration on physical properties of biodegradable films based on sugar palm (arenga pinnata) starch for food packaging. J Food Sci Technol 53(1):326–336. https://doi.org/10.1007/s13197-015-2009-7

Sariningsih N, Handayani DS, Kusumaningsih T (2019) Development and characterization of the mechanical properties of edible film from ginger starch, chitosan with glycerin as plasticizer to food packaging. IOP Conf Ser: Mater Sci Eng

Sen C, Das MJ (2018) Biodegradability of starch based self-supporting antimicrobial film and its effect on soil quality. J Polym Environ 26:4331–4337. https://doi.org/10.1007/s10924-018-1304-6

Şen F, Uzunsoy İ, Baştürk E, Kahraman MV (2017) Antimicrobial agent-free hybrid cationic starch/sodium alginate polyelectrolyte films for food packaging materials. Carbohyd Polym 170:264–270. https://doi.org/10.1016/j.carbpol.2017.04.079

Sganzerla WG, Rosa GB, Ferreira ALA, da Rosa CG, Beling PC, Xavier LO et al (2020) Bioactive food packaging based on starch, citric pectin and functionalized with Acca sellowiana waste by-product: characterization and application in the postharvest conservation of apple. Int J Biol Macromol 147:295–303. https://doi.org/10.1016/j.ijbiomac.2020.01.074

Shanbhag C, Shenoy R, Shetty P, Srinivasulu M, Nayak R (2023) Formulation and characterization of starch-based novel biodegradable edible films for food packaging. J Food Sci Technol 60(11):2858–2867. https://doi.org/10.1007/s13197-023-05803-2

Sharma S, Barkauskaite S, Jaiswal AK, Jaiswal S (2021) Essential oils as additives in active food packaging. Food Chem 343:128403

Shi C, Jia L, Tao H, Hu W, Li C, Aziz T et al (2024) Fortification of cassava starch edible films with Litsea cubeba essential oil for chicken meat preservation. Int J Biol Macromol 276:133920

Singh AK, Lee M, Jang D, Lee YS (2024) Non-conventional starch nanoparticles: Novel avenues towards improving sustainability of the food packaging sector. Trends Food Sci Technol 143. https://doi.org/10.1016/j.tifs.2023.104273

Su C-Y, Li D, Wang L-J, Wang Y (2023) Biodegradation behavior and digestive properties of starch-based film for food packaging–a review. Crit Rev Food Sci Nutr 63(24):6923–6945 https://doi.org/10.1080/10408398.2022.2036097

Sun X, Li Q, Wu H, Zhou Z, Feng S, Deng P et al (2023) Sustainable starch/lignin nanoparticle composites biofilms for food packaging applications. Polymers 15(8):1959. https://doi.org/10.3390/polym15081959

Suryavanshi VR, Santhosh R, Singhi H, Thakur R, Ahmed J, Gaikwad KK et al (2025) Preparation and characterization of kodo millet starch/gum tragacanth/copper oxide nanoparticles-based antimicrobial food packaging films. Mater Today Commun 42:111443. https://doi.org/10.1016/j.mtcomm.2024.111443

Syafiq R, Sapuan S, Zuhri M, Ilyas R, Nazrin A, Sherwani S, Khalina A (2020) Antimicrobial activities of starch-based biopolymers and biocomposites incorporated with plant essential oils: a review. Polymers 12(10):2403. https://doi.org/10.3390/polym12102403

Tavassoli M, Alizadeh Sani M, Khezerlou A, Ehsani A, Jahed-Khaniki G, McClements DJ (2022) Smart biopolymer-based nanocomposite materials containing pH-sensing colorimetric indicators for food freshness monitoring. Molecules 27(10). https://doi.org/10.3390/molecules27103168

Tavassoli M, Bahramian B, Abedi-Firoozjah R, Ehsani A, Phimolsiripol Y, Bangar SP (2024a) Application of lactoferrin in food packaging: a comprehensive review on opportunities, advances, and horizons. Int J Biol Macromol 273:132969. https://doi.org/10.1016/j.ijbiomac.2024.132969

Tavassoli M, Bahramian B, Abedi-Firoozjah R, Jafari N, Javdani H, Sadeghi SM et al (2024b) Comprehensive review on polyvinyl alcohol-based electrospun nanofibers for food packaging: applications, developments, and future horizon. Food Bioprocess Technol 18:3223–3244. https://doi.org/10.1007/s11947-024-03637-0

Todhanakasem T, Jaiprayat C, Sroysuwan T, Suksermsakul S, Suwapanich R, Maleenont KK et al (2022) Active thermoplastic starch film with watermelon rind extract for future biodegradable food packaging. Polymers 14(16). https://doi.org/10.3390/polym14163232

Tongdeesoontorn W, Mauer LJ, Wongruong S, Sriburi P, Rachtanapun P (2020) Physical and antioxidant properties of cassava starch-carboxymethyl cellulose incorporated with quercetin and TBHQ as active food packaging. Polymers 12(2). https://doi.org/10.3390/polym12020366

Travalini AP, Lamsal B, Magalhães WLE, Demiate IM (2019) Cassava starch films reinforced with lignocellulose nanofibers from cassava bagasse. Int J Biol Macromol 139:1151–1161. https://doi.org/10.1016/j.ijbiomac.2019.08.115

Trotta F, Da Silva S, Massironi A, Mirpoor SF, Lignou S, Ghawi SK, Charalampopoulos D (2024) Advancing food preservation: sustainable green-AgNPs bionanocomposites in paper-starch flexible packaging for prolonged shelf life. Polymers 16(7). https://doi.org/10.3390/polym16070941

Vianna TC, Marinho CO, Júnior LM, Ibrahim SA, Vieira RP (2021) Essential oils as additives in active starch-based food packaging films: a review. Int J Biol Macromol 182:1803–1819. https://doi.org/10.1016/j.ijbiomac.2021.05.170

Wang L, Yang C, Deng X, Peng J, Zhou J, Xia G et al (2023a) A pH-sensitive intelligent packaging film harnessing Dioscorea zingiberensis starch and anthocyanin for meat freshness monitoring. Int J Biol Macromol 245:125485. https://doi.org/10.1016/j.ijbiomac.2023.125485

Wang W, Zhang W, Li L, Deng W, Liu M, Hu J (2023b) Biodegradable starch-based packaging films incorporated with polyurethane-encapsulated essential-oil microcapsules for sustained food preservation. Int J Biol Macromol 235. https://doi.org/10.1016/j.ijbiomac.2023.123889

Wang P, Qin H, He D, Zou Z, Xu L, Tang Q (2024a) Developing colorimetric ammonia-sensing nanocomposite films based on potato starch/PVA and ZnCu-BTC nanorods for real-time monitoring food freshness. Int J Biol Macromol 277:134376. https://doi.org/10.1016/j.ijbiomac.2024.134376

Wang Z, Chen Q, Cai J (2024b) Research progress on starch-polyphenol composite film: from functional properties to food packaging application. Shipin Kexue/Food Sci 45(4):336–343. https://doi.org/10.7506/spkx1002-6630-20230903-016

Weligama Thuppahige VT, Moghaddam L, Welsh ZG, Wang T, Karim A (2023) Investigation of critical properties of Cassava (Manihot esculenta) peel and bagasse as starch-rich fibrous agro-industrial wastes for biodegradable food packaging. Food Chem 422. https://doi.org/10.1016/j.foodchem.2023.136200

Wu Y, Zhang J, Hu X, Huang X, Zhang X, Zou X, Shi J (2024) Preparation of edible antibacterial films based on corn starch/carbon nanodots for bioactive food packaging. Food Chem 444:138467. https://doi.org/10.1016/j.foodchem.2024.138467

Xia Y, Wang S, Meng F, Xu Z, Fang Q, Gu Z et al (2024) Eco-friendly food packaging based on paper coated with a bio-based antibacterial coating composed of carbamate starch, calcium lignosulfonate, cellulose nanofibrils, and silver nanoparticles. Int J Biol Macromol 254. https://doi.org/10.1016/j.ijbiomac.2023.127659

Xu F, Yan Y, Yong H, Yun D, Chen D, Liu J (2022) Comparison of the physical and functional properties of food packaging films containing starch and polyphenols from different varieties of wolfberry. J Food Meas Charact 16(6):4444–4456. https://doi.org/10.1007/s11694-022-01534-8

Yang Q, Zheng F, Cheng Y, Xu R, Yuan C, Wu Z et al (2024a) Improvement of the properties of essential oil-starch film with whey protein isolated fibril and sodium caseinate. Ind Crops Prod 222:119975. https://doi.org/10.1016/j.indcrop.2024.119975

Yang Z, Chen Q, Wei L (2024b) Active and smart biomass film with curcumin Pickering emulsion stabilized by chitosan-adsorbed laurate esterified starch for meat freshness monitoring. Int J Biol Macromol 275:133331. https://doi.org/10.1016/j.ijbiomac.2024.133331

Yang Z, Zhang D, Cai J (2024c) Robust design of starch composite nanofibrous films for active food packaging: towards improved mechanical, antioxidant, and antibacterial properties. Int J Biol Macromol 260:129329. https://doi.org/10.1016/j.ijbiomac.2024.129329

YousefniaPasha H, Mohtasebi SS, Tabatabaeekoloor R, Taherimehr M, Javadi A, Soltani Firouz M (2021) Preparation and characterization of the plasticized polylactic acid films produced by the solvent-casting method for food packaging applications. J Food Process Preserv 45(12):e16089. https://doi.org/10.1111/jfpp.16089

Zannou O, Beşir A, Awad NMH, Oussou KF, Odouaro OBO, Chabi IB et al (2023) Starch-based nanofibres for food packaging application. In: Starch based nanomaterials for food packaging: perspectives and future prospectus, pp 83–102

Zhang W, Azizi-Lalabadi M, Jafarzadeh S, Jafari SM (2023) Starch-gelatin blend films: a promising approach for high-performance degradable food packaging. Carbohyd Polym 320:121266. https://doi.org/10.1016/j.carbpol.2023.121266

Zhang R, Su Y, Xia F, Liu J, Xiao G, Sun D et al (2025) Study on oil resistance of different kinds of grinding starch used in paper food packaging. Chung-kuo Tsao Chih/China Pulp Pap 44(1):62–68, 84. https://doi.org/10.11980/j.issn.0254-508X.2025.01.008

Zhao Y, Teixeira JS, Gänzle MM, Saldaña MDA (2018) Development of antimicrobial films based on cassava starch, chitosan and gallic acid using subcritical water technology. J Supercrit Fluids 137:101–110. https://doi.org/10.1016/j.supflu.2018.03.010

Zhao Y, Teixeira JS, Saldaña MD, Gänzle MG (2019) Antimicrobial activity of bioactive starch packaging films against Listeria monocytogenes and reconstituted meat microbiota on ham. Int J Food Microbiol 305:108253. https://doi.org/10.1016/j.ijfoodmicro.2019.108253

Zhou X, Yang R, Wang B, Chen K (2019) Development and characterization of bilayer films based on pea starch/polylactic acid and use in the cherry tomatoes packaging. Carbohyd Polym 222:114912. https://doi.org/10.1016/j.carbpol.2019.05.042

Zhu W, Zhang D, Liu X, Ma T, He J, Dong Q et al (2022) Improving the hydrophobicity and mechanical properties of starch nanofibrous films by electrospinning and cross-linking for food packaging applications. LWT 169:114005.https://doi.org/10.1016/j.lwt.2022.114005

Zhu F (2025) Starch based films and coatings for food packaging: interactions with phenolic compounds. Food Res Int 204:115758. https://doi.org/10.1016/j.foodres.2025.115758

Zubair M, Rauf Z, Nawaz H, Shahzad S, Ullah A (2024) A review of recent advances in starch derived bionanocomposites for food packaging applications. Nano-Struct Nano-Objects 39:101204. https://doi.org/10.1016/j.nanoso.2024.101204

Chapter 4
Pectin-Based Edible Films: A Natural Approach to Food Packaging

Sadiya Anjum and Neetu Goel

Abstract With various environmental concerns and growing demands to search an ideal candidate for food packaging applications, biopolymers have garnered immense attraction to replace the traditional plastic-based materials used in the food packaging. These biopolymers consist of favourable properties i.e. biodegradable, biocompatible and non-toxicity and thus they are extensively applied in the design and development of food packaging materials. Among the large pool of biopolymers, pectin has started to evolve as a prime candidate for this application. Pectin is a low-cost polysaccharide having good film forming ability, bioactivity and biocompatibility with a distinctive oxygen barrier capacity. These properties help to increase the expiration date of foods. Furthermore, addition of bioactive agents in the pectin films focuses on the development of prolong shelf life as well as secure quality and safety of foods by limiting contamination of bacteria. In this chapter, we will review state-of-the-art methods of embodying bioactive materials into pectin-based films and development for food packaging applications. The addition of these bioactive materials in pectin films improves microbial resistance and also influences the functional properties of the packaging films. The efficacy of different classes of bioactive materials e.g., metal nanoparticles, phenolic compounds, essential oils and other bio-polymers have been discussed in the later sections of the chapters.

Keywords Pectin films · Biopolymers · Food packaging · Antimicrobial · Bioactive compounds · Microbial contamination · Shelf life

S. Anjum (✉)
Department of Applied Sciences & Humanities, IMS Engineering College, Ghaziabad, Uttar Pradesh, India
e-mail: sadiya2203@gmail.com

N. Goel
Department of Applied Sciences and Humanities, Inderprastha Engineering College, Ghaziabad, Uttar Pradesh, India

S. Roy and M. Shahid (eds.), *Biomaterials for Sustainable Food Packaging*, Environmental Footprints and Eco-design of Products and Processes,
https://doi.org/10.1007/978-3-032-08027-1_4

4.1 Introduction

Food packaging is a vital tool to preserve its quality, purity and extend food shelf life by acting as a protective barrier to the surrounding environment. The essential goal of the packaging materials is to facilitate the smooth transportation of food with its intact quality, shelf life and non-messy (Farrell et al. 2024). A range of options are available for this purpose such as plastics, paper, glass and metal, however, they all have advantages and limitations in their respective areas. Among the plethora of materials, plastics has gained enormous potential and are globally employed in food packaging industry. This is attributed to the ease to follow the shape, lightweight, robustness, transparent, easy to label, economical, color tuning and mass-production (Soltani Firouz et al. 2021). Irrespective to these benefits, it has a grave environmental concern i.e., discharging toxins that remain a fear for living organisms (Cheng et al. 2022; Klaiman et al. 2016; Sani et al. 2021). While other options i.e. glass and metals both are heavy in weight, expensive and not easy to use for food packaging. The unpleasant environmental impact of plastic-based packaging materials and on the living organisms have raised concerns to develop new materials which could overcome with previous challenges (Sani et al. 2021).

The basic material properties strongly influence the packaging ability of the material by determining its performance. The new material should be comprised of various smart features such as, barrier, permeability, mechanical properties, preserving food quality, appearance and hygiene with strong potential of degradation (Fig. 4.1) (Farrell et al. 2024). Figure 4.1 give information about the five critical properties which need to be considered when selecting/designing a suitable packaging material. The design and fabrication of archetype edible films for food packaging industry is revolutionary task with full of challenges. Although several articles are available to explain the fabrication of biodegradable materials for this purpose yet each material discussed had its own challenges which need to be overcome and propose an ideal candidate for this application (Hanani et al. 2014; Cheng et al. 2021; Birania et al. 2022; Amin et al. 2022; Cazón and Vázquez 2021; Nilsen-Nygaard et al. 2021; Westlake et al. 2022). Researchers have started to take a serious interest in edible films and coatings that could prolong the shelf life and promote the quality of foods (Nilsen-Nygaard et al. 2021).

A wide range of biopolymers i.e. chitosan, gelatin, alginate, cellulose, pectin, zein, gluten, collagen, whey protein, cellulose, starch, agar, carrageenan and their blends were used to develop various edible food packaging materials (Sani et al. 2021; Cheng et al. 2021; Amin et al. 2022; Cazón and Vázquez 2021; Akhter et al. 2024; Sadadekar et al. 2023; Bayer 2021). Although these biopolymer-based films lack of inherent mechanical strength, poor degradability and antimicrobial activity. These issues were overcome by blending with synthetic polymers, plasticizers, crosslinkers, fillers and bioactive agents (Bharadwaj et al. 2022). From the huge class of biopolymers, pectin has gained enormous interest due to its abundance (origin from plant sources) and favourable physical and functional properties (Singaram et al. 2024; Roy et al. 2023). Pectin is a plant-based polysaccharide occurring in vegetables

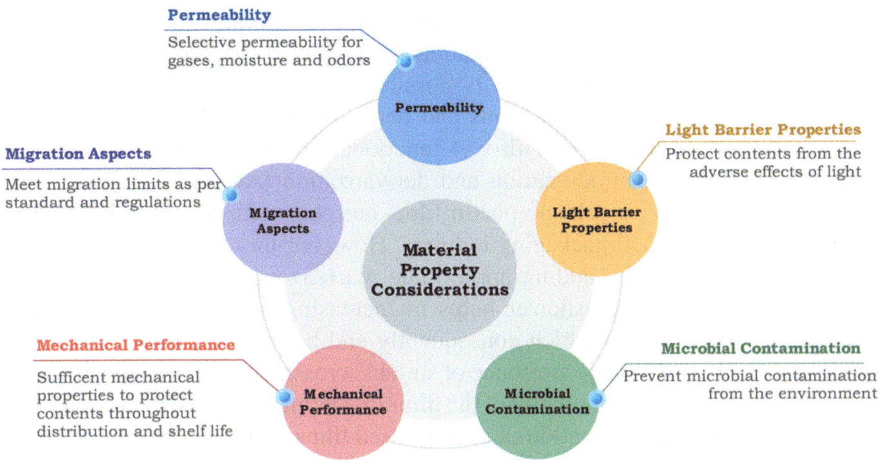

Fig. 4.1 Important key points for selecting food packaging materials. Reproduced with permission from Ref. Farrell et al. (2024)

and fruits which entails linear chains of *D*-galacturonic acid at $\alpha(1 \rightarrow 4)$ linkages (Anjum et al. 2023). It was widely used as a gelling, thickening and stabilizing agents. Pectin-based films are economic, sustainable, and bio-degradable and is classified as "green packaging" materials and generally regarded as safe (GRAS) by the Food and Drug Administration (FDA) (Singaram et al. 2024; Eça et al. 2015). In this book chapter, we will discuss the recent advancement in the design of pectin-based active food packaging materials. The possible methods for pectin film formation, their physical properties detailed investigation on the types of bio-active agents and their incorporation on pectin films, improved physical properties of packaging films were systemically deliberate in later section of the chapter.

4.2 Pectin and Its Processing

Pectin is a naturally occurring hetero-polysaccharides in the cell walls and lamellae (intercellular junction layer between the adjoining cells) of various plants and can be separated through different fruits and vegetables (Kumar et al. 2020; Li et al. 2021). Generally, the cell wall of plants contains different polysaccharides like cellulose, hemicellulose, pectin, lignin and various compounds (Roy et al. 2023). The characteristic features of pectin such as film-forming ability, solubility, gelation etc. primarily rely on the origin source and degree of esterification of its chains (Kumar et al. 2023). There are numerous sources available for pectin extraction but only a few were commercially exploited. Commercially, citrus fruits and apple pomace peels are largely exploited to produce pectin which is due to their high extraction yield without compromising its properties (Roy et al. 2023; Kumar et al. 2023; Huang et al. 2021).

These properties makes pectin-based materials as outstanding and viable choice to develop materials for packaging applications (Huang et al. 2021).

The film formation of polymers is possible by ionic, electrostatic, hydrogen bonding and hydrophobic interactions and depends on the polymer chain. Pectin contains various carboxyl and hydroxyl functional groups, which could easily be exploited for structural modification and derivatization (Anjum et al. 2023). The mechanical properties of pristine pectin films are poor with hydrophilicity which inhibit its use in the food packaging industry. However, these bottlenecks could be overcome or improved by adding appropriate plasticizers, fillers and crosslinking agents. The addition of plasticizer helps in increasing the internuclear distances between polymeric chains which consequently and bestow flexibility by reducing firmness and cohesion. While, presence of suitable crosslinkers in between the pectin chains improves the hydrophobicity of the films (Suderman et al. 2018). Various techniques are used for the fabrication of pectin-based films such as wet (solvent casting and spraying) and dry processes (compression moulding or extrusion) as demonstrated in Fig. 4.2 (Ghosh et al. 2025). The dry processes are environment-friendly alternative to the wet processes as it does not need harmful solvent(s).

A common method to prepare pectin films is solvent casting method which entails three steps. Initially, pectin is dissolved in a solvent of choice followed by addition of plasticizers and functional active components. The mixture is finally poured into a mould and left for drying at the required temperature (Huang et al. 2021; Ding et al. 2020). The spraying process is also a three-step process; (i) dissolution of material capable of forming film in the appropriate solvent, (ii) spraying the solutions onto a uniform surface and (iii) evaporation of the solvent with drying. Both aforementioned methods require low cost and skills and provides stronger molecular interaction with

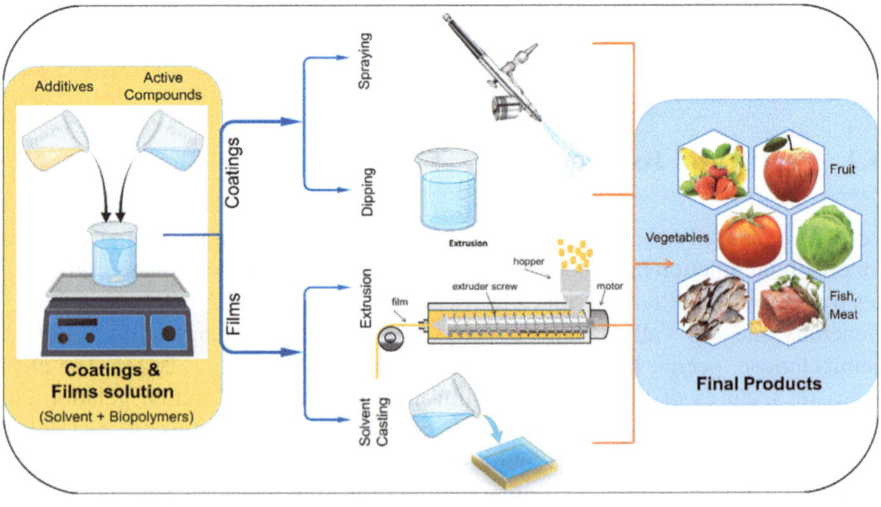

Fig. 4.2 A schematic diagram showing different methods to develop edible biopolymer-based films and coatings. Reproduced with permission from Ref. Ghosh et al. (2025)

closely packed structure, however, extended drying time of solution restricts their application on a large scale. The physical and chemical properties of the films solely depends on the solvent evaporation and drying conditions (Ortiz et al. 2017).

Extrusion is most commonly available dry process and is widely used to develop polymeric films (Gurram et al. 2018). It is a three-step process which includes (i) a feeding zone, where the presence of rotating screw helps in the mixing of polymers and bioactive agents at high pressure, (ii) a kneading zone where all ingredients were mechanically compressed and plasticized and (iii) heating zone, where the material is heated above the glass transition temperature to facilitate plasticization of polymeric chains (Gurram et al. 2018; Agarwal et al. 2021). This method has several advantages over the solvent casting method, it requires minimum time, energy, and solubility. But it's used is largely restricted due to high cost of the equipment. In conclusion, the above-stated processes are widely used for the designing of bioactive films of pectin for food packaging applications and will be discussed in subsequent sections.

4.3 Pectin as Bioactive Packaging Material: *State-of-the-Art*

Pectin films are utilized in the packaging of unprocessed and lightly processed food including fruits and vegetables (Eça et al. 2015). These films are considered as "*Green*" packaging materials due to several benefits over the conventional plastics, e.g., cost-effective, bio-compatibility and bio-degradability (Singaram et al. 2024; Eça et al. 2015). But, pristine pectin films are not very efficient to inhibit microbial growth at the desired site and hence these films are needed to be functionalized by adding suitable bioactive agents (Moeini et al. 2021). This functionalization leads to inherent bioactive pectin films which protect the quality, taste and shelf-life of the packed food items. Generally, *Escherichia coli (E. coli), Staphylococcus aureus (S. aureus), Bacillus cereus, Aspergillus, Rhizopus, Clostridium perfringens, Torulopsis, Candida* (yeasts) and *Salmonella* class of microorganisms are responsible for the food spoilage (Ahmed et al. 2017). The selection of bioactive agents is influenced by the presence of these microorganisms, therefore several bioactive agents are available and used for the designing of bioactive food packaging materials and show their activity against various pathogens. Natural compounds, essential oils, metal nanoparticles and blend with bioactive polymers are the prominent bioactive agents which are widely used for this purpose and show their efficacy by inhibiting microbial growth, free-radical scavenging, oxygen and moisture absorption capacity (Fig. 4.3) (Eça et al. 2015; Zhou et al. 2025; Chaichi et al. 2023). The addition of the bio-active agents in the pectin-based packaging films is found to propitious in reducing the microbial contamination of food. These bioactive agents work by different ways either they inhibit the metabolic pathways of microbes or destroy the structure of bacterial cell wall (Sofi et al. 2018). The addition bio-active compounds in pectin films and their different modes of action are discussed in the subsequent sub sections.

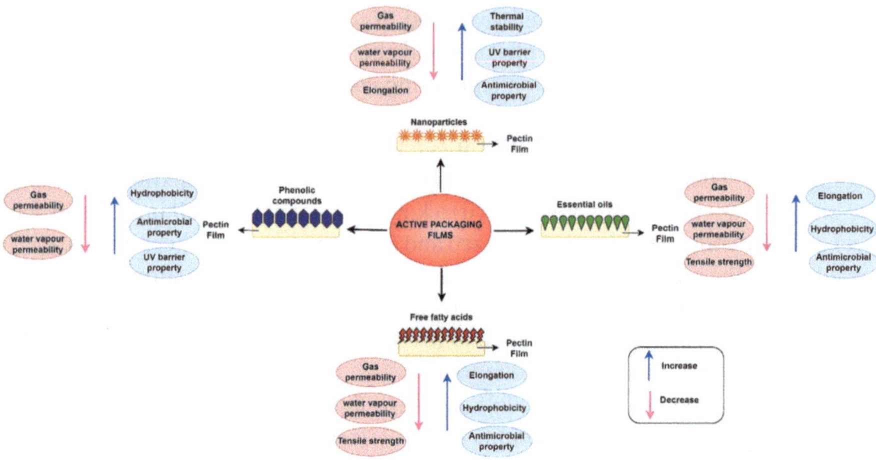

Fig. 4.3 Pectin-based films comprising different bio-active agents. Reproduced with permission from Ref. Singaram et al. (2024)

4.3.1 Essential Oils (EOs) Loaded Pectin Films

From several bio-active agents, EOs are one of the commonly used options in the designing of bioactive packaging materials (Seow et al. 2014; Carpena et al. 2021). These oils are low molecular weight distillation product of various plants which contains strong odour and flavour of the sources from which it is extracted. Most commonly used EOs are obtained from rosemary, oregano, clove, thyme, lemongrass, mandarin and cinnamon, they possess antifungal, antibacterial and antioxidant properties due to the presence of flavonoids, terpenes and terpenoids (Chaiwarit et al. 2018; Rodriguez-Garcia et al. 2016; Nisar et al. 2018). The hydrophobic nature of these oils plays a crucial role in altering the cellular metabolism of the cellular membrane of pathogenic bacteria and inhibit their growth. According to studies, EO shows high efficiency against gram-positive bacteria due to the thin peptidoglycan layer whereas, these oils are less capable to inhibit the growth of gram-negative bacteria due to the complex structure of the outer phospholipid layer (Norcino et al. 2020).

Almasi and co-workers reported the synthesis of nanodroplets of *Marjoram* EO loaded in pectin films for food packaging applications. The developed films exhibited good mechanical and water barrier properties. This was attributed highly dense and less permeable structure. Moreover, it displayed substantial inhibition in the growth of *S. aureus* and *E. coli* as vary the concentration of loaded EO in the films (Almasi et al. 2020). In a recent study, clove oil stabilized with nanocellulose crystals was loaded in pectin-konjac glucomannan films and was used for the packing of grapes. Results concluded that these films potentially delay the post-harvest aging of the grapes with its intact quality for 15 days at 25 °C. The 10 wt% clove oil doped film exhibited exceptional tensile strength (23.58 MPa) with the lowest water vapor permeability

(1.05×10^{-10} g/m s Pa). The composite film showed effective antimicrobial activity by zone of inhibition method against *E. coli* and *S. aureus* (Wang et al. 2025). Rehana et al. investigated the fabrication of bio-nanocomposite films of chitosan, gelatin, and pectin incorporated with rosemary oil. It was observed that these films showed an increased hydrophobic nature resulting in enhanced water vapor transmission rate and tensile strength as compared to the control films. Incorporation of rosemary oil exhibited the highest inhibitory activity against the tested pathogenic strains *Bacillus subtilis (B. subtilis)* and *E. coli* (Akhter et al. 2024).

Tanzeela et al. reported the development of antimicrobial composite films by incorporating different concentrations of clove bud EO into the citrus pectin to modify its inherent properties. The presence of oil in the composite films significantly improved the water barrier properties and were more resistant to breakage and were found to be more flexible than the control film. The inhibitory effect of pectin films with clove bud oil was also evaluated on three common foodborne bacteria and results revealed that EO has a good potential to inhibit the growth of bacterial and make antimicrobial edible films or coatings for various food applications (Nisar et al. 2018). Another study investigated and evaluated antibacterial activity against *S. aureus* when orange oil is directly mixed into pectin polymer films. It was noted that concentration of the oil played a pivotal role in introducing the mechanical properties (Chaiwarit et al. 2018). Recent research focused on the use of cassava peel induced cellulose nanofiber for the development of functionalized pectin/pullulan/olive oil nanocomposite films. These films were used for the cling-wrapping of chicken meat and their storage study was investigated. A substantial decrease in psychrophilic bacterial counts was observed in cling-wrapped chicken meat compared to unwrapped (Afzia et al. 2025). Sadadekar et al. reported the encapsulation of these potato peel extract and fennel oil in pectin-nanochitosan films for the development of bioactive packing material with improved antioxidant and antimicrobial properties of the material. Designed films showed enhanced biological and antioxidant properties of the functional films with the extended shelf life of the packed food. These active biodegradable films could be an alternative to replace plastic packing for food (Sadadekar et al. 2023). Another study reported the formulation of bioactive pectin-nanocellulose-based films comprising a blend of shirazi thyme oil, cinnamon oil and clove oil. These EO-loaded film showed maximum antioxidant and antimicrobial effects against *S. aureus*, *E. coli*, and *Pseudomonas aeruginosa, (P. aeruginosa)* fluorescent bacteria. While the presence of nanocellulose in between pectin led to improve the water-resistant property of the bioactive films which made them potential candidates for food packaging (Chaichi et al. 2023).

Due to the presence of active functional groups in EO's they have potential to decrease the growth rate of fungal pathogens. These functional groups are able to interact with the primary metabolic activities of fungi and thereby restrict the function of enzymes and cell metabolism resulting in cell death. Rodriguez-Garcia et al. investigated the effect of the addition of *Oregano (Lippia graveolens)* EO within pectin edible films/coatings specifically for tomatoes. The results concluded that these bioactive system inhibits fungal growth and increased the antioxidant potential of treated tomatoes (Rodriguez-Garcia et al. 2016). Akachat et al. reported lemon oil encapsulated pectin films for bioactive food packaging. These bioactive films exhibited efficient antibacterial activity against *S. aureus* and *E. coli*, whereas, these films shows strong inhibition in growth of *Botrytis cinerea* and reasonable for *alternata* (Akachat et al. 2025).

4.3.2 Nanoparticle (NPs) Loaded Pectin Films

NPs are a prime candidate due to their excellent antimicrobial activity against wide range of microbes (Anjum and Gupta 2018). Research showed that NPs of silver (Ag NPs), gold (Au NPs), Copper (Cu NPs), Titanium dioxide (TiO_2 NPs) and Zinc oxide (ZnO NPs) greatly hold the ability to slow-down the growth of microorganisms (Anjum et al. 2015, 2016a, b, 2018). This is primarily due to small (nano) size which increased the surface area and thus could easily cross the outer membranes of the bacterial strains and restrict the metabolic activities (Anjum and Gupta 2018; Anjum et al. 2019). These NPs are either encapsulated or grow within the pectin chains and is widely available in the literature. The examples of metallic NPs that have been utilized in tandem with pectin-based food packaging films are outlined below.

AgNPs is the safest bioactive material due to their effectiveness against wide range of microbes since it targets multiple functions in bacterial cells and inhibit their activity. AgNP gain a substantial interest towards the development of functional material in comparison to other bioactive molecules (Anjum and Gupta 2018). There are several reports which discussed the design and development of AgNPs loaded bioactive materials for food packaging applications. Li et al. reported AgNPs and curcumin (Cur)-loaded antibacterial multifunctional pectin-gelatin films for intelligent food packaging applications. It was noted that both Cur and AgNPs were to homogeneously distributed in the films and strongly interacted with pectin and gelatin. These composite films exhibited good antioxidant activity, mechanical performance, hydrophobicity and antibacterial ability. The films also demonstrated excellent water vapor barrier qualities and complete reduction in bacterial growth of both gram negative and gram-positive bacterial strains (Li et al. 2024). Another study investigated the incorporation of AgNPs with tannic acid in pectin-gelatin films and used them as ultravoilet (UV)-blocking nanocomposite (NC) films for strawberry preservation. The NC film displayed strong antibacterial activity against *S. aureus* and *E. coli* including complete blocking of UV rays to preserve strawberries and improve its shelf life (Li et al. 2025). An in-situ approach was used for the growth of AgNPs within the pectin matrix using gamma irradiation subsequently these films were used for the development of bioactive food packaging material. These NC films showed improve mechanical properties, water permeability and exhibited potent antibacterial activity against *Salmonella Typhimurium* and *E. coli* (Ardjoum et al. 2021). Recently, nano-architectonics of AgNPs and pectin chains had been investigated for the development of antibacterial bio-packaging films. The low methoxyl pectin with AgNPs showed improved functional properties such as effective antibacterial activity against both *E. coli* and *B. subtilis*. In addition to the antibacterial activity the developed film preserved the cherry tomatoes for ten days (Nguyen et al. 2025). Another group investigated the development of bioplastics based on pectin-starch-chitosan chains in which AgNPs were embedded and excellent antibacterial action allowed their use in the food industry. These biodegradable bioplastics become a promising candidate to replace plastics for food packaging purposes (Nogueira et al. 2025).

In an interesting study, Zhou et el. reported composite film embedding Ag nanosheets within the pectin chains. The developed composite film exhibited oriented structure which was beneficial for improving the mechanical and UV blocking properties. The composite film showed long term antimicrobial effect due to the controlled release of the Ag ions because of stretching orientation effect and depends on the stretching ratio of the composite films (Fig. 4.4a). The authors further investigated and evaluated the correlation between the microscopic orientation effect and the macroscopic antimicrobial effect. Figure 4.4b and c show the antimicrobial effect of the composite films against typical foodborne pathogens including *E. coli, S. aureus* and *Botrytis cinerea (B. cinerea).* Ag nanosheets embedded pristine composite film exhibited significantly improved antimicrobial properties than the pristine pectin film. The sharp edges of Ag nanosheets directly penetrated the cell membrane of bacteria and caused physical damage, however, bacterial inhibition rate slowly decreased as the degree of orientation increased. The results of the bacterial and fungal inhibition align very well with antibacterial activity as shown in Fig. 4.4c. As can be seen from Fig. 4.4c there was no antifungal activity for the pristine film while the Ag embedded composite film exhibited 95% antifungal rate. The resulting films shows noticeable delay in the discoloration and spoilage of *Agaricus bisporus* (Zhou et al. 2025).

Zinc oxide (ZnO) NPs are the most commonly FDA-approved bioactive agent used in topically applied creams, lotions and antibacterial coatings for various skin problems (Xing et al. 2019). These nanoparticles show bactericidal mechanisms for the prevention of bacterial growth rather than bacteriostatic properties. Some references are available which describe the uses of ZnO NPs for the development of bioactive packaging films are discussed. In recent study, the crystalline nanocellulose reinforced and ZnO NPs loaded novel pectin films were designed for food packing films. The optimized films exhibited enhanced UV-blocking, physical properties as compared to pristine pectin films. These films used as wrapping materials for artificially contaminated cheese samples and demonstrated their efficacy against *Salmonella enterica (S. enterica)* and *S. aureus.* A noticeable reduction was observed in bacterial growth after 5 days of storage of wrapped cheese sample (Sharaby et al. 2022). Another study investigated the fabrication of pectin-alginate aerogel containing ZnO NPs for the development of packaging material. These aerogel films shows potent antimicrobial activity against *E. coli* and *S. aureus* strains using low concentration of NP's (Mottola et al. 2025). The potential of ZnO NPs loaded pectin with hydroxyethyl cellulose-based biofilm was investigated for food packaging applications. The developed films show sharp improvement in tensile strength and flexibility after addition of 1% ZnO NPs from 1.2 to 2.6 MPa and 4.5 to 13 MPa, respectively. Excellent reduction in bacterial growth was observed against *E. coli, P. aeruginosa, S. enterica, S. aureus, Typhimurium, Listeria monocytogenes* and *Candida albicans* (Habbasha et al. 2025). Cur and sulphur NPs were utilized as antimicrobial

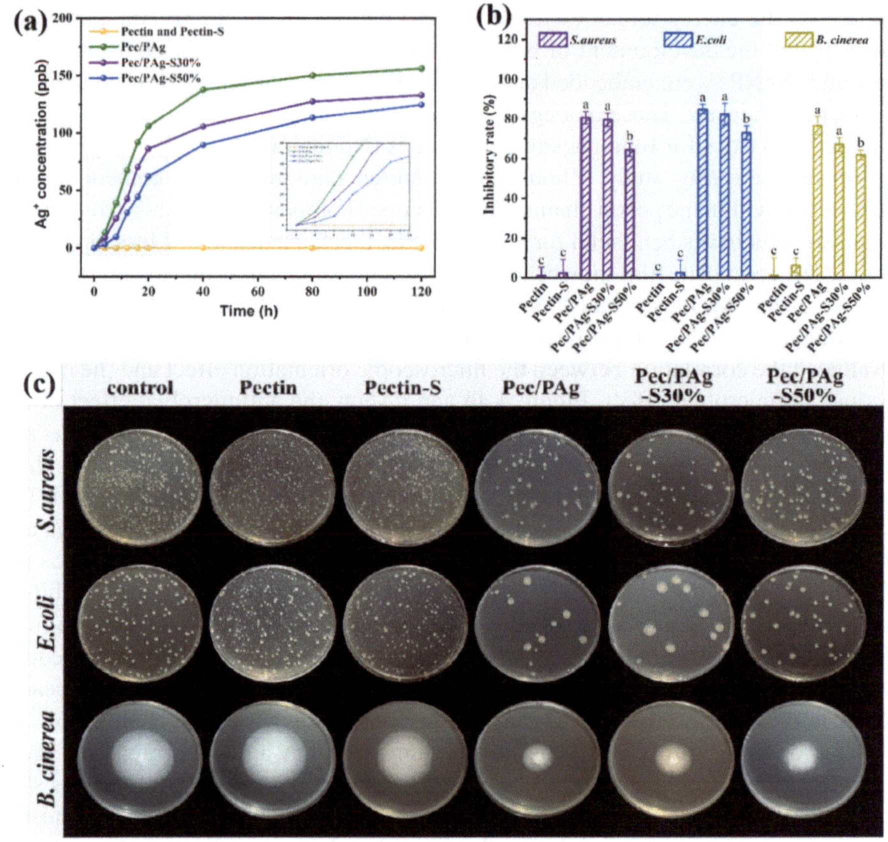

Fig. 4.4 Cumulative release of Ag^+ ions from composite films (**a**); In vitro antimicrobial activity (c) and inhibition rate of composite films against bacteria (*S. aureus* and *E. coli*) and fungi (*B. cinerea*) (**b**). Reproduced with permission from Ref. Zhou et al. (2025)

agents in the designing of pectin films for food packaging. The presence of Cur and NPs significantly affected the colour of surface and UV-blocking properties of the developed films. The composite films exhibited noticeable antioxidant properties and antibacterial activity against *E. coli* and *Listeria monocytogenes*. The composite film exhibited pH-responsive properties together with change in colour i.e., yellow to orange as the quality of packed shrimp changed (Ezati and Rhim 2020).

4.3.3 Phenolic Compounds Bioactive Agents

Various other bioactive molecules are also reported for the development of food packaging materials and discussed in this section. Polyphenols and polypeptides are the prime options for this application due to their well-recognized antioxidants and antimicrobial activities. They exhibit concentration-dependent properties, as they can damage the outer cell wall, restrict the nucleic acid synthesis and lead bacterial cell lysis. Bermudez-oria et al. developed pectin-gelatin composite films by incorporating hydroxytyrosol and 3,4-dihydroxy phenyl glycol for food packing. The antimicrobial potential of the films was tested on the perishable behaviour of the strawberries against mould and the significant delay was observed (Bermúdez-Oria et al. 2017). A bio-active compound tea polyphenols was casted pectin and chitosan films by Gao et al. and utilized as food packing material. Films showed high antioxidant and antimicrobial properties and strongly inhibited microorganisms growth as well as withhold its colour to a greater extent (Gao et al. 2019). Nisin, is a naturally occurring food preservative with antibacterial properties. Nisin was incorporated in the pectin-gellan films and used to determine the bioactivity and antimicrobial property. The film showed good antimicrobial resistance towards *S. aureus* and *Bacillus cereus*. Nisin-loaded pectin-gellan films displayed better release of the bioactive compound into the food which alleviated microbial contamination (Rivera-Hernández et al. 2021).

4.4 Conclusions and Future Perspectives

Bio-engineering of pectin films is the most feasible option for the development of edible packaging materials with advanced features to improve the quality and shelf life of food and consumer products. This chapter discusses the enormous potential of pectin-based active packaging systems with various bioactive agents. Pectin, being a multifunctional heteropolysaccharide has greater bio-compatibility and degradability and originate from the renewable sources which could be easily modified by integrating different additives such as AgNPs, ZnO NPs with desired functional properties. Pristine pectin films have some disadvantages which restrict their uses therefore, we need to overcome this problem. The addition of suitable compounds such as plasticizers, crosslinkers, fillers and bioactive agents make them prime candidates after enhancing the structural and functional properties of the inherent pectin films. After modification, the pectin films become more flexible, hydrophobic and having an excellent ability to inhibit bacterial growth and makes them suitable candidate for food packaging applications. These functionalized pectin films show excellent properties and significantly improve hygiene, shelf life and quality of packaged foods. Discussed bioactive agents (i.e. essential oils, nanoparticles, phenolic compounds and many more) exhibit different modes of action to inhibit or kill the propagation of food-borne microbes. The presence of bioactive agents also prevents the occurrence of food spoilage because of microbial growth. The future designing of packaging materials should focus on controlled architecture with specific functionality of the films by giving smart features like to detect the quality and taste of the food. The primary target should be on the advancement and enhancing materials properties keeping this in mind that the packaged food retains their quality until it reaches the consumer door step. It is also important that the developed material(s) must be economical so that it is commercially feasible and promising for widespread use. In conclusion, we believe the joint efforts of biotechnologist and material scientists will indeed continue to achieve the marvels in the development of innovative and smart edible packaging materials.

Declaration Artificial intelligence (AI) tools were used solely to improve the language, grammar, and readability of this manuscript. No content was generated or modified in a way that affects the scientific results, interpretations, or conclusions. The authors take full responsibility for the integrity and originality of the content.

References

Afzia N, Bora S, Ghosh T (2025) Utilization of cassava peel based cellulose nanofiber for developing functionalized pectin/pullulan/olive oil nanocomposite film for cling wrapping of chicken meat. Int J Biol Macromol 305:140879

Agarwal T, Costantini M, Maiti TK (2021) Extrusion 3D printing with Pectin-based ink formulations: recent trends in tissue engineering and food manufacturing. Biomed Eng Adv 2:100018

Ahmed I, Lin H, Zou L, Brody AL, Li Z, Qazi IM et al (2017) A comprehensive review on the application of active packaging technologies to muscle foods. Food Control 82:163–178

Akachat B, Himed L, Salah M, D'Elia M, Rastrelli L, Barkat M (2025) Development of pectin-based films with encapsulated lemon essential oil for active food packaging: improved antioxidant activity and biodegradation. Foods 14:353

Akhter R, Masoodi FA, Wani TA (2024) Chitosan, gelatin and pectin based bionanocomposite films with rosemary essential oil as an active ingredient for future foods. Int J Biol Macromol 272:132813

Almasi H, Azizi S, Amjadi S (2020) Development and characterization of pectin films activated by nanoemulsion and Pickering emulsion stabilized marjoram (Origanum majorana L.) essential oil. Food Hydrocoll 99:105338

Amin U, Khan MKI, Maan AA, Nazir A, Riaz S, Khan MU et al (2022) Biodegradable active, intelligent, and smart packaging materials for food applications. Food Packag Shelf Life 33:100903

Anjum S, Gupta B (2018) Bioengineering of functional nanosilver nanogels for smart healthcare systems. Global Chall 2:1800044

Anjum S, Ilmi R (2019) Functional nanomaterials for smart healthcare applications. In: Bhat AH, Khan I, Jawaid M, Suliman FO, Al-Lawati H, Al-Kindy SM (eds) Nanomaterials for healthcare, energy and environment. Springer, Singapore, pp 31–49

Anjum S, Sharma A, Tummalapalli M, Joy J, Bhan S, Gupta B (2015) A novel route for the preparation of silver loaded polyvinyl alcohol nanogels for wound care systems. Int J Polym Mater Polym Biomater 64:894–905

Anjum S, Gupta A, Sharma D, Gautam D, Bhan S, Sharma A et al (2016a) Development of novel wound care systems based on nanosilver nanohydrogels of polymethacrylic acid with Aloe vera and curcumin. Mater Sci Eng C 64:157–166

Anjum S, Gupta A, Sharma D, Kumari S, Sahariah P, Bora J et al (2016b) Antimicrobial nature and healing behavior of plasma functionalized polyester sutures. J Bioact Compat Polym 32:263–279

Anjum S, Singh S, Benedicte L, Roger P, Panigrahi M, Gupta B (2018) Biomodification strategies for the development of antimicrobial urinary catheters: overview and advances. Global Chall 2:1700068

Anjum S, Hasan N, Ilmi R (2023) Chapter 14—Pectin-based nanoformulations for therapeutic applications. In: Hasnain MS, Nayak AK, Aminabhavi TM (eds) Polymeric nanosystems. Academic Press, pp 405–439

Ardjoum N, Shankar S, Chibani N, Salmieri S, Lacroix M (2021) In situ synthesis of silver nanoparticles in pectin matrix using gamma irradiation for the preparation of antibacterial pectin/silver nanoparticles composite films. Food Hydrocoll 121:107000

Bayer IS (2021) Zein in food packaging. In: Sustainable food packaging technology, pp 199–224

Bermúdez-Oria A, Rodríguez-Gutiérrez G, Vioque B, Rubio-Senent F, Fernández-Bolaños J (2017) Physical and functional properties of pectin-fish gelatin films containing the olive phenols hydroxytyrosol and 3,4-dihydroxyphenylglycol. Carbohyd Polym 178:368–377

Bharadwaj S, Dhand V, Kalyana Lakshmi Y (2022) Biodegradable composite for smart packaging applications. In: Biodegradable materials and their applications, pp 593–625

Birania S, Kumar S, Kumar N, Attkan AK, Panghal A, Rohilla P et al (2022) Advances in development of biodegradable food packaging material from agricultural and agro-industry waste. J Food Process Eng 45:e13930

Carpena M, Nuñez-Estevez B, Soria-Lopez A, Garcia-Oliveira P, Prieto MA (2021) Essential oils and their application on active packaging systems: a review. Resources 10:7

Cazón P, Vázquez M (2021) Bacterial cellulose as a biodegradable food packaging material: a review. Food Hydrocoll 113:106530

Chaichi M, Badii F, Mohammadi A, Hashemi M (2023) Novel bioactive composite films based on pectin-nanocellulose-synergistic triple essential oils: development and characterization. Food Bioprocess Technol 16:1794–1805

Chaiwarit T, Ruksiriwanich W, Jantanasakulwong K, Jantrawut P (2018) Use of orange oil loaded pectin films as antibacterial material for food packaging. Polymers 10:1144

Cheng H, Chen L, McClements DJ, Yang T, Zhang Z, Ren F et al (2021) Starch-based biodegradable packaging materials: a review of their preparation, characterization and diverse applications in the food industry. Trends Food Sci Technol 114:70–82

Cheng H, Xu H, McClements DJ, Chen L, Jiao A, Tian Y et al (2022) Recent advances in intelligent food packaging materials: principles, preparation and applications. Food Chem 375:131738

Ding H, Cui SW (2020) Pectin Bioactivity. In: Kontogiorgos V (ed) Pectin: technological and physiological properties. Springer International Publishing, Cham, pp 165–188

Eça KS, Machado MT, Hubinger MD, Menegalli FC (2015) Development of active films from pectin and fruit extracts: light protection, antioxidant capacity, and compounds stability. J Food Sci 80:C2389–C2396

El Habbasha ES, Abouzeid R, Ibrahim FM, Youssef AM, Mahdy SZA, El-Liethy MA (2025) Developing a novel, low-cost, antimicrobial, and biodegradable pectin/HEC/ZnO biofilm for edible food packaging applications. Biomass Convers Biorefinery 15:6377–6388

Ezati P, Rhim J-W (2020) PH-responsive pectin-based multifunctional films incorporated with curcumin and sulfur nanoparticles. Carbohyd Polym 230:115638

Farrell R, Cortese YJ, Devine DM, Gately N, Rueda M, Rodriguez L et al (2024) The function and properties of common food packaging materials and their suitability for reusable packaging: the transition from a linear to circular economy. Curr Res Green Sustain Chem 100429

Gao H-X, He Z, Sun Q, He Q, Zeng W-C (2019) A functional polysaccharide film forming by pectin, chitosan, and tea polyphenols. Carbohyd Polym 215:1–7

Ghosh S, Ahmed M, Roy S (2024) Chapter 12—Nanotechnology in edible films and coatings. In: Ghosh T, Roy S, Łopusiewicz Ł (eds) Advancements in nanotechnology for food and packaging. Elsevier, pp 349–76

Gurram R, Souza Filho PF, Taherzadeh MJ, Zamani A (2018) A solvent-free approach for production of films from pectin and fungal biomass. J Polym Environ 26:4282–4292

Hanani ZN, Roos YH, Kerry JP (2014) Use and application of gelatin as potential biodegradable packaging materials for food products. Int J Biol Macromol 71:94–102

Huang J, Hu Z, Hu L, Li G, Yao Q, Hu Y (2021) Pectin-based active packaging: a critical review on preparation, physical properties and novel application in food preservation. Trends Food Sci Technol 118:167–178

Klaiman K, Ortega DL, Garnache C (2016) Consumer preferences and demand for packaging material and recyclability. Resour Conserv Recycl 115:1–8

Kumar M, Tomar M, Saurabh V, Mahajan T, Punia S, del Mar CM et al (2020) Emerging trends in pectin extraction and its anti-microbial functionalization using natural bioactives for application in food packaging. Trends Food Sci Technol 105:223–237

Kumar S, Reddy ARL, Basumatary IB, Nayak A, Dutta D, Konwar J et al (2023) Recent progress in pectin extraction and their applications in developing films and coatings for sustainable food packaging: a review. Int J Biol Macromol 239:124281

Li D-Q, Li J, Dong H-L, Li X, Zhang J-Q, Ramaswamy S et al (2021) Pectin in biomedical and drug delivery applications: a review. Intl J Biol Macromol 185:49–65

Li S, Wei N, Wei J, Fang C, Feng T, Liu F et al (2024) Curcumin and silver nanoparticles loaded antibacterial multifunctional pectin/gelatin films for food packaging applications. Int J Biol Macromol 266:131248

Li S, Fang C, Wei N, Wei J, Feng T, Liu F et al (2025) Antimicrobial, antioxidative, and UV-blocking pectin/gelatin food packaging films incorporated with tannic acid and silver nanoparticles for strawberry preservation. Int J Biol Macromol 308:142445

Moeini A, Germann N, Malinconico M, Santagata G (2021) Formulation of secondary compounds as additives of biopolymer-based food packaging: a review. Trends Food Sci Technol 114:342–354

Mottola S, Viscusi G, Oliva G, Vigliotta G, Cardea S, Gorrasi G et al (2025) Pectin/alginate aerogel containing ZnO produced from beetroot extract mediated green synthesis for potential applications in food packaging. J CO_2 Utiliz 91:103003

Nguyen CTK, Tran KQ, Nguyen ATL, Dang LLP, Le HN, Do KD et al (2025) Nanoarchitectonics of antibacterial bio-packaging film from grapefruit peel-derived low methoxy Pectin integrated with silver nanoparticles. J Polym Res 32:90

Nilsen-Nygaard J, Fernández EN, Radusin T, Rotabakk BT, Sarfraz J, Sharmin N et al (2021) Current status of biobased and biodegradable food packaging materials: impact on food quality and effect of innovative processing technologies. Compr Rev Food Sci Food Saf 20:1333–1380

Nisar T, Wang Z-C, Yang X, Tian Y, Iqbal M, Guo Y (2018) Characterization of citrus pectin films integrated with clove bud essential oil: physical, thermal, barrier, antioxidant and antibacterial properties. Int J Biol Macromol 106:670–680

Nogueira BR, Backx BP, Delazare T (2025) Starch, pectin and chitosan-based bioplastics with silver nanoparticles: an eco-friendly alternative for the food industry. Sustain Chem Environ 10:100246

Norcino L, Mendes J, Natarelli C, Manrich A, Oliveira J, Mattoso L (2020) Pectin films loaded with copaiba oil nanoemulsions for potential use as bio-based active packaging. Food Hydrocoll 106:105862

Ortiz CM, de Moraes JO, Vicente AR, Laurindo JB, Mauri AN (2017) Scale-up of the production of soy (Glycine max L.) protein films using tape casting: formulation of film-forming suspension and drying conditions. Food Hydrocoll 66:110–117

Rivera-Hernández L, Chavarría-Hernández N, López Cuellar MdR, Martínez-Juárez VM, Rodríguez-Hernández A-I (2021) Pectin-gellan films intended for active food packaging: release kinetics of nisin and physico-mechanical characterization. J Food Sci Technol 58:2973–2981

Rodriguez-Garcia I, Cruz-Valenzuela MR, Silva-Espinoza BA, Gonzalez-Aguilar GA, Moctezuma E, Gutierrez-Pacheco MM et al (2016) Oregano (Lippia graveolens) essential oil added within pectin edible coatings prevents fungal decay and increases the antioxidant capacity of treated tomatoes. J Sci Food Agric 96:3772–3778

Roy S, Priyadarshi R, Łopusiewicz Ł, Biswas D, Chandel V, Rhim J-W (2023) Recent progress in pectin extraction, characterization, and pectin-based films for active food packaging applications: a review. Int J Biol Macromol 239:124248

Sadadekar AS, Shruthy R, Preetha R, Kumar N, Pande KR, Nagamaniammai G (2023) Enhanced antimicrobial and antioxidant properties of nano chitosan and pectin based biodegradable active packaging films incorporated with fennel (Foeniculum vulgare) essential oil and potato (Solanum tuberosum) peel extracts. J Food Sci Technol 60:938–946

Sani MA, Azizi-Lalabadi M, Tavassoli M, Mohammadi K, McClements DJ (2021) Recent advances in the development of smart and active biodegradable packaging materials. Nanomaterials 11:1331

Seow YX, Yeo CR, Chung HL, Yuk H-G (2014) Plant essential oils as active antimicrobial agents. Crit Rev Food Sci Nutr 54:625–644

Sharaby MR, Soliman EA, Abdel-Rahman AB, Osman A, Khalil R (2022) Novel pectin-based nanocomposite film for active food packaging applications. Sci Rep 12:20673

Singaram AJV, Guruchandran S, Ganesan ND (2024) Review on functionalized pectin films for active food packaging. Packag Technol Sci 37:237–262

Sofi S, Singh J, Rafiq S, Ashraf U, Dar B, Nayik GA (2018) A comprehensive review on antimicrobial packaging and its use in food packaging. Curr Nutr Food Sci 14:305–312

Soltani Firouz M, Mohi-Alden K, Omid M (2021) A critical review on intelligent and active packaging in the food industry: research and development. Food Res Int 141:110113

Suderman N, Isa M, Sarbon N (2018) The effect of plasticizers on the functional properties of biodegradable gelatin-based film: a review. Food Biosci 24:111–119

Wang X, Li S, Zeng M, Gong H, Zhang Z, Yuan X et al (2025) Preparation, characterization and application of antimicrobial pectin-konjac glucomannan composite films incorporating cellulose nanocrystals stabilized clove essential oil Pickering emulsion. LWT 225:117855

Westlake JR, Tran MW, Jiang Y, Zhang X, Burrows AD, Xie M (2022) Biodegradable active packaging with controlled release: principles, progress, and prospects. ACS Food Sci Technol 2:1166–1183

Xing Y, Li W, Wang Q, Li X, Xu Q, Guo X et al (2019) Antimicrobial nanoparticles incorporated in edible coatings and films for the preservation of fruits and vegetables. Molecules 24:1695

Zhou Y, Wu W, Hileuskaya K, Shao P (2025) Oriented structure design of pectin/ag nanosheets film with improved barrier and long-term antimicrobial properties for edible fungi preservation. Food Chem 144451

Chapter 5
Cellulose Nanofiber-Based Innovation in Sustainable Food Packaging

Gurvendra Pal Singh, Krishna Aayush, and Vikas Bansal

Abstract The increasing environmental concerns seen with plastic pack waste have accelerated the demand for sustainable food packaging solutions. Among bio-based alternatives, cellulose nanofibers (CNFs) have developed as favourable materials due to their exceptional mechanical properties, biodegradability, and renewability. CNFs are derived from plant biomass and processed into nanoscale fibers with high strength, flexibility, and excellent barrier properties, making them suitable for food packaging applications. This chapter explores the production techniques of CNFs, including mechanical, chemical, and enzymatic methods, and their impact on material functionality. The structural characteristics and physicochemical properties of CNFs, such as surface area, water retention capacity, and film-forming ability, are discussed with their packaging potential. Integrating CNFs with biopolymers and active compounds enables the development of functional packaging materials with antimicrobial and oxygen-scavenging properties, enhancing food quality and shelf life. Despite their advantages, CNFs face challenges such as high production costs, moisture sensitivity, and regulatory barriers, which limit their large-scale commercialization. However, advancements in surface modifications, hybrid material formulations, and innovative processing techniques, including nanotechnology and 3D printing, are facilitating the way for improved CNF applications. Collaborative efforts between academics, industry, and policymakers are crucial for addressing these challenges and facilitating the widespread adoption of CNF-based packaging.

Gurvendra Pal Singh and Krishna Aayush—Equal contribution of the author.

G. P. Singh (✉)
Department of Food Technology and Nutrition, Lovely Professional University, Jalandhar 144411, India
e-mail: gurvendranohwar@gmail.com

K. Aayush
Department of Food Science and Technology, Graphic Era (Deemed to be University), Dehradun, Uttarakhand, India

V. Bansal
Department of Food Technology, School of Engineering and Technology, Jaipur National University, Jaipur, Rajasthan, India

S. Roy and M. Shahid (eds.), *Biomaterials for Sustainable Food Packaging*, Environmental Footprints and Eco-design of Products and Processes,
https://doi.org/10.1007/978-3-032-08027-1_5

Keywords Cellulose nanofibers · Sustainable packaging · Biodegradability · Food preservation · Bio-based materials · Smart packaging

5.1 Introduction

Cellulose nanofibers (CNFs) are nanoscale fibers obtained from cellulose, a naturally occurring polymer discovered in the cell walls of plants, algae, and bacteria (Talegaonkar et al. 2017). CNFs are yielded through mechanical, chemical, and enzymatic treatment of cellulose-rich materials, which reduces the fiber size to nanoscale dimensions (Talegaonkar et al. 2017). These fibers typically have diameters of 5–20 nm and lengths that can extend to several micrometers. CNFs possess a high aspect ratio, enhanced mechanical strength, and enlarged surface area, making them multipurpose materials for various applications. Their renewable origin, biodegradability, and non-toxicity further enhance their appeal as sustainable alternatives to synthetic materials. CNFs offer several advantages in the context of sustainable materials. Their abundance and renewability stem from cellulose being the abundant organic polymer on Earth, accounting for an estimated annual production of 1.5 trillion tons (Shogren et al. 2019). Moreover, CNFs are compatible with other biopolymers, making them suitable for use in hybrid material systems. The combination of these properties makes CNFs a main component in the production of environmentally friendly solutions across different industries, including food packaging.

The global support for conventional plastic packaging has created significant environmental challenges. Plastics derived from petroleum-based polymers are highly durable and nonbiodegradable. This durability, which is advantageous for packaging, poses a severe threat to ecosystems owing to the accumulation of plastic waste in soil, waterways, and the ocean (Zhang et al. 2021). An estimate, over 3 hundred million tons of non-biodegradable plastic waste are generated per year, with a substantial part used for single-use packaging (Macheca et al. 2024). The consequences of plastic pollution have been far-reaching. Marine and terrestrial ecosystems are particularly vulnerable because plastic debris disrupts habitats and threatens wildlife. Additionally, microplastics, which are small fragments of degraded plastic, have been detected in water sources, soil, and even in the food chain, raising concerns about human health and safety (Jadhav et al. 2021). Plastic production also contributes significantly to greenhouse gas emissions and exacerbates the global climate crisis. These environmental effects have prompted a global push for alternatives to conventional plastics. Biodegradable and bio-based materials such as cellulose nanofibers have emerged as promising substitutes. Unlike synthetic plastics, CNFs are derived from renewable resources and decompose naturally, thereby reducing their environmental footprint. The adoption of CNF-based materials aligns with circular economy principles that prioritize resource efficacy, waste cut, and sustainability. Consequently, CNFs are getting attention as a potential solution to the environmental challenges presented by traditional packaging materials.

The unique properties of cellulose nanofibers make them suitable for food packaging applications (Joshi et al. 2024). One of their most notable characteristics is their exceptional mechanical strength, which rivals that of synthetic fibers, such as Kevlar (Park and Rutledge 2017). This strength allows CNFs to reinforce packaging materials, thereby improving their durability and resistance to damage during transportation and storage. Additionally, CNFs exhibit excellent barrier properties, effectively restricting the transfer of gases (oxygen and carbon dioxide) (Nazrin et al. 2020). These barrier capabilities are crucial in preserving the purity and quality of perishable food products. Another advantage of CNFs is their transparency, which enables the production of visually appealing packaging that allows consumers to view the content. CNFs are also lightweight, which reduces transportation costs and energy consumption (Talegaonkar et al. 2017). Moreover, their natural origin and biodegradability address consumer demand for environmentally responsible packaging solutions.

In food packaging, CNFs can be utilized in many forms, including films, coatings, and composites. CNF-based films are particularly advantageous due to their flexibility, transparency, and high tensile strength. These films can serve as standalone packaging materials or as coatings to enhance the properties of existing substrates. For example, CNF coatings can enhance the barrier properties of paper-based packaging, making it more resistant to moisture and grease. CNFs also have the possibility to be integrated into active and smart packaging systems (Asgher et al. 2020). Active packaging incorporates functional agents, such as antioxidants and antimicrobial compounds, to increase the shelf life of food products (Aayush et al. 2024). On the other hand, Smart packaging includes sensors or indicators that check the condition of the packaged food, which provides real-time information on freshness, temperature, or humidity levels (Aayush et al. 2022). The versatility of CNFs in accommodating these functionalities further underscores their value in innovative food packaging solutions.

Although cellulose nanofibers offer numerous benefits for food packaging, their widespread adoption encounters several challenges. High production costs and scalability issues remain significant barriers to commercialization. Additionally, CNFs are hygroscopic, meaning they absorb moisture from the environment, which can compromise their performance under high humidity conditions. Overcoming these challenges through research and technological advancement is crucial to explaining the full potential of CNFs in sustainable packaging.

CNFs are a groundbreaking innovation in the chase of sustainable food packaging. This chapter explores their distinctive properties, including renewability, biodegradability, and exceptional mechanical and barrier performance, which make them a viable alternative to conventional plastics. As the global demand for environment friendly packaging solutions continues to rise, CNFs are set to play a critical role in transforming the food packaging industry and reducing its environmental impact.

5.2 Sources and Production of Cellulose Nanofibers

CNFs are obtained from a variety of lignocellulosic biomass sources. Wood pulp is main source for CNF production due to its excessive cellulose content and widespread availability (Joshi et al. 2024). Wood pulp is sourced from both hardwoods and soft-woods, making it a reliable and scalable option. It is highly available, especially in regions with established forestry industries. By-products of agricultural activities, such as rice husks, wheat straw, corn stover, and sugarcane bagasse, offer a sustainable alternative to wood pulp. These residues are abundant, cost-effective, and contribute to waste valorization. Abundant in agricultural regions. These residues are often underutilized or discarded, making them an attractive resource for CNF production (Bilal et al. 2020). Utilizing agricultural residues reduces waste and promotes a circular economy. However, competition with other uses, such as animal feed or bioenergy, must be considered. Fruit peels, such as those from citrus fruits, bananas, and apples, are rich in cellulose and hemicellulose (Meral et al. 2022). They represent a valuable resource for CNF extraction, particularly in regions with high agricultural output. Readily available as by-products of food processing industries. Valorizing fruit peels contributes to waste reduction and resource efficiency. The regional and seasonal nature of these materials may affect their consistent supply. Other ligno-cellulosic biomass is non-wood plants, including bamboo, kenaf, and hemp, as well as algae and bacterial cellulose, which serve as additional sources for CNFs. These materials often have unique properties that can be tailored for specific applications. Variable, depending on the source. Non-wood plants and algae are more accessible in specific geographic regions. These sources often have a lower environmental footprint compared to wood. However, large-scale cultivation of non-wood plants may require significant land and water resources. By evaluating the availability and sustainability of these sources, researchers and industry stakeholders can select the most appropriate feedstock for CNF production, aligning with regional resource availability and environmental goals.

5.3 Extraction Methods for Cellulose Nanofibers (CNFs)

5.3.1 Mechanical Methods

Mechanical methods primarily utilize physical forces to crush cellulose fibers into the nanoscale. These methods are widely used due to their scalability and ability to produce uniform nanofibers, as shown in Table 5.1.

Table 5.1 Different extraction methods for cellulose nanofibers

Extraction method	Modifications (if any)	Characteristics of obtained cellulose	Source of cellulose	References
Homogenization	Chemical pretreatment (e.g., enzymatic or alkali treatment) to reduce rigidity	Uniform nanofibers with good dispersion, energy-intensive process	Wood	Rajinipriya et al. (2018)
			Plant fibers	Zulaikha et al. (2022)
			Agricultural residues	Padhi et al. (2023)
Microfluidization	Pre-treatment required for improved efficiency	Fine, uniform nanofibers with high dispersibility, scalable but costly	Wood pulp	Zulaikha et al. (2022)
			Plant fibers	Padhi et al. (2023)
			Bacterial cellulose	Yahya et al. (2023)
Ultrasonication	Used in combination with other methods for better efficiency	Disaggregated nanofibers, useful for small-scale applications, energy-intensive	Cotton	Jordan et al. (2019)
			Wood	Jordan et al. (2019)
			Cellulose pulp	Padhi et al. (2023)
Acid hydrolysis	Controlled acid concentration to avoid fiber degradation	Cellulose nanocrystals (CNCs) with high crystallinity, environmental concerns due to acid waste	Wood pulp	Wulandari et al. (2016)
			Cotton	Jordan et al. (2019)
			Agricultural residues	Guo et al. (2020)
TEMPO-mediated oxidation	Introduction of carboxyl groups to enhance dispersibility	Negatively charged CNFs with improved aqueous suspension stability	Wood pulp	Hop et al. (2022)
			Plant cellulose	Sanchez-Salvador et al. (2025)
Alkali treatment	NaOH treatment to remove non-cellulosic components	Purified cellulose fibers, improved mechanical processing suitability	Plant	Barbash and Yaschenko (2021)
			Fibers, wood	Rajinipriya et al. (2018)
			Agricultural waste	García et al. (2017)
Enzymatic hydrolysis	Use of cellulase enzymes under mild conditions	Nanofibers with minimal structural damage, eco-friendly but slow process	Wood	Ribeiro et al. (2019)
			Cotton	Ribeiro et al. (2019)

5.3.1.1 Homogenization

Homogenization is a widely employed method wherein a cellulose slurry is forced through with narrow valve or nozzle at extremely elevated pressure. The intense shear and impact forces generated during this process disrupt the cellulose fibers, breaking them into nanoscale fibrils. This method is particularly advantageous for large-scale production due to its scalability and ability to produce uniform nanofibers (Kumara Behera and Varma 2017). However, it has significant energy requirements and often necessitates pre-processing treatments, such as chemical modifications, to lower the rigidity of the cellulose fibers and better processing efficiency. Without such treatments, the high energy input can lead to excessive wear on equipment and suboptimal yields. This method is particularly advantageous for large-scale production due to its scalability and ability to produce uniform nanofibers. However, it has significant energy requirements and often necessitates pre-processing treatments, such as chemical modifications, to reduce the rigidity of the cellulose fibers and improve processing efficiency. Without such treatments, the high energy input can lead to excessive wear on equipment and suboptimal yields (Salama and Abdel Aziz 2020).

5.3.1.2 Microfluidization

Microfluidization employs a high-pressure fluid system to pass cellulose fibers through a series of microchannels with progressively smaller diameters. This setup creates high shear and impact forces that effectively separate cellulose into nanoscale fibrils. The process is known for producing fine and uniform nanofibers with excellent dispersibility in aqueous media, making it ideal for applications requiring homogenous suspensions (Otoni et al. 2018). However, like homogenization, microfluidization is energy-intensive and demands durable equipment to withstand the rigorous forces generated during operation, which can result in increased operational costs. The process is known for producing fine and uniform nanofibers with excellent dispersibility in aqueous media, making it ideal for applications requiring homogenous suspensions. However, like homogenization, microfluidization is energy-intensive and demands durable equipment to withstand the rigorous forces generated during operation, which can result in increased operational costs (Otoni et al. 2018).

5.3.1.3 Ultrasonication

Ultrasonication utilizes high-intensity ultrasonic waves to create cavitation within a cellulose slurry. Cavitation involves the formation and collapse of microbubbles, generating localized high-pressure regions that disrupt the hydrogen bonding within cellulose fibers. This process effectively disaggregates cellulose fibers into nanofibers and is particularly suitable for small-scale production or laboratory settings (Adewuyi

2018). While ultrasonication is a powerful tool for disaggregation and provides precise control over the process, it is less efficient and cost-prohibitive for production due to its significant energy utilization. Ultrasonication is a powerful tool for disaggregation and provides precise control over the process, making it particularly suitable for small-scale production or laboratory settings (Wang et al. 2020).

5.3.2 Chemical Methods

Chemical methods are employed to modify or weaken the cellulose structure, facilitating mechanical processing and enabling the production of CNFs with specific characteristics. These methods often involve the selective removal of non-cellulosic components or the introduction of functional groups to enhance CNF properties.

5.3.2.1 Acid Hydrolysis

It is a chemical process that involves treating cellulose fibers with concentrated acids (sulfuric acid or HCl) During acid hydrolysis, hydrogen ions from the acid preferentially penetrate the less ordered, amorphous regions of cellulose, cleaving the β-1,4-glycosidic bonds and effectively removing these disordered domains. The crystalline regions, due to their tightly packed and ordered structure, resist acid attack and remain intact, resulting in the formation of highly crystalline cellulose nanocrystals (CNCs). The resulting CNCs exhibit high crystallinity, uniform size, and remarkable mechanical properties which making them suitable for various applications (Dussán et al. 2014). However, this method requires careful control of reaction conditions to prevent excessive degradation of the cellulose fibers. Additionally, the use of strong acids raises environmental concerns due to the potential for hazardous waste generation, necessitating appropriate waste management and recycling strategies. However, this method requires careful control of reaction conditions to prevent excessive degradation of the cellulose fibers. Additionally, the use of strong acids raises environmental concerns due to the potential for hazardous waste generation, necessitating appropriate waste management and recycling strategies (Singh et al. 2024a, b).

5.3.2.2 Oxidation (TEMPO-Mediated Oxidation)

It is a selective chemical process that employs the TEMPO (2,2,6,6-Tetramethylpiperidine-1-oxyl radical) catalyst with sodium hypochlorite and sodium bromide (Bragd et al. 2004). This process oxidizes the primary hydroxyl groups in cellulose to carboxylic groups, producing CNFs with negatively charged surfaces. The introduction of these charged groups significantly enhances the dispersibility of CNFs in aqueous systems and facilitates further functionalization for specific applications (Baron and Coseri 2020). Despite its effectiveness, TEMPO-mediated oxidation involves costly reagents and generates chemical waste that poses environmental challenges. Efforts to develop greener and more cost-effective alternatives are ongoing. The introduction of these charged groups significantly enhances the dispersibility of CNFs in aqueous systems and facilitates further functionalization for specific applications. Despite its effectiveness, TEMPO-mediated oxidation involves costly reagents and generates chemical waste that poses environmental challenges. Efforts to develop greener and more cost-effective alternatives are ongoing (Joshi et al. 2024).

5.3.2.3 Alkali Treatments

Alkali treatments involve the use of sodium hydroxide (NaOH) to eliminate impurities (lignin and hemicellulose) from cellulose fibers (Hassan et al. 2018). This process exposes the cellulose microfibrils and prepares them for subsequent mechanical processing. Alkali treatments are relatively simple, inexpensive, and widely accessible. However, careful handling is essential to prevent cellulose degradation and ensure consistent quality (Mukhtar et al. 2020). The effectiveness of this process depends on optimizing parameters such as temperature, concentration, and treatment duration. Alkali treatments are relatively simple, inexpensive, and widely accessible. However, careful handling is essential to prevent cellulose degradation and ensure consistent quality (Bharti et al. 2019).

5.3.3 Enzymatic Methods

Enzymatic methods utilize cellulase enzymes to degrade the amorphous sites of cellulose without affecting the crystalline sites. A cellulase enzyme is added to a cellulose slurry under controlled temperature and pH conditions (Ahamed and Vermette 2008). The enzymes break down glycosidic bonds in amorphous cellulose, releasing individual nanofibers (Kumari et al. 2019). Eco-friendly and operates under mild conditions, reducing energy consumption. Produces nanofibers with minimal structural damage. Relatively slow process compared to mechanical and chemical methods. Requires specific enzyme formulations, increasing costs. By combining these methods (e.g., enzymatic pretreatment followed by mechanical processing), it is possible to achieve high-quality CNFs with optimized properties for diverse applications such as packaging, composites, and biomedical products. This method is eco-friendly and operates under mild conditions, reducing energy consumption while producing nanofibers with minimal structural damage. However, it is a relatively slow process compared to mechanical and chemical methods and requires specific enzyme formulations, increasing costs. This format allows you to seamlessly integrate the advantages and disadvantages into each respective section while maintaining continuity (Adewuyi 2018).

5.4 Properties of Cellulose Nanofibers

Cellulose nanofibers (CNFs) exhibit a unique combination of properties such as high mechanical strength, biodegradability, and a large specific surface area that render them highly suitable for a wide range of applications are shown in Table 5.2.

5.4.1 *Mechanical Strength*

One of their standout features is their mechanical strength, with high tensile strength and modulus that often surpass synthetic fibers. The strength of cellulose nanofibers (CNFs) stems from their highly ordered crystalline structure and the extensive network of hydrogen bonds between fibrils, making them ideal for applications requiring durable materials (Joshi et al. 2024). Alongside this, CNFs exhibit excellent flexibility due to their nanoscale dimensions, allowing them to bend and fold without breaking (Huang et al. 2020). This combination of strength and flexibility makes CNFs ideal for producing films, coatings, and composites with adaptable mechanical properties. CNFs display exceptional mechanical performance, attributed to the high crystallinity and molecular orientation of cellulose chains. Their tensile strength ranges from 1 to 3 GPa and Young's modulus can reach up to 150 GPa, comparable to conventional materials like glass fiber and Kevlar. This makes them suitable for reinforcing polymer matrices and biodegradable composites (Klemm et al. 2011).

5.4.2 *Barrier Capabilities*

In addition to their mechanical properties, CNFs have remarkable barrier capabilities, particularly against oxygen and water vapor. Their compact, dense structure with minimal porosity provides superior oxygen barrier performance, making them highly effective in inhibiting oxidative spoilage and enhancing the shelf life of food products (Joshi et al. 2024). While they exhibit moderate water vapor barrier properties under dry conditions, their hydrophilic nature can reduce performance under high humidity (Sharma et al. 2020). However, this limitation can be mitigated through surface modifications with hydrophobic polymers, enabling their use in advanced packaging solutions for food, pharmaceuticals, and electronics. CNF-based films form dense and tortuous networks, which impede the diffusion of gases like oxygen, carbon dioxide, and water vapor. This makes them promising candidates for food packaging applications aimed at shelf-life extension (Amara et al. 2021).

Table 5.2 Properties of cellulose nanofibers (CNFs)

Property	Film composition	Description	Significance in applications	References
Mechanical strength	Casssava starch and modified microcrystalline cellulose/ nanocellulose	High tensile strength and modulus due to crystalline structure and hydrogen bonding	Enhances durability in composites, coatings, and films	Chen et al. (2020)
Flexibility	Poly(vinyl alcohol) and cellulose nanocrystals or cellulose nanofibers	Nanoscale dimensions allow bending and folding without breaking	Enables adaptable and flexible packaging materials	Chou et al. (2023)
Barrier properties	Cellulose nanomaterials	Excellent oxygen barrier, moderate water vapor barrier	Prevents oxidative spoilage; can be improved for humidity resistance	Wang et al. (2018)
Thermal stability	Chitosan/lignin-and cellulose nanofibrils	Resistant to moderate temperatures, but degrades at high temperatures (>200 °C)	Suitable for food packaging with limited heat exposure	Xu et al. (2022)
Transparency	Nanocellulose, dialdehyde-modified cellulose and polyvinyl alcohol	High transparency in thin films	Useful for clear food packaging and coatings	Kim et al. (2024)
Biodegradability	Nanocellulose	Fully degradable in soil, water, and compost	Supports sustainable packaging and reduces plastic waste	Shanmugam et al. (2019)
Surface functionality	Nanocellulose	Functional groups enable chemical modifications (e.g., carboxylation, acetylation)	Enhances compatibility with other biomaterials and allows active packaging applications	Ghasemlou et al. (2021)

(continued)

Table 5.2 (continued)

Property	Film composition	Description	Significance in applications	References
Hydrophilicity	Cellulose nanomaterials	Absorbs moisture due to hydroxyl groups	Can be a limitation in humid environments but improved through hydrophobic modifications	Wang et al. (2018)
Reinforcement ability	Chitosan, cellulose and nanocellulose	Acts as a reinforcing agent when blended with polymers	Increases mechanical and barrier properties in biocomposites	Abdul Khalil et al. (2016)
Antimicrobial potential	Zinc oxide nanoparticles, bacterial nanocellulose	Can be functionalized with antimicrobial agents	Suitable for active food packaging to prevent microbial contamination in chicken	Soares Silva et al. (2023)

5.4.3 Biodegradability

Another notable property of CNFs is their biodegradability. Derived from natural cellulose, they are fully degradable in soil, water, and compost, carbon dioxide, and biomass through microbial action (Harindintwali et al. 2020). This characteristic significantly reduces plastic waste and landfill burden, offering a sustainable alternative to conventional synthetic materials. CNFs are particularly suitable for single-use items, including food packaging, disposable utensils, and medical products, supporting the shift towards circular and eco-friendly material systems. Being naturally sourced, CNFs are renewable, biodegradable, and have a significantly lower environmental impact than petroleum-based materials. They decompose through enzymatic and microbial activity in soil and aquatic environments, making them suitable for sustainable packaging and green product development (Trache et al. 2020).

5.4.4 Compatibility with Other Biomaterials

CNFs exhibit excellent compatibility with other biomaterials, making them versatile for use in biocomposites. They can easily integrate with biopolymers like polylactic acid (PLA), starch, and chitosan (Kumari et al. 2022). When blended with these materials, CNFs enhance their thermal stability, mechanical strength, and barrier properties while maintaining biodegradability. For instance, CNFs improve the film-forming properties of starch-based materials and reinforce chitosan films, adding antimicrobial and mechanical benefits. Their ability to act as reinforcing agents in biocomposites further broadens their application in sustainable packaging, coatings, and bioplastics. This unique combination of properties features the potential of CNFs as a renewable, high-performance material for addressing modern environmental and material challenges.

5.5 Applications of Cellulose Nanofibers (CNFs) in Food Packaging

The unique properties of cellulose nanofibers (CNFs) make them an excellent material for innovative food packaging solutions, offering sustainability, enhanced performance, and functional versatility are shown in Fig. 5.1. Below are detailed applications in food packaging.

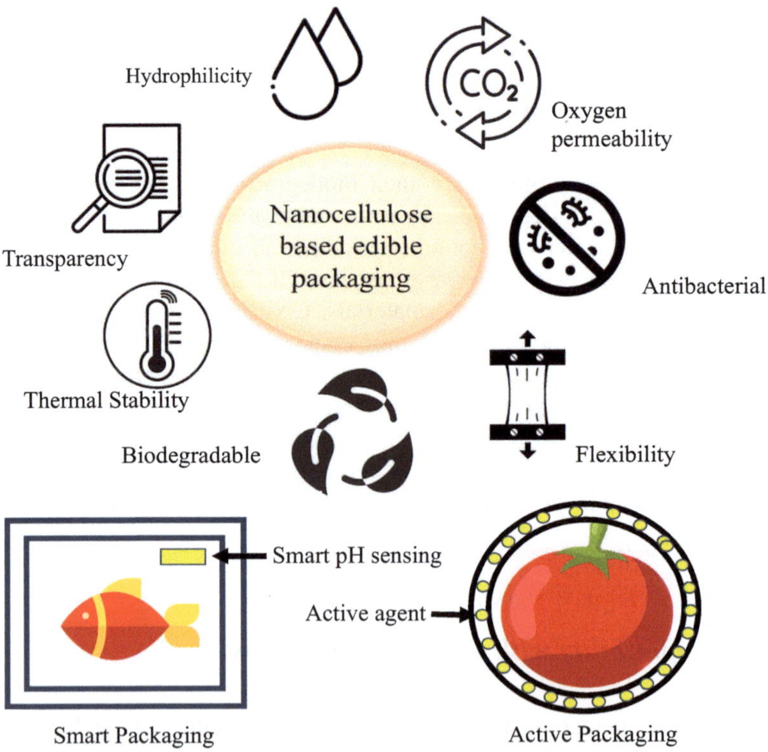

Fig. 5.1 Different properties of nanocellulose-based edible packaging

5.5.1 Edible Coatings and Films

CNF-based edible films are transparent, flexible, and exhibit excellent barrier properties, making them highly appropriate for food packaging applications. These films act as a protective layer over food products, preventing moisture loss and reducing exposure to oxygen, which can cause spoilage (Joshi et al. 2024). Their transparency ensures that the appearance of the food is not obscured, while their flexibility allows them to conform to irregular shapes. Moreover, these films are biodegradable and safe for consumption, providing a sustainable alternative to synthetic materials. Applications include fresh produce coatings, baked goods, and minimally processed foods, where shelf-life extension and quality preservation are essential (Chavan et al. 2023a, b; Singh et al. 2024a, b).

5.5.2 Composite Packaging Materials

When combined with biopolymers such as polylactic acid (PLA), starch, or chitosan, CNFs significantly improve the mechanical and thermal properties of packaging materials. CNFs act as reinforcing agents, enhancing tensile strength, elasticity, and durability (Talegaonkar et al. 2017). Additionally, they reduce the brittleness of biopolymers, making the resulting composites more robust and suitable for handling and transportation. The thermal stability of these materials is also improved, which is crucial for applications that require heat-resistant packaging (Joshi et al. 2024). These composite materials are ideal for biodegradable and compostable packaging solutions, addressing the demand for sustainable alternatives in the food industry.

5.5.3 Active Packaging

Active packaging involves incorporating functional components into packaging to enhance food safety and extend shelf life. CNF matrices can be infused with antimicrobial compounds such as essential oils, silver nanoparticles, or natural extracts (Aayush et al. 2024, 2025; Joshi et al. 2024). These agents release antimicrobial compounds in a controlled manner, effectively inhibiting the growth of bacteria, molds, and yeasts. This feature is especially beneficial for perishable food products like fresh fruits, vegetables, dairy, and meat, where microbial contamination is a significant concern (Aayush et al. 2022; Ban et al. 2020). By enhancing food safety and reducing spoilage, CNF-based active packaging helps minimize food waste.

5.5.4 Smart Packaging

Smart packaging integrates advanced functionalities to monitor and communicate the quality and freshness of food products. Functionalized CNFs can be embedded with sensors or indicators that respond to environmental changes such as temperature, pH, gas composition (Moazami Goodarzi et al. 2020; Talegaonkar et al. 2017). For instance, CNF-based smart packaging can include color-changing indicators that detect spoilage gases like ammonia or carbon dioxide, providing a visual cue to consumers or retailers (Ezati et al. 2021; Moazami Goodarzi et al. 2020). These systems help ensure food safety, improve inventory management, and reduce food waste by identifying products nearing spoilage. This innovative approach aligns with modern consumer demands for transparency and convenience in food packaging. Cellulose nanofibers enable the development of cutting-edge food packaging solutions that are sustainable, functional, and adaptable to various needs. Whether as edible films, reinforced composites, antimicrobial active packaging, or intelligent smart packaging, CNFs play a pivotal role in addressing current challenges in food preservation, safety, and sustainability.

5.6 Challenges in Utilizing Cellulose Nanofibers (CNFs)

Despite their promising properties and applications, the use of CNFs in food packaging faces several challenges that need to be addressed for wider adoption and commercialization. Below is a detailed explanation of these challenges.

5.6.1 High Production Costs

One of the most significant barriers to the large-scale utilization of CNFs is their high production cost. The processes involved in extracting CNFs, such as mechanical treatments (e.g., homogenization and ultrasonication), chemical methods (acid hydrolysis and TEMPO oxidation), enzymatic treatments require substantial energy and resources (Djafari Petroudy et al. 2021). These processes often involve expensive equipment, harsh chemicals, or long processing times, making scalability economically challenging. Additionally, the cost of raw materials and the need for extensive purification steps further add to the expense (Ul-Islam et al. 2020). Research into energy-efficient production methods, low-cost raw materials, and process optimization is crucial to address this challenge.

5.6.2 *Processing Challenges*

During the fabrication of CNF-based materials, several processing issues arise. CNFs had a strong tendency to agglomerate due to the high surface area and presence of hydroxyl groups, which promote hydrogen bonding in fibers (Parveen et al. 2013). This agglomeration can lead to uneven dispersion within polymer matrices or coatings, resulting in non-uniform properties. Furthermore, the compatibility of CNFs with hydrophobic polymers (e.g., PLA or polyethylene) is limited due to hydrophilic nature (Joshi et al. 2024). To overcome these issues, surface modifications, chemical functionalization, or compatibilizers are often required. However, these additional steps can complicate the manufacturing process and increase costs.

5.6.3 *Water Sensitivity*

CNFs exhibit hygroscopic behavior, meaning they readily absorb moisture from the near atmosphere. This water sensitivity can compromise the barrier properties, mechanical strength, and overall performance of CNF-based packaging under high humidity conditions. For instance, their oxygen barrier performance, which is excellent in dry conditions, significantly deteriorates when exposed to moisture (Asgher et al. 2020). This limitation restricts the use of CNFs in packaging applications that involve wet or humid environments. Strategies to address this challenge include surface hydrophobic modifications, blending with water-resistant polymers, or incorporating moisture-resistant coatings, but these solutions may introduce complexity and increase production costs.

5.6.4 *Regulatory and Standardization Issues*

The absence of standardized protocols for evaluating CNF-based packaging materials poses a significant challenge to their commercialization. Regulatory bodies have yet to establish comprehensive guidelines for the production, characterization, and safety assessment of CNFs, particularly for food contact applications. This lack of standardization leads to inconsistencies in the quality, performance, and safety of CNF-based products, making it difficult for manufacturers to gain regulatory approval and market acceptance. Furthermore, concerns about the potential toxicity of CNF-based materials, especially when chemically modified or combined with additives, require thorough investigation. Collaborative efforts among researchers, regulatory bodies, and industry stakeholders are needed to develop standardized protocols and ensure compliance with food safety regulations. While cellulose nanofibers are promising for advancing sustainable packaging, challenges related to high production costs, processing difficulties, water sensitivity, and regulatory issues must be addressed.

Innovations in production methods, material design, and regulatory frameworks will be critical to showing the potential of CNFs and achieving their widespread adoption in food packaging and other applications.

5.7 Recent Advancements in Cellulose Nanofibers (CNFs)

The cellulose nanofibers (CNFs) field has witnessed several advancements aimed at overcoming existing challenges and broadening their applications. These innovations enhance CNF functionality, improve processability, and align with sustainability goals.

5.7.1 Hybrid Materials

Hybrid materials that combine CNFs with other reinforcements, such as graphene, clay nanoparticles, or metallic oxides, have emerged as a promising area of research. By integrating these additives, the mechanical, thermal, and barrier properties of CNF-based materials can be significantly enhanced. For example, incorporating graphene improves electrical conductivity, opening opportunities for smart packaging with sensing capabilities (Idumah et al. 2020). Similarly, clay nanoparticles enhance oxygen and water vapor barrier properties, making these hybrid materials highly effective for food packaging (Zabihzadeh Khajavi et al. 2020). These composites leverage the strengths of multiple materials to create multifunctional packaging solutions.

5.7.2 Surface Modifications

Surface modifications, both physical and chemical, have been developed to address the hydrophilic nature of CNFs and improve their compatibility with non-polar polymers.. Techniques such as acetylation, silanization, and grafting hydrophobic functional groups onto CNF surfaces increase hydrophobicity and reduce water sensitivity (Sun et al. 2020). These treatments also enhance the dispersion of CNFs within hydrophobic polymer matrices, resulting in more uniform and high-performing composites. Physical modifications like plasma treatment or coating with hydrophobic agents also provide a simpler, eco-friendly alternative for improving CNF properties (Ranjha et al. 2023). These advancements expand the range of applications for CNF-based materials in packaging and beyond.

5.7.3 3D Printing Applications

The use of CNF-based inks in 3D printing is a novel advancement that enables the design of custom packaging solutions. CNFs can be formulated into printable hydrogels or composite inks, offering high precision and versatility in creating complex structures (Iglesias-Mejuto et al. 2024). This approach allows for the fabrication of lightweight, biodegradable, and mechanically robust packaging tailored to specific products. For instance, CNF-based 3D printing can produce custom food trays, protective packaging, or decorative elements with minimal waste from nanocomposite bio gel (Heggset et al. 2019). This innovation enhances packaging design and aligns with sustainable manufacturing practices.

5.7.4 Circular Economy Integration

Incorporating CNFs into circular economy models is a key advancement aimed at promoting sustainability. Researchers are developing fully recyclable or compostable CNF-based packaging systems that minimize environmental impact. These systems are designed to be compatible with existing recycling infrastructure or to decompose safely in natural environments. For instance, blending CNFs with biodegradable polymers like polylactic acid (PLA) or starch results in materials that can be industrially composted (Ezati et al. 2021). Additionally, CNFs derived from agricultural or industrial waste feedstocks contribute to waste valorization and resource efficiency. This approach supports the transition to a circular economy by reducing reliance on fossil-based materials and promoting sustainable material lifecycles. Recent advancements in cellulose nanofibers, including hybrid materials, surface modifications, 3D printing applications, and circular economy integration, demonstrate their potential to revolutionize sustainable packaging. These innovations address existing limitations while opening new opportunities for high-performance, environmentally friendly packaging solutions tailored to modern consumer and industrial needs (Heggset et al. 2019).

5.8 Future Prospects of Cellulose Nanofiber (CNF)-Based Food Packaging

The potential of CNFs in transforming the food packaging industry is vast, owing to their sustainability, excellent material properties, and versatility. However, the future success of CNF-based food packaging hinges on overcoming existing challenges and harnessing advancements across interdisciplinary domains. With strategic investments, policy support, and innovative technologies, CNF-based materials can become mainstream in the global packaging market.

One of the primary future goals for CNF-based food packaging is to address challenges like high production costs, water sensitivity, and processing difficulties. Innovations in production processes can significantly reduce costs, making CNFs economically viable for large-scale applications. For example, developing energy-efficient mechanical treatments or using low-cost biomass feedstocks can lower expenses while maintaining quality (Djafari Petroudy et al. 2021). Enzymatic methods, which are environmentally friendly and selective, could also be scaled up with advancements in biotechnology. To mitigate water sensitivity, future research will likely focus on advanced surface modifications and blending techniques. Hydrophobic coatings or functionalization of CNFs to repel moisture can enhance their performance in humid environments. Additionally, the development of hybrid materials combining CNFs with moisture-resistant biopolymers or nanoparticles will play a crucial role in creating robust and durable packaging materials. Processing challenges, such as agglomeration and compatibility with hydrophobic polymers, can be addressed through improved dispersion techniques and novel compatibilizers. For instance, advancements in nanotechnology can help produce uniformly dispersed CNF-polymer composites, ensuring consistent performance. These developments will be essential in making CNFs suitable for diverse applications, from edible films to composite packaging.

Integrating CNFs with cutting-edge technologies like nanotechnology, biotechnology, and artificial intelligence (AI) promises to unlock new possibilities in food packaging. Nanotechnology can enhance the performance of CNF-based packaging by enabling the incorporation of functional nanoparticles (Chavan et al. 2023a, b; Singh et al. 2022, 2023, 2024a, b). For instance, embedding silver or zinc oxide nanoparticles can impart antimicrobial properties, extending the shelf life of perishable foods (Youssef and El-Sayed 2018). Similarly, integrating graphene or conductive nanomaterials into CNF matrices can create smart packaging solutions with sensors to monitor freshness, temperature, or gas levels (Mianehrow et al. 2022). Biotechnology offers opportunities to refine CNF production and functionalization. Genetic engineering of microorganisms or plants could lead to more efficient and sustainable cellulose production, while bio-based additives or enzymes can be used to tailor CNF properties for specific applications (Dadwal et al. 2020; Misra et al. 2019). This approach aligns with the broader goal of minimizing environmental impact throughout the production process. Artificial Intelligence and machine learning can optimize the design and manufacturing of CNF-based materials (Choudhary et al. 2024). By analyzing vast datasets on material performance, AI can predict optimal formulations, processing conditions, and functionalization techniques. Additionally, AI-powered quality control systems can ensure the consistency and reliability of CNF-based products, facilitating their commercialization.

The future of CNF-based food packaging will also depend on supportive policies and increased investments in research, development, and infrastructure. Governments and regulatory bodies must establish clear standards and guidelines for the safety and evaluation of CNF-based materials, particularly for food contact applications. Such standardization will build consumer trust and facilitate market acceptance. Investment in pilot-scale and industrial-scale production facilities is crucial for bridging the gap between research and commercialization. Public–private partnerships and funding initiatives can accelerate the scaling-up process, reducing costs and improving accessibility. Policies promoting the use of biobased and biodegradable materials, such as tax incentives or bans on single-use plastics, will further drive the adoption of CNFs in the packaging industry (Zhang et al. 2021).

The role of CNFs in advancing the circular economy is a critical aspect of their prospects. As fully biodegradable and renewable materials, CNFs align perfectly with the principles of sustainability and waste reduction. Future packaging systems will likely focus on integrating CNFs into fully recyclable or compostable designs, minimizing environmental impact. Additionally, CNFs derived from agricultural or industrial byproducts can contribute to waste valorization, turning waste streams into high-value materials.

5.9 Conclusion

Cellulose nanofibers (CNFs) represent a transformative innovation in the pursuit of sustainable food packaging solutions, offering a viable alternative to conventional petroleum-based plastics. Derived from renewable plant biomass, CNFs possess a unique combination of desirable attribute including high mechanical strength, excellent barrier properties, biodegradability, and film-forming ability—which make them especially attractive for packaging applications aimed at preserving food quality and extending shelf life. The versatility of CNFs lies not only in their basic properties but also in their compatibility with other biopolymers and functional agents, enabling the development of active and intelligent packaging systems. Such innovations can impart antimicrobial, antioxidant, and oxygen-scavenging functionalities, thus contributing to food safety and waste reduction. Various extraction and processing methods such as mechanical, chemical, and enzymatic play a significant role in tailoring the morphology and functionality of CNFs, thereby enhancing their performance in diverse packaging formats. Despite their potential, CNF-based packaging materials face notable challenges. High production costs, sensitivity to moisture, and the lack of regulatory clarity hinder their commercial scalability. Addressing these limitations requires continued advancements in processing technologies such as surface modification, hybrid composites, and the integration of nanotechnology and additive manufacturing techniques like 3D printing. Looking ahead, the future of CNF-based food packaging will depend on collaborative efforts among researchers, industry stakeholders, and policymakers. Strategic partnerships and supportive policy frameworks can accelerate innovation and commercialization, ensuring that CNF-based materials become a practical, eco-friendly solution for the global food packaging industry. By leveraging scientific advancements and sustainable practices, cellulose nanofibers have the potential to revolutionize the packaging sector and contribute meaningfully to environmental conservation and circular economy goals.

Declaration Authors declare that AI-based language tools were used only to improve the grammar, wording, and clarity of this chapter. All the contributors have carefully reviewed the final content and take full responsibility for its accuracy, meaning, and overall presentation.

References

Aayush K, McClements DJ, Sharma S, Sharma R, Singh GP, Sharma K, Oberoi K (2022) Innovations in the development and application of edible coatings for fresh and minimally processed Apple. Food Control 109188

Aayush K, Sharma K, Singh GP, Chiu I, Chavan P, Shandilya M, Roy S, Ye H, Sharma S, Yang T (2024) Development and characterization of edible and active coating based on xanthan gum nanoemulsion incorporating betel leaf extract for fresh produce preservation. Int J Biol Macromol 132220

Aayush K, Singh GP, Chiu I, Joshi M, Sharma K, Gautam S, Chavan P, Jha N, Singh AK, Babaei A, Sharma S, Yang T (2025) Development and characterization of sodium alginate and β-cyclodextrin nanoemulsions encapsulating betel leaf (Piper betle L.) extract for enhanced antimicrobial efficacy against foodborne pathogen. Food Chem 463:141227. https://doi.org/10.1016/J.FOODCHEM.2024.141227

Abdul Khalil HPS, Saurabh CK, Adnan AS, Nurul Fazita MR, Syakir MI, Davoudpour Y, Rafatullah M, Abdullah CK, Haafiz MKM, Dungani R (2016) A review on chitosan-cellulose blends and nanocellulose reinforced chitosan biocomposites: properties and their applications. In: Carbohydrate polymers, vol 150. Elsevier, pp 216–226. https://doi.org/10.1016/j.carbpol.2016.05.028

Adewuyi YG (2018) Intensification of enzymatic hydrolysis of cellulose using high frequency ultrasound. RSC Green Chem 2018-January(55):166–187. https://doi.org/10.1039/9781788010320-00166

Ahamed A, Vermette P (2008) Culture-based strategies to enhance cellulase enzyme production from Trichoderma reesei RUT-C30 in bioreactor culture conditions. Biochem Eng J 40(3):399–407. https://doi.org/10.1016/j.bej.2007.11.030

Amara C, El Mahdi A, Medimagh R, Khwaldia K (2021) Nanocellulose-based composites for packaging applications. Curr Opin Green Sustain Chem 31:100512. https://doi.org/10.1016/j.cogsc.2021.100512

Asgher M, Qamar SA, Bilal M, Iqbal HMN (2020) Bio-based active food packaging materials: sustainable alternative to conventional petrochemical-based packaging materials. Food Res Int 137. https://doi.org/10.1016/J.FOODRES.2020.109625

Ban Z, Zhang J, Li L, Luo Z, Wang Y, Yuan Q, Zhou B, Liu H (2020) Ginger essential oil-based microencapsulation as an efficient delivery system for the improvement of Jujube (Ziziphus jujuba Mill.) fruit quality. Food Chem 306. https://doi.org/10.1016/j.foodchem.2019.125628

Barbash V, Yaschenko O (2021) Preparation, properties and use of nanocellulose from non-wood plant materials. In: Novel nanomaterials. IntechOpen. https://doi.org/10.5772/intechopen.94272

Baron RI, Coseri S (2020) Preparation of water-soluble cellulose derivatives using TEMPO radical-mediated oxidation at extended reaction time. React Funct Polym 157:104768. https://doi.org/10.1016/J.REACTFUNCTPOLYM.2020.104768

Bharti I, Singh S, Saxena DC (2019) Influence of alkali treatment on physicochemical, pasting, morphological and structural properties of mango kernel starches derived from Indian cultivars. Int J Biol Macromol 125:203–212

Bilal M, Rasheed T, Nabeel F, Iqbal HMN (2020) Bionanocomposites from biofibers and biopolymers. In: Biofibers and biopolymers for biocomposites: synthesis, characterization and properties. Springer International Publishing, pp 136–157. https://doi.org/10.1007/978-3-030-40301-0_6

Bragd PL, Van Bekkum H, Besemer AC (2004) TEMPO-mediated oxidation of polysaccharides: survey of methods and applications. Top Catal 27(1–4):49–66. https://doi.org/10.1023/B:TOCA.0000013540.69309.46/METRICS

Chavan P, Lata K, Kaur T, Jambrak AR, Sharma S, Roy S, Sinhmar A, Thory R, Singh GP, Aayush K (2023) Recent advances in the preservation of postharvest fruits using edible films and coatings: a comprehensive review. Food Chem 135916

Chavan P, Sinhmar A, Thory R, Sharma S, Sukhija S, Singh GP, Aayush K, Singh J, Kumar D (2023) Acid hydrolyzed pearl millet starch nanoparticles: synthesis and characterization. Starch-Stärke 2300172

Chen Q, Shi Y, Chen G, Cai M (2020) Enhanced mechanical and hydrophobic properties of composite cassava starch films with stearic acid modified MCC (microcrystalline cellulose)/NCC (nanocellulose) as strength agent. Int J Biol Macromol 142:846–854. https://doi.org/10.1016/j.ijbiomac.2019.10.024

Chou CT, Shi SC, Chen TH, Chen CK (2023) Nanocellulose-reinforced, multilayered poly(vinyl alcohol)-based hydrophobic composites as an alternative sealing film. Sci Progr 106(1). https://doi.org/10.1177/00368504231157142

Choudhary N, Tomar A, Bhardwaj S, Ćwiertnia J, Just D, Janas D, Chandra R, Maji PK (2024) Advancements in biomass-derived cellulose composite electrodes for supercapacitors: a review. J Mater Chem A. https://doi.org/10.1039/D4TA05470A

Dadwal A, Sharma S, Satyanarayana T (2020) Progress in ameliorating beneficial characteristics of microbial cellulases by genetic engineering approaches for cellulose saccharification. In: Frontiers in microbiology, vol 11. Frontiers Media S.A., p 552376. https://doi.org/10.3389/fmicb.2020.01387

Djafari Petroudy SR, Chabot B, Loranger E, Naebe M, Shojaeiarani J, Gharehkhani S, Ahvazi B, Hu J, Thomas S (2021) Recent advances in cellulose nanofibers preparation through energy-efficient approaches: a review. Energies 14(20):6792 (Multidisciplinary Digital Publishing Institute). https://doi.org/10.3390/en14206792

Dussán KJ, Silva DDV, Moraes EJC, Arruda PV, Felipe MGA (2014) Dilute-acid hydrolysis of cellulose to glucose from sugarcane bagasse. Chem Eng Trans 38:433–438. https://doi.org/10.3303/CET1438073

Ezati P, Priyadarshi R, Bang YJ, Rhim JW (2021) CMC and CNF-based intelligent pH-responsive color indicator films integrated with shikonin to monitor fish freshness. Food Control 126:108046. https://doi.org/10.1016/J.FOODCONT.2021.108046

García A, Labidi J, Belgacem MN, Bras J (2017) The nanocellulose biorefinery: woody versus herbaceous agricultural wastes for NCC production. Cellulose 24(2):693–704. https://doi.org/10.1007/s10570-016-1144-2

Ghasemlou M, Daver F, Ivanova EP, Habibi Y, Adhikari B (2021) Surface modifications of nanocellulose: From synthesis to high-performance nanocomposites. Progr Polym Sci 119:101418 (Pergamon). https://doi.org/10.1016/j.progpolymsci.2021.101418

Guo Y, Zhang Y, Zheng D, Li M, Yue J (2020) Isolation and characterization of nanocellulose crystals via acid hydrolysis from agricultural waste-tea stalk. Int J Biol Macromol 163:927–933. https://doi.org/10.1016/j.ijbiomac.2020.07.009

Harindintwali JD, Zhou J, Yu X (2020) Lignocellulosic crop residue composting by cellulolytic nitrogen-fixing bacteria: a novel tool for environmental sustainability. Sci Total Environ 715:136912 (Elsevier). https://doi.org/10.1016/j.scitotenv.2020.136912

Hassan A, Isa MRM, Ishak ZAM, Ishak NA, Rahman NA, Salleh FM (2018) Characterization of sodium hydroxide-treated kenaf fibres for biodegradable composite application. High Perform Polym 30(8):890–899. https://doi.org/10.1177/0954008318784997

Heggset EB, Strand BL, Sundby KW, Simon S, Chinga-Carrasco G, Syverud K (2019) Viscoelastic properties of nanocellulose based inks for 3D printing and mechanical properties of CNF/alginate biocomposite gels. Cellulose 26(1):581–595. https://doi.org/10.1007/S10570-018-2142-3/METRICS

Hop TT, Mai DT, Cong TD, Nhi TT, Loi VD, Huong NT, Tung NT (2022) A comprehensive study on preparation of nanocellulose from bleached wood pulps by TEMPO-mediated oxidation. Results Chem 4:100540. https://doi.org/10.1016/j.rechem.2022.100540

Huang Y, Gu C, He S, Zhu D, Liu X, Chen Z (2020) Development and characterization of an edible chitosan–whey protein nano composite film for chestnut (Castanea mollissima Bl.) preservation. J Food Sci 85(7):2114–2123

Idumah CI, Zurina M, Ogbu J, Ndem JU, Igba EC (2020) A review on innovations in polymeric nanocomposite packaging materials and electrical sensors for food and agriculture. Compos Interfaces 27(1):1–72. https://doi.org/10.1080/09276440.2019.1600972

Iglesias-Mejuto A, Malandain N, Ferreira-Gonçalves T, Ardao I, Reis CP, Laromaine A, Roig A, García-González CA (2024) Cellulose-in-cellulose 3D-printed bioaerogels for bone tissue engineering. Cellulose 31(1):515–534. https://doi.org/10.1007/S10570-023-05601-1/FIGURES/9

Jadhav EB, Sankhla MS, Bhat RA, Bhagat DS (2021) Microplastics from food packaging: an overview of human consumption, health threats, and alternative solutions. Environ Nanotechnol Monitor Manag 16:100608. https://doi.org/10.1016/J.ENMM.2021.100608

Jordan JH, Easson MW, Dien B, Thompson S, Condon BD (2019) Extraction and characterization of nanocellulose crystals from cotton gin motes and cotton gin waste. Cellulose 26(10):5959–5979. https://doi.org/10.1007/s10570-019-02533-7

Joshi M, Aayush K, Sharma K, Bose I, Khan AA, Atanassova M, Yang T, Murariu OC, Sharma S, Caruso G (2024) Fiber and nanofiber based edible packaging for enhancing the shelf life of food: a review. Food Biosci 59:103970. https://doi.org/10.1016/J.FBIO.2024.103970

Kim M, Lim T, Park H, Baek KY, Jeong Y, Sohn D, Cho KY, Cho S (2024) Nanocellulose-based optical and radio frequency transparent barrier coating for food packaging. Cellulose 31(8):5185–5197. https://doi.org/10.1007/s10570-024-05915-8

Klemm D, Kramer F, Moritz S, Lindström T, Ankerfors M, Gray D, Dorris A (2011) Nanocelluloses: a new family of nature-based materials. Angew Chem Int Ed 50(24):5438–5466. https://doi.org/10.1002/anie.201001273

Kumara Behera B, Varma A (2017) Downstream processing. In: Microbial biomass process technologies and management, pp 109–214. https://doi.org/10.1007/978-3-319-53913-3_3

Kumari P, Pathak G, Gupta R, Sharma D, Meena A (2019) Cellulose nanofibers from lignocellulosic biomass of lemongrass using enzymatic hydrolysis: characterization and cytotoxicity assessment. DARU J Pharm Sci 27(2):683–693. https://doi.org/10.1007/s40199-019-00303-1

Kumari SVG, Pakshirajan K, Pugazhenthi G (2022) Recent advances and future prospects of cellulose, starch, chitosan, polylactic acid and polyhydroxyalkanoates for sustainable food packaging applications. Int J Biol Macromol 221:163–182 (Elsevier). https://doi.org/10.1016/j.ijbiomac.2022.08.203

Macheca AD, Mutuma B, Adalima JL, Midheme E, Lúcas LHM, Ochanda VK, Mhlanga SD (2024) Perspectives on plastic waste management: challenges and possible solutions to ensure its sustainable use. Recycling 9(5):77. https://doi.org/10.3390/RECYCLING9050077

Meral R, Kose YE, Ceylan Z, Cavidoglu İ (2022) The potential use of agro-industrial by-products as sources of bioactive compounds: a nanotechnological approach. In: Studies in natural products chemistry, vol 73. Elsevier, pp 435–466. https://doi.org/10.1016/B978-0-323-91097-2.00014-5

Mianehrow H, Li L, Olsén P, Berglund L (2022) Moisture effects on mechanical behavior of CNF-RGO nanocomposites showing electrical conductivity. Compos A Appl Sci Manuf 163:107235. https://doi.org/10.1016/j.compositesa.2022.107235

Misra P, Shukla PK, Prasada Rao K, Ramteke PW (2019) Genetic engineering applications to improve cellulase production and efficiency: part II. In: New and future developments in microbial biotechnology and bioengineering: from cellulose to cellulase: strategies to improve biofuel production. Elsevier, pp 227–260. https://doi.org/10.1016/B978-0-444-64223-3.00014-X

Moazami Goodarzi M, Moradi M, Tajik H, Forough M, Ezati P, Kuswandi B (2020) Development of an easy-to-use colorimetric pH label with starch and carrot anthocyanins for milk shelf life assessment. Int J Biol Macromol 153:240–247. https://doi.org/10.1016/J.IJBIOMAC.2020.03.014

Mukhtar I, Leman Z, Zainudin ES, Ishak MR (2020) Effectiveness of alkali and sodium bicarbonate treatments on sugar palm fiber: mechanical, thermal, and chemical investigations. J Nat Fibers 17(6):877–889. https://doi.org/10.1080/15440478.2018.1537872

Nazrin A, Sapuan SM, Zuhri MYM, Ilyas RA, Syafiq R, Sherwani SFK (2020) Nanocellulose reinforced thermoplastic starch (TPS), polylactic acid (PLA), and polybutylene succinate (PBS) for food packaging applications. Front Chem 8:213 (Frontiers Media S.A.). https://doi.org/10.3389/fchem.2020.00213

Otoni CG, Carvalho AS, Cardoso MVC, Bernardinelli OD, Lorevice MV, Colnago LA, Loh W, Mattoso LHC (2018) High-pressure microfluidization as a green tool for optimizing the mechanical performance of all-cellulose composites. ACS Sustain Chem Eng 6(10):12727–12735. https://doi.org/10.1021/ACSSUSCHEMENG.8B01855/SUPPL_FILE/SC8B01855_SI_001.PDF

Padhi S, Singh A, Routray W (2023) Nanocellulose from agro-waste: a comprehensive review of extraction methods and applications. Rev Environ Sci Biotechnol 22(1):1–27. Springer. https://doi.org/10.1007/s11157-023-09643-6

Park JH, Rutledge GC (2017) 50th anniversary perspective: advanced polymer fibers: high performance and ultrafine. Macromolecules 50(15):5627–5642. https://doi.org/10.1021/ACS.MACROMOL.7B00864/ASSET/IMAGES/MEDIUM/MA-2017-00864M_0013.GIF

Parveen S, Rana S, Fangueiro R (2013) A review on nanomaterial dispersion, microstructure, and mechanical properties of carbon nanotube and nanofiber reinforced cementitious composites. J Nanomater 2013(1):710175. https://doi.org/10.1155/2013/710175

Rajinipriya M, Nagalakshmaiah M, Robert M, Elkoun S (2018) Importance of agricultural and industrial waste in the field of nanocellulose and recent industrial developments of wood based nanocellulose: a review. ACS Sustain Chem Eng 6(3): 2807–2828 (American Chemical Society). https://doi.org/10.1021/acssuschemeng.7b03437

Ranjha MMAN, Shafique B, Aadil RM, Manzoor MF, Cheng JH (2023) Modification in cellulose films through ascent cold plasma treatment and polymerization for food products packaging. Trends Food Sci Technol 134:162–176. https://doi.org/10.1016/J.TIFS.2023.03.011

Ribeiro RSA, Pohlmann BC, Calado V, Bojorge N, Pereira N (2019) Production of nanocellulose by enzymatic hydrolysis: Trends and challenges. Eng Life Sci 19(4):279–291 (Wiley). https://doi.org/10.1002/elsc.201800158

Salama HE, Abdel Aziz MS (2020) Optimized carboxymethyl cellulose and guanidinylated chitosan enriched with titanium oxide nanoparticles of improved UV-barrier properties for the active packaging of green bell pepper. Int J Biol Macromol 165:1187–1197. https://doi.org/10.1016/j.ijbiomac.2020.09.254

Sanchez-Salvador JL, Xu H, Balea A, Fuente E, Monte MC, Blanco A, Negro C, Sanchez-Salvador JL, Xu H, Balea A, Fuente E, Monte MC, Blanco A, Negro C (2025) Nanocellulose, a promising raw material: improving the scalability of TEMPO-mediated oxidation process. In: Cellulose—Biobased solutions for society [Working Title]. https://doi.org/10.5772/INTECHOPEN.1008755

Shanmugam K, Doosthosseini H, Varanasi S, Garnier G, Batchelor W (2019) Nanocellulose films as air and water vapour barriers: a recyclable and biodegradable alternative to polyolefin packaging. Sustain Mater Technol 22:e00115. https://doi.org/10.1016/j.susmat.2019.e00115

Sharma R, Jafari SM, Sharma S (2020) Antimicrobial bio-nanocomposites and their potential applications in food packaging. Food Control 112(September 2019):107086. https://doi.org/10.1016/j.foodcont.2020.107086

Shogren R, Wood D, Orts W, Glenn G (2019) Plant-based materials and transitioning to a circular economy. Sustain Prod Consum 19:194–215. https://doi.org/10.1016/J.SPC.2019.04.007

Singh GP, Bangar SP, Yang T, Trif M, Kumar V, Kumar D (2022) Effect on the properties of edible starch-based films by the incorporation of additives: a review. Polymers 14(10):1987

Singh GP, Aayush K, Yang T (2023) Nanotechnology enabled packaging materials. In: Reference module in food science. Elsevier. https://doi.org/10.1016/B978-0-12-822521-9.00154-4

Singh GP, Aayush K, Chavan P, Chiu I, Yan S, Verma R, Kuca K, Kumar D, Yang T (2024a) Valorization of agro-industrial wastes: Converting mango seed into starch nanoparticles-based edible coating for fresh produce preservation. Innov Food Sci Emerg Technol 103722

Singh GP, Bangar SP, Aayush K, Yang T, Verma R, Kuca K, Kumar D, Phimolsiripol Y (2024b) Value addition of mango kernel for development and characterization of starch with starch nanoparticles for packaging applications. Int J Biol Macromol 133185

Silva FA, de Carvalho TB, Dourado F, Gama M, Teixeira P, Poças F (2023) Performance of bacterial nanocellulose packaging film functionalised in situ with zinc oxide: migration onto chicken skin and antimicrobial activity. Food Packag Shelf Life 39:101140. https://doi.org/10.1016/j.fpsl.2023.101140

Sun L, Zhang X, Liu H, Liu K, Du H, Kumar A, Sharma G, Si C (2020) Recent advances in hydrophobic modification of nanocellulose. Curr Org Chem 25(3):417–436. https://doi.org/10.2174/1385272824999201210191041

Talegaonkar S, Sharma H, Pandey S, Mishra PK, Wimmer R (2017) Bionanocomposites: smart biodegradable packaging material for food preservation. In: Food packaging, no 4. Elsevier Inc. https://doi.org/10.1016/b978-0-12-804302-8.00003-0

Trache D, Tarchoun AF, Derradji M, Hamidon TS, Masruchin N, Brosse N, Hussin MH (2020) Nanocellulose: from fundamentals to advanced applications. Front Chem 8:392. https://doi.org/10.3389/fchem.2020.00392

Ul-Islam M, Ullah MW, Khan S, Park JK (2020) Production of bacterial cellulose from alternative cheap and waste resources: a step for cost reduction with positive environmental aspects. Korean J Chem Eng 37(6):925–937. https://doi.org/10.1007/S11814-020-0524-3/METRICS

Wang J, Gardner DJ, Stark NM, Bousfield DW, Tajvidi M, Cai Z (2018) Moisture and oxygen barrier properties of cellulose nanomaterial-based films. ACS Sustain Chem Eng 6(1):49–70 (American Chemical Society). https://doi.org/10.1021/acssuschemeng.7b03523

Wang H, Xu K, Ma Y, Liang Y, Zhang H, Chen L (2020) Impact of ultrasonication on the aggregation structure and physicochemical characteristics of sweet potato starch. Ultrason Sonochem 63:104868

Wulandari WT, Rochliadi A, Arcana IM (2016) Nanocellulose prepared by acid hydrolysis of isolated cellulose from sugarcane bagasse. IOP Conf Ser: Mater Sci Eng 107(1):012045. https://doi.org/10.1088/1757-899X/107/1/012045

Xu K, Li Q, Xie L, Shi Z, Su G, Harper D, Tang Z, Zhou J, Du G, Wang S (2022) Novel flexible, strong, thermal-stable, and high-barrier switchgrass-based lignin-containing cellulose nanofibrils/chitosan biocomposites for food packaging. Ind Crops Prod 179:114661. https://doi.org/10.1016/j.indcrop.2022.114661

Yahya EB, Elarbash SS, Bairwan RD, Mohamed MMI, Khan NB, Harlina PW, Abdul Khalil HPS (2023) The role of microorganisms in the isolation of nanocellulose from plant biomass. Forests 14(7):1457 (Multidisciplinary Digital Publishing Institute). https://doi.org/10.3390/f14071457

Youssef AM, El-Sayed SM (2018) Bionanocomposites materials for food packaging applications: concepts and future outlook. Carbohydr Polym 193:19–27. https://doi.org/10.1016/j.carbpol.2018.03.088

Zabihzadeh Khajavi M, Ebrahimi A, Yousefi M, Ahmadi S, Farhoodi M, Mirza Alizadeh A, Taslikh M (2020) Strategies for producing improved oxygen barrier materials appropriate for the food packaging sector. Food Eng Rev 12(3):346–363. https://doi.org/10.1007/S12393-020-09235-Y/METRICS

Zhang L, Yu D, Regenstein JM, Xia W, Dong J (2021) A comprehensive review on natural bioactive films with controlled release characteristics and their applications in foods and pharmaceuticals. Trends Food Sci Technol 112:690–707. https://doi.org/10.1016/j.tifs.2021.03.053

Zulaikha W, Hassan MZ, Ismail Z (2022) Recent development of natural fibre for nanocellulose extraction and application. Mater Today: Proc 66:2265–2273. https://doi.org/10.1016/j.matpr.2022.06.221

Chapter 6
Alginate in Food Packaging: Seaweed-Derived Solutions for Sustainability

I. K. Nishitha, Tintu Jose Manicketh, and Geo Joseph

Abstract The food industry and researchers are exploring biodegradable, renewable materials from natural sources to address the adverse environmental effects of synthetic packaging. The environmental impact of plastic waste is enormous, and the pressure to switch from conventionally polluting petrochemical polymers to sustainable ones is on the rise. The demand for innovative and healthier products has risen as a result of worldwide progress and growing knowledge of the health effects of food. Many natural polysaccharides possess valuable properties such as enhancing flavor, texture, shelf life, storage stability, water retention, and improving oil absorption in fried foods, making them highly suitable for food packaging. Among these, alginate stands out as a key polysaccharide. Alginate-derived food packaging materials have immense potential in the food industry due to their natural biodegradability, compatibility with the ecosystem, and capacity to prolong the shelf life of a variety of food products. Moreover, the alginates can have enhanced functionalities through various techniques. However, further study and advancement are required to improve their features and boost production for widespread use. The present chapter focuses on the alginate sources, extraction techniques, and alginate properties relevant to food packaging.

Keywords Alginates · Biodegradable · Biopolymers · Food packaging · Seaweed · Sustainability

I. K. Nishitha · T. J. Manicketh (✉)
Department of Botany, St. Teresa's College (Autonomous), Cochin, Kerala, India
e-mail: tintujose@tereasas.ac.in

I. K. Nishitha
e-mail: iknishitha@teresas.ac.in

G. Joseph
Department of Botany, Carmel College (Autonomous), Mala, Thrissur, Kerala, India
e-mail: geo@carmelcollegemala.ac.in

S. Roy and M. Shahid (eds.), *Biomaterials for Sustainable Food Packaging*,
Environmental Footprints and Eco-design of Products and Processes,
https://doi.org/10.1007/978-3-032-08027-1_6

6.1 Introduction

Food packaging is essential for maintaining the quality, safety, and shelf life of the food while ensuring its portability and protection from external factors (Theagarajan et al. 2019). Since plastics are most commonly used to package food, plastic pollution is one of the biggest risks to both the environment and human health (Saravanakumar et al. 2020). Plastics utilized in food packaging applications are regarded as "environmentally unfriendly materials" because they generate a lot of waste that enters the marine and terrestrial ecosystem and has a detrimental effect on wildlife and marine life (Andrady, 2011). Plastics accumulate in the environment because of their very slow rates of breakdown. These plastics decompose over time to form microplastics, which are far more persistent and pose more serious health risks to humans and other living things (Boone et al. 2023). Furthermore, a large amount of energy is needed to produce plastics, which leads to the release of greenhouse gases. Food waste, plastic pollution from packaging, and foodborne illness are major worldwide concerns that highlight the need for sustainable, cutting-edge food packaging solutions (Fengwei, 2024). Nowadays, people are more concerned about lowering the use of plastics and are now focusing on using natural food packaging alternatives that are safe for both the environment and human wellness (Huang et al. 2022).

An efficient, eco-friendly, and viable alternative to conventional food packaging materials is biopolymers (Metha et al. 2024). Biomaterials suitable for food packaging can be derived from various renewable resources. These consist of proteins such as whey, wheat, maize, and soy proteins; lipid compounds like waxes and triglycerides; and carbohydrates such as starch and its byproducts, carrageenan, and alginates (Embuscado and Huber 2009). Biopolymers exhibit unique qualities such as being biodegradable, edible, antioxidant, biologically compatible, and ecologically sustainable. Furthermore, the food's quality is consistently maintained in its packaging (Farris et al. 2009; Hoffman 2002; Rhim and Wang 2013). With rising consumer demand for eco-friendly alternatives and increasing environmental consciousness, the advent of biopolymer-based packaging enables brands to project an eco-friendlier image (González-López et al. 2023).

Alginate, a biopolymer made from seaweed and considered a "blue carbon", has recently attracted a lot of interest as an alternative for conventional polymers based on petroleum. Due to its widespread availability, seaweed does not compete with the quantity of land required for crop cultivation. Furthermore, growing seaweed reduces greenhouse gas emissions worldwide by sequestering CO_2. Alginate is non-toxic, biodegradable, and biocompatible, and it has exceptional gelation qualities. As a result, bulk alginate materials have demonstrated great promise for use in a variety of fields, including membrane separation, environmental cleanup, controlled release of nutrients or bioactive substances, food packaging, and tissue engineering (Xie et al. 2024). Longer shelf life is one of the many advantages of alginate-based food packaging for a variety of food items, such as dairy, meat, fish, and poultry. They were initially used as a protective layer to control respiration rate on perishable fresh produce (Kontominas 2020).

6.2 Alginate and Its Properties

Alginate, an anionic polysaccharide, occur in nature obtained from brown seaweeds. Chlorophylls a and c, which are coated in carotenes and xanthophylls, make up the pigments found in these algae. It is phycoxanthin (xanthophyll) that gives them their brown hue (Abka-khajouei et al. 2022).

Alginate can be commercially extracted from brown algae such as *Laminaria hyperborean, Macrocystis pyrifera, Ascophyllum nodosum, Eclonia maxima, Lesonia negrescens, Sargassum* sp. (Pawar and Edgar 2012; Senturk Parreidt et al. 2018; Bi et al. 2022). Alginate can also be extracted from other types of algae, including *Fucus vesiculosus* and *Macrocystis pyrifera.* Alginate is synthesized from certain bacteria, such as *Pseudomonas* and *Azotobacter*, for use in industry and research. Some fungi, such as *Saprolegnia* sp., can also synthesise alginate. However, this is less common (Xie et al. 2024). Alginate is primarily extracted by exploiting its high solubility in alkaline solutions and limited solubility in water. Sodium carbonate solution is employed to dissolve alginate from the algae's cell walls, after which a pH adjustment with acid facilitates its precipitation (Bi et al. 2022).

Alginate is made up of β-D-mannuronic acid (M) and α-L-guluronic acid (G), joined together by 1–4 bonds. These monomers are organized into blocks of alternating (MG), homopolymeric (MM or GG), or random sequences, which significantly affect the physical and chemical characteristics of the alginate. Alginate units are arranged at random in a linear chain with varying amounts of G and M, contingent on the source and treatment (Bannikova et al. 2018). The physical and chemical properties of alginate are significantly influenced by the relative abundance, distribution, and chemical changes of G and M units, which vary in amount and arrangement among seaweed species (Rhein-Knudsen et al. 2017). G-enriched alginate exhibits a stiff molecular structure, whereas M-rich alginate has a flexible structure (Xie et al. 2024). It is well known that alginate can create hydrogels when divalent cations are present. Alginate can create a gel matrix when combined with divalent cations, particularly Ca^{2+}. These characteristics make it possible to employ this polysaccharide to regulate the release of certain food ingredients, bioactive substances, and medicinal ingredients for a variety of goods (Bi et al. 2022).

Alginate is a very good adhesive substance for food surfaces because of its mucoadhesive qualities. It is a linear natural polymer that is non-toxic, environmentally benign, and does not harm public health. It can be beneficial in the food industry due to its firmness, gelling, stabilising, and water retention qualities as well as the elastic properties of the tissue and the sturdiness (Tammina et al. 2025). Nevertheless, they are extremely susceptible to water and oxygen due to their porous nature (Xiao et al. 2014). Alginate is very compatible with other biopolymers due to its powerful interactions between the hydrophilic chemical groups, especially bonding through hydrogen and electrostatic pull (Xie et al. 2024).

Alginate has exceptional gas and moisture-repellent properties. By creating a barrier around food products, alginate helps to maintain freshness and increases the shelf life. This feature reduces the wastage of food and enhances the total quality of the packaging, making it particularly beneficial for perishables (Tammina et al. 2025). The key advantage of alginate is its molding ability, which enables it to produce innovative packaging solutions according to certain specifications (Kaidi et al. 2022; Metha et al. 2024). Moreover, alginate's capacity to crosslink with different cations helps to produce hydrogels with precise mechanical qualities very rapidly. The longevity of food products can be increased by combining them with other biopolymeric materials to produce composite films with better functional qualities. Because of its low price, high nutritional content, sustainability, versatility and flexibility, as well as its low oxygen permeability, biodegradability and biocompatibility, alginate is highly preferable for creating active and intelligent food packaging solutions (Tammina et al. 2025).

Alginate has been approved by Food and Drug Administration of the United States and is widely employed in foods like desserts, sauces, and beverages. It has been discovered that coatings and films made of alginate enhance the mechanical and sensory qualities of a variety of foods (Heydari et al. 2015; Şen et al. 2015). The two commercially used alginates are calcium and sodium alginate (Hasnain et al. 2020).

6.3 Extraction and Purification of Alginates

Extraction involves the diffusion of a solvent into a matrix, facilitating the dissolution and subsequent segregation of bioactive compounds. This process typically comprises pre-treatment, cellular disruption, and the application of specific extraction techniques, which can be either conventional or novel. Novel methods are often favored due to their enhanced eco-friendliness and accelerated extraction kinetics (Ummat et al. 2021).

6.3.1 Pre-treatments

The cleaned algae are dried, powdered, and then treated with different solvents like ethanol, or combinations of solvents like methanol: chloroform: water (Behfar et al. 2022) or acetone: methanol: chloroform (Benslima et al. 2021). This process removes phenols, proteins, lipids, and pigments, ensuring the extraction of pure alginate, and removing non-targeted substances such as polyphenols, mannitol, and other readily extracted polysaccharides such as laminarins, fucoidan, as well as other low molecular weight substances. The Fenton reaction, used as a pre-treatment method, enhanced alginate extraction yield by 26% while reducing water consumption by 40% (Gao et al. 2018).

6.3.2 *Conventional Extraction*

6.3.2.1 Acid Treatment

In the first step, salts, organic solvent residue from the pre-treatment, non-target substances and low molecular weight compounds are eliminated by acid treatment. Acid treatments are a key step in alginate extraction, typically lasting from 30 min to 24 h, with duration and efficiency subjected to parameters such as temperature and acid content. These treatments enhance extraction by breaking down the cell wall, which makes it easier for the polysaccharides to be released. Lorbeer et al. (2015) found that temperature, treatment duration, and pH all significantly correlate, with lower pH levels producing higher extraction efficiencies. Optimal conditions for a 45% yield include a temperature of 55 °C for 160 min at a pH of 1. However, excessively low pH reportedly hydrolyzes alginate, resulting in products with large molecular weight and viscosity that are more unstable as well as less appropriate for industrial use.

6.3.2.2 Alkaline Extraction

Follows treatment of acid, with the medium adjusted to a pH of 9–10, typically using sodium carbonate (Na_2CO_3), though alternatives like NaOH, $NaHCO_3$, or NaCl may be employed (Bojorges et al. 2022; Yan et al. 2020; Davis et al. 2004). Water-soluble sodium alginate is created by this technique from the water-insoluble alginic acid found in the cell wall of algae. It is subsequently solidified, refined, and dried out. Temperature, alkaline concentration, time, pH, and their interactions all affect how effective this process is.

6.3.2.3 Precipitation

Three primary processes lead to alginate precipitating: (1) alcohol precipitating, which mostly uses ethanol; (2) calcium chloride ($CaCl_2$) precipitating; and (3) hydrochloric acid (HCl) precipitating. The alcohol method is the most widely used for the precipitation of sodium alginate due to its simplicity, high yield, and favourable rheological qualities. The $CaCl_2$ method produces calcium alginate, which is converted to alginic acid and can be neutralised into commercial alginate forms. The HCl route directly produces alginic acid but can degrade the polymer chain. Studies have found the ethanol route to be the most efficient, while the $CaCl_2$ and HCl routes often result in alginates with lower molecular weight or weak mechanical properties (Bojorges et al. 2023).

6.3.2.4 Purification and Drying

Alginate precipitates to create a solid, fiber-like paste, which is then dried and crushed to the appropriate size. In industrial settings, the drying process is typically done in a fluidised bed, followed by grinding with a turbine or a stationary hammer mill. For use in pharmaceutical or biomedical applications, alginate must pass more stringent purification procedures to get rid of biological and inorganic contaminants. In the literature, various purifying methods have been introduced (Bojorges et al. 2023).

6.3.3 Novel Extraction Methods

Novel extraction methods support sustainable development while improving the quality and functionality of alginate for diverse applications. These methods are environment-friendly, require little solvent, and have a high extraction rate.

6.3.3.1 Ultrasound-Aided Extraction (UAE)

Using ultrasonic waves to create cavitation, disrupt cell walls, and improve solvent absorption. In comparison to other extraction methods, it is an environmentally benign method because it uses little amounts of solvents, typically has quick extraction rates, is manageable, and can be scaled up industrially at a reasonable cost (Bojorges et al. 2023).

6.3.3.2 Microwave-Assisted Extraction (MAE)

Administering non-ionizing electromagnetic waves at different frequencies between 300 MHz and 300 GHz and wavelengths between 1 mm and 1 m is the basic principle of MAE, which disrupts or produces changes in the structure of the biomass-treated cell walls. Microwaves generate vibrations that raise the temperature of intracellular fluids. The internal heating leads to the breakdown of the cell wall, resulting in the emergence of compounds into the solvent (Gomez et al. 2020).

6.3.3.3 Pressurized Liquid Extraction (PLE)

Employs high-pressure conditions to enhance solvent diffusion and mass transfer. The technique relies on applying high pressures (1–20 MPa) and high temperatures (20–200 °C) to the solvent to maintain its liquid condition (Gullón et al. 2020).

6.3.3.4 Enzyme-Assisted Extraction

This method uses enzymes like cellulases and proteases, which causes the cell wall matrix disruption, releasing alginates. This method is advantageous as it can operate under mild conditions while preserving the quality of the extracted alginate. Alginate produced by EAE often has a lesser molecular size and an improved porous structure, which increases its antioxidant properties and qualifies it for use in biomedical fields (Bojorges et al. 2023).

6.4 Functional Properties and Enhancement Techniques of Alginates for Food Packaging

6.4.1 Edible Films and Coat-Forming Properties

The natural covering of food products can be strengthened or replaced with edible films and coatings, which are thin layers that are typically less than 0.3 mm thick. These can be removed before eating or consumed with the food (Guilbert et al. 1995; Pavlath and Orts 2009). Edible packaging consists of layers between sections or ingredients of the heterogeneous products or a thin layer on top of the food. These are designed to meet the particular requirements of the product and take into account the modifications that might take place through manufacturing, shipping, and storage. Alginate is frequently used in biofilms and coatings on account of its film-forming ability, biodegradability, and compatibility with various functional additives. The terms "edible coating" and "edible film" are often used synonymously, but they differ in how they are applied to food. Edible films are solid sheets that can be cut and wrapped around food, while edible coatings are directly applied to form a thin layer on the surface of the food item. Even when made from the same material, films and coatings can have distinct characteristics (Kramer 2009; Sánchez-Ortega et al. 2014). Biofilms and coatings made with alginate can be significantly enhanced through cross-linking technique.

6.4.1.1 Cross-Linking

One common technique of enhancing properties of alginate for varied applications in food packaging, is crosslinking. Crosslinking causes several changes that improve its durability for packaging applications by producing a material that is tough and resistant. Coatings and film food packaging were produced due to the crosslinking process, which involved the interaction of alginate with Ca^{2+} ions. These films and coatings have shown efficiency in prolonging the shelf life and improving the quality of food items (Senturk Parreidt et al. 2019). When crosslinking alginate for food packaging, it's important to take into account certain packaging requirements, including mechanical strength, biodegradability, and barrier qualities (Aguero et al. 2021). The procedure creates a hydrogel that is appropriate for packing by adding gelling ions to the alginate solution. This can be accomplished in two different ways of gelation, both internal and external gelation. This ensures that the finished product is both efficient and biodegradable, thereby ensuring an environmentally sustainable solution. In the traditional external gelation procedure, the alginate solution is exposed to a gelling ion solution. The carboxyl groups of the guluronic acid residues then react with the Ca^{2+} ions by ion diffusion to produce an irreversible hydrogel. During internal gelation, alginate solutions are treated with an insoluble source of gelling ions. Also, the addition of organic acids or the quick hydrolysis of lactones liberates the gelling ions, thereby lowering the pH (Metha et al. 2024).

6.4.2 *Water Absorption and Barrier Properties*

Alginates possess distinct hygroscopic properties, which are essential for food packaging. These films maintain the moisture content of food by absorbing water up to 200–300 times their weight (Senturk Parreidt et al. 2018). Nevertheless, swelling brought on by such absorption may result in increased water vapour uptake and transmission, which could affect the film's mechanical qualities (Gubitosa et al. 2023). Alginate films typically have poor water resistance, necessitating the development of various methods to enhance their water barrier properties (Akshaya and Nathaniel 2024). One common approach is cross-linking alginate with cations. Divalent and trivalent metal cations such as calcium, magnesium, manganese, aluminium, and iron help alginate solutions to form gels. One major drawback of using ionically cross-linked alginate is that it can be unstable in the presence of physiological fluids or solutions containing salts. In such environments, the divalent ions (like calcium) that hold the alginate network together can be replaced by other ions present in the solution, which can weaken or disrupt the gel structure and cause a gradual loss of mechanical strength (Vallée et al. 2009).

The water resistance of alginate can be improved by attaching hydrophobic groups onto its backbone. This modification enhances the molecular flexibility of alginate by reducing intramolecular hydrogen bonding, which in turn alters its activity. In a 2024 study, Paramabhorn et al. developed several hydrophobic alginate derivatives by chemically linking primary amines to alginate through an amidation process. The effectiveness of this modification depended on the length of the amine chain—a minimum of 12 carbon atoms was necessary to impart hydrophobicity to the alginate. When applied as surface coatings, these modified alginates showed strong water-repellent behavior, with paper proving to be the most effective surface.

When forming alginate-based biofilms, calcium ions play a key role by helping the alginate molecules stick together and form a gel. Calcium ions bind strongly with specific parts of alginate called the guluronate blocks. These blocks on different chains come into close alignment, allowing the calcium ions to act like bridges between them. The gel network forms when the calcium ions fit into spaces between two adjacent polymer chains that have guluronate blocks arranged in a helical shape. This structural alignment is known as the egg-box structure (Fig. 6.1) (Banks and Greenwood 1975; Hu et al. 2021). Therefore, alginate films have high water vapor permeability, impermeability to oils, and barrier property to oxygen (Xiao et al. 2014).

6.4.3 Enhancing Flexibility and Tear Resistance Using Plasticizers

Plasticizers, which typically blend well with the polymer and are non-volatile, can be used to make the polymer more flexible and enhance the mechanical characteristics of biodegradable packaging (Senturk Parreidt et al. 2018). The purpose of plasticizers is to increase the flexibility of the polymer. Plasticizers enable this by increasing the space between polymer chains and reducing the tight forces holding the chains together. This results in less brittleness, better resistance to tearing or impact, and improved flow when used in coatings. Ideally, plasticizers should also have a high boiling point so they don't evaporate easily, and they should be able to modify the material's physical and mechanical properties effectively and should have similar solubility characteristics as the polymer in the solvent system (Senturk Parreidt et al. 2018). Glycerol, sorbitol, acetylated monoglyceride, polyethylene glycol, sucrose, etc., have been employed as plasticizers in food coating research. In general, the addition of hydrophilic plasticizers increases water vapour permeability (WVP) and affects the mechanical characteristics of the coatings (Donhowe and Fennema 1994).

6.4.4 Adhesion and Uniform Coating Using Surfactants

Substances with low surface energy repel things, making the surface water resistant or less likely to stick with other materials. To overcome this, surfactants are incorporated into the coating material. Also, surfactants enable the coatings to spread more evenly by reducing the surface tension, allowing them to cover surfaces more smoothly and uniformly. Coat formulations are designed considering the hydrophobicity of the food, surface tension of the coating material, and its spreading coefficient (Senturk Parreidt et al. 2018).

6.4.5 Antimicrobial and Antioxidant Properties

Packaging with antimicrobial potential has gained significant interest in the packaging industry because of its ability to improve food safety, maintain quality, and

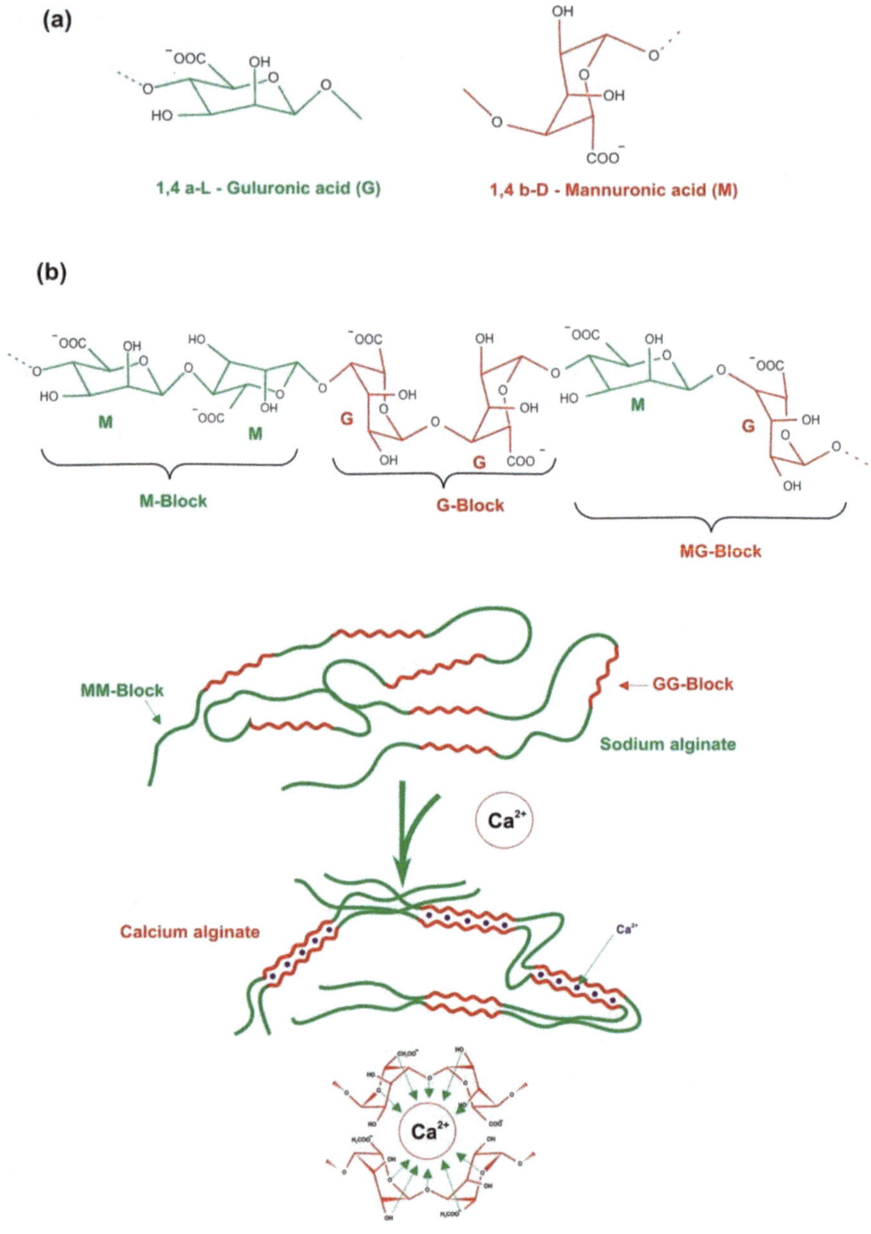

Fig. 6.1 Structure of alginate and egg box model for alginate gelation. *Source* Bojorges et al. (2023) with some modifications

extend shelf life. It is a great alternative method to lessen, limit, or prevent pathogenic and spoilage microorganisms in food products. Bacteriocins, essential oils, plant extracts, enzymes, organic acids and inorganic/metallic nanoparticles, etc. were employed as antimicrobial agents reviewed by Senturk Parriedt et al. (2018). Bioactive substances like flavonoids and phenolic compounds give alginate a naturally occurring antioxidant potential. These components can counteract free radicals and have antioxidant properties (Vengatesan et al. 2021). Functional groups present in alginate can combine with metal ions like iron and copper to form coordination complexes. These metal ions frequently serve as catalysts for food oxidative processes. Alginate-based packaging can help slow down food oxidation by effectively binding certain ions, which in turn enhances the preservation and extends the freshness of the food (Dawei and Jun 2022).

6.5 Innovations and Advances in Alginate Packaging

Alginate, a naturally derived polysaccharide obtained from brown seaweed, has gained increasing attention for its applications in sustainable and edible packaging systems. Its molecular structure, high in hydroxyl and carboxyl functional groups, enables strong chelation with divalent metal ions, which enhances its efficiency in binding heavy metals. In addition to its biodegradability and non-toxic nature, this underlines alginate's role as a suitable substance for environmentally friendly packaging materials (Zhang and Chen 2023). The use of essential oils (EOs) in alginate films is a recent development. EOs, which come from a wide range of plants and animals—such as lemongrass, oregano, thyme, rosemary, clove, cinnamon, sage, basil, garlic, and calendula—enhance the mechanical and functional qualities of packaging films in addition to acting as natural antimicrobial and antioxidant agents. They enhance thermal stability and provide effective barriers against UV radiation, both of which are critical for maintaining food quality and safety during storage and distribution (Saba et al. 2023). The incorporation of bioactive agents significantly broadens the functional scope of alginate-based films, enabling their utilization across diverse segments of the packaging sector. This approach supports current industry trends toward minimizing the use of synthetic additives and fossil fuel-derived polymers, thereby delivering a dual advantage of ecological sustainability and enhanced preservation efficacy for packaged goods.

However, while alginate-based films hold significant promise, there are economic and technical barriers that must be addressed before their widespread consumption. One of the most pressing concerns lies in the cost-effectiveness of producing bio-based films at an industrial scale. The extraction and purification of alginate, followed by the film-forming processes, can be capital- and energy-intensive. Additionally, the inclusion of natural additives, like essential oils or nanoparticles, can further increase production costs. To make these materials commercially viable, it is essential to focus on optimizing formulation techniques and production protocols. Strategies such as adjusting the concentration of active compounds, using cost-effective cross-linking agents (e.g., calcium chloride or natural polyphenols), and implementing efficient drying and casting technologies could significantly reduce manufacturing costs while maintaining product performance (Saba et al. 2023). The application of natural cross-linkers not only improves film stability and moisture resistance but also reduces the reliance on synthetic agents, making the process greener. Moreover, innovations in nanotechnology are opening up new avenues in the design of reinforced alginate-based composites, allowing for enhanced mechanical strength and barrier functionality with minimal material usage. These improvements are vital to creating a cost-effective solution that meets consumer expectations for performance and environmental impact. Policy incentives and increasing consumer awareness about the ecological benefits of biodegradable packaging could also help stimulate demand, further supporting the transition from conventional plastics to bio-based alternatives.

Functionally, alginate-based packaging exhibits numerous advantageous properties that make it highly appropriate for use in packaging fresh food products. Notably, their natural transparency allows consumers to easily inspect the visual quality of packaged goods, such as fruits and vegetables, without the need for opening the package—contributing to reduced food waste and better consumer satisfaction (Wang et al. 2025). This feature is especially important in retail environments where visual appeal can strongly influence purchasing decisions. Furthermore, recent studies have shown that combining alginate with other biopolymers or surface coatings, such as in tertiary blends with paper substrates, results in significantly improved packaging characteristics. These composite materials exhibit higher resistance to moisture and oil penetration, greater tensile strength, and enhanced hydrophobicity, making them suitable for diverse food packaging applications including greasy and moist foods (Shankar and Rhim 2018). Another major breakthrough in this domain is the development of polysaccharide-based nanocomposites. These next-generation materials integrate nano-fillers—such as cellulose nanocrystals, chitosan nanoparticles, or clay minerals—into the alginate matrix, leading to superior mechanical performance, heat resistance, and barrier properties. Such nanocomposites are particularly effective in preserving the safety and freshness of food products over longer storage periods, thereby supporting extended shelf life and food security (Yosri et al. 2024). Their optical clarity and antimicrobial potential further enhance their appeal as a smart packaging solution. As consumer demand shifts towards sustainable and high-performance packaging, these innovative materials offer a pathway to bridge the gap between environmental responsibility and industrial functionality. Ultimately, the adoption of alginate and its composites signals a transformative step toward reducing plastic dependency, ensuring food safety, and promoting a circular economy within the packaging sector.

6.6 Future Prospects of Alginates in Food Packaging

Antibacterial composite alginate packing films are opening new avenues in packaging by enhancing shelf life while being biodegradable, offering a sustainable alternative to traditional materials (Zhang and Chen 2023). In the food packaging industry, alginate-based biofilms ensure food safety by providing a protective barrier against contaminants and extending the freshness of packaged goods. As an edible packaging material, alginate not only preserves food but also serves as a dietary fiber source, aiding in human digestion and regulating lipid and glucose metabolism. Its water vapor permeability and water contact angle properties offer effective protection against rehydration and dehydration, maintaining the quality of food materials during storage and transport (Saba et al. 2023). These properties make alginate-based biofilms particularly effective in preserving the nutritional quality and safety of post-harvest fruits and vegetables (Yobal et al. 2021).

Innovative developments, such as incorporating copper oxide nanomaterials into biopolymers, have expanded the potential of alginate-based materials to include electronic food packaging with temperature and freshness indicators, enhancing consumer convenience and food monitoring (Ragab et al. 2024). Furthermore, treating alginate polymers with biobased acids like citric and tartaric acids has been shown to improve tensile strength, chemical resistance, and thermal properties, making them more robust and versatile for various packaging applications (Singh et al. 2020). These advancements demonstrate the versatility of alginate in meeting modern packaging needs by combining sustainability with functionality. With its biodegradability, nutritional benefits, and potential for technological integration, alginate-based materials represent a promising solution for reducing environmental impact while ensuring packaged food products' safety, quality, and longevity.

6.7 Conclusion

Food packaging is vital for preserving the quality, extending the duration of storage, and reducing waste. Alginate, a polysaccharide that occurs naturally derived from seaweed, holds potential as a renewable packaging material in addition to its variety of applications. Alginate serves as a feasible and sustainable packaging solution due to its natural biodegradable properties, compatibility with biological systems, and ability to prolong the life of a wide variety of products. Alginate-based solutions help to cut down plastic waste and promote a more environmentally responsible approach, especially in light of the rising demand for sustainable packaging.

Declaration During the preparation of this work, the authors used the AI tool to improve the language and readability of the content. After using these AI tools, the authors reviewed and edited the content as needed and take full responsibility for the content of the publication.

References

Abka-khajouei R, Tounsi L, Shahabi N, Patel AK, Abdelkafi S, Michaud P (2022) Structures, properties and applications of alginates. Mar Drugs 20(6):364. https://doi.org/10.3390/md20060364

Aguero L, Alpdagtas S, Ilhan E, Zaldivar-Silva D, Gunduz O (2021) Functional role of crosslinking in alginate scaffold for drug delivery and tissue engineering: a review. Eur Polym J 160:110807. https://doi.org/10.1016/j.eurpolymj.2021.110807

Akshaya S, Nathanael AJ (2024) A Review on hydrophobically associated alginates: approaches and applications. ACS Omega 9(4):4246–4262. https://doi.org/10.1021/acsomega.3c08619

Andrady AL (2011) Microplastics in the marine environment. Mar Pollut Bull 62(8):1596–1605. https://doi.org/10.1016/j.marpolbul.2011.05.030

Banks W, Greenwood CT (1975) Starch and its components. Edinburgh University Press, Edinburgh

Bannikova A, Evteev A, Pankin K, Evdokimov I, Kasapis S (2018) Microencapsulation of fish oil with alginate: in-vitro evaluation and controlled release. LWT-Food Sci Technol 90:310–315. https://doi.org/10.1016/j.lwt.2017.12.045

Behfar A, Aqajari N, Shushizadeh MR, Ramezani Z, Ghatrami ER (2022) Production of alginate from Persian gulf *Sargassum angustifolium* seaweeds: novel extraction and characterization methods. J Nat Pharm Prod 17(1):e106011. https://doi.org/10.5812/jjnpp.106011

Benslima A, Sellimi S, Hamdi M, Nasri R, Jridi M, Cot D, Li S, Nasri M, Zouari N (2021) The brown seaweed *Cystoseira schiffneri* as a source of sodium alginate: chemical and structural characterization, and antioxidant activities. Food Biosci 40:100873. https://doi.org/10.1016/j.fbio.2020.100873

Bi D, Yang X, Yao L, Hu Z, Li H, Xu X, Lu J (2022) Potential food and nutraceutical applications of alginate: a review. Mar Drugs 20(9):564. https://doi.org/10.3390/md20090564

Bojorges H, Fabra MJ, López-Rubio A, Martínez-Abad A (2022) Alginate industrial waste streams as a promising source of value-added compounds. Sci Total Environ 838:156394. https://doi.org/10.1016/j.scitotenv.2022.156394

Bojorges H, López-Rubio A, Martínez-Abad A, Fabra M (2023) Overview of alginate extraction processes: impact on alginate molecular structure and techno-functional properties. Trends Food Sci Technol 140:104142. https://doi.org/10.1016/j.tifs.2023.104142

Boone L, Préat N, Nhu TT, Fiordelisi F, Guillard V, Blanckaert M, Dewulf J (2023) Environmental performance of plastic food packaging: life cycle assessment extended with costs on marine ecosystem services. Sci Total Environ 894:164781. https://doi.org/10.1016/j.scitotenv.2023.164781

Davis TA, Ramirez M, Mucci A, Larsen B (2004) Extraction, isolation and cadmium binding of alginate from *Sargassum* spp. J Appl Phycol 16(4):275–284. https://doi.org/10.1023/B:JAPH.0000047779.31105.ec

Dawei Y, Jun L (2022) Recent advances on the development of food packaging films based on citrus processing wastes: a review. J Agric Food Res 9:100316. https://doi.org/10.1016/j.jafr.2022.100316

Donhowe IG, Fennema O (1994) Edible films and coatings: characteristics, formation, definitions, and testing methods. In: Krochta JM, Baldwin EA, Nisperos-Carriedo MO (eds) Edible coatings and films to improve food quality. Technomic Publ Co, Lancaster, NH, USA, pp 1–24

Embuscado ME, Huber KC (2009) Edible films and coatings for food applications. Springer, Berlin/Heidelberg, Germany

Farris S, Schaich KM, Liu L, Piergiovanni L, Yam KL (2009) Development of polyion-complex hydrogels as an alternative approach for the production of bio-based polymers for food packaging applications: a review. Trends Food Sci Technol 20(8):316–332. https://doi.org/10.1016/j.tifs.2009.04.003

Fengwei X (2024) Alginate-based nanocomposites for food preservation: recent progress showcasing heightened material properties and functionalities. Adv Nanocomp 1(1):248–274. https://doi.org/10.1016/j.adna.2024.07.002

Gao F, Liu X, Chen W, Guo W, Chen L, Li D (2018) Hydroxyl radical pretreatment for low-viscosity sodium alginate production from brown seaweed. Algal Res 34:191–197. https://doi.org/10.1016/j.algal.2018.07.017

González-López ME, Calva-Estrada SDJ, Gradilla-Hernández MS, Barajas-Álvarez P (2023) Current trends in biopolymers for food packaging: a review. Front Sustain Food Syst 7:1225371. https://doi.org/10.3389/fsufs.2023.1225371

Gubitosa J, Rizzi V, Marasciulo C, Maggi F, Caprioli G, Mustafa AM, Fini P, De Vietro N, Aresta AM, Cosma P (2023) Realizing eco-friendly water-resistant sodium-alginate-based films blended with a polyphenolic aqueous extract from grape pomace waste for potential food packaging applications. Int J Mol Sci 24:11462. https://doi.org/10.3390/ijms241411462

Gullón B, Gagaoua M, Barba FJ, Gullón P, Zhang W, Lorenzo JM (2020) Seaweeds as promising resource of bioactive compounds: overview of novel extraction strategies and design of tailored meat products. Trends Food Sci Technol 100:1–18. https://doi.org/10.1016/j.tifs.2020.03.039

Guilbert S, Gontard N, Cuq B (1995) Technology and applications of edible protective films. Packag Technol Sci 8(6):339–346. https://doi.org/10.1002/pts.2770080607

Hasnain MS, Jameel E, Mohanta B, Dhara AK, Alkahtani S, Nayak AK (2020) Alginates: sources, structure, and properties. In: Nayak AK, Hasnain, MS (eds) Alginates in drug delivery. Academic Press, pp 1–17. https://doi.org/10.1016/b978-0-12-817640-5.00001-7

Heydari R, Bavandi S, Javadian SR (2015) Effect of sodium alginate coating enriched with horsemint (*Mentha longifolia*) essential oil on the quality of bighead carp fillets during storage at 4 °C. Food Sci Nutr 3:188–194. https://doi.org/10.1002/fsn3.202

Hoffman AF (2002) Hydrogels for biomedical applications. Adv Drug Deliv Rev 54(1):3–12. https://doi.org/10.1016/S0169-409X(01)00239-3

Hu C, Lu W, Mata A, Nishinari K, Fang Y (2021) Ions-induced gelation of alginate: mechanisms and applications. Int J Biol Macromol 177:578–588. https://doi.org/10.1016/j.ijbiomac.2021.02.086

Huang S, Wang H, Ahmad W, Ahmad A, Ivanovich Vatin N, Mohamed AM, Deifalla AF, Mehmood I (2022) Plastic Waste management strategies and their environmental aspects: a scientometric analysis and comprehensive review. Int J Environ Res Pub Health 19(8):4556. https://doi.org/10.3390/ijerph19084556

Kaidi S, Bentiss F, Jama C, Khaya K, Belattmania Z, Reani A, Sabour B (2022) Isolation and structural characterization of Alginates from the kelp species *Laminaria ochroleuca* and *Saccorhiza polyschides* from the Atlantic Coast of Morocco. Colloids Interfaces 6(4):51. https://doi.org/10.3390/colloids6040051

Kontominas MG (2020) Use of alginates as food packaging materials. Foods 9(10):1440. https://doi.org/10.3390/foods9101440

Kramer ME (2009) Structure and function of starch-based edible films and coatings. In: Huber KC, Embuscado ME (eds) Edible films and coatings for food applications. Springer, New York, NY, USA, pp 113–134. https://doi.org/10.1007/978-0-387-92824-1_4

Lorbeer AJ, Lahnstein J, Bulone V, Nguyen T, Zhang W (2015) Multiple response optimization of the acidic treatment of the brown alga *Ecklonia radiata* for the sequential extraction of fucoidan and alginate. Bioresour Technol 197:302–309. https://doi.org/10.1016/j.biortech.2015.08.103

Gomez LP, Alvarez C, Zhao M, Tiwari U, Curtin J, Garcia-Vaquero M, Tiwari BK (2020) Innovative processing strategies and technologies to obtain hydrocolloids from macroalgae for food applications. Carbohydr Polym 248:116784. https://doi.org/10.1016/j.carbpol.2020.116784

Metha C, Pawar S, Vasanti S (2024) Recent advancements in alginate-based films for active food packaging applications. Sustain Food Technol 2:1246–1265. https://doi.org/10.1039/D3FB00216K

Paramabhorn T, Hannah SL, Christopher JC (2024) Developing hydrophobic alginate-amine derivatives as a coating for food packaging. ACS Appl Polym Mater 6(17):10669–10677. https://doi.org/10.1021/acsapm.4c01816

Pavlath AE, Orts W (2009) Edible films and coatings: why, what, and how? In: Huber KC, Embuscado ME (eds) Edible films and coatings for food applications. Springer, New York, USA, pp 1–23. https://doi.org/10.1007/978-0-387-92824-1_1

Pawar SN, Edgar KJ (2012) Alginate derivatization: a review of chemistry, properties and applications. Biomaterials 33(11):3279–3305. https://doi.org/10.1016/j.biomaterials.2012.01.007

Ragab HM, Diab NS, Aziz RA et al (2024) Development and characterization of HPMC/NaAlg-CuO bio-nanocomposites: enhanced optical, electrical, and antibacterial properties for sustainable packaging applications. Int J Biol Macromol 283(2):137774. https://doi.org/10.1016/j.ijbiomac.2024.137774

Rhein-Knudsen N, Ale MT, Ajalloueian F, Meyer AS (2017) Characterization of alginates from Ghanaian brown seaweeds: Sargassum spp. and Padina spp. Food Hydrocoll 71:236–244. https://doi.org/10.1016/j.foodhyd.2017.05.016

Rhim JW, Wang LF (2013) Mechanical and water barrier properties of agar/kappa carrageenan/konjac glucomannan ternary blend biohydrogel films. Carbohydr Polym 96(1):71–81. https://doi.org/10.1016/j.carbpol.2013.03.083

Saba E, Deblina B, Swarup R, David JM (2023) Incorporation of essential oils in edible seaweed-based films: a comprehensive review. Trend Food Sci Technol 135:43–56. https://doi.org/10.1016/j.tifs.2023.03.015

Sánchez-Ortega I, García-Almendárez BE, Santos-López EM, Amaro-Reyes A, Barboza-Corona JE, Regalado C (2014) Antimicrobial edible films and coatings for meat and meat products preservation. Sci World J 48935. https://doi.org/10.1155/2014/248935

Saravanakumar K, Sathiyaseelan A, Vijaya A, Mariadoss A, Xiaowen H, Wang M (2020) Physical and bioactivities of biopolymeric films incorporated with cellulose, sodium alginate and copper oxide nanoparticles for food packaging application. Int J Biol Macromol 153:207–214. https://doi.org/10.1016/j.ijbiomac.2020.02.250

Şen F, Uzunsoy İ, Başturk E, Kahraman MV (2015) Antimicrobial agent-free hybrid cationic starch/sodium alginate polyelectrolyte films for food packaging materials. Carbohydr Polym 170:264–270. https://doi.org/10.1016/j.carbpol.2017.04.079

Senturk Parreidt T, Lindner M, Rothkopf I, Schmid M, Müller K (2019) The development of a uniform alginate-based coating for cantaloupe and strawberries and the characterization of water barrier properties. Foods 8(6):203. https://doi.org/10.3390/foods8060203

Senturk Parreidt T, Müller K, Schmid M (2018) Alginate-based edible films and coatings for food packaging applications. Foods 7(10):170. https://doi.org/10.3390/foods7100170

Shankar S, Rhim JW (2018) Antimicrobial wrapping paper coated with a ternary blend of carbohydrates (alginate, carboxymethyl cellulose, carrageenan) and grapefruit seed extract. Carbohydr Polym 196:92–101. https://doi.org/10.1016/j.carbpol.2018.04.128

Singh P, Baisthakur P, Yemul OS (2020) Synthesis, characterization and application of crosslinked alginate as green packaging material. Heliyon 6(1):e03026. https://doi.org/10.1016/j.heliyon.2019.e03026

Tammina SK, Priyadarshi R, Khan A, Manzoor A, Abdel Rahman RSH, Banat F (2025) Recent developments in alginate-based nanocomposite coatings and films for biodegradable food packaging applications. Int J Biol Macromol 295:139480. https://doi.org/10.1016/j.ijbiomac.2025.139480

Theagarajan R, Dutta S, Moses JA, Chinnaswamy A (2019) Alginates for food packaging applications. In: Ahmad S (ed) Alginates: applications in the biomedical and food industries. Scrivener Publishing LLC. https://doi.org/10.1002/9781119487999.ch11

Ummat V, Sivagnanam SP, Rajauria G, O'Donnell C, Tiwari BK (2021) Advances in pre-treatment techniques and green extraction technologies for bioactives from seaweeds. Trend Food Sci Technol 110:90–106. https://doi.org/10.1016/j.tifs.2021.01.018

Vallée F, Müller C, Durand A, Schimchowitsch S, Dellacherie E, Kelche C, Cassel JC, Leonard M (2009) Synthesis and rheological properties of hydrogels based on amphiphilic alginate-amide derivatives. Carbohydr Res 344(2):223–228. https://doi.org/10.1016/j.carres.2008.10.029

Vengatesan M, Rambabu K, Banat F, Mittal V (2021) Natural antioxidants-based edible active food packaging: an overview of current advancements. Food Biosci 43:101251. https://doi.org/10.1016/j.fbio.2021.101251

Wang H, Yang L, Yang Y (2025) A review of sodium alginate-based hydrogels: structure, mechanisms, applications, and perspectives. Int J Biol Macromol 292:139151. https://doi.org/10.1016/j.ijbiomac.2024.139151

Xiao Q, Gu X, Tan S (2014) Drying process of sodium alginate films studied by two dimensional correlation ATR-FTIR spectroscopy. Food Chem 164:179–184. https://doi.org/10.1016/j.foodchem.2014.05.044

Xie F, Gao C, Avérous L (2024) Alginate-based materials: enhancing properties through multiphase formulation design and processing innovation. Mater Sci Eng: R: Rep 159:100799. https://doi.org/10.1016/j.mser.2024.100799

Yan J, Huang Y, Chrisey DB, Jorgensen M, Gibbons A, Sui K, Leigh Herran C, Chai W, Guo Y, Zhang S (2020) New extraction technology and characterization of sodium alginate. IOP Conf Ser: Earth Environ Sci 474(5):052092. https://doi.org/10.1088/1755-1315/474/5/052092

Yobal M, Oliva A., Saucedo-Rivalcoba V, Rivera-Armenta J, Hernández-Ramírez G, Salinas-Ruiz J, Herrera-Corredor A (2021) Evaluation of physicochemical properties of film-based alginate for food packing applications. e-Polym 21(1):82–95. https://doi.org/10.1515/epoly-2021-0011

Yosri N, Khalifa SAM, Attia NF et al (2024) Sustainability in the green engineering of nanocomposites based on marine -derived polysaccharides and collagens: a review. Int J Biol Macromol 274(1):133249. https://doi.org/10.1016/j.ijbiomac.2024.133249

Zhang M, Chen H (2023) Development and characterization of starch-sodium alginate-montmorillonite biodegradable antibacterial films. Int J Biol Macromol 233:123462. https://doi.org/10.1016/j.ijbiomac.2023.123462

Chapter 7
Chitosan in Food Packaging: Harnessing Crustacean Waste

Debabrata Nath, Pratap Kalita, Abhishek Parasar, and Dibyajyoti Das

Abstract Chitosan is a renewable, biodegradable, bioadvective, bioactive, sustainable and versatile biopolymer comprising unique characteristics such as crosslinking, film-forming properties, complexing, chelating, hydrophilicity, crystallinity, ionic conductivity, entrapment, adsorption, high viscosity etc. Crustacean waste contains higher levels of chitin, around 13–42% and deacetylation of chitin leads to chitosan. Chitosan is randomly distributed with β-(1–4) linkage or D units responsible for undergoing structural confirmation under acidic conditions to produce a wide range of functional groups within the chitosan skeleton and adopt several conformations with improved functional and physicochemical properties. Considering all these properties, chitosan is a biopolymer that can be applied in the fabrication of active, intelligent, and edible food packaging films that are used to maintain and improve the conditions of packaged foods by extending their shelf life and monitoring the packed foods. Adding the bioactive compound to the chitosan-based film-forming solution significantly improves the final film barrier and mechanical, thermal, and functional properties with potent antioxidant and antimicrobial activity. Therefore, in this chapter, considering all these factors, this attempt has been made to profile the crustacean waste sources of chitosan along with their extraction procedure and different applicability in food packaging.

Keywords Crustacean waste · Chitosan · Extraction · Food packaging · Functional properties

7.1 Introduction

Food packaging protects the food substances from physical, chemical and biological attacks and maintains the food chain and quality for an extended period. Food packaging is also essential to control the several influencing parameters such as odour,

D. Nath · P. Kalita (✉) · A. Parasar · D. Das
Pratiksha Institute of Pharmaceutical Sciences, Guwahati, Assam, India
e-mail: pratapkalita0@gmail.com

S. Roy and M. Shahid (eds.), *Biomaterials for Sustainable Food Packaging*, Environmental Footprints and Eco-design of Products and Processes,
https://doi.org/10.1007/978-3-032-08027-1_7

colour, humidity, light, pH, temperature, dust, moisture, etc, faced by the packaged food items (Cheng et al. 2022). In recent times, the utilization of conventional plastics has drastically increased due to their good properties, variability, and relative prices. Due to these properties, plastic replaces paper, glass, cartoons, etc. In addition, human lives depend on plastics, from clothing, toys, and cars to electronics, agriculture, healthcare products and packaging materials (Flórez et al. 2022). The serious drawbacks of plastic packaging materials are environmental issues such as air pollution, water pollution, and carbon footprint, and these problems arise due to their non-biodegradable characteristics, leading to resource depletion and massive waste accumulation (Cheng et al. 2022). An earlier report stated that an effort was made to recycle plastic packaging materials to reduce environmental damage and massive carbon dioxide and toxic compounds by incinerating petroleum-based plastic substances (Zhang et al. 2022). Alternatively, it was found that recycling suffers from various issues, such as recycling becoming waste when contamination of the plastic substances occurs with biological products, this is not a favourable approach all the time as it consumes more energy and time and leads to creating a carbon footprint. Considering all these factors, current researchers focus on developing biodegradable packaging materials that can replace plastic petroleum-based packaging materials and reduce pollution (Shlush and Davidovich-Pinhas 2022).

Biodegradable polymers such as chitosan have versatile properties, including antioxidant, antimicrobial, enzyme inhibiting and biodegradable nature that fit to apply in a packaging science field in the fabrication of active and intelligent food packaging materials (Liu et al. 2022). Active food packaging refers to maintaining and improving the conditions of the packaged foods by extending the shelf life and quality, and intelligent food packaging refers to monitoring the conditions or the environment surrounding the packaged foods. Nowadays, the active, intelligent and edible chitosan-based films have been generated by incorporating several bioactive compounds (such as essential oils, phenolic acids, flavonoids, etc.) in order to improve the antimicrobial and antioxidant properties along with improving both mechanical and barrier properties of chitosan films (Flórez et al. 2022). Considering all these properties of chitosan, recent studies have been conducted to obtain chitosan from chitin, which is abundantly available in several natural sources, including crustacean waste.

Chitin is an abundantly available second biopolymer on the earth after cellulose, made from N-acetyl-D-glucosamine units and obtained from natural sources like shrimp, krill, crab, other crustacean shells, insects, and microorganisms such as fungi, algae, and yeasts. Isolated chitin from crustacean waste, squid pens and fungi, a yeast called α-chitin (anti-parallel polymer chains), β-chitin (parallel polymer chains), γ-chitin (combination of α and β structures), respectively (Hajji et al. 2021; Pakizeh et al. 2021). Crustacean waste contains a higher level of chitin, around 13–42%, and proteins and mineral salts consist of 30–40% and 30–50%, respectively, and during the processing for human consumption, between 40 and 50% of the total mass is waste, and 40% of this waste is chitin (Muthu et al. 2021). The deacetylation of chitin leads to chitosan, which consists of co-polymers of D-glucosamine and N-acetyl-D-glucosamine, and the degree of deacetylation depends on these residues of the co-polymers (Pakizeh et al. 2021; Muthu et al. 2021). Chitosan is randomly distributed with β-(1–4) linkage or D units that are responsible for undergoing structural confirmation under acidic conditions to produce a wide range of functional groups within the chitosan skeleton and adopt several conformations with improved functional and physicochemical properties (Reshad et al. 2021).

Crustacean waste-derived chitosan has been used in developing food packaging materials that are used in packing several food items, and the isolated chitosan produced unique physicochemical attributes that make it fit to fabricate the active and intelligent packaging materials (Muthu et al. 2021). The extracted chitosan from natural sources is associated with excellent properties such as complexing, chelating, hydrophilicity, crystallinity, ionic conductivity, weak base (pKa: 6.3), high viscosity, insoluble in water and organic solvents, but soluble in dilute aqueous acid solution, the existence of reactive groups for chemical activation and crosslinking, forms salts with organic and inorganic acids, entrapment, adsorption, film forming ability, cationic biopolymer with high charge density (one positive charge per glucosamine residue) and flocculating agent (Amiri et al. 2022; Jiménez-Gómez and Cecilia 2020). Due to all these versatile characteristics of the chitosan obtained from sustainable crustacean waste, it is becoming a highlighting and functional material in food packaging. Therefore, this book chapter reveals several crustacean waste sources, their functional properties, and food packaging applications.

7.2 Sources, Chitosan Extraction and Its Chemistry

Because of its unique blend of antibacterial, biodegradable and biocompatible qualities, chitosan has become a promising material for food packaging (Pal et al. 2021). This natural polymer derived from chitin has shown promise in addressing important food preservation issues by preserving product quality, lowering microbiological contamination, and increasing shelf life (Liu et al. 2022). Chitosan can interact with negatively charged microbial cell membranes because of its cationic character, which is caused by the protonation of its amino groups in acidic environments (Jiang et al. 2023). Because of its antibacterial activity and ability to block oxygen and moisture, chitosan is a useful ingredient for making protective coatings and films. Furthermore, its eco-friendly and non-toxic qualities complement the growing need for packaging alternatives that are sustainable and biodegradable (Pal et al. 2021). To improve their mechanical strength, flexibility, and functional qualities, chitosan-based films are frequently mixed with additional natural polymers, bioactive substances, or nanoparticles (Liu et al. 2022; Flórez et al. 2022). In order to improve food preservation, these changes are intended to increase the packaging's resistance to water vapour, oxygen permeability, and microbial development (Jiang et al. 2023). In order to create multipurpose packaging materials that satisfy the various needs of the food industry, researchers are still working to optimize the qualities of chitosan for use in commercial food packaging applications. Some potential approaches include adding essential oils, antioxidants, and other bioactive substances (Pal et al. 2021; Liu et al. 2022; Flórez et al. 2022).

7.2.1 Sources

Chitosan is derived from the natural polymer chitin, which is found in many biological sources. These sources, which include crustaceans, fungi, insects, cephalopods, and other marine animals, are grouped according to their origin. The quality, yield, and characteristics of chitosan are strongly influenced by the source of chitin (Pellis et al. 2022).

7.2.1.1 Crustaceans

The most important commercial sources of chitin and chitosan are crustaceans, which include shrimp, crabs, lobsters, and prawns. Depending on the species and processing techniques, the seafood sector produces about 6–8 million tons of crustacean shell waste annually, with chitin contents ranging from 15 to 40% (Yan and Chen 2015; Aung et al. 2018). According to some studies, crab shells normally contain 18–32% chitin (Pambudi et al. 2018), whereas shrimp shells typically include 15–20% (Wahyuni 2022). Because crustacean shells are easily accessible as by-products, they are an economical and environmentally friendly source for large-scale chitosan manufacturing. However, some issues call for cautious processing, such as seasonal availability and worries about allergenic residues in chitosan obtained from crustaceans (Muzzarelli 2010).

7.2.1.2 Fungi

Another non-animal source of chitosan is the cell walls of fungi, especially those of species like *Rhizopus oryzae*, *Aspergillus niger* and *Mucor rouxii*. Fungal chitin concentration is lower than crustaceans, typically ranging from 5 to 20% of the dry biomass; nevertheless, the fungus may be grown year-round under regulated circumstances, guaranteeing a steady supply (Abduh et al. 2023; Erdoğmuş et al. 2023). Fungal-derived chitosan is also appropriate for particular consumer groups because it circumvents the problems of shellfish allergies and dietary or religious limitations. Since fungi may be grown on agro-industrial residues, they also support sustainable practices by helping to valorise waste (Crognale et al. 2022; Pochanavanich and Suntornsuk 2002; Huq et al. 2022).

7.2.1.3 Insects

Because of their great availability and agricultural efficiency, insect exoskeletons are becoming increasingly recognized as a potential source of chitin and chitosan. The dry weight of insects like mealworms, silkworm pupae, and grasshoppers contains 10–20% chitin. Insect exoskeletons, frequently seen as waste, offer an extra resource for the manufacture of chitosan as interest in insect farming for food and feed develops worldwide. According to research, for example, silkworm pupae produce chitosan that is acceptable for use in agriculture and biomedicine and has a degree of deacetylation about 83% (Battampara et al. 2020; Zainol Abidin et al. 2020).

7.2.1.4 Cephalopods

Although poorly studied, the inside shells of cephalopods like squid and cuttlefish are excellent sources of chitin. For example, squid pens are an effective source of chitin for specific applications since they can contain up to 40% chitin on a dry-weight basis. Compared to crustacean shells, cephalopods offer a purer form of chitin with a lower mineral content despite their small contribution to the production of chitosan worldwide (Messerli et al. 2019; Younes and Rinaudo 2015).

7.2.1.5 Other Marine Animals

Chitin is also found in marine animals such as sponges and other algae, although its application is restricted because of their limited availability and less effective extraction techniques. For example, it has been demonstrated that the sponge species *Hippospongia communis* contains chitin that may be converted into high-purity chitosan; nevertheless, the yields of this extraction are frequently below 10% (Younes and Rinaudo 2015; Kannappan et al. 2024; Souza et al. 2011).

7.2.2 Extraction Procedure of Chitosan

Chitosan is often separated from chitin using various techniques, each designed to maximize yield, purity, and functional qualities for particular uses.

7.2.2.1 Chemical Extraction Method

Chemical extraction is the most popular technique for obtaining chitosan since it is easy to use, economical, and scalable. Three main processes are involved in this process: demineralization, deproteinization, and deacetylation. Each stage is intended to gradually eliminate particular impurities to produce high-purity chitosan with regulated physiochemical properties (Younes and Rinaudo 2015; Pellis et al. 2022; Pakizeh et al. 2021).

Demineralization: The demineralization stage purges the raw material of calcium carbonate ($CaCO_3$) and other mineral components mostly found in crustacean shells. Because of its high reactivity and efficiency, hydrochloric acid (HCl) is the most widely used reagent (Younes and Rinaudo 2015; Pellis et al. 2022). An HCl solution ranging from 0.5 to 2 M is applied to the raw material, usually shrimp or crab shells. The reaction is conducted at room temperature for two to six hours or roughly 25–100 °C. According to studies, mineral removal can reach 95% when 1 M HCl is used for three hours (Pellis et al. 2022).

$$CaCO_3 + 2HCl \rightarrow CaCl_2 + CO_2 + H_2O \tag{7.1}$$

Elevating the temperature to 50 °C can accelerate the demineralization process. However, high acid concentrations (>2 M) can break down the structure of the chitin. As a result, it is crucial to optimize the acid content, temperature, and reaction time (Liyanage et al. 2023). Following the reaction, the material is carefully cleaned with deionized water to eliminate any remaining acid and reaction by-products until the wash water reaches a neutral pH (Younes and Rinaudo 2015; Varun et al. 2017).

Deproteinization eliminates proteins and other organic materials firmly attached to chitin's structure. Because of its potent alkalinity and capacity to efficiently hydrolyse proteinaceous materials, sodium hydroxide (NaOH) is the preferred reagent for this stage (Younes and Rinaudo 2015; Pellis et al. 2022). NaOH solutions, usually 1–3 M, are applied to the demineralized material at temperatures between 60 and 80 °C (Kandile et al. 2018). Usually, the reaction lasts between two and four hours. For example, shrimp shells treated with 2 M NaOH for three hours at 70 °C remove more than 90% of the protein (Soon et al. 2018). Protein peptide linkages are hydrolysed by NaOH, resulting in the formation of soluble amino acid salts. Reduced molecular weight can result from chitin structural degradation caused by higher alkali concentrations (>3 M) or longer reaction times (>4 h). Moderate conditions are preferred to balance structural integrity and efficiency (Novikov et al. 2023). After deproteinization, the material is periodically rinsed with water to eliminate any remaining alkali and neutralize the sample (Younes and Rinaudo 2015; Aldila et al. 2020).

Deacetylation: This process, which removes acetyl groups ($-COCH_3$) from the chitin polymer and leaves behind amino groups ($-NH_2$), is essential for turning chitin into chitosan. The degree of deacetylation (DDA), a crucial factor affecting the solubility, bioactivity, and functional characteristics of chitosan, is determined by this process (Mathaba and Daramola 2020; Reza et al. 2024). Concentrated NaOH solutions (40–50%) are applied to deproteinized and demineralized chitin for one to three hours at high temperatures (90–120 °C). For example, chitosan with a DDA of roughly 85–90% is produced by processing chitin with 50% NaOH at 100 °C for two hours (Pellis et al. 2022; Vicente et al. 2021; Younes and Rinaudo 2015). The reaction mechanism involves the hydrolysis of acetyl groups by NaOH, which produces water-soluble acetate ions (CH_3COO^-).

$$R\text{-}NH\text{-}COCH_3 + NaOH \rightarrow R\text{-}NH_2 + CH_3COONa \tag{7.2}$$

The final DDA depends on temperature, reaction duration, and alkali content. While inadequate deacetylation may result in poor stability in acidic conditions, excessive deacetylation might decrease molecular weight (Younes and Rinaudo 2015; Novikov et al. 2023). Following deacetylation, the material is dried at 50–60 °C until a consistent weight is reached after properly cleaning to eliminate any remaining alkali (Mathaba and Daramola 2020).

7.2.2.2 Enzymatic Extraction Method

Using particular enzymes to extract proteins and minerals from raw materials such as fungal biomass or crustacean shells, enzymatic extraction of chitosan is more environmentally friendly than traditional chemical procedures. Reduced chemical use, less influence on the environment, and improved preservation of chitosan's molecular weight and structural integrity are only a few benefits of this technology (Kaczmarek et al. 2019; Pratiwi et al. 2023; Pellis et al. 2022). According to Kaczmarek et al. (2019), enzymatic extraction usually consists of two main steps: enzymatic deproteinization and demineralization. Each process helps to purify chitin into a high-purity product.

Enzymatic Deproteinization: Deproteinization is the process of removing proteins from source materials' cell walls or exoskeletons that are attached to chitin. Proteins are frequently hydrolysed into smaller peptides or amino acids by proteolytic enzymes such as trypsin, papain, or alcalase, which are subsequently removed by washing (Pratiwi et al. 2023). For example, investigations have shown that over 90% protein removal can be achieved by treating shrimp shells with papain at a concentration of 10 U/ml at 50 °C for 24 h (Pérez et al. 2022). Compared to chemical approaches, enzymatic deproteinization offers a number of benefits. Because severe alkalis like NaOH can break down the chitin polymer and lower molecular weight, it does away with their need. Applications requiring high molecular weight chitosan, including those in the biomedical industry, benefit greatly from this strategy (Younes and Rinaudo 2015). Furthermore, because the by-products of enzymatic processes are biodegradable and can be utilized as fertilizer and animal feed, they reduce environmental pollution (Pratiwi et al. 2023; Kaczmarek et al. 2019).

Enzymatic Demineralization: In enzymatic extraction, demineralization is frequently accomplished by direct enzymatic action or using organic acids generated by lactic acid bacteria (LAB). Up to 95% demineralization efficiency has been demonstrated with LAB, such as *Lactobacillus plantarum*, at 30 °C for 48 h (Neves et al. 2017). Incorporating LAB fermentation into a biorefinery model improves process sustainability and removes the need for concentrated acids like HCl (Broquá et al. 2018). The use of enzymes such as carbohydrates for enzymatic demineralization has also been investigated. These enzymes facilitate the effective clearance of mineral deposits by dissolving the polysaccharide matrix that envelops them (Younes and Rinaudo 2015). These methods are becoming more popular due to their promise in completely enzymatic processes, notwithstanding their rarity (Sharma et al. 2020).

Combination Enzymatic Approach: A combination of LAB for demineralization and proteolytic enzyme for deproteinization has been used in certain investigations. While preserving the environmental advantages of enzymatic methods, this integrated approach produces chitin with excellent yields and purity (Kaczmarek et al. 2019). For example, chitin, with a purity of about 95%, has been produced by sequentially treating crab shells with alcalase for deproteinization and LAB fermentation for demineralization (Pohling et al. 2024). Chitosan with high levels of deacetylation (80–90%) and molecular weights appropriate for high-value applications is produced by further deacetylation with mild alkali treatments (Valdez-Peña et al. 2010; Rossi et al. 2024).

7.2.2.3 Microwave-Assisted Extraction of Chitosan

A contemporary method called microwave-assisted extraction (MAE) employs microwave radiation to help chitin deacetylate into chitosan. Compared to traditional procedures, this approach can drastically reduce reaction time, energy consumption, and chemical use, which has drawn much interest. The idea behind MAE is that microwaves' quick heating and effective energy transmission improve chitin's molecular structural disruption and quicken the reaction's kinetics during deacetylation (Sebastian et al. 2019; Kozma et al. 2022). In the presence of concentrated alkali, such as NaOH, MAE usually uses high-power microwave irradiation. According to research, chitosan with a high DDA can be produced by operating at a temperature between 90 and 120 °C and a power range of 400–800 W. One study, for example, showed that chitin obtained from shrimp shells treated with 50% NaOH under microwave irradiation at 800 W for only 15 min produced a DDA of almost 90%. This is a significant improvement compared to conventional techniques, which can take up to two or three hours to complete under comparable chemical conditions (Knidri et al. 2019; Sebastian et al. 2019). Optimising the microwave parameters, such as the alkali content, power level, and irradiation period, is essential to achieve the desired chitosan properties. For example, extending the irradiation period past 20 min may result in thermal degradation or excessive deacetylation, which lowers the molecular weight and compromises its mechanical qualities. Similarly, increasing power levels can boost reaction efficiency and raise the possibility of hot patches and uneven heating, so they must be carefully controlled (Kozma et al. 2022; Knidri et al. 2019; Li et al. 2023).

7.2.2.4 Ultrasonic-Assisted Extraction of Chitosan

A contemporary method called ultrasonic-assisted extraction (UAE) improves the extraction and processing of chitosan from chitin using ultrasonic waves. This approach has drawn attention compared to traditional methods since it can drastically reduce reaction times, chemical usage, and extraction efficiency. In the solvent medium, the ultrasonic waves produce cavitation bubbles that burst, producing localized high temperatures and pressures. By altering the chitin structure, these physical impacts improve chemical penetration and speed deacetylation (Singh et al. 2019; Buanasari et al. 2021). Ultrasound breaks down intermolecular connections in chitin by producing a cavitation effect at frequencies ranging from 20 to 40 kHz. The crystalline structure of chitin is broken down by the strong shear pressures created when cavitation bubbles collapse. The alkali solution (often NaOH) can interact with the acetyl groups more efficiently due to the increased surface area exposure, which speeds up the deacetylation process that forms chitosan (Martín-López et al. 2020; Shen et al. 2023).

According to research, the best outcomes are obtained with ultrasonic frequencies of about 20 kHz and power levels of 500 W. The molecular weight of chitosan is decreased by higher power levels, which produce more intense cavitation but may also lead to excessive polymer chain disintegration (Martín-López et al. 2020; Savic et al. 2022). Usually, the deacetylation agent is 40–50% NaOH. Compared to traditional heating techniques, ultrasonic waves improve the diffusion of NaOH into the chitin structure, increasing reaction efficiency (Singh et al. 2019; Buanasari et al. 2021). Temperatures between 80 and 100 °C are used for the process. Uniform deacetylation is ensured by keeping the bulk solution at the ideal temperature, even when ultrasonic waves produce localized heat (Martín-López et al. 2020; Savic et al. 2022). UAE drastically cuts down on processing time when compared to conventional techniques. According to studies, deacetylation of chitin under varying temperatures and durations resulted in chitosan yields of 50–65% (w/w) with a degree of deacetylation (DDA) ranging from 78 to 90%. Notably, ultrasonic-assisted deacetylation enhanced the DDA to $85.35 \pm 0.20\%$, compared to $80.24 \pm 0.19\%$ without ultrasound treatment (Singh et al. 2019; Buanasari et al. 2021).

7.2.2.5 Extraction from Fungal Sources

Because it is superior to chitosan derived from crustaceans, chitosan extracted from fungal sources has drawn more attention. The main source of fungal chitin and chitosan is the cell walls of fungi, especially those that are members of the genera *Rhizopus*, *Aspergillus*, and *Mucor*. Fungal sources, as opposed to crustacean shells, remove the allergic issues related to shellfish and fit in with vegetarian and vegan diets. Additionally, under controlled conditions, fungal cultivation can be carried out all year, guaranteeing a steady supply of raw materials (Dhillon et al. 2013; Pellis et al. 2022; Iber et al. 2022). Typically, fungal biomass is obtained by industrial fermentation procedures in which fungi are grown on agro-industrial wastes like maize steep liquor, sugarcane bagasse, or molasses. In addition to cutting waste, this biotechnological technology offers a sustainable way to obtain fungal cell walls high in chitin and chitosan. To avoid microbial contamination, the fungal biomass is dried at 50–60 °C after being cleaned with water to eliminate extracellular polysaccharides, colours, and contaminants (Pellis et al. 2022; Davis et al. 2024; Araújo et al. 2020).

An alkali treatment is used to deproteinize fungal biomass, which aids in the breakdown of protein complexes attached to the cell wall. NaOH solutions with concentrations between 1 and 3 M are frequently utilized. For example, a protein removal efficiency of 90% has been demonstrated when *Aspergillus niger* biomass is treated with 2 M NaOH at 90 °C for two to three hours. Enhancing the extracted chitin's purity and guaranteeing the manufacture of high-quality chitosan depends on this phase (Dhillon et al. 2013; Iber et al. 2022; Araújo et al. 2020). A demineralization phase is still required to eliminate traces of inorganic components, even if fungal cell walls have fewer minerals than crustacean shells. Usually, diluted HCl solutions (0.5–1 M) at room temperature are used for this. According to studies, leftover minerals are efficiently removed from fungal biomass after two hours of treatment with 1 M HCl, with no effect on the chitin's structure. Crucially, the technique is more ecologically benign since fungal-derived chitin requires less intense acid treatment than crustacean sources (Pellis et al. 2022; Davis et al. 2024; Iber et al. 2022). The deacetylation process turns chitin into chitosan by eliminating acetyl groups from the polymer chain. This is accomplished by applying concentrated NaOH (40–50%) to the demineralized and deproteinized fungal material for two to four hours at a high temperature (90–120 °C). According to research, depending on the temperature and reaction time, deacetylation of fungal chitin under these circumstances can produce chitosan with a DDA range from 80 to 90%. For example, a DDA of almost 85% is obtained by employing 50% NaOH at 100 °C for three hours, making it appropriate for use in pharmaceutical and biomedical applications (Dhillon et al. 2013; Iber et al. 2022; Mane et al. 2022). Additional purification procedures, including neutralization and washing with deionized water, are carried out to improve the quality of chitosan generated from fungi. In order to eliminate any remaining pigments and enhance the end product's whiteness, several procedures incorporate an optional bleaching phase that uses hydrogen peroxide (H_2O_2). Following purification, the chitosan is dried at 50–60 °C and milled into a fine powder for use and storage (Pellis et al. 2022; Davis et al. 2024; Araújo et al. 2020) (Fig. 7.1).

7.2.3 The Chemistry of Chitosan

Randomly distributed β-(1 → 4)-linked D-glucosamine (deacetylated units) and N-acetyl-D-glucosamine (acetylated units) make up chitosan, a linear polymer (Aranaz et al. 2021; Choi et al. 2016). According to studies, chitin, the second most abundant natural polymer after cellulose, is the source of its chemical structure and characteristics (Jiménez-Gómez and Cecilia 2020; Li et al. 2020a, b). Chitosan's distinctive reactivity and usefulness are attributed to its main amino group ($-NH_2$) at the C-2 position of the glucosamine units (Pellis et al. 2022; Alemu et al. 2023) (Fig. 7.2).

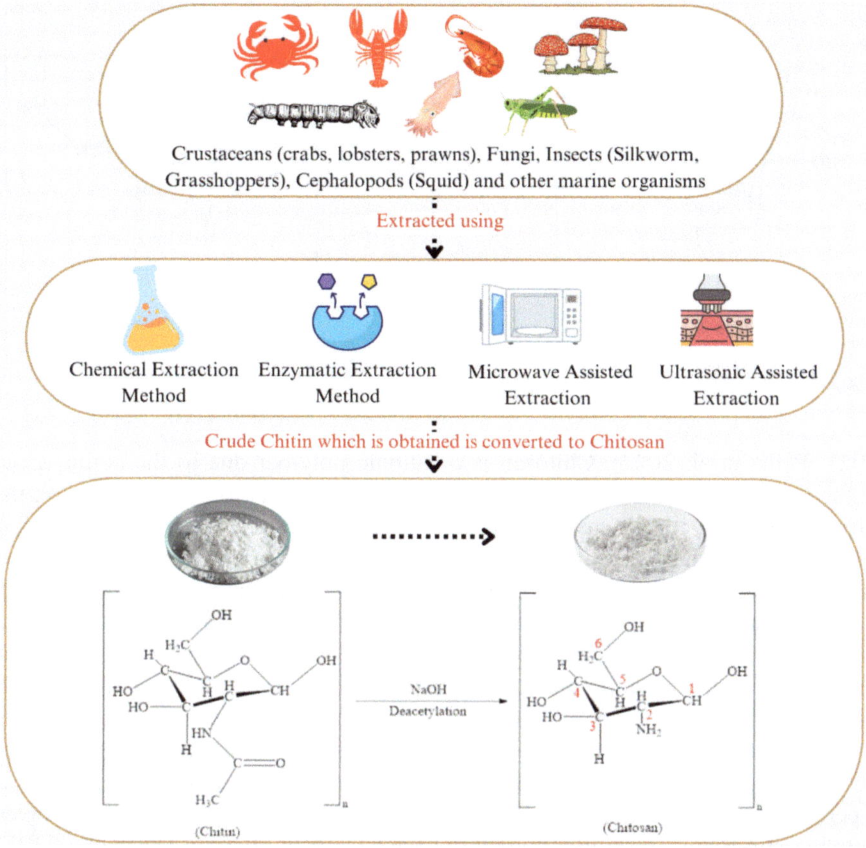

Fig. 7.1 Using chemical, enzymatic, microwave-assisted, or ultrasonic-assisted techniques, chitin can be extracted from a variety of natural sources, including crustaceans (such as crabs, lobsters, and prawns), fungi, insects (such as silkworms and grasshoppers), cephalopods (such as squids), and other marine species. After the crude chitin is removed, it is deacetylated using sodium hydroxide (NaOH) to produce chitosan. This process eliminates the acetyl groups ($-COCH_3$) from the chitin's structure, leaving the chitosan polymer with free amino groups ($-NH_2$). Because of its biocompatibility and useful qualities, the finished product, chitosan, is a white powder with a wide range of uses in industries like biomedicine, agriculture, food, and water treatment

The backbone of chitosan is made up of glucosamine units with acetylated units scattered throughout. The DDA, a crucial factor affecting its solubility, reactivity, and biological activity, is determined by the ratio of these units (Aranaz et al. 2021; Pellis et al. 2022). The chemical formula for chitosan is commonly expressed as $(C_6H_{13}NO_5)_n$, where "n" denotes the degree of polymerization that varies depending on the source and processing circumstances. Depending on the extraction and processing techniques, chitosan can have molecular weights ranging from 50 kDa to more than 1,000 kDa. Its mechanical characteristics, biological interactions, and viscosity are all influenced by its molecular weight (Aranaz et al.

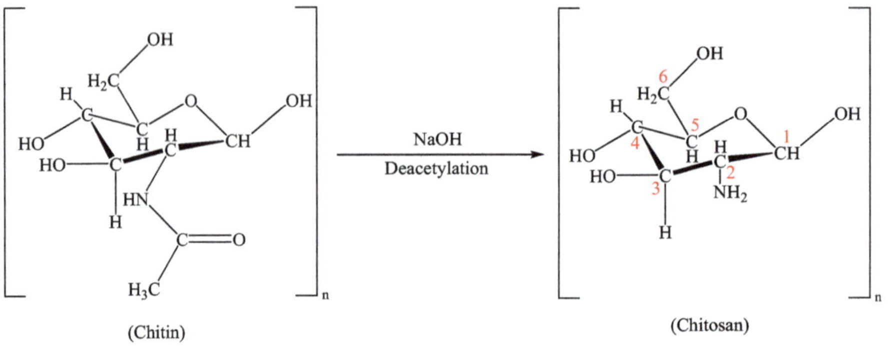

Fig. 7.2 Conversion of chitin to chitosan

2021; Pellis et al. 2022). Chitosan is a cationic polymer due to the amino group at the C-2 position of the glucosamine units. In acidic environments (pH < 6), this group is protonated, giving chitosan a positive charge that facilitates interactions with negatively charged molecules such as proteins, lipids, and polyanions. Another factor that makes chitosan soluble in acidic solutions is the protonation of amino groups (Aranaz et al. 2021). The structural flexibility and derivatization potential are enhanced by the participation of hydroxyl groups at the C-3 and C-6 positions in hydrogen bonding and chemical alterations (Pellis et al. 2022). The acetylated units contain the acetamide groups. These groups affect chitosan's crystallinity and hydrophobicity. The DDA has an inverse relationship with the percentage of these categories. The DDA defines the percentage of glucosamine units in the polymer as a percentage of all units. Higher DDA values indicate more solubility and biological activity; chitosan is generally thought to have a DDA of 80–95% (Yuan et al. 2011; Foster et al. 2015).

7.3 Properties of Chitosan

7.3.1 Physicochemical Properties

Molecular weight: The molecular weights of chitosan vary greatly, usually ranging from 50,000 to more than 1,000,000 Da. The molecular weight all directly influences its mechanics, solubility, and viscosity. Chitosan with a higher molecular weight is more viscous in solution and better suited for scaffold building and film production requiring strong mechanical properties. However, lower molecular weight chitosan is recommended in drug administration systems that require improved solubility and simpler biological interaction. By altering the reaction conditions or by enzymatic and chemical degradation, the molecular weight of chitosan can be regulated throughout its synthesis (Aranaz et al. 2021; Kose and Suvacı 2022; Román-Doval et al. 2023).

Degree of Deacetylation (DDA): The percentage of deacetylated glucosamine units in the chitosan polymer is known as the degree of deacetylation (DDA), and it normally falls between 50 and 95% (Mathaba and Daramola 2020). The DDA significantly influences the polymer's solubility, charge density, and reactivity (Pellis et al. 2022). Because free amino groups are more prevalent and protonated in acidic environments, chitosan with a greater DDA is more soluble in these situations (Potivas and Laokuldilok 2014; Nguyen et al. 2022). Furthermore, the DDA influences the polymer's capacity to combine with negatively charged materials to form complexes, impacting its applicability in tissue engineering, drug delivery, and water treatment (Novikov et al. 2023). Spectroscopic techniques such as Nuclear Magnetic Resonance (NMR) and Fourier-transform Infrared (FTIR) spectroscopy can be used to quantify the DDA (Pérez-Álvarez et al. 2018).

Solubility: At pH levels below 6.5, chitosan dissolves in aqueous solutions of mild organic acids such as citric, lactic, or acetic acids (Mathaba and Daramola 2020). The protonation of the amino groups in the polymer's glucosamine units is responsible for this solubility and for making it more hydrophilic (Potivas and Laokuldilok 2014). These amino groups deprotonate at higher pH values, which causes the polymer to become less soluble and precipitate (Nguyen et al. 2022). Chitosan's solubility is mostly determined by its molecular weight and DDA; a lower molecular weight and a higher DDA increase solubility (Pellis et al. 2022). Carboxymethylation or quaternization modifications can further enhance solubility in aqueous and neutral pH solutions, expanding its use in industrial and biological domains (Novikov et al. 2023).

Hydrophilicity: Because of the large number of hydroxyl and amino groups throughout its polymer chain, chitosan is naturally hydrophilic. Because it enables water interaction, chitosan is a great material for absorbing and retaining moisture (Piekarska et al. 2023). Its hydrophilicity further enhances its biocompatibility, allowing interaction with biological tissues without resulting in appreciable cytotoxicity (Cheung et al. 2015). This feature is helpful in applications such as wound dressing, where preserving a moist environment is essential for fostering healing (Aranaz et al. 2021). Its hydrophilic characteristic can be altered by crosslinking or blending with hydrophobic polymers, making it suitable for particular uses like packaging or controlled drug release (Thambiliyagodage et al. 2023).

pH Sensitivity: The ionization of chitosan's amino groups is the main factor influencing its notable pH sensitivity. The amino groups are protonated at acidic pH levels, which increases the polymer's solubility and gives it a positive charge (Alemu et al. 2023). According to Nilsen-Nygaard et al. (2015), chitosan's cationic nature allows it to interact with negatively charged molecules, including proteins, DNA, and anionic polymers, to create stable complexes (Nilsen-Nygaard et al. 2015). The amino groups lose protons when the pH gets closer to neutral or alkaline, which lowers the polymer's charge density and solubility (Piekarska et al. 2023). Because chitosan can form protective carriers that release their payload in reaction to pH changes, such as those experienced in different areas of the gastrointestinal tract, this pH-responsive behaviour is frequently used in drug delivery systems (Li et al. 2020a, b).

7.3.2 Barrier Properties

Gas Permeability: Because of its semi-crystalline structure and hydrogen bonds between polymer chains, chitosan has a limited gas permeability, especially for carbon dioxide and oxygen. Reducing respiration rates and oxidative processes make it a great barrier material for maintaining the quality of perishable goods such as fruits, vegetables, and meats (Shi et al. 2024). A number of variables, including humidity, molecular weight, DDA, and any further alterations, affect chitosan's gas barrier qualities (Butnaru et al. 2019). For instance, because water functions as a plasticizer and lessens intermolecular bonds, excessive humidity levels may weaken the polymer's gas barrier. Chitosan can be even more successful for some packaging applications by mixing it with other polymers or crosslinking it with substances like glutaraldehyde to increase its resistance to gas permeability (Hu et al. 2020).

Water Vapour Barrier Properties: According to some studies, chitosan has moderate water vapour barrier qualities that are especially useful for regulating moisture exchange in packaging applications (Cazón et al. 2022). Because of its hydrophilic character, Chitosan has a better vapour permeability than manufactured hydrophobic polymers like polyethylene. Nevertheless, this characteristic can be adjusted by adding hydrophobic compounds or making chemical changes. Chitosan's water vapour barrier function can be greatly enhanced by mixing it with substances like beeswax, essential oils, or other hydrophobic polymers (Shi et al. 2024). These changes expand its utility in biomedical applications like wound dressings, where controlled moisture retention is desired, and improve its suitability for food packaging (Butnaru et al. 2019). Furthermore, chitosan is a versatile material for applications where moisture control is crucial due to its capacity to create films with adjustable water vapour permeability (Cazón et al. 2022).

The Impact of Moisture Sensitivity on Barrier Properties: Because chitosan is naturally hydrophilic, its barrier qualities, especially against water vapour, are affected by ambient humidity. Swelling and enhanced permeability result from water molecules penetrating the chitosan matrix and breaking intermolecular hydrogen bonds at high relative humidity levels. This sensitivity may be a drawback for some applications, but it also offers chances to create responsive materials. For example, chitosan's sensitivity to moisture can be used in smart packaging systems, where the film's permeability varies according to the surrounding humidity, offering the contents dynamic protection (Aguirre-Loredo et al. 2016). Additionally, chitosan films can be made more water-resistant while maintaining their overall barrier qualities by undergoing chemical changes such as acetylation or the grafting of hydrophobic groups (Butnaru et al. 2019).

7.3.3 Mechanical Properties

Tensile strength: Chitosan is equivalent to certain synthetic polymers, often falling between 30 and 60 MPa. The molecular weight and DDA impact: Higher DDA and molecular weight result in stronger films and structures by increasing intermolecular hydrogen bonding (Gadgey and Sharma 2020). Reinforcing substances like cellulose nanofibers, graphene oxide, or nanoclays can also increase the tensile strength. By establishing robust connections with the chitosan matrix, these additions increase the material's strength and stiffness (Shi et al. 2024).

Elongation at Break: Relatively low elongation at break, usually between 2 and 20%, is a characteristic of chitosan. The material's limited elasticity and propensity to shatter under stress without undergoing significant deformation are reflected in this attribute (Alemu et al. 2023). Plasticizers like sorbitol, glycerol, or polyethylene glycol (PEG) can be added to improve flexibility. These substances improve the elongation at the break by reducing intermolecular contacts within the chitosan matrix and enabling free movement of the polymer chains (Habibie et al. 2016). Environmental factors, including humidity and moisture absorption, which function as a natural plasticizer, also affect the specific elongation value (El-Araby et al. 2024).

Modulus of Elasticity: A measurement of a material's stiffness, the modulus of elasticity is also called Young's modulus. Under normal circumstances, chitosan is a stiff material because of its relatively high modulus. Although this stiffness can be a drawback in applications that call for flexibility, it is beneficial in structural applications (Salavati 2023). Chitosan's stiffness can be adjusted by crosslinking it with substances like glutaraldehyde or genipin or mixing it with other polymers like polyvinyl alcohol (PVA) or polyethene oxide (PEO). By creating covalent links between chitosan chains, crosslinking improves the material's structural integrity by making it stiffer and less soluble in water (Shi et al. 2024).

Toughness and Flexibility: Chitosan's inherent stiffness may be a disadvantage when toughness and flexibility are essential. This can be lessened by creating composites with it and other biopolymers or synthetic materials. For example, chitosan can be combined with gelatin or alginate to create more flexible and less brittle films (Gadgey and Sharma 2020). Similarly, by more equally spreading stress throughout the material, the introduction of nanoparticles like zinc oxide (ZnO) or titanium dioxide (TiO_2) improves toughness in addition to mechanical strength (Salavati 2023).

Temperature and humidity, in particular, significantly impact chitosan's mechanical performance. When exposed to high humidity, Chitosan functions as a natural plasticizer, increasing elongation at break and decreasing tensile strength (El-Araby et al. 2024). The material becomes more brittle and likely to shatter at lower humidity levels. This sensitivity must be considered when creating applications for chitosan-based materials, mainly when using the materials in various environmental settings (Alemu et al. 2023).

7.3.4 Thermal Properties

Thermal Stability: Chitosan's DDA and molecular weight impact its good thermal stability. There are usually two phases to thermal deterioration. Because it is hydrophilic, the first stage is the loss of bound and adsorbed water below 150 °C. The second stage, which involves the breakdown of glycosidic linkages and depolymerization, occurs between 200 and 300 °C. The DDA has a significant impact on when thermal degradation occurs, with a greater DDA resulting in somewhat better thermal stability. The polymer degrades significantly above 300 °C, releasing volatile substances such as carbon dioxide, amines, and acetic acid. Because of these thermal characteristics, chitosan can be used in applications that do not require processing at high temperatures (Nainggolan et al. 2020; Zakaria et al. 2012).

Glass Transition Temperature (Tg): Chitosan does not show a sharp Tg because of its semi-crystalline structure. On the other hand, Tg values for chitosan are often found between 150 and 200 °C. A number of variables, including water content, DDA, and molecular weight, affect the Tg. Chitosan is plasticized by water, which increases molecular mobility and lowers its Tg. For example, hydrated chitosan has a much lower Tg than dry chitosan, which has a greater Tg. Depending on the needs of the application, the Tg can be further changed by the addition of crosslinking agents, fillers, or blending with other polymers, improving its flexibility or thermal performance (Eulalio et al. 2019; Dhawade and Jagtap 2012).

Melting Behaviour: Because of its strong hydrogen bonds within its molecular structure and semi-crystalline form, chitosan does not exhibit a clear melting point. Rather than melting completely, heating causes chitosan to change from an ordered crystalline structure to an amorphous one linked to decomposition. Although it may maintain structural integrity at moderate heat conditions, this behaviour restricts its applicability in thermoplastic applications (Gubanova et al. 2021; Zakaria et al. 2012).

Crystallinity and Thermal Modifications: Chitosan's thermal properties are significantly influenced by its degree of crystallinity. Greater thermal stability and slower deterioration are correlated with higher crystallinity. Processing circumstances, DDA, and preparation technique all impact crystallinity. Heat treatment or chemical changes like crosslinking or combining with other polymers can increase or decrease crystallinity. Altering the crystalline areas, for example, mixing with polyvinyl alcohol (PVA) or polyethene glycol (PEG) might increase thermal resistance. Similarly, crosslinked chitosan has improved thermal characteristics, making it appropriate for uses requiring moderate heat resistance, such as packaging and biomedical devices (Nainggolan et al. 2020; Gubanova et al. 2021).

Thermal Conductivity: Despite not being a prominent property, chitosan's poor thermal conductivity makes it an excellent insulator because of its organic structure. Applications needing thermal protection, such coating materials or food preservation, benefit from this feature. Enhancing its thermal conductivity through fillers or nanoparticle modifications expands its potential for advanced material applications (Eulalio et al. 2019; Zakaria et al. 2012).

7.3.5 *Biodegradability*

Due to its natural origin and structural resemblance to cellulose, chitosan is well known for its exceptional biodegradability. It is mainly broken down by enzymatic processes involving cellulase, lysozyme, and chitosanase. These enzymes break down the chitosan polymer into smaller oligosaccharides and glucosamine units by cleaving the polymer's β-(1 → 4) glycosidic linkages (Reshad et al. 2021). Several variables, such as the chitosan's molecular weight, DDA, and the surrounding environment, affect the rate and degree of enzymatic degradation (Islam et al. 2019). For example, because enzymes are less accessible to the polymer chain, larger molecular weight chitosan and lower DDA typically slow down the degradation process (Yarahmadi et al. 2024).

The DDA mostly determines the degrading behaviour of chitosan. Because it contains more free amino groups, which are more accessible to enzymatic action, chitosan with a high DDA—typically around 80%—tends to break down more easily (Jiménez-Gómez and Cecilia 2020). On the other hand, chitosan with a lower DDA has more acetyl groups, which can prevent enzymatic hydrolysis because of decreased solubility and steric hindrance (Reay et al. 2022). Furthermore, the pace at which chitosan degrades is influenced by its crystallinity. Because enzymes may more easily reach the polymer chains in amorphous portions of chitosan, they break down more readily than in highly crystalline sections, which are more resistant to enzymatic attack (Matica et al. 2017). The biodegradation of chitosan is also significantly influenced by environmental variables. Chitosan is broken down by naturally existing microorganisms that provide the required enzymes in natural settings, such as soil or composting environments (Casadidio et al. 2019). Variables like temperature, pH, and moisture content can significantly affect how quickly things break down. For example, moderate temperatures and neutral to slightly acidic environments promote enzymatic and microbial activity, speeding up biodegradation (Wrońska et al. 2023). Though often slower than enzymatic degradation, hydrolytic processes can also break down chitosan in aquatic environments (Reshad et al. 2021).

Glucosamine and N-acetylglucosamine, the main by-products of chitosan breakdown, are non-toxic and biodegradable in and of themselves. These by-products are organic substances that can be further broken down by microbes or incorporated into biological systems (Islam et al. 2019). Chitosan's environmentally benign disintegration process makes it desirable for biodegradable applications such as medication delivery systems, packaging, and agricultural films (Yarahmadi et al. 2024). Its adaptability for various applications is further increased by the capacity to customize its degradation rate by altering structural characteristics and environmental factors (Jiménez-Gómez and Cecilia 2020).

7.3.6 Biological Activities

Antimicrobial Activity: Because chitosan is efficient against a wide range of microorganisms, such as bacteria, yeasts, and mould, its antimicrobial activity is a focus of food preservation research. A number of variables, including molecular weight, degree of deacetylation, pH levels, and the presence of functional additives, affect this biopolymer's antibacterial activity (Ke et al. 2021; Yilmaz Atay 2019). Its primary antibacterial mechanism is the interaction of chitosan's positively charged amino groups with the negatively charged microbial cell membranes. The integrity of the membrane is compromised by this contact, which increases permeability, causes intracellular contents to seep out, and eventually results in cell death. Furthermore, critical metal ions can be chelated by chitosan, depriving microbes of essential nutrients for their growth (Goy et al. 2009). Its capacity to build films also produces a physical barrier that prevents microorganisms from accessing oxygen and nutrients (Dutta et al. 2012). In order to improve the antibacterial qualities of chitosan films, research has looked into adding metal oxides. For example, chitosan films containing silver nanoparticles, zinc oxide (ZnO), and titanium dioxide (TiO_2) have shown enhanced antibacterial action against foodborne pathogens such as *Listeria monocytogenes* and *Campylobacter jejuni*. To increase shelf life and guarantee safety, these composite films have the potential to be used as active food packaging materials (Mawazi et al. 2024; Liu et al. 2022). According to another study, moderately hydrolysed chitosan shows better antibacterial action than its natural form. By lowering its molecular weight, hydrolysis improves chitosan's solubility and ability to interact with microbial cells. However, the loss of antibacterial qualities can result from excessive degradation, underscoring the significance of regulated hydrolysis in creating potent chitosan-based preservatives (Hafdani and Sadeghinia 2011). However, creating edible films based on chitosan and enhanced with natural extracts, such as green tea extract, has demonstrated encouraging outcomes in lowering bacteria and foodborne viruses. Directly applying these films to food surfaces creates a barrier of protection that prevents microbiological growth and improves food safety (Ke et al. 2021; Mawazi et al. 2024; Aranaz et al. 2021).

Antioxidant Activity: A naturally occurring polysaccharide made from chitin, chitosan has strong antioxidant qualities that are especially advantageous for food preservation and medical applications. The main reason for its antioxidant activity is that it can scavenge free radicals and chelate metal ions, which prevents oxidative processes that can cause food to deteriorate and spoil (Liu et al. 2022; Muthu et al. 2021). Chitosan's molecular structure contains hydroxyl (−OH) and amino ($-NH_2$) groups, contributing significantly to its antioxidant effectiveness. In order to neutralize free radicals and stop chain reactions that result in oxidative damage to food products, these functional groups can donate hydrogen atoms or electrons (El-Araby et al. 2024; He et al. 2024). Furthermore, transition metals like Fe^{2+} and Cu^{2+}, which act as catalysts in oxidative reactions, can be bound by chitosan due to its potent metal-ion chelating properties. Chitosan efficiently lowers the rate of lipid oxidation in food matrices by capturing these metal ions (Hao et al. 2021).

The antioxidant capability of chitosan and its derivatives is supported by empirical research. For example, a study assessing the antioxidant activity of chitosan derived from shrimp exoskeletons showed that it was an efficient scavenger of free radicals. According to the study, Chitosan may be a natural antioxidant in food preservation (Muthu et al. 2021). According to a different study on the physicochemical characteristics and antioxidant activity of chitin and chitosan, both substances have strong antioxidant capabilities that may be advantageous for various uses, such as food preservation (El-Araby et al. 2024; Aranaz et al. 2021).

Film-forming and Coating Properties: Because of its potential to improve food safety and preservation, chitosan's film-forming and coating qualities have attracted much interest in the food industry. Chitosan effectively regulates moisture transfer, gas exchange, and microbiological contamination by forming a semi-permeable barrier on food goods' surface when applied as edible films or coatings. According to Muñoz-Tebar et al. (2023) and Khubiev et al. (2023), this barrier function plays a crucial role in prolonging the shelf life of perishable goods such as fruits, vegetables, meats, and seafood (Muñoz-Tebar et al. 2023; Khubiev et al. 2023). Chitosan-based films' mechanical and barrier qualities can be customised by altering variables, including the degree of deacetylation, molecular weight, and the addition of plasticizers or other biopolymers (Singh et al. 2015). For example, combining chitosan with polysaccharides, proteins, or lipids can make it more appropriate for specific food applications to improve the film's tensile strength, flexibility, and water vapour permeability (Nunes et al. 2023). A thorough analysis shows that composite films made of chitosan and other natural polymers have better mechanical strength and barrier performance than pure chitosan films, increasing their possible uses in food packaging (Luangapai et al. 2019). Chitosan coatings are effective in maintaining the quality of various food goods. For example, a study on strawberries coated with chitosan-based films found that the film's capacity to regulate gas exchange and prevent microbial development resulted in a notable decrease in weight loss and decay rate during storage (Šuput et al. 2015). Similarly, by minimizing moisture loss and enzymatic browning, chitosan coatings have proven successful in preserving fresh-cut fruits and vegetables' firmness and sensory qualities. Its antibacterial qualities further enhance Chitosan's usefulness as a food coating material. Chitosan coatings can lower the microbial load on food surfaces and improve food safety by breaking down the cell membranes of pathogens and spoilage microorganisms. According to studies, chitosan-based antimicrobial films are useful for prolonging the shelf life of perishable foods because they effectively stop the growth of prevalent foodborne infections. Films made of chitosan can be used for preservation and as vehicles for bioactive substances like natural extracts that have antibacterial or antioxidant qualities. These substances can be added to chitosan films to improve their functional qualities and add an extra layer of defence against oxidation and microbial deterioration. According to a review, chitosan films reinforced with natural extracts have better antibacterial and antioxidant qualities, making them helpful in preserving various food items (Shi et al. 2024; Suhag et al. 2020).

Anti-browning Effects: The anti-browning properties of chitosan in various food products, especially fresh-cut fruits and seafood, have been thoroughly investigated.

Its interaction with polyphenol oxidase (PPO), the enzyme that causes browning responses, is mainly responsible for its capacity to prevent enzymatic browning. According to a study, Chitosan coatings successfully inhibited microbial development and preserved firmness in fresh-cut apples stored at ambient temperature (22 °C) and in a refrigerator (5 °C). When chitosan coatings were applied, apple cubes' L values (lightness) were decreased, indicating less browning than in untreated samples. Chitosan coatings also reduced the loss of vitamin C, which helped maintain the quality of apples while they were being stored (Shyu et al. 2019). Chitosan and citric acid's effects on longan fruit's pericarp browning and PPO activity were assessed in a different study. According to Apai et al. (2009), combining treatments dramatically decreased browning. It inhibited PPO activity, indicating that chitosan and citric acid work together to limit enzymatic browning in longans (Apai et al. 2009). Chitosan coatings have been utilized in seafood applications to decrease shrimp browning and increase shelf life. The potential of chitosan-based coatings in seafood preservation was highlighted by a study that used chitosan-coated nanoliposomes loaded with liquorice root extract to show improved shelf life and enhanced anti-browning actions in shrimp (Hamdan et al. 2022). Additionally, it has been demonstrated that ascorbic acid and chitosan-based coatings can prevent browning, preserve the firmness of the flesh, and preserve phenolic components in chilled strawberry fruits. The quality of strawberries was maintained during storage thanks to this combination treatment, which successfully prevented enzymatic browning (Demircan and Velioglu 2024).

Prebiotic Activity: Chitosan oligosaccharides (COS), produced when chitosan is deacetylated, have attracted a lot of interest due to their prebiotic qualities, especially their capacity to alter the gut flora positively. COS provide a distinctive non-plant-derived substitute with various biological functions in contrast to conventional prebiotics obtained from plants. Studies have shown that COS can specifically encourage the development of good gut bacteria. In one study, for example, Lee et al. discovered that COS with a degree of polymerization (DP) between 2 and 8 significantly increased the growth of Lactobacillus species and Bifidobacterium bifidum in a dose-dependent way (Lee et al. 2002). Given that a higher population of these probiotics is linked to better gut health and stronger immune function, this targeted stimulation is essential (Guan and Feng 2022; Wang et al. 2021). Observations from animals further support these conclusions. Weaned piglets' ileal and colonic microbiota composition significantly improved after two weeks of food supplementation with 500 mg/kg/day of COS. In particular, there was a decrease in potentially hazardous bacteria like Clostridium species and Escherichia coli and an increase in beneficial bacteria like *Bifidobacterium* species, *Faecalibacterium prausnitzii*, and *Lactobacillus* species (Benchamas et al. 2021; Wang et al. 2021). Furthermore, COS supplementation decreased levels of toxic metabolites such as ammonia and increased concentrations of short-chain fatty acids (SCFAs), which are essential for colon health (Zhang et al. 2018). COS has prebiotic effects that go beyond the dynamics of bacterial populations. Research has demonstrated that COS can help mice with metabolic syndrome brought on by high-fat diets. In addition to preventing weight gain and the build-up of liver fat, using COS supplements also brought high blood glucose and fasting insulin levels back to normal (Wang et al. 2021; Lee et al. 2002). Beneficial changes in gut microbiota composition coincided with these physiological improvements, indicating a connection between changes in gut microbiota regulation and metabolic health (Zhang et al. 2018; Wang et al. 2021). Additionally, COS has been shown to improve aquatic species' digestive health. Immune responses, intestinal morphology, and growth performance were all enhanced by dietary COS supplementation in hybrid groups (male *Epinephelus lanceolatus* x female *Epinephelus fuscoguttatus*). The potential of COS as a functional feed addition in aquaculture was highlighted by these advantages, which were linked to the regulation of gut microbiota (Guan and Feng 2022).

Several different processes underlie COS's prebiotic activity. They can enter the colon undamaged because they are resistant to the digestive enzymes used by humans. Once there, they act as substrates for fermentation by particular gut bacteria. SCFAs, produced during this fermentation process, reduce intestinal pH and stop harmful bacteria from growing. Furthermore, by improving the function of the intestinal barrier and having anti-inflammatory properties, COS can alter the gut environment and support a balanced microbiota (Guan and Feng 2022; Zhang et al. 2018).

Emulsifying Properties and Fat Binding: Chitosan's ability to bind fat is ascribed to its cationic character, which enables it to interact with bile acids and negatively charged lipids in the gastrointestinal system. Because of this interaction, complexes that are difficult to absorb are formed, which lowers total fat absorption. Although the precise mechanisms are still being studied, in vitro studies have shown that chitosan can bind lipids and bile acids (Panith et al. 2016; Jin et al. 2017). The relationship between the physicochemical properties of chitosan and its ability to bind fat has been studied. For example, a study on crawfish chitosan discovered that viscosity changes affected fat-binding capacity, with lower viscosity being associated with lower fat-binding ability. The clinical relevance of chitosan's fat-binding qualities in weight management is still up for dispute, notwithstanding these findings. According to several reviews, chitosan can bind dietary lipids, but it has little effect on body weight loss and might not be worth using as a supplement for weight loss (Liu et al. 2021; Panith et al. 2016). Because of its amphiphilic nature, chitosan can stabilize oil-in-water emulsions by lowering the interfacial tension between immiscible phases (Jin et al. 2017). This characteristic is especially helpful for food compositions that call for stable emulsions, such as dressings. According to studies, adding chitosan to emulsions can improve their stability. For example, whey protein-stabilized emulsions with increased chitosan content had better creaming stability, higher viscosity, and smaller average particle sizes. Furthermore, chitosan's emulsifying qualities have been contrasted with other biopolymers. The potential of chitosan derivatives in emulsion stabilization is highlighted by research showing that, although chitin and chitosan may not form emulsions on their own, microcrystalline chitin has better emulsifying qualities than microcrystalline cellulose (Panith et al. 2016; Liu et al. 2021).

Adsorption of Toxins and pollutants: Chitosan's ability to effectively adsorb toxins and pollutants, especially mycotoxins and heavy metals, has been thoroughly investigated, underscoring its potential to improve food safety. When found in food and feed, mycotoxins—toxic secondary metabolites from fungi—pose serious health hazards.

It has been shown that chitosan may absorb many mycotoxins at once. In one study, for example, removal efficiency for palm kernel cake contaminated with different mycotoxins using chitosan ranged from 45.90% for aflatoxin B2 (AFB2) to 94.35% for aflatoxin B1 (AFB1) (Abbasi Pirouz et al. 2020). The ideal parameters for this adsorption were a pH of 4, an 8-h contact duration, and a temperature of 35 °C.

Additionally, the study found that temperature, pH, and contact time all affected the adsorption efficiency, with acidic conditions favouring the protonation of chitosan's amino groups and boosting electrostatic interactions with negatively charged mycotoxins (Abbasi Pirouz et al. 2020). The modification of chitosan to increase its adsorption ability has been the subject of additional investigation. According to one study, cross-linked chitosan polymers may remove up to 73% of specific mycotoxins, making them general adsorbents with simultaneous adsorption efficiencies (Zhao et al. 2015).

Chitosan has a great affinity for heavy metals, frequently found as contaminants in food products and mycotoxins. The chelation of metal ions is facilitated by the polymer's structure, which is abundant in amino and hydroxyl groups. The ability of chitosan to adsorb metals like copper, zinc, arsenic, and mercury has been shown in studies; the adsorption efficiency of chitosan is pH-dependent and typically increases under slightly acidic to neutral environments (pH 4–6). Chitosan has been modified with surfactants such as sodium dodecyl sulphate (SDS) to improve its adsorption capabilities. The resultant chitosan-SDS combination creates an insoluble matrix that can bind pollutants, such as heavy metals and anionic organic molecules. For uses in animal feed to stop the absorption of toxins, this complex has demonstrated enhanced stability across pH levels and maintained its adsorption capacity in various gastrointestinal circumstances (Delgado-Cedeño et al. 2022) (Table 7.1).

7.4 Chitosan Based Intelligent Packaging and Properties

Packaging involves the intersection of art, science, and technology to deliver products to consumers safely, efficiently, and with minimal cost. This is a critical process for maintaining both fresh and processed food. It plays a vital role in allowing people to consume the food grown in a specific region, sometimes days, weeks, or months later. With the growth of food production, storing food for future use became more convenient as the population grew and people needed shelter and housing. This gave rise to storage and packing technologies. Early materials used for this purpose were made of leaves, shells, woven grasses, hollow logs and animal organs (Berger 2003). The perfect food packaging material should preserve food quality for a n extended period, be light and easy to handle, be cheap enough to use, and be either renewable or biodegradable to prevent trash (Risch 2009).

The film-forming ability of chitosan is considered its most important characteristic from the perspective of food packaging and has received much research attention. Research has been done on using these films to increase food quality due to the barrier properties of chitosan and its natural antimicrobial properties. Considerable

Table 7.1 Properties of chitosan

Properties/ activity	Description	Influencing factors	Applications	References
Molecular weight	Varies from 50,000 to >1,000,000 Da; affects solubility, viscosity and mechanics	Reaction conditions, enzymatic/ chemical degradation	Scaffold building, film production, drug delivery systems	Aranaz et al. (2021), Kose and Suvacı (2022), Román-Doval et al. (2023)
Degree of deacetylation (DDA)	Percentage of deacetylated glucosamine units (50–95%); impacts solubility, charge density and reactivity	DDA levels, analytical methods (NMR, FTIR)	Tissue engineering, drug delivery, water treatment	Mathaba and Daramola (2020), Pellis et al. (2022), Novikov et al. (2023)
Solubility	Dissolves in acidic solutions ($pH < 6.5$); solubility influenced by molecular weight and DDA	Acidic pH, molecular weight, DDA, chemical modifications	Industrial uses, biological systems, neutral pH solutions (via modification)	Mathaba and Daramola (2020), Potivas and Laokuldilok (2014), Nguyen et al. (2022)
Hydrophilicity	Naturally hydrophilic due to hydroxyl and amino groups; enhances biocompatibility and moisture retention	Crosslinking blending with hydrophobic polymers	Wound dressing, packaging, controlled drug release	Piekarska et al. (2023), Cheung et al. (2015), Thambiliyagodage et al. (2023)
pH sensitivity	Responds to pH changes; protonated amino groups increase solubility and positive charge in acidic pH	pH levels, ionizations of amino groups	pH-responsive drug delivery systems, interaction with proteins, DNA and anionic polymers	Alemu et al. (2023), Nilsen-Nygaard et al. (2015), Li et al. (2020a, b)
Gas permeability	Limited permeability to gases like CO_2 and O_2 due to semi-crystalline structure and hydrogen bonding between polymer chains	Humidity, Molecular weight, DDA, modifications like crosslinking with glutaraldehyde or blending with other polymers	Food packaging, biomedical applications requiring gas regulation	Shi et al. (2024), Butnaru et al. (2019), Hu et al. (2020)
Water vapour barrier	Moderate barrier properties suitable for regulating moisture exchange	Hydrophilicity, additives, chemical modifications like acetylation or hydrophobic group grafting	Food packaging, wound dressings, biomedical applications	Shi et al. (2024), Cazón et al. (2022) Butnaru et al. (2019)

(continued)

Table 7.1 (continued)

Properties/activity	Description	Influencing factors	Applications	References
Impact of moisture sensitivity	Hydrophilic nature leads to sensitivity to ambient humidity, affecting water vapour barrier properties	High relative humidity, chemical modifications	Smart packaging	Aguirre-Loredo et al. (2016), Butnaru et al. (2019)
Tensile strength	Measures the ability to withstand pulling forces. Chitosan's tensile strength is typically 30–60 MPa	Molecular weight and DDA increases the strength through enhanced intermolecular hydrogen bonding, additives like cellulose or nanoclays improves strength	Biodegradable films, coatings and packaging materials	Gadgey and Sharma (2020), Shi et al. (2024)
Elongation at break	Indicates material's flexibility with values ranging between 2 and 20%	Addition of plasticizers like sorbitol, glycerol or PEG enhances flexibility by reducing intermolecular interactions. Environmental factors like humidity acts as natural plasticizers	Flexible films, coatings and biomedical scaffolds	Alemu et al. (2023), Habibie et al. (2016), El-Araby et al. (2024)
Modulus of elasticity	Represents the stiffness of the material. Chitosan is inherently stiff with a high modulus	Crosslinking with substances like glutaraldehyde or genipin increases stiffness. Blending with polymers like PVA or PEO adjusts flexibility and stiffness	Structural components, stiff films and materials requiring rigidity under mechanical stress	Salavati (2023), Shi et al. (2024)

(continued)

Table 7.1 (continued)

Properties/ activity	Description	Influencing factors	Applications	References
Toughness and flexibility	Combines strength and flexibility, critical for applications requiring durability	Composites with gelatin, alginate or synthetic materials reduce brittleness. Nanoparticles like ZnO or TiO_2 improve toughness and mechanical strength	Biomedical devices, packaging and applications requiring impact resistance and durability	Gadgey and Sharma (2020), Salavati (2023)
Environmental sensitivity	Chitosan's properties are influenced by humidity and temperature	High humidity increases elongation at break while reducing tensile strength; low humidity makes the material brittle	Design of chitosan-based materials for variable environmental conditions in packaging and coatings	Alemu et al. (2023), El-Araby et al. (2024)
Thermal stability	Chitosan has good thermal stability due to its DDA and molecular weight. Thermal degradation occurs in two phases: loss of water (<150 °C) and glycosidic linkage breakdown (200–300 °C). Significant degradation occurs above 300 °C	DDA, Molecular weight	Applications requiring moderate thermal stability such as coatings and biomedical devices	Nainggolan et al. (2020), Zakaria et al. (2012)
Glass transition temperature (Tg)	Chitosan has a Tg between 150 and 200 °C. Water acts as a plasticizer, lowering Tg. Crosslinking agents, fillers or blending can modify Tg	Water content, DDA, molecular weight, crosslinking agents	Applications requiring specific thermal performance or flexibility such as packaging and films	Eulalio et al. (2019), Dhawade and Jagtap (2012)
Melting behaviour	Does not exhibit a clear melting point due to strong hydrogen bonds and semi-crystalline structure. Heating transitions it to an amorphous state	Molecular structure, semi-crystallinity	Limited thermoplastic applications; suitable for moderate heat-resistant uses	Gubanova et al. (2021), Zakaria et al. (2012)

(continued)

Table 7.1 (continued)

Properties/ activity	Description	Influencing factors	Applications	References
Crystallinity and thermal modifications	Higher crystallinity enhances thermal stability. Crystallinity can be altered by heat treatment, crosslinking or blending with other polymers like PVA or PEG	DDA, processing conditions, preparation techniques	Packaging, biomedical devices and materials requiring enhanced thermal resistance	Nainggolan et al. (2020), Gubanova et al. (2021)
Thermal conductivity	Poor thermal conductivity makes chitosan a good insulator	Fillers, nanoparticle modifications	Thermal protection, food preservation and coating materials	Eulalio et al. (2019), Zakaria et al. (2012)
Biodegradability	Chitosan is enzymatically degraded into glucosamine and N-acetylglucosamine. Degradation depends on DDA, molecular weight, crystallinity and environmental conditions	DDA, molecular weight, crystallinity, environmental factors such as pH, temperature, moisture; structural modifications and microbial activity	Biodegradable packaging, drug delivery systems, agricultural films	Reshad et al. (2021), Islam et al. (2019), Jiménez-Gómez and Cecilia (2020), Yarahmadi et al. (2024), Wrońska et al. (2023), Casadidio et al. (2019), Matica et al. (2017)
Antimicrobial activity	Chitosan acts against bacteria, yeasts and mold by disrupting microbial cell membranes and chelating essential metal ions	Molecular weight, DDA, pH, functional additives	Active food packaging materials enhanced with Ag, ZnO or TiO_2 nanoparticles for improved antibacterial action	Ke et al. (2021), Yilmaz Atay (2019), Goy et al. (2009), Mawazi et al. (2024), Hafdani and Sadeghinia (2011)
Antioxidant activity	Exhibits free radical scavenging and metal ion chelating properties due to amino ($-NH_2$) and hydroxyl ($-OH$) groups	Molecular structure, derivation source and functional group availability	Natural antioxidant for food preservation; prevents lipid oxidation in food matrices	Liu et al. (2022), Muthu et al. (2021) Hao et al. (2021), El-Araby et al. (2024), He et al. (2024)
Film-forming properties	Forms semi-permeable edible films/coatings that control moisture transfer, gas exchange and microbial growth	DDA, Molecular weight, Addition of plasticizers and combination with other polymers	Food coatings, delivery of bioactive compounds in packaging	Muñoz-Tebar et al. (2023), Khubiev et al. (2023), Singh et al. (2015), Šuput et al. (2015), Luangapai et al. (2019)

(continued)

Table 7.1 (continued)

Properties/ activity	Description	Influencing factors	Applications	References
Anti-browning effects	Prevents enzymatic browning by inhibiting PPO activity and microbial growth; preserves firmness and vitamin C content	pH, combination with other agents such as citric acid, ascorbic acid; environmental temperature	Coatings for fresh-cut fruits and seafood; enhances visual and nutritional quality during storage	Shyu et al. (2019), Apai et al. (2009), Hamdan et al. (2022), Demircan and Velioglu (2024)
Prebiotic activity	Chitosan oligosaccharides (COS) promote growth of beneficial gut bacteria like *Lactobacillus* and *Bifidobacterium*	Degree of polymerization, dose and resistance to digestion in upper GI tract	Functional food ingredient; improves gut health and metabolic health; feed additive in aquaculture	Lee et al. (2002), Guan and Feng (2022), Zhang et al. (2018), Benchamas et al. (2021) Wang et al. (2021)
Emulsifying properties	Stabilizes oil-in-water emulsions by reducing interfacial tension; binds dietary fats and bile acids, forming non-absorbable complexes	Molecular weight, viscosity, amphiphilic nature	Stabilizes in dressing, sauces and emulsions; potential weight management supplements	Panith et al. (2016), Jin et al. (2017), Liu et al. (2021)
Adsorption of toxins	Binds mycotoxins and heavy metals through electrostatic interactions and chelation, improving food and feed safety	pH, contact time, temperature, chitosan modifications such as cross-linking, surfactant addition	Adsorbent in food/ feed to reduce contamination; potential applications in gastrointestinal detoxification	Solís-Cruzet et al. (2017), Abbasi Pirouz et al. (2020), Zhao et al. (2015), Delgado-Cedeño et al. (2022)

effort has occurred to develop pure chitosan films, chitosan blends, membranes, and edible coatings, which will be described in detail in theoretical discussion sections for food packaging applications.

Chitin is a cellulose-like biopolymer found in the exoskeletons of insects and crustaceans and in the fungi and yeast cell walls. Chitin has a structural role analogous to cellulose in plants. This polymer is mainly composed of (1–4)-linked 2-acetamido-2-deoxy-β-D-glucose monomers and is produced in metric ton quantities, about 1011 tons per year (Elieh-Ali-Komi and Hamblin 2016). However, only about 150,000 tons are available for commercial applications (Guan et al. 2019). Consequently, it is the second most prevalent biopolymer on the planet, after cellulose and the most abundant biopolymer derived from animal sources. Unlike other polysaccharides, chitin contains nitrogen. Chitosan characteristics, such as the degree of acetylation (DA), nitrogen content, N/C ratio, molecular size, and polydispersity, vary depending on the source of chitin. Typically, chitin has a DA above 90%, with a nitrogen content of 7% and an N/C ratio of 0.146. In contrast, chitosan has a DA of less than 40% and a nitrogen content greater than 7% (Purohit et al. 2024). The presence of amino groups in chitosan imparts unique functional properties; these electronegative groups can accept protons and acquire a positive charge, contributing to chitosan's diverse chemical, physical, and biological characteristics. Its solubility, film-forming ability, viscosity, ion binding, and antimicrobial properties render chitosan a versatile polymer suitable for various research applications, including food packaging (Wang et al. 2018).

7.4.1 Different Chitosan-Based Film

Chitosan can be used in food packaging as films or as coatings applied directly onto food products. Numerous researchers worldwide have investigated both methods. Chitosan-based films can be used as food packaging materials and extend the shelf life of food in the form of pure chitosan films, chitosan/biopolymer films, chitosan/synthetic polymer films, chitosan derivative films, etc.

Pure chitosan films: Chitosan dissolves in dilute acids like acetic and hydrochloric acid, enabling the creation of various products, including edible films. Research indicates that chitosan films effectively preserve foods such as bananas, pomegranates, carrots, mangoes, tomatoes, fish, papayas, and wolfberries by enhancing storage stability. These films can delay quality and nutritional changes, inhibit microbial growth, maintain antioxidant activity, and extend shelf life (Romanazzi et al. 2017).

Films made from traditional chitosan and submicron dispersions of 600 nm in diameter outperform those made from 1000 nm in food preservation (Ali et al. 2014). Adding plasticizers like glycerol and sorbitol improves the films strength and flexibility, while emulsifiers such as Tween 80 and Span 20 enhance film properties. Vacuum packaging is often used with chitosan films for fish preservation (Günlü et al. 2014). Despite their favourable characteristics, some films may face challenges like weight loss and the requirement for specific conditions, such as ultrasound, which could limit their mass application in food packaging (Mustafa et al. 2014).

Chitosan/Polysaccharides-Based Films: Polysaccharides are effective when blended with chitosan to create functional films. Starch, notable for its low cost, abundant availability, and biodegradability, is one of the key renewable polysaccharides sourced from plants. Both chitosan and starch possess strong film-forming abilities that facilitate the development of composite films (Talón et al. 2017). The resulting films demonstrated lower bacterial adhesion, enhanced antioxidant activity, and improved water vapour barrier properties, making them promising candidates for active packaging solutions (Xiao et al. 2013).

Alginate, a natural linear anionic polysaccharide, is a promising biopolymer due to its low toxicity and chemical adaptability. Stable composite films made from chitosan and alginate were created through electrostatic deposition of their opposing charges. These charges significantly affected the films' properties, including contact angle, microstructure, and thermal behaviour (Da Silva et al. 2013). The chitosan/alginate films exhibited excellent gas exchange and water vapour permeability, resulting in an appealing appearance and high microbiological and physicochemical quality in the treated foods. These chitosan/alginate films demonstrate significant potential for use in food packaging (Liu et al. 2013; Da Silva et al. 2013).

Pectin, a structural polysaccharide from plants, forms stable and uniform films through strong interactions with positively charged chitosan. This combination resulted in fresh-cut cantaloupe with high physicochemical and sensory qualities, making these composite films suitable alternatives to conventional food packaging (Hoagland and Parris 1996). Cyclodextrin has also been shown to enhance the tensile strength of chitosan films. When combined with active materials or photoactive agents, chitosan/cyclodextrin films exhibited improved antimicrobial properties and controlled release, positioning them as effective bioactive packaging materials (Higueras et al. 2014). Various polysaccharides, such as glucose, xylan, fucose, and konjac glucomannan, have been explored with chitosan for food packaging. For instance, chitosan/glucose films demonstrated enhanced antioxidant properties, effectively reducing decay and weight loss while preserving berry texture and sensory quality (Gao et al. 2013). Overall, these biodegradable chitosan/polysaccharide films offer promising enhancements for extending the shelf life of food products.

Chitosan/Protein-Based Films: Animal-derived proteins have garnered significant interest among researchers due to their film-forming capabilities, high nutritional value, and biocompatibility. Caseinate, known for its excellent thermoplastic and film-forming properties, can be combined with chitosan to create chitosan/caseinate films that improve water vapour permeability (Khwaldia et al. 2014). As a potential substitute for synthetic polymers, Collagen also forms composite films with chitosan, enhancing thermal stability, adhesion, and compatibility (Ahmad et al. 2016). Lysozyme-chitosan films have been shown to extend the freshness of eggs, strengthen their shells, and preserve internal quality (Yuceer and Caner 2014). Gelatin, a purified protein sourced from collagen, is often used in chitosan/gelatin films, demonstrating enhanced mechanical properties and effective barriers against water vapour and UV light (Kowalczyk et al. 2015). Foods like red bell peppers treated with chitosan-gelatin edible films exhibit reduced microbial decay and improved physicochemical and microbiological properties, leading to extended shelf life (Poverenov et al. 2014a, b).

Plant-derived proteins show great potential for food preservation due to their abundance and affordability. The chitosan/kidney bean protein isolate complex results in films that are less rigid and more flexible, featuring high surface hydrophobicity and low surface free energy, making them suitable for antimicrobial food packaging (Ma et al. 2013). Additionally, chitosan/quinoa protein composite films exhibit enhanced mechanical and water barrier properties, supporting their application in food packaging (Caro et al. 2016).

Microbial proteins can significantly contribute to packaging films. Nisin, an amphiphilic cationic peptide derived from microorganisms like *Lactococcus lactis*, can enhance the shelf life of packaged foods when combined with chitosan (Imran et al. 2014). When chitosan/ε-polylysine/nisin films were applied to fresh-cut carrots, they effectively reduced respiration rates, minimized ascorbic acid loss, inhibiting microbial growth, and suppressed the synthesis of white blush and lignin (Song et al. 2017).

Chitosan/Extracts-Based Films: Bee-derived extracts, such as beeswax and propolis, have been combined with chitosan to create food packaging films. Beeswax, known for its antimicrobial properties, offers an eco-friendly option for controlling pathogens and preserving food quality. When incorporated into chitosan films, beeswax enhances visual appeal and taste (Velickova et al. 2013). Additionally, extracts from bee carbodiimide, achieved through free-radical-initiated grafting, improve antioxidant capacity, antimicrobial effectiveness, tensile strength, and reduce water and oxygen permeability, making them suitable for multifunctional food packaging.

Essential oils extracted from plants also enrich chitosan films. For example, films infused with cinnamon oil exhibit strong antimicrobial properties. At the same time, carvacrol ensures food safety, especially when combined with techniques like gamma irradiation and modified atmosphere packaging (Wang et al. 2011; Kurek et al. 2013). Other essential oils, including olive, buriti, rosemary, and oregano, enhance the mechanical and barrier properties and the antioxidant activity of chitosan films, although some may impart a deep orange colour (Pelissari et al. 2009).

Plant extracts, being natural and non-toxic, have been utilized in food preservation for years. Adding plant extracts to chitosan films significantly alters their properties, enhancing antimicrobial (e.g., honeysuckle and citrus extracts) (Wang et al. 2017), antioxidant (e.g., clove eugenol and maqui berry extracts) (Genskowsky et al. 2015), barrier performance (e.g., thyme extract) (Talón et al. 2017), mechanical properties (e.g., tannic acid) (Rivero et al. 2012), thermal stability (e.g., young apple polyphenols) (Sun et al. 2017) and colour (e.g., carvacrol) (Kurek et al. 2013), resulting in a synergistic effect between chitosan and the extracts. Gallic acid, a promising functional material, can also be grafted onto chitosan (Guo et al. 2016).

Chitosan/Synthetic Polymer-Based Films: Wang et al. found that chitosan can be blended with synthetic polymers to create films with enhanced properties. Polyvinyl alcohol (PVA), a non-toxic and water-soluble synthetic polymer, forms strong intermolecular hydrogen bonds with chitosan, facilitating film fabrication. High concentrations of PVA improve elasticity, elongation, tensile strength, and barrier properties against water and oxygen, making these films suitable as universal antimicrobial food packaging materials (Wang et al. 2015a, b). Additionally, Feng et al. have incorporated various bioactive materials to impart unique characteristics, such as fire resistance and improved mechanical properties, thus expanding the applications of chitosan/PVA films (Feng et al. 2013). Chitosan can also be combined with synthetic acids, resulting in chitosan/polylactic acid films by blending chitosan with polylactic acid or incorporating lactic acid oligomer-grafted chitosan as a nanofiller. Chitosan films can incorporate synthetic acids, and chitosan/polylactic acid films can be created by blending chitosan with polylactic acid or adding lactic acid oligomer grafted-chitosan as a nanofiller. These composite films enhance properties like tensile strength and thermal resistance, making them suitable for food packaging (Pal and Katiyar 2016). Additionally, Zhang et al. found that films combining chitosan with synthetic acids (e.g., salicylic and fumaric acids) can reduce chilling injury and maintain food quality, such as in cucumbers (Zhang et al. 2015). Chitosan can also coat synthetic polymer films, like polyethylene, to impart antibacterial and hydrophilic properties. The films characteristics can be adjusted by varying the biopolymer's electrical charges and chitosan concentration. Chitosan is also effective with other synthetic polymers, enhancing antimicrobial and optical properties for food packaging applications (Wang et al. 2015a, b).

7.4.2 Fabrication Method for Chitosan-Based Film

Chitosan can be used in food packaging as films or as coatings applied directly onto food products. Numerous researchers worldwide have investigated both methods. This section outlines the techniques used to create chitosan films and coatings. Additionally, it reviews studies on the formulations and uses of pure chitosan films and chitosan films combined with other polymers or biopolymers, including polysaccharides, proteins, and synthetic plastics.

Direct casting: The direct casting method is a widely used and straightforward approach for preparing biopolymer films. The process involves several steps: (1) dissolving chitosan in a weak acidic solution with a specific pH; (2) adding active compounds, such as cross-linkers, plasticizers, and fillers; (3) blending with other polymers if necessary; (4) stirring to achieve a thick, uniform solution; (5) filtering or centrifuging to remove insoluble particles and air bubbles; (6) pouring the solution onto a flat, level surface coated with a non-stick material; (7) drying under controlled temperature and humidity for a specified duration; (8) peeling off the film; and (9) curing and storing it (Wang et al. 2018).

The two-step casting technique is also effective for film preparation. First, a fucose-rich exopolysaccharide solution is cast and dried on an adhesive surface, followed by the casting of a chitosan solution on top. This method produces dense and homogeneous films with excellent gas barrier properties, making them suitable for packaging low-moisture products (Ferreira et al. 2016). Additionally, Wu et al., have created green and immersing methods. This process involves: (1) casting a cellulose solution onto a glass plate; (2) immersing the resulting films in a coagulation bath of 5% (w/v) H_2SO_4 for 10 min, then into a 5 wt% Na_5lO_6 solution for varying duration; and (3) immersing the films in a 2.5 wt% chitosan solution to form composite films. These composite films offer better preservation than traditional polyethylene wraps (Wu et al. 2016).

Layer-by-layer assembly: Layer-by-layer (LBL) is a technique for electrostatic deposition that allows precise control over the properties and functions of biomaterial films. It can be used to create multi-component films without complex equipment. Additionally, the films produced are independent of the shape of the substrate (Li et al. 2013). pH plays a crucial role in forming LBL films, as it influences the electrical charge, affecting the amount of polymer deposition. A higher number of biopolymer molecules is needed to neutralize the previous layer (Acevedo-Fani et al. 2015). Additionally, the structure and characteristics of the multilayer films are significantly impacted by pH (Fernández et al. 2014). As pH decreases and the number of layers increases, the films' thickness, roughness, and elastic modulus also rise. Higher levels of photo orientation are achieved under these conditions. Some researchers found that films exhibiting optimal barrier performance were composed of chitosan (pH 5.5), polyacrylic acid (pH 3), chitosan (pH 5.5), and graphene oxide (48 nm). Chitosan becomes highly ionized at this pH by the polyacrylic acid counterion, allowing for greater incorporation of graphene oxide into the film. These multilayer films demonstrate combined properties of their components and show enhanced inhibition of E. coli and increased antioxidant activity with a higher bilayer count (Fernández et al. 2014; del Hoyo-Gallego et al. 2016).

Combining the LBL assembly technique with the immersion method is commonly utilized. In this process, films are created using oppositely charged materials, such as polyanion and polycation chitosan, by immersing objects multiple times in their respective solutions (Poverenov et al. 2014a, b). Various factors, including microbial quality, sensory analysis, pH, colour, moisture, texture and material growth, were assessed to determine that the resulting film could extend the shelf life of food. This combined method is relatively straightforward and provides an effective approach for fabricating functional materials that can enhance the shelf life and preserve the quality of treated food products. Additionally, under high humidity conditions, some of these films demonstrate excellent barrier properties (Tzeng et al. 2015).

Extrusion: Melt extrusion is a well-established technique for producing chitosan-based biodegradable active packaging films. The process includes several steps: (1) formulating materials with various compositions; (2) mixing and blending these materials; (3) extruding the blended material under controlled conditions; (4) cutting the extrudates into pellets using a pelletizer; (5) drying the pellets; and (6) either using a twin-screw extruder with a flat die to create flat sheets or employing a blown film extruder with an annular die to produce blown films (Martínez-Camacho et al. 2013).

The extrusion process often yields films with satisfactory mechanical properties and good thermal stability. During this process, low-density polyethene acts as the matrix polymer, while ethylene-acrylic acid copolymer is an adhesive to enhance adhesion between immiscible polymers (Martínez-Camacho et al. 2013). A two-step melt-compounding process is also commonly used for producing blown films. As the chitosan content increases, the breaking strength and elongation decrease while water vapour permeability improves (Wang et al. 2015a, b). Another film, created from polyethylene glycol, chitosan, and low-density polyethylene, was made using compression molding. The thermal stability of this film primarily relied on the low-density polyethylene. Although the films exhibited excellent transparency, adding polyethylene glycol reduced Young's modulus (Davidovich-Pinhas et al. 2014). A three-step extrusion process can be achieved using a coextrusion device, allowing for film production without degradation: (1) plasticizing starch or flour powder via twin-screw extrusion; (2) extruding thermosplasticized starches or flour granulates with polycaprolactone; and (3) co-extruding with polycaprolactone/chitosan (Alix et al. 2013). A blended matrix of polylactic acid/starch/chitosan was extruded into an antimicrobial film, with thicknesses ranging from 0.15 to 0.18 mm, making it suitable for protecting high-water-activity foods like fresh meat (Bie et al. 2013). Some researchers compared casting, extrusion, and coating films, finding that coated films performed best in preserving beefsteak. However, the presence of chitosan negatively impacted the mechanical properties of the extruded films, likely due to weak interactions between the polyamide matrix and chitosan, leading to poor affinity or miscibility and resulting in diminished mechanical performance and brittleness (Lago et al. 2014).

To sum up, extrusion efficiently produces films with strong mechanical, thermal, and antimicrobial properties. However, to optimize performance, it's important to consider the processing temperature and the interactions between the polymers involved (Alix et al. 2013).

7.4.3 Chitosan-Based Coating

The coating is commonly used to create chitosan-based thin films on surfaces, immobilize functional polymers for protective purposes, or impart unique properties. These biobased films, which are environmentally friendly, offer benefits such as low cost and biodegradability. They are attractive alternatives to synthetic materials for food packaging. Techniques like spread-coating and spray-coating show significant potential for enhancing food preservation.

Spread-Coating: Spread-coating is commonly employed to create chitosan films using brushes and spatulas. This method includes direct coating on food surfaces, such as vegetables, fruits, and meats, and indirect coating on packaging materials.

The direct coating method is crucial for reducing microbial growth on food surfaces. The process includes: (1) preparing a chitosan formulation that includes active ingredients like antioxidants, antimicrobials, or reinforcing agents; (2) preparing the food for coating through methods such as washing, cutting, pasteurizing, irradiating, or heating; (3) using a sterile tool to evenly apply the chitosan formulation on the food surface; (4) drying the coated food under specific conditions in a sterile environment like a laminar air-flow cabinet.; and (5) packaging and storing the food in a refrigerator or vacuum. For indirect coating, the focus is on creating multi-layered packaging materials that offer improved properties compared to single-layer options. Chitosan coatings on paper provide antimicrobial benefits and decrease moisture permeability without significantly impacting the material's mechanical properties (Khwaldia et al. 2014).

Spray-Coating: Spray-coating is typically performed on food using tools such as compressed air-assisted sprayers, knapsack sprayers, and copper backpacks. Combining spray-coating with other treatments offers several benefits. For example, using a combination of spray coating, gamma irradiation, and modified atmosphere packaging can enhance the effectiveness of all three methods, reducing microbial populations and ensuring the safety of green beans. Additionally, the combination of chitosan spray coating and heat treatment has been shown to improve the postharvest quality and acceptability of wolfberries significantly. Preharvest spraying with oligochitosan has also been effective in boosting the resistance of navel oranges to anthracnose. While chitosan film does not affect ripening or alcoholic fermentation, it can significantly influence the composition of berries and wines, including phenolic acids, flavonols, and nitrogenous compounds.

Overall, the coating process is straightforward and does not require complex equipment. It effectively protects food from spoilage and helps maintain its quality. In some cases, spray-coated films have been found to perform better than cast films in reducing lipid oxidation in food. (Ban et al. 2015; Nowzari et al. 2013).

Dip-coating: The dip-coating technique involves immersing food items in chitosan-based solutions, resulting in a uniform film on their surfaces. The formulation of this film is influenced by factors such as the wetting ability of the surface, processing time, and draining time. The process includes (1) preparing a chitosan solution with a specific pH, (2) selecting and preparing the food sample for coating, and (3) immersing the food sample in the solution for a set. duration, which can range from 30 s to 30 min depending on the type of food; (4) removing the food, draining off excess solution, and drying it; (5) placing the food on trays or plastic supports; and (6) packaging and storing the finished product. Dip-coated food samples have shown increased resistance to chilling injury and reduced colour variation and decay, improving quality and longer shelf life. Some studies recommend a double-dipping process, noting that cheese coated in this manner retains superior aroma and flavour. Overall, dip-coating is a straightforward, time-efficient method that does not require extensive equipment while effectively preserving food and maintaining its quality (Kanatt et al. 2013; Liu et al. 2016).

7.5 Application of Chitosan-Based Film or Coating in Food Packaging

Chitosan is an active food packaging material and can be used as a thin edible film or coating. The edible coating is created by applying a liquid chitosan solution to food, either by dipping the items or using an aerosol spray to apply a fine mist. In contrast, the chitosan film is a pre-made layer that can be placed on or between food items (Muxika et al. 2017). Using semipermeable chitosan films or coatings shows great potential for extending the shelf life of perishable fruits and vegetables as a post-harvest treatment. Chitosan coatings have been shown to preserve various fruits effectively. Compared to prochloraz-treated peaches, chitosan significantly reduced brown rot caused by *Monilina fructicola* and *Rhizopus stolonifera* while maintaining firmer texture, higher titratable acidity, and vitamin C levels and improving membrane integrity by reducing malondialdehyde formation and stimulating superoxide dismutase activity (Li and Yu 2001). Similarly, chitosan extended the shelf life of peeled litchi by minimizing weight loss, maintaining sensory quality, and suppressing polyphenol oxidase and peroxidase activity (Dong et al. 2004). For sliced mango, chitosan coatings reduced water loss, retained sensory quality, and enhanced soluble solids, titratable acidity, and vitamin C while inhibiting microbial growth (Chien et al. 2007).

The impact of chitosan coatings on tomatoes, mushrooms, and broccoli was studied, showing significant preservation benefits. In tomatoes, chitosan reduced respiration and ethylene production, maintained firmness, increased titratable acidity, minimized decay, and slowed red pigmentation during storage (Ghaouth et al. 1992). For fresh-cut mushrooms, chitosan delayed discolouration by reducing enzyme activity (e.g., polyphenol oxidase, peroxidase) and microbial counts (bacteria, yeast, mould) while lowering phenolic content (Eissa 2007). In minimally processed broccoli, chitosan-glycerol coatings controlled native microflora, inhibited *E. coli O157:H7*, reduced yellowing, and prevented floret. opening, and enhanced sensory quality (Moreira et al. 2011). Baby carrots were coated using spraying and dipping methods, with dipped samples showing slightly higher weight loss due to thicker coatings. Chitosan-based coatings effectively delayed microbial spoilage while maintaining colour, texture, and overall quality. Sensory analysis revealed similar acceptability levels for coated and uncoated carrots (Leceta et al. 2015).

Shahbazi et al., evaluated the impact of chitosan films enriched with cellulose nanoparticles (CN), ethanolic propolis extract (EPE), or both on the chemical, microbial, and sensory qualities of minced beef during 14 days of refrigeration. Treated samples showed a significant microbial reduction (1–5.5 log CFU/g) compared to the control. Films containing 2% EPE alone or combined with 1–2% CN effectively slowed microbial growth and oxidative spoilage without compromising odor, color, or overall acceptability (Shahbazi and Shavisi 2018). Chitosan films, with excellent oxygen-barrier properties, effectively protect meat fats from oxidation and control microbial growth. Using chitosan coatings or films is a promising, natural alternative to synthetic antimicrobial and antioxidant additives, enhancing fresh meat products shelf life and value (Bonilla et al. 2014a). Using chitosan films enriched with essential oils on pork fat and minced pork meat was studied (Bonilla et al. 2014b). While essential oils slightly reduced moisture barrier properties, the films prevented fat oxidation. In minced meat, reduced oxygen levels caused myoglobin to convert to metmyoglobin, darkening the meat and potentially affecting consumer acceptance. However, chitosan films enhanced safety and extended the product's shelf life.

Cooked pork sausages coated with chitosan, glycerol, and clove oil remained sensory acceptable for up to 20 days under refrigeration (Lekjing 2016). Similarly, dipping chicken breast meat in chitosan with oregano oil and modified atmosphere packaging inhibited microbial growth and extended shelf life to 21 days without significant colour changes, ensuring sensory acceptability throughout refrigerated storage (Chi et al. 2006) (Fig. 7.3).

Seafood products are highly prone to quality deterioration due to lipid oxidation, autolysis, microbial contamination, and loss of protein functionality (No et al. 2007). In addition to its antimicrobial properties, chitosan exhibits a strong antioxidant capacity, effectively slowing lipid oxidation in fresh fish muscle (Souza et al. 2010). This is achieved by chelating ferrous ions, reducing their prooxidant activity, or preventing their conversion to ferric ions (No et al. 2007). The effect of chitosan coating on salmon fillets stored at 0 °C showed a significant reduction in pH, ATP

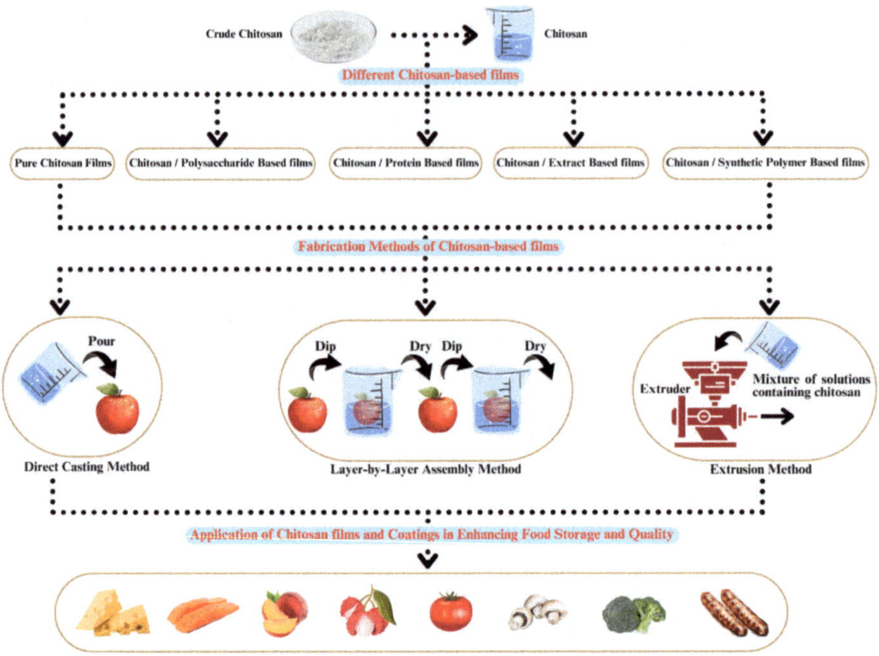

Fig. 7.3 A variety of films are made from crude chitosan, including pure films, films based on polysaccharides, films based on proteins, films based on extracts, and films based on synthetic polymers. These films are made by a variety of techniques, including the layer-by-layer assembly method, which involves dipping and drying layers of chitosan; the direct casting method, which involves pouring chitosan solution directly onto the surface; and the extrusion method, which involves combining chitosan solutions with other ingredients and creating films through extrusion. Various food products, including cheese, fish, fruits (like tomatoes, peaches, and litchi), vegetables (like broccoli and mushrooms), and processed meats (like sausages), benefit from the application of the resulting chitosan-based films and coatings, which improve food preservation and quality. These coatings offer a number of additional benefits in addition to extending shelf life

breakdown products after 6 days, and total volatile base nitrogen, trimethylamine, and thiobarbituric acid values after 9 days. Microbial growth was slower, extending the shelf life by 3 days (Souza et al. 2010).

Similarly, coating Mozzarella cheese with chitosan-lysozyme improves microbial safety by controlling post-processing contaminants, addressing the rising demand for safer cheese production. Chitosan films and coatings were evaluated for their ability to inhibit *Listeria monocytogenes, Escherichia coli, Pseudomonas fluorescens, Candida inconspicua*, and moulds, which can cause cheese-associated outbreaks, leading to flavour, texture, and appearance defects, economic losses, or severe illnesses. Chitosan-lysozyme films demonstrated the strongest antimicrobial effect due to their higher concentrations of chitosan and lysozyme than laminated films and coatings (Duan et al. 2007) (Table 7.2).

Table 7.2 Applications of chitosan films and coatings in enhancing food storage and quality

Food product	Chitosan application	Effects observed	References
Peaches	Chitosan coating	Reduced brown rot incidence, firmer texture, higher acidity, vitamin C retention, reduced malondialdehyde, stimulated superoxide dismutase activity, maintained membrane integrity	Li and Yu (2001)
Litchi fruit	Chitosan coating	Retarded weight loss, maintained sensory quality, preserved total soluble solids, titratable acidity, ascorbic acid and suppressed enzymatic activity	Dong et al. (2004)
Mango slices	Chitosan coating	Retarded water loss, improved sensory quality, increased soluble solids, titratable acidity, vitamin C and microbial inhibition	Chien et al. (2007)
Tomatoes	Chitosan coating	Reduced respiration and ethylene production, firmer texture, higher titratable acidity, reduced decay, less red pigmentation	Ghaouth et al. (1992)
Mushrooms	Chitosan coating	Delayed discoloration, reduced enzyme activities (polyphenoloxidase, peroxidase, catalase etc.), lower total phenolic content, reduced microbial colonies	Eissa (2007)
Broccoli	Chitosan-glycerol coating	Controlled microbial growth, improved sensory quality, inhibited yellowing and floret. opening	Moreira et al. (2011)
Baby carrots	Chitosan coating via spraying/dipping	Delayed microbial spoilage, maintained colour and texture, similar sensory acceptability compared to uncoated samples	Leceta et al. (2015)
Minced meat	Chitosan coating	Reduced oxidation and microbial growth, extended shelf life	Bonilla et al. (2014a, b)
Cooked pork sausages	Chitosan-glycerol clove oil coating	Sensory acceptability maintained for 20 days in refrigeration	Lekjing (2016)
Chicken breast	Chitosan-oregano oil dipping in modified atmosphere	Extended shelf life to 21 days, no significant colour changes	Chi et al. (2006)
Salmon fillets	Chitosan coating	Slower pH ad microbial growth, reduced spoilage markers (volatile base nitrogen, trimethylamine), extended shelf life by 3 days	Souza et al. (2010)

(continued)

Table 7.2 (continued)

Food product	Chitosan application	Effects observed	References
Mozzarella cheese	Chitosan-lysozyme coating	Improved microbial safety by reducing post-processing contaminants	Duan et al. (2007)
Seafood products	Chitosan films	Slowed lipid oxidation, antimicrobial properties	Souza et al. (2010)

7.6 Future Perspective and Conclusion

The massive use of plastic packaging materials in the food packaging field generates a huge amount of plastic waste that is directly involved in the damage to the planet in terms of air pollution, water pollution, and carbon footprint. Therefore, current trends focus on developing biodegradable packaging materials from chitosan, which can be obtained from renewable sources like crustacean waste. The literature revealed that chitosan is a biopolymer with multifarious properties and functions, fit to fabricate packaging materials with significant antioxidant and antimicrobial properties along with improved barrier, mechanical and functional properties. Further studies need to be conducted on chitosan know about the other radical scavenging potential, types of foodstuffs that can be candidates to package with chitosan films, and how the chitosan-based films will meet the practical criteria to compete with conventional petroleum-based packaging materials. In addition, how the D units or the linkage of the chitosan undergo different chemical modifications to adopt chitosan derivatives and how group modification occurred in the chitosan skeleton need to be studied in future investigations.

Declaration Authors used generative AI tools for the purpose of improving readability and language. The final content has been reviewed and verified by the authors after using these tools, and they take full responsibility for its accuracy and integrity.

References

Abbasi Pirouz A, Selamat J, Zafar Iqbal S, Iskandar Putra Samsudin N (2020) Efficient and simultaneous chitosan-mediated removal of 11 mycotoxins from palm kernel cake. Toxins 12(2):115

Abduh MY, Shafitri TR, Jamilah M, Firmansyah M, Manurung R (2023) Chemical and antibacterial properties of chitosan derived from Mucor spp., Rhizopus. Oryzae and Hermetia illucens. J Appl Biol Biotechnol 11(6):185–192

Acevedo-Fani A, Salvia-Trujillo L, Soliva-Fortuny R, Martín-Belloso O (2015) Modulating biopolymer electrical charge to optimize the assembly of edible multilayer nanofilms by the layer-by-layer technique. Biomacromolecules 16(9):2895–2903

Aguirre-Loredo RY, Rodríguez-Hernández AI, Morales-Sánchez E, Gómez-Aldapa CA, Velazquez G (2016) Effect of equilibrium moisture content on barrier, mechanical and thermal properties of chitosan films. Food Chem 196:560–566

Ahmad M, Nirmal NP, Danish M, Chuprom J, Jafarzedeh S (2016) Characterisation of composite films fabricated from collagen/chitosan and collagen/soy protein isolate for food packaging applications. RSC Adv 6(85):82191–82204

Aldila H, Fabiani VA, Dalimunthe DY, Irwanto R (2020) The effect of deproteinization temperature and NaOH concentration on deacetylation step in optimizing extraction of chitosan from shrimp shells waste. IOP Conf Ser: Earth Environ Sci 599(1):012003 (IOP Publishing)

Alemu D, Getachew E, Mondal AK (2023) Study on the physicochemical properties of chitosan and their applications in the biomedical sector. Int J Polym Sci 2023(1):5025341

Ali A, Zahid N, Manickam S, Siddiqui Y, Alderson PG (2014) Double layer coatings: a new technique for maintaining physico-chemical characteristics and antioxidant properties of dragon fruit during storage. Food Bioprocess Technol 7:2366–2374

Alix S, Mahieu A, Terrie C, Soulestin J, Gerault E, Feuilloley MG et al (2013) Active pseudo-multilayered films from polycaprolactone and starch based matrix for food-packaging applications. Eur Polym J 49(6):1234–1242

Amiri H, Aghbashlo M, Sharma M, Gaffey J, Manning L, Moosavi Basri SM et al (2022) Chitin and chitosan derived from crustacean waste valorization streams can support food systems and the UN sustainable development goals. Nat Food 3(10):822–828

Apai W, Sardsud V, Boonprasom P, Sardsud U (2009) Effects of chitosan and citric acid on pericarp browning and polyphenol oxidase activity of longan fruit. Songklanakarin J Sci Technol 31(6)

Aranaz I, Alcántara AR, Civera MC, Arias C, Elorza B, Heras Caballero A, Acosta N (2021) Chitosan: an overview of its properties and applications. Polymers 13(19):3256

Araújo D, Ferreira IC, Torres CA, Neves L, Freitas F (2020) Chitinous polymers: extraction from fungal sources, characterization and processing towards value-added applications. J Chem Technol Biotechnol 95(5):1277–1289

Aung KP, Win SZ, Thu SL (2018) Study on chitin extraction from crab shells waste. Int J Sci Eng Appl 7(11):437–441

Ban Z, Wei W, Yang X, Feng J, Guan J, Li L (2015) Combination of heat treatment and chitosan coating to improve postharvest quality of wolfberry (L ycium barbarum). Int J Food Sci Technol 50(4):1019–1025

Battampara P, Sathish TN, Reddy R, Guna V, Nagananda GS, Reddy N et al (2020) Properties of chitin and chitosan extracted from silkworm pupae and egg shells. Int J Biol Macromol 161:1296–1304

Benchamas G, Huang G, Huang S, Huang H (2021) Preparation and biological activities of chitosan oligosaccharides. Trends Food Sci Technol 107:38–44

Berger KR (2003) A brief history of packaging: ABE321/AE206, 12/2002. Edis 2003(17)

Bie P, Liu P, Yu L, Li X, Chen L, Xie F (2013) The properties of antimicrobial films derived from poly (lactic acid)/starch/chitosan blended matrix. Carbohydr Polym 98(1):959–966

Bonilla J, Fortunati ELENA, Atarés L, Chiralt A, Kenny JM (2014a) Physical, structural and antimicrobial properties of poly vinyl alcohol–chitosan biodegradable films. Food Hydrocoll 35:463–470

Bonilla J, Vargas M, Atarés L, Chiralt A (2014b) Effect of chitosan essential oil films on the storage-keeping quality of pork meat products. Food Bioprocess Technol 7:2443–2450

Broquá J, Zanin BG, Flach AM, Mallmann C, Taborda FGD (2018) Different aspects of chemical and biochemical methods for chitin production a short review. J Nanomed Nanosci: JNAN-145. 10:2577–1477

Buanasari B, Sugiyo W, Rustaman H (2021) Effect of ultrasonic assisted on the degree of deacetylation of chitosan extracted from portunus pelagicus. Jurnal Bahan Alam Terbarukan 10(1):17–23

Butnaru E, Stoleru E, Brebu MA, Darie-Nita RN, Bargan A, Vasile C (2019) Chitosan-based bionanocomposite films prepared by emulsion technique for food preservation. Materials 12(3):373

Caro N, Medina E, Díaz-Dosque M, López L, Abugoch L, Tapia C (2016) Novel active packaging based on films of chitosan and chitosan/quinoa protein printed with chitosan-tripolyphosphate-thymol nanoparticles via thermal ink-jet printing. Food Hydrocoll 52:520–532

Casadidio C, Peregrina DV, Gigliobianco MR, Deng S, Censi R, Di Martino P (2019) Chitin and chitosans: characteristics, eco-friendly processes, and applications in cosmetic science. Mar Drugs 17(6):369

Cazón P, Morales-Sanchez E, Velazquez G, Vázquez M (2022) Measurement of the water vapor permeability of chitosan films: a laboratory experiment on food packaging materials. J Chem Educ 99(6):2403–2408

Cheng H, Xu H, McClements DJ, Chen L, Jiao A, Tian Y et al (2022) Recent advances in intelligent food packaging materials: principles, preparation and applications. Food Chem 375:131738

Cheung RCF, Ng TB, Wong JH, Chan WY (2015) Chitosan: an update on potential biomedical and pharmaceutical applications. Mar Drugs 13(8):5156–5186

Chi S, Zivanovic S, Penfield MP (2006) Application of chitosan films enriched with oregano essential oil on bologna–active compounds and sensory attributes. Food Sci Technol Int 12(2):111–117

Chien PJ, Sheu F, Yang FH (2007) Effects of edible chitosan coating on quality and shelf life of sliced mango fruit. J Food Eng 78(1):225–229

Choi C, Nam JP, Nah JW (2016) Application of chitosan and chitosan derivatives as biomaterials. J Ind Eng Chem 33:1–10

Crognale S, Russo C, Petruccioli M, D'annibale A (2022) Chitosan production by fungi: current state of knowledge, future opportunities and constraints. Fermentation 8(2):76

da Silva MA, Iamanaka BT, Taniwaki MH, Kieckbusch TG (2013) Evaluation of the antimicrobial potential of alginate and alginate/chitosan films containing potassium sorbate and natamycin. Packag Technol Sci 26(8):479–492

Davidovich-Pinhas M, Danin-Poleg Y, Kashi Y, Bianco-Peled H (2014) Modified chitosan: a step toward improving the properties of antibacterial food packages. Food Packag Shelf Life 1(2):160–169

Davis D, Umesh M, Santhosh AS, Suresh S, Shanmugam S, Kikas T (2024) Extraction of Fungal Chitosan by leveraging pineapple peel substrate for sustainable biopolymer production. Polymers 16(17):2455

del Hoyo-Gallego S, Pérez-Álvarez L, Gómez-Galván F, Lizundia E, Kuritka I, Sedlarik V et al (2016) Construction of antibacterial poly (ethylene terephthalate) films via layer by layer assembly of chitosan and hyaluronic acid. Carbohydr Polym 143:35–43

Delgado-Cedeño A, Hernández-Martínez SP, Ramos-Zayas Y, Marroquín-Cardona AG, Méndez-Zamora G, Franco-Molina MA, Kawas JR (2022) Insoluble chitosan complex as a potential adsorbent for aflatoxin B1 in poultry feed. Front Mater 9:1044495

Demircan B, Velioglu YS (2024) Control of browning, enzyme activity, and quality in stored fresh-cut fruit salads through chitosan coating enriched with bergamot juice powder. Foods 13(1):147

Dhawade PP, Jagtap RN (2012) Characterization of the glass transition temperature of chitosan and its oligomers by temperature modulated differential scanning calorimetry. Adv Appl Sci Res 3(3):1372

Dhillon GS, Kaur S, Brar SK, Verma M (2013) Green synthesis approach: extraction of chitosan from fungus mycelia. Crit Rev Biotechnol 33(4):379–403

Dong H, Cheng L, Tan J, Zheng K, Jiang Y (2004) Effects of chitosan coating on quality and shelf life of peeled litchi fruit. J Food Eng 64(3):355–358

Duan J, Park SI, Daeschel MA, Zhao Y (2007) Antimicrobial chitosan-lysozyme (CL) films and coatings for enhancing microbial safety of mozzarella cheese. J Food Sci 72:355–362

Dutta J, Tripathi S, Dutta PK (2012) Progress in antimicrobial activities of chitin, chitosan and its oligosaccharides: a systematic study needs for food applications. Food Sci Technol Int 18(1):3–34

Eissa HA (2007) Effect of chitosan coating on shelf life and quality of fresh-cut mushroom. J Food Qual 30(5):623–645

El Ghaouth A, Ponnampalam R, Castaigne F, Arul J (1992) Chitosan coating to extend the storage life of tomatoes. HortScience 27(9):1016–1018

El-Araby A, Janati W, Ullah R, Ercisli S, Errachidi F (2024) Chitosan, chitosan derivatives, and chitosan-based nanocomposites: eco-friendly materials for advanced applications (a review). Front Chem 11:1327426

Elieh-Ali-Komi D, Hamblin MR (2016) Chitin and chitosan: production and application of versatile biomedical nanomaterials. Int J Adv Res 4(3):411

Erdoğmuş SF, Altıntaş ÖE, Çelik S (2023) Production of fungal chitosan and fabrication of fungal chitosan/polycaprolactone electrospun nanofibers for tissue engineering. Microsc Res Tech 86(10):1309–1321

Eulalio HYC, Rodrigues JFB, Kleilton O, Peniche C, Marcus V (2019) Characterization and thermal properties of chitosan films prepared with different acid solvents. Revista Cubana De Química 31(3):309–323

Feng X, Wang X, Xing W, Yu B, Song L, Hu Y (2013) Simultaneous reduction and surface functionalization of graphene oxide by chitosan and their synergistic reinforcing effects in PVA films. Ind Eng Chem Res 52(36):12906–12914

Fernández R, Ocando C, Fernandes SC, Eceiza A, Tercjak A (2014) Optically active multilayer films based on chitosan and an azopolymer. Biomacromolecules 15(4):1399–1407

Ferreira AR, Torres CA, Freitas F, Sevrin C, Grandfils C, Reis MA et al (2016) Development and characterization of bilayer films of FucoPol and chitosan. Carbohydr Polym 147:8–15

Flórez M, Guerra-Rodríguez E, Cazón P, Vázquez M (2022) Chitosan for food packaging: recent advances in active and intelligent films. Food Hydrocoll 124:107328

Foster LJR, Ho S, Hook J, Basuki M, Marcal H (2015) Chitosan as a biomaterial: influence of degree of deacetylation on its physiochemical, material and biological properties. Plos One 10(8):e0135153

Gadgey KK, Sharma GS (2020) Investigation of mechanical properties of chitosan based films prepared from narmada riverside crab shells. Int J Mech Eng Technol 11(6):21–28

Gao P, Zhu Z, Zhang P (2013) Effects of chitosan–glucose complex coating on postharvest quality and shelf life of table grapes. Carbohydr Polym 95(1):371–378

Genskowsky E, Puente LA, Pérez-Álvarez JA, Fernandez-Lopez J, Muñoz LA, Viuda-Martos M (2015) Assessment of antibacterial and antioxidant properties of chitosan edible films incorporated with maqui berry (Aristotelia chilensis). LWT-Food Sci Technol 64(2):1057–1062

Goy RC, Britto DD, Assis OB (2009) A review of the antimicrobial activity of chitosan. Polímeros 19:241–247

Guan Z, Feng Q (2022) Chitosan and chitooligosaccharide: the promising non-plant-derived prebiotics with multiple biological activities. Int J Mol Sci 23(12):6761

Guan G, Azad MAK, Lin Y, Kim SW, Tian Y, Liu G, Wang H (2019) Biological effects and applications of chitosan and chito-oligosaccharides. Front Physiol 10:516

Gubanova GN, Petrova VA, Kononova SV, Popova EN, Smirnova VE, Bugrov AN et al (2021) Thermal properties and structural features of multilayer films based on chitosan and anionic polysaccharides. Biomolecules 11(5):762

Günlü A, Sipahioğlu S, Alpas H (2014) The effect of high hydrostatic pressure on the muscle proteins of rainbow trout (Oncorhynchus mykiss Walbaum) fillets wrapped with chitosan-based edible film during cold storage (4 ± 1 °C). High Press Res 34(1):122–132

Guo P, Anderson JD, Bozell JJ, Zivanovic S (2016) The effect of solvent composition on grafting gallic acid onto chitosan via carbodiimide. Carbohydr Polym 140:171–180

Habibie S, Hamzah M, Anggaravidya M, Kalembang E (2016) The effect of chitosan on physical and mechanical properties of paper. J Chem Eng Mater Sci 7(1):1–10

Hafdani FN, Sadeghinia N (2011) A review on application of chitosan as a natural antimicrobial. Int J Pharmacol Pharm Sci 5(2):46–50

Hajji S, Kchaou H, Bkhairia I, Salem RBSB, Boufi S, Debeaufort F, Nasri M (2021) Conception of active food packaging films based on crab chitosan and gelatin enriched with crustacean protein hydrolysates with improved functional and biological properties. Food Hydrocoll 116:106639

Hamdan N, Lee CH, Wong SL, Fauzi CENCA, Zamri NMA, Lee TH (2022) Prevention of enzymatic browning by natural extracts and genome-editing: a review on recent progress. Molecules 27(3):1101
Hao W, Li K, Ma Y, Li R, Xing R, Yu H, Li P (2021) Preparation and antioxidant activity of chitosan dimers with different sequences. Mar Drugs 19(7):366
He C, Yuan L, Bi S, Zhou C, Yang Q, Gu J et al (2024) Modified chitosan-based coating/packaging composites with enhanced antibacterial, antioxidant, and UV-resistant properties for fresh food preservation. ACS Appl Mater Interfaces 16(36):48352–48362
Higueras L, López-Carballo G, Hernández-Muñoz P, Catalá R, Gavara R (2014) Antimicrobial packaging of chicken fillets based on the release of carvacrol from chitosan/cyclodextrin films. Int J Food Microbiol 188:53–59
Hoagland PD, Parris N (1996) Chitosan/pectin laminated films. J Agric Food Chem 44(7):1915–1919
Hu P, Raub CB, Choy JS, Luo X (2020) Modulating the properties of flow-assembled chitosan membranes in microfluidics with glutaraldehyde crosslinking. J Mater Chem B 8(12):2519–2529
Huq T, Khan A, Brown D, Dhayagude N, He Z, Ni Y (2022) Sources, production and commercial applications of fungal chitosan: a review. J Bioresour Bioprod 7(2):85–98
Iber BT, Kasan NA, Torsabo D, Omuwa JW (2022) A review of various sources of chitin and chitosan in nature. J Renew Mater 10(4):1097
Imran M, Klouj A, Revol-Junelles AM, Desobry S (2014) Controlled release of nisin from HPMC, sodium caseinate, poly-lactic acid and chitosan for active packaging applications. J Food Eng 143:178–185
Islam N, Dmour I, Taha MO (2019) Degradability of chitosan micro/nanoparticles for pulmonary drug delivery. Heliyon 5(5)
Jiang A, Patel R, Padhan B, Palimkar S, Galgali P, Adhikari A et al (2023) Chitosan based biodegradable composite for antibacterial food packaging application. Polymers 15(10):2235
Jiménez-Gómez CP, Cecilia JA (2020) Chitosan: a natural biopolymer with a wide and varied range of applications. Molecules 25(17):3981
Jin Q, Yu H, Wang X, Li K, Li P (2017) Effect of the molecular weight of water-soluble chitosan on its fat-/cholesterol-binding capacities and inhibitory activities to pancreatic lipase. PeerJ 5:e3279
Kaczmarek MB, Struszczyk-Swita K, Li X, Szczęsna-Antczak M, Daroch M (2019) Enzymatic modifications of chitin, chitosan, and chitooligosaccharides. Front Bioeng Biotechnol 7:243
Kanatt SR, Rao MS, Chawla SP, Sharma A (2013) Effects of chitosan coating on shelf-life of ready-to-cook meat products during chilled storage. LWT-Food Sci Technol 53(1):321–326
Kandile NG, Zaky HT, Mohamed MI, Nasr AS, Ali YG (2018) Extraction and characterization of chitosan from shrimp shells. Open J Organ Polym Mater 8(3):33–42
Kannappan N, Karuppasamy D, Muthusamy S (2024) Sources and applications of chitin: a review. Int J Pharm Sci 2(4):1152–1161
Ke CL, Deng FS, Chuang CY, Lin CH (2021) Antimicrobial actions and applications of chitosan. Polymers 13(6):904
Khubiev OM, Egorov AR, Kirichuk AA, Khrustalev VN, Tskhovrebov AG, Kritchenkov AS (2023) Chitosan-based antibacterial films for biomedical and food applications. Int J Mol Sci 24(13):10738
Khwaldia K, Basta AH, Aloui H, El-Saied H (2014) Chitosan–caseinate bilayer coatings for paper packaging materials. Carbohydr Polym 99:508–516
Knidri HE, Dahmani J, Addaou A, Laajeb A, Lahsini A (2019) Rapid and efficient extraction of chitin and chitosan for scale-up production: effect of process parameters on deacetylation degree and molecular weight. Int J Biol Macromol 139:1092–1102
Kose Y, Suvacı E (2022) Improving the properties of medium molecular weight chitosan/PVA composite polymers via designed ZnO particles. Eurasia Proc Sci Technol Eng Math 21:144–151
Kowalczyk D, Kordowska-Wiater M, Nowak J, Baraniak B (2015) Characterization of films based on chitosan lactate and its blends with oxidized starch and gelatin. Int J Biol Macromol 77:350–359

Kozma M, Acharya B, Bissessur R (2022) Chitin, chitosan, and nanochitin: extraction, synthesis, and applications. Polymers 14(19):3989

Kurek M, Moundanga S, Favier C, Debeaufort F (2013) Antimicrobial efficiency of carvacrol vapour related to mass partition coefficient when incorporated in chitosan-based films aimed for active packaging. Food Control 32(1):168–175

Lago MA, Sendón R, de Quirós ARB, Sanches-Silva A, Costa HS, Sánchez-Machado DI et al (2014) Preparation and characterization of antimicrobial films based on chitosan for active food packaging applications. Food Bioprocess Technol 7:2932–2941

Leceta I, Molinaro S, Guerrero P, Kerry JP, De la Caba K (2015) Quality attributes of map packaged ready-to-eat baby carrots by using chitosan-based coatings. Postharvest Biol Technol 100:142–150

Lee HW, Park YS, Jung JS, Shin WS (2002) Chitosan oligosaccharides, dp 2–8, have prebiotic effect on the Bifidobacterium bifidium and Lactobacillus sp. Anaerobe 8(6):319–324

Lekjing S (2016) A chitosan-based coating with or without clove oil extends the shelf life of cooked pork sausages in refrigerated storage. Meat Sci 111:192–197

Li H, Yu T (2001) Effect of chitosan on incidence of brown rot, quality and physiological attributes of postharvest peach fruit. J Sci Food Agric 81(2):269–274

Li F, Biagioni P, Finazzi M, Tavazzi S, Piergiovanni L (2013) Tunable green oxygen barrier through layer-by-layer self-assembly of chitosan and cellulose nanocrystals. Carbohydr Polym 92(2):2128–2134

Li B, Elango J, Wu W (2020a) Recent advancement of molecular structure and biomaterial function of chitosan from marine organisms for pharmaceutical and nutraceutical application. Appl Sci 10(14):4719

Li Q, Dunn ET, Grandmaison EW, Goosen MF (2020b) Applications and properties of chitosan. In: Applications of chitan and chitosan. CRC Press, pp 3–29

Li Z, Li MC, Liu C, Liu X, Lu Y, Zhou G et al (2023) Microwave-assisted deep eutectic solvent extraction of chitin from crayfish shell wastes for 3D printable inks. Ind Crops Prod 194:116325

Liu K, Lin X, Chen L, Huang L, Cao S, Wang H (2013) Preparation of microfibrillated cellulose/chitosan–benzalkonium chloride biocomposite for enhancing antibacterium and strength of sodium alginate films. J Agric Food Chem 61(26):6562–6567

Liu J, Meng CG, Wang XC, Chen Y, Kan J, Jin CH (2016) Effect of protocatechuic acid-grafted-chitosan coating on the postharvest quality of Pleurotus eryngii. J Agric Food Chem 64(38):7225–7233

Liu SH, Chen RY, Chiang MT (2021) Effects and mechanisms of chitosan and chitosanoligosaccharide on hepatic lipogenesis and lipid peroxidation, adipose lipolysis, and intestinal lipid absorption in rats with high-fat diet-induced obesity. Int J Mol Sci 22(3):1139

Liu T, Li J, Tang Q, Qiu P, Gou D, Zhao J (2022) Chitosan-based materials: an overview of potential applications in food packaging. Foods 11(10):1490

Liyanage CS, Gonapinuwala ST, Fernando CAN, de Croos MDST (2023) A simple and effective method to extract chitosan from crustacean shell waste. J Aquat Food Prod Technol 32(4):396–415

Luangapai F, Peanparkdee M, Iwamoto S (2019) Biopolymer films for food industries: properties, applications, and future aspects based on chitosan. Rev Agric Sci 7:59–67

Ma W, Tang CH, Yang XQ, Yin SW (2013) Fabrication and characterization of kidney bean (Phaseolus vulgaris L.) protein isolate–chitosan composite films at acidic pH. Food Hydrocoll 31(2):237–247

Mane S, Pathan E, Tupe S, Deshmukh S, Kale D, Ghormade V et al (2022) Isolation and characterization of chitosans from different fungi with special emphasis on zygomycetous dimorphic fungus benjaminiella poitrasii: evaluation of its chitosan nanoparticles for the inhibition of human pathogenic fungi. Biomacromolecules 23(3):808–815

Martínez-Camacho AP, Cortez-Rocha MO, Graciano-Verdugo AZ, Rodríguez-Félix F, Castillo-Ortega MM, Burgos-Hernández A et al (2013) Extruded films of blended chitosan, low density polyethylene and ethylene acrylic acid. Carbohydr Polym 91(2):666–674

Martín-López H, Pech-Cohuo SC, Herrera-Pool E, Medina-Torres N, Cuevas-Bernardino JC, Ayora-Talavera T et al (2020) Structural and physicochemical characterization of chitosan obtained by UAE and its effect on the growth inhibition of Pythium ultimum. Agriculture 10(10):464
Mathaba M, Daramola MO (2020) Effect of chitosan's degree of deacetylation on the performance of pes membrane infused with chitosan during amd treatment. Membranes 10(3):52
Matica A, Menghiu G, Ostafe V (2017) Biodegradability of chitosan based products. New Front Chem 26(1)
Mawazi SM, Kumar M, Ahmad N, Ge Y, Mahmood S (2024) Recent applications of chitosan and its derivatives in antibacterial, anticancer, wound healing, and tissue engineering fields. Polymers 16(10):1351
Messerli MA, Raihan MJ, Kobylkevich BM, Benson AC, Bruening KS, Shribak M et al (2019) Construction and composition of the squid pen from Doryteuthis pealeii. Biol Bull 237(1):1–15
Moreira MDR, Roura SI, Ponce A (2011) Effectiveness of chitosan edible coatings to improve microbiological and sensory quality of fresh cut broccoli. LWT-Food Sci Technol 44(10):2335–2341
Muñoz-Tebar N, Pérez-Álvarez JA, Fernández-López J, Viuda-Martos M (2023) Chitosan edible films and coatings with added bioactive compounds: antibacterial and antioxidant properties and their application to food products: a review. Polymers 15(2):396
Mustafa MA, Ali A, Manickam S, Siddiqui Y (2014) Ultrasound-assisted chitosan–surfactant nanostructure assemblies: towards maintaining postharvest quality of tomatoes. Food Bioprocess Technol 7:2102–2111
Muthu M, Gopal J, Chun S, Devadoss AJP, Hasan N, Sivanesan I (2021) Crustacean waste-derived chitosan: antioxidant properties and future perspective. Antioxidants 10(2):228
Muxika A, Etxabide A, Uranga J, Guerrero P, De La Caba K (2017) Chitosan as a bioactive polymer: processing, properties and applications. Int J Biol Macromol 105:1358–1368
Muzzarelli RA (2010) Chitins and chitosans as immunoadjuvants and non-allergenic drug carriers. Mar Drugs 8(2):292–312
Nainggolan I, Nasution TI, Ahmad KR (2020) Thermal properties of chitosan-roselle films
Neves AC, Zanette C, Grade ST, Schaffer JV, Alves HJ, Arantes MK (2017) Optimization of lactic fermentation for extraction of chitin from freshwater shrimp waste. Acta Scientiarum. Technol 39(2):125–133
Nguyen HTT, Tran TN, Ha AC, Huynh PD (2022) Impact of deacetylation degree on properties of chitosan for formation of electrosprayed nanoparticles. J Nanotechnol 2022(1):2288892
Nilsen-Nygaard J, Strand SP, Vårum KM, Draget KI, Nordgård CT (2015) Chitosan: gels and interfacial properties. Polymers 7(3):552–579
No HK, Meyers SP, Prinyawiwatkul W, Xu Z (2007) Applications of chitosan for improvement of quality and shelf life of foods: a review. J Food Sci 72(5):R87–R100
Novikov VY, Derkach SR, Konovalova IN, Dolgopyatova NV, Kuchina YA (2023) Mechanism of heterogeneous alkaline deacetylation of chitin: a review. Polymers 15(7):1729
Nowzari F, Shábanpour B, Ojagh SM (2013) Comparison of chitosan–gelatin composite and bilayer coating and film effect on the quality of refrigerated rainbow trout. Food Chem 141(3):1667–1672
Nunes C, Silva M, Farinha D, Sales H, Pontes R, Nunes J (2023) Edible coatings and future trends in active food packaging–fruits' and traditional sausages' shelf life increasing. Foods 12(17):3308
Pakizeh M, Moradi A, Ghassemi T (2021) Chemical extraction and modification of chitin and chitosan from shrimp shells. Eur Polym J 159:110709
Pal AK, Katiyar V (2016) Nanoamphiphilic chitosan dispersed poly (lactic acid) bionanocomposite films with improved thermal, mechanical, and gas barrier properties. Biomacromolecules 17(8):2603–2618
Pal K, Bharti D, Sarkar P, Anis A, Kim D, Chałas R et al (2021) Selected applications of chitosan composites. Int J Mol Sci 22(20):10968

Pambudi GBR, Ulfin I, Harmami H, Suprapto S, Kurniawan F, Ni'mah YL (2018) Synthesis of water-soluble chitosan from crab shells (Scylla serrata) waste. AIP Conf Proc 2049(1):020086 (AIP Publishing LLC)
Panith N, Wichaphon J, Lertsiri S, Niamsiri N (2016) Effect of physical and physicochemical characteristics of chitosan on fat-binding capacities under in vitro gastrointestinal conditions. LWT-Food Sci Technol 71:25–32
Pelissari FM, Grossmann MV, Yamashita F, Pineda EAG (2009) Antimicrobial, mechanical, and barrier properties of cassava starch-chitosan films incorporated with oregano essential oil. J Agric Food Chem 57(16):7499–7504
Pellis A, Guebitz GM, Nyanhongo GS (2022) Chitosan: sources, processing and modification techniques. Gels (Basel, Switzerland) 8(7):393
Pérez WA, Marín JA, López JN, Burgos MA, Rios LA (2022) Development of a pilot-ecofriendly process for chitosan production from waste shrimp shells. Environ Process 9(3):55
Pérez-Álvarez L, Ruiz-Rubio L, Vilas-Vilela JL (2018) Determining the deacetylation degree of chitosan: opportunities to learn instrumental techniques. J Chem Educ 95(6):1022–1028
Piekarska K, Sikora M, Owczarek M, Jóźwik-Pruska J, Wiśniewska-Wrona M (2023) Chitin and chitosan as polymers of the future—Obtaining, modification, life cycle assessment and main directions of application. Polymers 15(4):793
Pochanavanich P, Suntornsuk W (2002) Fungal chitosan production and its characterization. Lett Appl Microbiol 35(1):17–21
Pohling J, Ramakrishnan VV, Hossain A, Trenholm S, Dave D (2024) Optimization of enzymatic deproteination of northern shrimp (Pandalus borealis) shell chitin using commercial proteases. Mar Drugs 22(10):445
Potivas T, Laokuldilok T (2014) Deacetylation of chitin and the properties of chitosan films with various deacetylation degrees. Food Appl Biosci 13:559–568
Poverenov E, Danino S, Horev B, Granit R, Vinokur Y, Rodov V (2014a) Layer-by-layer electrostatic deposition of edible coating on fresh cut melon model: anticipated and unexpected effects of alginate–chitosan combination. Food Bioprocess Technol 7:1424–1432
Poverenov E, Zaitsev Y, Arnon H, Granit R, Alkalai-Tuvia S, Perzelan Y et al (2014b) Effects of a composite chitosan–gelatin edible coating on postharvest quality and storability of red bell peppers. Postharvest Biol Technol 96:106–109
Pratiwi RD, El Muttaqien S, Gustini N, Difa NS, Syahputra G (2023) Eco-friendly synthesis of chitosan and its medical application: from chitin extraction to nanoparticle preparation. ADMET DMPK 11(4):435–455
Purohit SD, Bhaskar R, Singh H, Priyadarshi R, Kim H, Son Y et al (2024) Chitosan-based electrospun fibers for bone-tissue engineering: recent research advancements. Int J Biol Macromol 136530
Reay SL, Jackson EL, Ferreira AM, Hilkens CM, Novakovic K (2022) In vitro evaluation of the biodegradability of chitosan–genipin hydrogels. Mater Adv 3(21):7946–7959
Reshad RAI, Jishan TA, Chowdhury NN (2021) Chitosan and its broad applications: a brief review. SSRN:3842055
Reza M, Gailani J, Susanti T, Zuhro SA, Sjaifullah A, Busroni B, Rahmawati I (2024) Effect of synthesis condition on degree of deacetylation of chitosan from shrimp waste for smart film applications. al Kimiya: Jurnal Ilmu Kimia dan Terapan 11(1):1–11
Risch SJ (2009) Food packaging history and innovations. J Agric Food Chem 57(18):8089–8092
Rivero S, García MA, Pinotti A (2012) Heat treatment to modify the structural and physical properties of chitosan-based films. J Agric Food Chem 60(1):492–499
Romanazzi G, Feliziani E, Baños SB, Sivakumar D (2017) Shelf life extension of fresh fruit and vegetables by chitosan treatment. Crit Rev Food Sci Nutr 57(3):579–601
Román-Doval R, Torres-Arellanes SP, Tenorio-Barajas AY, Gómez-Sánchez A, Valencia-Lazcano AA (2023) Chitosan: properties and its application in agriculture in context of molecular weight. Polymers 15(13):2867

Rossi N, Grosso C, Delerue-Matos C (2024) Shrimp waste upcycling: unveiling the potential of polysaccharides, proteins, carotenoids, and fatty acids with emphasis on extraction techniques and bioactive properties. Mar Drugs 22(4):153

Salavati M (2023) Mechanical properties of α-chitin and chitosan biocomposite: a molecular dynamic study. J Compos Sci 7(11):464

Savic IM, Savic Gajic IM, Milovanovic MG, Zerajic S, Gajic DG (2022) Optimization of ultrasound-assisted extraction and encapsulation of antioxidants from orange peels in alginate-chitosan microparticles. Antioxidants 11(2):297

Sebastian J, Rouissi T, Brar SK, Hegde K, Verma M (2019) Microwave-assisted extraction of chitosan from Rhizopus oryzae NRRL 1526 biomass. Carbohydr Polym 219:431–440

Shahbazi Y, Shavisi N (2018) A novel active food packaging film for shelf-life extension of minced beef meat. J Food Saf 38(6):e12569

Sharma A, Pranaw K, Singh S, Khare SK, Chandel AK, Nain PKS, Nain L (2020) Efficient two-step lactic acid production from cassava biomass using thermostable enzyme cocktail and lactic acid bacteria: insights from hydrolysis optimization and proteomics analysis. 3 Biotech 10:1–13

Shen L, Pang S, Zhong M, Sun Y, Qayum A, Liu Y et al (2023) A comprehensive review of ultrasonic assisted extraction (UAE) for bioactive components: principles, advantages, equipment, and combined technologies. Ultrason Sonochem 106646

Shi B, Hao Z, Du Y, Jia M, Xie S (2024) Mechanical and barrier properties of chitosan-based composite film as food packaging: a review. BioResources 19(2):4001

Shlush E, Davidovich-Pinhas M (2022) Bioplastics for food packaging. Trends Food Sci Technol 125:66–80

Shyu YS, Chen GW, Chiang SC, Sung WC (2019) Effect of chitosan and fish gelatin coatings on preventing the deterioration and preserving the quality of fresh-cut apples. Molecules 24(10):2008

Singh TP, Chatli MK, Sahoo J (2015) Development of chitosan based edible films: process optimization using response surface methodology. J Food Sci Technol 52:2530–2543

Singh A, Benjakul S, Prodpran T (2019) Ultrasound-assisted extraction of chitosan from squid pen: molecular characterization and fat binding capacity. J Food Sci 84(2):224–234

Solís-Cruz B, Hernandez-Patlan D, Beyssac E, Latorre JD, Hernandez-Velasco X, Merino-Guzman R et al (2017) Evaluation of chitosan and cellulosic polymers as binding adsorbent materials to prevent aflatoxin B1, fumonisin B1, ochratoxin, trichothecene, deoxynivalenol, and zearalenone mycotoxicoses through an in vitro gastrointestinal model for poultry. Polymers 9(10):529

Song Z, Li F, Guan H, Xu Y, Fu Q, Li D (2017) Combination of nisin and ε-polylysine with chitosan coating inhibits the white blush of fresh-cut carrots. Food Control 74:34–44

Soon CY, Tee YB, Tan CH, Rosnita AT, Khalina A (2018) Extraction and physicochemical characterization of chitin and chitosan from Zophobas morio larvae in varying sodium hydroxide concentration. Int J Biol Macromol 108:135–142

Souza BW, Cerqueira MA, Ruiz HA, Martins JT, Casariego A, Teixeira JA, Vicente AA (2010) Effect of chitosan-based coatings on the shelf life of salmon (Salmo salar). J Agric Food Chem 58(21):11456–11462

Souza CP, Almeida BC, Colwell RR, Rivera IN (2011) The importance of chitin in the marine environment. Marine Biotechnol 13:823–830

Suhag R, Kumar N, Petkoska AT, Upadhyay A (2020) Film formation and deposition methods of edible coating on food products: a review. Food Res Int 136:109582

Sun L, Sun J, Chen L, Niu P, Yang X, Guo Y (2017) Preparation and characterization of chitosan film incorporated with thinned young apple polyphenols as an active packaging material. Carbohydr Polym 163:81–91

Šuput DZ, Lazić VL, Popović SZ, Hromiš NM (2015) Edible films and coatings-sources, properties and application. Food Feed Res 42(1):11–22

Talón E, Trifkovic KT, Nedovic VA, Bugarski BM, Vargas M, Chiralt A, González-Martínez C (2017) Antioxidant edible films based on chitosan and starch containing polyphenols from thyme extracts. Carbohydr Polym 157:1153–1161

Thambiliyagodage C, Jayanetti M, Mendis A, Ekanayake G, Liyanaarachchi H, Vigneswaran S (2023) Recent advances in chitosan-based applications—A review. Materials 16(5):2073

Tzeng P, Stevens B, Devlaming I, Grunlan JC (2015) Polymer-graphene oxide quadlayer thin-film assemblies with improved gas barrier. Langmuir 31(21):5919–5927

Valdez-Peña AU, Espinoza-Perez JD, Sandoval-Fabian GC, Balagurusamy N, Hernandez-Rivera A, De-la-Garza-Rodriguez IM, Contreras-Esquivel JC (2010) Screening of industrial enzymes for deproteinization of shrimp head for chitin recovery. Food Sci Biotechnol 19:553–557

Varun TK, Senani S, Jayapal N, Chikkerur J, Roy S, Tekulapally VB et al (2017) Extraction of chitosan and its oligomers from shrimp shell waste, their characterization and antimicrobial effect. Vet World 10(2):170

Velickova E, Winkelhausen E, Kuzmanova S, Alves VD, Moldão-Martins M (2013) Impact of chitosan-beeswax edible coatings on the quality of fresh strawberries (Fragaria ananassa cv Camarosa) under commercial storage conditions. LWT-Food Sci Technol 52(2):80–92

Vicente FA, Huš M, Likozar B, Novak U (2021) Chitin deacetylation using deep eutectic solvents: ab initio-supported process optimization. ACS Sustain Chem Eng 9(10):3874–3886

Wahyuni E (2022) The use of shrimp shells in products that are not intended for consumption. Sci Res J Agric Vet Sci 1(1):1–8

Wang L, Liu F, Jiang Y, Chai Z, Li P, Cheng Y et al (2011) Synergistic antimicrobial activities of natural essential oils with chitosan films. J Agric Food Chem 59(23):12411–12419

Wang HS, Chen D, Chuai CZ (2015a) Mechanical and barrier properties of LLDPE/chitosan blown films for packaging. Packag Technol Sci 28(10):915–923

Wang H, Zhang R, Zhang H, Jiang S, Liu H, Sun M, Jiang S (2015b) Kinetics and functional effectiveness of nisin loaded antimicrobial packaging film based on chitosan/poly (vinyl alcohol). Carbohydr Polym 127:64–71

Wang L, Wang Q, Tong J, Zhou J (2017) Physicochemical properties of chitosan films incorporated with honeysuckle flower extract for active food packaging. J Food Process Eng 40(1):e12305

Wang H, Qian J, Ding F (2018) Emerging chitosan-based films for food packaging applications. J Agric Food Chem 66(2):395–413

Wang Y, Liu S, Tang D, Dong R, Feng Q (2021) Chitosan oligosaccharide ameliorates metabolic syndrome induced by overnutrition via altering intestinal microbiota. Front Nutr 8:743492

Wrońska N, Katir N, Nowak-Lange M, El Kadib A, Lisowska K (2023) Biodegradable chitosan-based films as an alternative to plastic packaging. Foods 12(18):3519

Wu Y, Luo X, Li W, Song R, Li J, Li Y et al (2016) Green and biodegradable composite films with novel antimicrobial performance based on cellulose. Food Chem 197:250–256

Xiao W, Xu J, Liu X, Hu Q, Huang J (2013) Antibacterial hybrid materials fabricated by nanocoating of microfibril bundles of cellulose substance with titania/chitosan/silver-nanoparticle composite films. J Mater Chem B 1(28):3477–3485

Yan N, Chen X (2015) Sustainability: don't waste seafood waste. Nature 524(7564):155–157

Yarahmadi A, Dousti B, Karami-Khorramabadi M, Afkhami H (2024) Materials based on biodegradable polymers chitosan/gelatin: a review of potential applications. Front Bioeng Biotechnol 12:1397668

Yilmaz Atay H (2019) Antibacterial activity of chitosan-based systems. In: Functional chitosan: drug delivery and biomedical applications, pp 457–489

Younes I, Rinaudo M (2015) Chitin and chitosan preparation from marine sources. Structure, properties and applications. Marine Drugs 13(3):1133–1174

Yuan Y, Chesnutt BM, Haggard WO, Bumgardner JD (2011) Deacetylation of chitosan: material characterization and in vitro evaluation via albumin adsorption and pre-osteoblastic cell cultures. Materials 4(8):1399–1416

Yuceer M, Caner C (2014) Antimicrobial lysozyme-chitosan coatings affect functional properties and shelf life of chicken eggs during storage. J Sci Food Agric 94(1):153–162

Zainol Abidin NA, Kormin F, Zainol Abidin NA, Mohamed Anuar NAF, Abu Bakar MF (2020) The potential of insects as alternative sources of chitin: an overview on the chemical method of extraction from various sources. Int J Mol Sci 21(14):4978

Zakaria Z, Izzah Z, Jawaid M, Hassan A (2012) Effect of degree of deacetylation of chitosan on thermal stability and compatibility of chitosan-polyamide blend. BioResources 7(4)

Zhang Y, Zhang M, Yang H (2015) Postharvest chitosan-g-salicylic acid application alleviates chilling injury and preserves cucumber fruit quality during cold storage. Food Chem 174:558–563

Zhang C, Jiao S, Wang ZA, Du Y (2018) Exploring effects of chitosan oligosaccharides on mice gut microbiota in in vitro fermentation and animal model. Front Microbiol 9:2388

Zhang M, Biesold GM, Choi W, Yu J, Deng Y, Silvestre C, Lin Z (2022) Recent advances in polymers and polymer composites for food packaging. Mater Today 53:134–161

Zhao Z, Liu N, Yang L, Wang J, Song S, Nie D et al (2015) Cross-linked chitosan polymers as generic adsorbents for simultaneous adsorption of multiple mycotoxins. Food Control 57:362–369

Chapter 8
Carrageenan-Based Biopolymer Films and Coatings for Sustainable Food Packaging

Debabrata Nath, Abhishek Parasar, Pratap Kalita, and Swarup Roy

Abstract Carrageenan is one of the seaweed-based biodegradable biopolymers that have gained significant interest in edible and sustainable food packaging. Three kinds of carrageenan, namely, kappa, iota, and lambda, are derived from red seaweed. Among them, kappa carrageenan is well-known for having good film-forming qualities. Carrageenan makes transparent and strong film and acts as a good carrier for bioactive ingredients; thus, it is suitable for developing sustainable, active, and intelligent packaging materials. The intrinsic limitations in carrageenan, such as lack of functionality, low barrier properties, and high hydrophilicity, can be tuned by adding functional active ingredients. Numerous studies on the possibility of carrageenan-based functional packaging for preserving and tracking food shelf life have been published recently. Therefore, this chapter covered the extraction of carrageenan using a variety of techniques, followed by sustainable films and coatings based on carrageenan and their essential characteristics, and finally, the use of carrageenan-based functional packaging in intelligent and active food packaging applications.

Keywords Seaweed · Algae · Edible polymers · Biodegradable · Packaging · Food

D. Nath · A. Parasar · P. Kalita (✉)
Pratiksha Institute of Pharmaceutical Sciences, Guwahati, Assam, India
e-mail: pratapkalita0@gmail.com

S. Roy
Department of Food Technology and Nutrition, School of Agriculture, Lovely Professional University, Phagwara, Punjab, India

S. Roy and M. Shahid (eds.), *Biomaterials for Sustainable Food Packaging*, Environmental Footprints and Eco-design of Products and Processes,
https://doi.org/10.1007/978-3-032-08027-1_8

8.1 Introduction

Petrochemical-derived plastics are still commonly used as food packaging materials, and as a result, there is an increasing concern about the impact of this non-renewable and non-biodegradable plastic on the environment, health, waste, and ecology. To address these challenges, there is a need for alternatives that can help our packaging be safe for living beings and the environment. In this connection, biopolymers can be used as a replacement for traditional packaging polymers. Biopolymers can be procured from natural and synthetic sources, and natural ones are superior to their synthetic counterparts. Polysaccharides are mostly used in developing biopolymer-based packaging, and seaweed polysaccharides have received a lot of attention in this regard (Roy et al. 2023a, b). Algae are mainly categorized into microalgae and macroalgae, and the latter is the source of biomass to make polysaccharides. Nowadays, there is growing attention to algae-generated renewable polysaccharides as they are abundant and sustainable materials. There are different types of macroalgae, such as red, green, and brown; among them, red is known as a source of carrageenan and agar (Roy and Rhim 2019). Carrageenan is one of the biopolymers extensively used in food and other applications. Carrageenan is a sulfated heteropolysaccharide composed of galactose and 3,6 anhydrogalactose unit (Cheng et al. 2022). Due to its excellent film-forming and gelling properties, carrageenan is a potential biopolymer used in developing biodegradable food packaging (Saedi et al. 2021). Even though there are advantages with carrageenan polymer in making film, the properties of this biopolymer are unmatched by conventional petroleum-derived plastics.

The weak barrier properties, low flexibility, and high hydrophilicity somewhat restrict its practical application, and thus, properties need to be improved (Sedayu et al. 2019; Zia et al. 2017). To this end, the properties of carrageenan film could be enhanced by mixing cross-linkers or reinforcing other biopolymers and bioactive fillers (Zia et al. 2017). Carrageenan-based packaging is commonly functionalized with nanofillers, polyphenols, essential oils, enzymes, natural colorants, etc., and the obtained functional film is found to be suitable for maintaining and monitoring the shelf life of preserved fresh or processed food (Dmitrenko et al. 2023; Oun et al. 2024; Roy and Rhim 2021; Roy et al. 2021). A number of studies on packaging films and coatings based on carrageenan, which are renewable and sustainable, have recently been published. While some of the work is focused on intelligent packaging, others have concentrated on active packaging (Mathew et al. 2024). Overall, carrageenan-based packaging shows much promise and potential in the packaging field. Therefore, this book chapter summarizes the recent development in carrageenan-based sustainable packaging and its application in the shelf-life improvement of food.

8.2 Carrageenan Extraction and Methods Optimization

8.2.1 Carrageenan Extraction Methods

Several methods used to optimize the yield, effectiveness, and quality of carrageenan extraction from red seaweeds have been investigated in recent studies. Conventional techniques, namely alkaline extraction, have been used extensively, but they frequently need strong alkali solutions, lengthy processing times, and significant energy consumption, all of which can have a deteriorating effect on the environment and the quality of the finished product (Mendes et al. 2024). As a result, scientists have been looking into cutting-edge and alternative extraction techniques that decrease processing time, energy consumption, and environmental effect while simultaneously improving extraction yields and the physicochemical characteristics of carrageenan. A number of new methods have demonstrated great potential for improving the extraction of carrageenan. These include microwave-assisted extraction (MAE), which uses microwave radiation to accelerate cell wall breakdown and enhance mass transfer (Álvarez-Viñas et al. 2023; Rupert et al. 2022); ultrasound-assisted extraction (UAE), which uses high-frequency ultrasonic waves to improve solvent penetration and disrupt cellular structures (Martín-del-Campo et al. 2021; Torres et al. 2021); eco-friendly calcium hydroxide extraction, which provides a sustainable alternative to conventional sodium hydroxide-based methods by reducing chemical waste and improving gel strength; ohmic heating extraction, which uses electrical resistance to generate uniform heating and minimize thermal degradation (Salengke et al. 2016; Hasizah et al. 2021); and response surface methodology (RSM), a statistical tool used to systematically optimize multiple extraction parameters for better yield and gel properties (Hasizah et al. 2021). The potential of each of these techniques to decrease processing time, energy consumption, and environmental effect while increasing carrageenan yield, viscosity, and gel strength has been investigated (Mendes et al. 2024; Salengke et al. 2016; Hasizah et al. 2021).

8.2.1.1 Ohmic Heating Extraction

With effectiveness and product quality benefits, ohmic heating has become a viable method for removing carrageenan from red seaweeds. According to these results, ohmic heating can produce extraction yields and product quality that are on par with or even better than those obtained using traditional heating techniques. The study also underlined how crucial each variable is to the extraction process. The extraction temperature, time, and solution-to-seaweed highly influenced the extraction yield. In particular, the yield was often improved by higher temperatures and longer extraction times until it reached an ideal point, after which it plateaued or decreased, probably due to heat deterioration. Likewise, improved extraction efficiency was made possible by a greater solution-to-seaweed ratio. KOH concentration and extraction temperature were important variables influencing gel strength and viscosity regarding product quality. Improved carrageenan quality was indicated by increased gel strength and viscosity at elevated temperatures and with the right KOH concentrations (Hasizah et al. 2021; Salengke et al. 2016).

Ohmic technology's quick and even heating eliminates thermal deterioration. It cuts down on processing time, producing higher-quality carrageenan that can be used in various ways, including environmentally friendly food packaging (Hasizah et al. 2021; Salengke et al. 2016). Ohmic heating is based on the electrical resistance principle, which states that heat is produced via resistance when an electric current flows through a conductive material. This enhances mass transfer throughout the extraction process and lowers the possibility of localized overheating by producing effective and consistent heating across the extraction medium.

A significant study on *Eucheuma spinosum* employed a pilot-scale static ohmic heating system to enhance extraction parameters. The procedure involves chopping the seaweed into tiny pieces, rinsing with tap water, and then soaking in a KOH solution at different concentrations. The soaked seaweed was subsequently put through ohmic heating, in which the RSM experimental design was used to adjust the temperature, duration, and solution-to-seaweed ratio. Following extraction, the carrageenan underwent filtering, ethanol precipitation, and drying. To systematically assess the impacts of four variables- extraction temperature, extraction length, potassium hydroxide (KOH) concentration, and solution-to-seaweed ratio—the study used a Central Composite Design (CCD). The ideal parameters were a solution-to-seaweed ratio of 45:1, an extraction temperature of 95 °C, an extraction duration of 240 min, and a KOH concentration 0.4 M. On a dry basis, the extraction yield was almost 61.59% under these circumstances. The resulting carrageenan had a viscosity of 284.51 mPa s. and a gel strength of 53.48 g/cm^2 (Hasizah et al. 2021). These findings highlight ohmic heating's potential as a competitive substitute for conventional extraction techniques, providing advantages regarding energy economy and product quality control. The rapid and uniform heating provided by ohmic technology reduces processing time and thermal deterioration, resulting in higher-quality carrageenan that may be used in a number of applications, such as eco-friendly food packaging (Hasizah et al. 2021; Salengke et al. 2016).

8.2.1.2 Microwave-Assisted Extraction

Compared to conventional procedures, microwave-assisted extraction (MAE) offers benefits such as shorter extraction times, decreased solvent consumption, and increased yield, making it a potential technology for separating carrageenan from red seaweeds. By heating the solvent and seaweed mixture with microwave energy, this technique breaks down cell walls and encourages the release of carrageenan into the solvent (Gomez et al. 2020). These results demonstrate MAE's promise as an effective and sustainable carrageenan extraction technique that preserves or improves the quality of the recovered polysaccharides while reducing processing time and energy usage (Mendes et al. 2024).

The principle involves using microwave energy to enhance extraction by rapidly heating the solvent and sample, improving mass transfer and reducing extraction time. Vázquez-Delfín et al. (2014) looked at the MAE of carrageenan derived from *Hypnea musciformis*. The researchers experimented with different extraction times (10 and 20 min) and temperatures (85, 95 and 105 °C). They discovered that the 3,6-anhydrogalactose level increased at 105 °C, suggesting a greater amount of κ-carrageenan. The quality of the extracted carrageenan was comparable to that of conventional methods despite the decreased yield, indicating that MAE is an effective alternative extraction approach (Vázquez-Delfín et al. 2014; Quitério et al. 2022). In a different study, Álvarez-Viñas et al. (2023) assessed the MAE of *Sarcopeltis skottsbergii* carrageenan. They operated at 110 or 160 °C for 5–7 min to obtain maximum extraction yields of up to 63–64%. At low shear rates (0.1 s^{-1}), the extracted carrageenan showed apparent viscosity values of about 10^3 mPa s, suggesting it is suitable for film formulation. Furthermore, extracts made below 140 °C showed gel-like qualities without needing ethanol precipitation, saving energy (Álvarez-Viñas et al. 2023).

The procedure involves cleaning the red seaweed to remove contaminants and drying it until its weight remains constant. It is then milled into a fine powder to enhance the surface area for extraction. A solvent usually distilled water or an alkaline solution, is combined in a predetermined ratio with a certain quantity of powdered seaweed. A microwave reactor is used to hold this combination. The reactor is exposed to microwave radiation at regulated temperatures and power levels (Kasim et al. 2023). The yield and quality of carrageenan are maximized by optimizing the parameters. The mixture is filtered after irradiation to separate the liquid extract from the solid residue. The final product can be obtained by precipitating the extract's carrageenan with ethanol and then drying it.

8.2.1.3 Ultrasound-Assisted Extraction

With benefits like shorter extraction times, higher yields, and enhanced functional qualities, ultrasound-assisted extraction (UAE) has become a viable method for separating carrageenan from red seaweeds. By breaking down cell walls with ultrasonic waves, this technique improves solvent penetration and makes it easier for intracellular chemicals to be released (Mendes et al. 2024; Carreira-Casais et al. 2021). These experiments demonstrate how effective the United Arab Emirates is at improving carrageenan extraction from red seaweeds. A useful method for the sustainable development of food packaging applications, UAE can increase yield and generate carrageenan with desired physicochemical and functional qualities by optimizing extraction parameters (Mendes et al. 2024; Shen et al. 2023).

The principle involves optimizing UAE conditions to enhance the yield and quality of carrageenan. To increase the quantity and quality of carrageenan, researchers in a study on the red seaweed *Mastocarpus stellatus* improved UAE conditions. The solid-to-liquid ratio, ultrasonic amplitude, and extraction duration were among the factors assessed by the experimental design. An extraction temperature of 80 °C for

35 min, an ultrasound frequency of 80 kHz, a solid-to-liquid ratio of 2 gm of seaweed per 100 gm of solvent, and an ultrasonic amplitude of 79% were found to be the ideal parameters (Torres et al. 2021). With a sulfate level of 46%, the carrageenan yield under these circumstances was roughly 40.3%. According to Shen et al. (2023), the extracted carrageenan demonstrated favourable mechanical characteristics with a viscoelastic modulus of 741.4 Pa and a gelling temperature of 39.4 °C. With EC_{50} values less than 51.9 μg/ml, the carrageenan also showed antiproliferative activities against human cancer cell lines (Martín-del-Campo et al. 2021). Another study looked into utilizing UAE in an alkaline environment to extract carrageenan from *Chondracanthus canaliculatus*. Compared to traditional approaches, the extraction yield was greatly enhanced to between 41 and 45% by using ultrasound, both alone and in conjunction with heat. Properties of the extracted carrageenan, including water retention capacity, syneresis behaviour, and odour, were similar to those of commercial carrageenan kinds. According to functional group analysis, both gelling and non-gelling fractions contained alginates and κ- and ι-carrageenan (Carreira-Casais et al. 2021; Martín-del-Campo et al. 2021).

The procedure includes washing the red seaweed to remove contaminants and drying it until its weight remains constant. It is then milled into a fine powder to improve the surface area for extraction. A solvent usually distilled water or an alkaline solution, combines a predetermined ratio with a certain quantity of powdered seaweed. This mixture is exposed to ultrasonic waves at a regulated frequency and amplitude in an ultrasonic reactor. According to the experimental design, the reactor runs at the ideal extraction temperature and duration (Torres et al. 2021). After ultrasonic treatment, the mixture is filtered to remove the solid residue and liquid extract. The final product can be obtained by precipitating the extract's carrageenan with ethanol and then drying it. Compared to traditional methods, the improved UAE parameters help improve carrageenan yield and functional characteristics, making it a more sustainable and effective extraction process.

8.2.1.4 Eco-Friendly Extraction Using Calcium Hydroxide

The potential use of calcium hydroxide [$Ca(OH)_2$] as a more ecologically friendly alternative to traditional sodium hydroxide (NaOH) methods has been highlighted by recent advancements in the extraction of carrageenan (Jiang et al. 2022). Though effective in dissolving seaweed cell walls and encouraging the production of 3,6-anhydro-D-galactose (3,6-AG), traditional alkaline methods mainly rely on NaOH, which raises environmental concerns because of its high chemical consumption, wastewater generation, and potential to degrade the quality of carrageenan (Lin et al. 2022). Red seaweed is usually boiled for long periods at high temperatures in an alkaline solution as part of the NaOH process, which raises processing expenses and energy consumption. A noteworthy work by Jiang et al. (2022) presented a technique for employing $Ca(OH)_2$ to extract ι-carrageenan from *Eucheuma denticulatum*, showing remarkable improvements in product quality and environmental effect. The efficient elimination of sulfate groups and the transformation of 6-sulfated molecules into 3,6-anhydro-D-galactose (3,6-AG) during the $Ca(OH)_2$ extraction procedure are responsible for these enhancements. The sulfate groups on the carrageenan molecular chains are balanced by calcium ions (Ca^{2+}), which promotes desulfation and improves gel characteristics. The procedure produces less wastewater with a more straightforward composition, facilitating recycling and treatment (Jiang et al. 2022; Mendes et al. 2024).

The principle involves using $Ca(OH)_2$ as an alternative alkali to improve carrageenan extraction efficiency and quality while minimizing environmental impact. Unlike sodium hydroxide (NaOH), which is commonly used in conventional methods, $Ca(OH)_2$ provides controlled alkalinity, reducing excessive degradation of carrageenan chains and preserving its gel-forming properties. Calcium ions (Ca^{2+}) play a crucial role in balancing sulfate groups, promoting desulfation, and enhancing the transformation of 6-sulfated molecules into 3,6-anhydro-D-galactose (3,6-AG), a key component responsible for improved gel strength (Jiang et al. 2022; Lin et al. 2022). The $Ca(OH)_2$ methodology showed a 17.6% increase in carrageenan yield, a 43.3% improvement in gel strength, indicating improved quality, and a reduction of 93.3% in alkali usage and 86.8% in water consumption when compared to the conventional NaOH extraction method. Furthermore, there was a 25% reduction in the overall extraction time. The physicochemical characteristics of the recovered ι-carrageenan were also evaluated in the study. The sulfate content of the $Ca(OH)_2$ and NaOH procedures did not differ significantly. Comparing the $Ca(OH)_2$ approach to the NaOH method, the 3,6-AG content increased by 13.6%. Carrageenan extracted with $Ca(OH)_2$ was more white (56.7%) than that extracted with NaOH (48.6%), suggesting that the pigment was removed more successfully. The difference in molecular weight brought on by the extraction conditions may be the reason why the viscosity of the $Ca(OH)_2$-extracted carrageenan was lower than that of the NaOH-extracted product (Jiang et al. 2022).

The procedure involves exposing the red seaweed to boiling water to break down the cell walls and allow the carrageenan to be released into the solution. After boiling, the mixture is mixed with a small amount of $Ca(OH)_2$ (0.5% w/w). After that, the extraction is conducted for four hours at 80 °C with a solid-to-liquid ratio of 1:25. To get rid of any remaining seaweed particles, the solution is filtered after extraction. The final product is then obtained by drying and grinding the carrageenan after it has been precipitated with alcohol (Jiang et al. 2022). Figure 8.1 shows the schematic representation of carrageenan extraction using various methods.

Carrageenan, a polysaccharide that finds extensive application in the food, pharmaceutical, and cosmetic industries, is extracted from red seaweed using a number of advanced techniques. These include microwave-assisted extraction, which increases yield and efficiency by rapidly heating, ultrasonic-assisted extraction, which uses ultrasonic waves to break down cell walls and improve extraction rates, eco-friendly extraction, which emphasizes sustainable and non-toxic solvents, and ohmic extraction, which uses electrical conductivity to save energy. The resulting carrageenan is useful for gelling, thickening, and stabilizing applications because it comes in three distinct structural forms: κ-carrageenan, λ-carrageenan, and ι-carrageenan. Each of these forms has unique chemical compositions and functional characteristics.

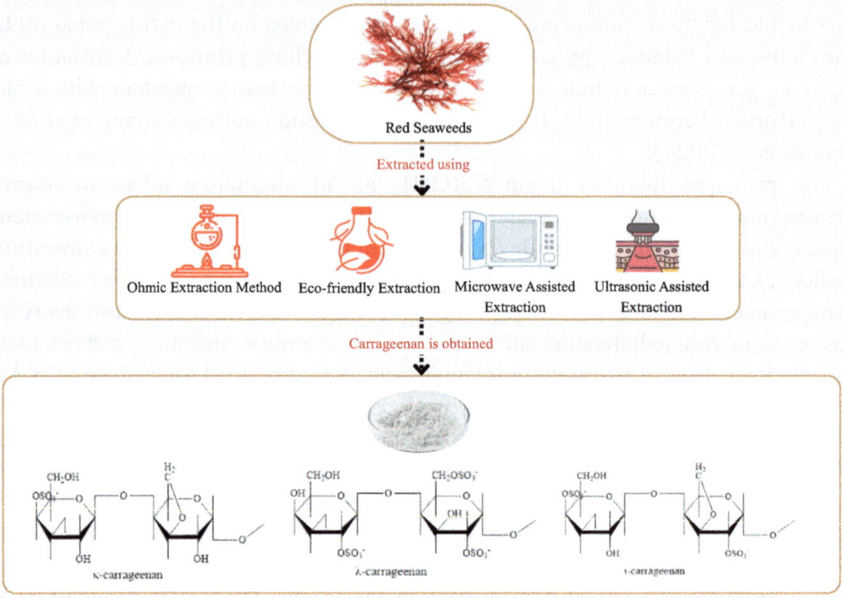

Fig. 8.1 Different extraction process of carrageenan

8.2.2 Carrageenan Methods Optimization

Response surface methodology is a statistical tool that systematically optimizes multiple extraction parameters for better yield and gel properties (Hasizah et al. 2021). Carrageenan extraction from *Kappaphycus alvarezii* has been successfully optimized using Response Surface Methodology (RSM), which focuses on variables including extraction temperature and time to improve yield, gel strength, and viscosity. Together, these experiments show how useful RSM is for methodically adjusting carrageenan extraction parameters, which improves yields and produces desired physicochemical characteristics. Researchers can more effectively determine ideal conditions by using RSM, which improves carrageenan extraction procedures' sustainability and financial feasibility (Webber et al. 2012; Hasizah et al. 2021).

RSM is a statistical and mathematical approach to optimize processes by evaluating the relationship between multiple independent variables and desired responses. It employs experimental designs like the Central Composite Design (CCD) to systematically vary factors, generate predictive models and create response surfaces that help identify optimal conditions. A CCD inside the RSM framework was used in a study by Webber et al. (2012) to systematically assess how temperature and time affect the extraction of carrageenan. The study found that while extraction time and temperature had an impact on gel strength, extraction temperature had a substantial impact on both yield and viscosity. It was found that 74 °C for four hours produced the best extraction results. The carrageenan extract was expected to have a yield of 31.17%, a gel strength of 158.27% g/cm^2, and a viscosity of 29.5 Cp under these circumstances. Results from experimental validation showed that the yield was 35 ± 4.68%, the gel strength was 112.50 ± 4.96 g/cm^2, and the viscosity was 16.02 ± 1.03 cP. These results highlight how effective RSM is in adjusting extraction parameters to raise the quality and productivity of carrageenan production (Webber et al. 2012). In order to maximize the extraction of carrageenan from *Eucheuma spinosum* utilizing ohmic heating, Hasizah et al. (2021) conducted additional research using RSM. Variables like extraction temperature, duration, KOH content, and solution-to-seaweed ratio were all examined in the study. The ideal parameters were determined to be a solution-to-seaweed ratio of 45:1, an extraction temperature of 95 °C, an extraction duration of 240 min, and a KOH concentration of 0.4 M. On a dry basis, the extraction yield under these circumstances was almost 61.59%. The resulting carrageenan had a viscosity of 284.51 mPa s and a gel strength of 53.48 g/cm^2. This study demonstrates how RSM may be combined with cutting-edge extraction methods like ohmic heating to improve the extraction efficiency of carrageenan (Hasizah et al. 2021).

The procedure involves pre-treating the seaweed by thoroughly washing it with distilled water to get rid of contaminants and then drying it at 60 °C until its weight remained consistent. After being dried, the seaweed was finely powdered and combined with KOH, an alkali solution, in different amounts. According to the RSM model, the mixture was cooked for varying amounts of time and at various temperatures. Following extraction, any remaining particles were filtered out of the solution, and acetic acid was added to the filtrate to neutralize it. After employing ethanol to precipitate the carrageenan, it was cleaned and dried (Webber et al. 2012).

The seaweed was sliced into small pieces, rinsed with tap water, and then soaked in a KOH solution at different concentrations in the Hasizah et al. (2021) investigation. The soaked seaweed was subsequently put through ohmic heating, in which the RSM experimental design was used to adjust the temperature, duration, and solution-to-seaweed ratio. Following extraction, the carrageenan underwent filtering, ethanol precipitation, and drying (Hasizah et al. 2021). These techniques successfully demonstrate how RSM may be used to optimize extraction conditions to increase yields and enhance the physicochemical characteristics of carrageenan.

8.3 Key Properties of Carrageenan-Based Films and Coatings

Because of its remarkable mechanical, barrier, antioxidant, and antibacterial qualities, carrageenan-based films and coatings have been thoroughly researched for their potential in sustainable food packaging.

8.3.1 Mechanical Properties

Because of their remarkable mechanical qualities, which are crucial for preserving integrity and durability during handling, storage, and transit, carrageenan-based films have drawn a lot of interest in sustainable food packaging (Cheng et al. 2022; Wang et al. 2024). To improve their mechanical qualities, additives, plasticizers, crosslinkers, and biopolymer blends are often used to improve the mechanical performance of pure carrageenan films (Kang and Yun 2022). Elongation at break (EAB) shows a material's ability to stretch before rupturing, whereas tensile strength (TS) gauges a material's resistance to breaking under stress. The TS of pure κ-carrageenan is about 39 MPa (Cheng et al. 2022). The addition of additives significantly improves these mechanical qualities. When the percentage of titanium dioxide (TiO_2) nanoparticles increases from 0 to 5% by weight, for instance, it has been demonstrated that adding TiO_2 nanoparticles to the carrageenan matrix increases the TS by 122% (Kokkuvayil Ramadas et al. 2024; Cheng et al. 2022).

In order to address the intrinsic brittleness and low mechanical strength of pure carrageenan films, research has investigated the use of various additives. For instance, it has been shown that adding pearl millet starch and carrageenan improves the TS of the resulting composite films. The composite film reportedly had a tensile strength of 28 MPa, higher than that of pure carrageenan film. Increased intermolecular interactions caused by the hydrogen bonds that develop between the starch and carrageenan molecules are thought to be the cause of this enhancement. In a different study, titanium dioxide (TiO_2) nanoparticles considerably raised the modulus and TS of carrageenan films. Specifically, adding 5% weight TiO_2 nanoparticles increased the composite films' TS and modulus by 39% and 122%, respectively. This enhancement results from a better physical contact between the filler and the carrageenan matrix. The goal of recent research has been to enhance the functional properties of carrageenan films, particularly by the addition of bioactive materials. One notable example is the addition of zinc oxide (ZnO) nanoparticles and chitosan to carrageenan films. The resulting composite film demonstrated enhanced antibacterial activity against Escherichia coli due to the synergistic action of ZnO and chitosan. Additionally, the film demonstrated greater mechanical strength and transparency when compared to films without these modifications. Natural extracts have also been studied to provide carrageenan films with antioxidant properties. For instance, adding olive oil extract (OOE) greatly enhanced the antioxidant activity of carrageenan films. The antioxidant capacity of the film increased from 0% in the control film to 32.55% in the OOE-containing film, as determined by the DPPH technique. This improvement is due to the phenolic compounds present in the OOE (Cheng et al. 2022).

Together with their mechanical and antibacterial benefits, films based on carrageenan have been developed for clever packaging applications. These films can be made into visual indicators of food freshness by adding anthocyanins and other naturally occurring pigments that are sensitive to pH. For example, a study made a double-layer indicator film that changed colour in response to pH variations using carrageenan and anthocyanin. By identifying colour changes that signify deterioration, this feature allows the film to monitor the freshness of meat (Rodriguez Carrillo et al. 2023; Pan et al. 2024).

Furthermore, combining κ-carrageenan with agar produces films with better mechanical qualities; κ-carrageenan films show higher TS than agar-based films, demonstrating the superior strength that carrageenan imparts (Sedayu et al. 2019). Biopolymer films are made more flexible by adding plasticizers, which lower intermolecular pressures. Their impact on mechanical characteristics varies, however. For example, because of its higher molecular mobility, glycerol lowers the TS of carrageenan films (Genecya et al. 2023), but sorbitol has a less noticeable effect (Wang et al. 2024). Therefore, to customize the mechanical properties of films based on carrageenan, it is essential to choose the right plasticizer and its concentration (Cheng et al. 2022). Carrageenan greatly improves mechanical performance when combined with biopolymers like alginate, gelatin, and carboxymethyl cellulose (CMC), making it possible to create composite films (Islamiyah et al. 2022). Crosslinkers, both chemical and physical, increase mechanical strength even further. While citric acid crosslinks CMC-HEC, boosting TS, Young's modulus (YM), and toughness, potassium chloride facilitates the gelation of κ-carrageenan (Kang and Yun 2022). Utilizing nanoparticles and bio-based fillers affects properties including EAB, puncture resistance (PR), TS and YM (Kokkuvayil Ramadas et al. 2024; Cheng et al. 2022; Asadzadeh et al. 2023). Carrageenan's mechanical and barrier qualities are improved by chemical changes like the addition of maleic anhydride (Zhou et al. 2021). This change decreases transparency and thermal stability while increasing water retention. Since more hydroxyl substitutions reduce intermolecular forces and increase molecular mobility and flexibility, modified carrageenan films likewise show lower TS. The incorporation of tapioca modified starch has an impact on mechanical and biodegradable qualities. While more modified starch improves tensile strength with little impact on EAB, more carrageenan content causes swelling (Islamiyah et al. 2022). Because of steric hindrance and a decrease in intermolecular hydrogen bonding, carrageenan combined with locust bean gum and guar gum changes the characteristics of the film by decreasing EAB and TS (Wang et al. 2024). The strength and stiffness of carrageenan films are increased by nanoparticles like cerium oxide (Ce_2O_3), silver (Ag), copper sulfide (CuS_2), and titanium dioxide (TiO_2). Ag nanoparticles lower TS, whereas Ce_2O_3 addition raises TS (Genecya et al. 2023). Without appreciably changing thickness, CuS_2 and limonene enhance mechanical and bactericidal qualities (Roy and Rhim 2020a, b). The addition of TiO_2 nanotubes increases film stiffness, and TS doped with cupric oxide (TNT-CuO), especially when TiO_2-TNT-CuO is combined (Ezati et al. 2021). Other polysaccharides' mechanical qualities are similarly enhanced by carrageenan. Research on biofilms with carrageenan, gelatin, and alginate indicates that while carrageenan has little effect on PR, it has a major effect on TS and EAB (Fathiraja et al. 2022). According to RSM analysis, carrageenan enhances TS more than agar does.

8.3.2 Barrier Properties

The barrier qualities of films based on carrageenan are being extensively researched since they are essential for food packaging applications. These characteristics, which assist preserve food quality and prolong shelf life, include water vapour permeability (WVP), gas barrier performance, and UV light blocking. However, because carrageenan is hydrophilic, it typically has low moisture resistance, requiring structural changes. Improvements can be made by chemical changes like carboxymethylation or physical therapies like irradiation, microwave therapy, or sonication (Kokkuvayil Ramadas et al. 2024; Cheng et al. 2022).

WVP is crucial in food packaging since it establishes how well a film can prevent moisture from transferring from food to its surroundings. Research indicates that adding nanomaterials and biopolymers to carrageenan films can dramatically lower WVP. Using glycerol as a plasticizer and κ-carrageenan and sweet potato starch as the fundamental polymer, Bharti et al. (2021) created edible films. The Water Vapour Transmission Rate (WVTR) of neat carrageenan films was lower than that of films containing starch (Bharti et al. 2021). Comparing carrageenan films to pure carrageenan, including pectin from the fruit peel of *Dillenia serrata*, resulted in a 25% decrease in WVP (Sedayu 2020). Carrageenan and pectin molecules combine to generate a denser polymer matrix that restricts moisture transport, resulting in this improvement (Sedayu et al. 2019; Cheng et al. 2022). WVP has also been impacted by additional changes made with nanoclay and *Zataria multiflora* extract (ZME). According to Nouri et al. (2020), κ-carrageenan films with 1% ZME showed a WVTR of 1.49×10^{-10} g m^{-1} s^{-1} Pa^{-1}, whereas films with 3% ZME showed a WVTR of 2.21×10^{-10} g m^{-1} s^{-1} Pa^{-1}. This confirmed that hydrogen bonding between the phenolic compounds in ZME, the hydroxyl groups in glycerol, and the functional groups in carrageenan alters permeability (Nouri et al. 2020). Furthermore, adding 1:1 carrageenan-alginate blends improved hybrid films' moisture resistance by lowering WVP by around 30% (Sedayu 2020). Gas permeability is another crucial component of food packaging, particularly for oxidation-sensitive items. It has been demonstrated that adding bio-based nanofillers, including cellulose nanocrystals (CNCs), improves the resistance of carrageenan films to gas diffusion. Compared to films without CNCs, carrageenan films with 9% CNC content showed a 40% decrease in oxygen permeability, according to Kokkuvayil Ramadas et al. (2024). The sinuous channel CNCs generate, which increases resistance to gas diffusion, is responsible for this improvement. These films work especially well for packaging goods prone to oxidative spoiling, like nuts and fatty foods (Kokkuvayil Ramadas et al. 2024; Cheng et al. 2022). Guo et al. (2022) conducted additional research to examine the effects of hydroxypropyl methylcellulose (HPMC) on gas barrier characteristics. Using a solvent-casting technique, they discovered that oxygen permeability decreased from 0.228 to 0.164 cm^3/m^2 24 h 0.1 MPa as HPMC concentration increased from 0 to 12%. It's interesting to note that water vapour permeability fluctuated, peaking at 12% HPMC (9.5×10^{-11} g Pa^{-1} s^{-1} m^{-1}) and falling to its lowest at 6% HPMC

(7.71×10^{-11} g Pa^{-1} s^{-1} m^{-1}). This suggests that the ideal HPMC concentration influences a higher gas barrier efficiency (Guo et al. 2022).

8.3.3 Thermal Properties

Because of its thermal characteristics, films and coatings based on carrageenan are becoming increasingly popular as environmentally friendly food packaging materials. In order to increase these qualities and their performance in food packaging applications, recent research has concentrated on adding different chemicals. Natural additions can greatly increase the thermal stability of carrageenan films. According to Sandhu et al. (2020), the melting point of carrageenan films rose from 98 to 118 °C when pearl millet starch was added. This alteration implies the development of a denser network of polymers, improving the film's thermal characteristics and general stability (Sandhu et al. 2020; Cheng et al. 2022). In the same way. It has been demonstrated that adding GFO to κ-carrageenan films increases their thermal resistance. In contrast to pure carrageenan films, which showed thermal degradation temperatures of 500 °C, GFO-incorporated films showed temperatures as high as 600 °C, according to thermogravimetric analysis (TGA). The crosslinking effect of polyphenolic chemicals in GFO improves thermal stability by raising the films' tensile strength and overall stability (Bhatia et al. 2024). These films are more resilient to changing environmental conditions because of their improved thermal stability. Furthermore, Gupta et al. (2021) investigated how amla pomace-derived cellulose nanocrystals (CNCs) affected carrageenan films. Degradation temperatures rose by about 20 °C when CNCs were added, improving the films' thermal stability compared to the control films. This improvement was brought about by creating new interfacial bonds between the carrageenan matrix and the CNCs' sulfate groups, strengthening the polymer network (Gupta et al. 2021). These improvements in thermal stability are essential for extending the shelf life and use of carrageenan-based food components. According to Meng et al. (2018), the thermal effects of adding liquified banana pseudostems (LBP) and polyvinyl alcohol (PVA) to carrageenan films have also been investigated. PVA/LBP/carrageenan films showed greater thermal breakdown temperatures than native ones. Because of the contacts and ions added during the film-forming process, PVA and LBP helped to boost thermal stability (Meng et al. 2018).

8.3.4 UV-Light Barrier and Transparency

Although carrageenan films are transparent by nature, they can effectively block UV rays, preventing the deterioration of foods sensitive to light. Several additives, such as plant-based extracts, nanoparticles, and polyphenol-rich substances, improve UV protection while preserving transparency, which is essential for consumer appeal

(Kim et al. 2014; Rapa et al. 2017). Research has shown that adding grapefruit essential oil (GFO) to carrageenan films greatly enhances their blocking of UV rays. According to Kokkuvayil Ramadas et al. (2024), coatings with 3% GFO improved light protection by reducing UV transmission at 280 nm by more than 70%. This enhancement is attributed to the crosslinking action of the polyphenolic chemicals found in GFO, which both absorb damaging UV rays and reinforce the film structure (Kokkuvayil Ramadas et al. 2024; Bhatia et al. 2024). Additionally, carrageenan films can be made more aesthetically pleasing by adding colour and UV protection, including plant-based pigments and extracts (Cheng et al. 2022). Yadav and Chiu (2019) investigated how cellulose nanocrystals (CNCs) affected the carrageenan films' UV transmittance. Using a solvent-casting technique, Glycerol was used as a plasticizer when CNCs were added in different concentrations (0–9%). Higher CNC concentrations resulted in a considerable decrease in UV transmittance, according to the UV/visible spectroscopy study. The UV transmittance of films containing 9% CNC was 24%, while that of films made entirely of carrageenan was 76%. This suggests that the UV-blocking action is enhanced by the dispersion of CNC in the polymer matrix (Yadav and Chiu 2019). Carbon dots (CDs) produced from Allium sativum were added to alginate/carrageenan composite films by Khan et al. (2023a, b) at concentrations ranging from 1 to 3%. By acting as reinforcing agents, these CDs increased hydrophobicity and mechanical strength. The discolouration of red meat under UV exposure (200–400 nm) was used to measure the effectiveness of UV blocking. In the visible light spectrum (400–800 nm), films containing 4% CDs maintained good transparency while offering 85–99% UV protection. According to LAB colour analysis, the films' brightness dropped as the CD concentration rose, although their redness and yellowness marginally increased (Khan et al. 2023a, b). Lignin was examined by Rukmanikrishnan et al. (2020) as a UV-blocking agent in films made of carrageenan. Because lignin and carrageenan interact, adding lignin (0–30%) dramatically decreased UV transmission from 84 to 9%, indicating strong UV protection. Higher lignin content, however, significantly decreased transparency; at 15% lignin concentration, light transmission dropped from 70 to 15%. The chromophore groups in lignin, which absorb visible light, are probably to blame for this decrease in clarity (Rukmanikrishnan et al. 2020).

8.3.5 Antioxidant Properties

Because of their antioxidant properties, films made from carrageenan have attracted a lot of attention in the field of sustainable food packaging, which assist in maintaining food quality by lowering oxidative process-induced spoiling. These coatings can improve food preservation and increase shelf life by adding beneficial chemicals. Numerous studies have looked into various natural additions to increase the carrageenan-based films' antioxidant activity. By using MAE to incorporate olive oil extract (OOE) into carrageenan films, Da Rosa et al. (2020) created antioxidant packaging films. Tests were conducted on four formulations: 0, 50, 100, and 200 mg OOE. The strength and thickness of the modified films were decreased, but their antioxidant qualities were much enhanced. Antioxidant activity rose from 0% in the control to 32.55% in OOE-loaded films using the DPPH assay. Additionally, their potential as antioxidant packing was confirmed by the increase in total phenolic content (TPC) from 3 mg gallic acid equivalents (GAE) to 68 mg GAE (Da Rosa et al. 2020). Another study looked into how GFO could increase the antioxidant activity of κ-carrageenan films. The DPPH assay improved from 12.5% in pure carrageenan films to 47.3% with 0.3% GFO, indicating that adding GFO considerably enhanced the films' radical scavenging activity. GFO's promise as a natural antioxidant in carrageenan-based films for active food packaging applications was also highlighted by the ABTS assay, which showed an increase from 18.7 to 52.8% with the same concentration of GFO (Bhatia et al. 2024). Guo et al. (2022) developed packaging that tracks the shelf life of pork by combining hydroxypropyl methylcellulose (HPMC) and cork bark extract (CBE) into carrageenan films. Ethanol was used to extract CBE, which was then put into films using a solvent-casting technique. The efficacy of films with 1.2% CBE in active packaging was demonstrated by their 58.8% antioxidant activity (DPPH assay), which was noticeably higher than that of control films (Guo et al. 2022). Carrageenan films combined with anthocyanin-rich Jaboticaba peel extract (JPE) were the subject of another study. As JPE concentrations rose, the films' total phenolic content rose, reaching 45.6 mg GAE/g film at maximal extract addition. According to Kokkuvayil Ramadas et al. (2024), this increase was associated with a notable increase in antioxidant activity determined by the DPPH assay. This suggests that JPE improves antioxidant qualities and adds value by using agro-industrial waste, which aligns with sustainable development goals (Kokkuvayil Ramadas et al. 2024). Using solvent casting, Lakshmi Balasubramaniam et al. (2021) improved carrageenan films with gallic acid and nanocellulose. Esterification and radical grafting were the two grafting techniques used; 95% DPPH activity was obtained with radical grafting. This study demonstrated how fabrication techniques affect mechanical characteristics and antioxidant capability (Lakshmi Balasubramaniam et al. 2021). Studies on films of κ-carrageenan activated with olive leaf extract (OLE) showed increased capacity to scavenge radicals. The effectiveness of OLE as a natural additive to enhance the antioxidant qualities of carrageenan-based films and prolong the shelf life of packaged goods was highlighted by the DPPH assay, which showed that films containing 15% OLE had about 70% antioxidant activity (Yin and Woo 2024). Water extract of germinated fenugreek seeds (WEGFS) was added to semi-purified carrageenan (SRC) films by Farhan and Hani (2020). Antioxidant activity increased with increasing WEGFS concentration, according to the DPPH study, rising from 6.52% (control) to 67.6% (25% WEGFS). TPC also showed promise in preserving fresh chicken, increasing from 5.89 mg GAE/g (control) to 38.3 mg GAE/g (Farhan and Hani 2020). Jaboticaba bark extract was incorporated into carrageenan films by Avila et al. (2022) using MAE. Because of the anthocyanin-rich extract, this alteration improved thickness, TS, barrier qualities, and antioxidant activities. Increased antioxidant capacity and better antibacterial activity against *Escherichia coli* were validated by the DPPH and TPC assays (Avila et al. 2022). Dordevic et al. (2023) used solvent casting to incorporate spent coffee ground oil (SCGO) into carrageenan films. Hexane was used to extract the oil, which was high in linoleic and palmitic acids, and Tween-20 and Tween-80 were used to emulsify it. The DPPH, FRAP, ABTS, and CUPRAC tests were used to evaluate the antioxidant capabilities. Carrageenan's appropriateness for active packaging was confirmed by the maximum TPC (7.00 mg GAE/g) and antioxidant activity displayed by films containing 1% SCGO and Tween-80 (Dordevic et al. 2023).

8.3.6 Antimicrobial Properties

By avoiding microbial contamination and prolonging shelf life, antibacterial substances are essential to food preservation. These ingredients are frequently used in processed foods, restaurant dishes, and home-cooked meals. Synthetic antibacterial agents previously dominated the industry, but growing consumer worries about their possible health dangers have resulted in a growing demand for natural alternatives (Valdés et al. 2017; Young et al. 2020). Of them, films and coatings based on carrageenan have drawn interest because of their antibacterial qualities, which improve food safety and preservation. Films based on carrageenan have potent antibacterial properties, especially when mixed with bioactive substances. Their efficiency in preventing foodborne infections and enhancing food packaging applications has been investigated in several research. Films containing κ-carrageenan and citric acid showed significant antibacterial action against both Gram-positive and Gram-negative bacteria, such as *Pseudomonas aeruginosa*, *Dickeya chrysanthemi*, *Proteus mirabilis*, and *Staphylococcus aureus*. In addition to improving the films' antibacterial qualities, citric acid increased their mechanical flexibility and solubility, making them more appropriate for use in packaging (El-Fawal 2014). In order to enhance the structural and optical characteristics of carrageenan-based films, Simona et al. (2021) treated them using trehalose and orange essential oil. Solvent casting was used to create the films, which had improved thickness and barrier qualities. Their antimicrobial activity shown significant suppression against *Staphylococcus aureus* but poor efficacy against Gram-negative bacteria and yeast when evaluated using the agar diffusion method (Simona et al. 2021). Clove essential oil (CEO) was added to carrageenan and alginate composite films by Prasetyaningrum et al. (2021) to enhance their functionality. Calcium chloride ($CaCl_2$) improved the films' physical characteristics. While the control film exhibited no antibacterial action, antibacterial testing revealed a significant inhibition zone of 113.14 mm^2 for *Escherichia coli* in films containing 3% CEO (Prasetyaningrum et al. 2021). Eugenol, the primary bioactive component in clove oil with strong bactericidal effects, was credited by another study that integrated CEO into edible films based on alginate/κ-carrageenan with inhibition zones of 20 mm against *Staphylococcus aureus* and 18 mm against *Escherichia coli* (Kokkuvayil Ramadas et al. 2024). Olive leaf extract (OLE)-infused biodegradable films based on κ-carrageenan showed antimicrobial effectiveness against major foodborne pathogens, with inhibition zones of 12 mm for *Staphylococcus aureus* and 14 mm for *Escherichia coli*. One of OLE's antibacterial properties is the disruption of microbial cell membranes caused by its phenolic components (Martiny et al. 2020a, b). Similarly, Martiny et al. (2020a, b) created carrageenan films that were filled with olive oil extract to increase the shelf life of lamb meat. According to an evaluation of these films' mechanical, barrier, and functional qualities, oil-infused films had CFU levels five times lower than those of commercial plastic packaging, suggesting that they could be a useful antimicrobial food packaging substitute (Martiny et al. 2020a, b). Incorporating polyoxometalates into carrageenan-based films was investigated further, and the results showed strong antibacterial action against foodborne microorganisms. According to Cheng et al. (2022), these results imply that carrageenan films can be designed to function as active antimicrobial packaging materials, improving food safety and shelf life (Cheng et al. 2022). Table 8.1 displayed key properties of carrageenan-based films and coatings with influencing factors.

8.4 Application of Carrageenan Based Film in Food Preservation

Fruits and vegetables are abundant sources of essential vitamins and minerals that contribute significantly to human health. However, they are highly perishable, with nearly 20% being discarded annually in developed nations, leading to substantial economic losses (Mostafidi et al. 2020). The primary factors influencing their spoilage include water content and respiration rate. Studies have shown that their high content (approximately 75–95%) and elevated respiration rate make them more vulnerable to bacterial and fungal growth, accelerating deterioration. Therefore, developing effective preservation techniques is crucial to prolong their shelf life (Otoni et al. 2017). Carrageenan film possesses outstanding mechanical and protective properties, making it a preferred choice for preserving fresh and vegetables. Its strong mechanical characteristics help maintain the film's integrity during transportation and distribution, preventing physical damage to the product. Furthermore, adding oxygen and ethylene scavengers to the film can lower the pace at which fruits and vegetables breathe while preventing the formation of aerobic bacteria. Although carrageenan has a relatively weak barrier against water vapour, other components can significantly enhance its resistance, minimizing moisture loss through surface pores and cracks (Thakur et al. 2018; Resende et al. 2018).

Carrageenan-zinc oxide nanocomposite coatings have been shown to effectively minimize both physical and biological damage to mangoes. By the 19th day, coated mangoes maintained their initial firmness, while slight discolouration and decay appeared by the 33rd day. This coating extended the shelf life of mangoes by 14 days compared to uncoated ones. Incorporating zinc oxide nanoparticles significantly enhanced the coating's tensile strength by 43%, reduced water vapour transmission by 9%, and inhibited the growth of *E. coli*, outperforming pure carrageenan coatings (Meindrawan et al. 2018). Unlike mangoes, strawberries are more susceptible to spoilage, highlighting the need for effective preservation methods. Carrageenan

Table 8.1 Key properties of carrageenan-based films and coatings with influencing factors

Property	Key findings	Influencing factors	References
Mechanical properties	κ-carrageenan films exhibit high TS (about 39 MPa), enhanced by additives, TiO_2 increases TS by 122%	Addition of plasticizers, crosslinkers, nanoparticles and biopolymer blends	Cheng et al. (2022), Wang et al. (2024), Kokkuvayil Ramadas et al. (2024)
Elongation at break (EAB)	EAB is reduced by crosslinkers and nanoparticles but improved with plasticizers like glycerol	Glycerol lowers TS but increases EAB; Sorbitol has a milder effect	Genecya et al. (2023), Wang et al. (2024)
Barrier properties	Nanomaterials and biopolymers reduce WVP and improve gas barrier properties. Alginate-carrageenan blends lower WVP by about 30%	Blending with alginate, CNCs, nanoclay and bio-based extracts like *Zataria multiflora*	Sedayu (2020), Nouri et al. (2020), Kokkuvayil Ramadas et al. (2024)
Thermal stability	Pearl millet starch increases melting point from 98 to 118 °C. CNCs improve thermal stability by about 20 °C	Addition of CNCs, polyphenols and crosslinking agents	Sandhu et al. (2020), Gupta et al. (2021)
UV-light barrier	3% GFO reduces UV transmission at 280 nm by 70%. CNCs and lignin enhances UV-blocking ability	Addition of GFO, cellulose nanocrystals, carbon dots and lignin	Kokkuvayil Ramadas et al. (2024), Yadav and Chiu (2019), Rukmanikrishnan et al. (2020)
Antioxidant activity	OOE increases antioxidant activity from 0 to 32.55% (DPPH assay). JPE enhances TPC to 45.6 mg GAE/g film	Incorporation of plant-based extracts rich in polyphenols	Da Rosa et al. (2020), Kokkuvayil Ramadas et al. (2024)
Bactericidal properties	CuS_2 and limonene enhance antimicrobial efficiency without altering film thickness. Ag nanoparticles decrease TS but improve antibacterial action	Incorporation of nanoparticles (Ag, Ce_2O_3, CuS_2) and essential oils	Roy and Rhim (2020a, b), Genecya et al. (2023)
Crosslinking effects	Citric acid crosslinks CMC-HEC, improving TS, Young's modulus and toughness. Potassium chloride (KCl) helps in κ-carrageenan gelation	Chemical and physical crosslinking (citric acid, KCl, Maleic anhydride)	Kang and Yun (2022), Zhou et al. (2021)

(continued)

Table 8.1 (continued)

Property	Key findings	Influencing factors	References
Biodegradability	Addition of modified tapioca starch increases biodegradability while maintaining TS	Biopolymer blends (alginate, gelatin, CMC), starch modifications	Islamiyah et al. (2022)
Optical transparency	CNCs at 9% reduce UV transmittance from 76 to 24%, while lignin at 15% decreases light transmission	Use of CNCs, carbon dots, lignin and plant pigments	Yadav and Chiu, (2019), Rukmanikrishnan et al. (2020), Khan et al. (2023a, b)

coatings have been applied to strawberry packaging, showing the lowest total soluble solids (9.50%) by the 12th day of storage, notably lower than the control group (12.20%). Additionally, the coating reduced the rot rate of strawberries by 14.29% by inhibiting metabolic activity and minimizing water loss (Wani et al. 2021). Figure 8.2 displays the detailed application of carrageenan-based films in food preservation.

Because of its useful qualities, carrageenan, a naturally occurring hydrocolloid obtained from red seaweeds, is widely used in the food sector. It creates a gel-like consistency when dissolved in water, improving the stability, texture, and shelf life

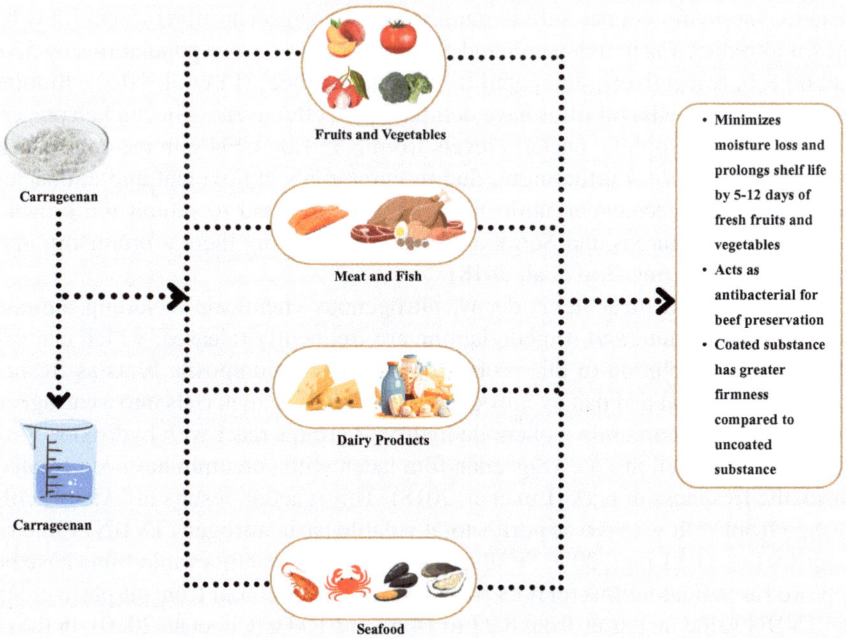

Fig. 8.2 Application of carrageenan-based films in food preservation

of a variety of perishable foods. Fruits and vegetables are frequently treated with it to prevent moisture loss and increase freshness by five to twelve days. Carrageenan is an antimicrobial and moisture-retaining agent in meat and fish products, especially helping preserve cattle. In comparison to uncoated substances, coated substances are firmer. It also preserves overall product quality, increases consistency, and improves texture in dairy and seafood products.

Vegetables can be preserved in storage using carrageenan film in a similar way. It has been demonstrated that tomatoes coated with an edible, biodegradable film composed of ι-carrageenan and arrowroot starch retain more firmness than tomatoes that are not coated (Abdillah and Charles 2021). Additionally, konjac glucomannan-SiO_2-carrageenan composite nanofilms help minimize weight loss and oxygen permeability in white mushrooms, showing their respiration and reducing browning. This preservation method effectively extends the shelf life of mushrooms stored at 4 °C by 5–12 days (Zhang et al. 2019).

Products made from fish and meat are prized for their high protein and unsaturated fatty acid content. However, they are highly perishable due to microbial contamination, fat oxidation, protein breakdown, and enzymatic activity (Ghimire et al. 2022). External conditions like temperature, oxygen, moisture, and light can significantly alter their colour, smell, texture, and flavour, making spoilage difficult to detect with the naked eye. Refrigeration and freezing are commonly used to preserve these products, but frozen fish and meat often have an inferior taste compared to chilled alternatives (Zhang et al. 2020). Carrageenan-based composite films have gained significant attention for extending the shelf life of chilled fish and meat products. For instance, applying copper sulfide nanoparticle-carrageenan films to beef has been shown to reduce Escherichia coli and *Staphylococcus aureus* populations by 52.6% and 69.8%, respectively, compared to unpackaged beef (Li et al. 2020). Similarly, carrageenan/camellia oil films have demonstrated effectiveness in chicken preservation, reducing cryophilic bacteria levels from 5.35 Log CFU/g in the control group to 3.86 Log CFU/g. Furthermore, due to curcumin's antioxidant and antibacterial properties, carrageenan/curcumin films have been found to inhibit the growth of Staphylococcus aureus and Serratia marcescens, making them a promising option for shrimp packaging (Liu et al. 2018).

When fish and meat items decay, nitrogenous chemicals including ammonia, dimethylammonium, and trimethylamine are frequently released, which can cause pH shifts. The solution to this problem is the use of composite films as freshness indicators for meat and fish by introducing pH-sensitive materials into a carrageenan matrix. Because curcumin's phenolic hydroxyl groups react with hydroxide ions to make phenolate anions, a carrageenan film laden with curcumin has been created to check the freshness of pork (Liu et al. 2018). This reaction results in a visible colour change from yellow to red as pork's total volatile basic nitrogen (TVBN) value rises from 4.91 to 31.11 mg/100 g. Similarly, κ-carrageenan/anthocyanin films have been explored as indicator films for pork freshness, changing colour from purple to green as the TVBN value increases from 8.23 to 14.63 mg/100 g (Chi et al. 2020). In the case of lard packaging, carrageenan/carboxymethylcellulose/plum sap composite films have demonstrated a colour shift from deep red to blue-grey as the pH rises from 3 to

13 (Sun et al. 2019). Additionally, natural dye-loaded carrageenan composite films have been applied to monitor fish's freshness (Sani et al. 2022) and shrimp (Liu et al. 2018), further enhancing their potential as intelligent food packaging materials.

Oxidation and microbial activity are key factors contributing to the spoilage of dairy products. In recent years, carrageenan-based composite films have gained increasing attention for their role in dairy product preservation. Research indicates that incorporating mulberry polyphenolic extract into a carrageenan matrix can serve as an indicator for monitoring milk freshness. When the film was placed over milk stored at 40 °C for 0–6 h, its colour shifted from grey to purple and eventually to dark pink. This color change occurred due to increased acidity resulting from microbial metabolism, which led to lactic acid production, a drop in pH, and ultimately, a visible alteration in the film's appearance (Liu et al. 2019a, b). During ice cream production, the whipping process accelerates lipid oxidation and microbial growth, making it essential to develop carrageenan-based composite films for preservation (Ullah et al. 2017). Studies have shown that an aloe vera gel/carrageenan composite film effectively inhibits the growth of various microbial pathogens, including *Staphylococcus aureus, Escherichia coli, Streptococcus agalactiae,* and *Klebsiella pneumoniae* in ice cream (Mahajan et al. 2021).

Additionally, carrageenan-based composite films have been applied in cheese preservation. Research has demonstrated that a carrageenan-black bean extract film, known for its strong antioxidant, mechanical, and water barrier properties, can effectively protect cheese. After 20 days of storage, the peroxide value of the coated cheese was 1.25 mEq O_2/kg, significantly lower than the 4.2 mEq O_2/kg recorded for uncoated samples, highlighting its potential in extending cheese shelf life (Pérez et al. 2021; Mortensen et al. 2002). Table 8.2 revealed the application of carrageenan-based film in food preservation.

8.5 Application of Carrageenan-Based Film in Intelligent Packaging

Intelligent or smart packaging is an advanced type designed to monitor internal conditions and effectively convey this information to consumers. It enhances consumer convenience and decision-making by allowing them to assess food freshness (Roy et al. 2023a, b). Furthermore, it helps preserve food quality by identifying dangerous substances or gases, warning customers of possible risks, and allowing distributors to spot tainted goods before they are distributed. Carrageenan is a suitable base material for smart packaging applications to monitor the freshness of various food items, with specific examples discussed in the following sections (Balbinot-Alfaro et al. 2019; Kalpana et al. 2019).

Liu et al. investigated the ability of curcumin, which is produced from turmeric, to change colour as a sign of freshness in shrimp and pork. The casting approach was used to evenly insert curcumin into carrageenan film, guaranteeing consistent distribution inside the container. Fresh pork and shrimp, which spoil easily because of

Table 8.2 Application of Carrageenan-based film in food preservation

Application	Base material	Additives	Function/mechanism	Results/effects	References
Fruits & vegetables	Carrageenan film	Oxygen & ethylene scavengers	Reduces respiration rate and inhibits bacterial growth	Minimizes moisture loss and prolongs shelf life of fresh fruits and vegetables	Otoni et al. (2017), Thakur et al. (2018)
Mango preservation	Carrageenan-ZnO nanocomposite coating	Zinc oxide nanoparticles	Enhances tensile strength, reduces water vapor transmission, and inhibits E. coli growth	Maintained firmness for 19 days; extended shelf life by 14 days compared to uncoated mangoes	Meindrawan et al. (2018)
Strawberry packaging	Carrageenan coating	None	Inhibits metabolic activity and minimizes water loss	Reduced total soluble solids (9.50% vs. 12.20%) and decreased rot rate by 14.29% after 12 days	Wani et al. (2021)
Tomato preservation	Arrowroot starch & ι-Carrageenan film	None	Maintains firmness	Greater firmness compared to uncoated tomatoes	Abdillah and Charles (2021)
Mushroom preservation	Konjac Glucomannan-SiO_2-Carrageenan composite film	None	Reduces weight loss and oxygen permeability	Extended shelf life by 5–12 days at 4 °C	Zhang et al. (2019)
Beef preservation	Carrageenan film	Copper sulfide nanoparticles	Antimicrobial activity against *E. coli* and *Staphylococcus aureus*	Reduced bacterial populations by 52.6% and 69.8%, respectively	Li et al. (2020)
Chicken preservation	Carrageenan/Camellia oil film	Camellia Oil	Reduces cryophilic bacteria levels	Bacteria reduced from 5.35 Log CFU/g to 3.86 Log CFU/g	Liu et al. (2018)

(continued)

Table 8.2 (continued)

Application	Base material	Additives	Function/mechanism	Results/effects	References
Shrimp packaging	Carrageenan/Curcumin film	Curcumin	Inhibits *S. aureus* and *S. marcescens* growth	Demonstrates antibacterial properties	Liu et al. (2018)
Fish freshness indicator	κ-Carrageenan/anthocyanin film	Anthocyanin extract	Monitors spoilage via pH-sensitive color change	Film changes color from purple to green as TVBN value increases	Chi et al. (2020)
Lard packaging	Carrageenan/carboxymethylcellulose/plum sap film	Plum sap	Indicates pH changes	Film color shifts from deep red to blue-gray as pH rises from 3 to 13	Sun et al. (2019)
Milk freshness indicator	Carrageenan film	Mulberry polyphenolic extract	Indicates milk freshness through pH-induced color changes	Film color shifted from gray to purple to dark pink as milk acidity increased	Liu et al. (2019a, b)
Ice cream preservation	Aloe vera gel/carrageenan composite film	Aloe vera gel	Antimicrobial activity	Inhibited growth of *S. aureus*, *E. coli*, *S. agalactiae*, and *K. pneumoniae*	Mahajan et al. (2021)
Cheese preservation	Carrageenan-black bean extract film	Black bean extract	Antioxidant, mechanical, and water barrier properties	Reduced peroxide value to 1.25 mEq O_2/kg (compared to 4.2 mEq O_2/kg for uncoated cheese)	Pérez et al. (2021), Mortensen et al. (2002)

their high moisture and nutrient content, were then packaged in this film. Pork protein decomposes during spoiling to produce total volatile base nitrogen (TVBN), which is high in amines and lowers the pH of the meat because it contains substances like trimethylamine and dimethylammonium. Curcumin changes colour in response to these pH shifts. Because of its high absorption in this range, the curcumin film initially appears yellow; but, after three days of storage, it turns red. The alkaline conditions inside the packaging brought on by the natural decomposition are responsible for this colour shift (Liu et al. 2018).

The freshness of chicken wings wrapped in films prepared from a mixture of carrageenan and arrowhead starch (*Sagittaria sagittifolia*) enhanced with extract from black chokeberries (*Aronia melanocarpa*) was examined by Wang et al. Anthocyanins, which are naturally occurring pH-sensitive indicators that change colour in response to pH changes, are abundant in this extract. To investigate its impact on film qualities, the extract was added to composite films at varying amounts using the solvent-casting technique. Protein breakdown is the main cause of chicken spoiling, just like it is for pork and shrimp. The production of amino compounds by spoilage bacteria during the hydrolysis of proteins causes the pH to rise to more basic levels. The film turns from pink to dark pink as a result of these pH shifts, which are recognized by anthocyanins. The degree of the shift is determined by the concentration of anthocyanins. A discernible hue shift in the film after three days of room temperature storage was a sign of obvious spoiling. The TVBN value, which differs by nation, was used to measure freshness. Changes in the film clearly show that the TVBN level should not be above 15 mg per 100 mg of chicken, which is the limit set by Chinese regulations (Wang et al. 2023).

Zhang et al. incorporated mulberry polyphenol extract into carrageenan-based films, which were then produced using solvent-casting. Initially, the film appeared dark blue, likely due to the interaction between anthocyanins in the mulberry extract and carrageenan. The extract was added at varying concentrations to analyze its impact on the film's colour change. Milk was used as a test medium to evaluate the film's effectiveness. When milk spoils, lactic acid bacteria become more active, leading to lactic acid accumulation and a drop in pH below 7, making it acidic. Upon contact with milk, the film turned purple. After six hours of storage at 40 °C—an approximate time frame for milk spoilage—the increasing acidity caused the film to change from purple to pink, signalling deterioration. These findings suggest that carrageenan films infused with mulberry polyphenol extract can be an effective tool for smart packaging applications (Zhang et al. 2020).

Liu et al., investigated κ-Carrageenan-based films incorporating mulberry phenolic extract have been effectively developed as intelligent packaging systems to monitor the freshness of milk. Mulberry phenolic extract is a natural source of anthocyanins that serve as pH-sensitive colour indicators. Initially, the fabricated film appears blue when combined with the extract. Upon contact with milk, the film turns purple. When the milk is stored at 40 °C for six hours, spoilage occurs due to the activity of lactic acid bacteria, leading to lactic acid accumulation and a reduction in the pH of the milk. This pH change shifts the film's colour from purple to pink, clearly indicating milk spoilage (Liu et al. 2019a, b).

κ-Carrageenan-based films combined with natural indicators such as curcumin, anthocyanins, and flavonoids have been extensively applied in intelligent food packaging systems to monitor spoilage. Incorporating curcumin powder into κ-carrageenan films helps identify when pork and shrimp are spoiling because microbial activity creates nitrogenous chemicals like ammonia, which raises the alkalinity and turns the film red by the third day of storage (Liu et al. 2018). Similarly, when milk deteriorates at ambient temperature due to pH shifts, a κ-carrageenan/gelatin composite film that contains propolis and shikonin transforms from purple to reddish pink (Roy and Rhim 2020a, b). In a different study, mulberry anthocyanin extracts infused κ-carrageenan/sodium carboxymethyl starch films detect spoilage in mirror carp fish by turning from red to dark blue, which reflects pH changes from slightly acidic to neutral (Zhang et al. 2020). Because the colour of the film changes from purple to brown as the pH of the surrounding environment changes, films that contain jaboticaba peel extract show promise as spoiling indicators for fish packaging (Avila et al. 2022). Additionally, grape skin powder containing κ-carrageenan/hydroxypropylmethylcellulose films efficiently track the deterioration of pork, changing colour from purple to green as nitrogen molecules build up and the meat's basicity increases (Chi et al. 2020).

When paired with other natural indications, films based on κ-carrageenan offer efficient ways to track food spoilage in applications including intelligent packaging. By going through a colour change, κ-carrageenan/polyvinyl alcohol films combined with purple sweet potato anthocyanins (PSA) and purple cabbage anthocyanins (PCA) can identify when shrimps are about to spoil. As the shrimp's pH rises from 6.2 to 8.3, the anthocyanin in the cabbage turns pale blue, and the anthocyanin in potato turns purple to brown (Yong et al. 2022). In the same way, κ-carrageenan/arrowhead starch films containing black chokeberry anthocyanins cause chicken wings to turn from pink to dark pink as they are about to spoil. This alteration is related to the build-up of volatile amines during the breakdown process, which raises the chicken's alkalinity (Wang et al. 2023).

Additionally, κ-carrageenan/sodium carboxymethyl cellulose (Na-CMC) films infused with bromothymol blue, a synthetic compound, are used for monitoring fresh-cut papaya. As the CO_2 concentration inside the packaging increases due to a CO_2-releasing tablet, the film colour changes from dark blue to green and eventually to yellow–green and yellow–brown, indicating the fruit's edibility and spoilage stages. These applications highlight the versatility and effectiveness of κ-carrageenan-based intelligent packaging systems (Rong et al. 2023).

κ-carrageenan films combined with TiO_2-doped carbon dots and sweet potato peel anthocyanin extract act as effective pH indicators. As shrimp decompose, the accumulation of volatile amines creates an alkaline environment, causing the film to change colour from red to dark brown, indicating spoilage (Riahi et al. 2024). Similarly, κ-carrageenan/gelatin/copper metal films incorporated with red cabbage anthocyanin extract respond to pH changes by transitioning from grey to reddish-yellow as the shrimp spoils. This colour change is influenced by the anthocyanin content in the films, providing a reliable visual cue to assess shrimp freshness (Khan et al. 2023a, b). These studies underscore the versatility and effectiveness of κ-carrageenan-based films in intelligent packaging applications. Table 8.3 explains the applications of carrageenan-based film in intelligent packaging.

Table 8.3 Application of Carrageenan-based film in intelligent packaging

Film composition	Indicator type	Tested food	Color changes	Spoilage mechanism	References
κ-carrageenan/ curcumin	Curcumin (pH-sensitive)	Pork, shrimp	Yellow to red	Alkalinity increase due to amine compounds	Liu et al. (2018)
κ-carrageenan/ arrowhead starch/black chokeberry	Anthocyanins	Chicken wings	Pink to dark pink	Alkalinity increases due to protein hydrolysis	Wang et al. (2023)
κ-carrageenan/ mulberry polyphenol	Anthocyanins	Milk	Dark blue to purple to pink	Acidic environment due to lactic acid accumulation	Zhang et al. (2020)
κ-carrageenan/ mulberry phenolic extract	Anthocyanins	Milk	Blue to purple pink	Acidic environment due to lactic acid accumulation	Liu et al. (2019a, b)
κ-carrageenan/ gelatin/ shikonin/ propolis	Shikonin & propolis (pH-sensitive)	Milk	Purple to reddish pink	pH drops due to spoilage	Roy and Rhim (2020a, b)
κ-carrageenan/ sodium carboxymethyl starch/mulberry	Anthocyanins	Mirror carp fish	Red to dark blue	pH shifts from slightly acidic to neutral	Zhang et al. (2020)
κ-carrageenan/ jaboticaba peel extract	Anthocyanins	Fish	Purple to brown	pH changes due to spoilage	Avila et al. (2022)
κ-carrageenan/ hpmc/grape skin powder	Anthocyanins	Pork	Purple to green	Basicity increases due to nitrogenous compound	Chi et al. (2020)
κ-carrageenan/ pva/purple sweet potato & cabbage	Anthocyanins	Shrimp	Dark blue to pale blue, purple to brown	pH increases from 6.2 to 8.3	Yong et al. (2022)
κ-carrageenan/ sodium carboxymethyl cellulose/ bromothymol blue	Bromothymol blue (synthetic indicator)	Fresh cut papaya	Dark blue to green to yellow brown	CO_2 concentration increase affecting pH	Rong et al. (2023)

(continued)

Table 8.3 (continued)

Film composition	Indicator type	Tested food	Color changes	Spoilage mechanism	References
κ-carrageenan/ TiO_2-doped carbon dots/ sweet potato peel	Anthocyanins	Shrimp	Red to dark brown	Alkalinity increases due to volatile amines	Riahi et al. (2024)
κ-carrageenan/ gelatin/ copper metal/red cabbage extract	Anthocyanins	Shrimp	Grey to reddish yellow	Alkalinity increases due to volatile amines	Khan et al. (2023a, b)

8.6 Summary and Future Outlook

Carrageenan, a seaweed-derived polysaccharide, can be suitable for making sustainable packaging, and it can be a good alternative to traditional synthetic polymers used in packaging. The readily availability, good film-forming properties, high transparency, compatibility with other polymers, and active ingredients are some of the advantages of carrageenan. Combining plasticizers, biopolymers, and active substances allows for the tuning of physical qualities like mechanical, barrier, and hydrophilicity limitations. Active and intelligent packaging based on carrageenan can extend the shelf life of fruits, vegetables, meat, and dairy products. It can also be used to track the freshness and deterioration of meat, fish, and milk in real time while they are being stored. However, the cost of carrageenan, lack of scalability, purity of carrageenan, and uniform extraction methods need more research. Further research on this area could improve the limitations and practical utility of carrageenan-based sustainable packaging.

Declaration Authors used generative AI tools for the purpose of improving readability and language. The final content has been reviewed and verified by the authors after using these tools, and they take full responsibility for its accuracy and integrity.

References

Abdillah AA, Charles AL (2021) Characterization of a natural biodegradable edible film obtained from arrowroot starch and iota-carrageenan and application in food packaging. Int J Biol Macromol 191:618–626. https://doi.org/10.1016/j.ijbiomac.2021.09.141

Álvarez-Viñas M, Rivas S, Torres MD, Domínguez H (2023) Microwave-assisted extraction of carrageenan from Sarcopeltis skottsbergii. Mar Drugs 21(2):83. https://doi.org/10.3390/md21020083

Asadzadeh N, Ghorbanpour M, Sayyah A (2023) Effects of filler type and content on mechanical, thermal, and physical properties of carrageenan biocomposite films. Int J Biol Macromol 253:127551. https://doi.org/10.1016/j.ijbiomac.2023.127551

Avila LB, Barreto ERC, Moraes CC, Morais MM, Rosa GSD (2022) Promising new material for food packaging: an active and intelligent carrageenan film with natural jaboticaba additive. Foods 11(6):792. https://doi.org/10.3390/foods11060792

Balbinot-Alfaro E, Craveiro DV, Lima KO, Costa HLG, Lopes DR, Prentice C (2019) Intelligent packaging with pH indicator potential. Food Eng Rev 11:235–244. https://doi.org/10.1007/s12393-019-09198-9

Bharti SK, Pathak V, Arya A, Alam T, Rajkumar V, Verma AK (2021) Packaging potential of Ipomoea batatas and κ-carrageenan biobased composite edible film: its rheological, physicomechanical, barrier and optical characterization. J Food Process Preserv 45(2):e15153. https://doi.org/10.1111/jfpp.15153

Bhatia S, Abbas Shah Y, Al-Harrasi A, Jawad M, Koca E, Aydemir LY (2024) Enhancing tensile strength, thermal stability, and antioxidant characteristics of transparent kappa carrageenan films using grapefruit essential oil for food packaging applications. ACS Omega 9(8):9003–9012. https://doi.org/10.1021/acsomega.3c07366

Carreira-Casais A, Otero P, Garcia-Perez P, Garcia-Oliveira P, Pereira AG, Carpena M et al (2021) Benefits and drawbacks of ultrasound-assisted extraction for the recovery of bioactive compounds from marine algae. Int J Environ Res Public Health 18(17):9153. https://doi.org/10.3390/ijerph18179153

Cheng C, Chen S, Su J, Zhu M, Zhou M, Chen T, Han Y (2022) Recent advances in carrageenan-based films for food packaging applications. Front Nutr 9:1004588. https://doi.org/10.3389/fnut.2022.1004588

Chi W, Cao L, Sun G, Meng F, Zhang C, Li J, Wang L (2020) Developing a highly pH-sensitive κ-carrageenan-based intelligent film incorporating grape skin powder via a cleaner process. J Clean Prod 244:118862. https://doi.org/10.1016/j.jclepro.2019.118862

da Rosa GS, Vanga SK, Gariepy Y, Raghavan V (2020) Development of biodegradable films with improved antioxidant properties based on the addition of carrageenan containing olive leaf extract for food packaging applications. J Polym Environ 28:123–130. https://doi.org/10.1007/s10924-019-01589-7

Dmitrenko M, Kuzminova A, Cherian RM, Joshy KS, Pasquini D, John MJ et al (2023) Edible carrageenan films reinforced with starch and nanocellulose: development and characterization. Sustainability 15(22):15817. https://doi.org/10.3390/su152215817

Dordevic D, Dordevic S, Abdullah FAA, Mader T, Medimorec N, Tremlova B, Kushkevych I (2023) Edible/biodegradable packaging with the addition of spent coffee grounds oil. Foods 12(13):2626. https://doi.org/10.3390/foods12132626

El-Fawal G (2014) Preparation, characterization and antibacterial activity of biodegradable films prepared from carrageenan. J Food Sci Technol 51:2234–2239. https://doi.org/10.1007/s13197-013-1255-9

Ezati P, Riahi Z, Rhim JW (2021) Carrageenan-based functional films integrated with CuO-doped titanium nanotubes for active food-packaging applications. ACS Sustain Chem Eng 9(28):9300–9307. https://doi.org/10.1021/acssuschemeng.1c01957

Farhan A, Hani NM (2020) Active edible films based on semi-refined κ-carrageenan: antioxidant and color properties and application in chicken breast packaging. Food Packag Shelf Life 24:100476. https://doi.org/10.1016/j.fpsl.2020.100476

Fathiraja P, Gopalrajan S, Karunanithi M, Nagarajan M, Obaiah MC, Durairaj S, Neethirajan N (2022) Response surface methodology model to optimize concentration of agar, alginate and carrageenan for the improved properties of biopolymer film. Polym Bull 79(8):6211–6237. https://doi.org/10.1007/s00289-021-03797-5

Genecya G, Adhika DR, Sutrisno W, Wungu TD (2023) Characteristic improvement of a carrageenan-based bionanocomposite polymer film containing montmorillonite as food packaging through the addition of silver and cerium oxide nanoparticles. ACS Omega 8(42):39194–39202. https://doi.org/10.1021/acsomega.3c04575

Ghimire A, Paudel N, Poudel R (2022) Effect of pomegranate peel extract on the storage stability of ground buffalo (Bubalus bubalis) meat. LWT 154:112690. https://doi.org/10.1016/j.lwt.2021.112690

Gomez L, Tiwari B, Garcia-Vaquero M (2020) Emerging extraction techniques: microwave-assisted extraction. In: Sustainable seaweed technologies. Elsevier, pp 207–224. https://doi.org/10.1016/B978-0-12-817943-7.00008-1

Guo J, Dong S, Ye M, Wu X, Lv X, Xu H, Li M (2022) Effects of hydroxypropyl methylcellulose on physicochemical properties and microstructure of κ-carrageenan film. Foods 11(19):3023. https://doi.org/10.3390/foods11193023

Gupta V, Yadav R, Tanwar R, Gaikwad KK (2021) κ-Carrageenan-based bio-nanocomposite film reinforced with cellulose nanocrystals derived from amla pomace for food packaging. Biomass Convers Biorefinery 1–10. https://doi.org/10.1007/s13399-021-02028-1

Hasizah A, Mahendradatta M, Laga A, Metusalach M, Salengke S (2021) Extraction of carrageenan from Eucheuma spinosum using ohmic heating: optimization of extraction conditions using response surface methodology. Food Sci Technol 41(4):928–937. https://doi.org/10.1590/fst.26220

Islamiyah HS, Alamsjah MA, Pujiastuti DY (2022) Application of modified starch in the Carragenan-based biodegradable packaging from Eucheuma Cottonii On biodegradablility and mechanical properties. IOP Conf Ser: Earth Environ Sci 1036(1):012072 (IOP Publishing). https://doi.org/10.1088/1755-1315/1036/1/012072

Jiang F, Liu Y, Xiao Q, Chen F, Weng H, Chen J et al (2022) Eco-friendly extraction, structure, and gel properties of ι-Carrageenan extracted using $Ca(OH)_2$. Mar Drugs 20(7):419. https://doi.org/10.3390/md20070419

Kalpana S, Priyadarshini SR, Leena MM, Moses JA, Anandharamakrishnan C (2019) Intelligent packaging: trends and applications in food systems. Trends Food Sci Technol 93:145–157. https://doi.org/10.1016/j.tifs.2019.09.008

Kang J, Yun SI (2022) Double-network hydrogel films based on cellulose derivatives and κ-Carrageenan with enhanced mechanical strength and superabsorbent properties. Gels 9(1):20. https://doi.org/10.3390/gels9010020

Kasim S, Alam G, Rifai Y, Utami RN (2023) Optimization of carrageenan extraction from Kappaphycus alvarezii red algae using microwave assisted extraction method with variation of solvent concentration. SCIREA J Phys 8(2):115–128. https://doi.org/10.54647/physics140518

Khan A, Priyadarshi R, Bhattacharya T, Rhim JW (2023a) Carrageenan/alginate-based functional films incorporated with Allium sativum carbon dots for UV-barrier food packaging. Food Bioprocess Technol 16(9):2001–2015. https://doi.org/10.1007/s11947-023-03048-7

Khan A, Riahi Z, Kim JT, Rhim JW (2023b) Gelatin/carrageenan-based smart packaging film integrated with Cu-metal organic framework for freshness monitoring and shelf-life extension of shrimp. Food Hydrocoll 145:109180. https://doi.org/10.1016/j.foodhyd.2023.109180

Kim YT, Min B, Kim KW (2014) General characteristics of packaging materials for food system. In: Innovations in food packaging. Academic Press, pp 13–35. https://doi.org/10.1016/B978-0-12-394601-0.00002-3

Kokkuvayil Ramadas B, Rhim JW, Roy S (2024) Recent progress of carrageenan-based composite films in active and intelligent food packaging applications. Polymers 16(7):1001. https://doi.org/10.3390/polym16071001

Lakshmi Balasubramaniam S, Singh Patel A, Nayak B (2021) Fabrication of antioxidative food packaging films using cellulose nanofibers, kappa-Carrageenan, and gallic acid. J Food Process Preserv 45(5):e15480. https://doi.org/10.1111/jfpp.15480

Li F, Liu Y, Cao Y, Zhang Y, Zhe T, Guo Z et al (2020) Copper sulfide nanoparticle-carrageenan films for packaging application. Food Hydrocoll 109:106094. https://doi.org/10.1016/j.foodhyd.2020.106094

Lin J, Jiao G, Kermanshahi-Pour A (2022) Algal polysaccharides-based hydrogels: extraction, synthesis, characterization, and applications. Mar Drugs 20(5):306. https://doi.org/10.3390/md20050306

Liu J, Wang H, Wang P, Guo M, Jiang S, Li X, Jiang S (2018) Films based on κ-carrageenan incorporated with curcumin for freshness monitoring. Food Hydrocoll 83:134–142. https://doi.org/10.1016/j.foodhyd.2018.05.012

Liu J, Wang H, Guo M, Li L, Chen M, Jiang S et al (2019a) Extract from Lycium ruthenicum Murr. Incorporating κ-carrageenan colorimetric film with a wide pH-sensing range for food freshness monitoring. Food Hydrocoll 94:1–10. https://doi.org/10.1016/j.foodhyd.2019.03.008

Liu Y, Qin Y, Bai R, Zhang X, Yuan L, Liu J (2019b) Preparation of pH-sensitive and antioxidant packaging films based on κ-Carrageenan and mulberry polyphenolic extract. Int J Biol Macromol 134:993–1001. https://doi.org/10.1016/j.ijbiomac.2019.05.175

Mahajan K, Kumar S, Bhat ZF, Naqvi Z, Mungure TE, Bekhit AEDA (2021) Functionalization of carrageenan based edible film using Aloe vera for improved lipid oxidative and microbial stability of frozen dairy products. Food Biosci 43:101336. https://doi.org/10.1016/j.fbio.2021.101336

Martín-del-Campo A, Fermín-Jiménez JA, Fernández-Escamilla VV, Escalante-García ZY, Macías-Rodríguez ME, Estrada-Girón Y (2021) Improved extraction of carrageenan from red seaweed (Chondracantus canaliculatus) using ultrasound-assisted methods and evaluation of the yield, physicochemical properties and functional groups. Food Sci Biotechnol 30:901–910. https://doi.org/10.1007/s10068-021-00935-7

Martiny TR, Pacheco BS, Pereira CM, Mansilla A, Astorga-España MS, Dotto GL et al (2020a) A novel biodegradable film based on κ-carrageenan activated with olive leaves extract. Food Sci Nutr 8(7):3147–3156. https://doi.org/10.1002/fsn3.1554

Martiny TR, Raghavan V, Moraes CCD, Rosa GSD, Dotto GL (2020b) Bio-based active packaging: carrageenan film with olive leaf extract for lamb meat preservation. Foods 9(12):1759. https://doi.org/10.3390/foods9121759

Mathew SS, Jaiswal AK, Jaiswal S (2024) Carrageenan-based sustainable biomaterials for intelligent food packaging: a review. Carbohyd Polym 342:122267. https://doi.org/10.1016/j.carbpol.2024.122267

Meindrawan B, Suyatma NE, Wardana AA, Pamela VY (2018) Nanocomposite coating based on carrageenan and ZnO nanoparticles to maintain the storage quality of mango. Food Packag Shelf Life 18:140–146. https://doi.org/10.1016/j.fpsl.2018.10.006

Mendes M, Cotas J, Gutiérrez IB, Gonçalves AM, Critchley AT, Hinaloc LAR et al (2024) Advanced extraction techniques and physicochemical properties of carrageenan from a novel Kappaphycus alvarezii Cultivar. Mar Drugs 22(11):491. https://doi.org/10.3390/md22110491

Meng F, Zhang Y, Xiong Z, Wang G, Li F, Zhang L (2018) Mechanical, hydrophobic and thermal properties of an organic-inorganic hybrid carrageenan-polyvinyl alcohol composite film. Compos B Eng 143:1–8. https://doi.org/10.1016/j.compositesb.2017.12.009

Mortensen G, Sørensen J, Stapelfeldt H (2002) Comparison of peroxide value methods used for semihard cheeses. J Agric Food Chem 50(18):5007–5011. https://doi.org/10.1021/jf0200220

Mostafidi M, Sanjabi MR, Shirkhan F, Zahedi MT (2020) A review of recent trends in the development of the microbial safety of fruits and vegetables. Trends Food Sci Technol 103:321–332. https://doi.org/10.1016/j.tifs.2020.07.009

Nouri A, Yaraki MT, Lajevardi A, Rahimi T, Tanzifi M, Ghorbanpour M (2020) An investigation of the role of fabrication process in the physicochemical properties of κ-carrageenan-based films incorporated with Zataria multiflora extract and nanoclay. Food Packag Shelf Life 23:100435. https://doi.org/10.1016/j.fpsl.2019.100435

Oun AA, Roy S, Hong SJ, Shin GH, Yoo S, Kim JT (2024) Development of smart colorimetric indicators for tracking kimchi freshness by loading aronia extract in agar, κ-carrageenan, and cellulose nanofiber films. Int J Biol Macromol 270:132343. https://doi.org/10.1016/j.ijbiomac.2024.132343

Otoni CG, Avena-Bustillos RJ, Azeredo HM, Lorevice MV, Moura MR, Mattoso LH, McHugh TH (2017) Recent advances on edible films based on fruits and vegetables—A review. Compr Rev Food Sci Food Saf 16(5):1151–1169. https://doi.org/10.1111/1541-4337.12281

Pan J, Li C, Liu J, Jiao Z, Zhang Q, Lv Z et al (2024) Polysaccharide-based packaging coatings and films with phenolic compounds in preservation of fruits and vegetables—A review. Foods 13(23):3896. https://doi.org/10.3390/foods13233896

Pérez MJ, Moreno MA, Martínez-Abad A, Cattaneo F, Zampini C, Isla MI et al (2021) Interest of black carob extract for the development of active biopolymer films for cheese preservation. Food Hydrocoll 113:106436. https://doi.org/10.1016/j.foodhyd.2020.106436

Prasetyaningrum A, Utomo DP, Raemas AFA, Kusworo TD, Jos B, Djaeni M (2021) Alginate/κ-carrageenan-based edible films incorporated with clove essential oil: physico-chemical characterization and antioxidant-antimicrobial activity. Polymers 13(3):354. https://doi.org/10.3390/polym13030354

Quitério E, Grosso C, Ferraz R, Delerue-Matos C, Soares C (2022) A critical comparison of the advanced extraction techniques applied to obtain health-promoting compounds from seaweeds. Mar Drugs 20(11):677. https://doi.org/10.3390/md20110677

Rapa M, Nita RND, Vasile C (2017) Influence of plasticizers over some physico-chemical properties of PLA. Mater Plast 54:73–78. https://doi.org/10.37358/MP.17.1.4789

Resende NS, Gonçalves GAS, Reis KC, Tonoli GHD, Boas EVBV (2018) Chitosan/cellulose nanofibril nanocomposite and its effect on quality of coated strawberries. J Food Qual 2018(1):1727426. https://doi.org/10.1155/2018/1727426

Riahi Z, Khan A, Rhim JW, Shin GH, Kim JT (2024) Carrageenan-based active and intelligent packaging films integrated with anthocyanin and TiO_2-doped carbon dots derived from sweet potato peels. Int J Biol Macromol 259:129371. https://doi.org/10.1016/j.ijbiomac.2024.129371

Rodriguez Carrillo SAA, Ruiz Olortino GP, Rafael Guevara DF (2023) bio-based carrageenan composite coatings for food packaging application. LACCEI 1(8). https://doi.org/10.18687/LACCEI2023.1.1.849

Rong L, Zhang T, Ma Y, Wang T, Liu Y, Wu Z (2023) An intelligent label using sodium carboxymethyl cellulose and carrageenan for monitoring the freshness of fresh-cut papaya. Food Control 145:109420. https://doi.org/10.1016/j.foodcont.2022.109420

Roy S, Rhim JW (2019) Carrageenan-based antimicrobial bionanocomposite films incorporated with ZnO nanoparticles stabilized by melanin. Food Hydrocoll 90:500–507. https://doi.org/10.1016/j.foodhyd.2018.12.056

Roy S, Rhim JW (2020a) Fabrication of copper sulfide nanoparticles and limonene incorporated pullulan/carrageenan-based film with improved mechanical and antibacterial properties. Polymers 12(11):2665. https://doi.org/10.3390/polym12112665

Roy S, Rhim JW (2020b) Preparation of gelatin/carrageenan-based color-indicator film integrated with shikonin and propolis for smart food packaging applications. ACS Appl Bio Mater 4(1):770–779. https://doi.org/10.1021/acsabm.0c01353

Roy S, Rhim JW (2021) Carrageenan/agar-based functional film integrated with zinc sulfide nanoparticles and Pickering emulsion of tea tree essential oil for active packaging applications. Int J Biol Macromol 193:2038–2046. https://doi.org/10.1016/j.ijbiomac.2021.11.035

Roy S, Ezati P, Rhim JW (2021) Gelatin/carrageenan-based functional films with carbon dots from enoki mushroom for active food packaging applications. ACS Appl Polym Mater 3(12):6437–6445. https://doi.org/10.1021/acsapm.1c01175

Roy S, Chawla R, Santhosh R, Thakur R, Sarkar P, Zhang W (2023a) Agar-based edible films and food packaging application: a comprehensive review. Trends Food Sci Technol 141:104198. https://doi.org/10.1016/j.tifs.2023.104198

Roy S, Deshmukh RK, Tripathi S, Gaikwad KK, Das SS, Sharma D (2023b) Recent advances in the carotenoids added to food packaging films: a review. Foods 12(21):4011. https://doi.org/10.3390/foods12214011

Rukmanikrishnan B, Rajasekharan SK, Lee J, Ramalingam S, Lee J (2020) K-Carrageenan/lignin composite films: biofilm inhibition, antioxidant activity, cytocompatibility, UV and water barrier properties. Mater Today Commun 24:101346. https://doi.org/10.1016/j.mtcomm.2020.101346

Rupert R, Rodrigues KF, Thien VY, Yong WTL (2022) Carrageenan from *Kappaphycus alvarezii* (rhodophyta, solieriaceae): metabolism, structure, production, and application. Front Plant Sci 13:859635

Saedi S, Shokri M, Roy S, Rhim JW (2021) Silver loaded aminosilane modified halloysite for the preparation of carrageenan-based functional films. Appl Clay Sci 211:106170. https://doi.org/10.1016/j.clay.2021.106170

Salengke S, Waris SA, Mochtar AH (2016) Design and optimization of pilot scale ohmic-based carrageenan extraction technology. J Nutr Food Sci 6(6):67

Sandhu KS, Sharma L, Kaur M, Kaur R (2020) Physical, structural and thermal properties of composite edible films prepared from pearl millet starch and carrageenan gum: process optimization using response surface methodology. Int J Biol Macromol 143:704–713. https://doi.org/10.1016/j.ijbiomac.2019.09.111

Sani MA, Tavassoli M, Salim SA, Azizi-lalabadi M, McClements DJ (2022) Development of green halochromic smart and active packaging materials: TiO_2 nanoparticle-and anthocyanin-loaded gelatin/κ-carrageenan films. Food Hydrocoll 124:107324. https://doi.org/10.1016/j.foodhyd.2021.107324

Sedayu BB (2020) Development of semi-refined k-carrageenan-based films for food packaging applications (Doctoral dissertation, Victoria University)

Sedayu BB, Cran MJ, Bigger SW (2019) A review of property enhancement techniques for carrageenan-based films and coatings. Carbohydr Polym 216:287–302. https://doi.org/10.1016/j.carbpol.2019.04.021

Shen L, Pang S, Zhong M, Sun Y, Qayum A, Liu Y et al (2023) A comprehensive review of ultrasonic assisted extraction (UAE) for bioactive components: principles, advantages, equipment, and combined technologies. Ultrason Sonochem 106646. https://doi.org/10.1016/j.ultsonch.2023.106646

Simona J, Dani D, Petr S, Marcela N, Jakub T, Bohuslava T (2021) Edible films from carrageenan/orange essential oil/trehalose—Structure, optical properties, and antimicrobial activity. Polymers 13(3):332. https://doi.org/10.3390/polym13030332

Sun G, Chi W, Zhang C, Xu S, Li J, Wang L (2019) Developing a green film with pH-sensitivity and antioxidant activity based on κ-carrageenan and hydroxypropyl methylcellulose incorporating Prunus maackii juice. Food Hydrocoll 94:345–353. https://doi.org/10.1016/j.foodhyd.2019.03.039

Thakur R, Pristijono P, Golding JB, Stathopoulos CE, Scarlett CJ, Bowyer M et al (2018) Development and application of rice starch based edible coating to improve the postharvest storage potential and quality of plum fruit (Prunus salicina). Sci Hortic 237:59–66. https://doi.org/10.1016/j.scienta.2018.04.005

Torres MD, Flórez-Fernández N, Dominguez H (2021) Ultrasound-assisted water extraction of mastocarpus stellatus carrageenan with adequate mechanical and antiproliferative properties. Mar Drugs 19(5):280. https://doi.org/10.3390/md19050280

Ullah R, Nadeem M, Imran M (2017) Omega-3 fatty acids and oxidative stability of ice cream supplemented with olein fraction of chia (Salvia hispanica L.) oil. Lipids Health Dis 16:1–8. https://doi.org/10.1186/s12944-017-0420-y

Valdés A, Ramos M, Beltrán A, Jiménez A, Garrigós MC (2017) State of the art of antimicrobial edible coatings for food packaging applications. Coatings 7(4):56. https://doi.org/10.3390/coatings7040056

Vázquez-Delfín E, Robledo D, Freile-Pelegrín Y (2014) Microwave-assisted extraction of the Carrageenan from Hypnea musciformis (Cystocloniaceae, Rhodophyta). J Appl Phycol 26:901–907. https://doi.org/10.1007/s10811-013-0090-8

Wang C, Cao J, Liu T, Jin L, Hang C, Zhang C et al (2023) Preparation and characterization of antioxidant and pH-sensitive films based on arrowhead (Sagittaria sagittifolia) starch, κ-carrageenan and black chokeberry (Aronia melanocarpa) extract for monitoring spoilage of chicken wings. Int J Biol Macromol 224:544–555. https://doi.org/10.1016/j.ijbiomac.2022.10.143

Wang X, Guo C, Guo H (2024) Progress of carrageenan-based films and coatings for food packaging applications. Packag Technol Sci 37(6):533–550. https://doi.org/10.1002/pts.2806

Wani SM, Gull A, Ahad T, Malik AR, Ganaie TA, Masoodi FA, Gani A (2021) Effect of gum Arabic, xanthan and carrageenan coatings containing antimicrobial agent on postharvest quality of strawberry: assessing the physicochemical, enzyme activity and bioactive properties. Int J Biol Macromol 183:2100–2108. https://doi.org/10.1016/j.ijbiomac.2021.06.008

Webber V, Carvalho SMD, Ogliari PJ, Hayashi L, Barreto PLM (2012) Optimization of the extraction of carrageenan from Kappaphycus alvarezii using response surface methodology. Food Sci Technol 32:812–818. https://doi.org/10.1590/S0101-20612012005000111

Yadav M, Chiu FC (2019) Cellulose nanocrystals reinforced κ-carrageenan based UV resistant transparent bionanocomposite films for sustainable packaging applications. Carbohydr Polym 211:181–194. https://doi.org/10.1016/j.carbpol.2019.01.114

Yin Y, Woo MW (2024) Transitioning of petroleum-based plastic food packaging to sustainable bio-based alternatives. Sustain Food Technol 2(3):548–566. https://doi.org/10.1039/D4FB00028E

Yong H, Liu J, Kan J, Liu J (2022) Active/intelligent packaging films developed by immobilizing anthocyanins from purple sweetpotato and purple cabbage in locust bean gum, chitosan and κ-carrageenan-based matrices. Int J Biol Macromol 211:238–248. https://doi.org/10.1016/j.ijbiomac.2022.05.046

Young E, Mirosa M, Bremer P (2020) A systematic review of consumer perceptions of smart packaging technologies for food. Front Sustain Food Syst 4:63. https://doi.org/10.3389/fsufs.2020.00063

Zhang R, Wang X, Li L, Cheng M, Zhang L (2019) Optimization of konjac glucomannan/carrageenan/nano-SiO_2 coatings for extending the shelf-life of Agaricus bisporus. Int J Biol Macromol 122:857–865. https://doi.org/10.1016/j.ijbiomac.2018.10.165

Zhang C, Sun G, Cao L, Wang L (2020) Accurately intelligent film made from sodium carboxymethyl starch/κ-carrageenan reinforced by mulberry anthocyanins as an indicator. Food Hydrocoll 108:106012. https://doi.org/10.1016/j.foodhyd.2020.106012

Zhou Y, Chen FQ, Chen S, Xiao Q, Weng HF, Yang QM, Xiao AF (2021) Preparation and characterization of κ-carrageenan modified with maleic anhydride and its application in films. Mar Drugs 19(9):486. https://doi.org/10.3390/md19090486

Zia KM, Tabasum S, Nasif M, Sultan N, Aslam N, Noreen A, Zuber M (2017) A review on synthesis, properties and applications of natural polymer based carrageenan blends and composites. Int J Biol Macromol 96:282–301. https://doi.org/10.1016/j.ijbiomac.2016.11.095

Chapter 9
Agar-Based Materials for Sustainable and Eco-Friendly Food Packaging Solutions

K. T. Remisha Jahan, P. Meenakshi, Gurmeet Kaur, Divya Mailk, Swarup Roy, and Deblina Biswas

Abstract Plastic-free agar obtained from marine seaweeds, specifically red macroalgae (Rhodophyta), is a sustainable food packaging option that protects food effectively and works in nature-friendly conditions. Natural packaging materials from agar contain two key components: agarose and agaropectin. These chemicals enable the red algae *Gelidium* and *Gracilaria* to create strong gels that hold moisture effectively. Agar packaging shows great potential as an eco-friendly alternative. Agar-based materials get more attention in sustainable packaging research for their use in films and coatings. These materials emerge from casting, extrusion, and molding processes and gain additional physical strength through blending with other biopolymer substances. The natural decomposition features of agar packaging maintain fresh food quality by preventing moisture loss and gas entry while staying non-harmful to consumers. The structural properties of agar packages provide necessary strength, flexibility, food preservation and water protection that allow these packages to serve many different kinds of food products. Food coating materials made from agar can stop harmful microorganisms from reaching the interior of food. Food packaging made from agar solves plastic pollution problems while keeping our food protected and safe. Future research on agar-based materials development will help us improve their use for preserving food and making sustainable packaging options.

Keywords Seaweed · Biopolymer · Film · Packaging · Food · Shelf life

---

K. T. Remisha Jahan · P. Meenakshi · G. Kaur · D. Mailk · S. Roy (✉)
Department of Food Technology and Nutrition, School of Agriculture, Lovely Professional University, Phagwara, Punjab, India
e-mail: swaruproy2013@gmail.com

S. Roy
Department of Food and Nutrition, Kyung Hee University, Seoul 02447, Republic of Korea

D. Biswas (✉)
Department of Instrumentation and Control Engineering, Dr B R Ambedkar National Institute of Technology Jalandhar, Jalandhar, Punjab, India
e-mail: deblinabi@gmail.com

S. Roy and M. Shahid (eds.), *Biomaterials for Sustainable Food Packaging*, Environmental Footprints and Eco-design of Products and Processes,
https://doi.org/10.1007/978-3-032-08027-1_9

9.1 Introduction

Packaging materials made of petroleum-based plastic are not biodegradable, which leads to massive environmental pollution (Moshood et al. 2022).Marine plastic pollution is estimated to be approximately 80–85% of the total (Jambeck et al. 2015), mainly derived from single use packaging as lightweight plastic bags and films accounting for almost half of all marine plastic waste (Foschi et al. 2019). Small plastics can change the immune system by harming body tissues and cause neurotoxicity as well as changes in acetylcholine esterase activity of marine fish. Taking in microplastics damaged the liver and gut, reduced the function of the gut wall, increased the wrong kinds of gut bacteria, influenced the metabolism of fish and lowered the activities of digestive enzymes, all which affected the fish's growth (Ghosh 2024).

The total production of plastic in 2018 was 359 million tons of which 90% was acquired from non-renewable resources. Around 95% of the current plastic packaging umbrella, which comprises polyolefins and polyethylene terephthalate are not reused and end up in recycling bin, on the other hand they are disposed to landfill as they cost the global economic value (Jones et al. 2021). People have been using 180 times more plastic today than they did in 1950. Experts estimate that worldwide plastic made in 2022 will be 400.3 million tons. The increase in plastic production is predicted to be highly rapid from now on (Rafey and Siddiqui 2023). As a result, we need an alternative to synthetic plastic which are biodegradable and renewable.

Natural origin biopolymeric macromolecules, such as polysaccharides and proteins, have emerged as the best candidates for the development of food packaging films (Zhang et al. 2021). In this context, agar based biodegradable films are one such innovation that elucidates immense potential in sustainable packaging. In recent times, agar has been obtained from seaweeds without negative impact on human beings and utilized as the biodegradable substitute to traditional plastic (Ebrahimzadeh et al. 2023; Jayasekara et al. 2022). Agar based film can be produced using water-based solvent and plasticizers like glycerol or sorbitol. Such material would be transparent, flexible, and suitable for use as a replacement of plastic food packaging films (Mostafavi and Zaeim 2020).

Agar is a polysaccharide-based biopolymer originated from red algae. Marine algae-derived agar consists of agarose and agaropectin substances (Lee et al. 2017; Surendran and Sherje 2022). Mainly, agar-based films are produced by the using solvent casting technique which has the advantage of flexibility, biodegradability, and film forming ability. Addition of fillers and other polymer helps overcome the challenges of low thermal stability and high-water permeability of agar. Agar have shown lot of promise in making biodegradable food packaging, (Roy and Rhim 2019; Wang et al. 2018). Therefore, in this chapter we covered the recent development in agar-based film and coating in food packaging application.

9.2 Sources and Chemistry of Agar

Agar, which is basically a phyco-colloid ("phyco" meaning algae and "colloid" meaning gel-like substances, derived from marine algae) is derived from the cell walls of specific types of algae mainly belonging to *Rhodophyceae* family. It is a hydrophilic colloid, meaning it has distinctive capability to interact with water molecules and form gel-like structures, thus providing a desirable consistency to various food products like jelly sweets, candies, cheese, marshmallows etc. (Fu and Kim 2010). This makes them widely recognized as substances used in food industry for gel formation, thickening and stabilization in order to provide the pleasant texture and consistency to the end product (Shah et al. 2024).

Marine reddish sea plant such as *Gelidium* and *Gracilaria*, is the main source of agar, *Gelidium* being the best among these two, owing to the superior quality of agar that it can produce. The seaweed *Gelidium cartilaganeum* grown in Mexico contains about 15–30% agar in its dried form. This amount however varies with season, for example, in July, i.e., summer season, the agar content is over 30% higher than in January, i.e., winter season. This variation is linked to the intensity of sunlight it gets, specifically ultraviolet (UV) rays, which are quite stronger in summer than in winters and seem to help seaweed produce more agar. Though it is harder to find and use *Gelidium* for commercial-based agar production. Therefore, *Gracilaria* is more preferred and popular source for commercial-based agar production because it's easier to grow and harvest. Three different red seaweed species were studied by Cebrián-Lloret et al. (2023), to produce agar-based hydrogels for food applications. The researchers evaluated the structure and gelling abilities of the agar extracted from *Gelidium corneum* and two *Gracilaria* species (*Agarophyton chilensis* and *Gracilariopsis longissima*). The researchers employed alkali to clean the agar and remove the unwanted impurities from it. Study revealed that *Gracilaria longissima* has exceptional potential to produce a stronger and compact gels, which also had higher melting and gelling temperatures compared to *G. corneum.* This is because Gracilaria agar contains semi-crystalline agaropectin which helps create tougher and more stable gel networks. Therefore, *Gracilariopsis longissima* can serve as a cost effective and efficient substitute to the frequently used *Gelidium corneum* for making high-quality agar-based hydrogels, for culinary uses. Apart from these two, agar can also be derived from other species of sea plants such as *Carragean moss, Chondrus* species like *C. crispus, C. ocellata, C. stellate, Gigantic stellate, Melanothamus somalensis, Pterocladia,* etc. (Roy et al. 2023).

Agar is a linear polysaccharide mainly composed of two basic polysaccharides, namely agarose and agaropectin, the chemistry and sources along with major reaction is illustrated in Fig. 9.1. Both agarose and agaropectin are made from similar building blocks but have certain crucial structural difference.

The first one, i.e., agarose, is the most important straight chain fraction in the whole composition of agar, that gives agar its special ability to form gels. This further makes it suitable for making films and coatings which can be used for manufacturing bio-based packaging in food industry. Agarose is a polysaccharide primarily made

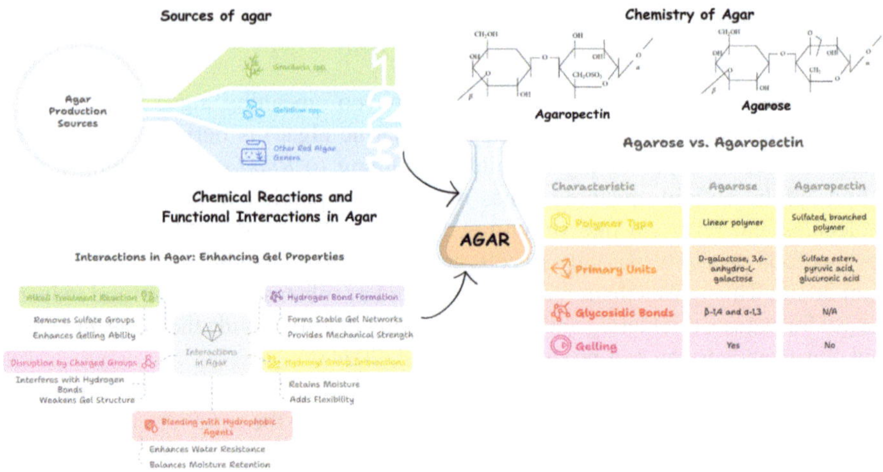

Fig. 9.1 Represents the chemistry and sources along with the functional interaction of agar

of recurring unit of agarobiose, which is a disaccharide made of two sugar units: Beta-D-galactose and 3,6-anhydro-alpha-L-galactose (3,6-AG), connected together through a special linkage called alpha-(1-3) and beta-(1-4) glycosidic linkage. It has no extra chemical groups like sulfate, thus is simple in its chemical composition. This simplified composition of agarose gives it strong gelling ability. The gelling happens because agarose molecules stick together through hydrogen bonds which are a type of weak but stable bonds. Inside the gel, agarose molecules form a network of double helices or twisted structures, which hold up a large amount of water within them. These water molecules act as plasticizers within the agarose network and aids in reducing rigidity and brittleness. This allows formation of flexible film which is less prone to crack under stress conditions and is more resistant to tearing and deformation. This moisture helps keep the agarose molecules closely linked and thus prevent the film from shrinking or developing weak spots on drying. Agarose molecules also have hydroxyl groups on them which makes the agar films hydrophilic in nature. These groups interact with water and allow film to retain moisture. However, while retaining moisture enhances the film's flexibility and strength, it also makes the film more sensitive to humidity and less water resistant, making it more prone to absorb moisture from environment, thus losing its mechanical strength. To address this issue and to attain a balance between strength and water resistance of the film, additional treatments are given, for example, blending with various hydrophobic materials such as waxes, oils, etc. can be done.

The second component, i.e., agaropectin, has similar backbone as agarose has, i.e., of repeating units of agrobiose but is slightly branched. While agarose is a neutral polysaccharide, agaropectin is a polysaccharide with sulphate groups, giving it acidic properties. Its sulfated groups make it more hydrophilic thus this fraction tends to reduce the gelling ability of agar, therefore further interfering with the bio-based film formation. Thus, to create a strong gel with good mechanical properties, it is better to use agar with less agaropectin. Agaropectin is even removed in case of commercial production of agar to ensure that the gel being formed is stronger and the film that can be formed from it is more resistant to external wear and tear and can be effectively used for various purposes. Agaropectin also has the same basic sugar units as agarose as mentioned earlier, though some of its 3,6-anhydro-alpha-L-galactose parts are substituted with extra chemical groups. These groups usually are negatively charged groups such as sulfate, methoxy, pyruvate or glycuronate. These groups make agaropectin less effective at forming strong gels as they interfere with the formation of strong and stable bonds because of repulsive forces between them. Further, they disrupt the hydrogen bonds that are mainly responsible for the integrity of film and film strength, thus weakening the gel network (Bertasa et al. 2020). Again, these negatively charged groups are hydrophilic in nature, therefore makes the film more prone to absorbing moisture. This weakens the film structure and makes it less durable. However, agaropectin also adds flexibility and elasticity to the film so as to prevent it from being too brittle. This allow its applications in various packaging materials and coatings where some stretch is required. Alkali treatment of agar usually done with Sodium hydroxide, enhances its gelling ability as it changes a part of the agar (L-galactose-6-sulphate) into another form (3,6-anhydro-L-galactose), which strengthens the gel (Mostafavi and Zaeim 2020). With fewer negatively charged groups like sulfates, the agar molecules can form compact, firm and more consistent gels or films. Though films made from alkali-treated agar are less elastic and have lower viscosity which can limit their applications.

9.3 Agar Based Materials in Sustainable Packaging

Agar is a jellylike substance consisting of polysaccharides and considered as a versatile substance with properties like stabilizing and gelling along with various other properties (Fu and Kim 2010). Agar does not dissolve in water whereas, it binds with boiling water molecules. This makes preparation of clear and isotropic solutions easy. The agar which cools and forms firm and stable gels under 35–45 °C, remains solid up to 80–85 (Zucca et al. 2016). Since edible films or coatings are a suitable alternative for plastic as moisture content is prevented by these films or coats and, transfer of atmosphere oxygen into food, use of these edible coating has been rapidly increased (Cazón et al. 2017). Agar gum has been considered as a good source of biodegradable film in the recent years and also, it is the first phyco-colloid utilised in the formulation of human food (Abdul Khalil et al. 2017; Guerrero et al. 2014).

Certain research indicates that the addition of AgNPs in the films plays major role in the structural properties, and a significant amount of AgNP improves the water vapour barrier and surface hydrophobicity without affecting the mechanical integrity. Strong antimicrobial action against *Listeria monocytogenes* and *Escherichia coli* is exhibited by the presence of >1 wt% AgNPs (Rhim et al. 2013). The development of GE/agar-based films, incorporated with copper-modified ZnO NPs (ZnO@Cu) and a clove oil Pickering emulsion (PE) upgraded the barrier against UV, mechanical strength, and both antimicrobial and antioxidant property. Wrapping pork belly meat in these films helped to increase its shelf life significantly (Roy et al. 2022). Incorporation of different metal composites into the film develops different properties which can be applied in food industry as well as several other industries.

9.3.1 Characteristics of Agar-Based Composite

Certain additives are incorporated to polymer matrix to form agar-based composites (ABCs), which enhance their properties (Shah et al. 2024). ABCs are environmentally degradable and renewable, which makes them green polymer. Addition of agar to the composite increases the mechanical strength, thermal stability, and biocompatibility of the final product (Jumaidin 2023; Kumar et al. 2022; Surendran and Sherje 2022). These films can be used as a substitute for the non-biodegradable plastic materials in most food packaging with good inhibitor properties to moisture and gases (Rukmanikrishnan et al. 2021; Surendran and Sherje 2022; Yin et al. 2018). Due to their exceptional flexibility these materials are highly effective in developing a sustainable environment (Betraoui et al. 2023; Hamimed et al. 2024).

9.3.2 Agar Based Materials

General techniques used within the industries for the agar gum extraction are (1) dissolving the agar–agar, present in cell wall which is detached by soaking the dry algae in boiling water (2) liquid extract is filtered, and (3) with the help of freezing and thawing, gum is separated from the water (Abdul Khalil et al. 2017; Xiao et al. 2019). When native agar was compared with the alkali modified agar, that is the agar treated with alkali, which improves its gelation ability by modifying structure responsible for gel strength, it was reported that the native agar offers both cost-effective extraction and enhanced safety compared to modified alternatives and also provided a higher viscosity (Sousa et al. 2015). Gelling property of the agar aids for the uninterrupted film-formation capability of agar.

In the film-forming process, the molding surface and the membrane-forming solution should maintain a temperature above the agar gelling temperature while casting agar solution. Premature gelation is prevented by following this. A continuous film network is formed by the interaction of molecules of non-ionic and linear structure of agarose through hydrogen bonds while drying the agar solution. Agar films connect effortlessly with different biologically active substances and form a protective layer over the surface of food products since agar films are biologically inert (da Rocha et al. 2018).

In both smart and active packaging, agar-based edible films have been utilized for packaging and this aided to enhance the shelf life and maintain the quality of food products (Roy et al. 2023). Coatings, labels, pads or sachets incorporated with active substances which have properties like antioxidant, antibacterial and anti-UV light is used in this technique (Rodrigues et al. 2021). It was stated that the addition of polymers/fillers in agar-based edible film/coating improved in various properties such as mechanical, barrier and antimicrobial activity (Bang et al. 2022; Priyadarshi et al. 2021; Uranga et al. 2020). Agar incorporated with different components have different applications in food packaging according to the components incorporated. For example, shelf-life of fresh potatoes were extended by using agar films incorporated with the extract of grape-fruit seed and silver nanoparticles (Wang and Rhim 2015). Prevention of accumulation of condensed moisture on the surface of the packaged film, and the change of potato's colour into green during storage along with its antimicrobial activity expanded the utilization of this film for highly respirating fresh agricultural produce (Mostafavi and Zaeim 2020). Table 9.1 shows the findings associated to the effects of integrating different polysaccharides in the agar-derived films for the functional properties.

9.4 Fabrication of Agar-Based Films and Coatings

Films and coatings are thin layer applied on food material to protect it from damage and spoilage. It preserves food and also acts as a barrier for environmental factor like gases, water vapour. Agar-based films are derived from red algae, are biodegradable and safe to eat. It can be used as separate films or surface coatings. These films improve food quality and shelf life by reducing moisture loss, oxidative spoilage, and microbial contamination (Roy et al. 2023).On the other hand, coating are thin layers applied directly to surfaces to provide improve functions like extending shelf life or protection against contaminations (Hernández et al. 2022).

9.4.1 Methods for Fabricating Agar-Based Films

9.4.1.1 Film Casting

Film casting is used most widely for producing agar-based films because of its simplicity and cost-effectiveness. For this, agar powder of concentration 1–5% (w/v) is dissolved in boiling water to form a homogenous gel. This gel is then cast on a plain surface, like a glass plate or a mould, and then left to dry under controlled conditions of temperature and humidity. The drying process is done at temperatures between 30 and 60 °C to make sure uniformity of film while preventing brittleness (Makhloufi et al. 2021; Hernández et al. 2022). Plasticizers like glycerol and sorbitol are mixed to the agar solution to enhance the flexibility and decrease brittleness. These added plasticizers cooperate with the polymer chains and increases their mobility as well as improve the film's mechanical properties. As the concentration

Table 9.1 Various biopolymer included agar-based composite and their key properties

Biopolymers and additives added to agar	Observations	References
Starch	Stiffness and strength of the blends were increased which resulted with promising barrier and tensile properties	Fekete et al. (2019)
Fish protein hydrolysate	It provided an antimicrobial property to the agar films	da Rocha et al. (2018)
Gelatin	When observed in SEM micrographs, changes were seen in different FTIR spectra as it suggested the protein-polysaccharide interactions were stabilized by hydrogen bonding	Jridi et al. (2019)
Pectin	The material's physical and mechanical properties, including UV protection was improved and anti-microbial characteristic also has been displayed	Roy and Rhim (2021c)
Alginate and collagen blends	Without compromising the transmittance, high UV screening was exhibited. Demonstrates anti-microbial property against both Gram positive and Gram-negative bacteria	Wang and Rhim (2015)
Locust bean gum (LBG)	Addition of LBG increases the flow resistance, decreases the gelation capability of the film-coating formulation	Sousa and Gonçalves (2015)
Corn starch (combined with halloysite clay)	Corn starch provides an antimicrobial activity and the incorporation of halloysite promote the strength of the film	Meira et al. (2017)
Chitosan	Since chitosan is hydrophobic, applications are allowed in products with high water activity	Contessa et al. (2021)
Lignin	Lignin increased the mechanical resistance and UV-light shield property of the film	Shankar and Rhim (2017)
Microcrystalline cellulose	Thermal stability and the water vapour barrier ability of the film improved	Harnkarnsujarit and Li (2017)
Arabinoxylan	Gels combined with arabinoxylan forms rapidly and are strong and stable at high temperature. They also contain a neutral flavour odour and colour and also have a dietary fibre nature	Mendez-Encinas et al. (2018)
Gelatin, copper doped-ZnO and clove essential oil	Films incorporated with fillers was observed to improve the mechanical and physical performance. It also enhanced the antioxidant property of the film	Roy et al. (2022)
AgNP	Packaging with AgNP incorporated films illustrated a bactericidal impact on foodborne contaminants like *E. coli,* and *L. monocytogenes*	Hong et al. (2021)

(continued)

Table 9.1 (continued)

Biopolymers and additives added to agar	Observations	References
TiO_2, anthocyanin and neem essential oil	Incorporation of anthocyanin extracted from red radish and the neem essential oil optimised the stabilizing and microbicidal and microbiostatic capacity of the coating. TiO_2 also enhanced the barrier and mechanical property	Yang et al. (2022)
CMC and summer savoury essential oil	Develop active films with antimicrobial properties against both Gram-positive and Gram-negative bacteria	Abdollahi et al. (2019)
Agar combined with alginate, glycerol and ascorbic acid	Slight difference was observed in the water vapour permeability (between 0.20 m^2 and 0.59 g mm/kPa h m^2). Tensile strength of the film exhibited highest value (56.43 MPa)	Sanyang et al. (2015)
Agar combined with alginate, glycerol and calcium chloride	Addition of $CaCl_2$ showed a water vapour permeability between 0.15 and 0.39 g mm/kPa h m^2	Amariei et al. (2022)
A combination of agar with thermoplastic corn starch and glycerol	Agar-thermoplastic corn starch provides a tensile strength of 10.7 ± 2.1 MPa along with enhanced barrier and low water penetration and dissolution	Perera et al. (2023)
Cassava starch	Water vapour permeability of the pure agar film was reduced by 54%	Abdollahi et al. (2019)
k-carrageenan and konjac glucomannan	The agar film became more transparent while the tensile strength of the film increased	Rhim and Wang (2013)
Green tree extract	Inhibition of microorganisms in the film due to the antimicrobial and antioxidant property of the extract	López de Lacey et al. (2014)

of plasticizer increases, improvement of elongation at break and reduction in tensile strength observed which make films more adaptable for flexible packaging applications (Guo et al. 2021). Other than plasticizers, active ingredients like essential oils and antioxidants can also be added during casting process to improve the functionality of the films. For example, adding grapefruit seed extract has been found to impart antimicrobial properties resulting in extending the shelf life of perishable food items (Roy and Rhim 2021b).

9.4.1.2 Melt Blending

Melt blending is another method used for fabrication purpose. It offers some advantages such as improving mechanical and barrier properties. In this methods agar is mixed other polymers like starch or pullulan, at high temperatures for the purpose of creating composite film with enhanced properties. This process forms the uniform distribution of additives and results in films having improved elasticity, water resistance, and durability (Roy et al. 2023). Addition of thermoplastic starch into agar films with help of melt blending process shows enhancement in the strength and flexibility of the materials. Furthermore, the addition of nanomaterials like graphene oxide can improve the films thermal stability and moisture resistance which making them more suitable for food packaging applications in humid environments (Belay et al. 2017).

9.4.1.3 Compression Molding

Compression molding is a viable fabrication technique for thick and durable films made of agar. Agar-based composite material is molten and subsequently compressed between two heated plates under high pressure. This method can be particularly advantageous in adding thermoplastic polymers and natural fibers because they improve the mechanical strength and water resistance of the films (Garrido et al. 2016). Pressure and temperature in compression are regulated meticulously to achieve homogeneity and to eliminate defects such as air bubbles. Compression molding can especially be applied to film for an application that demands high tensile strength and elasticity, so rather like in rigid food containers and multilayer packaging materials, this uses indeed high tensile strength and elasticity (Agusman et al. 2022; Garrido et al. 2016; Sousa and Gonçalves 2015).

9.4.1.4 Layer-by-Layer Assembly

Layer-by-layer (LbL) assembly is quite an advanced method that can be utilized in the preparation of multilayered films, based on agar. The technique comprises the deposition, to a substrate, of agar layers one after the other, incorporating some additives within each layer. All the deposited layers are allowed to dry before the next layer is applied, thereby giving total control over thickness as well as composition of the film. This allows for tailor-made films with improved functional properties such as barrier properties to oxygen and moisture by using hydrophobic or nanoparticle filled layers (Roy and Rhim 2022; Diop et al. 2023). For instance, the development of agar-based LbL structures combined with lipid-based or chitosan layers has been shown to markedly reduce water vapor transmission rates while preserving biodegradability. Such an approach is very promising for novel food packaging applications where active films exhibit multiple functions for protection against several environmental impacts (Lim et al. 2021; Ganjeh et al. 2024).

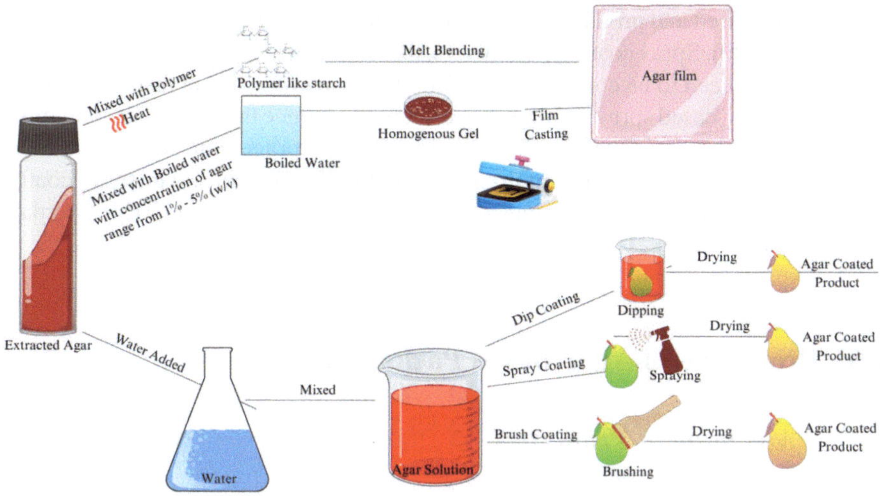

Fig. 9.2 Schematic representation of agar-based film and coating

9.4.2 Methods for Coating with Agar

9.4.2.1 Dip Coating

Dip coating technique is used widely for purpose of applying agar-based coatings onto the food products. Figure 9.2 schematically illustrates three methods of coating with agar: dip coating, spray coating, and brush coating. In this process, food items are dipped in agar solution. After withdrawal, a thin layer of agar solution adheres to the surface of product is observed. This coated food is then dried under controlled conditions to achieve a uniform layer. The thickness of the coating depends on factors like immersion time, withdrawal speed, and viscosity of solution (Contessa et al. 2021). Agar coatings applied using dip coating method decreases the moisture loss, slow down the oxidation process and also inhibits the microbial growth which results in longer shelf life of food products like fresh fruits and vegetables. Adding antimicrobial agents such as essential oils into the agar solution improves the protective capabilities of the coating and provide an added defence against spoilage (Wardana and Widyaningsih 2017; Carocho et al. 2019; Wei et al. 2017).

9.4.2.2 Spray Coating

Spray coating is an advanced technique that allows the more accurate application of agar-based coatings on the food surfaces. In this method, an agar solution is atomized and sprayed onto the food product to get uniform layers even on the product with irregular shape. This method is more advantageous to use where a thin, uniform layer of coating is required and product texture, appearance need to be maintained (Mostafavi and Zaeim 2020). Spray coating parameters, like nozzle size, spray pressure, and solution viscosity, need to be controlled carefully to achieve consistent results. Research shows that agar-based spray coatings can effectively reduce the microbial contamination which results in maintaining food quality during storage. Furthermore, addition of functional additives, such as antioxidants and flavour enhancers to spray coating allows improve in sensory attributes of the coated food (Gali et al. 2024; Giordano and Cremonesi 2021).

9.5 Properties of Agar-Based Packaging

9.5.1 Mechanical Properties

Agar-based films show moderate mechanical strength but are brittle in nature due to rigid polysaccharide structure (Perera et al. 2021). Pure agar films have a low elongation at break, which is a test of flexibility, and they tend to crack under stress. This restricts their use in flexible packaging unless plasticizers (glycerol or sorbitol) are included (Sudheer et al. 2023). Researchers have mixed agar with other natural biopolymers (e.g., starches, chitosan, or polyvinyl alcohol (PVA) to enhance their mechanical features. These blends make it better for elastic and tough (More et al. 2017). Furthermore, the mechanical strength is improved with less brittle nature of the gel through the crosslinking of calcium chloride or citric acid (Uranga et al. 2020). Agar-based films show increased tensile strength and resistance to breakage when reinforced with nanomaterials such as graphene oxide or cellulose nanofibers (Roy et al. 2023). The composition and additives may influence the mechanical strength of agar-based films. Agar films are known for their rigidity, inflexibility and brittleness, which inhibits their use in flexible packaging (Carina et al. 2021). Plasticizers (for example glycerol, sorbitol) are often included to increase flexibility, allowing a higher stretchability of the film without a loss of firmness. The selection of biopolymer blends like agar-starch composites or agar/chitosan blends increases mechanical durability and elongation at break (Guo et al. 2021).

9.5.2 Barrier Properties

Barrier properties of agar films quite relevant in food preservation, as they describe the ability of agar films to hinder the diffusion of gases and moisture. Due to its hydrophilic nature, pure agar films exhibit such low water vapor barrier properties that moisture can easily diffuse (Carina et al. 2021). This has limited use in terms of humid environments as humidity absorption weakens the film structure promoting food spoilage. The agar film has excellent barrier properties for oxygen, thus preventing the deterioration of food by oxidation. Agar films have shown to have higher oxygen transmission rates (OTR) than synthetic plastics such as polyethylene (Perera et al. 2021). Agar films have very good oxygen barrier property which minimizes oxidative spoilage in foods. Agar films exhibit low OTR in relation to synthetic plastics, enabling them to be utilized effectively in the preservation of products that are sensitive to oxygen, such as fresh meat or dairy. Agar films have the disadvantage of a high water vapor permeability (WVP) as a result of the hydrophilic properties of agar. That makes them easily soaked, which under humid conditions can lead to a weakened structure (Carina et al. 2021; Lee et al. 2019; Mostafavi and Zaeim 2020). Agar films can also be co-extruded with hydrophobic agents or lipid-based coatings (beeswax or stearic acid) to enhance moisture resistance. LbL assembly with other biopolymers such as alginate or cellulose has also been employed to increase moisture resistance and durability (Makaremi et al. 2019; Zhang et al. 2019). Researchers have explored lipid-based coatings (e.g., beeswax, stearic acid) and hydro phobic polymers (e.g., alginate, cellulose derivatives), intending to enhance moisture resistance. Multilayer films that combine agar and hydrophobic materials with improved barrier properties are more effective in humid areas (Mendez-Encinas et al. 2018; Devi et al. 2024).

9.5.3 Thermal Properties

The thermal delay of agar-based films is an average, because they ideal warmth without degradation. The glass transition temperature (Tg) of agar is about 40–50 °C, which means that films start to soften at these temperatures. However, their decomposition temperature is quite high, normally well over 200 °C, rendering them appropriate for food packaging applications where moderate heat is expected. Agar films are usually crosslinked or intermixed with biopolymers like chitosan or PVA to improve thermal resistance. Montmorillonite clay or silver nanoparticles have also been integrated within such matrices to enhance heat resistance and slow matrix degradation (Sudheer et al. 2023). Thermal properties improvement is an interesting approach but it is important to understand that only the gelling property can be enhanced by heating and that heating is not sufficient to mitigate thermal degradation and thermal stability can further be improved in the presence of calcium chloride or alginate as a crosslink agent which helps in enhancing polymer network before and after heating (Carina et al. 2021). Also, clay nanoparticles or metal oxides, which have been introduced into agar films to improve their mechanical properties without significant thermal deformation (Shah et al. 2024).

9.5.4 Optical Properties

Transparency in packaging materials is one of the essential factors for consumer attraction toward the product as well as in the product visibility. As agar-based films are highly transparent, consumers may have an opportunity to see food products inside the packaging without the need to open it. They are as clear as synthetic plastics, making them a good alternative for ecological packaging solution. One issue, however, is that agar films enable UV light penetration, resulting in oxidation and spoilage of some food products. To overcome this limitation, UV screeners like titanium dioxide, zinc oxide or natural pigments (i.e. anthocyanins, melanin) can be added to agar-based films in order to enhance UV protection. These changes in agar make the films more effective at blocking UV light and still remain clear (Amariei et al. 2022; Roy et al. 2021).

9.5.5 Edibility and Biodegradability

Agar was selected as a natural polysaccharide because one of the unique benefits of agar-based films is their edibility and biodegradability. In contrast with synthetic plastic films, the natural agar-based films decompose in soil or water, which has little environmental impact (Carina et al. 2021). Biodegradable plastics decompose in a range from few weeks to few months depending on the moisture content and microbial activity (Perera et al. 2021). These films are safe to consume, rendering them suitable for edible films on fresh produce, confectionery and ready-to-eat food They are a promising alternative in the quest for sustainable packaging solutions because of their ability to fully decompose without leaving behind microplastic residues.

9.5.6 Antimicrobial and Antioxidant Properties

Incorporation of antimicrobial agents can act as active packaging materials for agar-based films that prevent the growth of bacteria and fungi. Agar itself does not have innate antimicrobial activity, however, its benign character and ~90% compatibility with natural bioactive compounds make this polymer a suitable carrier for antimicrobial agents (Perera et al. 2021). Agar films containing essential oils (oregano, thyme and grapefruit seed extract) were developed with the aim of preventing microbial contamination and increasing the shelf life of food. Chitosan- or silver nanoparticle-doped agar films demonstrate a considerable antimicrobial activity against many pathogens such as *Escherichia coli* and *Staphylococcus aureus* (Carina et al. 2021). Moreover, the addition of essential oils in agar helps in improving the antioxidant performance of the film (Roy et al. 2023). Plant based materials are known for their natural antioxidant and the inclusion of this additive in agar-based film or coating provides functionality in packaging (Roy and Rhim 2021a). The addition of chitosan and natural plant extract or bioactive compounds also induce the free radical scavenging activity in agar film which during packaging retard the oxidation of lipid and protein rich food such as meat, fish, etc.

9.6 Application of Agar Based Ecofriendly Packaging in Food Preservation

In environmental aspects, food packaging film made up of agar which is also biodegradable, hold some applications and benefits of using the material in its natural form. First of all, agar is a natural gums substance considered generally safe to use for food packaging by independent governing body of most countries. To improve the quality, preservation and shelf life of food product researchers used agar-based films. Therefore, it indicates that other films such as the biodegradable food packaging films from agar could be used to accommodate variety of foods that include fresh fruits and vegetables, meat and bakery products. Figure 9.3 illustrates the major applications of agar-based edible functional packaging films in enhancing the preservation, shelf life, and quality of various food items through active and smart packaging (Dora et al. 2020; Mostafavi and Zaeim 2020). Flavors and odours that are undesirable and objectionable in foods are produced by fat and lipid oxidation, food spoilage and hosts of food pathogenic microorganisms. Oxidation reduces polyunsaturated fatty acids and generates toxic aldehydes beyond which they contribute to further deterioration of food. Edible functional packaging films based on agar have shown the potential for active and smart packaging for food products, with increased shelf life, and interactive quality monitoring (Tammina and Rhim 2023). Cherry tomato is common in culinary dishes worldwide due to the nutritional postharvest benefits and taste of this fruit unlike other fruits that can be kept for several weeks, cherry tomato has high water content and a short shelf life due to the tendency to rot (Álvarez-Hernández et al. 2021). Packaging helps in improvement in shelf life of cherry tomato. Table 9.2 represents the food preservation application of agar-based packaging materials.

A novel nanostructure was found by connecting an iron-based MOF with carbon quantum dots (CD@MOF). The formed CD@MOF crystallized in an octahedral structure and had positive antioxidant, antifungal and antibacterial actions. MOF and CD@MOF components are included in polymers made from agar/gelatin blends. To confirm CD@MOF compatibility and dispersion in the polymer, analysis with FTIR, XPS and SEM was conducted. The use of CDs loaded with active edges together with MOFs improved both antioxidant (71.9% DPPH, 100% ABTS) and antibacterial activities. The number of CD@MOF in the film led to better tensile strength, a greater gas barrier (water vapor and oxygen) and 100% UV blocking. Furthermore, tomato fruits were kept for 24 days in storage, with film packaging at 4 °C and the changes in their quality were investigated. From the experiments, it was suggested that the Agar/Gel/CD@MOF films helped keep tomato quality high and improve its shelf life (Riahi et al. 2024).

On the other hand, Rhim and Wang (2013) synthesize multicomponent hydrogel films based on agar, κ-carrageenan, konjac glucomannan powder and Cloisite® 30B nano-clay to improve its mechanical and water barrier characteristics. Present in these films, nano-clay contributed noticeable enhancements in terms of water resistance and mechanical strength whereby while water holding capacity of the individual component films range from 800% to up to 5488% that of the nanocomposite films.

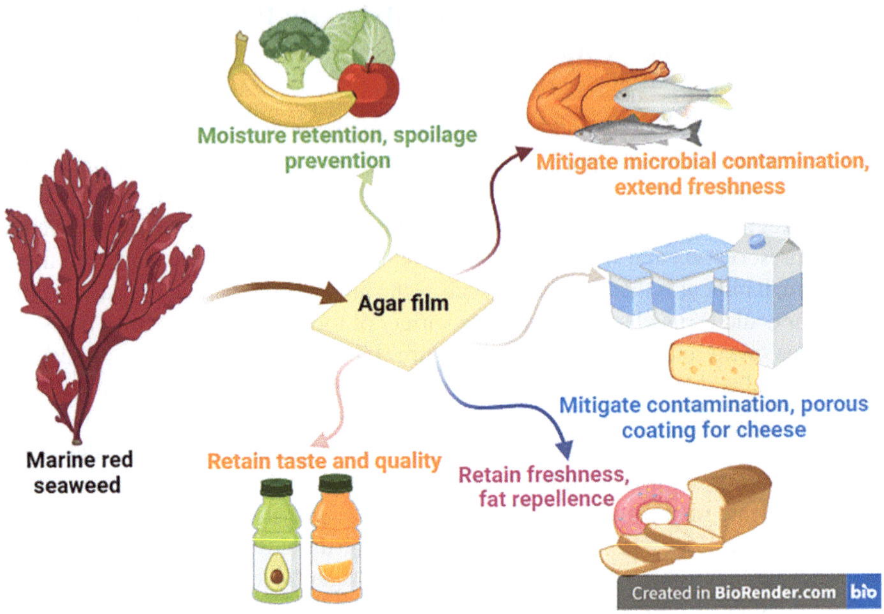

Fig. 9.3 Major applications of agar-based edible functional packaging films in enhancing the preservation, shelf life, and quality of various food items through active and smart packaging. Created in https://BioRender.com

In addition, the water vapor permeability, contact angle, solubility, and uptake ratios were improved, which means that the moisture resistance was improved as well. Furthermore, the films possessing good water vapor adsorption properties were also presented in the films. First trials of the films for fresh spinach packaging presented the usefulness of the films as antifogging packaging material for highly permeable horticultural crops. The ternary blend nanocomposite film also exhibited antibacterial characteristics against *Listeria monocytogenes*, which confirms its application in active food packaging (Rhim and Wang 2013). The functional composite films containing starch/agar were prepared by embedding green synthesized AgNPs that were capped by aqueous enoki mushroom extract. The characterization through FTIR and XPS clearly pointed to the nature of capping of the nanoparticles and the nanoparticles themselves were found to be of spherical nature and with an approximate size of 10 nm. The AgNPs were well embedded within the starch/agar polymer matrix and hence a great improvement in barrier properties and hydrophobicity of the films was observed. There was no significant influence of AgNPs on the mechanical properties of the starch/agar film but the films demonstrated good antibacterial properties against the food pathogens. These green synthesized AgNPs embedded starch/agar composite films with improved physical and functional characteristics, suggest the potential for application in active food packaging (Roy and Rhim 2022).

Table 9.2 Applications of agar-based films to enhance the quality, preservation, and storage life of food products

Film component	Food items	Characteristic features	References
Agar/gelatin + metal-organic framework (MOF) with carbon quantum dots (CD@MOF)	Tomato fruits	Strong antioxidant activity (71.9% DPPH, 100% ABTS), antibacterial and antifungal properties Moreover it also has improved tensile strength and gas/UV barrier properties	Riahi et al. (2024)
Agar, κ-carrageenan, konjac glucomannan powder, and Cloisite® 30B nanoclay hydrogel films	Spinach	Antifogging packaging material for highly permeable horticultural crops, antibacterial characteristics against *Listeria monocytogenes*, good water vapor adsorption properties and moisture resistance property	Rhim and Wang (2013)
Films containing starch/agar embedded by green synthesized silver nanoparticles (AgNPs)	Mushroom	Active packaging characteristics, films showed strong antibacterial qualities against food pathogens	Roy and Rhim (2022)
Agar-Nano zinc oxide (ZnO)	Green grapes	This film matrix was found to be directly related to preservation efficiency and less spoilage	Kumar et al. (2019)
Ternary blend bio-hydrogel films based on agar, alginate and collagen (A/A/C) incorporating silver nanoparticles (AgNPs) and grapefruit seed extract (GSE)	Fresh potato	Films reduced or eliminated surface fogging and greening during storage through the uses of UV-screening together with WVP functions and also may function as antifogging and antimicrobial packaging for extended storage period of the foods	Wang and Rhim (2015)
The active agar-titanium dioxide + active agar-red radish extract-neem essential oil (AG-TiO_2+AG-RRE-NEO) bilayer film	Banana	Film efficiently retarded weight diminution and firmness loss, suppressed microbial growth and senescent spotting in banana during storage. Film in regard to banana preservation and active packaging	Yang et al. (2022)
Copper-doped zinc oxide (ZnO@Cu) and gelatin/agar/ Pickering emulsion of clove essential oil (PEC)	Pork meat	Antimicrobial and antioxidant property	Roy et al. (2022)

(continued)

Table 9.2 (continued)

Film component	Food items	Characteristic features	References
Agar-based film with Aggregation-Induced Emission (AIE) properties that are photodynamic and bactericidal	Pork	Bacteriostatic activity against *E. coli, S. aureus, C. albicans* and other microorganisms and antimicrobial activity	Yang et al. (2024)
Guar gum (GG) and agar (AA) based biopolymer film incorporated with bee wax (BE)	Cheese	Water solubility was declined by 94.40% and the water vapor permeability was declined 53.70% owing to hydrophobic interactions to hinder the transport of water molecules. It can be easily replacing the traditional petroleum-based packaging materials	Pal and Agarwal (2024)

Wang and Rhim (2015) prepared agar ZnO nanocomposite films and tested for green grapes packaging applications with evident prolongation of shelf life. The control grapes packaged with a plastic film showed signs of mold, mildew, juice mark stains, or a total off smell after a week of storage. Grapes wrapped on control films had less spoilage and mould was observed on the top most part of them and few pale-yellow discolouration. Using 2% (w/w) Zinc oxide nanoparticles (ZnONPs) embedded in agar-ZnO films, grapes maintained their acceptable visual appeal for 14 days and grapes wrapped in films containing 4% (w/w) ZnONPs were fresh for up to 21 days. ZnONP content in the film matrix was found to be directly related to preservation efficiency. The enhancement of thermal stability, elongation at break and fruit preservation indicated the functionality of the incorporated ZnONPs in the shelf-life prolongation of fruits. Likewise, another study done on ternary blend bio-hydrogel films based on A/A/C were synthesized and improved by incorporating AgNPs and GSE for active food packaging. These films showed the test specimen to absorb up to 23.6 times their own weight in water while retaining water resistance. The enhancement of AgNPs or GSE had greatly enhanced on the UV-screening of the films, in which A/A/C(GSE) having a much superior UV shield without losing its transparency. The mechanical and barrier properties were also improved thereby making the films more suitable for practical applications. The antimicrobial effect against Gram positive (*L. montocytogenes*) and Gram negative (*E. coli*) food borne pathogen was evident in both A/A/C(AgNPs) and A/A/C(GSE) films. Safety tests using fresh potatoes for packaging showed that these films reduced or eliminated surface fogging and greening during storage through the uses of UV-screening together with WVP functions. From the films, the decrease of water condensation on the packaging surface was restrained so as to avoid spoiling effect. These, findings show that A/A/C incorporated with AgNPs or GSE can be used as functional packaging material for fresh horticultural produce especially for high respiratory produce. They may function as antifogging and antimicrobial packaging to increase the shelf life of the foods and also to inhibit post processing contamination and thereby retain food quality. These multifunctional films may offer a potential and efficient approach to environmentally friendly and smart structural food packaging materials (Wang and Rhim 2015).

The microstructure of functional agar bilayer films with bioactive and electrochemical characteristics was described to improve the bananas' postharvest quality. Decorated lower layer offered antioxidant and antimicrobial features with the help of red radish-derived extract and neem oil meanwhile the upper layer, containing TiO_2, contributed to the improvement of barrier and mechanical properties and total anthocyanin as well as the retention of neem essential oil. The bilayer film also enables one to write multicolour patterns on the surface of such a film. The AG-TiO_2+AG-RRE-NEO bilayer film efficiently retarded weight diminution and firmness loss, suppressed microbial growth and senescent spotting in banana during storage. The results provided in the present work present a very appealing application for the proposed multifunctional film in regard to banana preservation and active packaging (Yang et al. 2022). The shelf life of pork meat was increased by the agar-based functional film (gelatin/agar/PEC/copper doped zinc oxide (ZnO@Cu), which was found to be effective in reducing lipid oxidation and the microbial load. These elements include the essential antimicrobial qualities of the nanoparticles and antioxidant qualities that were proposed to be attributed to the high antioxidant activity of the essential oils and the functional features of edible packaging film based on gelatin and agar (Roy et al. 2022).

Films used in food packaging include agar films since they are cheap, edible and biodegradable but do not possess antibacterial qualities. In this regard, one newly designed light activated bactericidal Aggregation-Induced Emission characteristics (AIE) was embedded into the agar to generate a composite film with enhanced antimicrobial activity. The agar film doped with AIE exhibited bacteriostatic activity against *E. coli*, *S. aureus*, *C. albicans* and other microscopic organism and the pork wrapped in the film has a storage stability beyond 8 days. At the same time, the composite film had superior mechanical properties, water resistance, thermal stability, and light transmittance, and the function of photodynamic sterilization was added. Although the study focused solely on pork, it successfully demonstrated that AIE molecules enhance the antimicrobial and structural properties of agar films. This innovation offers useful suggestions for creating high-quality antimicrobial food packaging materials (Yang et al. 2024). In other research, hake fillets were treated with edible films made from bioactive agar, green tea extract, and specific probiotic strains, *Lactobacillus paracasei* L26 and *Bifidobacterium lactis* B94, to evaluate their effectiveness after up to 15 days of refrigeration. The study's film increased the fish's lactic acid bacteria while decreasing spoilage bacteria like *Shewanella putrefaciens* and *Photobacterium phosphoreum*. It also reduces the quality degradation parameter like total volatile basic nitrogen (TVB-N), trimethylamine nitrogen (TMA-N), and pH level. The coated hake fillets were within acceptable microbial and quality traits in the seven days treatment and the condition was sustained for an additional 8 days with the green tea + probiotic film which improved the probes of the fish (López de Lacey et al. 2014).

Pal and Agarwal (2024) synthesized a guar gum (GG) and agar (AA) bio-based polymer film incorporated with bee wax (BE) to preserve cheese. The improvements of tensile strength (TS) and elongation of break (EB) of the cross-linked composites based on GG and AA polymers can be associated with the hydrogen bonding with an increase of 61.24% and 87.37%, respectively. The employed GG/AA polymer matrix offered a lower free volume and much lower permeability evidenced from 99.5% reduction of permeability to oxygen in contrast to GG film. And introduction of BE into GG/AA matrix the water solubility was declined by 94.40% and the water vapor permeability was declined 53.70% owing to hydrophobic interactions to hinder the transport of water molecules. The GG/AA/BE20(50:50) The studied film exhibited 99% biodegradation within 28 days and kept cheese fresh by avoiding bacterial spoilage for up to 65 days. Technically, this composite film has demonstrated general characteristics enough as a substitute for conventional petroleum-based packaging materials (Pal and Agarwal 2024). The study conducted by Zhang et al. (2023) showed that incorporation of zinc nanoparticles (ZNPs) (0–5% w/w) safely acts as an antimicrobial agent for food packaging films (FPFs) made of polysaccharides and proteins. The use of ZNPs on films has benefits of increased barrier properties to UV light, gases and moisture, mechanical strength and water resistance. The extent to which the ZNPs affect the characteristic of resulting films varies with parameters such as concentration, size, surface chemistry of the NPs, distribution and aggregation state, and the properties of the biopolymer matrix. Typically, an optimal concentration of ZNP from 0–3% w/w is indicated in cases where consequent benefits are to be realized. Depending on the type of ZNPs incorporated into the FPFs, different types of perishable food products have been successfully maintained, such as fresh produce, animal proteins, and dairy product as well. Nevertheless, the monitoring of the ZNP transfer to foods is important to minimize the transfer of nanoparticles into the food chain (Zhang et al. 2023).

9.7 Conclusion

The global trend toward eco-friendly renewable packaging solutions supports agar-based materials as sustainable alternatives for food delivery packaging. The natural polysaccharide agar obtained from red seaweed demonstrates superior film-forming properties that qualify it as an environmental-friendly plastic replacement to reduce waste and pollution. Widespread usage of agar-based materials remains limited because of weak mechanical property, sensitivity to water, high production costs and challenges with regulatory standards. Enhanced agar utilization depends heavily on biopolymer blending advancement together with active agent inclusion and nanotechnology utilization and fabrication technique optimization. The effective utilization of agar depends on the advancement of biopolymer blending while utilizing active agents and nanotechnology and perfecting fabrication techniques.

The functional traits of agar-based materials experience enhanced improvements by integrating bioactive compounds which demonstrate antimicrobial and antioxidant properties. Agar-based materials function as shelf-life extenders by stopping microbial development along with oxidative processes. The improved features of agar-based packaging have increased its value for protecting fresh produce items and dairy products and meat products during packaging processes. Studies have proven the positive effects of nanotechnology applications with cellulose nanocrystals and clay nanoparticles on improving the mechanical performance and flexibility with enhanced barrier properties of agar-based films. For the successful deployment of agar-based food packaging both regulatory approval and consumer satisfaction stand as fundamental requirements. Extensive safety assessments together with full compliance with food regulations must be achieved since agar-based materials represent a fresh commercial packaging solution. Addressing consumer ignorance about biodegradable packaging by educating them about its benefits will help manufacturers gain wider adoption.

Future research should focus on optimizing formulations, enhancing functional performance, and ensuring cost-effectiveness for commercial viability. Conducting sustainability assessments and lifecycle analyses will further validate agar-based packaging as a viable alternative to synthetic plastics. With continued innovation and collaboration among researchers, industry stakeholders, and policymakers, agar-based food packaging can contribute to reducing plastic pollution, promoting a circular economy, and achieving global sustainability goals. By addressing these challenges through interdisciplinary research and technological advancements, agar-based materials have the potential to become a mainstream solution in the future of sustainable food packaging.

Declaration AI tools were used only for improving the language and readability. After using these tools, the authors have reviewed the final manuscript and take full responsibility for its content.

References

Abdollahi M, Damirchi S, Shafafi M, Rezaei M, Ariaii P (2019) Carboxymethyl cellulose-agar biocomposite film activated with summer savory essential oil as an antimicrobial agent. Int J Biol Macromol 126:561–568. https://doi.org/10.1016/J.IJBIOMAC.2018.12.115

Abdul Khalil HPS, Saurabh CK, Tye YY, Lai TK, Easa AM, Rosamah E, Fazita MRN, Syakir MI, Adnan AS, Fizree HM, Aprilia NAS, Banerjee A (2017) Seaweed based sustainable films and composites for food and pharmaceutical applications: a review. Renew Sustain Energy Rev 77:353–362. https://doi.org/10.1016/J.RSER.2017.04.025

Agusman FD, Murdinah WT, Irianto HE, Priambudi P, Fateha AAHD, Nissa RD, Firdiana B, Nurhayati (2022) Physical properties of bioplastic agar/chitosan blend. In: IOP conference series: earth and environmental science, vol 978, no 1, p 012046. https://doi.org/10.1088/1755-1315/978/1/012046

Álvarez-Hernández MH, Martínez-Hernández GB, Castillejo N, Martínez JA, Artés-Hernández F (2021) Development of an antifungal active packaging containing thymol and an ethylene scavenger. Validation during storage of cherry tomatoes. Food Packag Shelf Life 29:100734. https://doi.org/10.1016/J.FPSL.2021.100734

Amariei S, Ursachi F, Petraru A (2022) Development of new biodegradable agar-alginate membranes for food packaging. Membranes 12(6):576. https://doi.org/10.3390/MEMBRANES12060576

Bang YJ, Roy S, Rhim JW (2022) A facile in situ synthesis of resorcinol-mediated silver nanoparticles and the fabrication of agar-based functional nanocomposite films. J Compos Sci 6(5):124. https://doi.org/10.3390/JCS6050124/S1

Belay M, Nagarale RK, Verma V (2017) Preparation and characterization of graphene-agar and graphene oxide-agar composites. J Appl Polym Sci 134(33):45085. https://doi.org/10.1002/APP.45085

Bertasa M, Dodero A, Alloisio M, Vicini S, Riedo C, Sansonetti A, Scalarone D, Castellano M (2020) Agar gel strength: a correlation study between chemical composition and rheological properties. Eur Polym J 123:109442. https://doi.org/10.1016/J.EURPOLYMJ.2019.109442

Betraoui A, Seddiki N, Souag R, Guerfi N, Semlali A, Aouak T, Aliouche D (2023) Synthesis of new hydrogels involving acrylic acid and acrylamide grafted agar-agar and their application in the removal of cationic dyes from wastewater. Gels 9(6):499. https://doi.org/10.3390/GELS9060499

Carina D, Sharma S, Jaiswal AK, Jaiswal S (2021) Seaweeds polysaccharides in active food packaging: a review of recent progress. Trends Food Sci Technol 110:559–572. https://doi.org/10.1016/J.TIFS.2021.02.022

Carocho M, Heleno S, Rodrigues P, Barreiro MF, Barros L, Ferreira ICFR (2019) A novel natural coating for food preservation: effectiveness on microbial growth and physicochemical parameters. LWT 104:76–83. https://doi.org/10.1016/J.LWT.2019.01.031

Cazón P, Velazquez G, Ramírez JA, Vázquez M (2017) Polysaccharide-based films and coatings for food packaging: a review. Food Hydrocoll 68:136–148. https://doi.org/10.1016/J.FOODHYD.2016.09.009

Cebrián-Lloret V, Martínez-Abad A, López-Rubio A, Martínez-Sanz M (2023) Sustainable bio-based materials from minimally processed red seaweeds: effect of composition and cell wall structure. J Polym Environ 31(3):886–899

Contessa CR, da Rosa GS, Moraes CC (2021) New active packaging based on biopolymeric mixture added with bacteriocin as active compound. Int J Mol Sci 22(19):10628. https://doi.org/10.3390/IJMS221910628

da Rocha M, Alemán A, Romani VP, López-Caballero ME, Gómez-Guillén MC, Montero P, Prentice C (2018) Effects of agar films incorporated with fish protein hydrolysate or clove essential oil on flounder (Paralichthys orbignyanus) fillets shelf-life. Food Hydrocoll 81:351–363. https://doi.org/10.1016/J.FOODHYD.2018.03.017

Devi LS, Jaiswal AK, Jaiswal S (2024) Lipid incorporated biopolymer based edible films and coatings in food packaging: a review. Curr Res Food Sci 8. https://doi.org/10.1016/j.crfs.2024.100720

Diop CIK, Beltran S, Sanz MT, Garcia-Tojal J, Trigo-lopez M (2023) Designing bilayered composite films by direct agar/chitosan and citric acid-crosslinked PVA/agar layer-by-layer casting for packaging applications. Food Hydrocoll 144:108987. https://doi.org/10.1016/J.FOODHYD.2023.108987

Ebrahimzadeh S, Biswas D, Roy S, McClements DJ (2023) Incorporation of essential oils in edible seaweed-based films: a comprehensive review. Trends Food Sci Technol 135:43–56. https://doi.org/10.1016/J.TIFS.2023.03.015

Fekete E, Bella É, Csiszár E, Móczó J (2019) Improving physical properties and retrogradation of thermoplastic starch by incorporating agar. Int J Biol Macromol 136:1026–1033. https://doi.org/10.1016/j.ijbiomac.2019.06.109

Foschi E, Sciences AB-A, 2019, undefined (2019) The commitment of packaging industry in the framework of the European strategy for plastics in a circular economy. Adm Sci mdpi.Com 9(1). https://doi.org/10.3390/admsci9010018

Fu XT, Kim SM (2010) Agarase: review of major sources, categories, purification method, enzyme characteristics and applications. Marine Drugs 8(1):200–218. https://doi.org/10.3390/MD8010200

Gali SP, Reshma RG, Krishnamoorthy S, Meenatchi R (2024) Coat the flavor, preserve the quality: elevating food stability with cutting-edge coating materials. Food Coat Preserv Technol 211–258. https://doi.org/10.1002/9781394237623.ch7

Ganjeh AM, Saraiva JA, Pinto CA, Casal S, Gonçalves I (2024) Recent advances in nano-reinforced food packaging based on biodegradable polymers using layer-by-layer assembly: a review. Carbohydr Polym Technol Appl 7:100395. https://doi.org/10.1016/J.CARPTA.2023.100395

Garrido T, Etxabide A, Guerrero P, De La Caba K (2016) Characterization of agar/soy protein biocomposite films: effect of agar on the extruded pellets and compression moulded films. Carbohyd Polym 151:408–416. https://doi.org/10.1016/J.CARBPOL.2016.05.089

Ghosh T (2024) Microplastics bioaccumulation in fish: Its potential toxic effects on hematology, immune response, neurotoxicity, oxidative stress, growth, and reproductive dysfunction. Toxicol Rep 14:101854. https://doi.org/10.1016/J.TOXREP.2024.101854

Giordano A, Cremonesi P (2021) New methods of applying rigid agar gels: from tiny to large-scale surface areas. Stud Conserv 66:437–448. https://doi.org/10.1080/00393630.2020.1848272

Guerrero P, Etxabide A, Leceta I, Peñalba M, De La Caba K (2014) Extraction of agar from Gelidium sesquipedale (Rodhopyta) and surface characterization of agar based films. Carbohyd Polym 99:491–498. https://doi.org/10.1016/J.CARBPOL.2013.08.049

Guo Y, Zhang B, Zhao S, Qiao D, Xie F (2021) Plasticized starch/agar composite films: processing, morphology, structure, mechanical properties and surface hydrophilicity. Coatings 11. https://doi.org/10.3390/coatings11030311

Hamimed S, Ben Ammar NE, Slimi H, Asses N, Hamzaoui AH, Chatti A (2024) Innovative entrapped Yarrowia lipolytica within polyvinylpyrrolidone (PVP)/polyethylene glycol (PEG)/agar for improving olive mill wastewater bioremediation. J Clean Prod 449:141828. https://doi.org/10.1016/J.JCLEPRO.2024.141828

Harnkarnsujarit N, Li Y (2017) Structure–property modification of microcrystalline cellulose film using agar and propylene glycol alginate. J Appl Polym Sci 134(47). https://doi.org/10.1002/APP.45533

Hernández V, Ibarra D, Triana JF, Martínez-Soto B, Faúndez M, Vasco DA, Gordillo L, Herrera F, García-Herrera C, Garmulewicz A (2022) Agar biopolymer films for biodegradable packaging: a reference dataset for exploring the limits of mechanical performance. Materials 15(11):3954. https://doi.org/10.3390/ma15113954

Hong SI, Cho Y, Rhim JW (2021) Effect of Agar/AgNP composite film packaging on refrigerated beef loin quality. Membranes 11(10):750. https://doi.org/10.3390/MEMBRANES11100750

Jambeck JR, Geyer R, Wilcox C, Siegler TR, Perryman M, Andrady A, Narayan R, Law KL (2015) Plastic waste inputs from land into the ocean. Science 347(6223):768–771. https://doi.org/10.1126/SCIENCE.1260352

Jayasekara S, Dissanayake L, Jayakody LN (2022) Opportunities in the microbial valorization of sugar industrial organic waste to biodegradable smart food packaging materials. Int J Food Microbiol 377:109785. https://doi.org/10.1016/J.IJFOODMICRO.2022.109785

Jones H, Saffar F, Koutsos V, Ray D (2021) Polyolefins and polyethylene terephthalate package wastes: recycling and use in composites. Energies 14(21):7306. https://doi.org/10.3390/EN14217306

Jridi M, Abdelhedi O, Zouari N, Fakhfakh N, Nasri M (2019) Development and characterization of grey triggerfish gelatin/agar bilayer and blend films containing vine leaves bioactive compounds. Food Hydrocoll 89:370–378. https://doi.org/10.1016/J.FOODHYD.2018.10.039

Jumaidin R (2023) Agar based composite as a new alternative biopolymer 67–82. https://doi.org/10.1007/978-981-19-5327-9_3

Kumar S, Boro JC, Ray D, Mukherjee A, Dutta J (2019) Bionanocomposite films of agar incorporated with ZnO nanoparticles as an active packaging material for shelf life extension of green grape. Heliyon 5(6):e01867. https://doi.org/10.1016/J.HELIYON.2019.E01867

Kumar S, Jahan K, Verma A, Agarwal M, Chandraprakash C (2022) Agar-based composite films as effective biodegradable sound absorbers. ACS Sustain Chem Eng 10(26):8242–8253. https://doi.org/10.1021/ACSSUSCHEMENG.2C00168

Lee K, Baek S, Kim D, Seo J (2019) A freshness indicator for monitoring chicken-breast spoilage using a Tyvek® sheet and RGB color analysis. Food Packag Shelf Life 19:40–46. https://doi.org/10.1016/J.FPSL.2018.11.016

Lee WK, Lim YY, Leow ATC, Namasivayam P, Ong Abdullah J, Ho CL (2017) Biosynthesis of agar in red seaweeds: a review. Carbohyd Polym 164:23–30. https://doi.org/10.1016/J.CARBPOL.2017.01.078

Lim JW, Lim WS, Lee MH, Park HJ (2021) Barrier and structural properties of polyethylene terephthalate film coated with poly(acrylic acid)/montmorillonite nanocomposites. Packag Technol Sci 34(3):141–150. https://doi.org/10.1002/PTS.2547

López de Lacey AM, López-Caballero ME, Montero P (2014) Agar films containing green tea extract and probiotic bacteria for extending fish shelf-life. LWT 55(2):559–564. https://doi.org/10.1016/J.LWT.2013.09.028

Makaremi M, Yousefi H, Cavallaro G, Lazzara G, Goh CBS, Lee SM, Solouk A, Pasbakhsh P (2019) Safely dissolvable and healable active packaging films based on alginate and pectin. Polymers 11(10). https://doi.org/10.3390/POLYM11101594

Makhloufi N, Chougui N, Rezgui F, Benramdane E, Freire CSR, Vilela C, Silvestre AJD (2021) Bio-based sustainable films from the Algerian *Opuntia ficus-indica* cladodes powder: effect of plasticizer content. J Appl Polym Sci 138(20):e50450. https://doi.org/10.1002/app.50450

Meira SMM, Zehetmeyer G, Werner JO, Brandelli A (2017) A novel active packaging material based on starch-halloysite nanocomposites incorporating antimicrobial peptides. Food Hydrocoll 63:561–570. https://doi.org/10.1016/J.FOODHYD.2016.10.013

Mendez-Encinas MA, Carvajal-Millan E, Rascon-Chu A, Astiazaran-Garcia HF, Valencia-Rivera DE (2018) Ferulated arabinoxylans and their gels: functional properties and potential application as antioxidant and anticancer agent. Oxid Med Cell Longev 2018:2314759. https://doi.org/10.1155/2018/2314759

More A, Sen C, Das M (2017) Development of starch-polyvinyl alcohol (PVA) biodegradable film: effect of cross-linking agent and antimicrobials on film characteristics. J Appl Packag Res 9(3). https://repository.rit.edu/japr/vol9/iss3/1

Moshood TD, Nawanir G, Mahmud F, Mohamad F, Ahmad MH, AbdulGhani A (2022) Sustainability of biodegradable plastics: new problem or solution to solve the global plastic pollution? Curr Res Green Sustain Chem 5:100273. https://doi.org/10.1016/J.CRGSC.2022.100273

Mostafavi FS, Zaeim D (2020) Agar-based edible films for food packaging applications—a review. Int J Biol Macromol 159:1165–1176. https://doi.org/10.1016/J.IJBIOMAC.2020.05.123

Pal N, Agarwal M (2024) Development and characterization of eco-friendly guar gum-agar-beeswax-based active packaging film for cheese preservation. Int J Biol Macromol 277:134333. https://doi.org/10.1016/J.IJBIOMAC.2024.134333

Perera KY, Jaiswal AK, Jaiswal S (2023) Biopolymer-based sustainable food packaging materials: challenges, solutions, and applications. Foods 12:2422. https://doi.org/10.3390/FOODS12122422

Perera KY, Sharma S, Pradhan D, Jaiswal AK, Jaiswal S (2021) Seaweed polysaccharide in food contact materials (active packaging, intelligent packaging, edible films, and coatings). Foods 10(9):2088. https://doi.org/10.3390/FOODS10092088

Priyadarshi R, Ezati P, Rhim JW (2021) Recent advances in intelligent food packaging applications using natural food colorants. ACS Food Sci Technol 1(2):124–138. https://doi.org/10.1021/ACSFOODSCITECH.0C00039/ASSET/IMAGES/MEDIUM/FS0C00039_0005.GIF

Rafey A, Siddiqui FZ (2023) A review of plastic waste management in India–challenges and opportunities. Int J Environ Anal Chem 103(16):3971–3987. https://doi.org/10.1080/03067319.2021.1917560;JOURNAL:JOURNAL:GEAC20;REQUESTEDJOURNAL:JOURNAL:GEAC20;PAGE:STRING:ARTICLE/CHAPTER

Rajesh Kumar Dora T, Ghosh S, Damodar R (2020) Synthesis and evaluation of physical properties of Agar biopolymer film coating-an alternative for food packaging industry. Mater Res Express 7(9). https://doi.org/10.1088/2053-1591/ABB7AC

Rhim JW, Wang LF (2013) Mechanical and water barrier properties of agar/κ-carrageenan/konjac glucomannan ternary blend biohydrogel films. Carbohyd Polym 96(1):71–81. https://doi.org/10.1016/J.CARBPOL.2013.03.083

Riahi Z, Khan A, Rhim JW, Shin GH, Kim JT (2024) Synergistic effect of iron-based metal-organic framework hybridized with carbon quantum dots in agar/gelatin films for fruit preservation. Food Packag Shelf Life 45:101330. https://doi.org/10.1016/J.FPSL.2024.101330

Rodrigues C, Souza VGL, Coelhoso I, Fernando AL (2021) Bio-based sensors for smart food packaging—current applications and future trends. Sensors 21(6):2148. https://doi.org/10.3390/S21062148

Roy S, Chawla R, Santhosh R, Thakur R, Sarkar P, Zhang W (2023) Agar-based edible films and food packaging application: a comprehensive review. Trends Food Sci Technol 141:104198. https://doi.org/10.1016/J.TIFS.2023.104198

Roy S, Kim HJ, Rhim JW (2021) Effect of blended colorants of anthocyanin and shikonin on carboxymethyl cellulose/agar-based smart packaging film. Int J Biol Macromol 183:305–315. https://doi.org/10.1016/j.ijbiomac.2021.04.162

Roy S, Priyadarshi R, Rhim JW (2022) Gelatin/agar-based multifunctional film integrated with copper-doped zinc oxide nanoparticles and clove essential oil Pickering emulsion for enhancing the shelf life of pork meat. Food Res Int 160. https://doi.org/10.1016/J.FOODRES.2022.111690

Roy S, Rhim JW (2019) Agar-based antioxidant composite films incorporated with melanin nanoparticles. Food Hydrocoll 94:391–398. https://doi.org/10.1016/J.FOODHYD.2019.03.038

Roy S, Rhim JW (2021a) Carrageenan/agar-based functional film integrated with zinc sulfide nanoparticles and Pickering emulsion of tea tree essential oil for active packaging applications. Int J Biol Macromol 193:2038–2046. https://doi.org/10.1016/J.IJBIOMAC.2021.11.035

Roy S, Rhim JW (2021b) Fabrication of carboxymethyl cellulose/agar-based functional films hybridized with alizarin and grapefruit seed extract. ACS Appl Bio Mater 4(5):4470–4478. https://doi.org/10.1021/ACSABM.1C00214/SUPPL_FILE/MT1C00214_SI_001.PDF

Roy S, Rhim JW (2021c) Preparation of pectin/agar-based functional films integrated with zinc sulfide nano petals for active packaging applications. Colloids Surf, B 207. https://doi.org/10.1016/J.COLSURFB.2021.111999

Roy S, Rhim JW (2022) Starch/agar-based functional films integrated with enoki mushroom-mediated silver nanoparticles for active packaging applications. Food Biosci 49:101867. https://doi.org/10.1016/J.FBIO.2022.101867

Rukmanikrishnan B, Ramalingam S, Lee J (2021) Quaternary ammonium silane-reinforced agar/polyacrylamide composites for packaging applications. Int J Biol Macromol 182:1301–1309. https://doi.org/10.1016/J.IJBIOMAC.2021.05.060

Sanyang ML, Sapuan SM, Jawaid M, Ishak MR, Sahari J (2015) Effect of plasticizer type and concentration on tensile, thermal and barrier properties of biodegradable films based on sugar palm (Arenga pinnata) starch. Polymers 7(6):1106–1124. https://doi.org/10.3390/POLYM7061106

Shah M, Hameed A, Kashif M, Majeed N, Muhammad J, Shah N, Rehan T, Khan A, Uddin J, Khan A, Kashtoh H (2024) Advances in agar-based composites: a comprehensive review. Carbohydr Polym 346. https://doi.org/10.1016/j.carbpol.2024.122619

Shankar S, Rhim JW (2017) Preparation and characterization of agar/lignin/silver nanoparticles composite films with ultraviolet light barrier and antibacterial properties. Food Hydrocoll 71:76–84. https://doi.org/10.1016/J.FOODHYD.2017.05.002

Sousa AMM, Gonçalves MP (2015) Strategies to improve the mechanical strength and water resistance of agar films for food packaging applications. Carbohyd Polym 132:196–204. https://doi.org/10.1016/J.CARBPOL.2015.06.022

Sousa AMM, Souza HKS, Liu LS, Gonçalves MP (2015) Alternative plasticizers for the production of thermo-compressed agar films. Int J Biol Macromol 76:138–145. https://doi.org/10.1016/J.IJBIOMAC.2015.02.030

Sudheer S, Bandyopadhyay S, Bhat R (2023) Sustainable polysaccharide and protein hydrogel-based packaging materials for food products: a review. Int J Biol Macromol 248:125845. https://doi.org/10.1016/J.IJBIOMAC.2023.125845

Surendran G, Sherje AP (2022) Biomedical applications of agar and its composites: a mini-review. Nat Prod J 13(5):23–30. https://doi.org/10.2174/2210315512666220615113320

Tammina SK, Rhim JW (2023) Carboxymethylcellulose/agar-based functional film incorporated with nitrogen-doped polyethylene glycol-derived carbon dots for active packaging applications. Chemosphere 313:137627. https://doi.org/10.1016/J.CHEMOSPHERE.2022.137627

Uranga J, Nguyen BT, Si TT, Guerrero P, De la Caba K (2020) The effect of cross-linking with citric acid on the properties of agar/fish gelatin films. Polymers 12(2). https://doi.org/10.3390/POLYM12020291

Wang LF, Rhim JW (2015) Preparation and application of agar/alginate/collagen ternary blend functional food packaging films. Int J Biol Macromol 80:460–468. https://doi.org/10.1016/J.IJBIOMAC.2015.07.007

Wang X, Guo C, Hao W, Ullah N, Chen L, Li Z, Feng X (2018) Development and characterization of agar-based edible films reinforced with nano-bacterial cellulose. Int J Biol Macromol 118:722–730. https://doi.org/10.1016/J.IJBIOMAC.2018.06.089

Wardana AA, Widyaningsih TD (2017) Development of edible films from tapioca starch and agar, enriched with red cabbage (Brassica oleracea) as a sausage deterioration bio-indicator. IOP Conf Ser Earth Environ Sci 109(1):012031. https://doi.org/10.1088/1755-1315/109/1/012031

Wei D, Li Z, Wang H, Liu J, Xiao H, Zheng A, Guan Y (2017) Antimicrobial paper obtained by dip-coating with modified guanidine-based particle aqueous dispersion. Cellulose 24(9):3901–3910. https://doi.org/10.1007/S10570-017-1386-7/FIGURES/7

Xiao Q, Weng H, Ni H, Hong Q, Lin K, Xiao A (2019) Physicochemical and gel properties of agar extracted by enzyme and enzyme-assisted methods. Food Hydrocoll 87:530–540. https://doi.org/10.1016/j.foodhyd.2018.08.041

Yang P, Song YF, Qi H, Li LQ, Xie XB (2024) AIE-doped agar-based photodynamic sterilization film for antimicrobial food wrapping. LWT 202:116330. https://doi.org/10.1016/J.LWT.2024.116330

Yang Z, Zhai X, Zhang C, Shi J, Huang X, Li Z, Zou X, Gong Y, Holmes M, Povey M, Xiao J (2022) Agar/TiO_2/radish anthocyanin/neem essential oil bionanocomposite bilayer films with improved bioactive capability and electrochemical writing property for banana preservation. Food Hydrocoll 123. https://doi.org/10.1016/J.FOODHYD.2021.107187

Yin ZC, Wang YL, Wang K (2018) A pH-responsive composite hydrogel beads based on agar and alginate for oral drug delivery. J Drug Deliv Sci Technol 43:12–18. https://doi.org/10.1016/J.JDDST.2017.09.009

Zhang R, Wang W, Zhang H, Dai Y, Dong H, Hou H (2019) Effects of hydrophobic agents on the physicochemical properties of edible agar/maltodextrin films. Food Hydrocoll 88:283–290. https://doi.org/10.1016/j.foodhyd.2018.10.008

Zhang W, Sani MA, Zhang Z, McClements DJ, Jafari SM (2023) High performance biopolymeric packaging films containing zinc oxide nanoparticles for fresh food preservation: a review. Int J Biol Macromol 230:123188. https://doi.org/10.1016/J.IJBIOMAC.2023.123188

Zhang W, Zhang Y, Cao J, Jiang W (2021) Improving the performance of edible food packaging films by using nanocellulose as an additive. Int J Biol Macromol 166:288–296. https://doi.org/10.1016/J.IJBIOMAC.2020.10.185

Zucca P, Fernandez-Lafuente R, Sanjust E (2016) Agarose and its derivatives as supports for enzyme immobilization. Molecules 21(11). https://doi.org/10.3390/MOLECULES21111577

Chapter 10
Protein-Based Food Packaging Films Soy, Whey, and Beyond for Barrier Properties

Aswathi Raj K. V. and Swarup Roy

Abstract Protein-packaging films have the potential to substitute for conventional synthetic polymers as demand rises for sustainable and biodegradable packaging materials. The chapter reviews the potential of protein-based films, in the form of soy and whey proteins, as well as other candidate proteins, in food packaging. Proteins being structurally and functionally distinct entities provide superior film-forming attributes, mechanical strength, and barrier characteristics against gases, water, and lipids. Whey protein films are noted for their transparency, elasticity, and high barrier properties against fats and odors while soy protein films are noted for their richness, ease of biodegradation, and fair oxygen barrier properties. These films are obtained from different sources, which vary from plant proteins like soy, zein, and pea to animal proteins like whey and gelatin, each of which has unique functional properties. For the production of films with the best properties, processing methods like extrusion and solution casting are employed. These films have promising properties like mechanical strength, barrier performance against oxygen and water, and functional properties like antioxidant and antimicrobial. Protein-based films applications keep the food quality intact, increase the shelf life, and impart antimicrobial and antioxidant functionality. This chapter highlights the promise of protein films to alleviate the growing demand for green packaging within the food industry.

Keywords Protein · Biopolymer · Packaging · Film · Coating · Shelf life

Aswathi Raj K. V. · S. Roy (✉)
Department of Food Technology and Nutrition, School of Agriculture, Lovely Professional University, Punjab, India
e-mail: swaruproy2013@gmail.com

S. Roy
Department of Food and Nutrition, Kyung Hee University, Seoul 02447, Republic of Korea

S. Roy and M. Shahid (eds.), *Biomaterials for Sustainable Food Packaging*,
Environmental Footprints and Eco-design of Products and Processes,
https://doi.org/10.1007/978-3-032-08027-1_10

10.1 Introduction

Research on the development and application of bioactive packaging systems is imperative. Smart packaging systems are an effective tool for preventing food spoilage and reducing the risk of pathogenic microbial growth because they function as a barrier and have active ingredients at the product edge and packaging (Arkoun et al. 2018). The majority of the materials used in packaging today are made from petrochemical products. Alternative, environmentally friendly, and biodegradable polymers are in high demand due to the worldwide issue of environmental pollution (Kabir et al. 2020). The future food packaging should thus be constructed from innovative materials that are edible or biodegradable naturally. Apart from functional (water, vapours, oxygen, carbon dioxide, and ultraviolet (UV) light barrier), mechanical (strength), and physical (opacity and colour) properties, food-grade compound-only-made edible packaging systems should possess bioactive properties (antimicrobial, antioxidant, and anti-browning (Hamed et al. 2022).

Employing edible films in food packaging contains great prospects of maintaining the freshness, safety, transportation, storage, and display of several different foods. Most edible packaging films are made primarily from polysaccharides, proteins, and lipids in separate or in combination. Proteins are a great option for hydrocolloids because they have the good mechanical, optical (transparency), and physical (flexibility and resistance) qualities to create biodegradable films. They can also serve as barriers to organic compounds, oxygen, and scents. Casein, collagen, corn zein, ovalbumin, fish proteins, soy protein isolate (SPI), wheat gluten, and whey protein isolate (WPI) are among the protein materials that have been validated (Milani and Tirgarian 2020). These proteins are easily formulated into films and processed. They can be recognized as a more sustainable substitute for conventional synthetic plastics because they are manufactured from renewable resources. These benefits have led to an increase in the use of edible packaging as a substitute for both natural and synthetic biodegradable plastic food packaging (Shah et al. 2023).

Proteins are hetero-biopolymers composed of various types of amino acids, giving them distinctive three-dimensional network structures. Protein chains with different functional groups in their amino acids have great potential for improved mechanical and gas barrier qualities (Zubair and Ullah 2020). In addition, certain protein polymers possess amphiphilic characteristics that enable them to incorporate active agents such as various antimicrobial agents that can possibly preserve the nutritional quality and value of food (Bahrami et al. 2020). When making edible films, proteins are favoured over lipids and carbohydrates in order to get the required film qualities. Protein secondary, tertiary, and quaternary structures can be readily altered to produce improved film properties using a variety of techniques, such as thermal denaturation, chemical hydrolysis, enzymatic treatment, and chemical crosslinking (Shah et al. 2023). Protein films have been used to package fruits, vegetables, dairy products, meat products, and frozen foods. Protein biopolymers can be used to prolong the shelf life of fruits and vegetables by reducing the flow of gases and moisture between the environment and the product to be packaged (Assad et al. 2020). Because many microbiological and biochemical responses, including lipolysis, oxidation, proteolysis, and moisture, are taken into account during production, choosing the best packing material for milk products is challenging. The shelf life of dairy products provided by protein-based films is significantly influenced by the barrier qualities of the packaging materials (Sid et al. 2021). They prolong the shelf life of meat packing by preventing microbiological growth and lipid oxidation, preserving texture and color. They improve the safety and quality of meat by acting as transporters for active compounds like antioxidants (Figueroa-Enríquez et al. 2024).

10.2 Sources of Protein

There are numerous sources of protein and some are useful for the extraction biopolymers like SPI, WPI, etc. Various sources of protein are schematically represented in Fig. 10.1.

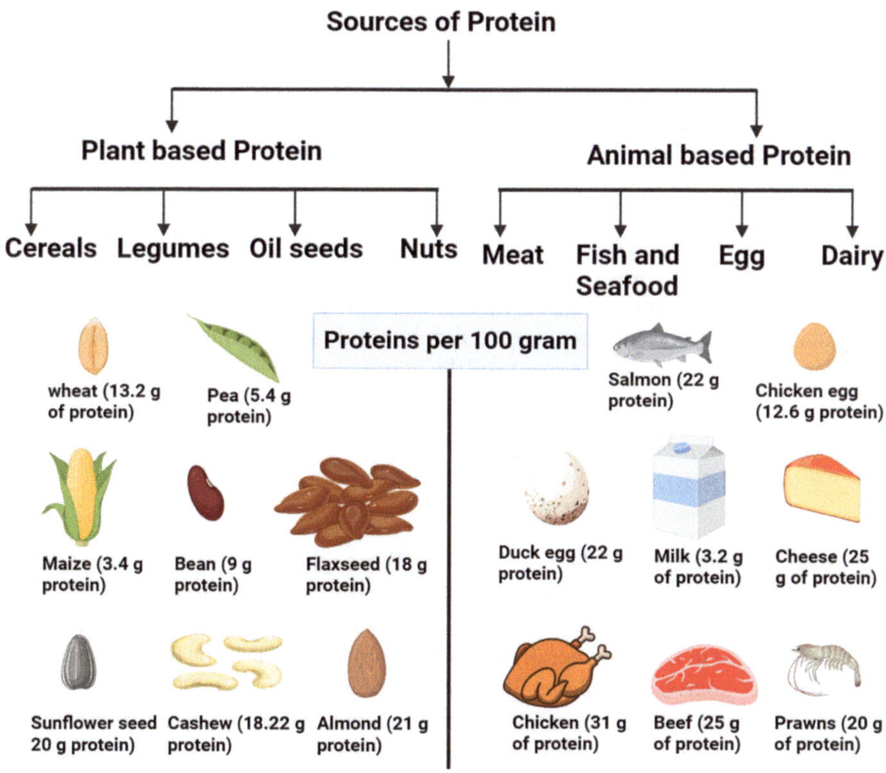

Fig. 10.1 Diagrammatic illustration of various sources of protein

10.2.1 Milk Proteins

Two types of milk proteins exist: casein protein and whey protein. Casein constitutes 80% of milk protein, they are composed of three main components such as alpha, beta and kappa which all combine to form colloidal micelles in milk (Chevalier et al. 2018). They have good film-forming ability, and at low relative humidity, it creates transparent and flexible films with excellent lipid and gas barrier performance (Sophia and Malar Retna 2018). Due to their random-coil nature and the abundance of intermolecular hydrogen, hydrophobic, and electrostatic bonds formed, casein can readily form films from water solutions without additional processing. Casein is a good protein to employ in biodegradable films because the protein possesses many features that include its stability at temperature, non-toxicity, ion and small molecule binding, and ability to form micelles. Caseins are good biomaterials to utilize in preparing edible films because they have high nutritional value, water solubility, and emulsification abilities.

Casein protein coagulates to produce whey protein. α-Lactalbumin, β-Lactoglobulin, bovine serum albumin immunoglobulins and protease peptones constitute whey protein contents (Mohamed et al. 2020). WPI and concentrates (WPC) can be employed to create whey-based films. The two materials have high cysteine, methionine and sulphur amino acids but contain different protein percentages by at least 90% and 50–80%, respectively. Whey-protein based coatings possess superior lipid and fragrance barrier properties. Whey protein films exhibit superior mechanical strength and gas barrier properties at low relative humidity. At increasing relative humidity, however, these properties become deteriorate (Cruz-Diaz et al. 2019). They possess superior oil, oxygen, and fragrance barriers too. However, the hydrophilic nature of whey-based films makes them poor moisture barriers. Lipid components like fat and oils are capable of improving the poor moisture barriers and also making whey-based films more hydrophobic. Plant oils, fatty acids waxes, and acetylated monoglycerides are some of the common lipid materials used.

10.2.2 Soy Protein

Soybeans are utilized to isolate soy protein. Three types of soy protein include soy isolates (90% protein and 2% carbohydrate), soy flour (56% protein and 34% carbohydrate), and soy concentrate (65% protein and 18% carbohydrate) (Mohamed et al. 2020). Soy protein film is generally produced from SPI. It is a cheap and biodegradable material source for edible film and coating production (Tian et al. 2018). Compared to synthetic films, soy protein films have lower mechanical and barrier properties due to their hydrophilicity and sensitivity to the environment (Tulamandi et al. 2016). Tensile strength and elongation improved with the increased molecular weight of soy proteins, but water vapor barrier properties were not affected. Many methods, including heating, extrusion, spinning, casting, and thermal compacting, have been reported for film-forming soy protein (Chen et al. 2019).

10.2.3 Gelatin

Gelatin is an animal protein derived from the hydrolysis of collagen and is a common constituent in (edible) food packaging. Collagen, which is used to produce gelatin, comes from a variety of sources, including fish, pork, and cows (Hanani et al. 2019). Proline, 4-hydroxyproline, and glycine are the major constituents of gelatin, but the percentages of these constituents differ with the source; for example, pigskin gelatin contains contents of 13%, 9%, and 33%, respectively (Chen et al. 2019). Pure and dry gelatin is pale yellow in colour and transparent, brittle, odourless, and glass-like. Gelatin is dissolved in hot water to form an edible film, which is then applied thinly on a plate and roasted to dryness (Mohamed et al. 2019) Gelatin-based edible films possess improved mechanical properties and a thicker film and is very hydrophobic and thus has a very low water vapor permeability (Hassan et al. 2018). Essential oils and lipids may be added to enhance water vapor permeability. Gelatin finds applications in the nutraceutical, pharmaceutical, and food industries due to its superior gelling capacity (Scartazzini et al. 2019). It serves as an effective stabilizer when formulating a film-forming solution. Essential oils and lipids may be added to enhance water vapor permeability (Shankar et al. 2019). Gelatin is a key raw material for manufacturing biodegradable films because it possesses good film-forming properties, is easily found and available, is cheap, biocompatible, and biodegradable. Gelatin films are reversible, oxygen-impermeable, and transparent. Because gelatin's melting point is close to body temperature, gelatin films are especially important raw materials for edible film production. Gelatin edible films exhibit low permeability to oxygen and can be formulated with antimicrobial and antioxidant properties through the incorporation of additives such as carvacrol and citrus essential oils (Mohamed et al. 2020).

10.2.4 Zein

Zein, corn's major protein, is responsible for creating adhesive film and carries hydrophobic, antioxidant, and antibacterial properties (Mohamed et al. 2020). They are prolamin proteins that become soluble in ethanol 70–80%. Naturally, they are hydrophobic in nature (Hassan et al. 2018). They are alcohol-soluble but water-insoluble based on the nonpolar nature of amino acids (Calva-Estrada et al. 2019). α-zein (MW = 21–25 k Da), β-zein (MW = 17–18 k Da), γ-zein (MW = 27 k Da), and δ-zein (MW = 9–10 k Da) are the four main fractions they possess (Zhang et al. 2020). Wet or solvent process can be employed in order to form zein-based films. Polyols or fatty acids are often employed as plasticizers in the development of film.

First, zein, plasticizers, and other agents are dissolved in the correct solvent. Then, the film-forming solution is coated on a nonstick, flat surface. A freestanding film can be taken off the surface after the solution has evaporated (Chen et al. 2019).

10.2.5 Gluten

Gliadin (34%) and glutenin (47%) are the two predominant types of gluten proteins. Although the former is soluble, the latter is insoluble (Calva-Estrada et al. 2019). The surface of gluten films is glossy, possesses good oxygen isolation properties, low resistance to water vapor, and poor mechanical properties (Rocca-Smith et al. 2016). They contain rich amounts of the amino acid leucine, threonine, and methionine. Since hydrated wheat gluten is a viscoelastic protein network by hydrogen bonding and hydrophobic interactions, it is an ideal biopolymer for film production (Hajikhani et al. 2020). The mechanical and barrier properties of gluten films are depended on the amount of protein content, pH, and ethanol. These qualities can be improved by the addition of binding or reinforcing agents such as lipids, polysaccharides, or phenols (Liu et al. 2018).

10.3 Protein-Based Packaging Film

The most common polymers employed to produce films for food packaging are polysaccharides, proteins, and lipids, either alone or in combination (Nur Hanani et al. 2014). The reason why these natural materials are used is that they are biodegradable and renewable. Besides that, the other advantages are expected when employing these materials for food packaging, like being able to act as a delivery vehicle of functional substances that confer wellness benefits, add flavourings and colourings, and enhance organoleptic properties (Trajkovska Petkoska et al. 2021). Recent advancements in food packaging have concentrated on increasing the sustainability, safety, and affordability of protein-based materials while also broadening their range of uses and enhancing their properties (Assad et al. 2020).

Protein is a heteropolymer consisting of more than 100 monomers, or amino acids, that are bonded together by peptide bonds. There are two kinds of naturally occurring proteins: fibrous and globular. Fibrous proteins repeat in a single unit to create chains that serve as connective tissue and closely connect with each other in parallel forms to impart strength and joint mobility, globular proteins are very long chains that have numerous branches and are wound up into elaborate spherical forms and stabilized by the combination of hydrogen, ionic, hydrophobic, and covalent bonds. Among such fibrous proteins that have drawn much interest as research subjects regarding food packaging material are collagen. In respect to globular proteins, significant research has focused on their functionality as edible packaging food. The globular proteins include casein, wheat gluten, corn zein, soy protein, whey protein, and mung bean protein (Said et al. 2019; Wittaya and Wittaya 2012).

Proteins are an excellent candidate among the hydrocolloids since they possess the good physical, optical (transparency), and mechanical qualities to form biodegradable films. They are excellent film-forming agents with excellent lipid barrier, O_2, and CO_2 properties along with good mechanical properties, especially under low relative humidity. Still, their inability to have better mechanical strength compared to synthetic polymers can restrict their application in food packaging (Azeredo and Waldron 2016). Protein-based film is biodegradable and compostable. Since they break down and emit nitrogen, they also impart a fertilizer advantage that other films that are not protein-based cannot. Lastly, there is considerable evidence that the bioactive peptides generated during the proteolytic digestion of proteins, more specifically the proteins of dairy origin, are of benefit to health in terms of antihypertensive and radical scavenging (Calva-Estrada et al. 2019). Many protein sources, i.e., corn zein, wheat gluten, SPI derived from plants and collagen, casein, fish protein, ovalbumin, WPI derived from animals have been utilized (Cinelli et al. 2014). Among them soy protein and whey protein have become increasingly popular because of their prime characteristics. Soy proteins were the focus of extensive research because of their remarkable capacity for film forming. Adhesiveness, cohesiveness, dough, emulsification, fat and water absorption, fiber formation, and texturizing capacity are some of the numerous beneficial properties of soy protein films. SPI, a high-purity soy protein with not less than 90% protein, is the foundation for most soy protein films (Tian et al. 2018). Whey protein films possess improved barrier properties as well as they are suitable as delivery systems for functional and bioactive components, improving the nutritional quality and sensory properties of foods and extending their shelf life and safety.

Protein, plasticizer, and solvent are the three main ingredients of the solutions utilized to form protein-based films. The inherent properties of the film constituents, as well as extrinsic processing conditions, determine the properties of the final film. Amino acid composition, crystallinity (of the protein and/or plasticizer), hydrophobicity/hydrophilicity, surface charge, pH, molecular size, and three-dimensional shape are some of the intrinsic protein properties. Temperature of processing, drying conditions, pH, ionic strength, salt type, relative humidity upon storage and processing, shear, and pressure are some of the extrinsic factors (Milani et al. 2020). Water or ethanol is commonly utilized as the solvent for preparing films since it is safe to be ingested. Plasticizers are the low molecular weight chemicals utilized in the film solution for increasing the mechanical and flexible characteristics of the matrix of the film. Most of the protein-based films are very tough but brittle unless plasticized; thus, a plasticizer needs to be added in order to enhance the protein-based films' potential for application (Nur Hanani et al. 2014).

The method most widely applied for the formulation of edible protein films is solution casting. This process is easy to prepare: the chosen materials are dissolved in a solvent, the polymer is gelatinized or swelled, the slurry is degassed or deaerated, the solution is cast into a thin film, the solution is poured into petri plates, and the solvent is evaporated, leaving the solid film behind (Deng et al. 2018). Solution casting is a versatile technique for the fabrication of thin films across numerous industries, given that it possesses numerous advantages, including uniformity, flexibility, microstructure control, high purity, ease on the wallet, silky-smooth finish, additive compatibility, conformal coating ability, scalability, and the possibility of being energy-efficient (Roy et al. 2024).

Andrade et al. (2021) investigated the formulation of whey protein concentrate (WPC) film using a Fucus vesiculosus extract to retard chicken breast lipid oxidation. Ten hydroethanolic F. vesiculosus extracts were prepared using the solvent casting method, which was utilized for film preparation. The incorporation of the *F. vesiculosus* extract seems to enhance the whey protein film's thickness, tensile strength, and elastic modulus. Moreover, it enhanced the permeability of the film to water vapor. At least 25 days of storage, the new film was capable of inhibiting lipid oxidation of stored poultry meat in packaged form as compared to the control whey protein film.

Rojas-Lema et al. (2021) prepared a transparent film from faba bean protein isolate (FBP) that was loaded with different concentrations of cellulose nanocrystals (CNCs) (1, 3, 5, and 7 weight percent). The CNCs were produced by acid hydrolysis of pine cones and glycerol as a plasticizer. The mechanical, thermal, barrier, optical, and morphological properties were found to be influenced upon the incorporation of different concentrations of CNC. On incorporating 7 weight percent of CNC into the FBP film, the break elongation was reduced by about 53.1%, whereas tensile strength and Young's modulus improved by 50.5 and 76.1%, respectively, when compared with the control film. Moreover, the OTR and WVTR values of the film were 36.2 and 22.7% less, respectively, than those of the unreinforced FBP film. Based on this research, CNCs prepared from pine cones play a vital role in improving the mechanical and physical properties of FBP films, which can ultimately lead to their use as edible food packaging material in the food packaging sector.

Fish gelatin edible films containing different loadings of Pomegranate seed juice by-product (PSP) residue as a reinforcing and antimicrobial agent were prepared and characterized. PSP was incorporated in the film (0, 10, 30 wt%) in three formulations by using the casting method. The film with 10 wt% was found to be rich in antioxidant activity and total phenolic content. PSP increased the stiffness of the films and reduced elongation at break, especially for 30% loading. When PSP was added, there were no discernible changes in the barrier qualities to oxygen and water vapor permeability when compared to the control; however, transparency values declined. Inhibition against *S. aureus* and high thermal stability of the films were noted. It was demonstrated that adding PSP at 10 wt% to the fish gelatin edible films could preserve the food material's integrity and shield food from lipid oxidation (Hanani et al. 2019).

One of the main polymer processing methods currently in use on a commercial scale is extrusion, which is also used to create polymeric films. This process modifies the materials' structure and enhances the extruded material's physiochemical characteristics. In general, there are three zones that make up the extrusion process: (i) the feeding zone, (ii) the kneading zone, and (iii) the heating zone at the machine's final part or exit (Calderón-Castro et al. 2018). First, the mixture of film components is moved into the feeding zone and compressed using air. This process is also known as a dry process since it works best when there is little water or solvent present. However, plasticizers are required to make the film more flexible. Plasticizers such as polyethylene glycol or sorbitol are frequently added by extrusion. As the ingredients move into the kneading zone, the mixture's density, temperature, and strain all rise (Suhag et al. 2020).

Different contents of PI (pea protein isolate) (10–50%) were blended with AS (acetylated cassava starch) and blown-extrusion was applied to produce the films. Mechanical strength lowered as PI dispersion decreased. The creation of a continuous protein network reinforcing starch matrices resulted in dramatic strength increase for AS-PI at a content of 20% PI. Excessive PI enhanced oxygen and water vapor permeability and barrier properties such as surface hydrophobicity. In addition, AS films exhibited significant storage-induced structural deformation, which PI hindered. Even though PI decreased glass transition and relaxation temperature, shrinkage of structure was averted successfully by limited protein's molecular mobility (Huntrakul et al. 2020).

10.4 Properties of Protein-Based Packaging Film

10.4.1 Mechanical Characteristics

Important functional characteristics of food packaging are mechanical characteristics of edible films, including tensile strength, elongation at break, and Young's modulus. YM and TS, which describe the stiffness of the films, and the EB, or flexibility index, are most commonly approximate mechanical parameters. Mechanical characteristics of films ultimately rely on intramolecular and intermolecular interactions of the film-forming components (Basiak et al. 2017; Jafarzadeh et al. 2022). The mechanical properties of protein-based films depend on the type of protein used, the processing treatment, and the composition of the films (Said et al. 2019). Another important ingredient that impacts the mechanical properties of protein-based films is plasticizer. The elongation at break of the polymer increases and its tensile strength reduces due to its reduction of intermolecular tensions caused by the plasticizer (Purewal et al. 2023).

This research revealed the application of kaolin (Kln) and silver-kaolin (Ag-Kln) in gelatin-composite films used for packaging active food products. Three forms of kaolin were produced through solution casting: raw Kln, Ag-Kln mixture (1:2), and Ag-Kln mixture (1:1) at different concentrations (15, 30, and 45%). Brittleness and elasticity of the film were utilized as measures to estimate its strength. Compared to Ag-Kln and kaolin films, the control film was found to be elastic in nature. Compared to raw kaolin and Ag-Kln (1:2) mix, incorporation of Ag-Kln (1:1) mix showed higher elasticity films, as the dramatic increase in YM (Youngs modulud) values between the films for active substances Ag-Kln and kaolin indicates. The incorporation of the silver particles brought a considerable change in the YM for the AgKln (1:1) and

Ag-Kln (1:2) blends, varying from 63.50 MPa to a prolonged height of 376.20 MPa and 63.35 to 115.46 MPa, respectively. Due to the flexibility of the gelatin, which was hard to crack due to its strong peptide bond, the TS value was maximized for control film when it was compared to kaolin and Ag-Kln films (Nur Amila Najwa et al. 2020).

Aqueous gelatin (GEL) and furcellaran (FUR) solutions containing varying amounts of nanoparticles (Se-AgNPs) added were employed to prepare nanocomposite films through solvent casting. As the concentration of Se-AgNPs increased, the modulus of elasticity of the film increases linearly. At loading concentrations of 5 and 10%, elongation at break of the nanocomposite film improved. This suggests that the FUR/GEL matrix became a strong matrix due to the assimilation of the nanoparticles. The matrix gets overloaded as NPs rise by an additional 15%. The extensibility of the polymer was likely increased by the addition of Se-AgNPs (with CMC), which further increased the mobility and stretching of the nanocomposite films. As per (Jamróz et al. 2019), the characteristics are likely the consequence of inter- and intra-molecular interaction within the film network.

10.4.2 Barrier Properties

The food film barrier property is its ability to control the passage of moisture or gasses in and out of food. Edible food films' capacity to limit water vapor flow into or from food is determined as water vapor permeability (WVP), whereas capacity to limit carbon dioxide or oxygen flow into or from food is determined as gas permeability (GP) (Benbettaïeb et al. 2019). Oxygen is the compound which can cause oxidation and thus initiate a variety of unwanted changes such as nutrient loss and changes in colour, flavour, and aroma (Sothornvit et al. 2007). CO_2 and O_2 permeabilities are only relevant when respiratory reactions can influence food quality. Some of the factors influencing the gas permeabilities of edible films include the integrity of the film, crystalline to amorphous zones ratio, hydrophilic–hydrophobic ratio, and polymeric chain mobility. How the plasticizer or additives interact with the film-forming polymers is another factor that influences the permeability of films (Kocira et al. 2021). Since protein-based films are hydrophilic in nature, they possess high water vapor permeability. Utilization of such packaging materials at low relative humidity is always recommended. At low relative humidity, protein-based films present a high gas barrier. With increasing humidity, the films will get wet more, which will decrease the oxygen barrier (Kumar et al. 2021).

Miriam Malherbi et al. (2019) formulated a bioactive edible film that can also be utilized as a sachet or packaging material for extra-virgin olive oil. The film was formulated using corn starch and activated gelatin from guabiroba pulp. Two of the numerous naturally occurring antioxidants present in guabiroba pulp are ascorbic acid and phenolic compounds. In the 2:1, 1:1, and 1:2 ratios, 2 and 4% native corn starch (NCS) and 5 and 10% gelatin (GEL) formulations were prepared. The film of the blend with 5% GEL and 2% NCS in 1:1 ratio was selected for incorporating GP

(10 and 20%). The WVP value of the control film was 3.88 g mm m^{-2} d^{-1} kPa^{-1}, while that of the active blend films varied between 6.75 and 12.95 g mm m^{-2} d^{-1} kPa1. These differences in the formulations were significant. Compared to the control film, the values of WVP increased as GP levels increased at both the 10 and 20% levels. The hydrophilic compound from the fruit pulp fibre and starch existing in the polymeric matrix of the films is accountable for the outcome. The total dietary fibre content of guabiroba fruit was 9.75 g 100 g^{-1}, out of which 0.28 g were soluble fibres. The fruit contained a greater number of insoluble fibres (9.47 g and 100 g^{-1}.100 g^{-1}), which may have influenced the WVP of active mix films.

Gelatin (Gel) and graphene oxide (GO)-functionalized cellulose nanofibrils (CNF) were blended to form nanocomposite films. Catechol-bearing dopamine self-polymerized to chemically reduce and dip-coat GO nanosheets in order to produce poly(dopamine) (PDA)-functionalized GO (PGO). One of the most significant functional properties of packaging films was their ability to impede moisture migration in complex conditions. The smooth film of gelatin exhibited relatively inferior water vapor barrier performance and the greatest WVP value. Because of the lower hydrophilicity and inferior permeability of GO nanosheets, the value of WVP for Gel/GO film was less than that of the control film after GO addition. Additionally, the WVP value of Gel/CNF-PGO film decreased even more, suggesting its better water vapor barrier properties. The primary cause of this were the good dispersion of the CNF-PGO nanohybrids in the gelatin matrix and the increased cohesiveness of the polymer due to filling of the open volume holes in the matrix. The lowest value of WVP is for Gel/CNF-PGO/Ag. Because the synergistic interaction between nanoparticles coupled together could create a more tortuous route for water vapor molecules to penetrate the gelatin-based composite film, this great decline in permeability was mainly associated with the decline in water vapor diffusivity (Li et al. 2019).

10.4.3 Functional Properties

10.4.3.1 Antimicrobial and Antioxidant Properties

Edible films have antimicrobial agents included to inhibit or delay the growth of microorganisms in the food item, and thus its shelf life and safety are increased. Whether volatile or non-volatile, these chemicals are emitted through diffusion or evaporation after they have stayed on the surface of the food with high concentrations during storage (Calva-Estrada et al. 2019). The main natural active substances considered in scientific literature are bacteriocins of microbial origin (nisin, natamycin), organic acids (ascorbic, sorbic, propionic, and citric acid), and animal-derived enzymes (lysozyme, lactoferrin) (Benbettaïeb et al. 2019). Membrane disruption with inhibition of ATPase activity, necessary biomolecule leakage from the cell, disruption of the proton motive force, and enzyme inactivation are some potential mechanisms of the active ingredient in antimicrobial agents (Pisoschi et al. 2018).

SPI-based active nanocomposite films reinforced with cellulose nanocrystals (CNC), zinc oxide nanoparticles (ZnONP), and CNC/ZnONP blends (physical blending) were fabricated. Whereas pure SPI film (control) and CNC film had no antibacterial activity, the three ZnO-containing films inhibited both *S. aureus* and *E. coli* readily. Also, due to the nature of the bacterial cell wall, ZnONP appears to be more effective against gram-negative *E. coli* than against gram-positive *S. aureus*. Gram-positive bacteria possess thick peptidoglycan multilayer cell walls that make it difficult for ZnONP to penetrate through them. Gram-negative bacteria contain an outer membrane covering their thin layer of peptidoglycan. Nanoparticles can utilize this layer for adhesion and adsorption. The three films above were applied to pork packaging. In comparison to the exposed sample, viable bacterial counts in film-packed samples reduced to different extends three- and six-days post-storage (Xiao et al. 2020).

Cui et al. (2020) prepared an active zein film with pomegranate peel extract (PE) incorporated in chitosan nanoparticles (CSNPs) for food packaging. Due to the fact that L. monocytogenes is a major food-borne pathogen, it was selected for this research. The control pork sample contained 3.40 Log CFU/g of L. monocytogenes cells at the start of the storage period, and on the fourteenth day, that value had risen to 7.38 Log CFU/g. It showed that L. monocytogenes, the challenge food pathogen, was able to grow rapidly without pomegranate peel extract. When CSNPs/PE matrix were incorporated into the zein film, the growth of L. monocytogenes was prevented. Since PE consists of hydrolysable polyphenols that cause lysis of bacterial cell membranes, it is antimicrobial in nature.

An antioxidant refers to any substance that is able to prevent, retard, or delay the development of rancidity or other spoilage reactions initiated by oxidation reactions. Antioxidants are incorporated into food items to avoid lipid oxidation, resulting in toxic aldehydes and off flavours as well as odours, making the food undesirable for human consumption (Parreidt et al. 2018). There exist many natural and synthetic antioxidants that are utilized. Artificial antioxidants such as thioester compounds and organophosphates have been employed; however, their application is in doubt due to their possible toxicity and migration into food products (Gómez-Estaca et al. 2014). Phenolic compounds, which are active plant constituents, and their secondary metabolites are some of the natural antioxidants (Adilah et al. 2018). There are two mechanisms through which antioxidant packaging can operate: either by releasing antioxidants into food or by scavenging unwanted substances that may develop at a specific stage in the food oxidation process (Ganiari et al. 2017). For the purpose of prolonging the shelf life of the product by protecting foods against oxidative rancidity, spoilage, and discoloration, antioxidants are incorporated into edible film-forming preparations. Antioxidant film production for food packaging must be done with great attention to active ingredient selection. To improve uniform antioxidant distribution and, thereby, encourage the release of antioxidants on all food-contacting surfaces, the film and antioxidant compounds should have compatibility (Benbettaïeb et al. 2019).

Active packaging was designed to track pork spoilage and to store Actinidia arguta by using SPI/trehalose (TRE) and Oxalis triangularis extract (OTE). Antioxidant capacity of OTE was strong as ABTS, and DPPH scavenging activities were more than 90% for SPI-OTE and ST-OTE (SPI +TRE+OTE). The antioxidant activity of the film reduced after storage, heat treatment, and irradiation by UV and visible light. The reason given was that anthocyanins were susceptible to degradation and loss of activity through oxygen oxidation in air, heat, and light. When ST-OTE was stored at 45 days, heated at 90 °C, and treated with UV and visible light, the highest activity of DPPH and ABTS scavenging was shown. This was because ST-OTE possessed a more compact network structure compared to SPI-OTE since the SPI film structure became denser following the addition of TRE. This may restrict the reduction in anthocyanin activity induced by adverse environmental conditions and preserve anthocyanins to some extent. In addition, the antioxidant activity of the film after visible light exposure for 14 days was also greater than after 48 h of exposure to UV radiation, indicating that UV irradiation was the key factor affecting the stability of the active film (Jiang et al. 2024).

To prepare gelatin-based nanocomposite films with varying MNP concentrations, i.e., 0, 0.25, 0.5, and 1.0 weight percent, melanin nanoparticles (MNP) were isolated from squid ink. Antioxidant activity of gelatin-based nanocomposite films was determined using the DPPH radical and ABTS+ radical scavenging assay, which is often employed in active food packaging studies. The rate at which hydrogen atoms are stripped away from the phenyl group and the stability of the produced radicals decide the capacity of antioxidants to scavenge radicals. In ABTS+ and DPPH testing, neat gelatin film showed a value of 7.8 ± 1.9 and $5.7 \pm 1.3\%$, respectively, implying some radical scavenging activity. Nevertheless, the antioxidant activity of the gelatin/MNP nanocomposite films as determined by both analysis methods significantly improved following mixing with MNP. The enhanced antioxidant activity of the gelatin/MNP nanocomposite films was affected by the MNP concentration. The ABTS+ and DPPH scavenging activities of the gelatin/MNP nanocomposite films were enhanced to 85.5 ± 2.5 and $56.5 \pm 4.2\%$, respectively, upon the incorporation of 1 weight percent of MNP. The primary factor for increasing the antioxidant activity of gelatin/MNP nanocomposite films was melanin, an effective antioxidant (Shankar et al. 2019). Table 10.1 shows the various types of protein-based packaging film and their important physical and functional properties.

10.5 Application of Protein-Based Packaging in Food

Applications of edible films in food are dependent upon a number of variables including cost, accessibility, the mechanical, optical, functional, and barrier properties of the packaging material, structural resistance, and sensory acceptability. The material of the film structural matrix, the operational circumstances of the film production process, and the kind and concentration of active agents added to the packaging all affect these requirements. Before the film is created, these elements must to be carefully considered (Ganiari et al. 2017). Dairy, meat, fish, poultry, fruits, and vegetables all use edible films. Edible packaging provides a selective oxygen and

Table 10.1 The different properties of various types of protein-based packaging film

Polymer	Key properties	References
SPI	Tensile strength was evidently enhanced in the mechanical properties of SPIs with CNC-containing films over the control film. It was found that the reinforcing of CNC enhanced the stiffness of SPI-based films. The enhancement of tensile strength was probably caused by the filling effect of CNC and the extensive interfacial contacts between CNC and the film matrix. However, the incorporation of reinforcing chemicals led to a sharp reduction in elongation at break, indicating that the flexibility of the film had been reduced. The dense cross-linked network structure might have reduced the molecular motility of the molecules that made up the film, and this is why the value of elongation was less. The incorporation of reinforcements did not have much effect on the thickness of the film	Xiao et al. (2020)
Zein and chitosan	Various ratios of zein to chitosan mix were employed to prepare various films. Initially, the zein and chitosan solutions were blended in different volume ratios such as 1:1 (C/Z-1), 3:1 (C/Z-3), 5:1 (C/Z-5), and 1:1 (C/Z-0). In comparison to the C/Z-0 film, the water vapor permeability of the C/Z-1 film was significantly lower. The C/Z-1 film also showed the lowest WVP, which reflects better moisture barrier properties compared to other films. Similar to dicarboxylic acid, which is able to cross-link chitosan molecules, hydrophobic zein can also hinder water vapor from diffusing through the mix layer. The WVP of the SPI film was considerably reduced by incorporating the corn zein film layer	Zhang et al. (2019)
Zein (Z)	Graphene oxide (GO) nanoparticles were suspended in maize protein zein through the solvent casting technique to form graphene oxide (GO)-zein nanocomposite films. The results show that elongation at break decreased with increased GO content, tensile strength, and Young's modulus. Compared to zein control films, Z-GO films had higher tensile strength and Young's moduli. With a Young's modulus of 145 MPa, 300% greater than the zein control (35 MPa), the Z–3%GO nanocomposite film possessed the highest value. The Z 3%GO nanocomposite film possessed an 82% greater tensile strength than the zein control film. As the GO content of the films increased, percentage EAB decreased. The control zein films possessed the greatest elongation (65%), while the Z–3%GO nanocomposite films possessed the least elongation (5.8%)	Rouf et al. (2019)

(continued)

Table 10.1 (continued)

Polymer	Key properties	References
SPI	Licorice residue extract (LRE) was incorporated into SPI films to produce antioxidant films. DPPH and ABTS radical scavenging assays were utilized to determine the release of antioxidant activity components of SPI-based films into the two food simulants. Incorporation of licorice extract into SPI films improved their antioxidant activity, as evident from the significant improvement in the film's DPPH and ABTS radical scavenging activities with the increase in LRE concentration. ABTS RSA and DPPH RSA were improved from 46.5% to 69.9% and from 3.0% to 18.9%, respectively. The control film had minimal DPPH RSA and 12.2% ABTS radical scavenging activity. DPPH RSA of the films was improved from 10.5% to 57.0% and their ABTS radical scavenging activity from 25.6 to 82.9% upon increasing the concentration of LRE from 10 to 70 g kg 1	Han et al. (2018)
WPI and Furcellaran (FUR)	Biopolymer films of FUR/WPI, FUR/WPI in combination with yerba mate extract (YM), and FUR/WPI in combination with white tea extract (WT) were successfully employed to package fresh soft rennet-curd cheese. The water vapour transition rate (WVTR) is determined by the film's crystallinity, the tortuosity of the path, surface and structural imperfections, and the hydrophobic to hydrophilic component ratio. The value of WVTR decreased when YM extract was incorporated into the FUR/WPI films. WT incorporation has no noticeable impact on the WVTR of FUR/WPI films. Following the incorporation of yerba mate into the film, solubility, water content, and WVTR decrease were induced by the hydrophobicity of the extract and the intense hydrogen bond interaction between the polar components of yerba mate and the furcellaran-whey protein groups; films prepared with cassava starch and yerba mate presented the same findings	Pluta-Kubica et al. (2020)

(continued)

Table 10.1 (continued)

Polymer	Key properties	References
SPI	*Cedrus deodara* pine needle extract and cellulose nanocrystals were employed to produce SPI film. Soy protein-based films were evaluated for antioxidant activity through the DPPH and ABTS free radical-scavenging assays. The soy protein films showed weak DPPH and ABTS free radical scavenging activities. The incorporation of CNCs enhanced the two free radicals' scavenging activity. In addition, a significant concentration-dependent enhancement of the antioxidant activity was observed upon the incorporation of PNE (2.5–10%) into the films, which indicates that PNE significantly enhanced the antioxidant activity of soy protein-based films. The soy protein film with 10% PNE exhibited the highest antioxidant activity among the fabricated film samples	Yu et al. (2018)
Mung bean protein	Depending on the protein content, pomegranate peel, a bioactive compound, was incorporated into mung bean protein films in different concentrations (0, 2.5, 12.5, and 25% w/w). Through the determination of the inhibitory zones, the antibacterial activity of the films against both Gram-positive (*L. monocytogenes*) and Gram-negative (*E. coli*) bacteria was assessed. Pomegranate peel-fortified film samples exhibited greater inhibition of *L. monocytogenes* and *E. coli* than the control mung bean protein film, which was not effective on the microorganisms under investigation. The antibacterial activity increased with increasing concentration of pomegranate peel. The mung bean protein film containing 25% pomegranate peel had the greatest inhibitory zone. The antibacterial activity of pomegranate peel was attributed to its content of tannins and polyphenols, especially ellagic acid, which possesses numerous antibacterial activities like membrane perforation, microbial enzyme inhibition, and reduced membrane fluidity	Moghadam et al. (2020)

(continued)

Table 10.1 (continued)

Polymer	Key properties	References
WPI	Films prepared with 5% glycerol, 5% WPI, and tarragon essential oil—both without and with heat treatment—were investigated in the present study. Micropaths of a film network microstructure are linearly related to its WVP; the greater the WVP, the more permeable the movie is to water. In UNT films, WVP ranged from 6.7 to 8.7 g mm/m2 day kPa, whereas in HT films, it ranged from 7.3 to 9.6 g mm/m2 day kPa. Heat treatment of the film-forming solution significantly affected the WVP and accounted for 19.9%. Its value reflected greater permeability to water in the HT film compared to the UNT film. The WVP was also significantly affected by the incorporation of tarragon essential oil (27.8% contribution), which increased as increasing amounts were added. Consequently, the ability of the film to prevent moisture was impaired	Socaciu et al. (2020)
Gelatin	Gelatin films were made through a covalent interaction with oxidized guar gum (OGG) containing green tea extract. Compared to pure gelatin film (3.47 $\pm$ 0.1 MPa), the TS of the gelatin-GG film at 36.24 $\pm$ 0.5 MPa is greater. The rise can be attributed to hydrogen bonds, van der Waals forces, and electrostatic force of attraction interactions between GG and the chains of gelatin within the film's matrix. Following the addition of OGG to the gelatin solution, the films obtained exhibited a significant decrease in EAB (4.48% $\pm$ 0.6) and an increase in TS (78.99 $\pm$ 0.4 MPa), which confirmed that cross-linking between the gelatin and OGG had occurred. Cross-linking, which hardened the film structure, may be behind the greater reduction in the EAB and rise in the TS factors in the gelatin-OGG films, particularly in the films with higher OGG content	Maroufi et al. (2020)
Soy protein	A biodegradable soy protein-based film supplemented with lemon peel essential oil displayed inhibitory potential against certain strains of bacteria. In soy protein films, there was a tendency for increasing diameter of the inhibition zones as there was more concentration of lemon peel essential oil added; however, the film that lacked lemon peel essential oil could not exhibit inhibitory action. The most susceptible bacterial species within the strains used was *Staphylococcus aureus*. The inhibitory diameter was 22 mm using 0.1 mL essential oil as an inhibitory agent in soy protein film. The least inhibiting *E. Coli* O157: H7 inhibitory zone was 18 mm. Lemon peel essential oil with D limonene and other compounds can inhibit film activity	Al-Sahlany (2017)

water barrier, prolonging the shelf life of food products in contrast to conventional plastic packaging (Ribeiro et al. 2021).

Meat, seafood, and poultry products have high water levels and thus tend to be extremely perishable. As a result, these are preserved by edible films. Under transportation and storage, they become contaminated with the environment, and this changes their nutritional composition and quality and also increases the potential for foodborne diseases. Besides, it minimizes their acceptability by the customers (Kumar et al. 2022). Meat, fish, and derivatives are some of the perishable commodities whose proper handling would help in prolonging their shelf life and avoiding deterioration, especially due to the development of spoiling microbes. Some of the major inherent parameters governing microbial deterioration of meat are pH, moisture content, water activity (aw), oxidation/reduction potential, nutritional composition, antimicrobial constituents, and biological structures. Most bacteria are reported to grow best at a pH of 7.0. The pH level of fresh meat is neutral, which is most suitable for most microorganisms. Similarly, aw, which is greater than 0.9 in meat, promotes the growth of microorganisms, especially pathogenic bacteria that do not grow below aw 0.91 (Putnik et al. 2019). The quality of meat, fish, and processed products can also be greatly affected by extrinsic factors, which are not affected by substrate characteristics. Storage temperature, surrounding air relative humidity, gases presence and concentration, and absence or presence of other microorganisms are some examples of relevant extrinsic factors (Fernandes et al. 2018). Through antioxidant activity, microbial growth suppression, and sensory property preservation, improved-functioning edible films have proven effective in halting quality degradation in fish, meat, and derivative products (Umaraw et al. 2020).

The antimicrobial films of SPI were created using EDTA (E), sodium lactate (NaL), nisin (N), and their optical combination (C). The TVB-N (total viable bacterial number) values of the meat samples exhibited an increasing trend as the storage time increased. The sample packaged with SPI antimicrobial film, regular wraps, and pure SPI film had TVB-N values of 14.23,14.20, and 10.83 mg/100 g on the third day. On the sixth day, however, samples wrapped with regular wraps and pure SPI film had values of 18.09 and 21.34 mg/100 g, respectively. The sample in the control group was still in second grade freshness on the ninth day (TVB-N value of 18.49 mg/100 g), while the sample wrapped in regular wraps was rotten with a TVB-N value of 21.15 mg/100 g. On the 12th day, the control group rotted, and its TVB-N value was 23.33 mg/100 g. As a result, SPI antibacterial film maintained its secondary freshness for 6 and 3 days longer than both pure SPI film and regular wraps (Liu et al. 2017).

The potential impact of WPI films on postponing the deterioration of minced lamb meat was investigated after they were prepared both alone and with added essential oils (WPI + EO) at varying concentrations. Fresh minced lamb meat's initial TVC value was 3.65 log CFU/g, indicating that the meat's quality was acceptable. On day 8 for WPI films, between days 8 and 11 for WPI + 0.5% EO, and on day 11 for WPI + 1% EO (ginger (GEO) and rosemary (REO) for both cases), the maximum allowable level for TVC was attained. WPI + 1% EO reaching the highest allowable level on day 11 suggests that the tested films may have antimicrobial properties. The TBARS values stayed low, suggesting a significant delay in the degree of oxidation of the meat samples, but the WPI film demonstrated a significant delay in the microbiological deterioration of minced lamb meat, particularly the films with 1% incorporated EO (both GEO and REO). The GEO and REO 1% films did not significantly differ, according to the results (Tsironi et al. 2022).

Dairy products like milk, yogurt, and cheese are a necessary component of a daily diet. The most varied and challenging category of dairy products is undoubtedly cheese. This is due to their primary components such as are lipids, proteins, minerals and vitamins. Cheese is made from a finely orchestrated series of consecutive and concomitant biochemical events that, if synchronized and balanced, lead to products with highly desirable aromas and flavours—but, when unbalanced, result in off-flavours and odours (Cruz-Diaz et al., 2019; Nunes et al., 2018). Even when cheese is produced and stored properly, gas exchange reactions triggered by microbes or enzyme activity result in biochemical events. Aside from biological activity, two important environmental factors that need to be taken into account during storage are exposure to light and oxygen. By creating efficient cheese packaging systems, scientists and technologists have investigated ways to get rid of or at least reduce some of these serious problems. Nowadays, it is theorized that packaging is essential to preserving and guaranteeing the quality of cheese (El-Sayed and Youssef 2024). Alternative biodegradable packaging options with negligible environmental effects are provided by edible films. Edible packaging materials can be used on foods to address issues with stable quality, market safety, nutritional value, and cost-effectiveness, even though they are not intended to completely replace synthetic packaging (Costa et al. 2018). Edible films have the following main benefits: (i) reduce the migration of moisture, lipids, and aromas between components and the outside; (ii) act as a gas barrier to regulate gas exchange between food and the surrounding atmosphere; (iii) limit the exchange of volatile compounds; and (iv) shield foods from physical damage brought on by pressure, vibrations, impact, and other mechanical factors, or vulnerable ingredients from oxidation—thereby increasing food quality and shelf life during storage (Bagheripoor et al. 2018).

Sorbitol-amended whey isolate-based film was treated with 1.5% (w/v) thyme and clove essential oils. Classic semi-hard Kashar cheese was coated with these films. The weight loss of Kashar cheese was positively impacted by the use of WPIOF (thyme-fortified film) and WPISF (clove-fortified film), which were produced by adding 1.5% (v/v) thyme and clove essential oils to sorbitol and WPI-based film. The application of a film made by sorbitol to WPI was found to reduce weight loss over a 60-day storage period, inhibit rind formation, and lessen moisture transition. Additionally, the cheese samples' fat ratio slightly increased after being coated with WPI-based essential oil-fortified films; however, this was determined to be statistically not significant. According to the study, the essential oils used had an antimicrobial effect; the samples with WPIOF film exhibited a higher level of this activity than the samples with WPISF film. During storage, the noncoated samples' microorganism counts rose. Therefore, it is suggested that extending the shelf life of kashar cheese may be possible with the WPI-based edible films made by adding 1.5% (v/v) thyme and clove essential oils (Kavas et al. 2015).

Edible films made from solutions based on whey protein concentrate were treated with varying concentrations of spirulina (sp.) (0.5, 1; 2, and 4% w/w). Films containing 2% sp. were chosen to be used as "kefalotyri" cheese surface packaging. On the surface of the cheeses without films, the deterioration caused by yeasts was visible from Day 30. Cheese samples covered with films containing 2% sp. did not exhibit any signs of spoiling bacteria until the end of the storage days. The broad range of bioactive phenolic compounds, including chlorogenic acid, and other compounds present in sp., including carotenoids, C-phycocyanin, and chlorophyll (a and b), may work in concert to produce the antimicrobial activity of sp. (Kontogianni et al. 2023).

Fruits and vegetables form, necessary part of the human diet due to being a storehouse of vitamins, vital minerals, bio-flavonoids, antioxidants, dietary fiber, and flavourings. They do, however, have a short postharvest life, due to being still living tissues at the time when consumed and subject to physiological and biochemical alterations that might have pathological or physical causes leading to significant financial losses. Transpiration makes the fruits and vegetables shed some weight when in post-harvest handling and storage processes that change the texture of these commodities and give the surface some shrinkage to decrease their shelf life (Raghav et al. 2016; Sapper et al. 2018). Fruit and vegetables have high respiration rate and excess free moisture thus being considered as perishable. Lowering respiration often helps edible films serve as a protective layer around the surface of fruits and vegetables, delaying spoilage (Dhaka et al. 2018). Since it is a thin protective layer, edible films help in extending food shelf life through phases of post-harvest, processing, transportation, storage, and consumption, halting dehydration, deterioration, and spoiling and preserving the colour, freshness, flavour, and nutrients of food. Further, edible films do not reduce or alter the nutritional value of fruits and vegetables (Liyanapathiranage et al. 2023).

Sajimon et al. (2023) develop a packing film with oregano essential oil and assess the film's physicochemical, mechanical, antibacterial, and antioxidant properties in relation to grape packaging. Following the addition of an essential oil nanoemulsion to a WPC-glycerol film-forming solution, the films were created using the casting method. The effects of various concentrations of 1, 2, 3, and 4% (w/w) of oregano essential oil (OEO) in WPC edible films were investigated. Grapes packed in F3 (WPC film with 3% OEO) and grapes packed in commercially available LDPE (L) packaging, were the two conditions under which samples were prepared. For the purpose of the shelf-life analysis, both sets of grapes were stored in a refrigerator for 21 days, or three weeks. For the first five days, there was no discernible difference in the appearance of the L and F3-wrapped grapes; however, after that, the former's color began to turn brown while the latter's stayed the same. F3-Grapes nearly kept their original characteristics. Both samples consistently lost weight during storage, regardless of the package. Even after ten days of storage, the grapes in WPC packaging did not lose any weight, but the grapes in LDPE did lose a small amount. Thus, when compared to commercially available pack aging material, the biodegradable film made from WPC is found to minimize the weight loss of fresh produce.

Strawberry's short shelf life and high perishability are mostly caused by microbial and physiological deterioration. Zein (Z) nanofiber encapsulated with thyme essential oil (TEO), a naturally occurring antioxidant and antimicrobial agent, may be electrospun to test the nanofiber's capacity to preserve strawberry quality. Zein nanofiber film and zein loaded with TEO nanofiber film (Z/TEO) were the names of the test groups. The findings demonstrated that the Z/TEO group's growth rate of bacteria, fungi, and yeast on strawberries was considerably impacted. When compared to the other groups, the Z/TEO group had the lowest total microbial load. The microbial load was probably reduced because the thymol was encapsulated in zein nanofiber, which released gradually into the fruit's storage container. It shows how TEO's antimicrobial qualities in active packaging affect things over time. The overall quality of the treated strawberries' appearance after 15 days of storage at 4 °C in various packages, such as control, contain zein, and Z/TEO nanofiber. On the

Table 10.2 Application protein-based packaging in preservation of fresh and processed food

Polymer	Food system	Remark	References
WPI and starch	Salami	WPI and WPI-starch films contained ethanolic fractions of rambutan peel extract (RPE) and cinnamon oil (CO). After ten days of storage, the salami's TVC (total viable count) was determined. WPI5 non-active films showed the highest TVC of up to 5.1 log cfu·g^{-1}, but WPI5 (starch ratios, specifically 5:0) active films offered the lowest microbial growth during storage. Active films prevented the growth of microorganisms since RPE and CO included various polyphenols and essential oils that improved their antibacterial qualities. As the amount of starch in WPI blend films grew, so did the microbial growth. The presence of more starch enhanced the release of polyphenols in aqueous environments	Chollakup et al. (2020)
SPI	Pork	Zinc oxide nanoparticles (ZnONP), cellulose nanocrystals (CNC), physical mixing of CNC/ZnONP, or CNC@ZnONP nanohybrids were utilized to improve SPI-based active nanocomposite films. Uncovered pork that was not covered with the ZnO-containing films formed evident bacterial colonies on the surface after storage for six days at 4 °C. Microbial growth and nutrient breakdown could be responsible for this. The viable counts of bacteria from film-packed samples were lowered to different extents in comparison to the exposed sample upon storage for three and six days, especially in the case of ZnO-containing film-packed samples	Xiao et al. (2020)

(continued)

Table 10.2 (continued)

Polymer	Food system	Remark	References
WPI and chitosan	Turkey	Whey and chitosan protein edible films, fortified with quince or cranberry juice, have been used on freshly cut turkey meat. From days one (24 h) to three (72 h) of storage at 5 °C, the population of bacteria in turkey meat covered with WPI/CHIT films and treated with cranberry juice quickly fell from 2.76 log cfu/g on day one to 2.23 log cfu/g. This shows inhibition of bacteria. By the sixth day, the number had increased somewhat to 3.04 log cfu/g. Almost the same results were found in turkey flesh samples coated with WPI/CHIT films and fortified with quince juice. Yet, on the sixth day (144 h) of storage, the overall bacterial count in the control samples (uncoated) rose continuously from 5.57 log cfu/g initially to 8.23 log cfu/g	Brink et al. (2019)
Zein	Iraninan white cheese	In order to enhance the shelf life of Iranian white cheese during storage, the morphology of electrospun zein nanofibers was altered to contain clove essential oil (CEO) at concentrations of 5, 7.5, and 10% w/w. The EO antibacterial activity was evaluated against gram-positive (*S. aureus* and *L. monocytogenes*) and gram-negative (*E. coli* O157: H7) bacteria; the gram-positive bacteria responded more positively to this EO. It was found that the samples of cheese were adequately shielded by the EO's long-lasting release throughout a storage period of 45 days at 4 °C when Iranian white cheese was stored in brine using zein films embedded with CEO. 7.5% CEO-treated samples had improved sensory characteristics and greater acceptance compared with the control sample and other samples. Electrospun zein films have the ability to form antimicrobial structures with appropriate bioactive properties as a result of their sustained-release properties, which may be employed as a bio-preservative in perishable food items	Tayebi-Moghaddam et al. (2021)

(continued)

Table 10.2 (continued)

Polymer	Food system	Remark	References
WPI	Kasar cheese	Kasar slices of cheese were coated with multiple layers of WPI films containing nisin (NIS), natamycin (NAT), garlic oil (GEO), and oregano oil (OEO). Although NIS-added WPI films exhibited a more intense bactericidal effect towards L. monocytogenes during Kasar cheese preservation, OEO-added WPI films exhibited greater antimicrobial effect against *E. coli* O157: H7. In addition, Penicillium spp. inhibition. In Kasar cheese samples was increased by NAT addition of WPI films. Although yeast and mold count in Kasar cheese samples that were uninoculated without film and Kasar cheese layered with NIS-added WPI film were comparable, NAT-added WPI films had higher inhibition levels against the naturally occurring mold and yeast in Kasar cheese samples that were not inoculated	Seydim et al. (2020)
Zein	Himalayan cheese	Zein films with various amounts of pomegranate peel extract (PPE) (0, 25, 50, and 75 mg/ml of film forming solution) were formulated as a novel package material for Himalayan cheese (kalari). The inhibitive ability of PPE-loaded films to prevent lipid and protein oxidation during storage of kalari cheese in them was assessed through MDA and carbonyl content assays. Throughout every storage period, the oxidation products of cheese stored in 25 mg PPE, 50 mg PPE, and 75 mg PPE films were less than the control films. Nevertheless, it was demonstrated that as PPE concentration in the films increased, the oxidation products decreased in all days of storage. During a period of 30 days, the cheese stored in film with highest PPE content recorded the least oxidation values when compared to C0 (Control film), C1 (25 mg PPE), and C2 (50 mg PPE)	Mushtaq et al. (2018)

(continued)

Table 10.2 (continued)

Polymer	Food system	Remark	References
WPC	Soft cheese	WPC and rosemary and sage aqueous extracts were employed for developing an active food packaging material. The application of WPC-based films with sage and rosemary infusion as a packaging material for soft cheese was successful in the preservation of the cheese from spoilage and harmful microorganisms. Microbiological developments of soft cheese covered in plant extract edible films were studied after storage at 4 °C for 60 days. All cheese samples lacked any deteriorating symptom or pathogenic microorganisms up to seventeenth-day storage. In addition to being eaten with cheese to save cheese wastage, it may also be applied in the process of regulating cheese nutrition content, flavor, and appearance. In the case of whey protein films, it can also boost protein consumption and worth	Kontogianni et al. (2022)
SPI and pectin	Strawberry	SPI and apple pectin (AP) were utilized to develop antimicrobial films packed with pomegranate punicalagin (PP). Punicalagin was produced using pomegranate peel, while apple pomace waste was employed in the production of AP. Starting from the left, strawberry groups were covered in various films during storage: control, SPI, SPI/AP, SPI/AP-0.1PP, SPI/AP-0.5PP, and SPI/AP-1.0PP. Strawberries coated with SPI and SPI/AP films revealed clear evidence of mildew and rotting after three days of storage. Strawberries covered by SPI/AP-PP films remained unaffected, however, and maintained their bright color. Strawberries treated with SPI/AP-0.1PP and SPI/AP-1.0PP films started deteriorating and revealing mildew after five days of storage. Strawberries, however, that were covered with the SPI/AP-0.5PP film were still brightly colored and did not start showing signs of spoilage until day seven	Guo et al. (2024)

(continued)

Table 10.2 (continued)

Polymer	Food system	Remark	References
Casein and chitosan	Cherry tomato	Casein, chitosan, and oregano essential oil (EO) edible films were prepared. Cherry tomatoes stored at 4 °C with and without essential oil research revealed that the bioactive activity of OEO preserves fruit quality. Compared to control fruits, the films can extend changes in titratable acidity (TA), percentage shrinkage, and weight loss during storage. Cold-temperature shelf life of cherry tomatoes was extended to thirty-two days. At different concentrations of OEO, the chitosan/casein edible films' antibacterial activity was evaluated against *Staphylococcus aureus* and *Escherichia coli*. For 0.5%, 1%, and 1.5% OEO-containing edible films, respectively, the *E. coli* exhibited an inhibitory zone with diameters of 14.7, 22.0, and 22.5 mm	Roshandel-hesari et al. (2022)
WPC and tamarind starch	Tomato	Thyme essential oil (TEO) nano-emulsion (5, 10, 15, and 20%) was incorporated into edible composite films of tamarind starch (TS) and WPC in specific ratios. The respiration rate of tomatoes was reduced by using edible films of nano-composite materials containing a 20% thyme essential oil nano-emulsion, reducing the ripening process. Because thymol and carvacrol within TEO nano-emulsion act as an anti-oxidant and anti-microbial agent, TEO was appropriately dissolved in the polymer film matrix and slowly released to produce an antimicrobial system and initiate an antimicrobial mechanism at the correct time and place. The film quality thus enhanced after post-harvest storage at room temperature apart from sustaining quality characteristics	Ghoshal and Shivani (2022)

ninth day of storage, the control group's strawberries displayed deterioration and a general decline in appearance, making them unfit for consumption (data not shown). However, since no noticeable deterioration was noticed at the 12th or 15th day of cold storage, respectively, the fruit packed in the zein nanofiber and Z/TEO nanofiber groups had a marketing-acceptable overall appearance (Ansarifar and Moradinezhad 2021). Table 10.2 showing the application of protein-based films in meat, seafood, poultry, dairy, fruits and vegetables. On the other hand, application of protein-based packaging in fruits and meat preservation is presented in Fig. 10.2.

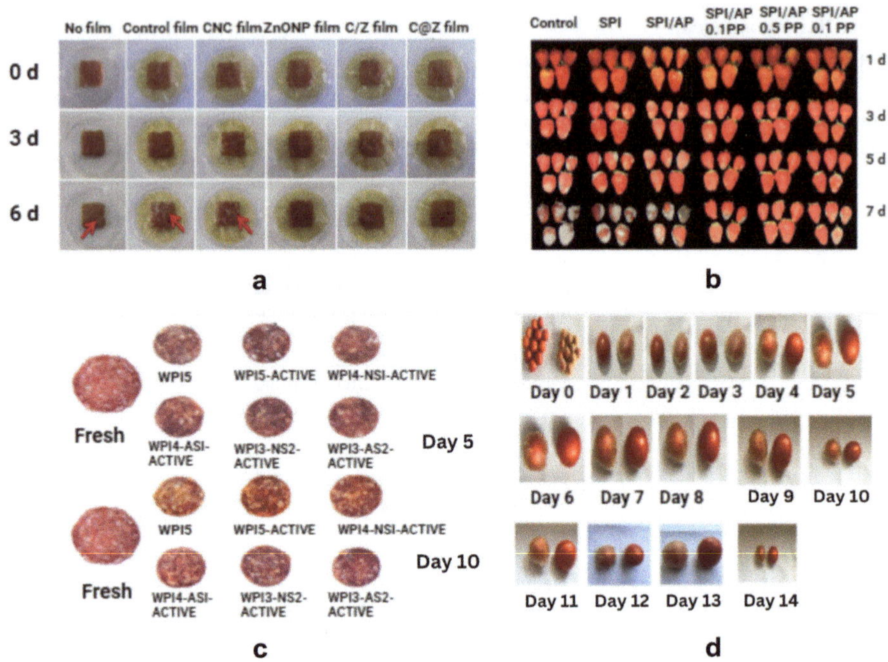

Fig. 10.2 Various types protein based active packaging film for food preservation application. **a** Active nanocomposite films based on soy protein isolate (SPI) reinforced with zinc oxide nanoparticles (ZnONP), cellulose nanocrystals (CNC), CNC/ZnONP mixture, or CNC@ZnONP nanohybrids. After six days of storage at 4 °C, visible bacterial colonies developed on the surface of the pork that was not covered with the ZnO-containing films (Xiao et al. 2020), **b** Development of antimicrobial films loaded with punicalagin using apple pectin (AP) and soy protein isolate (SPI). Five groups were submerged in various film solutions SPI, SPI/AP, SPI/AP-0.1PP (0.1%,), SPI/AP-0.5PP (0.5%,), and SPI/AP-1.0PP (1%,) (Guo et al., 2024), **c** WPI and WPI-starch films incorporated with rambutan peel extract (RPE) and cinnamon oil (CO). TVC in salami was measured after 10 days of storage. WPI5 (Starch-5:0) active films provided the lowest microbial growth during storage, whereas WPI5 non-active films displayed the highest TVC of up to 5.1 log cfu g^{-1} salami at 10 days (Chollakup et al., 2020), **d** It shows the impact of whey protein concentrate film with thyme essential oil (TEO) nano-emulsion (20%) incorporated edible composite films using tamarind starch (TS) on tomato at a storage period of 14 days (Ghoshal and Shivani 2022)

10.6 Summary

This chapter discusses the sources, development, characteristics, and uses of protein-based films as eco-friendly substitutes for synthetic packaging materials. Each of the natural protein sources employed to produce these films—soy, whey, zein, gelatin, and casein—has unique characteristics. Soy protein, which is derived from soybeans, is renowned for its excellent film-forming properties and biodegradability. Whereas whey protein, a dairy residue, is valued for its clarity and flexibility, they are abundant and affordable, and are often utilized for their outstanding mechanical toughness and oxygen barrier properties. Solvent casting or extrusion are most commonly used to produce these films, wherein proteins are dissolved in a solvent and then molded into thin layers. For the improvement of the properties of the film, crosslinkers, plasticizers, and bioactive agents serve as additives. The additives and the protein used determine the performance of protein-based films. These films are ideally used in food packaging because they have good barrier properties towards odour molecules, water, and oxygen. They also possess good mechanical properties such as elongation at break and tensile strength. Since oxygen and lipid barrier is crucial while packaging fresh foods such as fruit, cheeses, and meats, these films are particularly well-suited for this purpose. To further improve shelf life, they are also being researched for active packaging systems with antioxidant or antibacterial agents. Research on protein-based films has gained momentum with growing interest in environment-friendly plastic substitutes, and hence they are an interesting way to reduce pollution in the environment without compromising food quality and safety.

Declaration Authors used generative AI tools for the purpose of improving readability and language. The final content has been reviewed and verified by the authors after using these tools, and they take full responsibility for its accuracy and integrity.

References

Adilah AN, Jamilah B, Noranizan MA, Hanani ZN (2018) Utilization of mango peel extracts on the biodegradable films for active packaging

Al-Sahlany STG (2017) Production of biodegradable film from soy protein and essential oil of lemon peel and use it as cheese preservative. Basrah J Agricul Sci 30:27–35. https://doi.org/10.37077/25200860.2017.40

Andrade MA, Barbosa CH, Souza VGL et al (2021) Novel active food packaging films based on whey protein incorporated with seaweed extract: development, characterization, and application in fresh poultry meat. Coatings 11:229. https://doi.org/10.3390/COATINGS11020229

Ansarifar E, Moradinezhad F (2021) Preservation of strawberry fruit quality via the use of active packaging with encapsulated thyme essential oil in zein nanofiber film. Int J Food Sci Technol 56:4239–4247. https://doi.org/10.1111/IJFS.15130

Arkoun M, Daigle F, Holley RA et al (2018) Chitosan-based nanofibers as bioactive meat packaging materials. Packag Technol Sci 31:185–195. https://doi.org/10.1002/PTS.2366

Assad I, Bhat SU, Gani A, Shah A (2020) Protein based packaging of plant origin: fabrication, properties, recent advances and future perspectives. Int J Biol Macromol 164:707–716. https://doi.org/10.1016/J.IJBIOMAC.2020.07.140

Azeredo HMC, Waldron KW (2016) Crosslinking in polysaccharide and protein films and coatings for food contact—a review. Trends Food Sci Technol 52:109–122. https://doi.org/10.1016/J.TIFS.2016.04.008

Bagheripoor N, Khoshgozaran-Abras S, Sohrabvandi S et al (2018) Application of active edible coatings to improve the shelf-life of cheese. Food Sci Technol Res 24:949–962. https://doi.org/10.3136/FSTR.24.949

Bahrami A, Delshadi R, Assadpour E et al (2020) Antimicrobial-loaded nanocarriers for food packaging applications. Adv Colloid Interface Sci 278:102140. https://doi.org/10.1016/J.CIS.2020.102140

Basiak E, Lenart A, Debeaufort F (2017) Effect of starch type on the physico-chemical properties of edible films. Int J Biol Macromol 98:348–356. https://doi.org/10.1016/J.IJBIOMAC.2017.01.122

Benbettaïeb N, Debeaufort F, Karbowiak T (2019) Bioactive edible films for food applications: mechanisms of antimicrobial and antioxidant activity. Crit Rev Food Sci Nutr 59:3431–3455. https://doi.org/10.1080/10408398.2018.1494132

Brink I, Šipailienė A, Leskauskaitė D (2019) Antimicrobial properties of chitosan and whey protein films applied on fresh cut turkey pieces. Int J Biol Macromol 130:810–817. https://doi.org/10.1016/J.IJBIOMAC.2019.03.021

Calderón-Castro A, Vega-García MO, de Jesús Z-M et al (2018) Effect of extrusion process on the functional properties of high amylose corn starch edible films and its application in mango (Mangifera indica L.) cv. Tommy Atkins. J Food Sci Technol 55:905–914. https://doi.org/10.1007/S13197-017-2997-6

Calva-Estrada SJ, Jiménez-Fernández M, Lugo-Cervantes E (2019) Protein-based films: advances in the development of biomaterials applicable to food packaging. Food Eng Rev 11:78–92. https://doi.org/10.1007/S12393-019-09189-W

Chen H, Wang J, Cheng Y, et al (2019) Application of protein-based films and coatings for food packaging: a review. Polymers 11:2039. https://doi.org/10.3390/POLYM11122039

Chevalier E, Chaabani A, Assezat G et al (2018) Casein/wax blend extrusion for production of edible films as carriers of potassium sorbate—a comparative study of waxes and potassium sorbate effect. Food Packag Shelf Life 16:41–50. https://doi.org/10.1016/J.FPSL.2018.01.005

Chollakup R, Pongburoos S, Boonsong W et al (2020) Antioxidant and antibacterial activities of cassava starch and whey protein blend films containing rambutan peel extract and cinnamon oil for active packaging. LWT 130:109573. https://doi.org/10.1016/J.LWT.2020.109573

Cinelli P, Schmid M, Bugnicourt E et al (2014) Whey protein layer applied on biodegradable packaging film to improve barrier properties while maintaining biodegradability. Polym Degrad Stab 108:151–157. https://doi.org/10.1016/J.POLYMDEGRADSTAB.2014.07.007

Costa MJ, Maciel LC, Teixeira JA et al (2018) Use of edible films and coatings in cheese preservation: opportunities and challenges. Food Res Int 107:84–92. https://doi.org/10.1016/J.FOODRES.2018.02.013

Cruz-Diaz K, Cobos Á, Fernández-Valle ME et al (2019) Characterization of edible films from whey proteins treated with heat, ultrasounds and/or transglutaminase. Application in cheese slices packaging. Food Packag Shelf Life 22:100397. https://doi.org/10.1016/J.FPSL.2019.100397

Cui H, Surendhiran D, Li C, Lin L (2020) Biodegradable zein active film containing chitosan nanoparticle encapsulated with pomegranate peel extract for food packaging. Food Packag Shelf Life 24:100511. https://doi.org/10.1016/J.FPSL.2020.100511

Deng L, Kang X, Liu Y et al (2018) Characterization of gelatin/zein films fabricated by electrospinning vs solvent casting. Food Hydrocoll 74:324–332. https://doi.org/10.1016/J.FOODHYD.2017.08.023

Dhaka R, Upadhyay A (2018) Edible films and coatings: a brief overview. Pharma Innov J 7(7):331–333

El-Sayed SM, Youssef AM (2024) Emergence of cheese packaging by edible coatings for enhancing its shelf-life. J Food Measur Charact 18:5265–5280. https://doi.org/10.1007/S11694-024-02564-0/TABLES/2

Fernandes RPP, Trindade MA, Lorenzo JM, de Melo MP (2018) Assessment of the stability of sheep sausages with the addition of different concentrations of origanum vulgare extract during storage. Meat Sci 137:244–257. https://doi.org/10.1016/J.MEATSCI.2017.11.018

Figueroa-Enríquez CE, Rodríguez-Félix F, Ruiz-Cruz S et al (2024) Application of active packaging films for extending the shelf life of red meats: a review. Processes 12:2115. https://doi.org/10.3390/PR12102115

Ganiari S, Choulitoudi E, Oreopoulou V (2017) Edible and active films and coatings as carriers of natural antioxidants for lipid food. Trends Food Sci Technol 68:70–82. https://doi.org/10.1016/J.TIFS.2017.08.009

Ghoshal G, Shivani (2022) Thyme essential oil nano-emulsion/Tamarind starch/Whey protein concentrate novel edible films for tomato packaging. Food Control 138:108990. https://doi.org/10.1016/J.FOODCONT.2022.108990

Gómez-Estaca J, López-de-Dicastillo C, Hernández-Muñoz P, Catalá R, Gavara R (2014) Advances in antioxidant active food packaging. Trends Food Sci Technol
Guo H, Li A, Huang G et al (2024) Development of apple pectin/soy protein isolate-based edible films containing punicalagin for strawberry preservation. Int J Biol Macromol 273:133111. https://doi.org/10.1016/J.IJBIOMAC.2024.133111
Hajikhani M, Emam-Djomeh Z, Askari G (2020) Fabrication and characterization of gluten film reinforced by lycopene-loaded electrospun polylactic acid nano-fibers. Food Bioproc Tech 13:2217–2227. https://doi.org/10.1007/S11947-020-02561-3/FIGURES/7
Hamed I, Jakobsen AN, Lerfall J (2022) Sustainable edible packaging systems based on active compounds from food processing byproducts: a review. Compr Rev Food Sci Food Saf 21:198–226. https://doi.org/10.1111/1541-4337.12870
Han Y, Yu M, Wang L (2018) Preparation and characterization of antioxidant soy protein isolate films incorporating licorice residue extract. Food Hydrocoll 75:13–21. https://doi.org/10.1016/J.FOODHYD.2017.09.020
Hanani ZAN, Yee FC, Nor-Khaizura MAR (2019) Effect of pomegranate (Punica granatum L.) peel powder on the antioxidant and antimicrobial properties of fish gelatin films as active packaging. Food Hydrocoll 89:253–259. https://doi.org/10.1016/J.FOODHYD.2018.10.007
Hassan B, Chatha SAS, Hussain AI et al (2018) Recent advances on polysaccharides, lipids and protein based edible films and coatings: a review. Int J Biol Macromol 109:1095–1107. https://doi.org/10.1016/J.IJBIOMAC.2017.11.097
Huntrakul K, Yoksan R, Sane A, Harnkarnsujarit N (2020) Effects of pea protein on properties of cassava starch edible films produced by blown-film extrusion for oil packaging. Food Packag Shelf Life 24:100480. https://doi.org/10.1016/J.FPSL.2020.100480
Jafarzadeh S, Forough M, Amjadi S et al (2022) Plant protein-based nanocomposite films: a review on the used nanomaterials, characteristics, and food packaging applications. Crit Rev Food Sci Nutr 63:9667–9693. https://doi.org/10.1080/10408398.2022.2070721
Jamróz E, Kopel P, Juszczak L et al (2019) Development of furcellaran-gelatin films with Se-AgNPs as an active packaging system for extension of mini kiwi shelf life
Jiang L, Li M, Li Z et al (2024) Development, characterization and application of colorimetric active packaging based on soy protein isolate, trehalose and Oxalis triangularis extract. Ind Crops Prod 213:118456. https://doi.org/10.1016/J.INDCROP.2024.118456
Kabir E, Kaur R, Lee J et al (2020) Prospects of biopolymer technology as an alternative option for non-degradable plastics and sustainable management of plastic wastes. J Clean Prod 258:120536. https://doi.org/10.1016/J.JCLEPRO.2020.120536
Kavas G, Kavas N, Saygili D (2015) The effects of thyme and clove essential oil fortified edible films on the physical, chemical and microbiological characteristics of kashar cheese. J Food Qual 38:405–412. https://doi.org/10.1111/jfq.12157
Kocira A, Kozłowicz K, Panasiewicz K et al (2021) Polysaccharides as edible films and coatings: characteristics and influence on fruit and vegetable quality—a review. Agronomy 11:813. https://doi.org/10.3390/AGRONOMY11050813
Kontogianni VG, Kasapidou E, Mitlianga P et al (2022) Production, characteristics and application of whey protein films activated with rosemary and sage extract in preserving soft cheese. LWT 155:112996. https://doi.org/10.1016/J.LWT.2021.112996
Kontogianni VG, Kosma I, Mataragas M et al (2023) Innovative intelligent cheese packaging with whey protein-based edible films containing spirulina. Sustainability (Switzerland) 15:13909. https://doi.org/10.3390/SU151813909/S1
Kumar L, Ramakanth D, Akhila K, Gaikwad KK (2021) Edible films and coatings for food packaging applications: a review. Environ Chem Lett 20:1 20:875–900. https://doi.org/10.1007/S10311-021-01339-Z
Kumar L, Ramakanth D, Akhila K, Gaikwad KK (2022) Edible films and coatings for food packaging applications: a review. Environ Chem Lett 20:875–900. https://doi.org/10.1007/S10311-021-01339-Z

Li K, Jin S, Li J, Chen H (2019) Improvement in antibacterial and functional properties of mussel-inspired cellulose nanofibrils/gelatin nanocomposites incorporated with graphene oxide for active packaging. Ind Crops Prod 132:197–212. https://doi.org/10.1016/J.INDCROP.2019.02.011

Liu R, Cong X, Song Y et al (2018) Edible gum–phenolic–lipid incorporated gluten films for food packaging. J Food Sci 83:1622–1630. https://doi.org/10.1111/1750-3841.14151

Liu Y, Zhang H, Xu L et al (2017) Properties of soy protein isolate antimicrobial films and its application in preservation of meat. Emir J Food Agric 29:589–600. https://doi.org/10.9755/ejfa.2017-05-1070

Liyanapathiranage A, Dassanayake RS, Gamage A et al (2023) Recent developments in edible films and coatings for fruits and vegetables. Coatings 13:1177. https://doi.org/10.3390/coatings13071177

Maroufi LY, Ghorbani M, Tabibiazar M (2020) A gelatin-based film reinforced by covalent interaction with oxidized guar gum containing green tea extract as an active food packaging system. https://doi.org/10.1007/s11947-020-02509-7/Published

Milani JM, Tirgarian B (2020) An overview of edible protein-based packaging: main sources, advantages, drawbacks, recent progressions and food applications. J Packag Technol Res 4:103–115. https://doi.org/10.1007/S41783-020-00086-W

Miriam Malherbi N, Camila Scmitz A, Cristiani Grando R et al (2019) Corn starch and gelatin-based films added with guabiroba pulp for application in food packaging. 19:140–146. https://doi.org/10.1016/j.fpsl.2018.12.008

Moghadam M, Salami M, Mohammadian M et al (2020) Development of antioxidant edible films based on mung bean protein enriched with pomegranate peel. Food Hydrocoll 104:105735. https://doi.org/10.1016/J.FOODHYD.2020.105735

Mohamed SAA, El-Sakhawy M, El-Sakhawy MAM (2020) Polysaccharides, protein and lipid-based natural edible films in food packaging: a review. Carbohydr Polym 238:116178. https://doi.org/10.1016/J.CARBPOL.2020.116178

Mohamed SAA, El-Sakhawy M, Nashy ESHA, Othman AM (2019) Novel natural composite films as packaging materials with enhanced properties. Int J Biol Macromol 136:774–784. https://doi.org/10.1016/J.IJBIOMAC.2019.06.130

Mushtaq M, Gani A, Gani A et al (2018) Use of pomegranate peel extract incorporated zein film with improved properties for prolonged shelf life of fresh Himalayan cheese (Kalari/kradi). Innov Food Sci Emerg Technol 48:25–32. https://doi.org/10.1016/J.IFSET.2018.04.020

Nunes EM, Silva AI, Vieira CB et al (2018) Edible packaging for dairy products. Edible Food Packag 413–429. https://doi.org/10.1201/B19468-14

Nur Amila Najwa IS, Mat Yusoff M, Nur Hanani ZA (2020) Potential of silver-kaolin in gelatin composite films as active food packaging materials. Food Packag Shelf Life 26:100564. https://doi.org/10.1016/J.FPSL.2020.100564

Nur Hanani ZA, Roos YH, Kerry JP (2014) Edible films and coatings—sources, properties and application. Food Feed Res 71:11–22. https://doi.org/10.1016/J.IJBIOMAC.2014.04.027

Parreidt TS, Müller K, Schmid M (2018) Alginate-based edible films and coatings for food packaging applications. Foods 7(10):170. https://doi.org/10.3390/foods7100170

Pisoschi AM, Pop A, Cimpeanu C, Turcuş V, Predoi G, Iordache F (2018) Nanoencapsulation techniques for compounds and products with antioxidant and antimicrobial activity—a critical view. Eur J Med Chem 157:1326–1345

Pluta-Kubica A, Jamróz E, Kawecka A et al (2020) Active edible furcellaran/whey protein films with yerba mate and white tea extracts: preparation, characterization and its application to fresh soft rennet-curd cheese. Int J Biol Macromol 155:1307–1316. https://doi.org/10.1016/J.IJBIOMAC.2019.11.102

Purewal SS, Kaur A, Bangar SP et al (2023) Protein-based films and coatings: an innovative approach. Coatings 14:32. https://doi.org/10.3390/COATINGS14010032

Putnik P, Gabrić D, Roohinejad S, Barba FJ, Granato D, Mallikarjunan K, Lorenzo JM, Kovačević DB (2019) An overview of organosulfur compounds from Allium spp.: from processing and preservation to evaluation of their bioavailability, antimicrobial, and anti-inflammatory properties. Food Chem 276:680–691

Raghav PK, Agarwal N, Saini M (2016) Edible coating of fruits and vegetables: a review. Education 1(2):188–204

Ribeiro AM, Estevinho BN, Rocha F (2021) Preparation and incorporation of functional ingredients in edible films and coatings. Food Bioproc Tech 14:209–231. https://doi.org/10.1007/S11947-020-02528-4/TABLES/2

Rocca-Smith JR, Marcuzzo E, Karbowiak T et al (2016) Effect of lipid incorporation on functional properties of wheat gluten based edible films. J Cereal Sci 69:275–282. https://doi.org/10.1016/J.JCS.2016.04.001

Rojas-Lema S, Nilsson K, Trifol J et al (2021) Faba bean protein films reinforced with cellulose nanocrystals as edible food packaging material. Food Hydrocoll 121:107019. https://doi.org/10.1016/J.FOODHYD.2021.107019

Roshandel-hesari N, Mokaber-Esfahani M, Taleghani A, Akbari R (2022) Investigation of physicochemical properties, antimicrobial and antioxidant activity of edible films based on chitosan/casein containing Origanum vulgare L. essential oil and its effect on quality maintenance of cherry tomato. Food Chem 396:133650. https://doi.org/10.1016/J.FOODCHEM.2022.133650

Rouf TB, Schmidt G, Cakmak M, Kokini JL (2019) Design and mechanistic understanding of graphene oxide reinforced zein nanocomposites with improved mechanical, barrier and thermal properties. J Mater Sci 54:12533–12552. https://doi.org/10.1007/S10853-019-03817-W/TABLES/1

Roy S, Malik B, Chawla R et al (2024) Biocompatible film based on protein/polysaccharides combination for food packaging applications: a comprehensive review. Int J Biol Macromol

Said NS, Sarbon NM, Said NS, Sarbon NM (2019) Protein-based active film as antimicrobial food packaging: a review. Active Antimicrob Food Packag. https://doi.org/10.5772/INTECHOPEN.80774

Sajimon A, Edakkadan AS, Subhash AJ, Ramya M (2023) Incorporating oregano (Origanum vulgare L.) Essential oil onto whey protein concentrate based edible film towards sustainable active packaging. J Food Sci Technol 60:2408–2422. https://doi.org/10.1007/S13197-023-05763-7/FIGURES/4

Sapper M, Coatings AC (2018) Starch-based coatings for preservation of fruits and vegetables. Coatings. https://doi.org/10.3390/coatings8050152

Scartazzini L, Tosati JV, Cortez DHC et al (2019) Gelatin edible coatings with mint essential oil (Mentha arvensis): film characterization and antifungal properties. J Food Sci Technol 56:4045–4056. https://doi.org/10.1007/S13197-019-03873-9/TABLES/3

Seydim AC, Sarikus-Tutal G, Sogut E (2020) Effect of whey protein edible films containing plant essential oils on microbial inactivation of sliced Kasar cheese. Food Packag Shelf Life 26. https://doi.org/10.1016/J.FPSL.2020.100567

Shah YA, Bhatia S, Al-Harrasi A et al (2023) Mechanical properties of protein-based food packaging materials. Polymers 15:1724. https://doi.org/10.3390/POLYM15071724

Shankar S, Wang LF, Rhim JW (2019) Effect of melanin nanoparticles on the mechanical, water vapor barrier, and antioxidant properties of gelatin-based films for food packaging application. Food Packag Shelf Life 21:100363. https://doi.org/10.1016/J.FPSL.2019.100363

Sid S, Mor RS, Kishore A, Sharanagat VS (2021) Bio-sourced polymers as alternatives to conventional food packaging materials: a review. Trends Food Sci Technol 115:87–104. https://doi.org/10.1016/J.TIFS.2021.06.026

Socaciu MI, Fogarasi M, Semeniuc CA et al (2020) Formulation and characterization of antimicrobial edible films based on whey protein isolate and tarragon essential oil. Polymers 12:1748. https://doi.org/10.3390/POLYM12081748

Sophia A, Malar Retna A (2018) Stearic acid modified casein based nanocomposites with improved mechanical and thermal properties. Mater Today Proc 5:6247–6257. https://doi.org/10.1016/J.MATPR.2017.12.234

Sothornvit R, Pitak N (2007) Oxygen permeability and mechanical properties of banana films. Food Res Int 40(3):365–370

Suhag R, Kumar N, Petkoska AT, Upadhyay A (2020) Film formation and deposition methods of edible coating on food products: a review. Food Res Int 136:109582

Tayebi-Moghaddam S, Khatibi R, Taklavi S et al (2021) Sustained-release modeling of clove essential oil in brine to improve the shelf life of Iranian white cheese by bioactive electrospun zein. Int J Food Microbiol 355:109337. https://doi.org/10.1016/J.IJFOODMICRO.2021.109337

Tian H, Guo G, Fu X et al (2018) Fabrication, properties and applications of soy-protein-based materials: a review. Int J Biol Macromol 120:475–490. https://doi.org/10.1016/J.IJBIOMAC.2018.08.110

Trajkovska Petkoska A, Daniloski D, D'Cunha NM et al (2021) Edible packaging: sustainable solutions and novel trends in food packaging. Food Res Int 140:109981. https://doi.org/10.1016/J.FOODRES.2020.109981

Tsironi M, Kosma IS, Badeka AV (2022) The effect of whey protein films with ginger and rosemary essential oils on microbiological quality and physicochemical properties of minced lamb meat. Sustainability (Switzerland) 14:3434. https://doi.org/10.3390/SU14063434/S1

Tulamandi S, Rangarajan V, Rizvi SSH et al (2016) A biodegradable and edible packaging film based on papaya puree, gelatin, and defatted soy protein. Food Packag Shelf Life 10:60–71. https://doi.org/10.1016/J.FPSL.2016.10.007

Umaraw P, Munekata P, Verma A et al (2020) Edible films/coating with tailored properties for active packaging of meat, fish and derived products. Trends Food Sci Technol

Wittaya T, Wittaya T (2012) Protein-based edible films: characteristics and improvement of properties. Struct Funct Food Eng. https://doi.org/10.5772/48167

Xiao Y, Liu Y, Kang S et al (2020) Development and evaluation of soy protein isolate-based antibacterial nanocomposite films containing cellulose nanocrystals and zinc oxide nanoparticles. Food Hydrocoll 106:105898. https://doi.org/10.1016/J.FOODHYD.2020.105898

Yu Z, Sun L, Wang W et al (2018) Soy protein-based films incorporated with cellulose nanocrystals and pine needle extract for active packaging. Ind Crops Prod 112:412–419. https://doi.org/10.1016/J.INDCROP.2017.12.031

Zhang L, Liu Z, Wang X et al (2019) The properties of chitosan/zein blend film and effect of film on quality of mushroom (Agaricus bisporus). Postharvest Biol Technol 155:47–56. https://doi.org/10.1016/J.POSTHARVBIO.2019.05.013

Zhang W, Liu C, Qu M et al (2020) Characterization of a recombinant zein-degrading protease from Zea mays by Pichia pastoris and its effects on enzymatic hydrolysis of corn starch. Int J Biol Macromol 164:3287–3293. https://doi.org/10.1016/J.IJBIOMAC.2020.08.237

Zubair M, Ullah A (2020) Recent advances in protein derived bionanocomposites for food packaging applications. Crit Rev Food Sci Nutr 60:406–434. https://doi.org/10.1080/10408398.2018.1534800

Chapter 11
Application of Natural Wax in Biopolymer Food Packaging Film

Yiqin Zhang and Wanli Zhang

Abstract Natural wax-incorporated biopolymer films have emerged as promising sustainable packaging materials due to their enhanced functional properties and environmental benefits. This review comprehensively analyzes the properties and the effects of various natural waxes (beeswax, carnauba wax, candelilla wax, etc.) on the performance of polysaccharide-, protein-, and polyester-based packaging films. The incorporation of natural waxes significantly improves hydrophobicity, water vapor barrier properties, mechanical strength, and bioactive functionalities of biopolymer films. Furthermore, we critically evaluate the shelf-life extension capabilities of these composite films for food packaging applications. Despite these advantages, current challenges include insufficient standardization in wax extraction methods, limited understanding of wax-polymer interactions, and technological barriers in developing intelligent packaging systems. Future research should focus on optimizing wax processing techniques, elucidating structure–function relationships, and integrating smart sensing functionalities. Addressing these challenges will facilitate the development of high-performance, multifunctional food packaging materials that meet both sustainability and technological requirements.

Keywords Natural wax · Food packaging film · Preservation · Applications

11.1 Introduction

Food preservation is influenced by environmental factors, physiological/biochemical processes, and various enzymatic activities (Chu et al. 2024). Food packaging effectively isolates food products from external environments while prolonging shelf

Y. Zhang
College of Food Science, Sichuan Agricultural University, Ya'an, PR China
e-mail: scyszyq@163.com

W. Zhang (✉)
College of Food Science and Engineering, Hainan University, Haikou, PR China
e-mail: zwl@hainanu.edu.cn

S. Roy and M. Shahid (eds.), *Biomaterials for Sustainable Food Packaging*,
Environmental Footprints and Eco-design of Products and Processes,
https://doi.org/10.1007/978-3-032-08027-1_11

life for various foods (Otoni et al. 2017). Petroleum-based plastics have become the predominant packaging material owing to their superior flexibility, mechanical strength, chemical stability, and durability (Genovese et al. 2016). However, conventional plastics pose significant environmental challenges due to their non-recyclable, non-renewable, and non-biodegradable nature, threatening both ecosystem integrity and human health (Bayram et al. 2021; Nair et al. 2023). In alignment with sustainable development goals, biopolymer-based packaging films have emerged as promising alternatives, offering superior film-forming capacity, optimized water vapor barrier, and enhanced mechanical properties. These bio-based materials can additionally function as delivery systems for active compounds, enabling antimicrobial, antioxidant, and intelligent indicator functionalities (Bayram et al. 2021; Debeaufort 2021).

Biopolymers (including polysaccharides, proteins, and lipids) have gained extensive application in food packaging film development owing to their capacity to form continuous matrices (Dirpan et al. 2023). In particular, polysaccharides can establish three-dimensional networks through glycosidic bonding, hydrogen bonding, van der Waals interactions, and electrostatic forces (Zhao et al. 2021). From this perspective, polysaccharide-based food packaging demonstrates notable advantages in food preservation. For example, in the investigation of postharvest preservation of "Ninghaibai" loquats, Wang et al. (2016) demonstrated that application of a 0.5% konjac glucomannan coating film significantly enhanced superoxide dismutase (SOD) and catalase (CAT) activity while effectively maintaining total soluble solids (TSS), titratable acidity (TA), and ascorbic acid (AsA) content during 21-day storage at 4 °C. Nevertheless, there are still some limitations in the mechanical strength and barrier properties of polysaccharide-based food packaging films (Pan et al. 2024). In contrast, protein-based packaging films, derived from complex amino acid polymers linked by peptide bonds, exhibit superior mechanical strength and excellent barrier properties against vapors, liquids, and gases, although their water resistance remains suboptimal (Pan et al. 2024). Ghidelli et al. (Hassan et al. 2018) demonstrated that composite films incorporating soy protein isolate (SPI) with 0.5–1% cysteine effectively mitigated the weight loss and enzymatic browning in fresh-cut eggplants, thereby extending their marketable shelf life. Lipids, as natural hydrophobic polymers, demonstrate excellent water barrier properties when utilized in food packaging materials, significantly enhancing moisture resistance (Senturk Parreidt et al. 2018; Morillon et al. 2002). In addition, as the most representative substrate among lipids, waxes, composed of long-chain fatty alcohols and alkanes, represent highly hydrophobic compounds. Their incorporation into biopolymer films can effectively reduce the water vapor permeability (WVP) and the water solubility (Morillon et al. 2002; Rodrigues et al. 2014; Kowalczyk and Baraniak 2014; Navarro-Tarazaga et al. 2011). A report by Moyo et al. (2020) demonstrated that postharvest treatment combining fungicide application with 18% polyethylene-based wax coating film on citrus fruits (lemons and oranges) exhibiting black spot disease lesions completely suppressed spore release and significantly reduced latent infection under refrigerated storage conditions.

In fact, growing public concern over food safety has shifted research focus toward natural waxes as safer alternatives for lipid-based edible coating films. These wax mixtures offer abundant availability, cost-effectiveness, and inherent safety with minimal toxicity because they are mostly derived from renewable sources, including plant cuticles, seed coatings, and insect secretions (Winkler-Moser et al. 2019; Nurul Syahida et al. 2020). Therefore, compared to synthetic counterparts, natural waxes demonstrate superior consumer acceptance and research interest. On the other hand, these natural waxes exhibit excellent hydrophobic properties and moisture barrier performance, which can effectively inhibit moisture migration between the food and the environment, consequently delaying food deterioration processes. Oliveira et al. (2018) examined the efficacy of beeswax-incorporated cassava starch films on guava preservation under the storage conditions of 15 °C and 90–95% relative humidity (RH) for 15 days. Results showed that the coating film treatment maintained the superior postharvest quality parameters for the guava, including reducing physiological weight loss, promoting proper fruit ripening, retarding quality deterioration and chlorophyll degradation, etc. Furthermore, natural waxes also exhibit inherent antimicrobial/antioxidant properties that provide protection against microbial contamination and other food oxidation mechanisms. In a related study, Goncalves et al. (Gonçalves et al. 2010) reported that edible coating films containing 4.5–9% carnauba wax significantly suppressed *Monilinia fructicola* and *Rhizopus stolonifer* infections in nectarine and plum fruits by effectively inhibiting conidial germination.

In short, natural waxes have gained increasing attention as functional additives in biopolymer-based food packaging films. Their incorporation enhances crucial film properties, including water resistance, barrier performance, mechanical strength, and biodegradability. Thus, this chapter summarizes the characteristics of animal-derived beeswax, plant-based carnauba wax, and natural waxes from other sources, followed by a systematic review of their applications in food packaging (Fig. 11.1). By highlighting their role in improving biopolymer film's performance, this work can provide insights for future sustainable utilization of natural waxes in the food industry.

11.2 Commonly Used Natural Waxes in Biopolymer Food Packaging Films

Natural wax is a type of lipid-based material with complex compositions, primarily comprising long-chain esters, free fatty acids, alcohols, hydrocarbons (mostly 29–33 carbons), and resins such as triterpenes. And the minor constituents, including aldehydes, ketones, and sterols, further enhance their functional properties (Doan et al. 2017). Specifically, plant waxes are rich in oxygenated aliphatic compounds (e.g., alcohols, ketones, and fatty acids) with limited alkane content. In contrast, animal waxes lack triglycerides but contain sterols (e.g., cholesterol) and long-chain alcohols, imparting distinct physicochemical characteristics that differentiate them from plant and mineral waxes (Wang et al. 2025a). Figure 11.2 provides basic information on the chemical composition and property analysis of common natural waxes.

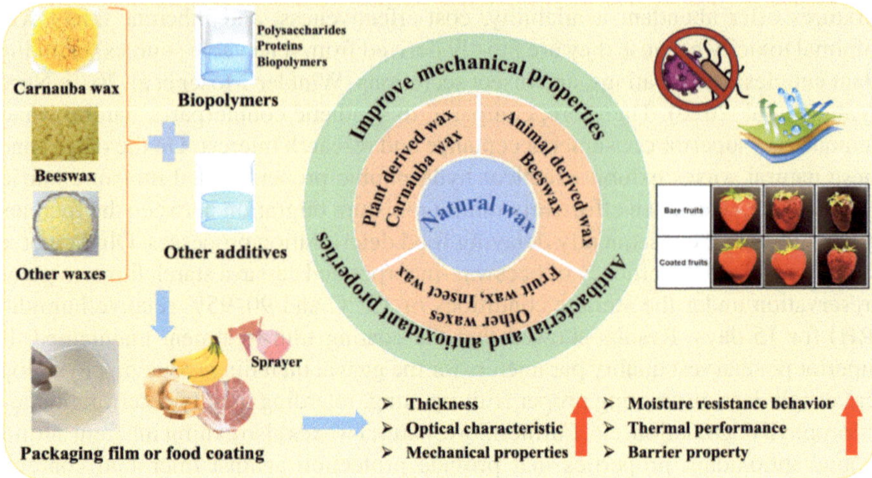

Fig. 11.1 Overview of characteristics of waxes and their applications in food packaging

For the application properties of film formation, natural waxes are characterized by chemical stability, ductility, and hydrophobicity (almost insoluble in water). They also exhibit remarkable resistance to saponification and chemical degradation at ambient temperatures, owing to their high saturation of fatty acids. Besides, with an HLB value of 9–10, they demonstrate excellent emulsifying capacity, facilitating stable emulsion formation. However, their crystallization behavior, including efficiency, crystal morphology, and melting point, varies significantly depending on the wax source and compositional differences (Gu et al. 2023). Subsequent sections will detail the properties of waxes derived from distinct origins.

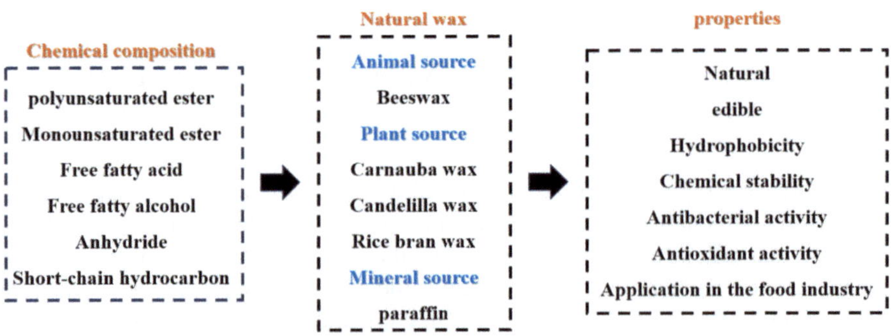

Fig. 11.2 Chemical composition and properties analysis of natural waxes

11.2.1 Animal Source: Beeswax

Beeswax, a viscous fatty substance secreted by worker bees' abdominal wax glands, serves as the primary material for hive construction, safeguarding honey, pollen, and larvae (Fratini et al. 2016). Initially liquid, it solidifies into scales upon air exposure, exhibiting a soft, solid state at room temperature (Kaluza et al. 2016). Freshly secreted beeswax appears nearly white but acquires a yellow–brown hue and aromatic fragrance due to pollen pigments (Fratini et al. 2016). Chemically, beeswax comprises about 300 components, predominantly hydrocarbons, free fatty acids, esters (mono-, poly-, and fatty acid esters), and a few unidentified compounds (Maia and Nunes 2013; Hosseini et al. 2023). Of course, exogenous residues (e.g., propolis, pollen, pollutants) and bioactive constituents (e.g., amino acids, flavonoids, sterols) are also present in beeswax (Winkler-Moser et al. 2019; Giampieri et al. 2022).

Notably, beeswax exhibits superhydrophobicity and low water solubility. Its chemical stability also ensures integrity across diverse environments without degradation. Additionally, with a low melting point of 62–67 °C, beeswax demonstrates excellent plasticity, enabling facile molding into intricate structures for film formation. Importantly, beeswax exhibits antimicrobial activity against both gram-positive (e.g., *Staphylococcus aureus, Streptococcus* spp.) and gram-negative bacteria (*Bacillus subtilis, Pseudomonas aeruginosa, Escherichia coli*), as well as yeast (*Candida albicans*) (Nevine 2011). A synergistic study demonstrates that a 1:1:1 (v/v) honey-beeswax-olive oil mixture inhibits *Staphylococcus aureus* and *Candida albicans*, producing 4.0 mm and 3.5 mm inhibition zones at 37 °C, respectively (Fratini et al. 2016). A beeswax-propolis mixture demonstrated an enhanced antimicrobial activity, producing inhibition zones of 20 and 22 mm against the aforementioned strains, respectively, and 13.5 mm/10.5 mm against *Staphylococcus epidermidis* and *Bacillus subtilis* (Castaldo and Capasso 2002; Al-Waili 2003). On the other hand, Zhao et al. (2020) demonstrated that beeswax exhibits significant antioxidant capacity, with an IC50 value of 5.91 mg/mL for DPPH, a FRAP value of 1.33 mg Trolox/g, and ferrous ion chelation of 0.38 mg EDTA/g. Similarly, Giampieri et al. (2018) also demonstrated that recycled beeswax protected Human Dermal Fibroblasts (HDF) from AAPH-induced oxidative damage, improving cellular antioxidant status, mitochondrial function, and wound healing. This antioxidant activity likely stems from the beeswax polyphenols, like quercetin, apigenin, and coniferin (Florido-Moreno et al. 2025). Thus, these antibacterial/antioxidant properties suggest the beeswax's potential as a functional additive for food packaging films.

In fact, beeswax is already effectively utilized in food packaging applications. For example, a hydroxypropyl methylcellulose-beeswax coating maintained the firmness and color of red guava "Pedro Sato" during 8-day room temperature storage by delaying ripening (Formiga et al. 2019). Similarly, cellulose nanocrystal-stabilized cornstarch-beeswax emulsions formed a sustainable edible coating that reduced spoilage for bananas, strawberries, and apple slices by limiting oxygen exposure and moisture loss, thereby preserving freshness and preventing oxidative/

enzymatic degradation (Trinh et al. 2022). Interestingly, a heat-resistant, edible superhydrophobic coating composed of beeswax and coffee lignin exhibited a micro/nanoscale hierarchical structure resembling a lotus leaf. This unique morphology enables liquid foods (e.g., milk, honey) to roll freely with minimal residue. Notably, the coating maintains super hydrophobicity even after 60 min at 120 °C, outperforming conventional edible coatings in thermal stability (Zhang et al. 2019; Aylanc et al. 2021; Fayaz et al. 2017; Moghtadaei et al. 2018). These properties, exceptional hydrophobicity, thermal resistance, and food safety, highlight the potential of beeswax for functional food packaging applications.

In summary, beeswax demonstrates significant potential for food industry applications due to its safety, thermal stability, and mechanical strength. In addition, emerging beeswax-based packaging materials exhibit notable antibacterial/antioxidant properties, effectively reducing oxidation and moisture loss in various foods (e.g., bread, dairy, honey, fresh produce), thereby extending shelf life while maintaining quality. These characteristics establish beeswax as a promising candidate for developing functional food packaging materials and inform future design strategies.

11.2.2 Plant Source: Carnauba Wax

Carnauba wax, the most economically significant plant-derived natural wax, forms a protective coating on Brazilian palm leaves to prevent desiccation and fungal infection while maintaining internal water balance (Romani et al. 2020). This saturated, water-insoluble lipid exhibits a characteristic yellow color and possesses the highest melting point among natural waxes. Notably, its exceptional hardness makes it the most rigid commercial wax available (Khosronia et al. 2023). Carnauba wax shares structural and physicochemical similarities with beeswax but differs in component proportions (Zhang et al. 2016). In detail, carnauba wax is primarily composed of esters (>80%, predominantly aliphatic esters and diester cinnamates), which also contain fatty acids, aromatic acids, free alcohols, hydrocarbons, ω-hydroxycarboxylic acids, and triterpene diols (Zhang et al. 2016; Susmita Devi et al. 2022). Environmental factors (precipitation, solar intensity, and soil properties) can alter tree biochemistry processes, subsequently modifying wax characteristics including viscosity, hardness, melting point, and solubility (De et al. 2013).

Carnauba wax exhibits a high melting point (82–86 °C) and limited solubility (slightly soluble with water at room temperature but readily dissolves in heated nonpolar solvents). Comparative studies demonstrate carnauba wax's superior hardness relative to palm oil and pure beeswax, enabling the formation of durable protective coatings that effectively resist scratches and surface wear (De et al. 2013). In addition, it exhibits selective permeability to O_2 and CO_2, along with notable waterproofing and UV-blocking properties that benefit plant protection (Campos et al. 2019). Moreover, the hexane and ether extracts of carnauba wax demonstrate antimicrobial and antioxidant activity, particularly against *L. infantum* amastigotes (IC_{50}

= 8–52 μM), attributed to bioactive constituents (Almeida et al. 2016). It is reported that carnauba wax contains defense proteins (chitinase, glucanase) that disrupt fungal growth by altering mycelial morphology (Cruz et al. 2002). Another study proved that carnauba wax effectively suppressed phytopathogens, with $\geq$1% concentrations inhibiting *Rhizopus stolonifer* and *Monilinia fructicola*. Notably, *Rhizopus stolonifer* showed complete germination inhibition at all concentrations, while 9% wax coating achieved 50 and 90% inhibition rates for *Monilinia fructicola* and *Rhizopus stolonifer* spore germination, respectively (Gonçalves et al. 2010). On the other hand, carnauba wax demonstrates notable antioxidant potential in a previous study, with water and ethanol extracts exhibiting DPPH radical IC_{50} values of 450 and 365 μg/mL, respectively, and ABTS radical IC_{50} values of 355 and 317 μg/mL. This scavenging activity was concentration-dependent below 1.0 mg/mL (Silva et al. 2018). These properties, along with its antimicrobial effects, enable carnauba wax to enhance the bioactivities of composite packaging materials. For instance, chitosan-casein films incorporating condensed tannins and carnauba wax demonstrated enhanced functionality, exhibiting a 40% higher DPPH radical scavenging capacity than controls. The composite also showed improved antibacterial activity, with inhibition zones of 6.2 mm (*Escherichia coli*) and 8.4 mm (*Staphylococcus aureus*) (Xu et al. 2024). In another research, a cassava starch/cellulose nanocrystal/carnauba wax nanoemulsion coating also demonstrated significant antioxidant and antimicrobial effects on strawberries. The coating preserved total phenolic content (TPC) by limiting gas exchange. After 12-day storage, DPPH scavenging capacity declined moderately (80% $\rightarrow$ 60%) versus controls (77.9% $\rightarrow$ 33.2%). The microbial counts were substantially reduced (3.59–4.03 log CFU g^{-1} for mesophilic aerobic bacteria and 4.45–5.22 log CFU g^{-1} for fungi and yeast) compared to uncoated samples (Oliveira Filho et al. 2022).

Owing to its physicochemical properties, carnauba wax is extensively employed in food applications. As an edible coating for fruits (oranges, apples, pomegranates) and vegetables (tomatoes, eggplants, guavas), it significantly prolongs shelf-life by reducing moisture loss and respiration rates while maintaining texture (firmness, brittleness) and marketability (Won and Min 2018; Santos et al. 2017; Caron et al. 2015; Barman et al. 2014). Take examples, a 9% carnauba wax nano emulsion with 1% peppermint oil reduced litchi decay incidence from 60 to 20% after 8-day storage (Fukuyama et al. 2024). Similarly, carnauba wax blended with grapefruit seed extract (1:1) or oregano oil (2:1) decreased *Penicillium italicum* infection on oranges by 23.6% and 25.0%, respectively. These coatings demonstrated enhanced colloidal stability in the coatings, reduced weight loss and suppressed respiration rates for fruits (Oliveira Filho et al. 2022). In baked goods, substituting oil gels with 6% carnauba wax or a 2% wax/4% adipic acid blend increased unsaturated fat content but reduced oxidation stability and hardness in chocolate sauces (Roufegarinejad et al. 2024). Additionally, carnauba wax encapsulated NaCl allowed a 35% salt reduction in bread without compromising sensory properties or dough rheology compared to controls (Beck et al. 2022). Interestingly, Filho et al. (2017) isolated a cinnamate ester derivative (PCOC) from carnauba wax, which demonstrated significant hypolipidemic and hypoglycemic effects at specific doses. Results showed that PCOC treatment effectively reduced total cholesterol and triglyceride-glucose levels, performing comparably or superior to γ-glutamate and reference agents in dyslipidemia management.

11.2.3 Other Natural Waxes: Insect Wax, Fruit Wax, Candelilla Wax, and Rice Bran Wax

Insect wax, mainly secreted by specific insects, is a type of chemically stable and glossy substance (Pan et al. 2024). Composed predominantly of fatty acid monoesters (C26–C30 hydrocarbons), it also contains minor constituents including free acids, wax alcohols, and resins (Ma et al. 2018). Currently, insect wax is primarily utilized in food packaging and as an edible coating for fruits (e.g., apples, lychees) to extend shelf life (Yan et al. 2020). Insect wax can be combined with antimicrobial agents (e.g., Ag^{+}/Cu^{2+} ions, quaternary ammonium salts, bacteriocins) to develop antibacterial packaging films that effectively inhibit microbial growth and extend food shelf life (Wang et al. 2025b). Additionally, its inherent hydrophobicity enhances waterproofing in packaging materials. For instance, insect wax-coated pulp bowls exhibit significantly reduced water absorption and improved self-cleaning properties (Ahuja and Kumar 2023).

Fruit wax, a natural epidermal coating biosynthesized by plants, serves as a protective barrier against water loss, nutrient leakage, mechanical damage, and pathogen invasion while mitigating environmental stresses like UV radiation (Rios et al. 2015; Mukhtar et al. 2014; Lara et al. 2019; Zhu et al. 2020; Shaheenuzzamn et al. 2021; Samuels et al. 2008). Composed of very-long-chain fatty acid derivatives (alkanes, alcohols, esters, etc.), this non-toxic fruit wax shows significant food industry potential (Khan et al. 2024). For instance, ash gourd peel wax extended strawberry shelf-life from blew 2–7 d at 25 °C while maintaining quality parameters (texture, color, acidity, microbial counts) (Sreenivas et al. 2011). Therefore, fruit wax-based edible films offer a sustainable alternative to chemical preservatives, aligning with green food preservation strategies.

Candelilla wax is a protective layer of stem wax produced by endemic plants in the arid regions of northern Mexico, endowing them with the ability to resist abiotic (extreme temperatures) and biological (insect) stressors (León-Zapata et al. 2018). When applied at $\leq$200 mg/kg, candlestick wax can be used as a multi-functional food coating to prevent microbial deterioration of fruits (such as citrus, apples, peaches, etc.) and extend their shelf life. Relevant study has demonstrated that composite films incorporating candelilla wax with pectin, aloe vera, and Larrea tridentata extract exhibited significant antifungal activity, reducing *Colletotrichum gloeosporioides* and *Alternaria alternata* damage by 22.0% and 24.5%, respectively (Aguirre-Joya et al. 2019).

Rice bran wax, extracted during rice bran oil refining, primarily consists of long-chain saturated fatty acids (C20-C26) and fatty alcohols (C30-C36). With a high melting point (78–81 °C) and thermal stability, it exhibits oil crystallization tendencies at room temperature (Thakur et al. 2023). Studies demonstrate its effectiveness in food preservation: A 0.6% wax/5.5% WPI coating optimally extended shelf-life for dry snacks in biodegradable packaging (Ganjizadeh Zavareh et al. 2021). Wax-treated cherry tomatoes showed reduced weight loss (13.54% vs the control's 16.02%) by inhibiting pectin chelation during refrigeration (Zhang et al. 2017).

This section provides a systematic overview of the sources, characteristics, and application value of various natural waxes. These natural waxes have characteristics such as hydrophobicity, plasticity, chemical stability, and antibacterial and antioxidant properties, which can effectively extend the shelf life of food and maintain the quality of fruits and vegetables. Natural waxes also demonstrate important value in food preservation, packaging material improvement, and other aspects (Fig. 11.3). Overall, natural wax has broad development prospects in the fields of sustainable packaging and food preservation due to its green, safe, and biodegradable characteristics.

Fig. 11.3 Overview of the characteristics of natural wax and its application scenarios in biobased food packaging films

11.3 Effect of Beeswax on Different Biopolymer Food Packaging Film

11.3.1 Polysaccharide Based Film

11.3.1.1 Thickness

The thickness of polysaccharide-based beeswax films is a crucial factor that significantly influences their performance, including mechanical strength, barrier properties, and cost-efficiency. An optimal thickness is essential to balance food protection and material economy. Studies have shown that the incorporation of beeswax increases film thickness by occupying space within the polysaccharide matrix (Maia and Nunes 2013; Sultan et al. 2021). However, microscopy analyses, such as scanning electron microscopy (SEM) and confocal laser scanning microscopy (CLSM), have revealed that higher beeswax concentrations tend to promote particle aggregation. This aggregation creates structural heterogeneity, which in turn further impacts the film thickness. In contrast, an appropriate beeswax content enhances barrier properties and mechanical stability, underscoring its functional value in edible film development (Lim et al. 2024).

11.3.1.2 Color and Light Transmittance

The addition of beeswax can significantly alter the optical properties of polysaccharide-based films. Specifically, it imparts a yellowish-green hue and reduces light transmittance. As a previous study has pointed out, increasing the beeswax content elevates the color difference (ΔE) and haze while decreasing transparency (Pirinc et al. 2021; Luchesi et al. 2024). These changes result from the hydrophobic particles of beeswax, which disrupt light transmission and create structural heterogeneity. For example, beeswax-chitosan composites exhibit a stronger yellow–brown coloration and matte surfaces compared to pure starch films (Velickova et al. 2013). Although these modifications may affect the visual appeal of the films, they can benefit food preservation. This is because the reduced light permeability helps to decrease photo-oxidation.

11.3.1.3 Water Resistance Behavior

Beeswax significantly enhances the water resistance of polysaccharide-based films through multiple mechanisms. Specifically, it improves moisture barrier properties by forming hydrophobic domains within the matrix, thereby reducing both WVP and overall water sensitivity. For example, gelatin/green algae films containing 5% beeswax not only maintained their antibacterial properties but also exhibited improved water resistance (Kazemi et al. 2023). Similarly, chitosan-beeswax composites demonstrated an increased contact angle (from 79.8° to 109.7°) and a substantial reduction in swelling (from 2172.09% in pure chitosan to 458.81%). Additionally, starch/gelatin-beeswax films exhibited superior hydrophobicity, with a contact angle of 102.6°, and enhanced thermal stability compared to other waxes (Sultan et al. 2021). These improvements are attributed to the small particle dispersion and hydrophobic surface properties of beeswax, which effectively limit water diffusion and absorption. These excellent water resistance properties not only extend the service life of these films, but also significantly broaden their potential applications in high moisture food packaging systems (Cheng et al. 2024, 2023).

11.3.1.4 Mechanical Properties

The incorporation of beeswax alters the mechanical properties of polysaccharide-based films in a formulation-dependent manner. While typically reducing tensile strength (as observed in starch-based systems), beeswax can simultaneously enhance film stiffness. For example, in chitosan-beeswax/pollen composites, Young's modulus doubled from 12.7 to 22.56 MPa (Sultan et al. 2021; Luchesi et al. 2024; Auras et al. 2009). The stiffening effect results from beeswax-matrix interactions that limit polymer chain mobility, but this also reduces flexibility. For instance, pure chitosan films demonstrated a two-fold decrease in elongation at break (from 35.81 to 14.09%) upon beeswax incorporation (Sultan et al. 2021). In contrast, cassava starch films showed increased maximum elongation with higher beeswax content, demonstrating system-specific effects. The ability to precisely tune these properties through controlled beeswax incorporation enables targeted optimization of packaging films for various applications, ranging from rigid protective layers to more flexible wrapping materials (Sultan et al. 2021; Luchesi et al. 2024; Auras et al. 2009).

11.3.1.5 Antibacterial and Antioxidant Activities

The incorporation of beeswax into food packaging films has demonstrated remarkable enhancements in both antibacterial and antioxidant properties, leading to significant improvements in food preservation performance. Experimental studies have revealed a synergistic antimicrobial effect when beeswax is combined with other natural components. For instance, composite films based on cassava starch incorporating both beeswax and propolis ethanol extract exhibited substantial antimicrobial efficacy, showing 51% growth inhibition against the food spoilage fungus *Aspergillus niger* (Betancur-D´Ambrosio et al. 2024). In addition, parallel investigations on tragacanth/hydroxypropyl methylcellulose films containing beeswax confirmed their antimicrobial potential. These composite materials displayed concentration-dependent antibacterial activity, with particularly strong effects against Gram-negative pathogenic bacteria (Bahrami et al. 2019). Beyond their antimicrobial effects, these beeswax-incorporated films also possess considerable antioxidant properties. Systematic studies on cellulose/glycerol/beeswax bioplastics documented sustained DPPH radical scavenging activity, maintaining a minimum inhibition rate of 3.3% even after prolonged 96 h exposure. Importantly, this antioxidant capacity showed a direct correlation with beeswax content (Giampieri et al. 2018). The observed antioxidant effects are primarily attributed to polyphenolic compounds naturally present in beeswax, such as kaempferol and apigenin, which contribute to oxidative stability. The synergistic integration of antimicrobial and antioxidant properties in beeswax-incorporated films establishes a dual-functional protective system for food packaging applications. These complementary functionalities provide comprehensive protection against microbial proliferation and oxidative degradation, significantly prolonging the shelf-life and maintaining the nutritional quality of packaged foods. These findings underscore the considerable potential of beeswax-based composite films as active packaging materials, offering enhanced preservation performance compared to conventional alternatives.

11.3.1.6 Other Performance

Coating adhesion serves as a critical parameter in food preservation applications, with optimized polysaccharide-beeswax ratios demonstrating significant improvements in film-substrate adherence. Recent advancements in this field reveal that ionically crosslinked chitosan-gellan gum composites exhibit substantially enhanced adhesion properties. These modified composites establish tighter contact with food surfaces, effectively minimizing air and moisture penetration while extending the shelf life of food (Cheng et al. 2024). Additionally, complementary thermal analysis studies provide further insight into these composition-property relationships. For example, cassava starch-beeswax blends display characteristic thermal degradation patterns, particularly exhibiting predominant weight loss within the 200–360 °C range at elevated component concentrations (Betancur-D´Ambrosio et al. 2024). Collectively, these findings demonstrate that both interfacial adhesion and thermal stability are fundamentally governed by the precise formulation of polysaccharide-beeswax composites. This strong dependence on compositional parameters highlights the essential role of component ratio optimization in developing advanced, high-performance food preservation coatings.

11.3.1.7 Application in Food Industry

Beeswax is a crucial functional additive in polysaccharide-based food coatings, significantly enhancing their preservation performance. Research has demonstrated that multilayer beeswax-chitosan coatings effectively regulate the respiration rates of strawberries. This regulation reduces weight loss and senescence while maintaining desirable color and texture attributes (Velickova et al. 2013). Similar preservation effects are observed in chitosan-beeswax/pollen coatings applied to Le Conte pears. These coatings exhibit minimal decay (7.15% after 75 d) and enhance firmness and total soluble solids retention (Sultan et al. 2021). The inherent hydrophobicity of beeswax facilitates a wide range of applications. For instance, bioinspired pectin-beeswax films that mimic the surface of lotus leaves have been developed for underwater food preservation (Wang et al. 2020). Additionally, heat-stable superhydrophobic coatings derived from beeswax-coffee lignin composites have been created. These coatings can withstand temperatures of 120 °C for 60 min while preventing the adhesion of liquid foods (Trinh et al. 2022). Collectively, these developments establish beeswax as a key component in advanced food packaging systems.

11.3.2 Protein Based Film

11.3.2.1 Thickness

The incorporation of beeswax into protein-based films generally leads to an increase in film thickness. This phenomenon is attributed to the physical filling of intermolecular voids and potential structural reorganization mediated by protein-wax interactions. Studies demonstrated this thickness enhancement across various protein matrices. Upon beeswax incorporation, the film thickness significantly increases across different protein matrices: corn zein films from 126.33 to 148.00 μm, soy protein isolate (SPI) films from 105.67 to 148.67 μm, and wheat gluten films from 110.67 to 148.00 μm. These changes are accompanied by observable microstructural modifications, which are revealed by SEM analysis to trend toward either rougher or more compact morphologies (Duan et al. 2023). However, when the concentrations of beeswax and glycerol in protein-based films are adjusted, the thickness variations of certain composite formulations are not statistically significant ($P > 0.05$), which highlights the composition-dependent nature of these structural modifications (Khanzadi et al. 2015).

11.3.2.2 Color and Light Transmittance

The incorporation of waxes significantly modifies the optical properties of films, with specific effects being highly dependent on the matrix composition. In gelatin-based films, increasing beeswax content results in higher L* values and pronounced yellowing, as indicated by elevated b* values. For instance, a wax concentration of 15% demonstrates significantly greater yellowness compared to lower concentrations, while simultaneously reducing transparency, as evidenced by values increasing from 3.90 to 14.0 ($P < 0.05$) (Zhang et al. 2018). Additionally, whey protein-pullulan composites show formulation-dependent ΔE variations, particularly with whey protein concentrate 30 (WPC30) displaying reduced ΔE at elevated beeswax levels (Khanzadi et al. 2015). Notably, zein films demonstrate unique optical responses.

The opacity of these films decreases from 9.35 to 6.51 Abs/mm upon beeswax addition, an effect that contrasts with other protein films. This reduction in opacity may be attributable to modified light scattering patterns resulting from altered particle size distribution and polymer chain mobility (Suwannasang et al. 2021).

11.3.2.3 Water Resistance Behavior

The incorporation of beeswax significantly enhances the hydrophobicity of protein-based films, although its impact on WVP varies depending on the matrix. The inherent hydrophobicity of beeswax (melting point 62–65 °C) notably reduces water solubility in fish gelatin films by 45.17% and decreases moisture content from 2.20 to 1.40% in zein films. These improvements are primarily attributed to hydrophobic interactions that limit water-polymer contact (Duan et al. 2023; Zheng et al. 2022). In terms of WVP reduction, elongated beeswax particles form tortuous diffusion pathways during low-temperature extrusion processing (<75 °C), resulting in a substantial decrease in WVP from 13.2×10^{-8} to 5×10^{-8} g mm/h cm^2 Pa in gelatin films (Zhang et al. 2018; Chevalier et al. 2018). However, certain protein matrices, such as whey protein, exhibit negligible changes in WVP, indicating insufficient crystalline barrier formation (Cortés-Rodríguez et al. 2020). In addition, contact angle measurements further reveal system-dependent behaviors. For instance, the contact angle increases from 25.47° to 27.17° in wheat protein films, while it decreases in SPI films (from 44.70° to 23.50°) due to glycerol migration during drying of protein-wax films (Maryam Adilah et al. 2018). These differential effects demonstrate that while beeswax consistently enhances overall film hydrophobicity, its effectiveness as a water barrier depends on specific polymer-wax interactions and the resulting morphological modifications.

11.3.2.4 Mechanical Properties

The influence of beeswax incorporation on the mechanical properties of protein-based films is highly matrix-dependent, with effects varying according to composition and molecular interactions. For instance, tensile strength reductions are observed in zein films (from 9.00 to 6.98 MPa) and SPI films (from 1.77 to 1.12 MPa), while wheat gluten protein (WGP) films show minimal changes (0.62 and 0.64 MPa). These changes are attributable to beeswax-induced disruption of hydrogen bonding networks and increased polymer chain free volume (Duan et al. 2023; Zhang et al. 2012; Li et al. 2019). However, optimized formulations can counteract these effects. For example, WPC30-Beeswax30 demonstrates enhanced tensile strength (10.73 MPa) through balanced component ratios (Khanzadi et al. 2015). Beeswax generally reduces brittleness while improving flexibility, as observed in corn zein films. Nevertheless, elongation behavior remains system-specific. SPI-arabic gum conjugates exhibit lower elongation (30.54%) compared to beeswax-modified counterparts (32.68–48.35%) (Xue et al. 2019). The plasticizing action of oleic acid combined with beeswax, which further increases film stretchability by altering melting characteristics during deformation (Monedero et al. 2009). Additionally, freezing treatments differentially affect mechanical performance, with beeswax-containing films showing reduced puncture strength due to matrix discontinuity (Soazo et al. 2013). These mechanical modifications primarily result from beeswax-mediated alterations in intermolecular bonding, polymer chain mobility, and microstructural heterogeneity. This allows for systematic tuning of the strength-flexibility balance through compositional optimization for targeted packaging applications.

11.3.2.5 Antibacterial and Antioxidant Activities

Beeswax enhances food preservation through establishing microbial barriers and providing antioxidant activity. The hydrophobicity of beeswax coatings forms a physical barrier that limits microorganism-food contact, thereby extending shelf life (Maia and Nunes 2013). Additionally, beeswax exhibits concentration—dependent antioxidant effects. For instance, beeswax-gelatin films demonstrate enhanced hydroxyl radical scavenging activity, with a 14.9% scavenging rate at 0.75 g/100 mL compared to 10.4% at 0.5 g/100 mL. This improvement results from the combined contributions of protein peptides and beeswax-derived aliphatic alcohols (Ravelo et al. 2011). Moreover, in composite systems, beeswax combined with spent coffee grounds achieves superior performance, with an 84.9% radical scavenging rate compared to 70% in control samples. This indicates synergistic interactions between the natural wax and plant-derived antioxidants (Papadaki et al. 2023).

11.3.2.6 Other Performance

The incorporation of beeswax induces matrix-specific modifications to the thermal and functional characteristics of protein-based films, revealing complex structure–property relationships. Specifically, beeswax decreases the denaturation temperatures from 77 to 72 °C in zein films and WGP films from 106 to 98 °C by weakening molecular interactions. In contrast, it enhances the thermal stability of gelatin films by raising the decomposition temperature range from 286–316 to 292–320 °C (Duan et al. 2023; Zhang et al. 2018). Notably, in SPI-arabic gum conjugate films, beeswax addition significantly increases the glass transition temperature from 85.25 to 142–148 °C, thereby demonstrating enhanced thermal resistance (Li et al. 2019). Beyond its impact on thermal properties, beeswax profoundly influences the functionality of these films. By increasing hydrophobicity, beeswax reduces lysozyme release rates by 2.5–17-fold, with the release kinetics further modulated by the melting behavior of the wax (Arcan and Yemenicioğlu 2013). Additionally, beeswax exerts predominant control over the stickiness of whey protein films compared to other additives. Meanwhile, the concentrations of 50–65% protein and 35–50% sorbitol have minimal effects on viscosity and appearance (Ozdemir and Floros 2008). Higher beeswax content increases particle size and solution viscosity during film formation in sodium-caseinate systems (Fabra et al. 2009). Importantly, these functional modifications do not compromise biodegradability. For example, controlled hydrolysis is achieved in fish gelatin composites (Zheng et al. 2022). Thus, beeswax emerges as a multifunctional modifier capable of precisely tuning packaging film properties while upholding sustainable material standards.

11.3.2.7 Application in Food Industry

The extrusion blow-molded hydrophobic starch/gelatin/beeswax (HSGBW) film has exceptional industrial potential for a wide range of food packaging applications. It is especially effective for high-moisture products such as dumpling fillings, high-fat items like milk powder, and oily foods such as chili oil. This is because it combines outstanding biocompatibility and storage stability with rapid high-temperature dissolution and complete biodegradability (Cheng et al. 2024). In a similar vein, spent coffee grounds/beeswax/whey protein films have been developed with similarly advantageous properties. These include low solubility and reduced water vapor permeability, which effectively prevent fruit dehydration while preserving crispness in baked goods through controlled moisture regulation (Papadaki et al. 2023). Collectively, these technological advancements demonstrate the growing adoption of beeswax-enhanced protein films in food packaging applications. These progresses have been realized through the strategic optimization of key material properties. These include enhanced hydrophobicity to improve moisture resistance, improved thermal stability to withstand processing and storage conditions, controlled optical characteristics to meet specific packaging needs, and tailored mechanical performance to ensure durability and functionality.

11.3.3 Biopolyester Based Film

11.3.3.1 Thickness

Beeswax incorporation typically enhances biopolymer film thickness due to its volumetric properties and phase-filling behavior within the polymer matrix. However, excessive beeswax content (such as 3 wt%) produces counterproductive effects: the increased solution viscosity promotes particle agglomeration, ultimately leading to non-uniform thickness distribution. This phenomenon is clearly evidenced in coated trays, where thickness variations escalate from 543 ± 37 to 755 ± 54 μm at higher wax concentrations (Lim et al. 2024; Reis et al. 2018).

11.3.3.2 Color and Light Transmittance

The inherent light-yellow coloration of beeswax causes yellowish discoloration and reduced transparency in bio-polyester films through chemical interactions with film constituents, potentially affecting color stability.

11.3.3.3 Water Resistance Behavior

The incorporation of beeswax into polyvinyl alcohol/polyacrylic acid (PVA/PAA) composite films significantly enhances moisture barrier properties, demonstrating three key improvements: (1) water absorption decreases from $11.91 \pm 0.11\%$ to $8.98 \pm 0.61\%$, (2) water solubility reduces from $23.98 \pm 0.23\%$ to $7.90 \pm 0.03\%$, and (3) WVP declines by up to 73.76% compared to pure PVA films (Lim et al. 2024; Peter et al. 2024). This enhancement primarily stems from beeswax's hydrophobic long-chain fatty alcohols and alkanes, which form hydrogen bonds with polymer matrices while simultaneously inhibiting water molecule interactions (Lee et al. 2018; Pereda et al. 2010). Advanced structural designs further optimize moisture resistance. In P-BC-P multilayer architectures, epitaxial crystallization between beeswax and rosin flakes creates highly ordered hydrophobic structures, achieving exceptionally low WVP values (5.6×10^{-13} g m/m^2 s Pa) (Cruces et al. 2021). Similarly, in coating applications, beeswax exhibits dose-dependent performance: a thin 5μm layer reduces polylactic acid (PLA) film permeability by 65%, while a thicker 20 μm coating achieves an 87% reduction (Martínez-Abad et al. 2014). These findings collectively establish beeswax as an effective modifier for developing high-performance barrier biopolymer packaging materials, with performance tunable through both compositional adjustment and structural design.

11.3.3.4 Mechanical Properties

The mechanical properties of polymer films are significantly altered by the incorporation of beeswax. The size of beeswax particles plays a crucial role in determining the tensile characteristics of these films. Small and uniformly dispersed particles enhance the interfacial interactions, while larger aggregates form heterogeneous structures that can compromise the integrity of the film. In PVA/PAA composites, the addition of beeswax leads to a reduction in tensile strength and Young's modulus. This is primarily due to weakened interactions between polymer chains, especially when the beeswax concentration exceeds 10%. At such higher concentrations, non-uniform dispersion of beeswax induces structural defects (Lim et al. 2024; Fasihi et al. 2019). However, a moderate amount of beeswax (5%) can increase fracture elongation from 274.27 ± 15.57 to $323.33 \pm 3.55\%$ through plasticization effects. This enhancement is observed before the properties start to diminish at higher beeswax concentrations (20%) (Lim et al. 2024). Similarly, when beeswax is applied as a surface coating (2–3 wt%) on TPS/PLA trays, it reduces tensile strength (with a Young's modulus range of 185–294 MPa) without significantly altering bulk elongation (21–27%). This modification maintains the practical flexibility of the trays while altering their surface properties (Reis et al. 2018). This outcome is mainly due to the non-polar nature of beeswax, which limits its interactions with polar polymers such as gum cordia. As a result, discontinuous phases are formed, which dilute the polymer matrix and consequently reduce the overall mechanical performance (Haq et al. 2016).

11.3.3.5 Antibacterial Activities

Core–shell polyethylene oxide/beeswax composite fibers have demonstrated broad-spectrum antimicrobial efficacy. They form inhibition zones measuring 3.6–3.8 cm against *Escherichia coli*, 1.9–3.8 cm against *Staphylococcus aureus*, and 4.3–5.0 cm against *C. albicans*. In contrast, blank controls show no antimicrobial activity (Kyuchyuk et al. 2022). Similarly, beeswax-coated PLA films maintain sustained antibacterial performance through controlled release mechanisms. These films effectively reduce *Salmonella* populations by 4–5 log units within 7 d, both in growth media and vegetable broth. The microbial counts ultimately reach 10–100 CFU/mL (Cruces et al. 2021).

11.3.3.6 Other Performance

The addition of beeswax significantly alters the gas barrier properties of polymer films. Specifically, incorporating 0.05 g/g beeswax increases oxygen permeability (OP) by 50-fold. This is due to the formation of non-polar diffusion pathways within the film structure. However, this effect levels off at higher beeswax concentrations, as the network becomes saturated (Haq et al. 2016). In PVA/PAA composites, OP rises significantly when beeswax content exceeds 5%. This increase is attributed to the development of hydrophobic oxygen channels and interfacial regions that facilitate diffusion. Additionally, enhanced surface hydrophobicity promotes the absorption of atmospheric oxygen (Winkler-Moser et al. 2019; Lim et al. 2024). Conversely, oil permeability decreases with increasing beeswax content. This is due to the formation of fat crystals within the polymer matrix, an effect that is further amplified by the dispersion of nanoparticles (Lim et al. 2024). Beeswax coatings at 1–3 wt% alter the thermal stability of glycerol without modifying the degradation patterns of starch/PLA (Reis et al. 2018). Additionally, beeswax regulates the release kinetics of silver ions in PLA films, which can be influenced by pH value (Peter et al. 2024).

11.3.3.7 Application in the Food Industry

Beeswax-modified polyvinyl alcohol (PVA-BW) shows enhanced performance in bread packaging. It delays crust drying by 4.33-fold and maintains bread softness 15% more effectively than pure PVA. This is a significant improvement compared to polyethylene, which is only 8% more efficient than PVA-BW. Pure PVA has a 23% deficit in this regard (Peter et al. 2024). Similarly, beeswax-coated starch/polylactic acid trays exhibit optimized water vapor permeability and mechanical properties. These properties make them suitable for fresh produce packaging (Reis et al. 2018).

These findings are systematically documented in Table 11.1, which provides a comprehensive summary of beeswax-polymer film applications across various food systems (Hosseini et al. 2023).

11.4 Effect of Carnauba Wax on Different Biopolymer Food Packaging Film

11.4.1 Polysaccharide Based Film

11.4.1.1 Thickness

Carnauba wax plays dual roles as a physical filler and a structural modifier within the polymer matrix, thereby systematically increasing the thickness of polysaccharide-based films. With its high-melting-point and hydrophobic properties, carnauba wax not only contributes to volumetric expansion but also simultaneously alters the arrangement and crystallization patterns of polysaccharide chains. This dual action ultimately leads to the formation of denser microstructures, with film thicknesses reaching up to 152.86 ± 7.56 μm (Zhang et al. 2023). Although film thickness generally correlates with wax concentration, this relationship becomes asymptotic at higher concentrations. This is due to increased molecular contacts between the polymer and the wax, which weaken chain aggregation forces and facilitate matrix unfolding (Zhang et al. 2023). In starch-based formulations, thickness variation is primarily governed by wax concentration and solid content rather than droplet size. Additional contributions come from wax-mediated reorganization of starch chains (Haq et al. 2016). This indicates that while wax content remains the primary determinant of thickness, its effects are ultimately regulated by complex, interdependent interactions between the polymer and the wax.

11.4.1.2 Color and Light Transmittance

The incorporation of carnauba wax induces significant modifications to the optical properties of polysaccharide-based films. Its hydrophobic particles reduce transparency through light scattering while simultaneously enhancing UV-blocking capacity across UVA, UVB, and UVC spectral regions (Corrêa-Filho et al. 2024). As the wax content increases progressively, visible light transmittance and lightness (L^* value) are systematically diminished, while redness (a^*) and yellowness (b^*) indices are elevated. This imparts a characteristic yellow-green coloration that stems

Table 11.1 The impact of beeswax and other different material-based food packaging films on food

Beeswax and other materials	Foods	Impacts	References
Coffee lignin and beeswax	Liquid foods, including honey, tea, Coca-Cola, wine, whole milk, coffee, and yoghourt	It has super-hydrophobicity, repulsion to water and many liquid foods, good thermal stability, coffee lignin can eliminate harmful free radicals to cells, and beeswax is non-toxic to mouse osteoblasts	Trinh et al. (2022)
Starch foam/5 wt% water hyacinth composite with beeswax coating	Water-rich fruit such as watermelons, strawberries, pomelos, cantaloupes, oranges, mangos and pineapples	The starch foam containing 5 wt% WH powder has the best mechanical properties, the lowest water content and the most uniform pore size distribution. The beeswax coating improves the hydrophobicity and low water solubility of starch foam composite, and can still maintain its shape after immersion in water	Chaireh et al. (2020)
5 wt% beeswax to glycerol-plasticized cellulose matrices	Pear slices	The addition of beeswax eliminates transparency and UV barrier, changes the main mechanical parameters, improves thermal stability and antioxidant capacity, as well as fluid dynamics and barrier properties. Has antibacterial activity and reduces browning by preserving freshly cut pear slices	Giampieri et al. (2018)
Beeswax and polyvinyl alcohol (PVA)/polyacrylic acid (PAA)	–	The addition of beeswax improves the barrier properties against oxygen, water, and oil. When the beeswax content increases to 10%, the tensile strength slightly decreases and the elongation rate increases	Lim et al. (2024)
Poly (lactic acid) and gelatin-beeswax emulsion	–	An increase in beeswax content significantly increases the color difference, opacity, and elongation at break of the film, but reduces the transmittance and tensile strength values	Pirinc et al. (2021)
Beeswax and modify maize starch, potato starch and other natural starch and glycerol	Climacteric fruits (bananas)	The composite lotion formed a coating on the surface of the banana. The weight loss rate of the coated banana decreased by 42%, the titratable acidity increased by 21%, and the vitamin C content increased by 20%, effectively maintaining the freshness of the fruit and extending its storage life	Sun et al. (2024)

(continued)

Table 11.1 (continued)

Beeswax and other materials	Foods	Impacts	References
Beeswax 10 (w/w, db) and cassava starch (13.3%, w/v)	Tomato	Delayed ripening, weight loss, firmness, and extended the storage life of fruits; preserved the antioxidant activity, total phenolic content, organoleptic properties and resisted enzymes activities	Osae et al. (2022)
Beeswax 1.7% (w/w), cornstarch (5%, w/v), CNCs (1–5%, w/v)	Bananas, strawberries, and apples	The formulated coating displayed magnificent color, freshness, and appearances preservation by reducing oxygen activities and preventing moisture losses	Formiga et al. (2019)
Beeswax 5 and 10% (w/v), cornstarch (3%, w/v), cassava starch (2%, w/v), Gelatin (5%, w/v)	Guava	Minimized weight loss, guaranteeing adequate ripening of the fruits, delaying mass and chlorophyll loss and keeping them in levels of high quality	Nurul Syahida et al. (2020)
Beeswax 5, 10 and 15% (w/v), CMC (1.0, 1.5 and 2.0%, w/v), Chitosan (0.5, 1.0 and 1.5%, w/v)	Kinnow mandarin fruit	Reduced weight loss, spoilage, juice content, firmness, pectin content, the activity of fruit softening enzymes; maintained higher levels of juice content, TSS, TA, ascorbic acid, and carotenoids	Baswal et al. (2020)

from the intrinsic pigmentation of wax (Zhang et al. 2023; Oliveira Filho et al. 2020). Cassava starch-carnauba wax blends invariably increase cashew gum film opacity independent of composition ratios (Morillon et al. 2002). However, the resulting color alteration (ΔE change minimal in KW20%) and compromised visual appeal restrict their utility in transparent packaging systems, notwithstanding their superior barrier performance (Haruna et al. 2019).

11.4.1.3 Water Resistance Behavior

Carnauba wax significantly enhances the hydrophobicity of biopolymer films, thanks to its unique composition of fatty esters and hydroxy acids. These components restrict starch–water interactions, thereby reducing water solubility and decreasing moisture content by 13.5% in chitosan-based films (Rocha et al. 2020; Wang et al. 2022). Generally, the inherent hydrophobicity of wax tends to lower the WVP value when its water-repellent effect outweighs the reduction in matrix cohesion. In particular, nanoemulsion films exhibit superior barrier performance compared to microemulsion systems (Oliveira Filho et al. 2020; Liu et al. 2022). This improvement is primarily attributed to wax-starch interactions that limit the accessibility of hydroxyl groups and establish tortuous diffusion pathways. However, structural defects may partially offset these benefits (Nur Hanani 2017). The observed increase in contact angle, which correlates with wax content but remains unaffected by emulsion droplet size, confirms the efficacy of carnauba wax in developing moisture-resistant food packaging, which can extend the shelf life of food (Oliveira Filho et al. 2020).

11.4.1.4 Mechanical Properties

The incorporation of carnauba wax significantly modifies the mechanical properties of biopolymer films. It leads to substantial reductions in tensile strength (49%) and Young's modulus (60%), while simultaneously increasing elongation at break and puncture resistance across both chitosan and corn starch matrices (Corrêa-Filho et al. 2024; Gomes et al. 2019). These softening effects are primarily attributed to wax-induced discontinuities in the polymer network and the inherent plasticizing properties of the wax. However, the addition of ZnO nanoparticles can partially mitigate this increased flexibility (Chiumarelli and Hubinger 2014; Santos et al. 2014). Although elevated wax concentrations (0.40%) may eventually constrain elongation capacity, the lipid phase generally enhances film ductility. This enhancement is primarily due to interfacial interactions that are predominantly governed by wax distribution within the matrix rather than droplet size (Winkler-Moser et al. 2019; Santos et al. 2014). As a result, the mechanical properties of the films can be precisely modulated by wax to meet specific packaging requirements.

11.4.1.5 Antibacterial and Antioxidant Properties

The chitosan-casein-condensed tannin-carnauba wax (CCTW) film demonstrates superior antioxidant and antimicrobial properties compared to conventional food packaging materials. It has a DPPH radical scavenging rate of 36%, which significantly exceeds those of control films. This enhanced performance is attributed to the synergistic interactions among condensed tannins, casein, chitosan, and carnauba wax (Silva et al. 2018; Yu et al. 2024). In addition to its antioxidant capabilities, the film also exhibits exceptional antimicrobial efficacy. It forms inhibition zones measuring 6.2 ± 0.37 mm for *Escherichia coli* and 8.4 ± 0.35 mm for *Staphylococcus aureus*. Furthermore, soy polysaccharide-carnauba wax composites have been shown to significantly reduce bacterial counts during food storage (Ma et al. 2023). Collectively, these findings highlight the dual functionality of carnauba wax in improving both oxidative stability and microbial protection in active food packaging systems.

11.4.1.6 Other Performance

The incorporation of carnauba wax significantly impacts the thermal and gas barrier properties of biopolymer films, although these effects vary depending on the matrix composition. For example, carnauba wax-ZnO nanocomposites exhibit reduced thermal stability, with degradation onset at 190 °C and a peak degradation temperature of 414 °C. This is due to weakened interactions within the polymer network (Santos et al. 2014). In contrast, CCTW films demonstrate superior thermal resistance, reaching a maximum mass loss at 426.7°C and retaining 24.53% of their weight at 800 °C. This enhanced thermal stability is attributed to the formation of a cross-linked network (Silva et al. 2018). In terms of gas barrier performance, carnauba wax typically boosts the resistance to gas permeation. This is evidenced by reduced O_2 and CO_2 permeability in chitosan films, which results from the creation of tortuous diffusion paths (Corrêa-Filho et al. 2024). However, excessive wax content in starch films may form interfacial channels that increase the respiration rates of apple slices (Chiumarelli and Hubinger 2014). Furthermore, the efficiency of the oxygen barrier is positively correlated with cohesive energy density and inversely correlated with the size of micro-ordered regions. This suggests that optimization can be achieved through the control of wax nanoparticle size (Liu et al. 2022). Importantly, these enhancements can be achieved while maintaining full biodegradability, which is crucial for environmentally sustainable food packaging applications.

11.4.1.7 Application in Food Packaging

Carnauba wax—enhanced composite films demonstrate significant efficacy in preserving various fruits and vegetables. For instance, CCTW films maintain banana freshness for 7 d by minimizing enzymatic browning and dehydration (Silva et al. 2018). Similarly, sodium alginate-carnauba wax coatings effectively preserve fresh-cut apples. These films achieve this by reducing soluble solid losses while simultaneously inhibiting microbial growth and polyphenol oxidase activity (Lin et al. 2023). In contrast, soybean polysaccharide-wax films primarily enhance grape shelf life through improved water retention (Ma et al. 2023). Furthermore, optimized carnauba wax formulations that incorporate poly ethylene glycol (PEG), sodium alginate, and sodium dodecyl sulfate (SDS) significantly maintain eggplant firmness ($P < 0.05$) and moisture content during ambient storage (Singh et al. 2016). These findings highlight the versatility of carnauba wax in enhancing polysaccharide-based films. These films prolong food quality through multifunctional mechanisms, including moisture regulation, texture preservation, and oxidative/microbial protection.

11.4.2 Protein Based Film

11.4.2.1 Thickness

The addition of carnauba wax significantly increases film thickness. This is because wax particles disrupt the organized molecular arrangement during film formation. At higher concentrations, this effect is further amplified. The enhanced interactions between the wax and gelatin weaken the forces that hold protein chains together. This results in a more expanded matrix structure and increases the thickness (Winkler-Moser et al. 2019; Tongnuanchan et al. 2015).

11.4.2.2 Color and Light Transmittance

Incorporating carnauba wax (5–15% w/w) into gelatin films significantly increases opacity and induces yellowing, while reducing brightness (L* from 98.9 to 94.1–97.2) and light transmittance. In particular, medium caseinate films show pronounced increases in opacity (from 2.28 to 10.55 Abs/mm) (Zhang et al. 2018; Galus et al. 2020). These optical changes may compromise applications that require high transparency. The modifications primarily result from the wax's intrinsic coloration and its impact on the miscibility of the polymer matrix. However, excessive concentrations of carnauba wax may ultimately impair film functionality due to substantial light scattering and significant color alteration.

11.4.2.3 Water Resistance Behavior

The addition of carnauba wax to sodium caseinate films significantly enhances their hydrophobic properties. This is evidenced by decreased WVP (from 4.75×10^{-10} g/m Pa s) and reduced water solubility (from 45.53% in control films to 44.46–45.42%) (Galus et al. 2020). These improvements are primarily due to the inherent hydrophobicity of wax. However, homogeneous wax layers are more effective than dispersed phases in enhancing these properties. Furthermore, higher concentrations of carnauba wax (1.5%) in sodium alginate-whey protein films exhibit superior moisture resistance, with values of 3.12 ± 0.31 g mm/kPa h m^2 (Butt et al. 2023). This underscores the potential of carnauba wax for use in coatings for lipid-rich or moisture-sensitive foods. However, it is noted that further optimization is necessary for specific applications.

11.4.2.4 Mechanical Properties

The mechanical properties of protein-based films respond in complex ways to the incorporation of carnauba wax. For instance, sodium caseinate films demonstrate increased tensile strength (from 1.65 to 1.91–2.60 MPa) and Young's modulus despite the typical strength-reducing effects of lipids. This phenomenon is attributed to wax-induced matrix compaction and molecular chain reorganization (Galus et al. 2020). In contrast, a 15% wax loading in gelatin films decreases stiffness (a 25% reduction in Young's modulus) while simultaneously enhancing strength (by 39%) and ductility (a 62% increase in elongation) (Winkler-Moser et al. 2019). However, these effects are modulated by emulsifiers such as Tween-80. This emulsifier can reverse the anti-plasticizing effects of wax by increasing chain mobility (from 63.7 to 114.5%

elongation), thanks to its small molecular size (Galus and Kadzińska 2015). These opposing trends arise from the wax's dual capacity. It can reinforce polymer networks through physical interactions while also creating discontinuities that reduce elasticity. This demonstrates that the mechanical influence of wax depends critically on its distribution, concentration, and interactions with other film components (Galus et al. 2020).

11.4.2.5 Antibacterial Properties

Carnauba wax-based edible coatings (4.5–9% concentration) exhibit moderate postharvest antifungal activity, particularly against *Monilinia fructicola* and *Rhizopus stolonifer*, in stone fruit preservation. However, supplemental antimicrobial agents are required to achieve complete disease control (Oliveira et al. 2018). When integrated with natural antibacterial treatments, the performance of carnauba wax-based coatings is significantly enhanced. For instance, during storage at 25 °C, the incidence of *Penicillium digitatum* in honey oranges was reduced from 72.2–96.0% in untreated fruit to minimal levels following combined applications of CaO cleaning and grapefruit seed extract/carnauba wax (GSE/CW) coatings (Choi et al. 2019). These findings confirm that carnauba wax can serve as an effective antimicrobial component in protein-based composite films when properly formulated with complementary agents.

11.4.2.6 Viscosity and Oxygen Permeability

The incorporation of carnauba wax into sodium alginate/whey protein systems significantly enhances solution viscosity by forming a biopolymer-mediated gel-like structure. Specifically, viscosity increases linearly with rising wax concentration, up to a 20% increase, due to larger particle formation and restricted flow (Butt et al. 2023). Similarly, casein-carnauba wax composites exhibit reduced oxygen permeability compared to controls. This is because the wax effectively fills surface micropores and increases film density (Silva et al. 2018). Therefore, carnauba wax demonstrates a dual capacity to optimize both rheological and barrier properties for improved food packaging applications, through viscosity modulation and pore-sealing mechanisms.

11.4.2.7 Application in Food

The application of GSE/CW coatings, especially in combination with CaO cleaning, effectively preserves mandarin quality during storage at both 4 and 25 °C. This is achieved by maintaining CO_2 production, firmness, phenolic content, pH, ascorbic acid levels, and antioxidant capacity, while also inhibiting the growth of *Aspergillus* and enhancing sensory attributes (Choi et al. 2019). Similarly, carnauba wax films significantly reduce weight loss and firmness decline in stored oranges compared to untreated controls. Although there are gradual decreases in vitamin C and acidity levels, these coatings still demonstrate efficacy in minimizing fruit decay over extended storage periods (40–80 d) (Babarabie et al. 2024). Collectively, these findings confirm that carnauba wax is a highly effective coating material for prolonging postharvest fruit quality through integrated physiological and microbial protection mechanisms.

11.4.3 Biopolyester-Wax Film

11.4.3.1 Water Resistance Behavior

The incorporation of carnauba wax significantly enhances the water barrier properties of biopolymer films. This enhancement results in values only twice that of low-density polyethylene (0.21), thereby achieving a "high-to-medium" barrier performance compared to most biodegradable polymers (Polunin et al. 2023; Wu et al. 2021). When combined with glycerol in PVA matrices, carnauba wax further improves hydrophobicity through optimized polymer dispersion and surface energy modification, achieving contact angles greater than 90° (Tijani et al. 2024). Additionally, PLA/carnauba wax composites demonstrate exceptional super hydrophobicity, with a water contact angle of 160.4° and a sliding angle of 6°. These composites also exhibit broad liquid repellency, with contact angles ranging from 149.1° to 152.3° for various aqueous solutions (Wang et al. 2023). Collectively, these findings confirm that carnauba wax is highly effective in developing moisture-resistant coatings. These coatings minimize water loss and microbial contamination in fruit and vegetable preservation applications.

11.4.3.2 Mechanical Properties

The incorporation of carnauba wax into various polymer matrices demonstrates significant potential for developing sustainable food packaging alternatives. For instance, babassu oil-based acrylic polymer (PBBM)-xylan-carnauba wax (PBBM-X/CW) blends containing 40% wax achieve mechanical properties (475 MPa storage modulus and 248 MPa Young's modulus) that approach those of low-density polyethylene while maintaining ductile plastic behavior, despite lower yield strength (Polunin et al. 2023). In PVA systems, optimal wax/glycerol ratios (1:3 and 2:6) enhance modulus. However, excessive carnauba wax content induces phase separation and reduces tensile strength due to weakened PVA-wax interactions and disrupted matrix integrity (Tijani et al. 2024). For PLA composites, carnauba wax exhibits dual functionality as both a plasticizer and a reinforcement, depending on frequency. It increases the elastic modulus (G') at low frequencies while decreasing it at high frequencies, except in PLA-CBA blends, where G' remains constant (Mendoza-Duarte et al. 2023). Thus, these findings establish carnauba wax as a versatile bio-based modifier capable of tailoring mechanical performance for specific packaging applications while promoting environmental sustainability.

11.4.3.3 Thermal Performance

The incorporation of carnauba wax into polymer matrices significantly enhances thermal stability. For example, CW-PVA-glycerol composites exhibit delayed thermal degradation above 450 °C due to strengthened intermolecular bonds (Tijani et al. 2024). Although pure carnauba wax coatings lose their microstructure at 120 °C, PLA/CW composites maintain super hydrophobicity. This is attributed to the higher melting point of PLA, which preserves the rough surface morphology essential for water repellency (Wang et al. 2023). Furthermore, thermogravimetric analysis reveals that the incorporation of CW does not affect the thermal degradation profile of PLA (T10% and Tmax) and processing window (<200 °C) (Mendoza-Duarte et al. 2023). Summary, these findings confirm that carnauba wax is effective in improving heat resistance for food packaging applications, ensuring thermal stability during common processing and storage conditions.

11.4.3.4 Application in Food Packaging

Carnauba wax-based coatings have demonstrated significant efficacy in postharvest preservation. For example, wax-essential oil composites have been shown to reduce cucumber weight loss during 15-day storage at 10 °C (Gutiérrez-Pacheco et al. 2020). Similarly, wax-glyceryl laurate formulations maintain fruit quality at 20 °C by delaying color changes, preserving chlorophyll and ascorbic acid content, and reducing respiration rates compared to controls (Chen et al. 2019). Furthermore, combined putrescine-carnauba wax treatments effectively mitigate chilling injury symptoms in pomegranates. These symptoms include skin discoloration, weight loss, and texture degradation. The treatments also suppress ethylene production and respiratory activity (Barman et al. 2011). These findings establish carnauba wax as a versatile coating material capable of extending shelf life through multiple preservation mechanisms. These mechanisms are comprehensively summarized in the overview of functional wax-composite packaging films (Susmita Devi et al. 2022) in Table 11.2.

Table 11.2 Use of carnauba wax in biopolymer-based composite films for fruits and vegetables packaging (Susmita Devi et al. 2022)

Biopolymers and others functional ingredients	Carnauba wax (%)	Effect of carnauba wax on composite films	References
Arrowroot starch 2 wt%	Carnauba wax 0–15 wt%	Carnauba wax increased hydrophobicity, reduced solubility, and improved moisture, light barrier and thermal stability of composite films	Oliveira Filho et al. (2020)
Cassava starch 2.5–3.5% w/w	Carnauba wax 0.1–0.4% wt%	Composite films showed good mechanical, thermal, physical and structural properties	Chiumarelli and Hubinger (2014)
Cashew tree gum 70–85% w/v Cassava starch 15–30% w/v	Carnauba wax 0–15% w/v	Addition of CW improved barrier, water resistance properties of the films	Morillon et al. (2002)
Gelatin 5% w/v	Carnauba wax 0.25–0.75% w/v	Addition of carnauba wax into gelatin films improved thermal stability, barrier to light and moisture	Zhang et al. (2018)
Myofibrillar proteins 3% w/v	Carnauba wax 1%, w/v	Carnauba wax resulted in increased tensile strength and barrier properties of the composite films	Moghtadaei et al. (2018)
Chitosan 0.8% w/v	Carnauba wax 10% v/v	Incorporation of CW decreased WVT of the composite n films	Gutiérrez-Pacheco et al. (2020)
Arrowroot starch: cellulose nanocrystals 5:95, w/w Essential oils 0.1–0.3%, v/v	Carnauba wax nano emulsion 15% w/w	Composite film shows excellent barriers against water vapor permeability, UV light, and fungal growth	Oliveira et al. (2021)

(continued)

Table 11.2 (continued)

Biopolymers and others functional ingredients	Carnauba wax (%)	Effect of carnauba wax on composite films	References
Konjac glucomannan solution 0.6% w/v	Carnauba wax 0, 4, 8, 12, 16 and 20% w/w	Increased concentration of carnauba wax, significantly increased composite films hydrophobicity, and their mechanical, barrier and functional properties	Haruna et al. (2019)
Sodium caseinate 8% w/v	Carnauba wax 0.5–1%	CW enhanced mechanical and water solubility of the composite film	Galus et al. (2020)

11.5 Other Natural Wax Functionalized Food Packaging Films

11.5.1 Candelilla Wax

Candelilla wax, a natural plant wax derived from *Euphorbia antisyphilitica* Zucc., is characterized by its excellent hydrophobic properties and environmental sustainability, making it a promising candidate for food packaging applications. When applied to fruit coatings at concentrations below 200 mg/kg, it effectively enhances the shelf life and quality of perishable produce (León-Zapata et al. 2018). The incorporation of candelilla wax into edible films significantly reduces water vapor permeability, with reductions ranging from 11.1 to 41.7% in starch/gelatin films. This improvement is achieved without compromising the mechanical properties of the films. In fact, an optimal concentration of 0.5% candelilla wax has been shown to markedly enhance the waterproof performance of these films (Rodrigues et al. 2014; Cheng et al. 2024). Furthermore, candelilla wax exhibits synergistic effects when combined with other ingredients. For instance, when blended with whey protein and glycerol, it forms effective barrier films that are particularly suitable for fresh produce. The addition of tar polyphenols to candelilla wax further boosts its antioxidant and antifungal activity, thereby providing additional protection against spoilage and microbial contamination (Alvarez-Perez et al. 2022). In addition to its barrier and antimicrobial properties, candelilla wax serves as an oil-gelling agent and carrier for food additives. This dual functionality enhances its versatility in edible coating formulations. By balancing biodegradability with enhanced preservation performance through hydrophobic modification and structural reinforcement, candelilla wax offers a sustainable and effective solution for extending the shelf life of perishable foods (Mert and Demirkesen 2016).

11.5.2 Rice Bran Wax

Rice bran wax, a byproduct of rice bran oil refining, is composed of long-chain fatty acids (C20–C26) and alcohols (C30–C36). It exhibits exceptional thermal stability with a melting point of 78–81 °C and a tendency to crystallize at ambient temperatures (Thakur et al. 2023). These properties make it a valuable material for food packaging applications. In practical applications, rice bran wax enables optimized formulations that significantly enhance product performance. For instance, a formulation containing 0.6% rice bran wax and 5.5% whey protein has been shown to extend the shelf life of popcorn cups (Ganjizadeh Zavareh et al. 2021). Additionally, coatings made with rice bran wax on cherry tomatoes reduce weight loss to 13.54%, compared to 16.02% in controls (Zhang et al. 2017). Furthermore, chitosan composites incorporating rice bran wax demonstrate a 1.73-fold increase in elongation and a 0.19-fold reduction in water vapor permeability (Liu et al. 2025). Beyond these conventional uses, rice bran wax also facilitates the development of intelligent packaging systems. It exhibits pH and ammonia-responsive color changes, which can be used for monitoring freshness. Simultaneously, it provides hydrophobic protection against microbial invasion and enzymatic browning. These multifunctional properties make rice bran wax suitable for a wide range of applications, from fresh-cut produce to dairy and meat products (Abhirami et al. 2020). Overall, rice bran wax demonstrates its value as an environmentally sustainable material that enhances both protective and smart packaging performance.

11.5.3 Sugarcane Wax

Sugarcane wax effectively functions as a key component in edible coatings for fruit and vegetable preservation by mitigating water evaporation and microbial contamination, thereby extending shelf life. Research has demonstrated that coatings combining sugarcane wax with other natural compounds significantly improve postharvest quality. Its inherent hydrophobicity enhances the water resistance of packaging films, particularly when incorporated into chitosan matrices. This results in films with superior hydrophobic and stable properties (Apicella et al. 2022). The addition of sugarcane wax not only improves film flexibility, increasing elongation at break by 1.73-fold, but also reduces water vapor permeability by 0.19 times. These properties make sugarcane wax particularly valuable in food packaging applications. It serves dual purposes as both a preservative agent and a component in intelligent packaging systems, while simultaneously supporting environmental sustainability objectives.

11.5.4 Insect Wax

Insect wax serves as an effective hydrophobic coating for paper-based packaging materials, such as cups and bowls. Optimized wax composition and annealing treatments significantly enhance hydrophobicity and moisture resistance while reducing water absorption and imparting self-cleaning and antifouling properties (Ahuja and Kumar 2023). When combined with biopolymers such as proteins, polysaccharides, or lipids, insect wax forms functional edible films with excellent film-forming capacity and safety. These films can be further modified through the incorporation of antimicrobial agents, including natural compounds, synthetic chemicals, or metal nanoparticles, to enhance antibacterial performance (Wang et al. 2025b). Moreover, synergistic combinations with silver/copper ions, quaternary ammonium salts, or bioactive peptides like lactoferrin produce antimicrobial packaging films that effectively inhibit microbial growth and extend food shelf life. In summary, these properties establish insect wax as a multifunctional component in sustainable packaging. It simultaneously improves waterproofing, mechanical strength, and antimicrobial functionality while supporting the development of eco-friendly packaging solutions.

11.5.5 Wax Extracted from Fruits

The incorporation of natural berry wax (5–25 wt%) into chitosan films significantly enhances their functional properties by reducing moisture content, swelling capacity, solubility, and water vapor permeability, while simultaneously improving opacity, thermal stability, and tensile strength. Notably, the composite films exhibit superior thermal stability and UV/visible light barrier properties compared to pure chitosan films (Khan et al. 2024). Similarly, mango seed wax has been shown to effectively plasticize PLA films, increasing their hydrophobicity by 26% and significantly enhancing barrier properties, as evidenced by a 55.6% reduction in water vapor transmission rate and a 10% decrease in oxygen transmission rate. Although minor reductions in glass transition (10%) and melting temperatures (3%) were observed, these trade-offs do not diminish the overall improvements in film performance (Thomas et al. 2022). In conclusion, these results demonstrate that various natural waxes, both plant-derived and animal-derived, can substantially improve the performance of food packaging films, reduce the weight loss and extend the shelf life of food (Fig. 11.4). This is particularly evident in terms of water resistance and mechanical properties, thereby establishing their potential as sustainable materials for food packaging applications.

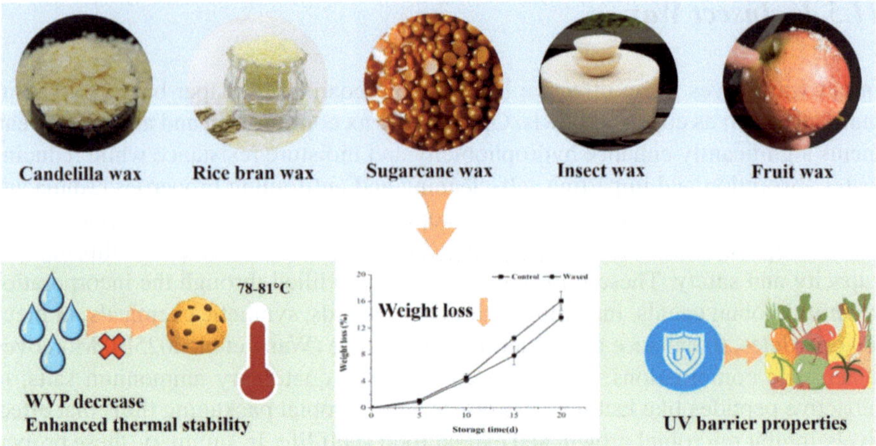

Fig. 11.4 The influence of other natural waxes on packaging films and their practical application in products

11.6 Conclusion and Future Prospects

Natural wax exhibits exceptional potential in biopolymer-based food packaging due to its inherent hydrophobicity, renewability, biodegradability, and bioactive properties, as evidenced by this review's systematic examination of how various waxes (beeswax, carnauba wax, etc.) enhance key film characteristics when incorporated into polysaccharide-, protein-, or polymer-based matrices, including mechanical strength, barrier properties, and antimicrobial/antioxidant activities. Moreover, practical applications confirm that wax-enhanced films effectively prolong the shelf life of perishables like strawberries and citrus by mitigating moisture loss and microbial growth. However, current limitations involve disjointed production processes between wax extraction and film fabrication, indicating that future research should prioritize process integration for cost efficiency. Beyond this, further development opportunities exist for creating multifunctional smart packaging through the incorporation of active compounds and spoilage indicators, though comprehensive safety assessments remain imperative despite waxes' natural origin.

Declaration AI tools were used only to improve language and readability. The authors have reviewed the final manuscript and take full responsibility for its content.

References

Abhirami P, Modupalli N, Natarajan V (2020) Novel postharvest intervention using rice bran wax edible coating for shelf-life enhancement of Solanum lycopersicum fruit. J Food Process Preserv 44(12):e14989

Aguirre-Joya JA, Cerqueira MA, Ventura-Sobrevilla J et al (2019) Candelilla wax-based coatings and films: functional and physicochemical characterization. Food Bioprocess Technol 12(10):1787–1797

Ahuja A, Kumar RV (2023) Spray coating of edible insect waxes for liquid food packaging. Appl Surf Sci 624:157150

Alvarez-Perez OB, Ventura-Sobrevilla JM, Torres-León C et al (2022) Development and characterization of whey protein films incorporated with tarbush polyphenols and candelilla wax. Food Biosci 45:101505

Al-Waili NS (2003) Topical application of natural honey, beeswax and olive oil mixture for atopic dermatitis or psoriasis: partially controlled, single-blinded study. Complement Ther Med 11(4):226–234

Apicella A, Barbato A, Garofalo E et al (2022) Effect of PVOH/PLA+ wax coatings on physical and functional properties of biodegradable food packaging films. Polymers 14(5):935

Arcan I, Yemenicioğlu A (2013) Development of flexible zein–wax composite and zein–fatty acid blend films for controlled release of lysozyme. Food Res Int 51(1):208–216

Auras R, Arroyo B, Selke S (2009) Production and properties of spin-coated cassava-starch-glycerol-beeswax films. Starch/Staerke 61(8):463–471

Aylanc V, Falcão SI, Ertosun S et al (2021) From the hive to the table: nutrition value, digestibility and bioavailability of the dietary phytochemicals present in the bee pollen and bee bread. Trends Food Sci Technol 109:464–481

Babarabie M, Sardoei A S, Jamali B et al (2024) Carnauba wax-based edible coatings retain quality enhancement of orange (Citrus sinensis cv. Moro) fruits during storage. Sci Rep 14(1):4133

Bahrami A, Rezaei Mokarram R, Sowti Khiabani M et al (2019) Physico-mechanical and antimicrobial properties of tragacanth/hydroxypropyl methylcellulose/beeswax edible films reinforced with silver nanoparticles. Int J Biol Macromol 129:1103–1112

Barman K, Asrey R, Pal RK (2011) Putrescine and carnauba wax pretreatments alleviate chilling injury, enhance shelf life and preserve pomegranate fruit quality during cold storage. Sci Hortic 130(4):795–800

Barman K, Asrey R, Pal R K et al (2014) Influence of putrescine and carnauba wax on functional and sensory quality of pomegranate (Punica granatum L.) fruits during storage. J Food Sci Technol 51(1):111–117

Baswal AK, Dhaliwal HS, Singh Z et al (2020) Influence of carboxy methylcellulose, chitosan and beeswax coatings on cold storage life and quality of Kinnow mandarin fruit. Sci Hortic 260:108887

Bayram B, Ozkan G, Kostka T et al (2021) Valorization and application of fruit and vegetable wastes and by-products for food packaging materials. Molecules 26(13):4031

Beck PHB, De Camargo Lima Beluci N, Monteiro ARG et al (2022) Carnauba wax utilization in salt encapsulation: application in bread. J Food Sci Technol 59(8):3307–3311

Betancur-D´Ambrosio MC, Pérez-Cervera CE, Barrera-Martinez C et al (2024) Antimicrobial activity, mechanical and thermal properties of cassava starch films incorporated with beeswax and propolis. J Food Sci Technol 61(4):782–789

Butt MS, Akhtar M, Maan AA et al (2023) Fabrication and characterization of carnauba wax-based films incorporated with sodium alginate/whey protein. J Food Measur Charact 17(1):694–705

Caron VC, Jacomino AP, Sarantópoulos CIGL et al (2015) Carnauba wax and modified atmosphere in refrigerated preservation of 'Tahiti' acid limes. Packag Technol Sci 28(7):647–656

Castaldo S, Capasso F (2002) Propolis, an old remedy used in modern medicine. Fitoterapia 73(Suppl 1):S1-6

Chaireh S, Ngasatool P, Kaewtatip K (2020) Novel composite foam made from starch and water hyacinth with beeswax coating for food packaging applications. Int J Biol Macromol 165:1382–1391

Chen H, Sun Z, Yang H (2019) Effect of carnauba wax-based coating containing glycerol monolaurate on the quality maintenance and shelf-life of Indian jujube (Zizyphus mauritiana Lamk.) fruit during storage. Sci Hortic 244:157–164

Cheng Y, Zhai X, Wu Y et al (2023) Effects of natural wax types on the physicochemical properties of starch/gelatin edible films fabricated by extrusion blowing. Food Chem 401:134081

Cheng Y, Wang W, Hou H et al (2024) A highly hydrophobic sustainable starch/gelatin-beeswax biodegradable film: easy to industrial scale-up, recyclable, and suitable for multiple packaging application scenarios. ACS Sustain Chem Eng 12(19):7351–7362

Chevalier E, Chaabani A, Assezat G et al (2018) Casein/wax blend extrusion for production of edible films as carriers of potassium sorbate—a comparative study of waxes and potassium sorbate effect. Food Packag Shelf Life 16:41–50

Chiumarelli M, Hubinger MD (2014) Evaluation of edible films and coatings formulated with cassava starch, glycerol, carnauba wax and stearic acid. Food Hydrocoll 38:20–27

Choi HY, Bang IH, Kang JH et al (2019) Development of a microbial decontamination system combining washing with highly activated calcium oxide solution and antimicrobial coating for improvement of mandarin storability. J Food Sci 84(8):2190–2198

Chu Z, Wang H, Dong B (2024) Research on food preservation based on antibacterial technology: progress and future prospects. Molecules 29(14):3318

Corrêa-Filho LC, Santos Junior JRD, Ramos AV et al (2024) Chitosan-based nanocomposite films with carnauba wax, rosin resin, and zinc oxide nanoparticles. Food Res Int 188:114475

Cortés-Rodríguez M, Villegas-Yépez C, Gil González JH et al (2020) Development and evaluation of edible films based on cassava starch, whey protein, and bees wax. Heliyon 6(9):e04884

Cruces F, García MG, Ochoa NA (2021) Reduction of water vapor permeability in food multilayer biopackaging by epitaxial crystallization of beeswax. Food Bioprocess Technol 14(7):1244–1255

Cruz MAL, Gomes VM, Fernandes KVS et al (2002) Identification and partial characterization of a chitinase and a β-1,3-glucanase from Copernicia cerifera wax. Plant Physiol Biochem 40(1):11–16

De Almeida BC et al (2016) Antiprotozoal activity of extracts and isolated triterpenoids of 'carnauba' (Copernicia prunifera) wax from Brazil. Pharm Biol 54(12):3280–3284

De Campos A, Claro PC, Luchesi BR et al (2019) Curaua cellulose sheets dip coated with micro and nano carnauba wax emulsions. Cellulose 26(13):7983–7993

De León-Zapata MA, Ventura-Sobrevilla JM, Salinas-Jasso TA et al (2018) Changes of the shelf life of candelilla wax/tarbush bioactive based-nanocoated apples at industrial level conditions. Sci Hortic 231:43–48

De Oliveira FJG, Albiero BR, Cipriano L et al (2021) Arrowroot starch-based films incorporated with a carnauba wax nanoemulsion, cellulose nanocrystals, and essential oils: a new functional material for food packaging applications. Cellulose 28(10):6499–6511

De S. Dantas AN, Magalhães TA, Matos WO et al (2013) Characterization of Carnauba wax inorganic content. J Am Oil Chem Soc 90(10):1475–1483

Debeaufort F (2021) Active biopackaging produced from by-products and waste from food and marine industries. FEBS Open Bio 11(4):984–998

Dirpan A, Ainani AF, Djalal M (2023) A review on biopolymer-based biodegradable film for food packaging: trends over the last decade and future research. Polymers 15(13):2781

Doan CD, To CM, De Vrieze M et al (2017) Chemical profiling of the major components in natural waxes to elucidate their role in liquid oil structuring. Food Chem 214:717–725

Duan J, Zhou Q, Fu M et al (2023) Research on properties of edible films prepared from zein, soy protein isolate, wheat gluten protein by adding beeswax. Food Bioprocess Technol 16(11):2443–2454

Fabra MJ, Jiménez A, Atarés L et al (2009) Effect of fatty acids and beeswax addition on properties of sodium caseinate dispersions and films. Biomacromol 10(6):1500–1507

Fasihi H, Noshirvani N, Hashemi M et al (2019) Antioxidant and antimicrobial properties of carbohydrate-based films enriched with cinnamon essential oil by Pickering emulsion method. Food Packag Shelf Life 19:147–154

Fayaz G, Goli S a H, Kadivar M et al (2017) Potential application of pomegranate seed oil oleogels based on monoglycerides, beeswax and propolis wax as partial substitutes of palm oil in functional chocolate spread. LWT 86:523–529

Filho ACVA, Rodrigues PAS, Benjamin SR et al (2017) Hypolipidemic activity of P-methoxycinnamic diester (PCO-C) isolated from copernicia pruńifera against Triton WR-1339 and hyperlipidemic diet in mice. Environ Toxicol Pharm, 56:198–203

Florido-Moreno P, Benítez JJ, González-Buesa J et al (2025) Plasticized cellulose bioplastics with beeswax for the fabrication of multifunctional, biodegradable active food packaging. Food Hydrocoll 162:110933

Formiga AS, Pinsetta JS, Pereira EM et al (2019) Use of edible coatings based on hydroxypropyl methylcellulose and beeswax in the conservation of red guava 'Pedro Sato.' Food Chem 290:144–151

Fratini F, Cilia G, Turchi B et al (2016) Beeswax: a minireview of its antimicrobial activity and its application in medicine. Asian Pac J Trop Med 9(9):839–843

Fukuyama CWT, Duarte LGR, Pedrino IC et al (2024) Effect of carnauba wax nanoemulsion associated with Syzygium aromaticum and Mentha piperita essential oils as an alternative to extend lychee post-harvest shelf life. Sustain Food Technol 2(2):426–436

Galus S, Kadzińska J (2015) Food applications of emulsion-based edible films and coatings[J]. Trends Food Sci Technol 45(2):273–283

Galus S, Gaouditz M, Kowalska H et al (2020) Effects of candelilla and carnauba wax incorporation on the functional properties of edible sodium caseinate films. Int J Mol Sci 21(24):9349

Ganjizadeh Zavareh S, Javanmard Dakheli M, Tajeddin B (2021) Optimization of biodegradable paper cup packaging coated with whey protein isolate and rice bran wax as potential popcorn package. Food Sci Nutr 9(12):6762–6775

Genovese L, Lotti N, Gazzano M et al (2016) Novel biodegradable aliphatic copolyesters based on poly(butylene succinate) containing thioether-linkages for sustainable food packaging applications. Polym Degrad Stab 132:191–201

Giampieri F, Gasparrini M, Forbes-Hernández TY et al (2018) Beeswax by-products efficiently counteract the oxidative damage induced by an oxidant agent in human dermal fibroblasts. Int J Mol Sci 19(9):2842

Giampieri F, Quiles JL, Cianciosi D et al (2022) Bee products: an emblematic example of underutilized sources of bioactive compounds. J Agric Food Chem 70(23):6833–6848

Gomes ÁVR, Gonçalves FCP, Silva Júnior MQD et al (2019) Effect of carnauba wax and coconut fiber contents on tensile properties of corn starch-based biocomposites. Mater Res 22

Gonçalves FP, Martins MC, Junior GJS et al (2010) Postharvest control of brown rot and Rhizopus rot in plums and nectarines using carnauba wax. Postharvest Biol Technol 58(3):211–217

Gu X, Du L, Meng Z (2023) Comparative study of natural wax-based W/O emulsion gels: microstructure and macroscopic properties. Food Res Int 165:112509

Gutiérrez-Pacheco MM, Ortega-Ramírez LA, Silva-Espinoza BA et al (2020) Individual and combined coatings of chitosan and carnauba wax with oregano essential oil to avoid water loss and microbial decay of fresh cucumber. Coatings 10(7):614

Haq MA, Hasnain A, Jafri FA et al (2016) Characterization of edible gum cordia film: effects of beeswax. LWT Food Sci Technol 68:674–680

Haruna MH, Wang Y, Pang J (2019) Konjac glucomannan-based composite films fabricated in the presence of carnauba wax emulsion: hydrophobicity, mechanical and microstructural properties evaluation. J Food Sci Technol 56(11):5138–5145

Hassan B, Chatha SAS, Hussain AI et al (2018) Recent advances on polysaccharides, lipids and protein based edible films and coatings: a review. Int J Biol Macromol 109: 1095–1107

Hosseini SF, Mousavi Z, Mcclements DJ (2023) Beeswax: a review on the recent progress in the development of superhydrophobic films/coatings and their applications in fruits preservation. Food Chem 424:136404

Kaluza BF, Wallace H, Heard TA et al (2016) Urban gardens promote bee foraging over natural habitats and plantations. Ecol Evol 6(5):1304–1316

Kazemi M, Jahanbin K, Ojagh SM et al (2023) Development of antimicrobial gelatin-ulvan-beeswax composite films: optimization of formulation using mixture design methodology. Int J Biol Macromol 231:123384
Khan S, Hashim SBH, Arslan M et al (2024) Berry wax improves the physico-mechanical, thermal, water barrier properties and biodegradable potential of chitosan food packaging film. Int J Biol Macromol 261:129821
Khanzadi M, Jafari SM, Mirzaei H et al (2015) Physical and mechanical properties in biodegradable films of whey protein concentrate–pullulan by application of beeswax. Carbohyd Polym 118:24–29
Khosronia A, Asadi G, Jafarpour A et al (2023) Feasibility of encapsulation of baking powder with carnauba and beeswax and investigating the physicochemical properties of produced powder. J Food Meas Charact 17(5):5163–5174
Kowalczyk D, Baraniak B (2014) Effect of candelilla wax on functional properties of biopolymer emulsion films—a comparative study. Food Hydrocoll 41:195–209
Kyuchyuk S, Paneva D, Karashanova D et al (2022) Core-sheath-like poly(ethylene oxide)/beeswax composite fibers prepared by single-spinneret electrospinning. Antibacterial, antifungal, and antitumor activities. Macromol Biosci 22(6):2200015
Lara I, Heredia A, Domínguez E (2019) Shelf life potential and the fruit cuticle: the unexpected player. Front Plant Sci 10
Lee MH, Kim SY, Park HJ (2018) Effect of halloysite nanoclay on the physical, mechanical, and antioxidant properties of chitosan films incorporated with clove essential oil. Food Hydrocoll 84:58–67
Li C, Wang L, Xue F (2019) Effects of conjugation between proteins and polysaccharides on the physical properties of emulsion-based edible films. J Am Oil Chem Soc 96(11):1249–1263
Lim WS, Kim MH, Park HJ et al (2024) Characterization of Polyvinyl Alcohol (PVA)/Polyacrylic Acid (PAA) composite film-forming solutions and resulting films as affected by beeswax content. Polymers 16(3):310
Lin X, Zhang H, Guo X et al (2023) A novel sodium alginate-carnauba wax film containing calcium ascorbate: structural properties and preservative effect on fresh-cut apples. Molecules 28(1):367
Liu S, Li L, Li B et al (2022) Size effect of carnauba wax nanoparticles on water vapor and oxygen barrier properties of starch-based film. Carbohyd Polym 296:119935
Liu T, Xue J, Qin R et al (2025) Bioinspired hydrophobic chitosan-based films for monitoring the visual freshness of pork. Food Hydrocoll 162:110932
Luchesi BR, Moreira FKV, Marconcini JM (2024) Scalable production of hydrophobic starch/beeswax films by continuous solution casting. J Appl Polym Sci 141(1):e54730
Ma J, Ma L, Zhang H et al (2018) Policosanol fabrication from insect wax and optimization by response surface methodology. PLoS ONE 13(5):e0197343
Ma Z, Xing Z, Zhao Y et al (2023) Lotus leaf inspired sustainable and multifunctional Janus film for food packaging. Chem Eng J 457:141279
Maia M, Nunes FM (2013) Authentication of beeswax (Apis mellifera) by high-temperature gas chromatography and chemometric analysis. Food Chem 136(2):961–968
Martínez-Abad A, Lagarón JM, Ocio MJ (2014) Antimicrobial beeswax coated polylactide films with silver control release capacity. Int J Food Microbiol 174:39–46
Maryam Adilah ZA, Jamilah B, Nur Hanani ZA (2018) Functional and antioxidant properties of protein-based films incorporated with mango kernel extract for active packaging. Food Hydrocoll 74:207–218
Mendoza-Duarte ME, Estrada-Moreno IA, López-Martínez EI et al (2023) Effect of the addition of different natural waxes on the mechanical and rheological behavior of PLA—a comparative study. Polymers 15(2):305
Mert B, Demirkesen I (2016) Reducing saturated fat with oleogel/shortening blends in a baked product. Food Chem 199:809–816

Moghtadaei M, Soltanizadeh N, Goli SAH (2018) Production of sesame oil oleogels based on beeswax and application as partial substitutes of animal fat in beef burger. Food Res Int 108:368–377
Monedero FM, Fabra MJ, Talens P et al (2009) Effect of oleic acid–beeswax mixtures on mechanical, optical and water barrier properties of soy protein isolate based films. J Food Eng 91(4):509–515
Morillon V, Frédéric D, Geneviève B et al (2002) Factors affecting the moisture permeability of lipid-based edible films: a review. Crit Rev Food Sci Nutr 42(1):67–89
Moyo P, Fourie PH, Masikane SL et al (2020) The effects of postharvest treatments and sunlight exposure on the reproductive capability and viability of phyllosticta citricarpa in citrus black spot fruit lesions. Plants 9(12):1813
Mukhtar A, Damerow L, Blanke M (2014) Non-invasive assessment of glossiness and polishing of the wax bloom of European plum. Postharvest Biol Technol 87:144–151
Nair SS, Trafiałek J, Kolanowski W (2023) Edible packaging: a technological update for the sustainable future of the food industry. Appl Sci 13(14):8234
Navarro-Tarazaga ML, Massa A, Pérez-Gago MB (2011) Effect of beeswax content on hydroxypropyl methylcellulose-based edible film properties and postharvest quality of coated plums (Cv. Angeleno). LWT—Food Sci Technol 44(10):2328–2334
Nevine GB (2011) Study on the antimicrobial activity of honey products and some Saudi Folkloric substances. Res J Biotechnol 6(4):38–43
Nurul Syahida S, Ismail-Fitry MR, Ainun ZMAA et al (2020) Effects of palm wax on the physical, mechanical and water barrier properties of fish gelatin films for food packaging application. Food Packag Shelf Life 23:100437
Oliveira VRL, Santos FKG, Leite RHL et al (2018) Use of biopolymeric coating hydrophobized with beeswax in post-harvest conservation of guavas. Food Chem 259:55–64
Oliveira Filho JGD, Bezerra CCDON, Albiero BR et al (2020) New approach in the development of edible films: the use of carnauba wax micro- or nanoemulsions in arrowroot starch-based films. Food Packag Shelf Life 26:100589
Oliveira Filho JGD, Albiero BR, Calisto ÍH et al (2022) Bio-nanocomposite edible coatings based on arrowroot starch/cellulose nanocrystals/carnauba wax nanoemulsion containing essential oils to preserve quality and improve shelf life of strawberry. Int J Biol Macromol 219:812–823
Osae R, Apaliya MT, Alolga RN et al (2022) Influence of shea butter, bee wax and cassava starch coatings on enzyme inactivation, antioxidant properties, phenolic compounds and quality retention of tomato (Solanum lycopersicum) fruits. Appl Food Res 2(1):100041
Otoni CG, Avena-Bustillos RJ, Azeredo HMC et al (2017) Recent advances on edible films based on fruits and vegetables—a review. Compr Rev Food Sci Food Saf 16(5):1151–1169
Ozdemir M, Floros JD (2008) Optimization of edible whey protein films containing preservatives for water vapor permeability, water solubility and sensory characteristics. J Food Eng 86(2):215–224
Pan J, Li C, Liu J et al (2024) Polysaccharide-based packaging coatings and films with phenolic compounds in preservation of fruits and vegetables—a review. Foods 13(23):3896
Papadaki A, Alexandri M, Kachrimanidou V et al (2023) Enhancing hydrophobic and antioxidant properties of whey protein films through the inclusion of oil and oleogel sourced from spent coffee grounds. Sustain Chem Pharm 36:101293
Pereda M, Aranguren MI, Marcovich NE (2010) Caseinate films modified with tung oil. Food Hydrocoll 24(8):800–808
Peter A, Pop L, Mihaly Cozmuta L et al (2024) Beeswax-poly(vinyl alcohol) composite films for bread packaging. Food Chem: X 24:102053
Pirinc FT, Dagdelen AF, Saricaoglu FT (2021) Mechanical, barrier, thermal, and microstructural properties of poly (lactic acid) and gelatin–beeswax emulsion bi-layer films. J Food Process Preserv 45(12):e16073
Polunin Y, Kirianchuk V, Mhesn N et al (2023) Tough bioplastics from babassu oil-based acrylic monomer, hemicellulose xylan, and carnauba wax. Int J Mol Sci 24(7):6103
Nur Hanani ZA (2017) Physicochemical characterization of kappa-carrageenan (Euchema cottoni) based films incorporated with various plant oils. Carbohydr Polym 157:1479–1487

Ravelo Y, Molina V, Carbajal D et al (2011) Evaluation of anti-inflammatory and antinociceptive effects of D-002 (beeswax alcohols). J Nat Med 65(2):330–335
Reis MO, Olivato JB, Bilck AP et al (2018) Biodegradable trays of thermoplastic starch/poly (lactic acid) coated with beeswax. Ind Crops Prod 112:481–487
Rios JC, Robledo F, Schreiber L et al (2015) Association between the concentration of n-alkanes and tolerance to cracking in commercial varieties of sweet cherry fruits. Sci Hortic 197:57–65
Rocha BAD, Francisco L, Almeida MD et al (2020) Antiinflammatory activity of carnauba wax microparticles containing curcumin. J Drug Deliv Sci Technol 59:101918
Rodrigues DC, Caceres CA, Ribeiro HL et al (2014) Influence of cassava starch and carnauba wax on physical properties of cashew tree gum-based films. Food Hydrocoll 38:147–151
Romani VP, Olsen B, Pinto Collares M et al (2020) Cold plasma and carnauba wax as strategies to produce improved bi-layer films for sustainable food packaging. Food Hydrocoll 108:106087
Roufegarinejad L, Habibzadeh Khiabani A, Konar N et al (2024) Carnauba wax and adipic acid oleogels as an innovative strategy for cocoa butter alternatives in chocolate spreads. J Food Sci Technol 61(2):331–339
Samuels L, Kunst L, Jetter R (2008) Sealing plant surfaces: cuticular wax formation by epidermal cells. Ann Rev Plant Biol 59:683–707
Santos TM, Pinto AMB, Deoliveira AV et al (2014) Physical properties of cassava starch–carnauba wax emulsion films as affected by component proportions. Int J Food Sci Technol 49(9):2045–2051
Santos FKGD, Silva KNDO, Xavier TDN et al (2017)Effect of the addition of carnauba wax on physicochemical properties of chitosan films. Mater Res 20
Senturk Parreidt T, Müller K, Schmid M (2018) Alginate-based edible films and coatings for food packaging applications. Foods 7(10):170
Shaheenuzzamn M, Shi S, Sohail K et al (2021) Regulation of cuticular wax biosynthesis in plants under abiotic stress. Plant Biotechnol Rep 15(1):1–12
Da Silva Andrade LB, Da Silva Julião MS, Carneiro Vera Cruz R et al (2018) Antioxidant and antifungal activity of carnauba wax powder extracts. Ind Crops Prod 125:220–227
Singh S, Khemariya P, Rai A et al (2016) Carnauba wax-based edible coating enhances shelf-life and retain quality of eggplant (Solanum melongena) fruits. LWT 74:420–426
Soazo M, Pérez LM, Rubiolo AC et al (2013) Effect of freezing on physical properties of whey protein emulsion films. Food Hydrocoll 31(2):256–263
Sreenivas KM, Chaudhari K, Lele SS (2011) Ash gourd peel wax: extraction, characterization, and application as an edible coat for fruits. Food Sci Biotechnol 20(2):383–387
Sultan M, Hafez OM, Saleh MA et al (2021) Smart edible coating films based on chitosan and beeswax–pollen grains for the postharvest preservation of Le Conte pear. RSC Adv 11(16):9572–9585
Sun J, Li J, Liu J et al (2024) Using natural starch granules to disperse solid beeswax into micron-sized droplets in emulsion. Int J Biol Macromol 282:137093
Susmita Devi L, Kalita S, Mukherjee A et al (2022) Carnauba wax-based composite films and coatings: recent advancement in prolonging postharvest shelf-life of fruits and vegetables. Trends Food Sci Technol 129:296–305
Suwannasang S, Thumthanaruk B, Zhong Q et al (2021) The improved properties of zein encapsulating and stabilizing Sacha Inchi oil by surfactant combination of lecithin and tween 80. Food Bioprocess Technol 14(11):2078–2090
Thakur D, Singh A, Suhag R et al (2023) Oleogelation based on plant waxes: characterization and food applications. J Food Sci Technol 60(12):2927–2944
Thomas T, Olongal M, Athiyanathil S (2022) Antioxidant activity of mango seed wax additive on the properties of poly(lactic acid) transparent films for food packaging application. J Vinyl Add Tech 28(2):305–320
Tijani AT, Ayodele T, Liadi M et al (2024) Mechanical and thermal characteristics of films from glycerol mixed emulsified carnauba wax/polyvinyl alcohol. Polymers 16(21):3024

Tongnuanchan P, Benjakul S, Prodpran T et al (2015) Emulsion film based on fish skin gelatin and palm oil: physical, structural and thermal properties. Food Hydrocoll 48:248–259
Trinh BM, Smith M, Mekonnen TH (2022) A nanomaterial-stabilized starch-beeswax Pickering emulsion coating to extend produce shelf-life. Chem Eng J 431:133905
Velickova E, Winkelhausen E, Kuzmanova S et al (2013) Impact of chitosan-beeswax edible coatings on the quality of fresh strawberries (Fragaria ananassa cv Camarosa) under commercial storage conditions. LWT Food Sci Technol 52(2):80–92
Wang Y, Shan Y, Chen J et al (2016) Comparison of practical methods for postharvest preservation of loquat fruit. Postharvest Biol Technol 120:121–126
Wang D, Huang J, Guo Z (2020) Tomato-lotus inspired edible superhydrophobic artificial lotus leaf. Chem Eng J 400:125883
Wang L, Xu J, Zhang M et al (2022) Preservation of soy protein-based meat analogues by using PLA/PBAT antimicrobial packaging film. Food Chem 380:132022
Wang X, Chen K, Liu Y et al (2023) Preparation and application of biodegradable and superhydrophobic polylactic acid/carnauba wax coating. Prog Org Coat 177:107434
Wang X, Wei X, Liang Y et al (2025a) Advances and perspectives in natural wax composite coatings on wood-based materials: Mechanisms and applications. Mater Today Commun 42:111377
Wang J, Wang X, Liu B et al (2025b) Shellac-based films/coatings: progress, applications and future trends in the field of food packaging. Food Chem 467:142326
Winkler-Moser JK, Anderson J, Felker FC et al (2019) Physical properties of beeswax, sunflower wax, and candelilla wax mixtures and oleogels. J Am Oil Chemists' Soci 96(10):1125–1142
Won MY, Min SC (2018) Coating Satsuma mandarin using grapefruit seed extract–incorporated carnauba wax for its preservation. Food Sci Biotechnol 27(6):1649–1658
Wu F, Misra M, Mohanty AK (2021) Challenges and new opportunities on barrier performance of biodegradable polymers for sustainable packaging. Prog Polym Sci 117:101395
Xu X, Deng S, Essawy H et al (2024) Chitosan-casein blended with condensed tannin and carnauba wax for the fabrication of antibacterial and antioxidant food packing films. Int J Biol Macromol 277:133784
Xue F, Gu Y, Wang Y et al (2019) Encapsulation of essential oil in emulsion based edible films prepared by soy protein isolate-gum acacia conjugates. Food Hydrocoll 96:178–189
Yan J, Zhang J, Hu C et al (2020) Use of carvacrol and thymol in shellac coating to control stem-end rot on 'Ruby Red' grapefruit and maintain fruit quality during simulated storage and marketing. Sci Hortic 272:109606
Yu J, Xu S, Chen R et al (2024) A promising bioactive chitosan film in strawberry fresh-keeping: plasticized with tomato processing by-product extract of deep eutectic solvent. Food Hydrocolloids 151:109859
Zhang C, Ma Y, Guo K et al (2012) High-pressure homogenization lowers water vapor permeability of soybean protein isolate-beeswax films. J Agric Food Chem 60(9):2219–2223
Zhang Y, Adams MJ, Zhang Z et al (2016) Plasticisation of carnauba wax with generally recognised as safe (GRAS) additives. Polymer 86:208–219
Zhang L, Chen F, Zhang P et al (2017) Influence of rice bran wax coating on the physicochemical properties and pectin nanostructure of cherry tomatoes. Food Bioprocess Technol 10(2):349–357
Zhang Y, Simpson BK, Dumont M-J (2018) Effect of beeswax and carnauba wax addition on properties of gelatin films: a comparative study. Food Biosci 26:88–95
Zhang Y, Bi J, Wang S et al (2019) Functional food packaging for reducing residual liquid food: thermo-resistant edible super-hydrophobic coating from coffee and beeswax. J Colloid Interface Sci 533:742–749
Zhang N, Gao C, Meng L et al (2023) Preparation and characterization of carnauba wax-based particle with hierarchical structure and its use as hydrophobic coating for chitosan films. Carbohyd Polym 319:121224
Zhao H, Zhu M, Wang K et al (2020) Identification and quantitation of bioactive components from honeycomb (Nidus Vespae). Food Chem 314:126052

Zhao Y, Li B, Li C et al (2021) Comprehensive review of polysaccharide-based materials in edible packaging: a sustainable approach. Foods 10(8):1845

Zheng H, Zhao M, Dong Q et al (2022) Extruded transglutaminase-modified gelatin–beeswax composite packaging film. Food Hydrocoll 132:107849

Zhu M, Ji J, Wang M et al (2020) Cuticular wax of mandarin fruit promotes conidial germination and germ tube elongation, and impairs colony expansion of the green mold pathogen, penicillium digitatum. Postharvest Biol Technol 169:111296

Chapter 12
Agro-Waste Materials in Sustainable Food Packaging

Ana Caroline Ferreira Carvalho, Thaís Costa Nihues, Camila Vespúcio Bis Souza, Amanda Alves Prestes, Karine Marafon, Caroline Meinert, Elane Schwinden Prudêncio, Júlia Prebianca, and Carolina Krebs de Souza

Abstract This chapter presents an approach using agricultural waste as sustainable alternative to conventional food packaging, such as petroleum-derived polymers. Agro-waste, derived from sources such as fruit peels, grain husks, and plant fibers, is defined along with its origins in agricultural residues and food processing by-products. The chapter highlights its environmental benefits, including actions to lower carbon emissions, reduce waste and reduce the need to use non-renewable resources. Key processing techniques are analyzed, such as biotransformation, molding, and treatments to enhance durability. The chapter explores recent technological advancements and innovative applications, presenting real-world case studies of successful implementations. Challenges such as cost, scalability, and durability are discussed, along with strategies to address these obstacles. The chapter also evaluates consumer acceptance, market trends, and regulatory frameworks, providing insights into the factors influencing adoption. It concludes by identifying future research opportunities and emerging trends, emphasizing agro-waste's transformative potential in driving a sustainable and eco-friendly packaging industry.

Keywords Sustainable packaging · Circular economy · Renewable resources

A. C. F. Carvalho · A. A. Prestes · K. Marafon · E. S. Prudêncio
Postgraduate Program in Food Engineering, Federal University of Santa Catarina, Technology Center, Trindade, Florianópolis, SC, Brazil

T. C. Nihues · C. V. B. Souza · C. Meinert · J. Prebianca · C. K. de Souza (✉)
Department of Chemical Engineering, University of Blumenau, Itoupava Seca, Blumenau, SC, Brazil

E. S. Prudêncio
Department of Food Science and Technology, Federal University of Santa Catarina, Itacorubi, Florianópolis, SC, Brazil

S. Roy and M. Shahid (eds.), *Biomaterials for Sustainable Food Packaging*, Environmental Footprints and Eco-design of Products and Processes,
https://doi.org/10.1007/978-3-032-08027-1_12

12.1 Introduction

The packaging industry innovations have been motivated by the continuous rise in the global population, leading to increased food demand and, consequently, higher packaging production (Khandeparkar et al. 2024). The sector also addresses the alarming global food and beverage waste of 1.3 billion tons annually. This waste happens throughout the production chain (Heydari 2024). Efficient packaging should protect, enhance, and maintain food quality during transportation, distribution, and storage, as well as minimize food waste (Brennan et al. 2021). Traditional packaging materials, such as synthetic polymers, have been widely used for many years due to their properties, such as flexibility, strength, and transparency. However, these materials raise environmental concerns as they are not biodegradable, renewable, or reusable, leading to ecosystem pollution (Abdulla et al. 2024). Given these concerns, there is growing pressure from governments, the packaging industry, and the food sector to develop new materials as cost-effective, abundant, and eco-friendly alternatives (Stark and Matuana 2021; Carvalho et al. 2025a).

One strategy for producing sustainable packaging involves the application of agro-industrial waste (Fig. 12.1), which includes organic materials such as crop leftovers, seeds, fruit peels, husks, grains, and vegetable stalks (Vigneswari et al. 2024).

These residues contain underutilized compounds, such as bioactive molecules, proteins, and polysaccharides, which add value to packaging and support the circular bioeconomy (Freitas et al. 2021; Ferraz and Pyka 2023). However, despite the growing interest and ongoing research, the possibilities of applying agro-industrial

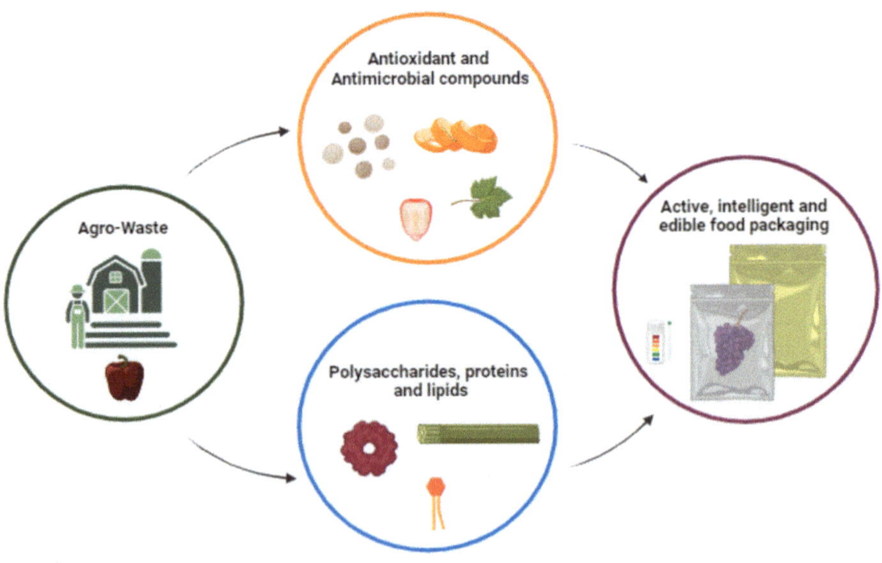

Fig. 12.1 Representative drawing of the use of agricultural waste in the development of sustainable materials for application in food packaging

waste in packaging are still underutilized, mainly due to the high production costs of bioplastics (Phiri et al. 2024). To make these sustainable materials more economically viable and widely available, further research and technological innovations are essential to reduce production expenses (Rosenboom et al. 2022). One of the main challenges of biodegradable packaging is to balance its durability with the shelf life of food products (Cheng et al. 2024; Finardi et al. 2021), which requires innovative solutions that ensure that this packaging remains functional and reliable throughout the product life cycle (Sabet 2025). A critical aspect is the evaluation and monitoring of thermal, mechanical, and barrier properties, which must be optimized to effectively prevent undesirable permeability, which may harm the quality of the packaged products (Andrew and Dhakal 2022).

Furthermore, these materials must be cost-effective, transparent, and mechanically robust to compete with traditional plastic packaging (Fang et al. 2021). Other unique advantages of these materials over conventional plastics include classifications such as edible, active, intelligent, and smart packaging (Babaremu et al. 2023; Trajkovska Petkoska et al. 2021). By integrating the packaging into the food, edible packaging increases sustainability and encourages more efficient use of resources (Yang et al. 2023).

Edible films and coatings are generally formulated from natural substances, such as polysaccharides, proteins, and lipids (Finardi et al. 2022), which are chosen for their safety, biodegradability, and ability to form thin, flexible films that can encapsulate food items (Gupta et al. 2024).

With continuous advancements in technology, the scalability and commercial feasibility of edible packaging are projected to improve significantly. These advancements could make edible packaging an increasingly affordable and practical option for both manufacturers and consumers (Trajkovska Petkoska et al. 2021).

Transforming disposable packaging into a consumable component aligns with the principles of innovative sustainability. It reduces waste, leverages renewable resources, and promotes fostering a circular economy (Zhu et al. 2022; Hamed et al. 2022).

Active packaging acts beyond the conventional function of protecting and maintaining food quality, extending shelf life and safety (Du et al. 2023). These materials can release or absorb antioxidants, antimicrobial agents, or moisture regulators, which reduce the need for artificial additives, allowing these foods to be healthier and more natural (Kuai et al. 2021). This is an interesting strategy to extend the shelf life of food, preserving the integrity of desirable nutritional and sensory aspects and meeting the demand for cleaner labels (Da Rocha et al. 2023; Lohita and Srijaya 2024).

Smart packaging features sensors and indicators that can provide real-time information about data on food quality (Hoffmann et al. 2022). This packaging can monitor factors such as temperature, humidity, and gas composition inside the package, alerting consumers to any changes that may impact food safety. These systems often include temperature sensors that track and record fluctuations in temperature during storage and transportation (Bhatlawande et al. 2024).

Smart packaging technologies can incorporate oxygen scavengers that actively monitor and reduce oxygen levels. This effectively extends shelf life and preserves food quality by minimizing oxidative damage (Thirupathi Vasuki et al. 2023). The embrace of innovative biopolymer-based packaging improves food preservation and provides added value, while meeting consumer expectations for safer and more environmentally friendly packaging solutions (Bhaskar et al. 2023). Some biopolymers, including polysaccharides and proteins, are obtained directly from biomass (Mueller et al. 2024), while others, such as polyhydroxyalkanoates are produced by microorganisms (de Souza et al. 2023), or synthesized from bio-based monomers, such as polylactic acid (Cywar et al. 2021). While increasing consumer awareness of environmental challenges, packaging industries must innovate, prioritize, and adopt more sustainable practices to meet these changes (Duarte et al. 2024).

12.2 Traditional Packaging for Food and Its Environmental Impact

Plastics-based packaging, primarily composed of synthetic material throughout fossil fuels, has significant environmental impacts. Versatility and low costs make it widely used but they pose serious environmental challenges, particularly regarding waste management and pollution. Plastics used in conventional packaging have persisted in the environment for centuries, contributing to pollution and environmental degradation. This evidence is a significant topic in discussions about sustainability and environmental conservation.

While food wrapping is indispensable to keep food products quality and preservation, it is also the key to avoiding its waste. From plastics to metals, paper, and other materials, conventional food packaging, in special day-use ones, has an enormous contribution to the pollution, affecting natural ecosystems, fauna, and flora, and, most importantly, environmental balance. The packaging industry is a major consumer of these materials, with food packaging plastics being an important portion of plastic disposal (Ncube et al. 2020). Ecosystems and their flora and fauna could be very damaged by this discarded material and its fractions broken down (Santos et al. 2024).

Packaging emerged as a practical solution for preserving and transporting food. However, with the advancement of industrialization and mass consumption, packaging has become more complex, durable, and often less recyclable. While glass, metal, and paper packaging have been used throughout time, plastics, especially disposable plastic, have grown exponentially since the mid-twentieth century. This transition to plastic and non-biodegradable packaging has brought a series of negative environmental impacts that are still felt today. Plastic is the material that has been associated with one of the greatest contemporary environmental crises. Plastic food packaging is widely used for its lightness, resistance, and low cost. However, these benefits also translate into serious environmental consequences (Santos et al. 2024).

Plastic, especially polyethylene, polypropylene, and PVC, when discarded without any treatment, has an estimated lifespan of hundreds of years, degrading extremely slowly. During this process, plastic breaks down into microplastics, particles so small that they can infiltrate soil and water bodies, posing a threat to aquatic and terrestrial ecosystems. The buildup of plastic in the oceans harms water quality, negatively affecting marine biodiversity. This accumulation can have consequences throughout the food chain, including for humans who eat seafood contaminated with microplastics. While plastic pollution is often linked to the oceans, its impact on soil and terrestrial ecosystems is also significant. Plastic packaging is often not recycled or reused efficiently and ends up in landfills. These landfills, over time, can leach harmful chemicals into soil and groundwater, polluting drinking water sources and affecting soil quality (Navarre et al. 2022). However, raising consumer awareness so that they have a greater connection with marine life could be a key to finding solutions and greater industry participation to reduce the environmental impact of waste (Nuojua et al. 2024).

Furthermore, plastic decomposition is still harming the environment. During this process, many of these materials release compounds, like bisphenol A (BPA) and phthalates, potentially harmful to human and animal health. Contamination by these substances can irreversibly affect fauna and flora, compromising biodiversity and essential ecosystem services, such as pollination and food production (Navarre et al. 2022).

Conventional wrapping materials production involves the extraction of non-renewable resources, which leads to resource depletion (Santos et al. 2024). Materials like polyethylene and PET are heavily reliant on finite fossil fuel resources, contributing further to both resource depletion and environmental degradation (Ncube et al. 2020). Life Cycle Assessments (LCAs) related that conventional packaging considerably impacts climate change, human toxicity, and resource depletion compared to some innovative packaging solutions. Steps such as the extraction and processing of these components lead to greenhouse gas emissions and various environmental issues (Santos et al. 2024). Conventional food packaging, especially plastics, contributes significantly to environmental pollution. Plastics has persisted in the environment for centuries, leading to ambiental dispersion of micro and nanoplastics, which pose serious ecosystem and body life safety risks (Ncube et al. 2020; Santos et al. 2024). Traditional packaging materials production like polyethylene and PET heavily relieve finite fossil fuel resources, which contributes to resource depletion and environmental degradation (Ncube et al. 2020). Unconscious disposal of conventional food wrapping significantly emits gases that promote global warming. Studies show that traditional packaging affects climate change more adversely than innovative packaging solutions (Jeswani et al. 2023).

The reducing of conventional packaging disposal is a significant barrier. Many conventional plastics are not biodegradable, leading to long-term environmental persistence. Even when recycling is possible, the process can be inefficient and limited by contamination and material quality issues (Kakadellis and Harris 2020).

Environmental concerns regarding conventional food packaging would be related to plastic waste and pollution, as it is a finite resource, and to life cycle impacts. Food packaging plastics are a major contributor to plastic waste, which pollutes terrestrial and marine environments (Table 12.1). The persistence of these materials in nature poses long-term ecological risks, including contamination of ecosystems with micro- and nanoparticles of plastic (Ncube et al. 2020; Santos et al. 2024). Reliance on petroleum-based plastics is unsustainable given that fossil fuel reserves are finite. Therefore, this calls for the studies of materials from different sources that provide the same duties as traditional ones without depleting environmental resources (Ncube et al. 2020; Maurizzi et al. 2022). The life cycle generates impacts on nature, from production to disposal. Life cycle assessments show that the choice of materials and the amount of packaging used have a significant influence on environmental outcomes. In fact, these factors often matter more than whether the materials are recyclable or compostable (Saeteros-Hernández et al. 2025; Molina-Besch et al. 2019).

Conventional storage materials, particularly plastics, cause substantial ambiental damage through their persistence, and resource-intensive manufacturing, even in its challenges in waste management. While alternative materials have presented great benefits and show new obstacles to being addressed to achieve eco-friendly companies. Efforts by improving the ecological output of packaging should focus on reducing resource use, enhancing recyclability, and minimizing pollution throughout the packaging life cycle. Bioplastics and other alternative materials are being developed, and they often come with their environmental trade-offs. For instance, bioplastics may reduce impacts in certain categories, like climate change, but can have higher impacts in others, such as eutrophication and acidification. Additionally, the production processes for some biodegradable alternatives can be resource-intensive and may not always result in a net environmental benefit (Ncube et al. 2020).

12.3 Benefits of Using Agro-Waste in Food Packaging

Improper management of waste generated by agricultural industries can contribute to environmental pollution, posing significant human and animal health risks. Much of this waste remains untreated and is often disposed of through unregulated landfilling or incineration. Such actions intensify environmental challenges, such as climate change, by amplifying greenhouse gas emissions into the atmosphere. However, agricultural waste holds substantial potential for transformation into value-added products. Most agricultural residues can be repurposed using various technologies, including their application in food packaging manufacturing. The unique composition of each type of agricultural waste determines its suitability for producing different value-added products, offering an eco-friendly substitute for conventional materials (Varghese et al. 2023).

Table 12.1 Environmental impact of conventional food packaging

Study title	Key findings	References
Environmental Impact of Food Packaging Materials	Highlights the significant environmental footprint of conventional packaging materials	Ncube et al. (2020)
Environmental Impacts of Plastic Packaging of Food Products	Analyzes the life cycle environmental impacts of plastic food packaging	Kan and Miller (2022)
Environmental Performance of Plastic Food Packaging: Life Cycle Assessment	Evaluates the environmental performance of plastic packaging using LCA	Boone et al. (2023)
A Review of Takeaway Packaging Waste: Types, Ecological Impact	Reviews the ecological impact of takeaway packaging waste, especially single-use plastics	(Zhuo et al. 2023)
Comparison of the Environmental Impact of Typical Packaging Materials	Compares the impacts of packaging materials, showing the burdens of conventional options	Meng et al. (2023)
Environmental Impacts Assessment in Packaging and Its Contribution to Food Waste Reduction	Assess packaging's role in reducing food waste and associated environmental effects	Paiva and Ugaya (2024)
The Green Era of Food Packaging: General Considerations and New Trends	Discusses the shift toward sustainable packaging and environmental implications	Maurizzi et al. (2022)
Recent and Emerging Food Packaging Alternatives: Chemical Safety and Environmental Concerns	Explores new alternatives, benefits, and safety concerns versus conventional materials	Lacourt et al. (2024)
Biotechnology in Food Packaging Using Bacterial Cellulose	Highlights the challenges of traditional packaging and the potential of biodegradable cellulose	Santos et al. (2024)
Development of Sustainable Food Packaging Material Based on Natural Polymers	Investigate natural polymers' environmental advantages for sustainable packaging	Andrade et al. (2022)

The widespread application of synthetic materials in food packaging, non-biodegradable and derived from non-renewable resources, has led to severe environmental problems. These issues include soil, water, and air contamination and the increasing burden on landfills. Additionally, synthetic polymers are difficult to break down, release harmful toxins when burned, and accelerate the exhaustion of fossil fuel reserves. Therefore, shifting to eco-friendly alternatives is crucial to replace these detrimental and non-degradable materials (Carvalho et al. 2025a).

Using agro-waste in food packaging offers significant environmental benefits such as pollution reduction and supports a circular economy by utilizing agricultural residues. Agrowaste-based packaging materials are biodegradable, which minimizes the ecological footprint linked to the disposal of conventional plastic packaging. This shift can significantly decrease pollution levels (Monye et al. 2023).

Various parts of crops, such as stems, pods, leaves, and straw, have been successfully utilized to produce packaging materials with significant potential. These materials can be transformed into edible, biodegradable, antioxidant, or antibacterial films, coatings, and agricultural films with various functional properties. This approach helps to repurpose waste and highlights its potential to create value-added products. Agricultural residues serve as a rich source of fibrous biomass, which could be effectively utilized in producing paper and biopolymers (Klai et al. 2021). Specific agricultural waste including lignocellulose can also be broken down by filamentous fungi rich in components like chitin, glucan, proteins, and lipids (Varghese et al. 2023). Utilizing agrarian waste to manufacture bioplastics addresses disposal challenges, reduces environmental pollution, reduces dependence on artificial, petroleum-derived polymers and generates affordable biodegradable plastics.

Straw, husks, ash, bran, and bagasse are the main solid waste generated throughout crop cycles. By-products from agriculture and the food industry serve as plentiful sources of biopolymers such as cellulose, starch, fiber, and protein. These biopolymers can be obtained through mechanical techniques like milling or chemical processes involving ethanol, acid, or alkaline treatments (Varghese et al. 2023).

Integrating reinforcements sourced from agricultural waste is especially beneficial for enhancing the properties of bio-based food packaging materials. These improvements include increased mechanical and thermal stability, lower water absorption, superior barrier performance, and biodegradability. Additionally, such reinforcements increase durability and adaptability for diverse packaging applications, ensuring greater functionality and sustainability. By utilizing agricultural residues, manufacturers can reduce reliance on synthetic polymers, address waste management challenges, and support eco-friendly practices (Carvalho et al. 2025a).

A research study explored adding phenolic compounds derived from agro-waste into chitosan films to improve their antibacterial and antioxidant capabilities in food packaging. Agrowaste-derived phenolic compounds were extracted from rice husk, sugarcane bagasse, and bamboo using dilute acid hydrolysis and hydrophobic resin. As a result, the agro-waste-derived phenolic compounds chitosan films exhibited significant antibacterial activity against *B. subtilis* and *P. aeruginosa*. The films also demonstrated enhanced antioxidant properties, including improved scavenging and reducing power. Additionally, the inclusion of agro-waste-derived phenolic compounds reduced film cytotoxicity, increasing cell viability. This research emphasizes the potential of phenolic compounds from agro-waste to enhance chitosan films, offering sustainable and functional solutions for active food packaging (Rai et al. 2021).

A study explored the extraction of phenolic compounds from sugarcane bagasse using various solvents and evaluated the aqueous extract as a natural preservative for chilled fresh meat. The water extract contained significant levels of total phenolic compounds (17.90 mg/g) and flavonoids (4.50 mg/g), with antioxidant activity of

45.90%. It exhibited antibacterial effects against Staphylococcus sp., reduced the total microbial count, and inhibited psychotropic bacteria. Meat treated by immersion in the sugarcane bagasse extract extended its shelf life to over 10 days, demonstrating its effectiveness as a natural and sustainable preservation solution (Mohamed et al. 2021).

Composite linear low-density polyethylene films incorporating orange peel were manufactured through blown film extrusion. The study indicates that films with up to 11.5% orange peel can be manufactured satisfactorily, maintaining water barrier properties, crystallinity, and melting temperatures like those of pure low-density polyethylene films. A decrease in tensile strength and elongation at break was observed. Additionally, the inclusion of orange peel enhanced the films' ability to block UV and visible light, making them appropriate for use in food packaging (Fehlberg et al. 2023).

The morphological, chemical, and thermal properties, along with the crystalline phases and chemical compositions, of cassava peel and bagasse were analyzed to assess their suitability as alternative materials for biodegradable food packaging. The residual starch content in cassava bagasse and peel closely resembled compared to commercial cassava starch. High potassium levels in the peel and calcium in the bagasse were identified. These advantageous characteristics underscore the potential of cassava peel and bagasse as promising matrices for developing biodegradable food packaging solutions (Weligama Thuppahige et al. 2023).

Optimizing extraction and manufacturing processes is critically important for making agro-waste packaging economically viable. Developing efficient, cost-effective methods for isolating valuable compounds, such as biopolymers, phenolic compounds, and other bioactive components, is essential to fully utilizing agricultural residues. Furthermore, advancements in scalable production techniques, including blown film extrusion, bioplastic synthesis, and other innovative technologies, must ensure these materials can compete effectively with conventional packaging options. Increased investments in research and development to refine these processes will lower production costs and boost industrial adoption and market competitiveness. By prioritizing these efforts, agro-waste packaging has the potential to transition from niche solutions to widespread applications, significantly mitigating environmental impacts, reducing dependency on fossil fuels, and fostering long-term sustainability in the food packaging industry (Carvalho et al. 2025a).

12.4 Sustainable Food Packaging from Agro-Waste

The use of agro-industrial waste as a raw material to produce packaging is a concern to make consumption more conscious and sustainable, appropriating the idea of the bio-cycle economy (Jain et al. 2024). Agricultural and biological waste is produced on a large scale, with the fruit, vegetable and wet meat processing industries being the main contributors and is usually dumped or incinerated to reduce its volume (Birania et al. 2022). Practices that cause significant environmental damage due to the emission of toxic gases, such as carbon monoxide and dioxide and methane, as well as the release of harmful substances, like slurry and heavy metals, contaminate soils and aquatic ecosystems (Panja et al. 2023). Active and intelligent food packaging can increase product shelf life through the antimicrobial and antioxidant properties derived from its by-products. This approach decreases waste and minimizes nature's impact while enhancing economic prospects for the packaging industry and food sector (da Silva et al. 2024; Perumal et al. 2025).

12.4.1 Types of Agro-Waste and Their Generation

Agro-waste refers to biodegradable residual components generated by agro-industries and their entire chain, from the field to their final consumers. It also includes waste from gardens, parks, wood and paper processing, scrap stores, and residential waste. Reusing these materials is an excellent waste management strategy (Jain et al. 2024).

Agro-waste can come from food waste, agriculture or even animal sources, as depicted in Fig. 12.2.

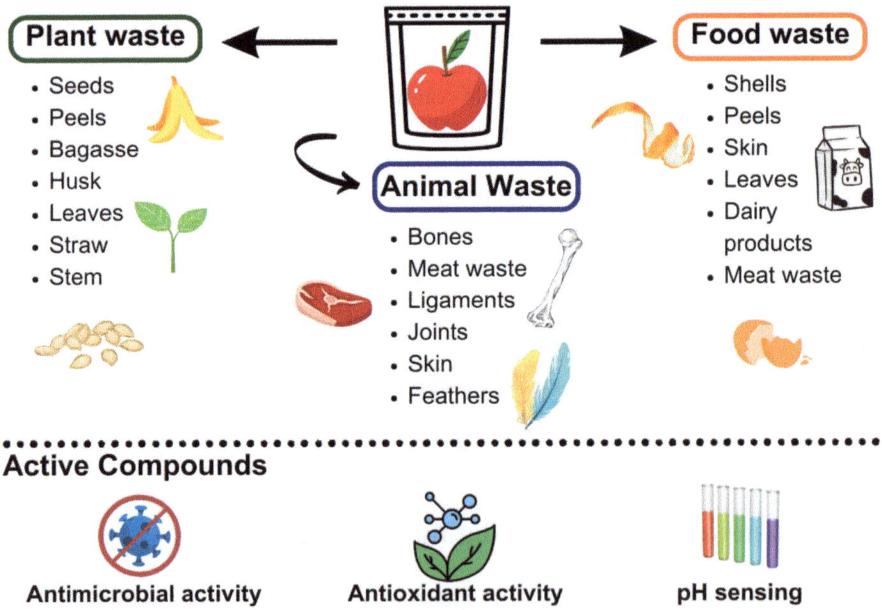

Fig. 12.2 Types of agro-waste and active properties for food packaging

12.4.1.1 Composts from Food Waste

Food disposal comes from cafeterias, restaurants, kitchens, processing waste, such as seafood, dairy, poultry feed, and waste. On average, one-third of all food becomes waste (Rajendran et al. 2024). Kitchen waste includes leftover leaves, seeds, vegetable peels, and fruit (bananas, apples, oranges, beans, rice, potatoes, tomatoes, and carrots). Food waste can include fish and meat, pasta, rice, milk, coffee grounds, etc. This food waste is rich in carbohydrates, about 50%, which favors the use of valuable by-products, one of which is lactic acid (LA), a precursor of polylactic acid (PLA). LA is produced naturally in household waste by microorganisms that grow and multiply there. The overall composition of the waste also includes starch (~30%), lipids (12–37.7%) and proteins (10–20%) (Swetha et al. 2023).

These components can be obtained from waste biomass, such as pectin from orange pomace and banana peel. The latter represents 40% of fruit waste and is also rich in hemicellulose and cellulose (Panja et al. 2023). Nazrin et al. (2021), proposed a film combining thermoplastic palm starch (TPS) with nanocellulose and PLA for food wrapping. Their results showed good water permeability when they developed a very uniform PLA/TPS composition with an ideal ratio of 60:40. Quilez-Molina et al. (2023), studied the use of pectin and its extraction obtained from orange peel to apply in food films. To extract the pectin, they partially hydrolyzed it with citric acid and applied a cross-linking structure in the pectin matrix structure between carboxylic acid groups and calcium present. The concentration of anionic polysaccharide in orange peel is about 30% and it has gelling properties. The application of pine nut flakes is another example of the use of a waste product with the potential for intelligent packaging, in which the pine nut flake extract gives the film the property of sensitivity to pH variations. By adjusting its color, the consumer is made aware that the product is not safe to consume (Carvalho et al. 2025b).

12.4.1.2 Agricultural Waste

Agricultural waste (agro-waste) is derived from both animal and plant sources, including dairy products, fruit processing, vegetables, meat, and fish (Varghese et al. 2023). Worldwide, around 1.3 billion tons of agro-waste is discarded annually (Thakur et al. 2025; Singh et al. 2024).

The main composition of waste generated by agriculture is cellulose (30–50%), followed by lignin (20–30%) and hemicellulose (15–30%). The stratification of waste compounds materials varies according to the composition of the material and the species responsible for its production.

Cellulose, the most abundant natural biopolymer, presents significant value for food wrappings. Its nanofibrils contribute to improved mechanical, chemical and thermal properties, turning it into an appealing component for sustainable packaging solutions (Birania et al. 2022; Stoica et al. 2024; Carvalho et al. 2025a). Zavaleta-Cavero et al. (2024) developed a cellulose-reinforced film with antibacterial properties incorporating copper nanoparticles. The formulation with the highest antimicrobial activity against E. coli was CMF20, which contained 79.20% microfibrillated cellulose (CMF). CMF enhances the structural properties of the films and improves their overall performance.

Starch is a highly versatile co-product used in bio-thermoplastics. It can be obtained from several supplies, including grains such as rice, corn, and soy, as well as from potatoes and other plants like quinoa, lentils, and beans. Starch is valuable for creating films, displays excellent barrier properties, and is biocompatible and non-toxic. This makes it a significant contributor to a circular economy (Stoica et al. 2024). Hasan et al. (2024) utilized nitrogen-doped carbon quantum dots (N-CQDs) derived from discarded pitahaya peels to enhance the properties of palm starch-based films reinforced with chitosan nanoparticles. In their study, they developed a material that exhibited improved thermal, mechanical and antioxidant qualities that help preserve the quality of packaged foods.

12.4.1.3 Animal Waste

The waste produced by slaughterhouses totals billions of tons yearly, primarily due to portions that do not fit in with the regulated cuts for human consumption. This waste is generated by agricultural slaughterhouses and processing industries that handle ruminants, monogastric animals, poultry, and aquatic animals (Khan et al. 2023). It is estimated that only 45–60% of an animal's weight is edible, with this figure varying by species (Chowdhury et al. 2022). Chitosan, a polymer of animal origin derived from chitin, can be produced from waste products of marine sources. When modified into carboxymethyl chitosan (CMC), it becomes particularly appealing for the food packaging sector for their film-forming properties, biocompatibility, and biodegradability. However, improvements are needed in its mechanical and barrier properties.

When it comes to applying whey Deng et al. (2025) valued the use of cheese and paper production by-products employing isolated protein from whey and lignin to create a film. The lignin incorporation in its alkaline form proved better in UV-blocking, antioxidant activity, opacity, and resistance.

In Rani et al. (2024) study, they combined chitosan, pectin and neem extract (*Azadirachta indica*) to explore the properties of this composition. The researchers developed an active and more resistant packaging material, demonstrating biodegradable potential for food packaging applications. In another study, Tang et al. (2025) investigated the combination of chitosan nanoparticles with konjac glucomannan to create a film exhibiting excellent mechanical properties, UV protection, antioxidant and antimicrobial activities.

While gelatin is a type of protein extracted from the breakdown of collagen by hydrolysis and is found in the bones, skin and cartilage of mammals and fish. The properties of gelatin are significantly influenced by the source of collagen. Skin and bones from pigs and cattle are the most common sources. However, there has been a growing trend ~~towards~~ extracting gelatin from infrequent sources, such as fish waste (Rather et al. 2022; Zuev et al. 2023). Wu et al. (2024) developed a smart film using gelatin combined with zein, exhibiting UV-blocking properties, sensitivity to pH changes, and antioxidant and antimicrobial activity. Collagen, a high molecular weight protein, is highly valued for film formation due to its biodegradability and potential tensile strength compared to polysaccharides (Gui et al. 2024). In addition to serving as an affordable bio-based co-product for food packaging applications (Stoica et al. 2024). Gui et al. (2024) created a film that exhibits antibacterial properties and can prolong the shelf life of pork beyond four days collated to traditional ones. To create this film, they combined collagen with histidine-modified ε-polylysine, finding that the optimal concentration of this compound was 5 wt%. Table 12.2 shows other recent examples of biodegradable packaging.

In conclusion, utilizing agricultural waste by-products is an effective waste management strategy. When these products are combined with bio-compounds that provide smart properties to food packaging, they depict a significantly reduced food waste. This is due to their ability to incorporate antioxidant, antibacterial, and free radical-absorbing agents, which help enlarge the shelf life of food (Das et al., 2024).

12.5 Processing Techniques for Agro-Waste Materials in Food Packaging

The environment may suffer unique impacts from every stage of a production's cycle, that could include manufacturing, consumer use, and disposal. Contributing to various objectives, such as minimizing biologically hazardous materials, repurposing waste into valuable products and preserving the environment can be achieved by adopting circular economic approaches to commercialize bio-waste for biomaterial production (Oumer et al. 2018). Enhancing ecological performance is facilitated by smart manufacturing and digital transformation. Significant concerns regarding the availability of agricultural land, crop production management, and the waste generated from these activities arise due to fast urbanization, industrialization, and accelerated population growth. Residual components derived from farmland and their processing facilities constitute agricultural waste biomass (Panja et al. 2023).

The incorporation of agro-industrial waste into packaging manufacturing significantly contributes to reducing waste. Reusing materials that would otherwise be discarded minimizes the amount of waste that ends up in landfills, reducing soil, water, and air pollution. In addition, replacing fossil fuel-based materials with bioplastics and other renewable materials reduces the carbon footprint of packaging production (Uysal-Unalan et al. 2024; Nath et al. 2024).

Reducing the use of plastic materials would significantly lower greenhouse gas emissions (Uysal-Unalan et al. 2024). Moreover, substituting fossil-based plastic materials with bioplastics in food packaging would help re-establish food systems

Table 12.2 Innovative combinations enhancing biomaterials for food packaging applications

Packaging composition	Fragility	Improvement	Application	Intelligent property	References
Thermoplastic reinforced with sugar palm crystalline nanocellulose and starch (TPS)/poly (lactic acid) (PLA) blend bio-nanocomposites	Poor mechanical properties and waterproof barrier	Inclusion of nanocellulose palm starch	–	–	Nazrin et al. (2021)
Biodegradable pine nutshell and rice husk polymer for food packaging	Production of intelligent food packaging	Introduction of pine nutshell	Cheese interfoliar	pH sensitivity	Carvalho et al. (2025b)
N-CQD loaded in starch film with chitosan nanoparticles	Poor mechanical and thermal properties	Incorporation of N-CQD	–	Antioxidant activity	Hasan et al. (2024)
Konjac glucomannan-based chitosan bio-nanocomposite film	Poor mechanical properties of Konjac	Incorporation of TP-FCNPs	Cherry tomato	UV resistance, and antioxidant and antibacterial	Tang et al. (2025)
Green walnut husk, chitosan, ethyl cellulose paper-based	Fragility in barrier properties	Addition of nanocellulose and green walnut husk extract	Fatty and high viscosity foods	Antioxidant and antibacterial activity	Liu et al. (2024a, b)
O-carboxymethyl chitosan/pectin/neem composite film	Poor mechanical strength and barrier property	Inclusion of pectin and *Azadirachta indica* extract	Suitable for various temperature-sensitive applications	Antimicrobial	Rani et al. (2024)
Bio-composed films based on kidney bean protein and pectin	Poor thermal, structural and barrier properties	Incorporation of kidney bean protein isolated and grape pomace pectin	–	Antibacterial and antioxidant activity	Amiri Samari et al. (2025)

(continued)

Table 12.2 (continued)

Packaging composition	Fragility	Improvement	Application	Intelligent property	References
Pectin film reinforced with calcium	Poor mechanical strength and water sensibility	Incorporation of calcium	Apple slices	Antioxidant activity	Quilez-Molina et al. (2022)
Film based on gelatin and pectin incorporated with tannin acid and silver nanoparticles	Limited barrier properties	Incorporation of tannin-acid and silver nanoparticles	Strawberry	UV-protection, antioxidant and antimicrobial activity	Li et al. (2025)
Chitosan/polyvinyl alcohol film added of quaternary ammonium lignin (QAL) and ZnO nanoparticles	Improve mechanical antibacterial and antioxidant activity	Utilization of ZnO nanoparticles and QAL	Grapes	UV-blocking, antibacterial and antioxidant activity	Zhang et al. (2025)
Cinnamaldehyde (CIN) with ZnO nanoparticles (NPs) in a gelatin film	Poor barrier properties	Incorporation of CIN and NPs	–	Antimicrobial activity	Sultan et al. (2024)
Film of anthocyanin and thymol into polymer fibers of gelatin and zein	Hydrophilicity and limited mechanical strength of gelatin	Combination of gelatin and zein	Shrimps	UV blocking, pH sensing, anti-microbial and antioxidant activity	Wu et al. (2024)
Fish protein concentrates from discards and combined application with gelatin	Management of fish by-products	Use of fish protein concentrate	–	–	Athanasopoulou et al. (2023)
Neoteric functional gelatin films	Poor barrier properties	Incorporation of bioactive compounds	Blueberries	Antioxidant activity and free radical scavenging	Liu et al. (2024a, b)

(continued)

Table 12.2 (continued)

Packaging composition	Fragility	Improvement	Application	Intelligent property	References
Collagen film containing histidine-modified ε-polylysine	Poor mechanical properties	Inclusion of histidine-modified ε-polylysine	Pork preservation	Antimicrobial activity	Gui et al. (2024)
Film based on sodium alginate, anthocyanin, curcumin and gellan gum	Poor barrier and mechanical properties	Prepare a double layer with strategic composition	Shrimp freshness	Ammonia sensing, antioxidant and antibacterial properties	Li et al. (2024)

within circular value chains. An increased need for food and essential goods arises due to the continuous growth of the global population, leading to a substantial rise in waste biomass production. This escalation results in environmental issues and higher costs associated with waste management. Throughout the year, excess crop residues are consistently produced by agriculture, one of the most common human activities, with a significant portion often remaining unused (Ezeorba et al. 2024).

Therefore, using agro-industrial waste in packaging production represents an essential strategy to promote sustainability, conserve natural resources, boost the circular economy, and stimulate innovation. By efficiently reusing this waste, a greener and more sustainable future can be created for everyone.

Biopolymers can be obtained from agro-food waste through processes like extraction or fermentation, including solid-state fermentation, to produce fermentable sugars. These biopolymers, derived from agro-food waste, possess diverse chemical and mechanical characteristics and biodegradability, bio-functionality, biostability, and biocompatibility, making them suitable for various uses (Gautam et al. 2022). Multiple applications including emulsions, edible films, food packaging, drug delivery systems, tissue scaffolds, and wound dressings in the pharmaceutical industry, can be achieved using recycled biopolymers. To lower the costs of biodegradable polymers-based products, whether natural or synthetic, they are typically composed of 70% renewable materials and 30% fillers, which may be natural or inorganic, such as talc or calcium carbonate (Nath et al. 2024).

The chemical complexity of bio-based feedstocks, the conditions of fermentation or extraction, and the downstream separation process for product recovery are factors linked to the higher costs of producing bioplastics compared to conventional fossil-based polymers (Naranjo et al. 2013; Lopez-Arenas et al. 2017; Tassinari et al. 2023). As a result of this, recent research initiatives within industrial environments, funded by the European Union and the Bio-based Industries Consortium, have demonstrated the technical feasibility of a novel product comprising a PLA-PHB blend plasticized with oligomeric lactic acid (OLA), suitable for specific food packaging requirements (De Carolis 2021; BBI JU 2022). All elements of this plastic blend film are obtained from renewable sources. Specifically, the production of PHB utilizes potato peel waste from agricultural and food processing sectors, while both PLA and OLA can be synthesized from starch (Kwan et al. 2015; Asgher et al. 2020; Tassinari et al 2023).

The performance of biopolymers is influenced by various factors such as the type, amount, and quality of the solvent used, as well as the processing methods that shape the final structure and the interacting components (Flórez et al. 2023). Primary strategies for producing biopolymers from agro-food waste include solvent casting, electrospinning, injection molding, extrusion, compression molding, and three-dimensional printing (Nath et al. 2024).

Cellulose (35–50%), hemicelluloses (15–35%), lignin (15–25%), and various inorganic materials comprise lignocellulosic biomass. High-cellulose content waste includes food waste like corn stoves, citrus peels, straws, and peanut husks (Tsang et al. 2019). Hydrogen bonds formed between cellulose and hydroxyl groups primarily influence the physical properties of cellulosic materials.

Potato waste, which is a common byproduct of potato chip manufacturing, is abundant in starch. Biopolymers, such as xanthan gum, have been derived from this waste through enzymatic hydrolysis (Bilanovic et al. 2010). Collagen, a protein-based biopolymer, is abundantly found in animals and is typically sourced from pig and cattle bones and skins, as well as from chicken processing waste, eggshells, and fish waste such as scales and bones (Ranganathan et al. 2020). Acid and enzymatic hydrolysis are commonly used to extract collagen from these residual materials.

Lignocellulose primarily originates from sugarcane bagasse, an affordable byproduct of the sugarcane industry. Its main components include ash, lignin, cellulose, and hemicelluloses (Sadh et al. 2023). The microbial breakdown process is slowed and complicated by the fibrous residues in sugarcane bagasse. To improve substrate availability and optimize fermentation, bagasse undergoes pretreatment. Alkaline methods are used to extract hemicelluloses from bagasse and cereal straw (Chen et al. 2022).

Adjusting the physical, chemical, and mechanical properties of materials to meet the specific needs of food products poses a significant challenge. Additionally, although sustainable waste management practices such as reusability, recyclability, biodegradability, and compostability are technically possible, they have not yet been fully adopted in practice. Harmonizing material use is essential to enhance sorting efficiency and optimize the economic aspects of the process (Uysal-Unalan et al. 2024).

Composites based on PLA, cellulose, and starch are currently being evaluated as potential substitutes for conventional PS or PET trays for MAP, PE films for wrapping, and PS foam trays/boxes (Uysal-Unalan et al. 2024). These materials have demonstrated promising oxygen barrier properties, making them suitable for MAP applications (Costa et al. 2023). For instance, the feasibility of replacing the conventional PVC lid film sealed on amorphous PET and PE trays with a PLA-based system was evaluated for red meat under a modified atmosphere (66% O2, 25% CO_2, and 9% N_2). The PLA solution maintained the red color of the meat and preserved it from oxidation (Panseri et al. 2018).

A similar study showed as a result that PLA and starch-based packaging material can be promising when used in replacing amorphous PET/PE and HDPE packaging with a modified atmosphere (60% CO_2 and 40% N_2) (Pettersen et al. 2011).

In a study, the use of glycerol and olive oil as plasticizers to improve the flexibility and moisture barrier of films, respectively, was investigated. It was found that glycerol content increased extensibility, while olive oil improved the moisture barrier. Films blended with 10% glycerol and 10% olive oil showed properties comparable to films with 20% of a single plasticizer. Furthermore, it was demonstrated that CMC derived from rice waste (rCMC) can replace up to 50% of commercial CMC without affecting mechanical and permeability properties. It was concluded that rCMC could form edible films and coatings as a renewable and environmentally friendly packaging material (Rodsamran and Sothornvit 2017).

The results of another study using agricultural waste as a substitute for biodegradable packaging indicated that biodegradable packaging produced from pitaya peel and palm oil stalks is efficient and versatile for different types of packaging. The density of the packaging ranged from 0.03 to 0.56 g/cm^3, proving to be comparable to other biodegradable packaging. The acceptability analysis revealed that most judges liked the appearance, resistance, and reliability of the packaging, with 78.26% of consumers stating that they would certainly buy the product. The comments highlighted innovation and environmental concern as positive factors. It is concluded that these residues are suitable to produce biodegradable packaging, with good acceptance by the public (Costa et al. 2023).

The potential valorization of cassava peels and bagasse as fibrous reinforcing agents and alternative starch-rich matrices for biodegradable food packaging was investigated in a study. The residual starch granules in cassava peels and bagasse were found to be morphologically similar to commercial starch granules. Slight differences in morphological structures, distinct chemical compositions, varying weight percentages and slightly different crystal lattice parameters were observed in the peels and bagasse. Although the peels underwent additional thermal decomposition and showed slight variations in mass losses, the thermal decomposition and stability of both materials were nearly identical. The presence of starch was suggested by the main crystalline phase, α-amylose dihydrate, while quartz (SiO_2) and whewellite ($CaC_2O_4 \cdot H_2O$) in the peels and bagasse indicated the presence of Si and Ca content in the samples. Crystalline phases were identified in nanocrystalline forms, potentially contributing to reinforcing effects. The study highlighted the potential for industrial cassava residue valorization in fiber- and starch-rich matrices for biodegradable food packaging, encouraging further research into their direct utilization and starch extraction processes for sustainable packaging solutions (Weligama Thuppahige et al. 2023).

12.6 Case Studies of Agrowaste Packaging Solutions

Converting agro-waste into value-added packaging materials supports a circular economy by providing two main benefits: waste valorization and the production of biodegradable, functional packaging. Researchers and industries are developing innovative materials with superior performance utilizing the distinctive characteristics of agricultural by products, including better mechanical strength, heat resistance, and active features such as antibacterial or antioxidant properties (Carvalho et al. 2025a). This sustainable approach minimizes waste and fosters environmental preservation by reducing reliance on non-renewable resources. Furthermore, biodegradable materials hold the potential to mitigate pollution and lessen the environmental impact of traditional plastic packaging. With the increasing demand for sustainable solutions, packaging materials made from agricultural waste present a viable substitute for conventional petroleum-based plastics, contributing to the shift towards a more sustainable and circular economy. Furthermore, developing these materials can contribute to reducing production costs for packaging while providing an opportunity for agro-industries to explore new revenue streams from their by-products (Monye et al. 2023).

12.6.1 Agro-Industrial Waste of Plant Origin Used in Packaging

Using by-products like seeds to extract valuable compounds and enhance their economic value has been extensively researched. An example is mango seeds, a fiber-rich agro-waste source, effectively utilized in sheet packaging applications through injection molding processes. Adding 20% fiber significantly enhanced the material's mechanical and barrier properties. This fiber incorporation resulted in a 38% increase in elastic modulus, indicating improved rigidity and strength of the packaging sheets. These advancements demonstrate the potential of mango seed fibers as a sustainable reinforcement material for creating high-performance, eco-friendly packaging solutions (Lima et al. 2021).

Similarly, tamarind seeds have been utilized to develop matrix films through casting, showcasing promising results for sustainable packaging. These films exhibit excellent light barrier properties and notable antioxidant activity attributed to interactions between polyphenols and biopolymers. Additionally, the flexible molecular structure of xyloglucan in tamarind seeds provides adequate elongation at break, ensuring the films maintain flexibility and mechanical integrity. These features emphasize the promise of tamarind seed-based films as a sustainable option, offering both functional and protective qualities for innovative packaging solutions (Thivya et al. 2021).

Research conducted by Zhao et al. (2022), silver nanoparticles were synthesized using grape seed extracts and incorporated into a chitosan-based coating to assess their ability to maintain the post-harvest quality of grapes. The findings demonstrated that grapes treated with this coating experienced a reduction in rot and loss in weight while maintaining higher titratable acidity levels compared to untreated samples. Furthermore, the coating effectively suppressed the total mold count during storage. These outcomes highlight the potential of chitosan coatings enriched with silver nanoparticles derived from grape seed waste to maintain fruit quality and significantly prolong the shelf life of grapes.

Hass avocado seeds (*Persea americana* Mill) were utilized to extract starch, which was then incorporated into biofilms. Adding glycerol to the formulation resulted in a favorable elongation value for the developed films. These results indicate that Hass avocado seed waste can effectively extract starch and produce biofilms, adding value to the byproduct (Jiménez et al. 2022).

Fruit and vegetable peels and residues have valuable applications in developing packaging materials. The cocoa pod husk (agricultural by-product) was used to develop a film-forming matrix for packaging applications using the solution casting method. The optimal formulation comprised 75% cellulose and 25% fiber, demonstrating superior performance in reduced water absorption and minimized water vapor permeability. These characteristics highlight the potential of cocoa pod husk-based films as effective and sustainable packaging materials with improved barrier properties (Azmin et al. 2020).

Sweet potato residues were incorporated into film production using injection and molding through compression techniques. Films incorporating up to 10% by weight of sweet potato residues demonstrated excellent mechanical properties, maintaining tensile strength above 30 MPa and strain exceeding 3.2%.

Polyphenols from apple peel were integrated into chitosan film. Incorporating apple peel polyphenols enhanced the film's thickness, density, and opacity while decreasing solubility, swelling, moisture content, and permeability. The film also exhibited increased antioxidants and antimicrobial activity. Apple peel polyphenols in chitosan film showed favorable mechanical properties and strong antimicrobial qualities, positioning it as a promising bio-composite material for sustainable food packaging applications (Riaz et al. 2018).

Bran and husks, composed of high concentrations of fibers and biopolymers, are effectively utilized as raw materials for developing biodegradable packaging solutions. It used soybean bran, a by-product of soybean processing, to produce a layer film through the casting method. The resulting films demonstrated complete biodegradation within 10 days, showcasing their environmentally friendly nature. Additionally, they exhibited no toxicity, as confirmed by successful cress seed growth during testing. However, when applied to tofu samples, the films did not display antibacterial or antioxidant properties, indicating their primary potential lies in biodegradable packaging rather than active preservation (Tkaczewska et al. 2023).

A study explored using Pinhão husks as a reinforcing material for creating edible films, utilizing Pinhão seed flour and bovine gelatin as base materials. The findings indicated that the films exhibited uniform and cohesive structures. Including Pinhão husk significantly enhanced the films' properties, improving tensile strength and reducing water vapor permeability. Additionally, thermal degradation analysis showed that the films maintain stability at temperatures up to 150 °C, demonstrating their potential for use in food packaging applications (De Santana Neto et al. 2024).

Researchers have extensively investigated the potential of bagasse for packaging applications, yielding promising and satisfactory results. Cassava bagasse produced packaging films using casting technique, leading to enhanced mechanical characteristics.

The films also demonstrated faster biodegradation and a reduced water absorption capacity (Abotbina et al. 2022). Travalini et al. (2019) also produced a film with cassava bagasse that significantly reduced opacity and water absorption values while tensile stress improved, indicating enhanced mechanical performance; additionally, water vapor permeability was reduced. The film was developed by Cardoso et al. (2019), demonstrated the maximum elongation achieved in biobased films, showcasing their improved flexibility and promising potential for sustainable packaging solutions.

Additionally, the study revealed that the formulation with cassava bagasse significantly influenced properties such as color, viscosity, solubility, thickness, and moisture content.

In another study, sugarcane bagasse and asparagus peel fibers were used as waste sources to develop eco-friendly foam trays using the thermo pressing method. Adding asparagus peel fibers improved the trays' properties, reducing porosity, lowering water absorption capacity, and enhancing tensile strength. Notably, higher concentrations of asparagus peel fibers (above 30%) produced more extensible foam trays, highlighting their potential for creating versatile and sustainable packaging materials (Cruz-Tirado et al. 2019).

The process of obtaining compounds from agro-industrial residues and by-products has been investigated. *Berberis crataegina* DC has developed chitosan-based edible films, seed oil and fruit extract into the chitosan matrix. The chitosan and fruit extract film demonstrated excellent thermal stability, along with notable antioxidant, antimicrobial, and anti-quorum sensing properties. Adding *B. crataegina* seed oil and fruit extract to the chitosan film notably decreased UV–vis transmittance but improved tensile strength (Kaya et al. 2018). In another study, pectin extracted from pineapple peels was utilized as a natural solvent and plasticizer to develop biopolymer films. Raising the concentration of pineapple peel extract improved the film's opacity, reduced its water vapor permeability, and increased its total phenolic content, thereby greatly enhancing its antioxidant capabilities (Rodsamran and Sothornvit 2019b). Films incorporated with natural antioxidants from lime peel residue extract and plasticized with coconut water were created and used as sachets to slow down the oxidation of soybean oil. The combination of coconut water and lime peel extract enhanced the film's water barrier properties, increased its total phenolic content, and boosted its antioxidant effectiveness (Rodsamran and Sothornvit 2019a).

12.6.2 Agro-Industrial Waste of Animal Origin Used in Packaging

In the animal processing industry, shells, skin, feathers, and other body parts excluded from primary process are classified as by-products. Researchers have recently explored their potential utilization for a cleaner industry concept. Animal-derived agricultural waste offers promising opportunities as sustainable alternatives for food packaging and containers (Varghese et al. 2023).

Chicken skin was used as a waste source for gelatin extraction to produce films using the casting method. The incorporation of virgin coconut oil significantly influenced the film's properties. As the amount of virgin coconut oil increased, the tensile strength and glass transition temperature decreased, indicating reduced rigidity. However, this modification enhanced the film's flexibility, as evidenced by improved elongation at break. The films exhibited increased antioxidant capacity (Jusoh et al. 2022).

Chicken feathers, a sustainable waste source, were used to extract keratin to produce biodegradable films via casting. The resulting films exhibited enhanced mechanical properties and water stability, rendering them appropriate for packaging applications. Additionally, they demonstrated eco-friendly biodegradability, with 24% degradation achieved within 12 days (Oluba et al. 2021).

Chicken eggshells were processed into powder and incorporated as a filler to produce rigid foam via compression molding. The eggshell filler significantly enhanced the rigid foam's dimensional stability and inhibited microbial growth, making it more durable. However, adding the filler decreased the foam's glass transition temperature (Leszczyńska et al. 2020). Also, eggshell waste was repurposed using a coating bar method to apply a coating film. The resulting material exhibited enhanced hydrophobicity, improved adhesion, and notable anticorrosive and antimicrobial properties (Patil et al. 2019).

An eggshell membrane was used as a source for gelatin to produce composite films through casting. The incorporation of eggshell membranes resulted in films with decreased water solubility and water vapor permeability compared to films made from a single material. The microstructure analysis revealed that gelatin and chitosan were utterly miscible. Additionally, adding chitosan enhanced the elongation at the break of the composite film, improving its flexibility (Mohammadi et al. 2018).

Poultry feathers were processed into a powder and utilized to produce films through compression molding. A feather-to-propylene glycol ratio of 30:70 demonstrated optimal mechanical properties, highlighting the potential of feather-based films for sustainable packaging solutions. The researchers present results so that the basic technique for implementing feather-based polymers can be implemented in the polymer industry. In addition, an estimate was made of the cost generation of the polymer, with a saving of 56% when compared to pure polypropylene (McGauran et al. 2021).

Carp skin was used by Tkaczewska et al. (2021), as a source for gelatin extraction to create films through casting, combined with furcellaran algae, the resulting gelatin film effectively delayed lipid oxidation and inhibiting microbial growth, extending the shelf life by 2 days on Atlantic mackerel under refrigeration conditions. In addition, according to consumers, the results showed that the coating application did not negatively affect the product's attractiveness parameters.

Egg and shrimp shells were utilized as a powder to enhance starch-based foam in packaging applications. Incorporating eggshell powder improved the Izod impact strength and reduced the foam's density, resulting in a more efficient material. Adding the filler also led to a distribution in the foam, though the degradation temperature decreased with filling (Kaewtatip et al. 2018).

A biodegradable coating made from poultry waste gelatin (derived from chicken feathers) and lotus stem starch has been developed to extend the shelf life of cherry tomatoes. This coating effectively preserved the firmness, pH, and minimized weight loss of the tomatoes during a 15-day storage period (Rather et al. 2022).

12.6.3 Challenges and Opportunities in Scaling Sustainable Packaging Solutions from Agro-Industrial Waste

Transitioning from the laboratory to the industrial scale is a crucial challenge for developing sustainable and innovative processes. Optimizing experimental conditions and testing materials on a small scale is possible in the laboratory phase. Still, when moving to large-scale production, issues such as economic viability, process uniformity, and the adaptation of technologies to larger volumes arise. Furthermore, challenges like cost reduction, maintaining quality, and environmental sustainability become more complex. Overcoming these barriers requires technological innovations, investments, and a thorough analysis of processes and impacts (Xie et al. 2024).

Agricultural sectors produce large volumes of waste and by-products during food production, management and processing, which has become a growing concern financially and ecologically. Improper disposal of these wastes can cause severe environmental damage, such as soil, water, and air contamination, and contribute to greenhouse gas emissions (Carvalho et al. 2025a). However, these wastes can be effectively valorized by adopting innovative strategies to produce sustainable packaging.

The development of composite materials based on agro-industrial waste provides a low-cost alternative with excellent properties that can be applied to packaging, offering an efficient solution to the waste problem and meeting the growing demand for environmentally friendly options in the packaging market (Fig. 12.3). For this to become a reality, governments and authorities must promote legislative support and encourage research and development of new technologies for utilizing these wastes. Measures aimed at large-scale use of agro-industrial waste help minimize environmental impact and result in more economical and sustainable solutions for industry (Klai et al. 2021; Xie et al. 2024).

12.7 Comparative Analysis of Agrowaste Packaging and Conventional Materials

The packaging sector is essential to the global economy, playing a vital role in the protection, transportation, and storage of products. However, growing awareness of the environmental impacts caused by conventional materials, such as glass, paper, metal, and plastic, has driven the search for more sustainable alternatives. In this context, packaging made from agricultural waste has been gaining prominence as an innovative and environmentally responsible solution. A comparative analysis between these two approaches reveals challenges and opportunities in environmental, economic, and functional terms, offering important support for decision-making by industries and consumers.

Conventional packaging is widely used due to its versatility, low cost, and functional efficiency. Plastic, for example, is the most common material, being used in

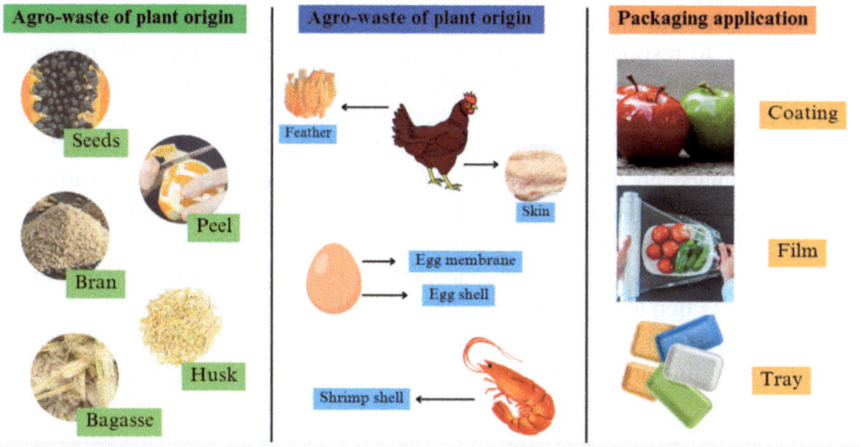

Fig. 12.3 Application of agro-industrial waste from plant and animal sources in packaging (coatings, films, and trays)

applications ranging from the food sector to electronic products. However, degradation is extremely slow in the environment and the low recycling rate in many countries has led to serious environmental consequences, such as marine pollution and the increase in the volume of solid waste. Studies indicate that, globally, only 9% of all plastic produced is recycled, while the majority is discarded in landfills or in the environment, worsening the waste crisis (Hopewell et al. 2009; Geyer et al. 2017).

Materials such as glass and metal have better recyclability indicators, but they are not free from impact. Glass, although 100% recyclable, requires large amounts of energy in its manufacturing and recycling process, in addition to being heavier, which increases the cost and carbon emissions during transportation. Metals, such as aluminum, have the advantage of being infinitely recyclable without losing quality, but their extraction involves intensive activities in greenhouse gas emissions and natural sources (Ndlovu et al. 2022). Paper and cardboard, often seen as ecological alternatives, also have environmental implications, including deforestation and high consumption of water and energy during production (Pereira et al. 2020).

On the other hand, packaging made from agricultural waste offers an innovative and sustainable solution. These materials include peels, fibers, pomace, and other byproducts of agricultural crops, which would normally be discarded or burned. Transforming waste into packaging helps reuse resources and reduce waste, ultimately reducing the environmental impact of the manufacturing process. Because these packages are biodegradable and, in some cases, compostable, they can re-enter the natural cycle without significantly damaging the environment (Sharma et al. 2021). Current studies have investigated the potential of packaging produced from agricultural waste as eco-friendly options to conventional packaging, generating valuable insights into their characteristics, advantages, and challenges. These investigations seek to respond to the growing demand for environmentally responsible solutions, especially considering the substantial ecological effects linked to plastic usage and other traditional materials (Pereira et al. 2023; Sasikumar et al. 2023; D'Almeida and de Albuquerque 2024). Studies show that packaging made from sugarcane pomace, coconut fiber or rice husks have much faster decomposition rates in comparison to petroleum-derived plastics, they are perfect for applications with short lifespans. Moreover, utilizing agricultural waste for their production supports the circular economy, promoting resource efficiency and decreasing reliance on new raw materials (Ellen MacArthur Foundation 2019).

In economic terms, this packaging can enhance the value of agro-industrial by-products, providing an extra revenue stream for farmers and industries.

Although packaging made from agricultural waste has clear environmental advantages, its functionality still faces limitations compared to conventional materials. One example is resistance to moisture and impacts, properties that limit its application in sectors that demand greater durability. However, recent advances in the development of natural coating technologies have shown significant improvements in this aspect, allowing these materials to become more competitive (Mendes et al. 2023). Researchers have also explored combinations of agricultural waste with biodegradable polymers, leading to composite materials with enhanced characteristics (Xu et al. 2022).

In relation to logistics and transport, plastic remains the lightest and most economical material, which reduces costs and carbon emissions associated with distribution. However, packaging made from agricultural waste, such as molded cellulose, also stands out for its lightness and for offering adequate protection to fragile products, such as eggs and household utensils (Júnior et al. 2021). In addition, the aesthetic and sustainable appeal of this packaging has attracted consumers who value responsible practices, which can generate competitive advantages for companies that adopt these alternatives. Studies that have been successful in various regions globally demonstrate the potential of these packages.

In Brazil, the use of sugarcane pomace to manufacture disposable plates, trays, and cups is on the rise, especially as a replacement for Styrofoam, which is widely criticized for its environmental impact. In India, farmers have been using banana leaves to create biodegradable packaging, a low-cost and highly sustainable alternative. Startups in the United States are developing packaging molded from corn and apple waste, offering innovative solutions to replace plastic in fast-moving consumer products (Sharma et al. 2021; Silva et al. 2023).

While agricultural by-products packaging offers numerous benefits, its widespread adoption encounters significant challenges. A major obstacle is the insufficient infrastructure for composting and recycling, especially in developing nations. Additionally, the high initial costs of research and development create significant barriers for small and medium-sized companies trying to enter the market. Consumer acceptance is also essential, as many people still equate sustainability with higher costs or lower quality. Public policies, such as tax incentives and stricter regulations on the use of conventional plastics, are essential to drive this transition.

The comparative analysis between packaging made from agricultural waste and conventional materials highlights the challenges and opportunities of each approach. While traditional materials continue to dominate the market due to their efficiency and scale of production, the environmental impacts associated with them reinforce the need for more sustainable alternatives. Packaging made from agricultural waste represents a promising solution, however, requires continued investment in technology, infrastructure, and awareness to become viable on a large-scale. Through collaborative actions among governments, businesses, and individuals, it is achievable to drive substantial progress toward a more sustainable system in line with the values of the circular economy.

12.8 Regulations and Standards for Agro-Waste Packaging

Globally, food packaging has become a critical factor in environmental degradation, primarily due to its persistent accumulation in solid waste streams, limited recyclability, and inefficiencies in recycling processes. The reliance on single-use plastics worsens waste management challenges, as the packaging sector discards over 95% of its annual plastic output, amounting to 150 million tons per year (Varghese et al. 2023). This issue contributes to marine pollution, soil degradation, and waterway obstructions, particularly impacting developing nations. An estimated100 million metric tons of food are wasted annually within the European Union, with the misinterpretation of date labels identified as a principal factor contributing to premature disposal. As a result, sustainable packaging is essential for minimizing food waste and encouraging responsible practices consumption (Cristofoli et al. 2023; Thapliyal et al. 2024).

In sustainability, food packaging is expected to contribute to resource optimization and reduce material intensity, recycling, and preservation without compromising safety or functionality. Recycling reduces the environmental impact of packaging, contingents on efficient collection and sorting systems. Utilizing agro-waste materials like fruit peels, grain husks, and plant fibers offers renewable and biodegradable alternatives that support the circular economy. Nonetheless, regulatory uncertainties hinder their market introduction, especially regarding toxicological safety (Carnaval et al. 2024). Food labeling regulations differ around the globe, reflecting various approaches to food safety, public health, and consumer information. The European Union (EU) has a robust set of regulations, such as *(EC) No. 1935/2004* and *GMP (EC) No. 2023/2006*, which stipulate that food contact materials must maintain sufficient inertness to prevent the transfer of harmful substances (Cherel-Bonnemaison et al. 2022). *Good Manufacturing Practice (GMP)* standards require quality assurance and traceability throughout the supply chain. Regulation *(EC) No. 282/2008* asserts that recycled plastics used in food packaging must come from approved processes and comply with safety standards. Validated scientific methods are necessary to demonstrate contaminant elimination (Cherel-Bonnemaison et al.

2022; Thapliyal et al. 2024). The *Directive (EU) 2019/904* focuses on single-use plastics and aims to mitigate environmental impacts through labeling, recycling goals, and integrating recycled materials (Cherel-Bonnemaison et al. 2022).

Within the framework of the Federal Food, Drug, and Cosmetic Act (1938), the FDA oversees the safety of food contact materials. As part of efforts to support sustainable packaging practices, the agency released guidance in 2021 detailing methods for evaluating chemical safety in recycled plastics used for food applications. The guidance highlights that exposure to contaminants under certain thresholds is generally safe (Versino et al. 2023). The FDA also advises that recycled plastics, when used in multilayer packaging with a reliable barrier, do not pose migration risks. US regulations vary by state, particularly for plastic and sustainable food packaging (Thapliyal et al. 2024).

India has various packaging regulations. FSSAI, as India's central regulatory agency, is tasked with ensuring compliance in all aspects of food safety, including the approval and monitoring of materials intended for direct food contact. The 2022 draft regulatory update introduced by FSSAI supports the integration of recycled plastics in food-grade packaging, provided these materials are proven to meet safety benchmarks and maintain functional stability during storage and transit. Under the current Indian regulatory framework, packaging compliance must be validated through NABL-accredited certification, ensuring safety and material standards conformity.

The updated Plastic Waste Management Rules (2021) further refine these measures by restricting recycled plastic in direct food-contact applications and prescribes structural specifications for single-use plastic bags (Varghese et al. 2023). The use of specific single-use plastics, including polystyrene, has been prohibited, while entities engaged in recycling operations must obtain registration from the respective *State Pollution Control Boards. Under the Plastic Waste Management Rules (2022), Extended Producer Responsibility (EPR)* mandates that producers, importers, and brand owners establish quantifiable targets for recycling and final disposal of plastic waste. These stakeholders must also submit annual reports detailing the volume of waste collected and processed. The EPR framework enhances circularity by promoting the development of decentralized and accessible plastic waste collection systems (Thapliyal et al. 2024).

The Food Standards Australia New Zealand Act (1991) provides the legal basis for establishing FSANZ, the regulatory agency overseeing the development and implementation of comprehensive food standards, including those governing ingredient composition and labeling obligations. Under a co-regulatory agreement with the Australian government, APCO plays a central role in promoting sustainable packaging systems by aligning industry practices with national waste reduction targets. Australia's National Plastic Plan (2021) sets targets for packaging, including eliminating single-use plastics and increasing recycling (Pauer et al. 2019). The National Environment Protection Measure (2011) supports waste reduction and recycling. The Sustainable Packaging Guidelines (2020) assist businesses in achieving packaging sustainability goals aligned with national objectives (Pauer et al. 2019).

In Japan, the oversight of food packaging materials falls under the jurisdiction of the *Ministry of Health, Labor, and Welfare (MHLW)*, which enforces safety through the implementation of technical standards and routine compliance inspections. The *Food Sanitation Law (1947)* prohibits harmful packaging materials and mandates safety certifications. The *Containers and Packaging Recycling Act (1995)* advances recycling and minimization of packaging waste through a system coordinated by the Japan Containers and Packaging Recycling Association. The *Plastic Resource Circulation Act (2021)* encourages sustainable plastic use, design for recycling, and reducing single-use plastics. Manufacturers, retailers, and municipalities collect and recycle plastic waste to support a circular economy (Nemes et al. 2020; Thapliyal et al. 2024).

The Ministry of Food and Drug Safety (MFDS) oversees the implementation of safety regulations related to the food supply in South Korea. The 2011 Food Sanitation Act restricts packaging materials that may pose chemical or toxicological risks when in contact with food. Additionally, the Resource Recycling Act of 2002 requires producers to assess the recyclability of their packaging.

The *2021 Standards and Specifications for Utensils, Containers, and Packages* outline the criteria necessary to ensure the safety of materials intended for direct food contact. Recycled plastics may be used in packaging areas that do not meet food but must meet strict requirements. This includes the use of chemical recycling or mechanical recycling of PET to prevent harmful substances from migrating into food (Thapliyal et al. 2024).

The *Ministry of Health*, in coordination with the *National Health Commission*, is responsible for implementing national standards concerning both food safety and compliance with packaging materials.

Packaging manufacturers must obtain a quality certificate from the AQSIQ. One of the foundational documents in China's regulatory framework for biodegradable packaging, *GB/T 38,082–2019*, defines essential specifications regarding the labeling, structural integrity, and environmental decomposition performance of plastic bags, and *GB 4806.6–2016* governing polymers in contact with food, with detailed limits on chemical migration and standardized benchmarks for material safety. Other regulations, like *GB/T 20,197–2006* and *GB/T 27,868–2011*, specify the classification, degradability, and test procedures for degradable plastics, including starch-based biodegradable resins (Cristofoli et al. 2023; Thapliyal et al. 2024).

Market integration of environmentally sustainable materials in food packaging will likely accelerate as economic barriers decrease and supply chains become more established (Santi et al. 2022). Increasing consumer preference for environmentally responsible packaging is expected to drive innovation and accelerate the integration of sustainable materials within circular economy frameworks (Trotta et al. 2023).

12.9 Consumer Perception and Acceptance

Due to growing consumer demand for more sustainable options, interest in the circular economy has increased significantly, encouraging the responsible use of raw materials, as well as the reuse, recovery and recycling of by-products, contributing to value generation throughout the entire production chain. The reintegration of by-products into their original chain requires a rigorous and comprehensive process assessment to ensure its feasibility, guaranteeing consumer safety and health. Biodegradable films produced from agro-industrial waste are gaining increasing strength and recognition in the food production chain. These materials represent an innovative solution to minimize environmental impacts by allowing the reuse of by-products that would otherwise be discarded. However, using these materials in direct contact with food requires special attention. It is essential to conduct thorough analyses and continuous monitoring to prevent the migration of toxic, carcinogenic, or harmful compounds, thereby ensuring food safety and consumer health. With increasing pressure to align industrial practices with sustainability, it is crucial that innovation is accompanied by detailed research and strict regulations, promoting a balance between environmental benefits and the protection of human health (Cristofoli et al. 2023; Donkor et al. 2023; Maraveas, 2020).

The consumer's perception of a product is crucial for its commercial success, as the value recognized by the customer influences various purchasing decisions. When it comes to products resulting from reuse cycles, such as those packaged with films derived from agricultural waste, significant concerns arise regarding perceived value and food safety. The frequent association of these industrial wastes with trash is one of the main barriers to the acceptance of these products by consumers, leading to a reduction in price and purchase intent. Furthermore, this negative association may raise doubts about the product's quality and safety, leading to the belief that it may be contaminated in some way (Aschemann-Witzel and Stangherlin 2021). The perception that waste-based products are of inferior quality is influenced by several factors, including consumers' sociodemographic characteristics, social environment, purchasing power, and environmental context. In this regard, the use of transparent information about production and sourcing methods, necessary certifications, and studies that prove quality and, above all, ensure food safety becomes a powerful strategy for transforming negative opinions consumers may hold (Cristofoli et al. 2023; Donkor et al. 2023).

The safety of using agro-industrial waste in food products is still a concern for consumers. This uncertainty often results in a devaluation of these materials. There is also a notable lack of studies assessing the toxicity of the components involved, especially regarding food packaging. It is crucial to conduct detailed investigations to determine whether residues used in films might migrate into the food, compromising safety, quality, and, most importantly, the health of the final consumer. Providing clear and scientific information about the safety of these films, along with data on the methods used in their development, can transform consumer perception. Emphasizing that such technologies are both safe and environmentally friendly can foster a more positive attitude toward the use of packaging made from agro-industrial waste. When consumers understand the benefits and safety of these solutions, they are likely to feel confident and open to adopting products that help recirculate the economy and ensure and preserve food safety (Aschemann-Witzel and Stangherlin 2021; Grasso and Asioli 2020; Zhang et al. 2020).

Consumer preferences and perspectives on sustainability significantly impact cost and competition. However, it is essential to acknowledge that not all consumers will adjust their purchasing choices based solely on ecological considerations. For this audience, it is essential that the sustainable product delivers performance equivalent to a conventional product; consumers need to perceive value in this exchange, whether through functionality, quality, or some other perceived benefit. Another relevant factor is generational differences. Millennial consumers (born between 1980 and 1994) show greater support for the production and use of environmentally conscious products.

However, this does not mean their purchasing intent is guided exclusively by these values, as other factors, such as price, also play a significant role. Generation X (born between 1965 and 1979) tends to show lower acceptance and perception of quality regarding recycled products. Baby Boomers (born between 1944 and 1964) demonstrate higher acceptance of such products, possibly due to associations with values of reuse and frugality that are already familiar to them. Finally, consumers who are concerned about environmental issues, regardless of their generation, tend to show greater acceptance and perceive value in products derived from agro-industrial waste. This consumer profile values tangible actions that demonstrate a commitment to Sustainability (Aschemann-Witzel and Stangherlin 2021; Maraveas 2020; Zhang et al. 2020).

In summary, the acceptance and purchase intentions for products that use agro-industrial waste in their packaging are influenced by individual, contextual, and product-specific factors. To increase acceptance and encourage consumer purchases, it is essential to emphasize environmental aspects and benefits. Clearer and more detailed information about the positive environmental impacts can significantly improve consumer perception of these products. Additionally, highlighting health benefits and environmental gains in an accessible and transparent manner is an effective strategy. Comprehensive and clear communication that reinforces the sustainability, quality, and food safety of these products can help build trust and promote a more positive image. Showing how these products help society by minimizing environmental harm and encouraging more responsible practices can significantly influence consumer choices (Donkor et al. 2023; Grasso and Asioli, 2020; Zhang et al. 2020).

12.10 Challenges and Solutions in the Implementation of Agricultural Waste Packaging

Growing global concern about environmental sustainability and the pressing need to minimize plastic waste have sparked significant interest in the use of agro-waste co-products by-products (Kumar Gupta et al. 2024). The worldwide food sector is currently dealing with a significant issue stemming from the vast amount of waste produced across its supply chain. Data from the Food and Agriculture Organization of the United Nations (FAO) indicate that approximately 1.3 billion tons of edible food are thrown away each year. Exacerbating environmental pollution and accelerating the exhaustion of natural resources (Prokic et al. 2022).

In emerging countries, food production and waste disposal encounter difficulties intensified by inadequate storage facilities, ineffective transportation systems, and consumer-related waste. Fruit waste, rich in bioactive compounds, presents itself as a viable substrate for producing metal oxide nanoparticles, which are then used to manufacture active packaging materials for food (Leta et al. 2024). In this way, the use of organic materials to create new products with lower costs and different properties represents a promising opportunity (Engel et al. 2022).

The production of by-products can vary throughout the year, which requires industrial plants to store large quantities of materials to keep production stable. The application of biotechnology with different raw materials, each with its own seasonality, represents an essential advance in making biofilm more competitive, reducing operating costs and the final price. To optimize the transport of this waste, the product must be affordable and easily available to the consumer (Tassinari et al. 2023). Considering economic factors and the context of large-scale development, converting agricultural waste into novel biomaterials typically demands a centralized and effective system for collection and transportation (Ren et al. 2024).

Studies on cellulose-based packaging indicate that these materials are cost-competitive, have outstanding ability to biodegrade and strong resistance to tension under elevated temperatures, while materials made from PLA (polylactic acid) offer advantages such as greater waterproofing and mechanical resistance in humid environments. However, there are still challenges to be overcome in the use of these materials, which direct future research toward the advancement of innovative technologies aimed at the extraction and preparation of lower-cost antimicrobial and biodegradable materials, as well as the implementation of more sustainable production routines, with reduced water consumption and carbon footprint (Ma et al. 2023).

To use these by-products in packaging, it is crucial to evaluate the risk of harmful substance release, ensuring that the material is completely free from agents that could contaminate the packaged food. Therefore, various tests must be performed to ensure safe application, as demonstrated in studies involving nanocellulose (Ren et al. 2024).

Research using fruit waste to synthesize metal oxide nanoparticles for edible packaging and coatings shows great potential. However, there are still significant challenges, including concerns related to toxicity from nanoparticle migration, challenges in scaling up production, and the absence of standardized procedures. In addition, a robust commercial infrastructure is needed to make large-scale production viable. These obstacles require interdisciplinary collaboration, drawing on nanotechnology, food science and environmental stewardship. To overcome these limitations and expand applications, it is essential to stimulate technological innovations in synthesis and processing techniques, with a view to improving scalability and reproducibility (Leta et al. 2024).

Studies of PHAs (polyhydroxyalkanoates), biodegradable polymers produced by microorganisms from food waste, highlight their strong promise as an eco-friendly substitute for traditional plastics. The fact that these materials are biodegradable, combined with their suitability for current plastic processing methods, positions them as strong options for diverse uses in food storage systems. Nevertheless, aspects like high production costs, scalability issues and specific material characteristics continue to hinder their widespread use in the packaging industry. Addressing these challenges is essential for PHAs to move beyond niche applications and become a widely adopted packaging alternative (Kusuma et al. 2024).

The utilization of food waste encounters several obstacles, but the recovery of these materials transforms waste into a valuable resource, reducing environmental impact and encouraging the circular economy. This progress makes it possible to create sustainable packaging, lowering the impact of plastic waste and reducing dependence on finite materials. Moreover, incorporating agricultural by-products enhances the overall value within the supply chain and economically benefits farmers and processors (Gupta et al. 2024). This establishes an alternative market for waste products, which can generate economic benefits for farmers and processors. This dual benefit, which combines waste reduction with value generation, is especially groundbreaking because of its combined and environmentally conscious methodology (Kumar and Goel 2019). Therefore, the current presents a prime opportunity for a major shift in the food packaging sector toward solutions grounded in biotechnology. Achieving this transformation requires a comprehensive strategy that considers every component of the food packaging system (Uysal-Unalan et al. 2024).

12.11 Future Trends and Innovations

Conventional food storage, designed with plastics and other non-biodegradable materials, poses one of the greatest environmental threats. The food packaging environmental impact involves the entire production cycle until the disposal of these materials. The production of plastics, for example, demands large amounts of natural resources and energy, furthermore emitting polluting chemicals into the environment.

After use, many of these packages finish up in the oceans or in landfills, aggravated by their centuries-long, decomposition in these places. Furthermore, they crumble into microplastics, which spread in the ambient water and soil, affecting fauna and flora. In this way, the improper disposal of non-degradable materials pollutes nature, causing irreparable damage to marine life, damaging biodiversity, and the health of aquatic ecosystems. The search for more sustainable alternatives has been driven by concern caused by environmental shock due to food storage disposal. Developing biodegradable and compostable packaging from agro-waste co-products is a promising area of innovation in food packaging. Agricultural waste, also known as agricultural by-products, is the leftover raw material generated after the harvesting and processing of agricultural products. This waste includes peels, stalks, seeds, and leaves. Agricultural waste used in food packaging means biodegradability, whereas petroleum-based plastics and synthetic materials are known for their negative environmental impact. Innovation in using this waste aims not only to replace plastic materials but also to give a new purpose to waste that would otherwise be underused, contributing to reducing waste.

Using agricultural waste in packaging production offers several environmental and economic benefits. Key advantages include reducing the amount of waste generated in agriculture, decreasing reliance on petroleum sources, and allowing a more circular and ecological production cycle. Thus, agricultural waste serves as a significant source of discarded materials repurposed as co-products for manufacturing packaging. Furthermore, many of these packages are compostable, simplifying organic waste management. Consequently, producing packaging from agricultural waste has the advantage of absorbing carbon and minimizing its footprint, differing from traditional plastics, which depend on petroleum. Manufacturing biodegradable packaging from agro-waste is less energy intensive and produces fewer greenhouse gas emissions. Packaging made from agricultural waste is biodegradable, decomposing naturally in the environment without leaving toxic or polluting residues. Furthermore, many of these packages can be composted, making organic waste easier. Thus, using agricultural waste to produce packaging would further promote innovation in the packaging sector, creating new technologies and materials that meet the growing demand for more environmentally friendly and sustainable solutions.

Various types of agricultural waste have been studied and used to produce food packaging. The processing of agricultural waste to create packaging involves multiple techniques tailored to the specific type of material being used, including extrusion, blow molding, and injection molding. The production of bottles and food packaging has also been used in bioplastics. Compared to conventional packaging, packaging made from agricultural waste has several advantages, especially regarding sustainability. However, packaging made from agricultural waste also faces challenges, such as durability and resistance to weathering. Agricultural materials may not be as resistant as conventional plastics, and their large-scale production is still in development, which may hinder their widespread adoption. Several countries are actively developing regulations to make food packaging from agro-waste. Establishing standards for biodegradable and compostable products, such as compostable certification, is crucial for ensuring that packaging is environmentally friendly and safe for human consumption. For example, in Brazil, the regulatory authorities, the National Health Surveillance Agency (ANVISA) and the National Institute of Metrology, Quality, and Technology (INMETRO) have implemented safety guidelines and regulations for quality and safety assurance of packaging for food, even those made from biodegradable materials. Another relevant point is the production large-scale of these packaging made with agricultural by-products. This packaging faces various hurdles, such as the readiness of raw materials, production costs, and the need for adequate infrastructure. For this reason, investing in research, innovation, and creating public policies that encourage using these alternatives is necessary. Finally, using agricultural waste in food packaging represents an innovative and sustainable solution, potentially significantly reducing the environmental impacts associated with using conventional plastics. However, its widespread adoption will depend on technological advances, greater consumer awareness, and a regulatory framework that favors sustainable innovation.

Acknowledgements The authors are grateful to the Foundation for Research Support and Innovation of Santa Catarina State (FAPESC) for financial support (2022TR000162), to the National Council for Scientific and Technological Development (CNPq n° 302867/2023-6) (CNPq n° 303069/2022-8) (Mai/Dai CNPq n° 68/2022), and the Ministry of Science, Technology, Innovation and Communications (MCTIC). This study was financed in part by the Coordenação de Aperfeiçoamento de Pessoal de Nível Superior e Brasil (CAPES) Finance Code 001. We would also like to thank All4Labels Global Packaging Group for funding our line of research into the development of packaging materials.

Declaration AI tools were used to enhance the language and readability of this chapter. The authors have reviewed and verified the final content and take full responsibility for the final content.

References

Abdulla SF, Shams R, Dash KK (2024) Edible packaging as sustainable alternative to synthetic plastic: a comprehensive review. Env Sci Poll Res. https://doi.org/10.1007/s11356-024-32806-z

Abotbina W, Sapuan SM, Ilyas RA et al (2022) Preparation and characterization of black seed/cassava bagasse fiber-reinforced cornstarch-based hybrid composites. Sustainability 14:12042. https://doi.org/10.3390/su141912042

Amiri Samani S, PourvatanDoust S, Savarolyia M et al (2025) Valorization of red grape pomace for sustainable food packaging: development of pectin/kidney bean protein based biocomposite films enriched with grape pomace polyphenols. Food Hydrocol 160:110806. https://doi.org/10.1016/j.foodhyd.2024.110806

Andrade MS, Ishikawa OH, Costa RS et al (2022) Development of sustainable food packaging material based on biodegradable polymer reinforced with cellulose nanocrystals. Food Pac Shelf Life 31:100807. https://doi.org/10.1016/j.fpsl.2021.100807

Andrew JJ, Dhakal HN (2022) Sustainable biobased composites for advanced applications: recent trends and future opportunities—a critical review. Compos Part C: Open Access 7:100220. https://doi.org/10.1016/j.jcomc.2021.100220

Aschemann-Witzel J, Stangherlin IDC (2021) Upcycled by-product use in agri-food systems from a consumer perspective: a review of what we know, and what is missing. Technol Forecast Soc Change 168:120749. https://doi.org/10.1016/j.techfore.2021.120749

Asgher M, Qamar SA, Bilal M, Iqbal HMN (2020) Bio-based active food packaging materials: sustainable alternative to conventional petrochemical-based packaging materials. Food Res Int 137:109625. https://doi.org/10.1016/j.foodres.2020.109625

Athanasopoulou E, Michailidi A, Ladakis D et al (2023) Extraction of fish protein concentrates from discards and combined application with gelatin for the development of biodegradable food packaging. Sustainability 15:12062. https://doi.org/10.3390/su151512062

Azmin SNHM, Hayat NA, Binti M, Nor MSM (2020) Development and characterization of food packaging bioplastic film from cocoa pod husk cellulose incorporated with sugarcane bagasse fibre. J Bioresources Biop 5(4):248–255. https://doi.org/10.1016/j.jobab.2020.10.003

Babaremu K, Oladijo OP, Akinlabi E (2023) Biopolymers: a suitable replacement for plastics in product packaging. Adv Ind Eng Polym Res 6:333–340. https://doi.org/10.1016/j.aiepr.2023.01.001

BBI JU (2022) Development of new competitive and sustainable bio-based plastics. Available at: http://www.bbeu.org/pilotplant/newpack/

Bhaskar R, Zo SM, Narayanan KB et al (2023) Recent development of protein-based biopolymers in food packaging applications: a review. Polym Test 124:108097. https://doi.org/10.1016/j.polymertesting.2023.108097

Bhatlawande AR, Ghatge PU, Shinde GU et al (2024) Unlocking the future of smart food packaging: biosensors, IoT, and nano materials. Food Sci Biotech 33:1075–1091. https://doi.org/10.1007/s10068-023-01486-9

Bilanovic D, Malloy D, Remeta P (2010) Solid and semi-solid fermentation of xanthan on potato or potato waste. US Patent 7:B2

Birania S, Kumar S, Kumar N et al (2022) Advances in development of biodegradable food packaging material from agricultural and agro-industry waste. J Food Process Eng 45. https://doi.org/10.1111/jfpe.13930

Boone L, Préat N, Nhu TT et al (2023) Environmental performance of plastic food packaging: life cycle assessment extended with costs on marine ecosystem services. Sci Tot Env 894:164781. https://doi.org/10.1016/j.scitotenv.2023.164781

Brennan L, Langley S, Verghese K et al (2021) The role of packaging in fighting food waste: a systematised review of consumer perceptions of packaging. J Clean Prod 281:125276. https://doi.org/10.1016/j.jclepro.2020.125276

Cardoso T, Esmerino LA, Bolanho BC et al (2019) Technological viability of biobased films formulated with cassava by-product and Spirulina platensis. J Food Process Eng 42. https://doi.org/10.1111/jfpe.13136

Carnaval L de SC, Jaiswal AK, Jaiswal S (2024) Agro-food waste valorization for sustainable bio-based packaging. J Compos Sci 8:41. https://doi.org/10.3390/jcs8020041

Carvalho ACF, Bose I, Ghosh S et al (2025a) Valorization of agro-waste derived materials for food packaging application. In: Sust Mat for Food Pack and Pres. Elsevier, pp 199–221

Carvalho ACF, Hoffmann TG, Helm CV, Prudêncio ES, Roy S, de Souza CK (2025b) Smart biopolymer based on rice husk extracts and Pinhão failure applied as an interlayer for sliced mozzarella cheese. J Vinyl Add Tech. https://doi.org/10.1002/vnl.22222

Chen W, Dong T, Bai F et al (2022) Lignin–carbohydrate complexes, their fractionation, and application to healthcare materials: a review. Int J Biol Macromol 203:29–39. https://doi.org/10.1016/j.ijbiomac.2022.01.132

Cheng J, Gao R, Zhu Y, Lin Q (2024) Applications of biodegradable materials in food packaging: a review. Alex Eng J 91:70–83. https://doi.org/10.1016/j.aej.2024.01.080

Cherel-Bonnemaison C, Feber D, Leger S et al (2022) Sustainability in packaging: global regulatory development across 30 countries. McKinsey: Paris, France. https://www.mckinsey.com/industries/packaging-and-paper/our-insights/sustainability-in-packaging-global-regulatory-development-across-30-countries. Accessed 2 Jan 2025

Chowdhury MW, Nabi MN, Arefin MA et al (2022) Recycling slaughterhouse wastes into potential energy and hydrogen sources: an approach for the future sustainable energy. Bioresour Technol Rep 19:101133. https://doi.org/10.1016/j.biteb.2022.101133

Costa A, Encarnação T, Tavares R et al (2023) Bioplastics: innovation for green transition. Polymers (Basel) 15:517. https://doi.org/10.3390/polym15030517

Cristofoli NL, Lima AR, Tchonkouang RDN et al (2023) Advances in the food packaging production from agri-food waste and by-products: market trends for a sustainable development. Sustainability 15:6153. https://doi.org/10.3390/su15076153

Cruz-Tirado JP, Vejarano R, Tapia-Blácido DR et al (2019) The addition of sugarcane bagasse and asparagus peel enhances the properties of sweet potato starch foams. Packag Technol Sci 32:227–237. https://doi.org/10.1002/pts.2429

Cywar RM, Rorrer NA, Hoyt CB et al (2021) Bio-based polymers with performance-advantaged properties. Nat Rev Mater 7:83–103. https://doi.org/10.1038/s41578-021-00363-3

Da Rocha J, Mustafa SK, Jagnandan A et al (2023) Development of active and biodegradable film of ternary-based for food application. Potravinarstvo Slovak J Food Sci 17:148–158. https://doi.org/10.5219/1853

De Carolis C (2021) Life cycle assessment. In: Biobased products from food sector waste. Springer International Publishing, pp 365–377 da Silva ACP, Barbosa JR, da Silva AC et al (2024) A new edible coating of fish gelatin incorporated into açaí oil to increase the post-harvest shelf life of tomatoes. Food Chem 438:138047. https://doi.org/10.1016/j.foodchem.2023.138047

D'Almeida AP, de Albuquerque TL (2024) Innovations in food packaging: from bio-based materials to smart packaging systems. Processes 12(10):2085. https://doi.org/10.3390/pr12102085

Das PP, Prathapan R, Ng KW (2024) Advances in biomaterials based food packaging systems: current status and the way forward. Biomater Adv 164:213988. https://doi.org/10.1016/j.bioadv.2024.213988

De Santana Neto DC, Paiva TS, Dos Santos RWS et al (2024) Potential of Pinhão husk (Araucaria angustifolia) as a structural reinforcement agent in the properties of edible films of Pinhão flour and gelatin. J Appl Polym Sci 141. https://doi.org/10.1002/app.55233

de Souza CK, Ghosh T, Lukhmana N et al (2023) Pullulan as a sustainable biopolymer for versatile applications: a review. Mater Today Commun 36:106477. https://doi.org/10.1016/j.mtcomm.2023.106477

Deng Y, Kolodjski S, Lewis G et al (2025) Physical and functional characterization of whey protein-lignin biocomposite films for food packaging applications. Future Foods 11:100554. https://doi.org/10.1016/j.fufo.2025.100554

Donkor L, Kontoh G, Yaya A et al (2023) Bio-based and sustainable food packaging systems: relevance, challenges, and prospects. Appl Res 3:100356. https://doi.org/10.1016/j.afres.2023.100356

dos Santos MR, Durval IJB, de Medeiros ADM et al (2024) Biotechnology in food packaging using bacterial cellulose. Foods 13:3327. https://doi.org/10.3390/foods13203327

Du H, Sun X, Chong X et al (2023) A review on smart active packaging systems for food preservation: applications and future trends. Trends Food Sci Technol 141:104200. https://doi.org/10.1016/j.tifs.2023.104200

Duarte P, Silva SC, Roza AS, Dias JC (2024) Enhancing consumer purchase intentions for sustainable packaging products: an in-depth analysis of key determinants and strategic insights. Sustain Futures 7:100193. https://doi.org/10.1016/j.sftr.2024.100193

Ellen MacArthur Foundation (2019) Completing the picture: how the circular economy tackles climate change. Ellen MacArthur Foundation. https://ellenmacarthurfoundation.org/completing-the-picture

Engel JB, Luchese CL, Tessaro IC (2022) Making the reuse of agro-industrial wastes a reality for starch-based packaging sector: a storage case study of carrot cake and cherry tomatoes. Int J Biol Macromol 206:740–749. https://doi.org/10.1016/j.ijbiomac.2022.03.074

Ezeorba TPC, Okeke ES, Mayel MH et al (2024) Recent advances in biotechnological valorization of agro-food wastes (AFW): optimizing integrated approaches for sustainable biorefinery and circular bioeconomy. Bioresour Technol Rep 26:101823. https://doi.org/10.1016/j.biteb.2024.101823

Fang D, Yu H, Dirican M, Tian Y, Xie J, Jia D, Yan C, Liu Y, Li C, Liu H, Wang J, Tang F, Chen G, Zhang X, Tao J (2021) Disintegrable, transparent and mechanically robust high-performance antimony tin oxide/nanocellulose/polyvinyl alcohol thermal insulation films. Carbohyd Polym 266:118175. https://doi.org/10.1016/j.carbpol.2021.118175

Fehlberg J, McKay S, Matuana LM, Almenar E (2023) Use of orange juice processing waste to produce films using blown film extrusion for food packaging. J Food Eng 341:111337. https://doi.org/10.1016/j.jfoodeng.2022.111337

Ferraz D, Pyka A (2023) Circular economy, bioeconomy, and sustainable development goals: a systematic literature review. Environ Sci Pollut Res. https://doi.org/10.1007/s11356-023-29632-0

Finardi S, Hoffmann TG, Angioletti BL et al (2022) Development and application of antioxidant coating on fragaria spp. Stored under isothermal conditions. J Microbiol Biotechnol Food Sci 11:e5432. https://doi.org/10.55251/jmbfs.5432

Finardi S, Hoffmann TG, Schmitz FRW et al (2021) Comprehensive study of Light-Emitting Diodes (LEDs) and ultraviolet-LED lights application in food quality and safety. J Pure Appl Microb 15:1125–1135. https://doi.org/10.22207/JPAM.15.3.54

Freitas LC, Barbosa JR, da Costa ALC et al (2021) From waste to sustainable industry: how can agro-industrial wastes help in the development of new products? Resour Conserv Recycl 169:105466. https://doi.org/10.1016/j.resconrec.2021.105466

Flórez M, Cazon P, Vazquez M (2023) Selected biopolymers' processing and their applications: a review. Polymers 15:641. https://doi.org/10.3390/polym15030641

Gautam K, Vishvakarma R, Sharma P et al (2022) Production of biopolymers from food waste: constrains and perspectives. Bioresour Technol 361:127650. https://doi.org/10.1016/j.biortech.2022.127650

Grasso S, Asioli D (2020) Consumer preferences for upcycled ingredients: a case study with biscuits. Food Qual Prefer 84:103951. https://doi.org/10.1016/j.foodqual.2020.103951

Geyer R, Jambeck JR, Law KL (2017) Production, use, and fate of all plastics ever made. Sci Adv 3(7):e1700782. https://doi.org/10.1126/sciadv.1700782

Gui N, Zhang X, Yang C et al (2024) A high-strength collagen-based antimicrobial film grafted with ε-polylysine fabrication by riboflavin-mediated ultraviolet irradiation for pork preservation. Food Chem 461:140889. https://doi.org/10.1016/j.foodchem.2024.140889

Gupta D, Lall A, Kumar S et al (2024) Plant-based edible films and coatings for food-packaging applications: recent advances, applications, and trends. Sust Food Technol 2:1428–1455. https://doi.org/10.1039/D4FB00110A

Hamed I, Jakobsen AN, Lerfall J (2022) Sustainable edible packaging systems based on active compounds from food processing byproducts: a review. Compr Rev Food Sci Food Saf 21:198–226. https://doi.org/10.1111/1541-4337.12870

Hasan M, Nurfathinah D, Gani A et al (2024) Thermomechanical properties of bio-nanocomposite film based on sugar palm starch/chitosan nanoparticles matrices loaded with N-CQD from dragon fruit peel extract. Case Stud Chem Environ Eng 10:101026. https://doi.org/10.1016/j.cscee.2024.101026

Heydari M (2024) Cultivating sustainable global food supply chains: a multifaceted approach to mitigating food loss and waste for climate resilience. J Clean Prod 442:141037. https://doi.org/10.1016/j.jclepro.2024.141037

Hoffmann TG, Angioletti BL, Bertoli SL, de Souza CK (2022) Intelligent pH-sensing film based on jaboticaba peels extract incorporated on a biopolymeric matrix. J Food Sci Technol 59:1001–1010. https://doi.org/10.1007/s13197-021-05104-6

Hopewell J, Dvorak R, Kosior E (2009) Plastics recycling: Challenges and opportunities. Philos Trans Roy Soc B: Biol Sci 364(1526):2115–2126. https://doi.org/10.1098/rstb.2008.0311

Jain A, Sarsaiya S, Gong Q et al (2024) Bioresources from biowaste for advancing manufacturing of packaging materials with modern circular economy towards a sustainable approach. Environ Dev Sustain. https://doi.org/10.1007/s10668-024-05734-8

Jeswani HK, Perry MR, Shaver MP, Azapagic A (2023) Biodegradable and conventional plastic packaging: Comparison of life cycle environmental impacts of poly(mandelic acid) and polystyrene. Sci Total Environ 903:166311. https://doi.org/10.1016/j.scitotenv.2023.166311

Jiménez R, Sandoval-Flores G, Alvarado-Reyna S et al (2022) Extraction of starch from Hass avocado seeds for the preparation of biofilms. Food Sci Technol 42. https://doi.org/10.1590/fst.56820

Júnior AS, Andrade DF, Silva LC (2021) Aplicações da celulose moldada em substituição a plásticos rígidos: avanços e limitações. Polímeros: Ciência e Tecnologia 31(2):45–58. https://doi.org/10.1590/0104-1428.20210045

Kaewtatip K, Chiarathanakrit C, Riyajan S-A (2018) The effects of egg shell and shrimp shell on the properties of baked starch foam. Powder Technol 335:354–359. https://doi.org/10.1016/j.powtec.2018.05.030

Kakadellis S, Harris ZM (2020) Don't scrap the waste: the need for broader system boundaries in bioplastic food packaging life-cycle assessment—a critical review. J Clean Prod 274:122831. https://doi.org/10.1016/j.jclepro.2020.122831
Kan M, Miller SA (2022) Environmental impacts of plastic packaging of food products. Resour Conserv Recycl 180:106156. https://doi.org/10.1016/j.resconrec.2022.106156
Kaya M, Ravikumar P, Ilk S et al (2018) Production and characterization of chitosan based edible films from Berberis crataegina's fruit extract and seed oil. Innov Food Sci Emerg Technol 45:287–297. https://doi.org/10.1016/j.ifset.2017.11.013
Khan RL, Khraibi AA, Dumée LF, Corridon PR (2023) From waste to wealth: repurposing slaughterhouse waste for xenotransplantation. Front Bioeng Biotechnol 11. https://doi.org/10.3389/fbioe.2023.1091554
Khandeparkar AS, Paul R, Sridhar A et al (2024) Eco-friendly innovations in food packaging: a sustainable revolution. Sustain Chem Pharm 39:101579. https://doi.org/10.1016/j.scp.2024.101579
Klai N, Yadav B, El Hachimi O et al (2021) Agro-industrial waste valorization for biopolymer production and life-cycle assessment toward circular bioeconomy. In: Biomass, biofuels, biochemicals. Elsevier, pp 515–555
Kuai L, Liu F, Chiou B-S et al (2021) Controlled release of antioxidants from active food packaging: a review. Food Hydrocoll 120:106992. https://doi.org/10.1016/j.foodhyd.2021.106992
Kumar Gupta R, Ali E, Abd El Gawad F et al (2024) Valorization of fruits and vegetables waste byproducts for development of sustainable food packaging applications. Waste Manage Bull 2:21–40. https://doi.org/10.1016/j.wmb.2024.08.005
Kumar N, Goel N (2019) Phenolic acids: natural versatile molecules with promising therapeutic applications. Biotechnol Rep 24
Kusuma HS, Sabita A, Putri NA et al (2024) Waste to wealth: polyhydroxyalkanoates (PHA) production from food waste for a sustainable packaging paradigm. Food Chem Mol Sci 9. https://doi.org/10.1016/j.fochms.2024.100225
Kwan TH, Pleissner D, Lau KY, Venus J, Pommeret A, Lin CSK (2015) Techno-economic analysis of a food waste valorization process via microalgae cultivation and co-production of plasticizer, lactic acid and animal feed from algal biomass and food waste. Biores Technol 198:292–299. https://doi.org/10.1016/j.biortech.2015.09.003
Jusoh NAM, Isa MIN, Sarbon NM (2022) Physical, mechanical and antioxidant properties of chicken skin gelatin films incorporated with virgin coconut oil. Biocatal Agric Biotechnol 45:102525. https://doi.org/10.1016/j.bcab.2022.102525
Lacourt C, Mukherjee K, Garthoff J et al (2024) Recent and emerging food packaging alternatives: chemical safety risks, current regulations, and analytical challenges. Comp Rev in Food Sci Food Saf 23(6):e70059. https://doi.org/10.1111/1541-4337.70059
Leszczyńska M, Ryszkowska J, Szczepkowski L et al (2020) Cooperative effect of rapeseed oil-based polyol and egg shells on the structure and properties of rigid polyurethane foams. Polym Test 90:106696. https://doi.org/10.1016/j.polymertesting.2020.106696
Leta TB, Adeyemi JO, Fawole OA (2024) Utilizing fruit waste-mediated nanoparticles for sustainable food packaging materials to combat food loss and waste. Food Biosci 59
Li C, Song A, Wu Y et al (2024) Intelligent double-layer film based on gellan gum/modified anthocyanin/curcumin/sodium alginate/zinc oxide for monitoring shrimp freshness. Int J Biol Macromol 274:132724. https://doi.org/10.1016/j.ijbiomac.2024.132724
Li S, Fang C, Wei N et al (2025) Antimicrobial, antioxidative, and UV-blocking pectin/gelatin food packaging films incorporated with tannic acid and silver nanoparticles for strawberry preservation. Int J Biol Macromol 308:142445. https://doi.org/10.1016/j.ijbiomac.2025.142445
Lima EMB, Middea A, Neumann R et al (2021) Biocomposites of PLA and mango seed waste: potential material for food packaging and a technological alternative to reduce environmental impact. Starch Stärke 73. https://doi.org/10.1002/star.202000118

Liu B, Sun F, Zhu P et al (2024a) Preparation of multi-barrier and multi-functional paper-based materials by chitosan, ethyl cellulose and green walnut husk biorefinery products for sustainable food packaging. Int J Biol Macromol 278:134557. https://doi.org/10.1016/j.ijbiomac.2024.134557

Liu S, Qiao S, Zhu J et al (2024b) Enhanced barrier and antioxidant properties of gelatin films by structural-colored bioactive materials for food packaging. Food Hydrocoll 150:109744. https://doi.org/10.1016/j.foodhyd.2024.109744

Lohita B, Srijaya M (2024) Novel technologies for shelf-life extension of food products as a competitive advantage: a review, pp 285–306

Lopez-Arenas T, González-Contreras M, Anaya-Reza O, Sales-Cruz M (2017) Analysis of the fermentation strategy and its impact on the economics of the production process of PHB (polyhydroxybutyrate). Comput Chem Eng 107:140–150. https://doi.org/10.1016/j.compchemeng.2017.03.009

Ma J, He J, Kong X et al (2023) From agricultural cellulosic waste to food delivery packaging: a mini-review. Chin Chem Lett 34. https://doi.org/10.1016/j.cclet.2022.04.005

Maraveas C (2020) Production of sustainable and biodegradable polymers from agricultural waste. Polymers (Basel) 12:1127. https://doi.org/10.3390/polym12051127

Maurizzi E, Bigi F, Quartieri A et al (2022) The green era of food packaging: general considerations and new trends. Polymers (Basel) 14:4257. https://doi.org/10.3390/polym14204257

McGauran T, Harris M, Dunne N et al (2021) Development and optimisation of extruded bio-based polymers from poultry feathers. Eur Polym J 158:110678. https://doi.org/10.1016/j.eurpolymj.2021.110678

Meng T, Wang Y, Fu L et al (2023) Comparison of the environmental impact of typical packaging systems for food cold chain express based on life cycle assessment. J Clean Prod 430:139756. https://doi.org/10.1016/j.jclepro.2023.139756

Mendes AM, Silva PR, Oliveira JP (2023) Resíduos agroindustriais: potencial e desafios para embalagens sustentáveis. Revista Brasileira de Inovação e Sustentabilidade 12(1):45–58. https://doi.org/10.1590/rbis.20230012

Mohamed RM, Ali MR, Smuda SS, Abedelmaksoud TG (2021) Utilization of sugarcane bagasse aqueous extract as a natural preservative to extend the shelf life of refrigerated fresh meat. Braz J Food Technol 24. https://doi.org/10.1590/1981-6723.16720

Mohammadi R, Mohammadifar MA, Rouhi M et al (2018) Physico-mechanical and structural properties of eggshell membrane gelatin-chitosan blend edible films. Int J Biol Macromol 107:406–412. https://doi.org/10.1016/j.ijbiomac.2017.09.003

Molina-Besch K, Wikström F, Williams H (2019) The environmental impact of packaging in food supply chains—does life cycle assessment of food provide the full picture? Int J Life Cycle Assess 24:37–50. https://doi.org/10.1007/s11367-018-1500-6

Monye SI, Afolalu SA, Ogedengbe TS (2023) suitability of agro wastes for biodegradable food packaging—a review. In: 2023 2nd international conference on multidisciplinary engineering and applied science (ICMEAS). IEEE, pp 1–6

Mueller E, Hoffmann TG, Schmitz FRW et al (2024) Development of ternary polymeric films based on cassava starch, pea flour and green banana flour for food packaging. Int J Biol Macromol 256:128436. https://doi.org/10.1016/j.ijbiomac.2023.128436

Naranjo JM, Posada JA, Higuita JC, Cardona CA (2013) Valorization of glycerol through the production of biopolymers: The PHB case using Bacillus megaterium. Bioresour Technol 133:38–44. https://doi.org/10.1016/j.biortech.2013.01.129

Nath PC, Sharma R, Debnath S et al (2024) Recent advances in production of sustainable and biodegradable polymers from agro-food waste: applications in tissue engineering and regenerative medicines. Int J Biol Macromol 259:129129. https://doi.org/10.1016/j.ijbiomac.2023.129129

Navarre N, Mogollón JM, Tukker A, Barbarossa V (2022) Recycled plastic packaging from the Dutch food sector pollutes Asian oceans. Resour Conserv Recycl 185:106508. https://doi.org/10.1016/j.resconrec.2022.106508

Nazrin A, Sapuan SM, Zuhri MYM et al (2021) Water barrier and mechanical properties of sugar palm crystalline nanocellulose reinforced thermoplastic sugar palm starch (TPS)/poly(lactic acid) (PLA) blend bionanocomposites. Nanotechnol Rev 10:431–442. https://doi.org/10.1515/ntrev-2021-0033

Ndlovu S, Farrokhpay S, Bradshaw DJ (2022) Environmental implications of metal production: a circular economy approach. J Clean Prod 340:130728. https://doi.org/10.1016/j.jclepro.2022.130728

Ncube LK, Ude AU, Ogunmuyiwa EN et al (2020) Environmental impact of food packaging materials: a review of contemporary development from conventional plastics to polylactic acid based materials. Materials 13:4994. https://doi.org/10.3390/ma13214994

Nemes SA, Szabo K, Vodnar DC (2020) Applicability of agro-industrial by-products in intelligent food packaging. Coatings 10:550. https://doi.org/10.3390/coatings10060550

Nuojua S, Pahl S, Wyles KJ, Thompson RC (2024) The impact of virtual reality exposure on ocean connectedness and consumer responses to single-use packaging. J Environ Psychol 99:102450. https://doi.org/10.1016/j.jenvp.2024.102450

Oluba OM, Osayame E, Shoyombo AO (2021) Production and characterization of keratin-starch bio-composite film from chicken feather waste and turmeric starch. Biocat Agric Biotech 33:101996. https://doi.org/10.1016/j.bcab.2021.101996

Oumer AN, Hasan MM, Baheta AT et al (2018) Bio-based liquid fuels as a source of renewable energy: a review. Renew Sustain Energy Rev 88:82–98. https://doi.org/10.1016/j.rser.2018.02.022

Paiva CL, Ugaya CML (2024) Environmental impacts assessment in packaging and its contribution to reducing food waste. Clea and Circ Bio 8:100083. https://doi.org/10.1016/j.clcb.2024.100083

Panja A, Paul S, Jha P et al (2023) Waste and their polysaccharides: are they worth bioprocessing? Bioresour Technol Rep 24:101594. https://doi.org/10.1016/j.biteb.2023.101594

Panseri S, Martino PA, Cagnardi P et al (2018) Feasibility of biodegradable based packaging used for red meat storage during shelf-life: a pilot study. Food Chem 249:22–29. https://doi.org/10.1016/j.foodchem.2017.12.067

Patil CK, Jirimali HD, Paradeshi JS et al (2019) Functional antimicrobial and anticorrosive polyurethane composite coatings from algae oil and silver doped egg shell hydroxyapatite for sustainable development. Prog Org Coat 128:127–136. https://doi.org/10.1016/j.porgcoat.2018.11.002

Pauer E, Wohner B, Heinrich V, Tacker M (2019) Assessing the environmental sustainability of food packaging: an extended life cycle assessment including packaging-related food losses and waste and circularity assessment. Sustainability 11:925. https://doi.org/10.3390/su11030925

Perumal T, Krebs de Souza C, Nihues TC et al (2025) A review on biopolymer-based oil and water-resistant functional paper coating for food packaging. Food Biosci 63:105656. https://doi.org/10.1016/j.fbio.2024.105656

Pettersen MK, Bardet S, Nilsen J, Fredriksen SB (2011) Evaluation and suitability of biomaterials for modified atmosphere packaging of fresh salmon fillets. Packag Technol Sci 24:237–248. https://doi.org/10.1002/pts.931

Pereira KY, Jaiswal AK, Jaiswal S (2023) Biopolymer-based sustainable food packaging materials: challenges, solutions, and applications. Foods 12(12):2422. https://doi.org/10.3390/foods12122422

Pereira RS, Souza TA, Almeida FC (2020) Impacto ambiental na produção de papel e alternativas sustentáveis. Ambiente and Sociedade 23(2):102–117. https://doi.org/10.1590/ambsoc2020.102

Phiri R, Mavinkere Rangappa S, Siengchin S (2024) Agro-waste for renewable and sustainable green production: A review. J Clean Prod 434:139989. https://doi.org/10.1016/j.jclepro.2023.139989

Prokic D, Curcic L, Stepanov J, Stojic N, Pucarevic M (2022) The role of circular economy in food waste management in fulfilling the United Nations' sustainable development goals. In: AlimentAriA, vol 15

Quilez-Molina AI, Mazzon G, Athanassiou A, Perotto G (2022) A novel approach to fabricate edible and heat sealable bio-based films from vegetable biomass rich in pectin. Mater Today Commun 32:103871. https://doi.org/10.1016/j.mtcomm.2022.103871

Quilez-Molina AI, Oliveira-Salmazo L, Amezúa-Arranz C et al (2023) Evaluation of the acid hydrolysis as pre-treatment to enhance the integration and functionality of starch composites filled with rich-in-pectin agri-food waste orange peel. Ind Crops Prod 205:117407. https://doi.org/10.1016/j.indcrop.2023.117407

Rai S, Dutta PK, Mehrotra GK (2021) Agrowaste derived phenolic compounds as additives to chitosan film for foodpackaging applications: antibacterial and antioxidant study. J Ind Chem Soc 93:767–774

Rajendran DS, Venkatraman S, Rahul R et al (2024) Manifesting sustainability toward food waste into bioenergy: biorefinery in a circular economic approach, pp 431–460

Ranganathan S, Dutta S, Moses JA, Anandharamakrishnan C (2020) Utilization of food waste streams for the production of biopolymers. Heliyon 6:e04891. https://doi.org/10.1016/j.heliyon.2020.e04891

Rani S, Lal S, Kumar S et al (2024) Utilization of marine and agro-waste materials as an economical and active food packaging: antimicrobial, mechanical and biodegradation studies of O-Carboxymethyl chitosan/pectin/neem composite films. Int J Biol Macromol 254:128038. https://doi.org/10.1016/j.ijbiomac.2023.128038

Rather JA, Akhter N, Ashraf QS et al (2022) A comprehensive review on gelatin: understanding impact of the sources, extraction methods, and modifications on potential packaging applications. Food Pack Shelf Life 34:100945. https://doi.org/10.1016/j.fpsl.2022.100945

Ren H, Huang Y, Yang W et al (2024) Emerging nanocellulose from agricultural waste: Recent advances in preparation and applications in biobased food packaging. Int J Biol Macromol 277

Riaz A, Lei S, Akhtar HMS et al (2018) Preparation and characterization of chitosan-based antimicrobial active food packaging film incorporated with apple peel polyphenols. Int J Biol Macromol 114:547–555. https://doi.org/10.1016/j.ijbiomac.2018.03.126

Rodsamran P, Sothornvit R (2017) Rice stubble as a new biopolymer source to produce carboxymethyl cellulose-blended films. Carbohydr Polym 171:94–101. https://doi.org/10.1016/j.carbpol.2017.05.003

Rodsamran P, Sothornvit R (2019a) Preparation and characterization of pectin fraction from pineapple peel as a natural plasticizer and material for biopolymer film. Food Biop Proc 118:198–206. https://doi.org/10.1016/j.fbp.2019.09.010

Rodsamran P, Sothornvit R (2019b) Lime peel pectin integrated with coconut water and lime peel extract as a new bioactive film sachet to retard soybean oil oxidation. Food Hydrocoll 97:105173. https://doi.org/10.1016/j.foodhyd.2019.105173

Rosenboom J-G, Langer R, Traverso G (2022) Bioplastics for a circular economy. Nat Rev Mater 7:117–137. https://doi.org/10.1038/s41578-021-00407-8

Sasikumar Nair S, Trafiałek J, Kolanowski W (2023) Edible packaging: A technological update for the sustainable future of the food industry. Appl Sci 13(14):8234. https://doi.org/10.3390/app13148234

Sabet M (2025) Exploring biodegradable polymer composites for sustainable packaging: a review on properties, manufacturing techniques, and environmental impacts. Iran Polym J 34:123–142. https://doi.org/10.1007/s13726-024-01365-y

Sadh PK, Chawla P, Kumar S et al (2023) Recovery of agricultural waste biomass: a path for circular bioeconomy. Sci Total Environ 870:161904. https://doi.org/10.1016/j.scitotenv.2023.161904

Saeteros-Hernández A, Chalen-Moreano F, Zurita-Gallegos R et al (2025) Quantitative and qualitative characterization of food waste for circular economy strategies in the restaurant sector of Riobamba, Ecuador: a case study approach. Biomass 5:18. https://doi.org/10.3390/biomass5020018

Santi R, Garrone P, Iannantuoni M, Del Curto B (2022) Sustainable food packaging: an integrative framework. Sustainability 14:8045. https://doi.org/10.3390/su14138045

Silva TF, Costa MA, Almeida RV (2023) Embalagens biodegradáveis: um estudo sobre o impacto dos resíduos agropecuários no mercado sustentável. Sustain Packag J 15(3):89–105. https://doi.org/10.1016/spj.2023.15089
Singh AK, Itkor P, Lee M et al (2024) Synergistic integration of carbon quantum dots in biopolymer matrices: an overview of current advancements in antioxidant and antimicrobial active packaging. Molecules 29:5138. https://doi.org/10.3390/molecules29215138
Sharma S, Basu S, Gupta R (2021) Agricultural waste: a new resource for sustainable packaging. J Clean Prod 278:124–132. https://doi.org/10.1016/j.jclepro.2020.124132
Stark NM, Matuana LM (2021) Trends in sustainable biobased packaging materials: a mini review. Mater Today Sustain 15:100084. https://doi.org/10.1016/j.mtsust.2021.100084
Stoica M, Bichescu CI, Crețu C-M et al (2024) Review of bio-based biodegradable polymers: smart solutions for sustainable food packaging. Foods 13:3027. https://doi.org/10.3390/foods13193027
Sultan M, Ibrahim H, El-Masry HM, Hassan YR (2024) Antimicrobial gelatin-based films with cinnamaldehyde and ZnO nanoparticles for sustainable food packaging. Sci Rep 14:22499. https://doi.org/10.1038/s41598-024-72009-7
Swetha TA, Ananthi V, Bora A et al (2023) A review on biodegradable polylactic acid (PLA) production from fermentative food waste—its applications and degradation. Int J Biol Macromol 234:123703. https://doi.org/10.1016/j.ijbiomac.2023.123703
Tang J, Huang C, Liu W et al (2025) Preparation and characterization of a konjac glucomannan-based bio-nanocomposite film and its application in cherry tomato preservation. Food Hydrocoll 159:110689. https://doi.org/10.1016/j.foodhyd.2024.110689
Tassinari G, Bassani A, Spigno G et al (2023) Do biodegradable food packaging films from agro-food waste pay off? A cost-benefit analysis in the context of Europe. Sci Total Environ 856:159101. https://doi.org/10.1016/j.scitotenv.2022.159101
Thakur I, Sharma RP, Kumar S (2025) Agricultural Wastes: Their Implications and Management. In: Organic production of vegetable crop. Apple Academic Press, pp 267–288
Thapliyal D, Karale M, Diwan V et al (2024) Current status of sustainable food packaging regulations: global perspective. Sustainability 16:5554. https://doi.org/10.3390/su16135554
Thirupathi Vasuki M, Kadirvel V, Pejavara Narayana G (2023) Smart packaging—an overview of concepts and applications in various food industries. Food Bioeng 2:25–41. https://doi.org/10.1002/fbe2.12038
Thivya P, Bhosale YK, Anandakumar S et al (2021) Exploring the effective utilization of shallot stalk waste and tamarind seed for packaging film preparation. Waste Biomass Valorization 12:5779–5794. https://doi.org/10.1007/s12649-021-01402-4
Tkaczewska J, Jamróz E, Zając M et al (2023) Antioxidant edible double-layered film based on waste from soybean production as a vegan active packaging for perishable food products. Food Chem 400:134009. https://doi.org/10.1016/j.foodchem.2022.134009
Tkaczewska J, Kulawik P, Jamróz E et al (2021) One- and double-layered furcellaran/carp skin gelatin hydrolysate film system with antioxidant peptide as an innovative packaging for perishable foods products. Food Chem 351:129347. https://doi.org/10.1016/j.foodchem.2021.129347
Trajkovska Petkoska A, Daniloski D, D'Cunha NM et al (2021) Edible packaging: Sustainable solutions and novel trends in food packaging. Food Res Int 140:109981. https://doi.org/10.1016/j.foodres.2020.109981
Travalini AP, Lamsal B, Magalhães WLE, Demiate IM (2019) Cassava starch films reinforced with lignocellulose nanofibers from cassava bagasse. Int J Biol Macromol 139:1151–1161. https://doi.org/10.1016/j.ijbiomac.2019.08.115
Trotta F, Da Silva S, Massironi A et al (2023) Silver bionanocomposites as active food packaging: recent advances & future trends tackling the food waste crisis. Polymers (Basel) 15:4243. https://doi.org/10.3390/polym15214243
Tsang YF, Kumar V, Samadar P et al (2019) Production of bioplastic through food waste valorization. Environ Int 127:625–644. https://doi.org/10.1016/j.envint.2019.03.076

Uysal-Unalan I, Sogut E, Realini CE et al (2024) Bioplastic packaging for fresh meat and fish: Current status and future direction on mitigating food and packaging waste. Trends Food Sci Technol 152:104660. https://doi.org/10.1016/j.tifs.2024.104660

Varghese SA, Pulikkalparambil H, Promhuad K et al (2023) Renovation of agro-waste for sustainable food packaging: a review. Polymers (Basel) 15:648. https://doi.org/10.3390/polym15030648

Versino F, Ortega F, Monroy Y et al (2023) Sustainable and bio-based food packaging: a review on past and current design innovations. Foods 12:1057. https://doi.org/10.3390/foods12051057

Vigneswari S, Kee SH, Hazwan MH et al (2024) Turning agricultural waste streams into biodegradable plastic: A step forward into adopting sustainable carbon neutrality. J Environ Chem Eng 12:112135. https://doi.org/10.1016/j.jece.2024.112135

Weligama Thuppahige VT, Moghaddam L, Welsh ZG et al (2023) Investigation of critical properties of Cassava (Manihot esculenta) peel and bagasse as starch-rich fibrous agro-industrial wastes for biodegradable food packaging. Food Chem 422:136200. https://doi.org/10.1016/j.foodchem.2023.136200

Wu X, Liu Z, Liu J et al (2024) Development of novel intelligent and edible dual-functional electrospun nanofiber films for shrimp preservation and freshness monitoring in real-time. Food Biosci 61:104828. https://doi.org/10.1016/j.fbio.2024.104828

Xie D, Ma H, Xie Q et al (2024) Developing active and intelligent biodegradable packaging from food waste and byproducts: a review of sources, properties, film production methods, and their application in food preservation. Compr Rev Food Sci Food Saf 23. https://doi.org/10.1111/1541-4337.13334

Xu W, Zhang X, Li Y (2022) Hybrid bioplastics from agricultural residues: balancing sustainability and functionality. Int J Biol Macromol 203:764–774. https://doi.org/10.1016/j.ijbiomac.2021.203764

Yang M, Chen L, Wang J et al (2023) Circular economy strategies for combating climate change and other environmental issues. Environ Chem Lett 21:55–80. https://doi.org/10.1007/s10311-022-01499-6

Zavaleta-Cavero DA, Chumpitaz-Principe DA, Gutarra-Espinoza AA et al (2024) From Bolaina Blanca wood fibers to antimicrobial films: characterization and application in the food industry using copper nanoparticles. J Nat Fibers 21. https://doi.org/10.1080/15440478.2024.2431314

Zhang J, Wen C, Zhang H et al (2020) Recent advances in the extraction of bioactive compounds with subcritical water: a review. Trends Food Sci Technol 95:183–195. https://doi.org/10.1016/j.tifs.2019.11.018

Zhang X, Li G, Chen C et al (2025) Chitosan/PVA composite film enhanced by ZnO/lignin with high-strength and antibacterial properties for food packaging. Int J Biol Macromol 306:141658. https://doi.org/10.1016/j.ijbiomac.2025.141658

Zhao X, Tian R, Zhou J, Liu Y (2022) Multifunctional chitosan/grape seed extract/silver nanoparticle composite for food packaging application. Int J Biol Macromol 207:152–160. https://doi.org/10.1016/j.ijbiomac.2022.02.180

Zhu Z, Liu W, Ye S, Batista L (2022) Packaging design for the circular economy: a systematic review. Sustain Prod Consum 32:817–832. https://doi.org/10.1016/j.spc.2022.06.005

Zhuo Y, He J, Li W et al (2023) A review on takeaway packaging waste: types, ecological impact, and disposal route. Environ Pollut 337:122518. https://doi.org/10.1016/j.envpol.2023.122518

Zuev YF, Derkach SR, Bogdanova LR et al (2023) Underused marine resources: sudden properties of cod skin gelatin gel. Gels 9:990. https://doi.org/10.3390/gels9120990

Chapter 13
Edible Films and Coatings: Towards Zero-Waste Packaging

Łukasz Łopusiewicz

Abstract In light of the growing environmental and economic challenges associated with the linear economy model (based on the extract–produce–dispose pattern), circular strategies focusing on the efficient use of resources, reuse, and waste management are becoming increasingly important. Within this framework, edible films and coatings emerge as a viable substitute for traditional packaging materials, particularly in relation to circular economy principles and the zero-waste approach. This chapter examines recent scientific developments and novel strategies involving the utilization of agri-food by-products in the formulation of environmentally sustainable edible films and coatings. The potential of waste materials, both plant and animal, as sources of biopolymers (including cellulose, starch, pectins, alginates, carrageenans, agar, chitosan, whey proteins, and waxes) used to develop sustainable packaging matrices is discussed. Furthermore, the role of these raw materials as sources of functional additives that significantly influence the performance of finished materials is highlighted. The chapter emphasizes the potential for aligning waste management practices with the development of innovative, biodegradable packaging solutions that contribute to a more sustainable food system and minimize the environmental footprint of packaging.

Keywords Natural compounds · Proteins · Polysaccharides · Packaging materials · Circular economy · Bio-based materials

---

Ł. Łopusiewicz (✉)
School of Medical and Health Sciences, Vizja University, Warsaw, Poland
e-mail: l.lopusiewicz@vizja.pl

Institute of Pharmacy, Department Pharmaceutical Biology, Greifswald University, Greifswald, Germany

S. Roy and M. Shahid (eds.), *Biomaterials for Sustainable Food Packaging*,
Environmental Footprints and Eco-design of Products and Processes,
https://doi.org/10.1007/978-3-032-08027-1_13

13.1 Introduction

The currently dominant economic model, the linear model (Fig. 13.1), is based on a framework inherited from the Industrial Revolution. It assumes a continuous supply of short-lived products, necessitating increased production to meet the ever-growing demands of consumers. The linear economy (extraction, production, and disposal) leads to uncontrolled exploitation of finite natural resources, potentially resulting in severe ecological and economic crises (Osorio et al. 2021; Costa et al. 2023; Hashimoto et al. 2025). While this model has facilitated rapid industrial and cultural development, its structure—centered around extraction, consumption, and disposal—adversely affects the availability of natural resources required to meet the needs of the global population. Over the past five decades, the global population has increased from 3.7 billion to 7.9 billion. Projections from a recent United Nations report indicate that this number will rise to 8.6 billion by 2030, 9.8 billion by 2050, and may reach 11.2 billion by the year 2100 (Berenguer et al. 2023; Yadav et al. 2024). Population growth drives increased consumption of natural resources and food. However, current resources are limited and insufficient to provide food for the growing population while simultaneously generating significant amounts of waste. The greatest challenge will be ensuring high-quality food while protecting and enhancing the environment (Kuttiyatveetil et al. 2019; Bilal et al. 2025). In response, the United Nations has identified 17 Sustainable Development Goals (SDGs) to mitigate major global challenges, including the eradication of poverty, zero hunger, good health and well-being, clean water and sanitation, innovative industry and infrastructure, responsible consumption and production, and climate action. In recent years, there has been increasing recognition that minimizing food loss and waste plays a pivotal role in achieving sustainable development for the global population. In fact, Goal 12.3 of the United Nations' 2030 Agenda for Sustainable Development aims to halve global food waste by 2030. Achieving this reduction would significantly support SDGs such as "zero hunger" and substantially improve the environmental impact of food production chain (Ben-Othman et al. 2020). Improving the food production chain from farm to table will be a key factor in meeting these objectives, given that the food supply system is among the least sustainable due to its complexity and interconnectedness (Tonolo et al. 2023). Despite ongoing initiatives and policies, an estimated 870 million people suffer from chronic undernourishment, while 1.3–1.6 billion tons of food waste are generated globally each year—equivalent to one-third of total food production (Wang et al. 2020; Rakesh and Mahendran 2024). Each year, this waste ends up in landfills, reducing available land for agriculture and causing health and environmental damage. Food waste and food loss have been widely addressed across various stages of the supply chain. The term "food waste" refers to both edible items and inedible components that are either deliberately discarded or inadvertently lost due to spoilage or expiration, with households representing a major source alongside other contributing sectors. Food loss is characterized as a "reduction in the quantity or quality of food, primarily due to the functioning of the food production and supply system or institutional and legal frameworks" (Melini et al.

2020; Rakesh and Mahendran 2024). The food processing industry—encompassing sectors such as fruits and vegetables, dairy, meat, poultry, seafood, brewing, and grain processing—generates substantial amounts of residue (waste) as by-products. Global estimates indicate that approximately 30% of food products are considered waste (Sadh et al. 2018; Ben-Othman et al. 2020; Osorio et al. 2021; Yadav et al. 2024). In the early stages of the Food Value Chain (FVC), about 13.3% of produced food is lost (e.g., during harvest, slaughter, post-harvest, transformation, or manufacturing), while 17% is discarded at the consumer level (Ortiz-Sanchez et al. 2023). It is estimated that as much as 30–40% of food waste occurs during food processing, indirectly increasing anthropogenic greenhouse gas emissions by 6–8% (Rakesh and Mahendran 2024; Bilal et al. 2025). The most wasted food categories include cereals (30%), fruits and vegetables (45%), and fish and seafood (35%) (Ben-Othman et al. 2020). In 2020, the total global generation of solid waste reached 2.24 billion tons, equating to an average of around 0.79 kg per person per day. By 2050, this figure is projected to rise by approximately 73%, resulting in 3.88 billion tons of waste generated annually worldwide (Faria et al. 2023; Bühlmann et al. 2023). To effectively manage by-products, a significant transformation of the agri-food system is required. The circular economy offers a viable and effective strategy for the medium to long term, facilitating the prevention, reuse, and recovery of natural resources and by-products, as illustrated in Fig. 13.1 (Ancuţa and Sonia 2020; Raţu et al. 2023). The circular economy is an economic model aimed at fostering human activities that result in no negative environmental or societal impacts. This model emphasizes collaboration between businesses and consumers to establish closed material loops, minimizing the materials and raw resources used in production while reducing waste through rational management of production processes (Morseletto 2020; Carvalho et al. 2025). The concept of a circular economy also involves extending the lifecycle of products by ensuring that by-products or waste can become valuable components of new production processes (Morseletto 2020; Rashwan et al. 2023). It is founded on the application of the 5Rs principles of sustainable development—reduce, reuse, recycle, recovery, and restore—which aim to replace the conventional linear model of "take, make, use, and dispose" (Lee et al. 2024). At its core, the circular economy focuses on efficient resource management and promotes a global economic model that seeks "win–win" solutions—both economically and environmentally beneficial (Yang et al. 2023). Closing the material loop is expected to yield numerous advantages, from the procurement of raw materials and auxiliary inputs for production to the resale and reuse of products, by-products, and waste (Salas-Millán and Aguayo 2024). The circular economy promotes the efficient utilization of resources across every phase of a product's lifecycle—from initial design and raw material extraction to processing, consumption, waste collection, and eventual repurposing or reintegration into the production cycle. By-products or waste generated during production are treated as potential raw materials, allowing the same resources to be reused without further environmental intervention (Rashwan et al. 2023). This approach fundamentally reshapes the prevailing culture and systems of extraction, production, and consumption by emphasizing the renewal and regeneration of natural resource value. It aims to curb excessive resource and energy consumption while

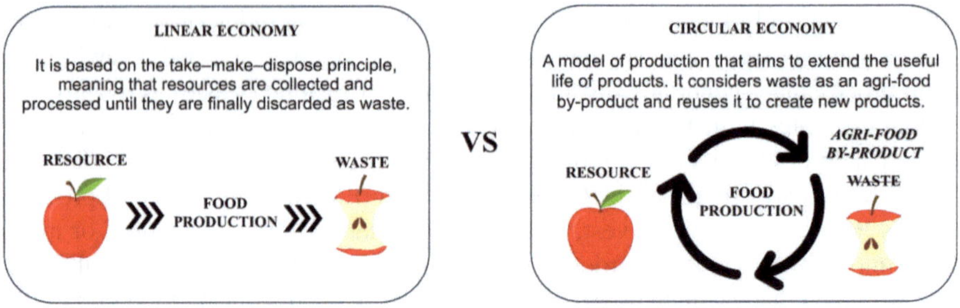

Fig. 13.1 Comparison of linear economy and circular economy models

avoiding unnecessary waste generation. In the circular economy framework, products are designed to minimize or entirely eliminate waste (Fraga-Corral et al. 2021). In addition, this approach fosters a culture where products are created with the intention of enabling a second life with added value, while reducing energy and raw material consumption. In the food industry, the circular economy primarily pertains to the secondary use of various types of biomass (Osorio et al. 2021). Biomass sources include plants (fruits, tubers, roots, leaves, stems, bark, flowers, etc.) and animals (milk and dairy products, meat, bones, blood, internal organs, glands, skin, wool, hair, etc.) (Rakesh and Mahendran 2024). Significant contributions also come from water-based products, such as fish and seafood. Recently, algae, lichens, and other aquatic or semi-aquatic plants have gained increasing importance. By-products can be repurposed for numerous applications, including animal and plant nutrition, biogas production, food, oil extraction, bioactive substances, organic fertilizers and the production of biomaterials. Additionally, the reuse of by-products from industrial processing as potential edible films and coatings represents an innovative aspect of the circular economy (Hamed et al. 2022; Jorge et al. 2023; Matloob et al. 2023; Gupta et al. 2024). However, certain legislative changes are also necessary to facilitate the implementation of sustainable approaches and innovations, as well as to increase the flow of proposed solutions from academia to industry. In fact, Socas-Rodríguez et al. highlighted the lack of specific regulations ensuring the safety of valorized by-products (Socas-Rodríguez et al. 2021).

13.2 Agri-Food Waste and By-Products Generated Across Different Branches of the Agricultural and Food Processing Industries—Examples

Biomass resulting from food loss, waste, and non-commercially viable portions contains considerable nutritional value, encompassing proteins, carbohydrates, lipids, minerals, and bioactive compounds such as polyphenols, dietary fiber, glucosinolates, and vitamins (Castrica et al. 2019; Álvarez-Castillo et al. 2021; Costa et al.

2023). These bioactive compounds have been linked to a lower risk of degenerative disorders, such as specific types of cancer, cardiovascular diseases, diabetes, and age-related ailments. Predominantly found in fruits and vegetables, phytochemicals exhibit physiological effects that surpass their fundamental nutritional functions (Sorrentino et al. 2022; Rojas-García et al. 2023; Farid et al. 2025). Agricultural by-products, derived from plants and animal production and processing, represent a valuable resource for repurposing rather than disposal (Sharma et al. 2016). For example, crop production by-products frequently result from the improper use of fruits and vegetables, which are often seasonally overproduced and largely discarded during processing. In the case of vegetables, common by-products include peels, seeds, stems, and leaves, whereas for fruits, they typically consist of peels, pulp, and pits (Rațu et al. 2023). Numerous by-products generated during food processing serve as valuable sources of dietary, functional, and emerging fibers. These fibers, whether used in their native state or following modifications for extraction or purification, can be integrated into a diverse range of food products (Sharma et al. 2016; Yadav et al. 2024). This integration not only maximizes the use of by-products but also enhances the nutritional and functional value of food, contributing to sustainable food systems and waste reduction (Sharma et al. 2016). Conversely, the potential of these compounds can also be harnessed to add value in the creation of functional and innovative packaging materials (Jorge et al. 2023).

The waste and by-products generated in fruit and vegetable processing rank among the richest organic materials within agriculture and the processing industry. The total organic content in these materials is substantial, ranging from 50 to 94% Available reports highlight their potential as a source of valuable bioactive compounds, predominantly accumulated in peels (Panzella et al. 2020; Roy et al. 2023b, c; Salas-Millán and Aguayo 2024). The biomass of such residues holds significant potential not only due to its nutritional value but also for energy and fertilizer applications, making it of keen interest to the food, biotechnology, pharmaceutical, and energy industries, among others. One major challenge is their perishability and susceptibility to microbial growth, necessitating rapid processing and presenting logistical difficulties in handling. Due to their high fiber content, fruit and vegetable residues are used in the production of dietary fiber preparations. These residues are ideal food and material additives, enriching products with proteins, vitamins, polyphenols, organic acids, and minerals, that also give new functionalities to packaging materials (Panzella et al. 2020). The composition of these compounds is influenced by factors such as the type of raw material, storage time, plant species, and the technological processes employed (Fig. 13.2). Furthermore, residues and other production by-products are sources of flavors and aromatic compounds, as well as natural colorants (Sales et al. 2018; Wang et al. 2020). Key pigments extracted include anthocyanins, betalains, carotenoids, and chlorophylls (Panzella et al. 2020; Roy et al. 2023b). Peels and pomace are particularly suited for pigment extraction, as they accumulate the highest concentration of colorants. Extracted pigments are available in liquid or dried forms with varying degrees of concentration and are widely used to color foods and to develop pH-sensitive intelligent films (Kokkuvayil Ramadas et al. 2024; Thakur et al. 2024). Pomace can be utilized either in its dried and milled form or through the incorporation of specific compounds extracted from it. Adding these components to foods can qualify for nutritional or health claims on product labels. Depending on the purpose of the addition, pomace or extracts can enhance the product's structure, improve antioxidant properties, or extend shelf life.

The growing manufacture of cereal- and pseudocereal-based food products leads to the accumulation of substantial quantities of milling by-products (Sabater et al. 2020; Rashwan et al. 2023; Rațu et al. 2023). These by-products are commonly used as animal feed; however, their rich content of nutrients and bioactive compounds also makes them well-suited for the development of functional food products. Antioxidant compounds in cereal grains are primarily concentrated in the outer layers and the germ, rendering milling by-products like bran and germ especially high in antioxidant capacity. Cereal bran, produced as a milling by-product, is valued for its abundance of health-promoting constituents, including dietary fiber, essential minerals, vitamins, polyphenols, and phytosterols (Nemes et al. 2022). Its chemical composition, low cost, and wide availability give it substantial potential as a nutrient source for functional foods as well as fiber source for packaging films (Zhang et al. 2024a). Bran can be successfully used as an ingredient in baked goods to enrich them with protein and dietary fiber. The utility of cereal bran in food production can be enhanced through various processing techniques, including mechanical, enzymatic, and thermal treatments.

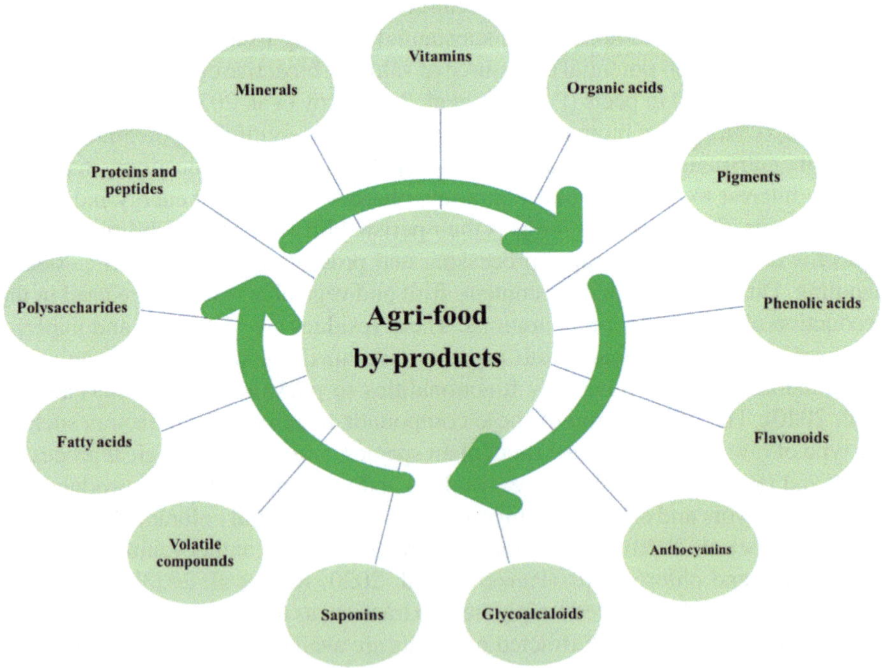

Fig. 13.2 Agri-food by-products as a source of valuable compounds

Brewer's spent grain (BSG) represents a lignocellulosic biomass by-product generated in significant volumes during the beer brewing process. Beer production yields approximately 25–30% waste in different forms, with BSG comprising 80–85% of this amount, leading to considerable quantities of waste on a global scale. Despite its rich composition, including carbohydrates, fiber, proteins, fatty acids, lipids, minerals, and vitamins, the development of efficient and economically viable utilization methods remains a challenge. These products find applications in energy production, the food industry, healthcare, and biomaterials, providing an innovative and sustainable solution for managing this abundant waste resource (Ahuja et al. 2024).

Coffee production represents a major global economic sector, with forecasts indicating a harvest of 171.4 million bags for the 2023/2024 season, primarily driven by leading producers such as Brazil, Colombia, and Ethiopia. Spent coffee grounds (SCG), a prominent by-product of coffee beverage preparation, are generated in vast quantities—estimated at around 60 million kilograms per day worldwide. This plentiful biomass holds significant potential for valorization, enabling sustainable applications across diverse fields including energy, agriculture, materials science, food, and cosmetics (Hashimoto et al. 2025; Vallejos-Jiménez et al. 2025).

The large quantities of by-products generated during oil production, such as oilseed meals and oilcakes, have driven efforts to find sustainable ways to utilize these resources (Ancuța and Sonia 2020; Petraru and Amariei 2023). Oilseed by-products, after appropriate processing, are commonly used as high-protein feed for livestock, significantly reducing the need for imported soybean meal. They are also employed as biomass for energy production (Ancuța and Sonia 2020). Cold-pressed oils, which are gaining popularity among consumers, can be extracted from seeds (e.g., sunflower, rapeseed, flax, camelina, evening primrose, borage, amaranth, safflower, poppy, black cumin, hemp), kernels (e.g., palm, black currant), fruits (e.g., olives, sea buckthorn, avocado), nuts (e.g., hazelnuts, walnuts, Brazil nuts), and cereal germ (e.g., wheat) (Cerveró et al. 2010; Ancuța and Sonia 2020). The cold-pressing process generates oil cakes as by-products, which, due to their chemical composition, are similarly utilized as animal feed components. However, given the larger scale of oilseed meal production compared to oil cakes, the economic feasibility of using oil cakes for feed is reduced. The chemical composition of cold-pressed oil cakes makes them a promising resource for functional food production and source of proteins and mucilages/gums that can be used for preparation of packaging films and coatings (Petraru and Amariei 2023; Pirsa et al. 2023; Akhila et al. 2023; Rani et al. 2023). Oil cakes serve as readily available protein sources for the formulation of plant-based protein products and are abundant in bioactive compounds with notable health-promoting properties. For example, flaxseed cakes are particularly rich in lignans, powerful antioxidants with significant anti-inflammatory activity (Łopusiewicz et al. 2023), and their properties can also be used in films and coatings formulations (Pedreiro et al. 2021). Some other by-products, such as fruit and tomato pomace, contain seeds that offer potential for the extraction of oils and valuable bioactive constituents. Depending on the plant species, seeds may contain 10–30% fat. The oil is extracted through pressing or supercritical CO_2 extraction. Studies have shown that oils from seeds are rich in valuable components, such as polyunsaturated fatty acids (PUFAs), squalene, vitamin E, and phytosterols. For instance, soy oil has been shown to effectively preserve the original quality characteristics and extend the shelf life of apple slices when incorporated into sago starch-based coatings (Zhang et al. 2024b). Challenges remain in the broader utilization of oil cakes, such as rapeseed cakes, for commercial production of proteins and bioactive compounds. Key issues include the presence of antinutritional or harmful substances, such as cyanogenic compounds, glucosinolates and phytates, which necessitate further detailed research to ensure safety and effectiveness in various applications (Ancuța and Sonia 2020).

Whey, the main by-product of the dairy industry, serves as both a low-cost substrate for diverse applications and a resource with a valuable chemical profile (Ryan and Walsh 2016; Kandasamy et al. 2021; Uribarrena et al. 2024). Products that can be obtained from whey or developed using it as a base include animal feed, microbial biomass, edible fermented goods, baker's yeast, organic and amino acids, enzymes, flavor compounds, colorants, microbial gums, polysaccharides, and proteins. However, the economic viability of producing some of these products often remains a challenge, hindering their commercialization. Another approach to whey utilization involves employing specific microorganisms capable of transforming its components into valuable products, primarily for applications in the food and pharmaceutical industries (Ryan and Walsh 2016; Rochín-Medina et al. 2018). Owing to its substantial lactose content, whey represents an appropriate substrate for lactic acid fermentation, during which lactose is transformed into lactic acid by lactic acid bacteria (Łopusiewicz et al. 2022a). While the high mineral and lactic acid content in acid whey limits its applications, technological advancements have enabled the production of valuable products from this by-product. Whey powder is produced through nanofiltration followed by spray drying. A popular processing pathway for acid whey is protein recovery using membrane separation techniques. Demineralization of whey, achieved through electrodialysis as an intermediate step before spray drying, provides valuable components for the production of infant formula. Whey protein has been characterized as containing essential amino acids, bioactive peptides, antioxidants, and immunomodulatory compounds, significantly contributing to its importance in functional food production. In addition, whey proteins possess distinctive characteristics—such as solubility over a broad pH spectrum, a well-balanced amino acid composition, multifunctionality, and cost-effective production—that render them highly suitable as raw materials for numerous food applications. Commercially, whey proteins are available as various protein preparations, including whey protein concentrate (WPC) containing 35% to over 80% protein, whey protein isolate (WPI) with protein content exceeding 90%, and highly purified single fractions such as β-lactoglobulin, which contains over 98% protein. These characteristics make whey proteins versatile and highly valuable for a range of applications, from functional and fortified foods to specialized nutritional products such as sports supplements, infant formula, medical dietary solutions, and packaging materials (Ryan and Walsh 2016; Kandasamy et al. 2021; Uribarrena et al. 2024).

13.3 Redefining Packaging: The Shift from Petrochemical Plastics to Eco-Designed Biodegradable Alternatives

The role of packaging has significantly increased in food production and sales. Packaging often has a decisive influence on the acceptability and desirability of the product it contains. Today's consumers are increasingly aware and have higher expectations when it comes to food packaging. The wide range and nature of food products are closely linked to the rapidly growing packaging market. Before a food product is sold, it is usually packaged (Jorge et al. 2023; Carvalho et al. 2025). This is done to preserve quality and minimize nutritional loss. Increasing consumer expectations, a growing diversity of food products, and the surge in waste generation have driven substantial transformations in the packaging sector, resulting in the development of numerous novel, and at times contentious, packaging solutions. Currently, two major trends can be observed in the packaging market. The first concerns eco-friendly packaging, which is made from environmentally friendly materials based on natural and synthetic compounds (González-Arancibia et al. 2024; Carvalho et al. 2025). The second trend involves intensifying product protection and the informational function of packaging. In addition, there is a growing trend towards innovative functional packaging, which includes smart and active packaging. In addition to protecting food, smart packaging allows its condition and environment to be monitored, while active packaging improves its quality and shelf life. Currently, the packaging market is dominated by plastics (causing excessive waste pollution), which are also the most criticized type of packaging (Kocira et al. 2021; Roy et al. 2023c; Jorge et al. 2023). Plastics possess limited or nonexistent biodegradability and are prone to flammability, safety concerns, reduced strength and stability, and substance migration; coupled with the urgent need for improved recycling and energy recovery, these drawbacks underscore the necessity of developing alternative solutions. Hybrid materials, which combine the advantages of organic polymers (plasticity, elasticity, processability, and ease of forming) with the desirable properties of inorganic components (hardness, chemical, mechanical, and thermal resistance), may be an alternative to plastics. Organic–inorganic hybrids have a unique structure and a wide range of applications, such as gels, thin films, fibers, and powders (Zhu et al. 2023; Ghosh et al. 2025).

The ongoing development of packaging has also contributed to the increased functionality of innovative food packaging materials, including active and smart packaging. There is a growing demand for packaging solutions that are environmentally sustainable and aligned with the principles of the circular economy and zero-waste practices (Carvalho et al. 2025). Proper management of packaging waste can significantly reduce the amount of waste sent to landfills. In response to the climate crisis and the influx of data on environmental degradation, legislation is forcing changes that will enable climate neutrality by 2050. As an example, under the European Green Deal, the European Commission has introduced initiatives aimed at replacing plastic packaging with alternatives that are either reusable or recyclable. By 2030, packaging is expected to be bio-based or biodegradable. Additionally, there is growing emphasis on the role of active packaging in reducing food waste. In addition to packaging made from traditional materials, biopolymer-based materials are also noteworthy, as they are biodegradable and fit in with the objectives of both action plans (circular economy and European Green Deal), which aim to prevent packaging waste, improve food quality, and reduce product losses.

In this regard, the advancement of renewable, biodegradable, and edible materials presents a promising approach to addressing the complex challenges confronting modern packaging systems (Lee et al. 2024). Therefore, a promising solution in the field of packaging is the production of edible packaging in the form of films and coatings. Nowadays, biobased polymers derived from sustainable and renewable natural sources are of great interest as an alternative to control the severe damage already caused by petro-chemical-based polymers (Carvalho et al. 2025). At the research stage, most attention is focused on edible films and coatings that are directly applied on food products. Food coating is used to extend the shelf life of many perishable raw materials and food products. In recent years, food coatings have gained growing attention from both manufacturers and consumers, as edible films and coatings offer an effective means of prolonging the shelf life of various food products. Minimizing dependence on conventional packaging materials by utilizing edible coatings contributes to a decrease in packaging waste, thereby supporting sustainable practices and environmental preservation initiatives (Pillai et al. 2024). Food coating is a technology that uses protective coatings made from renewable, biodegradable, and edible ingredients, and is therefore expected to reduce environmental pollution, even if they are not reused. Food industry residues are increasingly being used as materials for edible films and coatings (Hamed et al. 2022; Jorge et al. 2023; González-Arancibia et al. 2024). These materials can help inhibit alterations in color and texture, suppress the growth of undesirable microorganisms, and reduce the loss of essential vitamins and minerals. They should minimize the adverse effects of sunlight and moisture. Thus, edible coatings and films can contribute to improving food quality and safety (Hamed et al. 2022; Matloob et al. 2023). The development of edible packaging may be a beneficial and, over time, a necessary alternative to synthetic packaging, for economic and health reasons, among others. The growing application of biodegradable edible coatings across various food sectors is intended to lessen the reliance on individual packaging and curb the escalating volume of environmental waste. Within the field of food coating, ongoing efforts focus on developing formulations that exhibit optimal functional properties (Gupta et al. 2024). The introduction of antimicrobial substances or nanomaterials may increase the scope of application of edible coatings and determine the acceptance of such products by consumers (Hamed et al. 2022; Ghosh et al. 2025). Intensive research efforts are currently focused on developing efficient techniques for producing edible films and coatings, as well as on tailoring their functional characteristics. The origins of edible coatings date back to the twelfth century in China, where they were used to protect citrus fruits and minimize weight loss. In the early fifteenth century, Japan developed the first edible packaging from soy milk proteins, while in 19th-century England, coatings were applied to safeguard meat products. In the fruit and vegetable industry, coatings were introduced in the 1930s, enabling respiration and minimizing moisture loss during transport. Edible coatings are defined as thin layers of material applied directly to the surface of food products (Matloob et al. 2023). The performance of edible coatings is primarily influenced by the nature of the coating materials and the processing conditions under which they are produced. Key factors include the choice of solvent, pH, temperature, and the concentration of the formulation components. Edible films, defined as thin, consumable layers placed on or between food products, help protect products from mechanical damage as well as physical, chemical, and microbiological deterioration. Ongoing research seeks to develop biodegradable alternatives that reduce dependence on conventional synthetic packaging. The principal role of edible coatings is to act as selective barriers to water vapor, oxygen, carbon dioxide, volatile compounds, and lipids. When applied by dipping or spraying, these coatings can enhance the quality and shelf life of fresh produce such as fruits and vegetables. It is also important to be able to modify the atmosphere, e.g. around the fruit, which can improve the overall appearance of the coated product. The right choice of coating also reduces the activity of enzymes that cause browning and discoloration of fruit. The use of edible coatings is a relatively inexpensive method, usually used as an additional step in the production process to maintain the quality and freshness of products. Edible coatings also increase the attractiveness of the product, e.g., by making it shiny, but should not affect the taste of the coated product. This will ensure the expected sensory acceptability of the products (Matloob et al. 2023).

Food products that are susceptible to oxidation and enzymatic browning during storage may exhibit visible defects such as discoloration or moisture loss. The application of edible coatings can minimize undesirable changes and enhance the visual appeal of food products. These coatings contribute to maintaining appropriate texture, often through the inclusion of agents like calcium ions. Moisture loss, a significant factor contributing to economic losses, is mitigated by incorporating water-insoluble or low-permeability compounds into the coating formulation. In the storage of fruits and vegetables, edible coatings are employed to alter the internal gas composition by reducing oxygen levels and limiting ethylene production, which in turn helps preserve freshness. Additionally, they play a role in retaining bioactive compounds during storage. Coatings also offer protection against surface discoloration, inhibit the development of harmful microflora, and help maintain the natural appearance of products (Hamed et al. 2022). It has been found that the use of edible coatings for perishable and easily spoiling fruits and vegetables is an effective and efficient method due to the small amount of coating solution in relation to the weight of the coated raw materials. This method reduces the cost of preserving products compared to the use of low temperatures and other methods of protecting raw materials from spoilage during storage and transport. In the meat industry, edible coatings are used on fresh and frozen products derived from meat, poultry, and fish. They take the form of coatings placed, among other things, on the surface of kabanos sausages or other sausages (Song et al. 2021). They are also an alternative to their conventional counterparts and chemical preservatives. Edible coatings used for meat are carriers of antimicrobial agents, prevent moisture loss during storage, and also allow the product to remain juicy after packaging and absorb foreign odors (Song et al. 2021; Pires et al. 2024).

Biodegradable edible coatings are also utilized to shield food products from microbial contamination and to prevent mechanical, physical, and chemical degradation. Moreover, they can act as carriers for various functional additives, including antimicrobial agents, antioxidants, and anti-browning compounds, thereby contributing to the preservation of food quality and safety. Beyond their barrier properties that enhance product stability, these coatings may also interact beneficially with both the food surface and its surrounding environment, resulting in the formation of active coatings with targeted functionalities. The active properties of the coating enable the slow release of, for instance, antimicrobial or antioxidant active compounds, slowing down food degradation processes or removing undesirable compounds that contribute to its decomposition.

13.4 Edible Films and Coatings in the Circular Economy: Harnessing Agri-Food By-Products and Wastes for Both Structural and Functional Roles in Zero-Waste Packaging

Recently, biological refining, i.e., the sustainable processing of biomass into high value-added products or energy, has gained importance as one of the key technological innovations in the field of sustainable development (Berenguer et al. 2023; Yadav et al. 2024; Rakesh and Mahendran 2024). It is based on the belief that diverse biomass resources, including food industry by-products, can be efficiently broken down into basic molecular components such as polysaccharides, proteins, phenolic compounds, and triglycerides. These valuable fractions can then be used as raw materials for the production of a wide range of high-value materials, including bioplastics (Lee et al. 2024). Among the substances employed in the formulation of edible films and coatings, particular emphasis is placed on polysaccharides, proteins, and lipids, as well as on systems composed of two or more of these components in combination (Jorge et al. 2023; Pashova 2023; Pan et al. 2024).

Film and coating materials also include multi-component combinations of carbohydrates, proteins, and fats. For instance, multi-component protective layers for vegetables and fruits should contain compounds similar to natural waxes (Pashova 2023). The multi-component structure of edible coatings allows for the development of layers with tailored functional characteristics. An appropriately designed composition can effectively minimize moisture loss, support the formation of a beneficial internal atmosphere, and help maintain the nutritional and bioactive qualities of the food product. In order to increase the functionality of coatings, small amounts of edible chemical additives (including plasticizers, emulsifiers, plant extracts, and fragrances) are added to the membrane or coating matrix (Kocira et al. 2021; Liu et al. 2024; Pan et al. 2024; Ghosh et al. 2025). These can influence the mechanical properties, increase adhesion and stability, as well as improve food safety and quality. Plasticizers are low molecular weight, non-volatile substances that are added directly to coating formulations to influence their elasticity. Examples include glycerol, sorbitol, sucrose, polyethylene glycol, fatty acids, and monoacylglycerols, all characterized by low molecular weight. Depending on the type of coating, the plasticizer content ranges from 10 to 60%. The incorporation of plasticizers weakens the intermolecular interactions between polymer chains, leading to coatings that exhibit increased flexibility and extensibility. The possibility of shrinkage during transport and storage and cracking of the coatings is reduced, but the inclusion of too much of these substances in the structure can lead to phase separation and crystallization on the surface of the final product (Sothornvit and Krochta 2005). To further enhance the sustainability of edible films and coatings, plasticizers used in film-forming formulations can be derived from waste products and by-products. Other attractive routes for obtaining plasticizers are also based on biotechnological processes such as fermentation or biotransformation using a wide range of microorganisms (Yang et al. 2012; Ribeiro et al. 2017; Chilakamarry et al. 2021; Phannarangsee et al. 2024). Besides offering barriers to gases and moisture, edible films and coatings can also act as carriers for antimicrobial agents that inhibit the growth of pathogenic bacteria, as well as for antioxidants, flavor compounds, colorants, enzymes, and nutrients, components that enhance product attractiveness from the consumer's perspective. Antimicrobial substances are usually added directly to food, which may reduce the functional activity of the added substance. Numerous studies and reports indicate that a wide range of by-products and waste from many branches of the agri-food industry can serve as an attractive sustainable source of a wide variety of compounds with multiple effects, including antimicrobial and antioxidant activity (Hamed et al. 2022; Matloob et al. 2023; Moura-Alves et al. 2023; Berenguer et al. 2023; Gupta et al. 2024; Yadav et al. 2024). Therefore, the use of a coating may be more effective than the direct addition of such a substance to the product, as these substances can migrate selectively to the surface of the food. All ingredients used to prepare coating solutions should have as little impact as possible on the sensory quality of the products. It is important that the color of the coated product is maintained, as a change in color may suggest that the product has lost its freshness or is unripe.

At present, considerable research is focused on identifying and developing novel materials that possess appropriate film-forming capabilities for application in protective edible coatings and biodegradable packaging films.

13.4.1 Polysaccharides

Edible polysaccharide materials are characterized by low oxygen permeability, which has a beneficial effect on reducing oxidation reactions in food products. Because of their hydrophilic nature, such coatings generally provide limited resistance to water vapor; however, reducing their solubility can be achieved through cross-linking reactions. Coatings obtained from polysaccharides are usually colorless, characterized by good stability and biodegradability (González-Arancibia et al. 2024; Gupta et al. 2024; Pan et al. 2024). The main polysaccharides used for coating include cellulose, starch, pectins, carrageenans, chitosan, alginates, agar, and pullulan. Table 13.1 presents selected zero-waste strategies and the utilization of by-products in polysaccharide-based films and coatings.

13.4.1.1 Plant Polysaccharides

Cellulose

Cellulose, a widely available, renewable, and biocompatible raw material, has found application in the packaging of food products. Cellulose is the most abundant polymer in the world, with an estimated natural production of 10^{10}–10^{11} tons per year, of which only about 6×10^9 tons are used in various industries, including the paper, textile, chemical, and materials industries (Nargotra et al. 2023; González-Arancibia et al. 2024). Cellulose and its derivatives are extensively utilized in the food packaging sector, with biowaste and residues from various food sources serving as the primary feedstock. Polymers derived from cellulose are valued for their excellent biocompatibility, biodegradability, mechanical strength, thermal stability, water absorption capacity, porosity, crystallinity, solubility, sensory attributes, and other beneficial biological characteristics. Cellulose-based fibers offer a range of functional advantages. Although cellulose itself shows relatively low antioxidant and antimicrobial activity, these limitations can be addressed by incorporating natural plant extracts (Liu et al. 2021). For instance, Nemazifard et al. used ethanol extract from pomegranate seeds to modify cellulose films. They found a significant increase in antioxidant properties, but the ethanol extract from pomegranate seeds caused a decrease in tensile strength, elastic modulus, and swelling, as well as an increase in elongation at break and water permeability through the film, and also affected the morphology of the film (Nemazifard et al. 2017). In recent years, cellulose nanocrystals (CNC) have gained attention as highly promising biomaterials for the development of sustainable active packaging. Their appeal stems from a combination of advantageous features such as availability, consistency, low density, outstanding mechanical strength, biocompatibility, biodegradability, renewability, large surface area, tunable surface chemistry, and non-toxic nature (Chaka 2022; Nargotra et al. 2023; Ghosh et al. 2024, 2025). CNC are typically derived from a variety of lignocellulosic sources, including wood, plant fibers, and agricultural residues. It is indicated that cellulose nanocrystals as nanofillers influence the properties of packaging films, and their extraction from waste raw materials such as husks, cobs, fibers, seeds, shells, bagasse, and straw is one of the promising avenues in sustainable development (Chaka 2022; Thongsomboon et al. 2023; Ghosh et al. 2024). Other by-product sources include garlic stems, corn cobs, and chopped sugarcane stalks, from which Rovera et al. obtained cellulose yields of 35.73, 37.15, and 39.10% (Rovera et al. 2023). In another study, Lee et al. obtained cellulose from kimchi production. Optical analyses showed that films containing cellulose extracted from kimchi had better surface dispersion than films made from commercially available cellulose. Analyses of physical properties demonstrated the hydrophilic nature of the films, as indicated

Table 13.1 Zero-waste strategies and by-product applications in polysaccharides-based films and their impact on packaging performance

Polysaccharide biopolymers	Zero-waste approaches, by-product use, and impact on material properties	References
Cellulose	Source: ethanol extract from pomegranate seeds (by-product), cellulose derivatives (carboxymethylcellulose, hydroxyethylcellulose, hydroxypropylmethylcellulose, methylcellulose); process: incorporation of extract into cellulose films under mild conditions; properties: increased antioxidant activity, decreased tensile strength, elastic modulus, and swelling, increased elongation at break and water permeability, altered morphology with coarse microstructure and extract aggregates	Nemazifard et al. (2017)
Cellulose	Source: Brachiaria decumbens grass biomass (by-product); process: mild alkaline hydrolysis (1 M NH_3, 24 h) without separating cellulose, lignin, or hemicellulose; ε-polylysine as plasticizer and antimicrobial agent; properties: tensile strength ~10–21 MPa, Young's modulus ~700–2000 MPa, improved flexibility, water resistance, antimicrobial and antioxidant activity, significantly lower carbon footprint than PLA and PET	Estrada-Sotomayor et al. (2025)
Cellulose	Source: cellulose from grapevine pruning waste (by-product); process: alkaline and bleaching extraction, $ZnCl_2$ solubilization, Ca^{2+} crosslinking, glycerol plasticization; properties: transparent (~84%), tensile strength 15–18 MPa, low water solubility (14–19%), reduced water vapor permeability, full biodegradation in 17 days	Paudel et al. (2025)
Cellulose	Source: cellulose extracted from kimchi production waste (by-product); process: hot synthetic method with sorbitol and citric acid, followed by solvent casting; properties: better surface dispersion compared to films made with commercial microcrystalline cellulose, hydrophilic character (wetting angle < 20°), thermal stability (glass transition temperature) comparable to films with commercial microcrystalline cellulose	Lee et al. (2024)

(continued)

Table 13.1 (continued)

Polysaccharide biopolymers	Zero-waste approaches, by-product use, and impact on material properties	References
Cellulose	Source: broccoli stems (by-product); process: extraction of cellulose and carboxymethylation to produce carboxymethyl cellulose (CMC); properties: adjustable tensile strength and flexibility depending on glycerol content, good water vapor permeability	Sayanjali et al. (2024)
Cellulose	Source: rice straw, corncob, pineapple leaves/peels (by-products); process: bleaching and alkaline extraction for cellulose isolation; properties: highest yield and purity from corncob, high crystallinity, excellent thermal stability	Romruen et al. (2022)
Cellulose	Source: pumpkin seeds, broken rice, orange pectin, quince peel, potato peel/pulp, orange peel, pumpkin peel (by-products); process: edible film formation with various agro-waste components; properties: best mechanical and barrier performance with orange pectin, broken rice, pumpkin peel; good water contact angle, tensile strength, and elongation; promising substitutes for commercial starch and glycerol	Jorge et al. (2023)
Starch	Source: corn starch, moringa gum, pine cone extract (forest waste, by-product); process: composite film formation with bioactive extract; properties: increased elongation at break, reduced tensile strength, strong UV-blocking capacity, suitable for packaging light-sensitive foods	Nayak et al. (2024)
Starch	Source: arrowroot powder, corn starch, refined wheat flour; low water vapor permeability; thermal stability up to 190 °C (~5% weight loss); partial biodegradation in 22 days, complete by 60 days; refined wheat flour films showed uneven texture due to large particle size	Shanbhag et al. (2023)
Starch	Source: ripe banana peel flour (by-product), corn starch, glycerol; process: casting technique with varying cornstarch concentration and heating time; properties: enhanced mechanical strength with higher cornstarch, increased water vapor permeability due to starch hydrophilicity, uniform chroma	Arquelau et al. (2019)

(continued)

Table 13.1 (continued)

Polysaccharide biopolymers	Zero-waste approaches, by-product use, and impact on material properties	References
Starch	Source: potato waste and peel starch (by-product); process: hydrothermal and acid-alcohol treatments, plasticization with glycerol and sorbitol; properties: highest tensile strength and elongation with hydrothermal treatment, improved solubility with acid-alcohol treatment, best tensile strength with 35% sorbitol, increased water vapor transmission with 55% glycerol, significant biodegradability by soil microorganisms	Karki et al. (2020)
Starch	source: starch from potato processing slurries, powdered onion peel (by-products); process: direct incorporation of onion peel into starch-based films (zero-waste approach); properties: enhanced Young's modulus, improved gas barrier and water resistance, natural pigmentation, strong antioxidant activity (up to 93% ABTS inhibition), scalable circular economy solution for biodegradable packaging	Vallejo et al. (2025)
Pectin	Source: fruit and vegetable residues (sweet orange, passion fruit, watermelon, zucchini, lettuce, carrot, spinach, mint, yam, cucumber, arugula), methoxylated pectin; process: biodegradable film formation with different granulometry; properties: reduced solubility and elongation, increased tensile strength, uniform yellowish appearance, high water solubility, potential for tailored applications based on particle size and composition	Brito et al. (2019)
Pectin	Source: sweet potato starch, pectin from lemon waste (by-product), titanium oxide nanoparticles; process: casting of starch-pectin-TiO_2 nanocomposite films; properties: reduced moisture content, solubility, and moisture uptake, improved mechanical strength and thermal stability, UV-protective capacity	Dash et al. (2019)
Pectin	Source: pectin from coffee mucilage and pulp (by-products), bacterial cellulose, coffee oil; process: pectin isolation by citric acid hydrolysis, composite film formation; properties: improved tensile strength and elasticity, moderate water solubility, spontaneous soil degradation in 16 days	Vallejos-Jiménez et al. (2025)

(continued)

Table 13.1 (continued)

Polysaccharide biopolymers	Zero-waste approaches, by-product use, and impact on material properties	References
Pectin	Source: pectin, polyvinylpyrrolidone, bael fruit peel extract (by-product), titanium nanoparticles; process: casting method to create active nanocomposite films; properties: enhanced tensile strength, antioxidant activity comparable to gallic acid, antibacterial effects against *E. coli* and *L. monocytogenes*, reduced transparency, UV protection	Deshmukh et al. (2022)
Pectin	Source: pectin, chitosan, red prickly pear waste (by-product); process: incorporation of powdered or aqueous extracts into edible films; properties: substantially enhanced antioxidant activity, minimal impact on physicochemical characteristics, color changes depending on polymer	Kurek et al. (2021)
Pectin	Source: pectin and betacyanins from pitaya (*Hylocereus* spp.) peel waste (by-product), montmorillonite; process: composite film formation with colorimetric and functional additives; properties: enhanced UV shielding and antioxidant activity, reduced mechanical strength and thermal stability, high sensitivity to ammonia, effective freshness indicator for shrimp via progressive color change	Jiang et al. (2023)
Carrageenan	Source: carrageenan, starch, cellulose nanocrystals from pineapple leaves (by-product); process: nanocomposite film formation with glycerol and aloe vera gel; properties: enhanced mechanical strength and surface hydrophilicity, stable up to 200 °C, reduced water vapor and oil permeability, promising as biodegradable edible packaging for fruits and vegetables	Dmitrenko et al. (2023)
Carrageenan	Source: κ-carrageenan, cellulose nanocrystals from Indian gooseberry (amla) pomace (by-product); process: solution casting of bio-nanocomposite films; properties: improved mechanical strength and barrier performance at 5% CNC, reduced water vapor permeability, increased crystallinity, agglomeration at higher CNC levels, potential for high-barrier food packaging	Gupta et al. (2023)

(continued)

Table 13.1 (continued)

Polysaccharide biopolymers	Zero-waste approaches, by-product use, and impact on material properties	References
Carrageenan	Source: κ-carrageenan, hydroxypropyl methylcellulose, grape skin powder from wine production (by-product); process: composite film formation with anthocyanin-rich grape skin as pH indicator; properties: high pH sensitivity, color change indicating pork freshness, decreased light transmittance and mechanical strength with more grape skin, smart packaging solution for monitoring meat spoilage	Chi et al. (2020)
Alginate	Source: alginate, cellulose nanofibrils from cocoa by-products; process: edible coating formation with calcium chloride crosslinking; properties: decreased transparency and water vapor permeability, improved visual quality and firmness of blueberries, reduced respiration rate and moisture loss during 21 days of cold storage	Medina-Jaramillo et al. (2020)
Alginate	source: sodium alginate, grape pomace extract (by-product); process: internal and external gelation with $CaCl_2$, incorporation of polyphenolic extract; properties: reduced hydrophilicity, thermal and pH stability, UV shielding, antioxidant activity, swelling in concentrated salt solutions, suitable for protecting food from light-induced spoilage	Gubitosa et al. (2023)
Alginate	Source: alginate, mangaba pulp (Hancornia speciosa, by-product), Saccharomyces boulardii; process: bioactive edible film formation with probiotics; properties: improved water resistance, mechanical strength, thermal stability, antioxidant activity, UV and visible light barrier, probiotic viability during storage	Oliveira Filho et al. (2023)
Alginate	Source: sodium alginate, carboxymethylcellulose, gluten, onion peel and stalk extracts (by-products); process: blend film formation with bioactive extracts; properties: improved water barrier and tensile strength, increased phenolic content and antioxidant activity, reduced microbial load on peeled shallot onions, extended storage at 4 °C for 21 days	Thivya et al. (2022)

(continued)

Table 13.1 (continued)

Polysaccharide biopolymers	Zero-waste approaches, by-product use, and impact on material properties	References
Alginate	Source: alginate, melanin from watermelon seeds (*Citrullus lanatus*, by-product), silver and zinc oxide nanoparticles; process: composite film formation with melanin-stabilized nanoparticles; properties: high antioxidant and antimicrobial activity, improved mechanical and barrier properties, UV blocking, no cytotoxicity, potential for food packaging and biomedical applications	Łopusiewicz et al. (2022b)
Pullulan	Source: pectin, pullulan, grape seed extract (by-product); process: binary blend film formation for edible coatings; properties: antioxidant and antibacterial activity, UV blocking, high gas barrier, 75% reduction in peanut rancidity after 30 days, extended shelf life of coated peanuts	Priyadarshi et al. (2022)
Pullulan	Source: pullulan, essential oil from *Cistus ladanifer* (weed by-product); process: incorporation of essential oil into bioactive films; properties: antioxidant and antibacterial activity (stronger against Gram-positive), quorum sensing inhibition, antibiofilm effects, potential for packaging oxidation-prone foods	Luís et al. (2020)
Agar	Source: agar, bioactive extracts from pea husks (by-product) obtained with natural deep eutectic solvents; process: agar-glycerol film formation with incorporated extracts; properties: improved tensile strength and hydrophobicity, reduced water solubility, antibacterial activity against E. coli and S. aureus, enhanced cheese preservation	Akbaş and Erdoğan (2025)
Agar	Source: rice cooking water (by-product), agar, plant extracts from Ocimum tenuiflorum and Murraya koenigii, green silver nanoparticles; process: starch-agar composite film formation with plant extract and AgNPs; properties: antimicrobial activity, improved mechanical strength and barrier properties, extended grape shelf life to 12 days	Sarma et al. (2024)
Agar	source: agar, cellulose nanocrystals from onion skins (by-product); process: composite film formation with nanofiller reinforcement; properties: improved tensile strength and Young's modulus, reduced water vapor permeability, enhanced mechanical and barrier performance with 3% nanocrystal content	Rhim et al. (2015)

(continued)

Table 13.1 (continued)

Polysaccharide biopolymers	Zero-waste approaches, by-product use, and impact on material properties	References
Agar	Source: unpurified agar extracts from *Gelidium sesquipedale* (with natural proteins and polyphenols); process: film formation without added plasticizers; properties: improved moisture resistance, enhanced antioxidant and UV-blocking properties, flexible structure due to natural admixture	Martínez-Sanz et al. (2019)
Agar	Source: agar, grapefruit seed extract; process: solvent casting of antimicrobial films; properties: enhanced antimicrobial activity *against L. monocytogenes*, *B. cereus*, and *E. coli*, reduced tensile strength and elastic modulus, increased water vapor permeability	Kanmani and Rhim (2014)
Chitosan	Source: chitosan from crab-uçá (*Ucides cordatus*) waste (by-product); process: chitosan extraction and edible film formation with glycerol; properties: good optical and mechanical properties, satisfactory water vapor permeability and solubility, antimicrobial activity	do Vale et al. (2021)
Chitosan	Source: chitosan from blue crab waste, pectin from orange peel (by-products); process: casting method with glycerol as plasticizer; properties: increased elasticity and flexibility in blend films, higher solubility and moisture with more pectin, less stiffness compared to single-polymer films, brighter and less opaque control films	Baron et al. (2017)
Chitosan	Source: chitosan, pomegranate peel extract (by-product); process: incorporation of extract into edible films; properties: increased thickness, tensile strength, opacity, water and gas barrier, antioxidant and antimicrobial activity; reduced thermal stability, transparency, solubility, and swelling at higher extract levels	Kumar et al. (2021)
Chitosan	Source: chitosan, banana peel extract (by-product); process: composite film formation with antioxidant and cross-linking functions; properties: reduced hydrophilicity, improved mechanical strength, excellent antioxidant activity, enhanced postharvest quality of coated apples	Zhang et al. (2020)

by wetting angles below 20°. Thermal analysis revealed comparable glass transition temperatures (Tg), confirming similar thermal stability between films made with commercially available microcrystalline cellulose and those incorporating cellulose extracted from kimchi (Lee et al. 2024). In the work of Sayanjali et al., broccoli stalks (which are usually discarded) were used as a source of cellulose, which the researchers used then to produce carboxymethylcellulose in a carboxymethylation reaction using chloroacetic acid and sodium hydroxide (Sayanjali et al. 2024). Romruen et al. used alkaline extraction with a prior bleaching process of rice straw, corn cobs, pineapple leaves, and pineapple peels to obtain cellulose. They found that thanks to the prior bleaching processes, which allow for a higher content of cellulose free from impurities, these raw materials can be an attractive and sustainable source of cellulose as a biodegradable material (Romruen et al. 2022). Jorge et al. investigated the potential of plant by-products such as pumpkin seeds, broken rice, and orange pectin, as well as waste materials including quince peel, potato peel, potato pulp, orange peel, and pumpkin peel. Their findings indicated that films with elevated sugar content displayed reduced mechanical strength, enhanced flexibility, and lower onset temperatures for thermal degradation (Jorge et al. 2023). Estrada-Sotomayor et al. developed bioplastic films from *Brachiaria decumbens* grass biomass through full conversion without separating cellulose, lignin, or hemicellulose. They applied mild alkaline hydrolysis (1 M NH_3, 24 h), minimizing solvent use and waste generation, aligning with zero-waste principles. ε-Polylysine was incorporated as a plasticizer and antimicrobial agent, enhancing the films' flexibility, water resistance, and bioactivity. The resulting materials showed good mechanical properties (tensile strength ~10–21 MPa, Young's modulus ~700–2000 MPa) and effective antimicrobial activity against *E. coli* and *S. aureus*. Overall, the process significantly reduced the carbon footprint compared to PLA and PET, highlighting its potential for sustainable packaging applications (Estrada-Sotomayor et al. 2025). Paudel et al. developed biodegradable films from grapevine pruning waste, extracting cellulose through alkaline and bleaching treatments and processing it with $ZnCl_2$, Ca^{2+} crosslinking, and glycerol plasticization. This zero-waste strategy valorized abundant agricultural residues, transforming them into transparent (~84%) and mechanically strong films (15–18 MPa tensile strength) that biodegraded fully within 17 days in soil. The films exhibited low water solubility (~14–19%) and reduced water vapor permeability, making them suitable for food packaging applications (Paudel et al. 2025).

Starch

Starch is among the most abundant plant-derived raw materials, functioning as a storage substance in plants and accumulating primarily in tubers, roots, and seeds (Kocira et al. 2021; González-Arancibia et al. 2024). It is a natural biopolymer that plays a key role in food technology, especially in determining the rheological properties of food. It can also be used as a component of packaging material because it has beneficial functional properties, is cheaper than synthetic polymers, biodegradable, and easy to process. Starches contain varying amounts of amylopectin and amylose. Cereal starch contains approximately 25% amylose and 75% amylopectin. The ability to form gels depends on the amount of amylose (Kocira et al. 2021; Dutta and Sit 2024). High-amylose starch from corn seeds contains up to 85% amylose. Starches with a high (over 70%) amylose content are used to produce edible films. Starch possesses a strong capacity for forming edible films with satisfactory mechanical properties and can be applied to food products to minimize moisture loss or shield them from oxidative deterioration. Additionally, it enables the formation of transparent, tasteless, and odorless films that offer excellent oxygen barrier performance, making it highly advantageous for food preservation (Dash et al. 2019). Starch films are formed by evaporating aqueous solutions of gelatinizing amylose. Moisture significantly increases permeability. For instance, hydroxypropyl starch derivatives are more soluble and produce films that are more permeable to water vapor and gases. Edible coatings and films made exclusively from starch are brittle, so lipids and gums are added to modify their mechanical properties. In addition, their mixtures give food new textural properties, including increased water retention capacity.

In a context of sustainable solutions, Nayak et al. developed edible films based on corn starch/moringa gum with the addition of pine cone extract (PCE)—an underappreciated forest waste that contains bioactive compounds acting as UV filters. The study revealed that the incorporation of PCE markedly enhanced the elongation at break from 3.27 to 35.2%, albeit with a notable reduction in tensile strength. Nonetheless, starch-based films enriched with PCE extract demonstrated considerable UV-blocking capacity, highlighting their suitability as sustainable packaging materials for protecting light-sensitive food products (Nayak et al. 2024). Shanbhag et al. investigated the formulation and characterization of novel starch-based edible films utilizing ingredients such as arrowroot powder, corn starch, and refined wheat flour. Their findings indicated that the films exhibited low water vapor permeability, making them suitable candidates for food packaging applications. Partial biodegradation of all film types was observed within 22 days, with complete degradation occurring by day 60. The films maintained good thermal stability up to 190 °C, showing approximately 5% weight loss. However, those prepared with refined wheat flour displayed uneven surface textures, attributed to the larger particle size of the flour (Shanbhag et al. 2023). In another study, Arquelau et al. formulated and analyzed films composed of ripe banana peel flour, corn starch, and glycerol. A higher proportion of corn starch enhanced the mechanical strength of the films, attributed to the polymer's capacity to intensify interactions within the matrix. However, this modification also led to an undesirable rise in water vapor permeability, stemming from the inherently hydrophilic nature of starch polymer chains (Arquelau et al. 2019). Potatoes and their processing by-products are one of the richest sources of starch (Arias et al. 2024). For instance, Karki et al. developed biodegradable films utilizing starch extracted from potato waste and peelings, incorporating glycerol and sorbitol at varying concentrations (35, 45, and 55% w/w of dry starch). They applied two modification techniques—hydrothermal treatment (HTT) and acid-alcohol treatment (AAT)—to assess the impact of both plasticizer concentration and modification method. Their findings indicated that films produced with HTT starch exhibited superior tensile strength (10 N/m^2) and elongation (9.47%), while those made with AAT starch demonstrated enhanced solubility (25.8%). Elongation (8.91%) and solubility (29.98%) were significantly better in films containing 35 and 55% glycerol, but tensile strength (13.02 N/m^2) was better in films containing 35% sorbitol. Water Vapor Transmission Rate (WVTR) was higher at 91% relative humidity in films containing 55% glycerol (999 g/m^2/d) (Karki et al. 2020). In a context of utilization of potato by-products, Othman et al., studying films made from starchy potato peel waste, pointed out important factors affecting the properties of the material, which may also influence the potential area of application, especially of films and semi-films developed from hydrophilic polymers. Relative humidity and temperature significantly influence the water sorption behavior and permeability of edible films. Research findings demonstrated that both the water vapor transmission rate and water vapor permeability increased with rising humidity and temperature. To minimize moisture uptake by edible films and the packaged food products (thereby prolonging shelf life), Starch is among the most abundant plant-derived raw materials, functioning as a storage substance in plants and accumulating primarily in tubers, roots, and seeds (Kocira et al. 2021; González-Arancibia et al. 2024). It is a natural biopolymer that plays a key role in food technology, especially in determining the rheological properties of food. It can also be used as a component of packaging material because it has beneficial functional properties, is cheaper than synthetic polymers, biodegradable, and easy to process. Starches contain varying amounts of amylopectin and amylose. Cereal starch contains approximately 25% amylase and 75% amylopectin. The ability to form gels depends on the amount of amylose (Kocira et al. 2021; Dutta and Sit 2024). High-amylose starch from corn seeds contains up to 85% amylose. Starches with a high (over 70%) amylose content are used to produce edible films. Starch possesses a strong capacity for forming edible films with satisfactory mechanical properties and can be applied to food products to minimize moisture loss or shield them from oxidative deterioration. Additionally, it enables the formation of transparent, tasteless, and odorless films that offer excellent oxygen barrier performance, making it highly advantageous for food preservation (Dash et al. 2019). Starch films are formed by evaporating aqueous solutions of gelatinizing amylose. Moisture significantly increases permeability. For instance, hydroxypropyl starch derivatives are more soluble and produce films that are more permeable to water vapor and gases. Edible coatings and films made exclusively from starch are brittle, so lipids and gums are added to modify their mechanical properties. In addition, their mixtures give food new textural properties, including increased water retention capacity.

In a context of sustainable solutions, Nayak et al. developed edible films based on corn starch/moringa gum with the addition of pine cone extract (PCE)—an underappreciated forest waste that contains bioactive compounds acting as UV filters. The study revealed that the incorporation of PCE markedly enhanced the elongation at break from 3.27 to 35.2%, albeit with a notable reduction in tensile strength. Nonetheless, starch-based films enriched with PCE extract demonstrated considerable UV-blocking capacity, highlighting their suitability as sustainable packaging materials for protecting light-sensitive food products (Nayak et al. 2024). Shanbhag et al. investigated the formulation and characterization of novel starch-based edible films utilizing ingredients such as arrowroot powder, corn starch, and refined wheat flour. Their findings indicated that the films exhibited low water vapor permeability, making them suitable candidates for food packaging applications. Partial biodegradation of all film types was observed within 22 days, with complete degradation occurring by day 60. The films maintained good thermal stability up to 190 °C, showing approximately 5% weight loss. However, those prepared with refined wheat flour displayed uneven surface textures, attributed to the larger particle size of the flour (Shanbhag et al. 2023). In another study, Arquelau et al. formulated and analyzed films composed of ripe banana peel flour, corn starch, and glycerol. A higher proportion of corn starch enhanced the mechanical strength of the films, attributed to the polymer's capacity to intensify interactions within the matrix. However, this modification also led to an undesirable rise in water vapor permeability, stemming from the inherently hydrophilic nature of starch polymer chains (Arquelau et al. 2019). Potatoes and their processing by-products are one of the richest sources of starch (Arias et al. 2024). For instance, Karki et al. developed biodegradable films utilizing starch extracted from potato waste and peelings, incorporating glycerol and sorbitol at varying concentrations (35, 45, and 55% w/w of dry starch). They applied two modification techniques—hydrothermal treatment (HTT) and acid-alcohol treatment (AAT)—to assess the impact of both plasticizer concentration and modification method. Their findings indicated that films produced with HTT starch exhibited superior tensile strength (10 N/m^2) and elongation (9.47%), while those made with AAT starch demonstrated enhanced solubility (25.8%). Elongation (8.91%) and solubility (29.98%) were significantly better in films containing 35 and 55% glycerol, but tensile strength (13.02 N/m^2) was better in films containing 35% sorbitol. Water Vapor Transmission Rate (WVTR) was higher at 91% relative humidity in films containing 55% glycerol (999 $g/m^2/d$) (Karki et al. 2020). In a context of utilization of potato by-products Othman et al., studying films made from starchy potato peel waste, pointed out important factors affecting the properties of the material, which may also influence the potential area of application, especially of films and semi-films developed from hydrophilic polymers. Relative humidity and temperature significantly influence the water sorption behavior and permeability of edible films. Research findings demonstrated that both the water vapor transmission rate and water vapor permeability increased with rising humidity and temperature. To minimize moisture uptake by edible films and the packaged food products (thereby prolonging shelf life) it is advisable to store such films in environments with low relative humidity and reduced temperature (Othman et al. 2017).

Other starch sources, such as cassava waste, may also be attractive sources of biopolymers for the development of starch-based edible films (Arias et al. 2024). Cassava peel, the main waste product generated during cassava tuber peeling, accounts for more than 5% of the cassava root mass and is discarded during the production of cassava flour or starch (Silveira et al. 2025). For example, Tafa and Engida prepared films using starch extracted from cassava peel and reinforced with silica nanoparticles (SNP) extracted from rice husks and bamboo leaves, also using glycerol as a plasticizer and essential oil from rosemary leaves (REO). The addition of SNP increased tensile strength and elongation at break. However, when the concentration of SNP exceeded a certain threshold, a decline in mechanical properties was observed, likely resulting from agglomerate formation that disrupted effective interactions between the filler and the polymer matrix. Films with REO showed some inhibitory activity against common food pathogens (*Escherichia coli* and *Staphylococcus aureus*) (Tafa and Engida 2022). In the work of Silveira et al., periderm and cassava bark were bleached with NaClO or H_2O_2 before water starch extraction. In addition, the films were modified with clove essential oil as a potential antibacterial agent. The films showed some antibacterial activity against foodborne pathogens: *S. aureus*, *Listeria monocytogenes*, and *Salmonella enterica* Typhimurium. The films showed an average water vapor permeability of 0.14 g mm/m^2 h kPa, an average tensile strength of 0.17 MPa, and an elongation at break of 32.90% (Silveira et al. 2025). Vallejo et al. developed starch-based films by directly incorporating powdered onion peel (OP) at 5–20 wt% into starch recovered from potato processing slurries, eliminating extraction steps and fully valorizing agro-industrial by-products. This zero-waste approach enhanced the films' mechanical properties, gas barrier, and antioxidant activity, while providing natural pigmentation and improved water resistance. At 10 wt% OP, Young's modulus increased to ~1500 MPa, although tensile strength and elongation slightly decreased, and water vapor permeability dropped by up to 60%. Notably, 20 wt% OP films achieved ~93% ABTS radical inhibition, showcasing potent antioxidant capacity. The study demonstrated a scalable, circular economy strategy converting two major food waste streams into multifunctional, biodegradable packaging materials (Vallejo et al. 2025).

Pectin

Pectins are plant-derived polysaccharides (Roy et al. 2023c). Pectins come from the middle layer of plant cells. These polysaccharides are composed of D-galacturonic acid polymers with varying degrees of methyl esterification and are primarily derived from natural sources, particularly by-products from fruit and vegetable processing. Such materials (commonly discarded peels) constitute a sustainable and readily available raw material for biopolymer extraction. In industrial contexts, pectins are primarily extracted from citrus peels, including lemon, orange, lime, and grapefruit, which are particularly rich in this compound. Beyond citrus, numerous other agricultural residues, such as apple pomace, sugar beet pulp, banana peels, pumpkin waste, and sisal by-products, have also been investigated as potential sources. The molecular composition of pectin varies depending on its origin, leading to distinct structural and functional characteristics across different sources. As a result, the physicochemical properties of extracted pectin are closely influenced by the choice of raw material (Roy et al. 2023c). Pectin membranes are produced by evaporating water from a pectin gel. They are mainly used for packaging foods with low water content because they dissolve easily in water. They form gels in the presence of calcium salts. However, pectin is most often used in combination with other polymers (e.g. starch, soy proteins, gelatin, alginate), in which case it exhibits much better mechanical and barrier properties than when used alone. They are finding increasingly wider application in the food sector due to their excellent barrier properties against oxygen, lipids, and odor compounds. When combined with other polymers, the coatings are more stable, but there is still a need for research to develop more effective pectin membranes for food packaging on a laboratory and industrial scale (Roy et al. 2023c).

In the study conducted by Brito et al., edible films were produced using various fruits (sweet orange, passion fruit, watermelon) and vegetables (zucchini, lettuce, carrot, spinach, mint, yam, cucumber, and arugula), incorporating methoxylated pectin at concentrations of 0.25, 0.5, and 1% (w/w). These films exhibited a uniform, yellowish appearance and plastic-like characteristics, along with high water solubility. The inclusion of pectin led to a 50% reduction in solubility and elongation, while tensile strength increased, which was attributed to pectin's ability to reinforce the polymer network through hydrophobic interactions between methyl ester groups or hydrogen bonding between carboxyl and hydroxyl groups (Brito et al. 2019). In another research, Dash et al. analyzed the properties of edible films based on sweet potato starch and pectin from lemon waste with the addition of titanium oxide nanoparticles (0.5–4% w/w). The solid substance contained sweet potato starch and pectin from lemon waste in a ratio of 3:1 (w/w). Glycerol in a 30% (w/v) aqueous solution was used as a plasticizer. Increasing the TiO_2-NPs content resulted in a significant decrease in moisture content, solubility, and moisture absorption by the films. The addition of low concentrations of TiO_2-NPs to starch-pectin films significantly improved their mechanical and barrier properties, and good particle dispersion was achieved even at low TiO_2-NPs contents (0.5–2% by weight). With increasing TiO_2 concentration, light transmittance also decreased, demonstrating that, thanks to their UV-protective properties, the developed films can be used as biodegradable packaging materials for food contact (Dash et al. 2019). Vallejos-Jiménez et al. used coffee waste (slime and pulp) to develop a pectin-based film. Pectin was isolated by hydrolysis with citric acid. The researchers also used bacterial cellulose and coffee oil as film components, which improved the physical and mechanical properties, elasticity, and functionality (tensile strength of 3.4 MPa, elongation at break of 24.5%, water solubility of 40%, and spontaneous degradation within 16 days in soil under uncontrolled conditions) (Vallejos-Jiménez et al. 2025). Deshmukh et al. used a developed composite nanocomposite packaging material based on a pectin-polyvinylpyrrolidone system, which they further enriched by impregnating it with bioactive ingredients. The role of an antimicrobial agent was achieved by adding titanium nanoparticles, while the role of an antioxidant was played by bael fruit peel extract, which is a good source of phenolic antioxidants that can be used in the food industry. The researchers developed a flexible nanocomposite film characterized by a smooth surface. Incorporating bael fruit peel extract notably influenced the film's properties: the addition of 5% or 20% extract decreased the film's transparency relative to the control, yet markedly enhanced its tensile strength—from 5.57 MPa in the pure film to 30.02 MPa with 20% extract. Furthermore, the films exhibited strong antioxidant activity on par with gallic acid and demonstrated considerable antibacterial effects against foodborne pathogens, including *E. coli* and *L. monocytogenes* (Deshmukh et al. 2022). Kurek et al. also examined the impact of incorporating waste derived from prickly pear into pectin- and chitosan-based edible packaging films. Their findings indicated that, while the addition of this waste material had minimal influence on the physicochemical characteristics of the films, it substantially enhanced their antioxidant activity (Kurek et al. 2021). In a more recent investigation, a smart packaging film based on pectin was engineered by incorporating betacyanins extracted from pitaya peel waste. Although the inclusion of these pigments adversely affected the film's mechanical strength, water resistance, and thermal stability, it notably enhanced its UV shielding capacity and antioxidant potential. Furthermore, the film demonstrated high sensitivity to volatile ammonia, allowing it to function effectively as a freshness indicator for shrimp. The progressive color transition (from red to reddish-brown and eventually to brown) correlated with the extent of spoilage (Jiang et al. 2023).

13.4.1.2 Microbial Polysaccharides

Carrageenans

Carrageenans are natural, water-soluble polysaccharides obtained through extraction from red algae, primarily species such as *Chondrus crispus*, *Kappaphycus alvarezii*, and *Eucheuma denticulatum*. Chemically, they are a linear polymers composed of partially sulphated galactose units, forming a complex mixture of sulphated galactans (Kocira et al. 2021; Cheng et al. 2022). This structural composition endows carrageenans with significant attributes, including bioactivity, biocompatibility, and biodegradability, which contribute to its broad applicability across various industries. In the food industry, they are commonly employed as a gelling, stabilizing, and emulsifying agent (Dmitrenko et al. 2023). Moreover, their membrane-forming ability enables their application in edible films and coatings. There are three major types of carrageenan: κ (kappa), ι (iota), and λ (lambda), each differing in sulphation degree and gel-forming capacity. The solubility of carrageenan in water is influenced by its sulphate ester content and the type of associated cations (Farhan and Hani 2017). While carrageenan is more commonly applied as a coating material, the development of edible films based on this biopolymer remains less explored. To broaden its potential use in packaging applications, further research is needed to improve its mechanical strength and water vapor barrier properties. Strategies such as the incorporation of hydrophobic additives and cross-linking agents are recommended to enhance film performance and durability (Kokkuvayil Ramadas et al. 2024).

In a context of circular economy, Dmitrenko et al. modified carrageenan membranes using a by-product of pineapple production. They used pineapple leaves as a raw material to obtain cellulose nanocrystals. The study revealed that increasing the nanocellulose content in the membranes enhanced both the mechanical strength and surface hydrophilicity, as evidenced by improved wetting angle values and mechanical parameters. The optimal formulation was identified as a film with a carrageenan to starch ratio of 1.5:1, supplemented with nanocellulose and glycerol as a plasticizer. This film exhibited a water vapor permeability of 6.4×10^{-10} g/(s m Pa), oil permeability of 2%, water solubility of 42%, and moisture absorption of 29% (Dmitrenko et al. 2023). In a similar study, Gupta et al. modified κ-carrageenan by incorporating cellulose nanocrystals (CNC) derived from Indian gooseberry (amla pomace) at varying concentrations (1, 3, 5, and 7%) and produced the films using the solution casting method. The CNC-reinforced biocomposite films exhibited enhanced barrier and mechanical properties compared to the control. Specifically, the addition of 5% CNC reduced water vapor permeability from 3.21 to 2.36 g mm/m^2 day kPa and increased tensile strength from 23.28 to 39.75 MPa. The incorporation of CNC also led to an increase in film crystallinity; however, at higher concentrations (7%), agglomeration effects were observed (Gupta et al. 2023). In another study, Chi et al. used grape skin, a by-product of wine production, to modify κ-carrageenan-hydroxypropyl methylcellulose films. Grape skin is abundant in anthocyanins, which, due to their color change in acidic and alkaline environments, are used as pH indicators, providing an alternative to pH-sensitive synthetic dyes in the preparation of pH-sensitive films. The resulting carrageenan-based composite films were highly sensitive to pH changes. The film turned blue-green when the pH reached 7, indicating that it is highly sensitive to pH, and when TVB-N (Total Volatile Basic Nitrogen) was 14.63 mg/100 g, it changed color from purple to green, suggesting that it is a smart solution for monitoring the freshness of pork meat (Chi et al. 2020).

Alginates

Alginates are water-soluble polysaccharides of natural origin obtained from brown seaweed. They are commercially extracted primarily in the form of sodium alginate. The annual global production of brown seaweed, calculated on a dry weight basis, is estimated at around 85,000 tons, from which approximately 23,000 tons of alginates are obtained (Bojorges et al. 2023). Thanks to their excellent film-forming abilities, alginates are commonly utilized in the formulation of edible coatings and films. They produce uniform, transparent, and glossy membranes that effectively block fats and oils. However, due to their hydrophilic character, they exhibit high water vapor permeability—a feature that can be adjusted through modifications in the film's composition and structural design, e.g., by adding calcium ions (chlorides, lactates, acetates, oxalates) or by using appropriate gelation conditions, it is possible to obtain a material with customized physicochemical properties. In addition, factors such as pH, temperature, cross-linking agent exposure time, and the use of various fillers influence the structure and functionality of the films (Kocira et al. 2021; Bojorges et al. 2023). Alginates have unique colloidal properties: they act as thickening, stabilizing, and suspending agents, and also have the ability to form gels and emulsions. Based on these features, alginate coatings effectively reduce water evaporation, reduce food shrinkage, retain color and aroma, and reduce oil absorption and the rate of oxidative rancidity. In the case of vegetables and fruits, this coating first undergoes dehydration, protecting the inner layers of the plant from loss of turgor. Moreover, alginates act as efficient oxygen barriers, thereby inhibiting lipid oxidation, limiting the proliferation of surface microflora, and minimizing weight loss and product deterioration during storage. Coatings based on alginates are especially advantageous for highly perishable foods, including fruits, vegetables, and various types of meat such as poultry, beef, pork, and lamb. Their application supports prolonged shelf life and enhanced sensory attributes of food products.

To increase the efficiency of alginate extraction in line with the zero waste concept, green extraction methods, including traditional alginate extraction processes, are being replaced by more environmentally friendly techniques such as ultrasound-assisted extraction (UAE), microwave-assisted extraction (MAE), enzyme-assisted extraction (EAE) and extrusion (ExAE). These methods allow for increased extraction efficiency, shorter processing times, and reduced energy and chemical consumption. Alternative sources of biomass and its production are also being sought to achieve a further green approach in production (Bojorges et al. 2023). Combining alginates with other biopolymers derived from waste sources is also one of the directions for introducing greater sustainability, as demonstrated by de Oliveira Filho, who obtained bioactive films with, among other things, antimicrobial activity enriched with protein hydrolysates obtained by the enzymatic hydrolysis of underutilized cottonseed proteins (Oliveira Filho et al. 2019). In a study by Medina-Jaramillo et al., edible coatings from alginate modified with cellulose nanofibrils (CNFs) isolated from cocoa by-products were analyzed. The coatings were tested on Andean blueberries (*Vaccinium meridionale*), and their effects on fruit freshness were monitored over 21 days of refrigerated storage (4 °C, 90% relative humidity). The study revealed that incorporating cellulose nanofibers (CNFs) into alginate films notably decreased their transparency and water vapor permeability. When applied to the blueberries, both alginate and alginate/CNF coatings improved the visual quality and firmness of the fruits. Additionally, these coatings effectively lowered the respiration rate and reduced moisture loss during the entire storage duration (Medina-Jaramillo et al. 2020). Gubitosa et al. developed water-resistant films for food packaging using sodium alginate. The materials showed a decrease in hydrophilicity and high stability at elevated temperatures and varying pH. However, the film swelled on exposure to concentrated salt solutions. In addition, grape pomace waste extract was used to improve UV light shielding and antioxidant properties. The materials were proposed to be used to protect food from light-induced spoilage while being waterproof and stable (Gubitosa et al. 2023). Oliveira Filho et al. investigated edible bioactive films composed of alginate, mangaba pulp (*Hancornia speciosa*), and the probiotic yeast Saccharomyces boulardii. The inclusion of mangaba pulp, a by-product from fruit processing, had a pronounced impact on the characteristics of alginate-based films. It enhanced their water resistance, mechanical strength, and thermal performance—evidenced by reduced water solubility and vapor permeability, along with increased thermal stability, tensile strength, and elongation at break. As the concentration of mangaba pulp in the film-forming solution increased, the films became darker, exhibiting intensified yellow hues and higher saturation. Additionally, the incorporation of bioactive constituents such as carotenoids, vitamin C, and phenolic compounds improved the antioxidant activity and strengthened the films' barrier functions against ultraviolet and visible light (Oliveira Filho et al. 2023). Thivya et al. prepared sodium alginate/carboxymethylcellulose/gluten blends modified with onion waste (from peel and stalk) extracts. The materials were applied on peeled and shallot onions. The modified films were found to possess improved water barrier properties and tensile strength compared to SA-CMC film and reduced microbial load of packed food products (Thivya et al. 2022). Another by-product of the fruit industry, watermelon (*Citrullus lanatus*) seeds, were used as a natural source of melanin to prepare composite alginate-based films (Łopusiewicz et al. 2022b). Melanin, a natural compound with reducing properties, was used to synthesize silver and zinc oxide nanoparticles. The addition of melanin significantly increased the antioxidant activity of the film, achieving high free radicals scavenging capacity (ABTS and DPPH). In addition, melanin acted as a stabilizer for silver nanoparticles (AgNP) and zinc oxide nanoparticles (ZnONP), which contributed to the improvement of the mechanical and barrier properties of the film. It was also found that the incorporation of AgNP and ZnONP into alginate films, using melanin as a stabilizer, imparted antimicrobial properties to the materials, effectively inhibiting the growth of *E. coli* and *S. aureus.* Furthermore, incorporating ZnO nanoparticles (ZnONP) enhanced the UV-blocking capacity of the films while preserving satisfactory transmittance in the visible light range. In contrast, the addition of silver nanoparticles (AgNP) led to nearly complete UV blockage but also diminished visible light transmittance. Cytotoxicity assays indicated that the developed films did not negatively affect cell viability, suggesting their potential suitability for applications not only in the food industry but also in biomedical fields, such as wound healing materials (Łopusiewicz et al. 2022b).

Pullulan

Pullulan is a polysaccharide obtained through aerobic extracellular fermentation, mainly by yeasts of the species *Aureobasidium pullulans* (Farris et al. 2014; Singh et al. 2019). This polymer has a linear structure, is soluble in water, non-ionic and chemically neutral. Its structure is based on repeating maltotriose units linked by α-(1,6) and α-(1,4) bonds. Pullulan is not degraded by human digestive enzymes but undergoes enzymatic decomposition in the environment, making it biodegradable. It is also non-toxic, non-mutagenic, and non-carcinogenic (Dewan and Islam 2024; Rashid et al. 2024). This compound is commercially available as a white or slightly creamy powder, tasteless and odorless, that dissolves easily in both cold and hot water. Due to its unique properties, such as high mechanical strength, good adhesion, film- and fiber- forming ability, as well as very good oxygen barrier properties, pullulan is considered a promising material for packaging applications in the food industry (Luís et al. 2020). In 2002, it was recognized by the US FDA as a GRAS (Generally Recognized As Safe) substance, and in 2004 it was approved for use in the European

Union as a food additive labeled E1204 (Rashid et al. 2024). Edible films and coatings based on pullulan are particularly effective in protecting products with a short shelf life, such as fresh or minimally processed fruits and vegetables. In addition, these films can act as a barrier against light and oxygen, preventing vitamin degradation and slowing down oxidation processes. Due to the low viscosity of aqueous solutions, high stability over a wide pH range, and lack of gelation, pullulan is ideal for thin-film coating applications. It can also be used as a starch substitute in low-calorie foods and as a packaging material for nuts, pasta, confectionery, vegetables, and meat (Dewan and Islam 2024; Rashid et al. 2024).

Among the by-products of the agri-food industry that have been reported as alternative sources of nutrients for microorganisms in the production of pullulan, the following have been reported: sugarcane bagasse and molasses, potato and sweet potato residues (after enzymatic hydrolysis), grape residues, rice and wheat bran, coconut and palm kernels, soybean meal, coconut water and coconut milk, jaggery, corn liqueur, whey, and oilcakes (Wani et al. 2021; Mishra et al. 2023). Furthermore, various fermentation strategies (solid state and submerged) were investigated to find alternative solutions to reduce the production costs of pullulan (Singh et al. 2019). In the context of sustainable approach, Priyadarshi et al. prepared edible pectin/pullulan coatings with the addition of antioxidant grape seed extract and applied them to peanuts to prevent rancidity. They found that coatings containing grape seed extract exhibited some antibacterial activity against *E. coli* and *L. monocytogenes*, delaying bacterial growth. In addition, the film with the extract showed antioxidant potential due to the ability of the polyphenolic compounds in the extract to scavenge free radicals. Furthermore, the films had significant UV-blocking properties, which they owed to pectin and grape seed extract, as well as a high gas barrier provided by pullulan, and significantly reduced the rancidity of peanuts (75% decrease in peroxide value compared to uncoated peanuts after 30 days under ambient conditions) (Priyadarshi et al. 2022). In another study, Luis et al. developed pullulan-based films containing essential oil from the expansive weed (*Cistus ladanifer*), grown on agricultural land and meadows in the mountainous regions of the Iberian Peninsula. The films showed antioxidant properties (indicating the possibility of using the developed films for packaging food susceptible to oxidation and rancidity) and antibacterial activity (stronger against Gram-positive than Gram-negative bacteria). Interestingly, the films also showed promising quorum sensing inhibitory activity, as they were able to disrupt intercellular communication by inhibiting the production of violacein in the *Chromobacterium violaceum* model (Luís et al. 2020).

Agar

Agar, obtained mainly from red algae, is a mixture of two polysaccharide fractions: gelling agarose and non-gelling agaropectin (Kocira et al. 2021; Roy et al. 2023a). Owing to its unique structure, based on galactose and its isomers, agar has the ability to form transparent, flexible, and mechanically stable films that are biodegradable, non-toxic, and economical (Guerrero et al. 2014; Rhim et al. 2015;

Sousa and Gonçalves 2015; Martínez-Sanz et al. 2019; Thiviya et al. 2023). Its thermoreversible properties, stability at low pH, and high oxygen barrier properties make it a promising candidate for packaging materials, but its hydrophilicity limits its resistance to water vapor and requires modification (Kanmani and Rhim 2014; Sousa and Gonçalves 2015; Roy et al. 2023a; Baskaran and Bose 2024). From the point of view of the circular economy, innovative approaches using waste and by-products in the production of agar films are particularly interesting. An example is the work of Akbaş and Erdoğan, in which extracts from pea husks obtained using green solvents were used, which improved the strength (TS up to 9.03 MPa), hydrophobicity (contact angle 75.75°) and antibacterial properties of the film (Akbaş and Erdoğan 2025). Sarma et al. used rice cooking water (often treated as waste) to prepare a starch-agar composite enriched with green silver nanoparticles, which extended the shelf life of grapes to 12 days (Sarma et al. 2024). Rhim et al. presented the reinforcement of agar films with cellulose nanocrystals from onion skins, improving mechanical (TS 50 MPa) and barrier (WVP 1.78×10^{-9} g/m^2 s Pa) parameters (Rhim et al. 2015). Martínez-Sanz et al. showed that the use of unpurified agar extracts from *Gelidium sesquipedale*, containing natural admixtures of proteins and polyphenols, avoids the addition of plasticizers, increases the moisture resistance of the film, and improves its antioxidant properties (Martínez-Sanz et al. 2019). In addition, Kanmani and Rhim showed that enriching agar films with grapefruit seed extract increases their antimicrobial activity against *L. monocytogenes*, *B. cereus*, and *E. coli*, although it also reduces mechanical strength and increases water vapor permeability (Kanmani and Rhim 2014). Roy et al., on the other hand, emphasized the importance of the stability of agar films under low pH conditions, which makes them promising candidates for packaging products requiring an acidic environment (Roy et al. 2023a). Such zero-waste approaches significantly reduce production costs and energy consumption, which is in line with the ideas of sustainable development. In summary, although agar requires modification to meet the requirements of commercial food packaging, its combination with agricultural by-products, green solvents, nanomaterials, and the use of simplified extraction methods open up new perspectives for the development of ecological, active, and intelligent packaging materials.

13.4.1.3 Animal-Derived Polysaccharides

Chitin and Chitosan

Chitin and its deacetylated derivative, chitosan, exhibit excellent film-forming capabilities, making them suitable materials for edible films and coatings that serve as effective barriers against carbon dioxide and oxygen. Additionally, these biopolymers are recognized for their notable antibacterial activity. Chitin naturally occurs in the cell walls of a wide range of fungi and in the exoskeletons of various invertebrates, such as crustaceans and insects. Globally, chitin ranks as the second most abundant natural polysaccharide, surpassed only by cellulose in overall mass (Kozma et al. 2022). One thousand billion tons of chitin are generated annually, much of it by food-related organisms including shellfish, mollusks, fungi and insects (Knorr 2025). About 30% of chitin is obtained from shrimp shells (Kocira et al. 2021). Chitosan is obtained by alkaline deacetylation of chitin. It belongs to non-toxic polycationic polymers composed of D-glucosamine residues linked by β-1,4-glycosidic bonds. Chitosan-based films are typically transparent, colorless, or exhibit a slight yellow hue, depending on both the chitosan source and film thickness. In acidic aqueous environments, these films tend to swell or even dissolve, whereas at a pH near 6, their solubility diminishes and is largely determined by the polymer's degree of deacetylation. The mechanical characteristics of chitosan films are influenced by the molecular weight of the polymer and the nature and concentration of the acid employed during solubilization. As a high-molecular-weight, non-toxic, and biologically active polymer, chitosan is valued for its fungicidal properties and its ability to stimulate defense responses in plant tissues. Given these biochemical attributes, chitosan serves as an effective preservative coating for fresh fruits and vegetables. Its application contributes to prolonged shelf life and decreased incidence of postharvest decay, underscoring its suitability for use in the preservation and storage of horticultural produce (Kocira et al. 2021; Kozma et al. 2022; Knorr 2025).

The utilization of chitin derived from crustacean shells remains an active area of research, with considerable attention focused on establishing environmentally sustainable processes that enable the extraction of high-purity chitin while minimizing waste generation. Conventional approaches commonly involve chemical and biological methods; however, recent investigations have explored the use of ionic liquids and deep eutectic solvents as alternative, potentially greener extraction systems (Kozma et al. 2022). Some local crustacean species of minor economic importance are also considered alternative sources of chitin for the production of chitosan used in the manufacture of edible films. For instance, do Vale et al. obtained chitin and chitosan from the remains of the crab-uçá (*Ucides cordatus*), a species characteristic of tropical and subtropical regions and the most important crustacean inhabiting the mangrove forests of the Brazilian coast (do Vale et al. 2021). Combining chitosan with other biopolymers derived from waste materials is also one way to introduce greater sustainability, as demonstrated by Baron et al., who obtained chitosan-based films from blue crab (*Callinectes sapidus*) waste and orange peel pectin (Baron et al. 2017). Kumar et al. enhanced chitosan-based edible films by incorporating pomegranate peel extract. As the extract volume fraction rose, the films displayed greater thickness, tensile strength, moisture content, opacity, and improved water- and gas-barrier performance, alongside elevated phenolic levels and antioxidant capacity. Conversely, higher extract concentrations led to diminished thermal stability, reduced transparency, increased solubility and swelling, and noticeable color changes (Kumar et al. 2021). Zhang et al. adopted an alternative strategy, producing composite films from commercial chitosan and an agricultural by-product—banana peel. Although bananas are widely consumed worldwide, their peels are typically discarded as industrial waste, despite being rich in bioactive compounds such as carotenoids, phenolics, and biogenic amines that offer pro-vitamin A and substantial antioxidant potential.

Incorporating banana peel extract into the chitosan matrix yielded an active packaging film exhibiting antioxidant activity, making it suitable for fruit packaging applications (Zhang et al. 2020).

13.4.2 *Proteins*

To tailor edible films for particular applications, the molecular structure of proteins (macromolecules with defined amino acid sequences) can be modified via enzymatic, chemical, or physical methods (Purewal et al. 2024; Liu et al. 2024). The resulting films are often significantly weaker than synthetic materials. Protein films used for coating or packaging food provide an excellent barrier against oxygen, carbon dioxide, and lipids, but are sensitive to water vapor and moisture (Gupta et al. 2024; Khin et al. 2024). Given the wide range of polymers employed in edible films and membranes, the most commonly used are described below. Protein films are mainly obtained from animal proteins (collagen, gelatin, casein, and whey proteins) and plant proteins (including soy and zein proteins). Table 13.2 presents selected zero-waste strategies and the utilization of by-products in protein-based films and coatings.

13.4.2.1 Animal Proteins

Caseinates and Whey Proteins

The dairy industry generates significant amounts of liquid waste, one of the main by-products is whey—a product obtained after the coagulation of casein during cheese production. Dairy whey is a light yellow, water-soluble liquid rich in protein, formed after the separation of casein. It comes from various types of milk (cow, goat, sheep, or camel), but in Western countries, cow whey is most commonly used. It accounts for 85–95% of the volume of the original milk and contains about 55% of its nutrients (Kandasamy et al. 2021; Uribarrena et al. 2024). Whey proteins, which are the main component of the whey protein fraction, have beneficial properties: they are colorless, odorless, and neutral in taste. This makes them ideal for use in edible coatings, which are flexible, transparent, and have good barrier properties against gases and volatile compounds, including aromas. Importantly, whey protein films are biodegradable, making them an attractive alternative to conventional plastics (Ramos et al. 2012). However, the presence of lactose in the raw material can lead to crystallization during film formation, resulting in a non-uniform structure and reduced adhesion to surfaces. In addition to whey proteins, casein is also a popular milk raw material. Packaging based on casein, the main protein in milk, has excellent barrier and mechanical properties, as well as high water solubility. As a result, it effectively prolongs the freshness of products and can also enrich their nutritional value. Casein films are used in the packaging of fruit, vegetables, dairy products, meat, fish, and snacks such as

Table 13.2 Zero-waste strategies and by-product applications in protein-based films and their impact on packaging performance

Protein biopolymers	Zero-waste approaches, by-product use, and impact on material properties	References
Whey protein	Source: whey protein, antioxidant extract from spent coffee grounds (by-product); process: edible film formation with pH adjustment and antioxidant addition; properties: enhanced UV-blocking, increased tensile strength and elastic modulus, high antioxidant activity, increased water vapor permeability	Papadaki et al. (2022)
Whey protein	Source: whey protein isolate, lignin from pulp and paper industry (by-product); process: film formation with heat-induced protein denaturation and lignin addition; properties: enhanced UV shielding, improved antioxidant activity, increased glass transition temperature and opacity, reduced flexibility	Deng et al. (2025)
Whey protein	Source: whey protein concentrate and isolate, melanin from watermelon seeds (*C. lanatus*, by-product); process: film modification with antioxidant pigment; properties: improved UV shielding, water vapor barrier, swelling behavior, mechanical strength, enhanced antioxidant activity, no cytotoxicity, suitable for food packaging and biomedical use	Łopusiewicz et al. (2020)
Casein	Source: casein, melanin from watermelon seeds (*C. lanatus*, by-product); process: film modification with antioxidant additive, exposed to UV aging; properties: improved UV barrier, enhanced antioxidant activity, protection of sulfhydryl bonds and proteins, preserved mechanical properties	Macieja et al. (2023)
Collagen	Source: collagen, tannins from pomegranate by-products; process: cross-linking with pomegranate extracts; properties: improved mechanical strength, increased cross-linking density, retained collagen helix structure, enhanced thermal stability	El Moujahed et al. (2024)
Gelatin	Source: chicken gelatin from poultry by-products, caffeic acid, rutin (by-products); process: cross-linking with plasticizer; properties: increased tensile strength, improved thermal stability, reduced water solubility, enhanced opacity	Erge et al. (2024)
Gelatin	Source: gelatin, cactus mucilage from *Cylindropuntia fulgida* stems (by-product), Euphorbia caducifolia extract; process: biocomposite film formation; properties: improved elongation at break, enhanced water vapor barrier, strong antibacterial and antifungal activity due to bioactive flavonoids	Kumar et al. (2022)

(continued)

Table 13.2 (continued)

Protein biopolymers	Zero-waste approaches, by-product use, and impact on material properties	References
Gelatin	Source: gelatin, bioactive compounds from acerola cherry pomace (by-product); process: active films applied on beef cutlets; properties: controlled color changes, reduced lipid and protein oxidation, effective up to 4% extract addition, prolonged shelf life of frozen beef patties	da Nóbrega Santos et al. (2022)
Gelatin	Source: gelatin, melanin from mushroom waste (*Agaricus bisporus* by-product); process: solution casting of composite films; properties: improved oxygen and UV–visible light barrier, enhanced antioxidant activity, no change in mechanical or water vapor barrier properties, no antimicrobial effect, altered color characteristics	Łopusiewicz et al. (2018a)
Gelatin	Source: gelatin, fungal melanin from *A. bisporus* waste; process: modified gelatin coatings applied on pork lard; properties: reduced oxidative rancidity during storage, effective prevention of lard oxidation, enhanced oxidative stability compared to control coatings	Łopusiewicz et al. (2018b)
Zein	Source: zein, flax mucilage from flaxseed cake (by-product); process: one-pot glycation, solvent casting, injection molding; properties: low water solubility, thermoplastic and elastic behavior, reduced crystallinity with DMSO, slightly lower mechanical strength but enhanced plasticity	Tallawi et al. (2023)
Soy protein	Source: soy protein, nanocellulose and phenolic compounds from date palm leaves (by-product); process: edible film formulation with reinforcing and bioactive agents; properties: improved mechanical strength and barrier properties, enhanced antioxidant and antibacterial activity	Mostafa et al. (2023)
Soy protein	Source: soy protein, mango seed extract (by-product); process: antioxidant films for active packaging; properties: reduced lipid oxidation in mayonnaise by up to 65%, prolonged shelf life by 14 days, effective without added preservatives	Maryam Adilah et al. (2023)
Soy protein	Source: chitosan, soy protein, anthocyanins and carbon nanodots from pistachio shell waste (by-product); process: biopolymer film blending for smart packaging; properties: UV protection, antibacterial and antioxidant activity, color change with pH and ammonia, extended fish shelf life, visual freshness indicator	Hadavifar et al. (2025)

(continued)

Table 13.2 (continued)

Protein biopolymers	Zero-waste approaches, by-product use, and impact on material properties	References
Soy protein	Source: soy protein, methylcellulose, betacyanins and carbon nanodots from *Mirabilis jalapa* petal waste (by-product); process: biofilm fabrication with natural color indicators; properties: improved UV-blocking, wettability, tensile strength, reduced water vapor permeability, pH and ammonia sensitivity, enabled visual monitoring of shrimp freshness	Ebrahimi et al. (2025)
Soy protein	Source: soy protein, cinnamon and paprika oleoresins (by-products from spice industry); process: incorporation into films in free and encapsulated forms; properties: reduced light transmission, antifungal activity against *Penicillium paneum*, unchanged moisture and permeability, decreased tensile strength, enhanced preservative potential	Procopio et al. (2024)
Soy protein	Source: soy protein, chitin nanofibers from shrimp shells (by-product), Clitoria ternatea flower extract; process: biodegradable pH-sensitive films; properties: good tensile strength with chitin, color change depending on pH, enabled freshness monitoring of food	Koshy et al. (2022)
Soy protein	Source: soy protein, keratin from bovine horns (by-product), blueberry extract; process: compression molding with deep eutectic solvent; properties: improved tightness, heat-sealing ability, reduced water uptake, pH sensitivity, ammonia detection for food spoilage indication	Uribarrena et al. (2025)
Soy protein	Source: soy protein, microcrystalline cellulose from corn husk (by-product); process: chemical denaturation with covalent and non-covalent cross-linking; properties: improved tensile strength, elastic modulus, water resistance, and thermal stability; higher microcrystalline cellulose content slightly reduced elongation at break	Zhang et al. (2024a)
Gluten	Source: wheat gluten, apple pectin, blueberry anthocyanin extract; process: smart active films for ammonia detection; properties: antioxidant activity, good mechanical strength, color stability, sensitive pH and ammonia response; enabled real-time monitoring of shrimp freshness via color change	Hu et al. (2024)

crisps and nuts (Ramos et al. 2012; Schmid and Müller 2019; Pires et al. 2024). Casein salts (caseinates) are produced by precipitating casein from skimmed milk at a pH of 4.6 and a temperature of 20 °C, followed by dissolving the precipitate in solutions of sodium, potassium, calcium or magnesium hydroxides to obtain a pH close to natural (approx. 6.7). The casein films obtained in this way are usually odorless and tasteless, transparent or translucent, and their properties depend on the purity of the protein used, the type of fraction, and the method of formation. Like other milk protein-based films, they are fully soluble in water, which is an advantage in the context of sustainable food packaging systems (Schmid and Müller 2019).

In line with zero waste trends, Papadaki et al. provided evidence that the utilization of spent coffee grounds is a suitable approach to obtain antioxidants s to tune physical and functional properties of whey films. The application of coffee by-products resulted in increase of UV-blocking properties, mechanical strength and antioxidant activity. However, the modified films showed increased water vapor permeability (Papadaki et al. 2022). Similarly, Deng et al. reported that whey protein-based films modified with by-product from pulp and paper industries (lignin) enhanced significantly UV shielding and antioxidative properties of the materials (Deng et al. 2025). Łopusiewicz et al. incorporated melanin extracted from watermelon (Citrullus lanatus) seeds into whey-protein-concentrate and whey-protein-isolate films (WPC and WPI). Adding this pigment enhanced the films' UV shielding, water–vapor barrier efficiency, swelling behaviour, and mechanical strength, along with changing their colour. The melanin-enriched WPC/WPI films also demonstrated substantial antioxidant activity and showed no cytotoxicity (Łopusiewicz et al. 2020). In a similar study, melanin was used to modify casein films exposed to artificially accelerated UV aging, and due to strong antioxidant activity showed protective properties for sulfhydryl bonds and proteins, thus acted as a natural plant-biomass stabilizer for polymeric films (Macieja et al. 2023). An interesting approach was reported by Hernández-Carranza et al. who reported edible films made from potato starch and whey fermented with lactic acid bacteria (LAB). The films could be an interesting carrier of beneficial microflora as well as shown antimicrobial activity against food pathogen *E. coli* (Hernández-Carranza et al. 2023).

Collagen

Collagen is an animal protein that is the main component of connective tissue in vertebrates and invertebrates. It is found in the skin, tendons, cartilage, bones, connective tissue membranes, and blood vessels. Its physicochemical properties, such as elasticity, thermal stability, and mechanical strength, depend on the animal species, its environment, and the type of collagen, which varies in molecular weight, amino acid composition, and spatial structure (El Moujahed et al. 2024; Ezati et al. 2025). Collagen is widely used in the meat industry, especially in the form of coatings used in sausage production. Insoluble collagen films are also used to protect smoked meat products, such as ham, from the penetration of elastic nets into the product during heat treatment. After removal of the net, the collagen coating can be safely consumed with the product. From a technological point of view, collagen has excellent film-forming properties, especially in combination with other hydrophilic biopolymers (El Moujahed et al. 2024; Ezati et al. 2025). Incorporating additives such as chitosan, pectin, or gelatin into collagen composites enables the development of materials with enhanced barrier effectiveness and mechanical strength. Thanks to their biodegradability, biocompatibility, and edibility, these composites offer a promising alternative to conventional synthetic packaging, supporting the drive toward sustainable, zero-emission practices in the food industry (Salim et al. 2024).

In order to increase the possibilities of obtaining collagen from waste sources, intensive research is being conducted to develop effective strategies for the extraction of this biopolymer. Sources such as cattle, sheep, pigs, poultry, as well as their by-products and waste, are the most important sources of collagen and gelatin (Munasinghe et al. 2015; Ben Slimane and Sadok 2018). In recent years, there has been growing interest in collagen derived from fish, as well as from sea sponges and jellyfish, which is in line with the idea of using unconventional resources and waste materials (Salim et al. 2024). Other waste materials can also be a source of biopolymers used to modify collagen films. For example, in a study by El Moujahed, tannins from pomegranate by-products were analyzed as a cross-linking agents in collagen films, showing a significant influence on the mechanical properties of the materials (El Moujahed et al. 2024).

Gelatin

Gelatin is a natural, water-soluble blend of animal-derived polymers produced through denaturation and partial hydrolysis of collagen, the fibrous protein present in skin, bones, tendons, and other connective tissues of both vertebrates and invertebrates. Its structure consists mainly of glycine, proline, and hydroxyproline, arranged in a specific amino acid sequence. In aqueous solutions, gelatin takes the form of an odorless polymer with a random configuration of peptide chains. At a temperature of around 40 °C, it forms a sol, which upon cooling transforms into a physical, thermoreversible gel. Thanks to properties such as biocompatibility, biodegradability, ease of film formation, and low melting and gelling temperatures, gelatin is widely used in the production of edible films and coatings (Kumnerdsiri et al. 2024). The US Food and Drug Administration (FDA) has granted it GRAS (Generally Recognized as Safe) status, which allows this biopolymer to be used safely in contact with food. Gelatin films exhibit effective oxygen barrier performance under low to moderate relative humidity conditions and possess adequate mechanical strength (Łopusiewicz et al. 2018a, b). Despite its beneficial properties, the hydrophilic nature of gelatin means that films containing it have poor resistance to water penetration. Therefore, various strategies are used to improve barrier properties, such as combining gelatin with hydrophobic components or using cross-linking agents. In industrial practice, gelatin is used, among other things, to coat meat in order to reduce water loss, inhibit the growth of microflora, slow down the oxidation of fat components, and reduce fat absorption during frying. In addition, the use of gelatin coatings can help extend the shelf life and improve the sensory quality of food products (Łopusiewicz et al. 2018b). In the case of gelatin, as with collagen, alternative sources and methods of extraction from waste generated during the processing of animal-derived raw materials are being sought (Usman et al. 2022).

Protein-based films are often hindered by their relatively high permeability to water vapour and modest mechanical strength. Chemical cross-linking is the most common strategy for improving their mechanical integrity, thermal stability, and barrier properties. In this context, Erge et al. found that gelatin films cross-linked with caffeic acid and rutin obtained from by-products showed increased tensile strength and improved thermal stability (Erge et al. 2024). In another study, Kumar et al. used cactus mucilage (*Cylindropuntia fulgida*) to modify gelatin films. To obtain the mucilage, the cactus must be cut. Therefore, cactus stems are a waste product containing mucilage, which has many applications. In addition, *Euphorbia caducifolia* (EC) extract was added at concentrations of 0, 1, 5, 10, and 20%. The authors found that the modified films showed a significant improvement in mechanical properties (elongation at break improved from 2 to 60.59%), water vapor barrier properties (from 3.34 to 2.59 10^{-4} g mm/mm^2 day kPa), and antibacterial properties. High antibacterial activity was attributed to bioactive compounds in EC, mainly flavonoids, including luteolin and luteolin 7-o-β-glucoside present in EC extract, which exhibit significant biological activity, such as antibacterial and antifungal properties (Kumar et al. 2022). Similarly, acerola cherry pomace served as a source of bioactive compounds for modifying gelatin films applied to beef cutlets. The films effectively controlled color changes and oxidation reactions (by reducing the formation of malonaldehyde and the total content of carbonyl compounds) in lipids and proteins (da Nóbrega Santos et al. 2022). Łopusiewicz et al. used melanin from mushroom waste as a modifier to increase the value of films and coatings gelatin-based materials (Łopusiewicz et al. 2018b, a). Incorporating fungal melanin into gelatin films did not alter their mechanical strength or resistance to water vapour, but it markedly enhanced their barriers to oxygen and UV–visible light. The treated films displayed notable antioxidant capacity, although they showed no antimicrobial activity against pathogenic microorganisms (Łopusiewicz et al. 2018a). In another study, modified gelatin-melanin coatings were applied to the surface of pork lard and the potential of the coatings to reduce oxidative rancidity was investigated. The authors concluded that lard coated with modified coatings showed reduced oxidative rancidity, and therefore, modified coatings containing fungal melanin were effective in preventing lard oxidation (Łopusiewicz et al. 2018b).

13.4.2.2 Plant Proteins

Zein

Zein is a protein fraction isolated from corn endosperm, classified as a prolamin, which is soluble in 70–80% ethyl alcohol (Lan et al. 2023). It is a natural, readily available, biodegradable material with excellent film-forming properties. Thanks to its hydrophobic nature, zein is an effective barrier against oxygen and water vapor, also limits fat migration (Lawton 2002). Its shiny, flexible coatings are particularly valued in the food industry for their aesthetics and functionality; they help preserve the color and firmness of products while protecting them from adverse oxidative changes (Lan et al. 2023). Zein can be obtained not only from food-grade corn, but also from corn processing by-products (e.g., high-starch industrial corn varieties), which is in line with the principles of the circular economy and zero waste strategy. Due to its properties, it is used in the production of films for separating food layers (e.g., cheese slices), coating fresh and cooked products such as fruits, vegetables, nuts, sweets, and meat. Zein films are particularly useful as protective coatings for deep-fried foods (Lawton 2002). The simplest method of obtaining zein films is to dry an alcoholic solution with the addition of a plasticizer (e.g., glycerol) (Lan et al. 2023). However, due to their limited resistance to water and water vapor, these films require modification, for example by adding lipids to increase their hydrophobicity. Zein could also be used in the green synthesis of nanoparticles with antimicrobial properties, making it particularly attractive in the context of new-generation active packaging materials (Tivano and Chiono 2023). Tallawi et al. developed bioplastic films by conjugating zein protein with flax mucilage extracted from flaxseed cake, a by-product of flax oil production, applying a simple, non-toxic one-pot glycation process. This zero-waste approach valorized agricultural residues without competing with food resources, yielding conjugates with Young's modulus of 1–3 GPa, glass transition temperatures between 49–103 °C, and water solubility below 1%. The resulting films showed excellent processability via solvent casting, injection molding, and foam techniques, offering promising thermoplastic and elastic properties. Notably, DMSO-treated samples exhibited reduced crystallinity and enhanced plasticity, although with slightly lower mechanical strength (Tallawi et al. 2023).

Soy Protein Isolate

Soy protein, composed mainly of globulins and albumins, is increasingly used in the development of environmentally friendly packaging materials and carriers of bioactive compounds. Such films are yellow tinged, flexible, offer excellent oxygen barrier performance and exhibit limited resistance to water vapor (Purewal et al. 2024; Chen et al. 2025). Due to its properties—high nutritional value, gelation, emulsification, and film-forming abilities, as well as biocompatibility and biodegradability—it is an attractive alternative to synthetic materials (Ahmadzadeh et al. 2023; Peydayesh et al. 2023; Jahangiri et al. 2024; Chen et al. 2025; Pawde et al. 2025). Nevertheless, its natural hydrophilicity and limited mechanical strength require the use of strategies to improve its performance parameters, such as cross-linking (physical, chemical, enzymatic) or reinforcement with nanofibers, polysaccharides, and lipids (Peydayesh et al. 2023; Khin et al. 2024; Granados-Carrera et al. 2025).

It is worth noting the growing importance of the use of soy protein obtained from agri-food industry waste, such as post-extraction meal, okara, soy whey, and lecithin (Singh and Krishnaswamy 2022; Chen et al. 2025; Karim et al. 2025). For example, Mostafa et al. developed films from soy protein isolate (SPI) with the addition of nanocellulose and phenols obtained from date palm leaves, which increased both mechanical strength and antioxidant properties (Mostafa et al. 2023). In turn, Mayram

Adilah et al. showed that enriching SPI films with mango kernel extract reduced lipid oxidation in mayonnaise by as much as 65% compared to the control film (Maryam Adilah et al. 2023). In line with the zero waste trend, Hadavifar et al. used anthocyanins and carbon nanodots obtained from pistachio shell waste to develop smart films made of chitosan and SPI, which not only extended the shelf life of fish but also signaled the degree of freshness based on color change (Hadavifar et al. 2025). Ebrahimi et al. used waste from *Mirabilis jalapa* petals as a source of betacyanins and carbon nanodots to create SPI-methylcellulose biofilms for monitoring the freshness of shrimp (Ebrahimi et al. 2025). Systems enriched with waste oleoresins are also being developed. Procopio et al. introduced oleoresins from the spice industry (cinnamon, paprika) into SPI films, obtaining materials with reduced light transmission and delayed fungal growth (*Penicillium paneum*), which increased their preservative potential (Procopio et al. 2024). Koshy et al. developed indicator SPI films modified with chitin nanofibers (from shrimp shells) and *Clitoria ternatea* flower extract, which changed color depending on the pH (Koshy et al. 2022). Uribarena et al. proposed another approach, combining SPI with keratin from bovine horns dissolved in a eutectic solvent, which improved tightness, enabled heat sealing, and, thanks to the addition of blueberry extract, also enabled the detection of ammonia as an indicator of food spoilage (Uribarrena et al. 2025). By-products from other branches of the agri-food industry can also be an attractive source of compounds for modifying soy protein films, thereby closing production loops in various areas. For instance, Zhang et al. obtained microcrystalline cellulose from a by-product of the cereal industry (corn husk) used to modify soy protein films to enhance mechanical properties by covalent and non-covalent cross-linking (Zhang et al. 2024a, b). In summary, the integration of soy protein isolate with additives derived from agri-food industry by-products is in line with the ideas of the circular economy, reducing raw material waste, and limiting environmental impact. Research indicates that these modifications not only improve physicochemical properties but also allow for the creation of active and intelligent packaging systems tailored to the needs of the modern food market.

Gluten

Gluten, a mixture of gliadin and glutenin, is a plant protein obtained mainly from wheat, as well as corn and barley. Its structure, rich in hydrophobic amino acids and its ability to form disulfide bonds, gives it unique film-forming properties (Khin et al. 2024; Karnwal et al. 2025). Gluten consists of approximately 30% gliadin and 40% glutenin, which determines its elasticity and ability to form disulfide bonds (Jahangiri et al. 2024). Gluten is characterized by excellent barrier properties against oxygen and carbon dioxide, high elasticity, and good thermal stability, making it an attractive material for edible films and. However, its limitations include low resistance to water vapor transmission, structural rigidity, and the fact that it is a strong allergen, which requires caution in food applications coatings (Fabra et al. 2015; Dong et al. 2022; Pawde et al. 2025). In practice, gluten films are obtained by solvent casting or thermoplastic processing, often with the addition of plasticizers such as glycerol, which improves their mechanical properties. However, the differences between films obtained by casting (greater flexibility) and thermoforming (greater tensile strength) were reported (Chen et al. 2019). It was reported that gluten coatings extend the shelf life of perhisable products such as fruits and vegetables. Moreover, gluten has very low oxygen permeability (0.1–0.4 cm^3/m^2 d atm), although its high water vapor permeability (4–8 g mm/m^2 d kPa) limits its use to dry products or requires modification (Jahangiri et al. 2024; Karnwal et al. 2025).

From a zero-waste perspective, gluten has enormous potential. It is a by-product of the starch and bioethanol industries, and is also increasingly derived from wet rice milling, with glutenins and gliadins performing different functions in processing techniques—glutenin films have greater strength, while gliadin films have greater elasticity; in addition, corn and rice gluten are increasingly being studied as raw materials for biopolymers. Research indicates the possibility of direct use of gluten flours from corn or wheat, without costly purification steps, as components of biodegradable composites (Peydayesh et al. 2023). Examples of innovations include the use of gluten in combination with PBAT, PCL, or wood to improve the biodegradability of materials, as well as the use of mineral oils to reduce water permeability by 25% (Chen et al. 2019). In addition, Hu et al. showed that gluten-pectin films combined with blueberry extract enables the creation of smart ammonia-detecting films, paving the way for applications in active and indicator packaging (Hu et al. 2024). Reports described the mechanical properties of such films, and emphasized the compatibility of gluten with other biopolymers, which is crucial for the creation of biodegradable composites with sustainable performance parameters. To sum up gluten is a cheap, widely available biopolymer with great potential in the development of edible, active, and smart packaging materials. Solutions based on the use of its waste streams fit into the circular economy and open up new opportunities for environmentally friendly food packaging technologies (Peydayesh et al. 2023; Pawde et al. 2025).

13.4.2.3 Lipids and Waxes

Edible lipid-based films and coatings, when compared to those based on proteins have excellent moisture barrier properties (Pashova 2023; Gupta et al. 2024). However, lipid films are extremely brittle and, due to their hydrophobicity, are also thicker. Films and coatings based on lipid compounds are mainly made from beeswax, candelilla wax, carnauba wax, and shellac. Vegetable oils can be used to produce emulsifiable edible coatings (e.g., in combination with polysaccharides), and enable the production of transparent coating materials. An exception is the use of oil from the fruit of the oil palm (*Elaeis guineensis*), which gives the coating a characteristic orange-red color due to its high carotenoid content (Rodrigues et al. 2016; Łopusiewicz 2024). The homogeneous dispersion of smaller lipid droplets in the emulsion has a beneficial effect on the formation of coatings with vegetable oils, as it increases the resistance of the coating to the effects of water vapor. Table 13.3

presents selected zero-waste strategies and the utilization of by-products in lipid- and wax-based films and coatings.

Beeswax

Beeswax is most commonly used to coat fruits and vegetables. It protects against water loss and the penetration of microorganisms, thereby maintaining high product quality for as long as possible. Waxing prevents weight loss by up to 50%. At the same time, the beeswax coating improves the appearance of the product by increasing its shine. Due to its hydrophobic properties, beeswax is an excellent barrier against moisture in edible coatings (Pashova 2023).

Vergel-Alfonso et al. developed coatings based on pectins and beeswax enriched with anthocyanins extracted from purple carrots. The film coatings were characterized by continuous areas of pectin with embedded wax crystals, exhibiting antioxidant activity as well as antibacterial activity against the pathogenic fungus *Colletotrichum siamense* (Vergel-Alfonso et al. 2025). Sultan et al. developed smart edible films based on chitosan and beeswax that were further modified with pollen grains. The pollen-coated film exhibited a higher water contact angle than the unmodified chitosan film. Incorporating both beeswax and pollen reduced the WVTR to roughly half that of the chitosan control and tended to increase the film's stiffness. The authors also analyzed the effect of coatings on the quality of Le Conte pears stored in a cold room for 105 days during their shelf life after a simulated sales period of 7 days and found that all coated fruits showed a reduction in weight loss, rotting, and softening rate (Sultan et al. 2021). Tedeschi et al. proposed a novel hydrophobic, cuticle-mimicking coating formulated from sodium alginate, fatty acid fractions derived from tomato processing waste, and beeswax. The ingredients were blended in environmentally friendly solvents (water and ethanol) and subjected to heat treatment at 150 °C for 8 h to polymerize the tomato-derived unsaturated and polyhydroxy fatty acids. Incorporating beeswax enhanced the coating's hydrophobicity, mechanical strength, and barrier performance, while also lowering water uptake (Tedeschi et al. 2018).

Carnauba and Candellila Waxes

Carnauba wax, also known as the "queen of vegetable waxes", is obtained from the surface of the leaves of the Brazilian palm tree *Copernicia prunifera* (syn. *Copernicia cerifera*). It is one of the hardest natural waxes, with a melting point of approximately 82.5–83 °C, making it a desirable hardening agent in various wax blends. Its unique chemical composition includes long-chain alcohols and esters, including unesterified alcohols (12%), hydroxyl esters (14%), and hydroxycinnamic acid esters (30%) (Susmita Devi et al. 2022; Pashova 2023). In edible coating formulations, incorporating carnauba wax substantially decreases water–vapor permeability and strengthens the films' barriers against gases and light. The wax also lowers film

Table 13.3 Zero-waste strategies and by-product applications in lipid- and wax-based films and their impact on packaging performance

Lipid and wax polymers	Zero-waste approaches, by-product use, and impact on material properties	References
Beeswax	Source: pectin, beeswax, anthocyanins from purple carrot (by-product); process: filmogenic solutions for active coatings; properties: antioxidant and antifungal activity, continuous pectin matrix with wax crystals, acidic pH, reddish color; effective against Colletotrichum siamense, suitable for extending food shelf life	Vergel-Alfonso et al. (2025)
Beeswax	Source: chitosan, beeswax, pollen grains; process: smart edible films applied as coatings on pears; properties: increased water contact angle, reduced water vapor transmission rate, higher stiffness, reduced elongation at break; coatings decreased weight loss, decay, and softening during 105-day cold storage	Sultan et al. (2021)
Beeswax	Source: sodium alginate, fatty acids from tomato pomace (by-product), beeswax; process: blending in green solvents with thermal polymerization; properties: enhanced hydrophobicity, mechanical strength, and water barrier, reduced water uptake; beeswax addition improved overall performance, creating cuticle-mimicking coatings	Tedeschi et al. (2018)
Candelilla wax	Source: whey protein, candelilla wax, Tarbush polyphenols; process: edible film casting with glycerol as plasticizer; properties: reduced water vapor permeability, improved barrier function, polyphenols enhanced antioxidant and antifungal activity, slight effects on solubility and transparency; suitable for coating fresh or minimally processed fruits and vegetables	Alvarez-Perez et al. (2022)
Carnauba wax	Source: arrowroot starch, carnauba wax; process: film casting with micro- and nanoemulsions; properties: increased hydrophobicity, reduced water solubility and permeability, improved light barrier and thermal stability; nanoemulsions provided better tensile strength and smoother microstructure; enhanced barrier and elongation	Oliveira Filho et al. (2020)
Shellac	Source: shellac (natural resin by-product), cellulose nitrate, olive leaf extract, grape pomace extract (by-products); process: active coating solutions applied for antioxidant packaging; properties: provided antioxidant activity in food simulants, stronger release from cellulose nitrate, antioxidant levels increased with higher extract concentration	Licciardello et al. (2015)

(continued)

Table 13.3 (continued)

Lipid and wax polymers	Zero-waste approaches, by-product use, and impact on material properties	References
Shellac	Source: pectin, shellac (natural resin by-product), plasticizers (polyethylene glycol 400, diethyl phthalate); process: film casting for composite films; properties: reduced hydrophilicity and water vapor permeability, improved moisture protection with 50% shellac, but weaker mechanical strength; plasticizers enhanced mechanical performance with slight effect on moisture barrier	Luangtana-anan et al. (2017)
Shellac	Source: alginate, shellac (natural resin by-product), plasticizers (propylene glycol, glycerol, polyethylene glycol 400); process: edible composite coatings applied on passion fruit; properties: preserved nutritional quality, reduced weight loss and color change, antimicrobial and antioxidant effects; best performance with 10% propylene glycol; maintained fruit quality over 56 days	Ha et al. (2025)
Shellac	Source: gelatin, shellac (with or without PEG 400); process: composite edible coatings applied by dipping; properties: delayed banana ripening, reduced weight loss, softening, acid and sugar loss; best performance with 60% shellac, 40% gelatin; effectively maintained post-harvest quality over 30 days	Soradech et al. (2017)

solubility in water, thereby improving stability in humid environments. Carnauba wax is widely utilised in the food sector, especially for confectionery glazes, and is likewise common in cosmetic products and furniture finishes. In combination with biopolymers such as starch, chitosan, gelatin, fish protein or glucomannan (some of which may come from by-products), it enables the creation of functional composite films with increased mechanical and thermal resistance (Pashova 2023). Coatings with carnauba wax effectively extend the shelf life of fruits such as oranges, apples, pomegranates, tomatoes, guavas, and eggplants. They work by limiting water loss, reducing respiration intensity, and maintaining the firmness, crispness, and sensory appeal of products for longer periods of time (Pashova 2023).

For comparison, candelilla wax, obtained from the plant *Euphorbia antisyphilitica*, is also used in the production of biodegradable, edible coatings. Its hardness is between that of carnauba wax and beeswax. It has excellent moisture and gas barrier properties and improves the flexibility of coatings by reducing brittleness by weakening intermolecular forces in polymer chains (Aranda-Ledesma et al. 2022).

In order to increase the efficiency of wax extraction from natural sources and waste materials, approaches based on, for instance, the sequential use of solvents or multi-solvent systems are also suggested (Susmita Devi et al. 2022; Pashova 2023). Alvarez-Perez et al. used candelilla wax to modify whey-based films. Furthermore, Tarbush polyphenols were added as a bioactive agent. It was noticed that the addition of candelilla wax decreased the permeability of the edible films, whereas polyphenol compounds of Tarbush improved the barrier properties of the films (Alvarez-Perez et al. 2022). In another study, de Oliveira Filho modified arrowroot starch-based films with micro- and nano-emulsified carnauba wax. The authors found that carnauba wax improved the water vapor and light barrier properties of the films. Moreover wax nanoemulsions promoted less change in tensile strength and microstructure of the films (Oliveira Filho et al. 2020).

Shellac

Shellac is a natural lipid polymer with highly hydrophobic properties and very good resistance to moisture, mainly due to the presence of a wax fraction in its composition. This substance is insoluble in water, edible, and classified as a food additive with the number E904. Due to its properties, shellac is widely used as an ingredient in glazes for confectionery and pharmaceutical products, as well as a protective coating applied to the surface of fruit to extend its shelf life (Ahuja and Rastogi 2023). Products coated with shellac gain an attractive, shiny appearance that enhances their visual and consumer value. Importantly, the gloss effect is maintained even after prolonged exposure to UV radiation, with only minimal loss. Shellac coatings have good adhesion to product surfaces, which promotes their mechanical stability during storage. However, it should be noted that shellac also has certain technological limitations. Coatings based on it have relatively low elasticity and mechanical strength, which may make them susceptible to cracking. In addition, increasing the amount of material applied may lead to adverse taste changes in the food product. Despite these limitations, shellac remains an important ingredient in edible coating formulations, especially where an effective barrier to water vapor and an attractive surface appearance are required.

To maintain the integrity of the packaged product, shellac was tested in active packaging together with olive leaf/grape pomace extract obtained from by-products as a coating solution (Licciardello et al. 2015). Luangtana-anan et al. modified pectin films with shellac and recorded reductions in stress and strain of 66.6% and 66%, respectively. The decrease in deformation stemmed from the intrinsic brittleness of shellac, while the lower film stress was attributed to weaker hydrogen bonding, since shellac is less polar than pectin. The reduced polarity of the blend, declining from 46 to 32.68%, enhanced the film's water barrier performance by 54.5% (Luangtana-anan et al. 2017). In a recent study, Ha et al. analyzed the effect of alginate-shellac coatings with three different plasticizers (propylene glycol, glycerol, and polyethylene glycol 400 at concentrations of 5–20%) on the quality of purple passion fruit (*Passiflora edulis*) during a storage period of 56 days. They found that the coatings (especially those with 10% propylene glycol) effectively preserved the quality and nutritional properties of the fruit, such as vitamin C content, total soluble solids, titratable acidity, total phenolic content, total flavonoid content, and DPPH radical scavenging capacity (Ha et al. 2025). Similarly, gelatin coatings with shellac showed a beneficial effect on delaying the ripening process of banana fruit, as reported by Soradech et al.

The coatings effectively acted as a physical barrier around the fruit, causing a slow decrease in weight, firmness, acid, and sugar content, and maintaining fruit quality after harvest for over 30 days compared to uncoated fruit (Soradech et al. 2017).

13.5 Conclusions and Future Perspectives

Food waste and loss pose a serious challenge to both the environment and the economy, not only because of inefficient waste management systems, but also because of the enormous waste of resources such as land, water, and energy. This problem is exacerbated by the dominance of packaging made from synthetic polymers, which are among the main sources of persistent environmental pollution and non-degradable waste. In view of these challenges, it is necessary to switch as quickly as possible to sustainable packaging materials that are environmentally friendly and resource-efficient. Biopolymers derived from natural, renewable sources are an attractive alternative. Unlike petroleum-based plastics, they are biodegradable, non-toxic and compatible with ecosystems. In this context, edible films and coatings are a particularly promising group of materials, especially when designed using bioactive compounds and structural components derived from waste and by-products of the agri-food industry. However, the large-scale implementation of these innovations still faces significant challenges. The variable composition and limited and uneven availability of waste and by-products pose a serious obstacle to their standardisation and industrial use. Differences in origin, collection time, contamination levels and the dispersed nature of production make it difficult to effectively valorize these raw materials. In many low- and middle-income countries, waste collection systems are fragmented or underdeveloped, hindering integration with circular economy models. Overcoming these limitations requires investment in infrastructure, the development of standardised classification and processing protocols, and the creation of a clear regulatory framework. Decentralised local recovery models are a promising direction, particularly in small and medium-sized food processing plants. Methods such as fermentation, enzymatic treatment, drying and extraction can be used on site to transform by-product streams into valuable raw materials for packaging production. These approaches unlock the hidden potential of by-products, especially in regions where centralised waste management systems are lacking. Despite growing interest from researchers, there is still a significant technology gap in the development of effective, scalable and economically viable conversion methods. Filling this gap requires further investment not only in advanced processing technologies (such as microbial biotransformations or enzymatic syntheses), but also in systems that integrate waste valorization with biopolymer production, particularly for applications in edible films and coatings.

To fully exploit the potential of bio-based packaging, future research should focus on replacing traditional plastics with functional, biodegradable alternatives. It will be necessary to optimise the composition of edible films and coatings and make full use of underappreciated by-products as structural and active components. Such strategies can contribute to the implementation of a truly circular and waste-free approach in the packaging sector—supporting the extension of food shelf life, reducing dependence on fossil fuel-based plastics, and supporting global food security goals. Equally important will be building consumer awareness and acceptance, especially for edible packaging. This requires targeted education, transparent labelling and appropriate marketing communication that clearly communicates the environmental and functional benefits of these materials. Closing the loop between food production, waste generation and packaging innovation can make the food and packaging industries key players in the transition to a more circular, resilient and sustainable bioeconomy.

Declaration AI-based tools were used solely for language proofreading and editing purposes. After using these tools, the authors have reviewed and verified the final content and take full responsibility for the final content.

References

Ahmadzadeh S, Hettiarachchy N, Luthra K et al (2023) Effects of polyphenol-rich grape seed and green tea extracts on the physicochemical properties of 3D-printed edible soy protein films. Food Packag Shelf Life 40:101184. https://doi.org/10.1016/j.fpsl.2023.101184

Ahuja A, Rastogi VK (2023) Shellac: from isolation to modification and its untapped potential in the packaging application. Sustainability 15. https://doi.org/10.3390/su15043110

Ahuja V, Chauhan S, Purewal SS et al (2024) Microbial alchemy: upcycling of brewery spent grains into high-value products through fermentation. Crit Rev Biotechnol 44:1367–1385. https://doi.org/10.1080/07388551.2023.2286430

Akbaş P, Erdoğan H (2025) Production and characterization of agar based biodegradable food films developed from deep eutectic pea hull (Pisum sativum) extract. Int J Biol Macromol 319:145231. https://doi.org/10.1016/j.ijbiomac.2025.145231

Akhila K, Ramakanth D, Rao LL, Gaikwad KK (2023) UV-blocking biodegradable film based on flaxseed mucilage/pectin impregnated with titanium dioxide and calcium chloride for food packaging applications. Int J Biol Macromol 239:124335. https://doi.org/10.1016/j.ijbiomac.2023.124335

Álvarez-Castillo E, Felix M, Bengoechea C, Guerrero A (2021) Proteins from agri-food industrial biowastes or co-products and their applications as green materials. Foods 10. https://doi.org/10.3390/foods10050981

Alvarez-Perez OB, Ventura-Sobrevilla JM, Torres-León C et al (2022) Development and characterization of whey protein films incorporated with tarbush polyphenols and candelilla wax. Food Biosci 45:101505. https://doi.org/10.1016/j.fbio.2021.101505

Ancuța P, Sonia A (2020) Oil press-cakes and meals valorization through circular economy approaches: a review. Appl Sci 10. https://doi.org/10.3390/app10217432

Aranda-Ledesma NE, Bautista-Hernández I, Rojas R et al (2022) Candelilla wax: prospective suitable applications within the food field. LWT 159:113170. https://doi.org/10.1016/j.lwt.2022.113170

Arias LV, Silva VD, Vieira JM et al (2024) Plant-based films for food packaging as a plastic waste management alternative: potato and cassava starch case. Polymers 16. https://doi.org/10.3390/polym16172390

Arquelau PB de F, Silva VDM, Garcia MAVT et al (2019) Characterization of edible coatings based on ripe "Prata" banana peel flour. Food Hydrocoll 89:570–578. https://doi.org/10.1016/j.foodhyd.2018.11.029

Baron RD, Pérez LL, Salcedo JM et al (2017) Production and characterization of films based on blends of chitosan from blue crab (Callinectes sapidus) waste and pectin from Orange (Citrus sinensis Osbeck) peel. Int J Biol Macromol 98:676–683. https://doi.org/10.1016/j.ijbiomac.2017.02.004

Baskaran P, Bose VG (2024) Development and characterisation of agar/silver nanoparticles-based film from Gracilaria canaliculata for food packaging. IFRJ 31:737–747. https://doi.org/10.47836/ifrj.31.3.17

Ben Slimane E, Sadok S (2018) Collagen from cartilaginous fish by-products for a potential application in bioactive film composite. Mar Drugs 16. https://doi.org/10.3390/md16060211

Ben-Othman S, Jõudu I, Bhat R (2020) Bioactives from agri-food wastes: present insights and future challenges. Molecules 25. https://doi.org/10.3390/molecules25030510

Berenguer CV, Andrade C, Pereira JAM et al (2023) Current challenges in the sustainable valorisation of agri-food wastes: a review. Processes 11. https://doi.org/10.3390/pr11010020

Bilal M, Niu D, Wang Z (2025) Biotechnological strategies and perspectives for food waste treatment: the role of lactic acid and microbial biomass. Waste Biomass Valorization 16:547–569. https://doi.org/10.1007/s12649-024-02705-y

Bojorges H, López-Rubio A, Martínez-Abad A, Fabra MJ (2023) Overview of alginate extraction processes: impact on alginate molecular structure and techno-functional properties. Trends Food Sci Technol 140:104142. https://doi.org/10.1016/j.tifs.2023.104142

Brito TB, Carrajola JF, Gonçalves ECBA et al (2019) Fruit and vegetable residues flours with different granulometry range as raw material for pectin-enriched biodegradable film preparation. Food Res Int 121:412–421. https://doi.org/10.1016/j.foodres.2019.03.058

Bühlmann CH, Mickan BS, Tait S et al (2023) Waste to Wealth: The power of food-waste anaerobic digestion integrated with lactic acid fermentation. Front Chem Eng 5:1285002. https://doi.org/10.3389/fceng.2023.1285002

Carvalho ACF, Ghosh S, Hoffmann TG et al (2025) Valuing agro-industrial waste in the development of sustainable food packaging based on the system of a circular bioeconomy: a review. Clean Waste Syst 11:100275. https://doi.org/10.1016/j.clwas.2025.100275

Castrica M, Rebucci R, Giromini C et al (2019) Total phenolic content and antioxidant capacity of agri-food waste and by-products. Ital J Anim Sci 18:336–341. https://doi.org/10.1080/1828051X.2018.1529544

Cerveró JM, Skovgaard PA, Felby C et al (2010) Enzymatic hydrolysis and fermentation of palm kernel press cake for production of bioethanol. Enzyme Microb Technol 46:177–184. https://doi.org/10.1016/j.enzmictec.2009.10.012

Chaka KT (2022) Extraction of cellulose nanocrystals from agricultural by-products: a review. Green Chem Lett Rev 15:582–597. https://doi.org/10.1080/17518253.2022.2121183

Chen H, Wang J, Cheng Y et al (2019) Application of protein-based films and coatings for food packaging: a review. Polymers 11. https://doi.org/10.3390/polym11122039

Chen L, Ramezan Y, Pourramezan H et al (2025) Soy protein isolate (SPI)-based films/coatings for food packaging: research progress on properties and applications. Compr Rev Food Sci Food Saf 24:e70181. https://doi.org/10.1111/1541-4337.70181

Cheng C, Chen S, Su J et al (2022) Recent advances in carrageenan-based films for food packaging applications. Front Nutr 9:1004588. https://doi.org/10.3389/fnut.2022.1004588

Chi W, Cao L, Sun G et al (2020) Developing a highly pH-sensitive κ-carrageenan-based intelligent film incorporating grape skin powder via a cleaner process. J Clean Prod 244:118862. https://doi.org/10.1016/j.jclepro.2019.118862

Chilakamarry CR, Sakinah AMM, Zularisam AW, Pandey A (2021) Glycerol waste to value added products and its potential applications. Syst Microbiol Biomanuf 1:378–396. https://doi.org/10.1007/s43393-021-00036-w

Costa JM, Strieder MM, Saldaña MDA et al (2023) Recent advances in the processing of agri-food by-products by subcritical water. Food Bioprocess Technol 16:2705–2724. https://doi.org/10.1007/s11947-023-03071-8

da Nóbrega SE, de Albuquerque C, Sousa T, de Santana C, Neto D et al (2022) Edible active film based on gelatin and Malpighia emarginata waste extract to inhibit lipid and protein oxidation in beef patties. LWT 154:112837. https://doi.org/10.1016/j.lwt.2021.112837
Dash KK, Ali NA, Das D, Mohanta D (2019) Thorough evaluation of sweet potato starch and lemon-waste pectin based-edible films with nano-titania inclusions for food packaging applications. Int J Biol Macromol 139:449–458. https://doi.org/10.1016/j.ijbiomac.2019.07.193
Deng Y, Kolodjski S, Lewis G et al (2025) Physical and functional characterization of whey protein-lignin biocomposite films for food packaging applications. Future Foods 11:100554. https://doi.org/10.1016/j.fufo.2025.100554
Deshmukh RK, Kumar P, Tanwar R, Gaikwad KK (2022) Pectin-polyvinylpyrrolidone based antimicrobial and antioxidant nanocomposite film impregnated with titania nanoparticles and bael shell extract. Food Bioprocess Technol 15:2839–2853. https://doi.org/10.1007/s11947-022-02922-0
Dewan MDF, Islam MdN (2024) Pullulan-based films: unveiling its multifaceted versatility for sustainability. Adv Polym Technol 2024:2633384. https://doi.org/10.1155/2024/2633384
Dmitrenko M, Kuzminova A, Cherian RM et al (2023) Edible carrageenan films reinforced with starch and nanocellulose: development and characterization. Sustainability 15. https://doi.org/10.3390/su152215817
do Vale DA, Vieira CB, Vidal MF et al (2021) Chitosan-based edible films produced from crab-uçá (ucides cordatus) waste: physicochemical, mechanical and antimicrobial properties. J Polym Environ 29:694–706. https://doi.org/10.1007/s10924-020-01913-6
Dong M, Tian L, Li J et al (2022) Improving physicochemical properties of edible wheat gluten protein films with proteins, polysaccharides and organic acid. LWT 154:112868. https://doi.org/10.1016/j.lwt.2021.112868
Dutta D, Sit N (2024) Comprehensive review on developments in starch-based films along with active ingredients for sustainable food packaging. Sustain Chem Pharm 39:101534. https://doi.org/10.1016/j.scp.2024.101534
Ebrahimi A, Kiani-Salmi N, Tavassoli M et al (2025) Monitoring food spoilage using smart plant-based packaging materials: methylcellulose/soy protein films loaded with betacyanin and carbon dots. Future Foods 11:100610. https://doi.org/10.1016/j.fufo.2025.100610
El Moujahed S, Errachidi F, Morosanu A-M et al (2024) Sustainable collagen film preparation with tannins extracted from moroccan pomegranate byproduct varieties: thermal, structural, and nanoscaled studies. ACS Omega 9:27428–27437. https://doi.org/10.1021/acsomega.4c02321
Erge A, Güler BZ, Eren Ö (2024) Optimization and characterization of biodegradable films from chicken gelatin crosslinked with oxidized phenolic compounds. Food Chem 438:137923. https://doi.org/10.1016/j.foodchem.2023.137923
Estrada-Sotomayor JD, Łopusiewicz Ł, Lizundia E et al (2025) Full conversion of grass biomass into sustainable functional antimicrobial bioplastics. Green Chem 27:4587–4602. https://doi.org/10.1039/D5GC00643K
Ezati P, Khan A, Bhattacharya T et al (2025) Recent advances in collagen and collagen-based packaging materials: a review. ACS Food Sci Technol. https://doi.org/10.1021/acsfoodscitech.4c00938
Fabra MJ, Lopez-Rubio A, Lagaron JM (2015) Effect of the film-processing conditions, relative humidity and ageing on wheat gluten films coated with electrospun polyhydryalkanoate. Food Hydrocoll 44:292–299. https://doi.org/10.1016/j.foodhyd.2014.09.032
Farhan A, Hani NM (2017) Characterization of edible packaging films based on semi-refined kappa-carrageenan plasticized with glycerol and sorbitol. Food Hydrocoll 64:48–58. https://doi.org/10.1016/j.foodhyd.2016.10.034
Faria DJ, Carvalho AP, Conte-Junior CA (2023) Valorization of fermented food wastes and byproducts: bioactive and valuable compounds, bioproduct synthesis, and applications. Fermentation 9. https://doi.org/10.3390/fermentation9100920
Farid MS, Shafique B, Xu R et al (2025) Potential interventions and interactions of bioactive polyphenols and functional polysaccharides to alleviate inflammatory bowel disease—a review. Food Chem 462:140951. https://doi.org/10.1016/j.foodchem.2024.140951

Farris S, Unalan IU, Introzzi L et al (2014) Pullulan-based films and coatings for food packaging: present applications, emerging opportunities, and future challenges. J Appl Polym Sci 131. https://doi.org/10.1002/app.40539

Fraga-Corral M, Otero P, Echave J et al (2021) By-products of agri-food industry as tannin-rich sources: a review of tannins' biological activities and their potential for valorization. Foods 10. https://doi.org/10.3390/foods10010137

Ghosh S, Mandal RK, Mukherjee A, Roy S (2025) Nanotechnology in the manufacturing of sustainable food packaging: a review. Discover Nano 20:36. https://doi.org/10.1186/s11671-025-04213-x

Ghosh T, Roy S, Khan A et al (2024) Agricultural waste-derived cellulose nanocrystals for sustainable active food packaging applications. Food Hydrocoll 154:110141. https://doi.org/10.1016/j.foodhyd.2024.110141

González-Arancibia F, Mamani M, Valdés C et al (2024) Biopolymers as sustainable and active packaging materials: fundamentals and mechanisms of antifungal activities. Biomolecules 14. https://doi.org/10.3390/biom14101224

Granados-Carrera CM, Castro-Criado D, Jiménez-Rosado M et al (2025) Reinforcement of soy protein-based bioplastics as potential sustainable packaging solutions. Future Foods 11:100524. https://doi.org/10.1016/j.fufo.2024.100524

Gubitosa J, Rizzi V, Marasciulo C et al (2023) Realizing eco-friendly water-resistant sodium-alginate-based films blended with a polyphenolic aqueous extract from grape pomace waste for potential food packaging applications. Int J Mol Sci 24. https://doi.org/10.3390/ijms241411462

Guerrero P, Etxabide A, Leceta I et al (2014) Extraction of agar from gelidium sesquipedale (rodhopyta) and surface characterization of agar based films. Carbohyd Polym 99:491–498. https://doi.org/10.1016/j.carbpol.2013.08.049

Gupta D, Lall A, Kumar S et al (2024) Plant-based edible films and coatings for food-packaging applications: recent advances, applications, and trends. Sustain Food Technol 2:1428–1455. https://doi.org/10.1039/D4FB00110A

Gupta V, Yadav R, Tanwar R, Gaikwad KK (2023) κ-Carrageenan-based bio-nanocomposite film reinforced with cellulose nanocrystals derived from amla pomace for food packaging. Biomass Convers Biorefinery 13:16899–16908. https://doi.org/10.1007/s13399-021-02028-1

Ha PTT, Trang PT, Van Khoi N et al (2025) Alginate/shellac composite films—novel materials for preservation of the purple passion fruits (Passiflora edulis Sims). ChemistrySelect 10:e202404383. https://doi.org/10.1002/slct.202404383

Hadavifar S, Abedi-Firoozjah R, Bahramian B et al (2025) Multifunctional performance of chitosan/soy protein isolation-based films impregnated carbon dots/anthocyanin derived from purple hull pistachio for tracking and extending the shelf life of fish. Food Hydrocoll 159:110678. https://doi.org/10.1016/j.foodhyd.2024.110678

Hamed I, Jakobsen AN, Lerfall J (2022) Sustainable edible packaging systems based on active compounds from food processing byproducts: a review. Compr Rev Food Sci Food Saf 21:198–226. https://doi.org/10.1111/1541-4337.12870

Hashimoto EH, de Cassia Campos Pena A, da Cunha MAA, et al (2025) Fermentation-mediated sustainable development and improvement of quality of plant-based foods: from waste to a new food. Syst Microbiol Biomanuf 5:69–100. https://doi.org/10.1007/s43393-024-00292-6

Hernández-Carranza P, Fierro-Corona G, Tapia-Maruri D et al (2023) Bioactive edible films based on LAB-fermented whey solution and potato starch: characterization and storage behavior. Food Bioprocess Technol 16:3045–3056. https://doi.org/10.1007/s11947-023-03115-z

Hu F, Song Y-Z, Thakur K et al (2024) Blueberry anthocyanin based active intelligent wheat gluten protein films: Preparation, characterization, and applications for shrimp freshness monitoring. Food Chem 453:139676. https://doi.org/10.1016/j.foodchem.2024.139676

Jahangiri F, Mohanty AK, Misra M (2024) Sustainable biodegradable coatings for food packaging: challenges and opportunities. Green Chem 26:4934–4974. https://doi.org/10.1039/d3gc02647g

Jiang H, Zhang W, Pu Y et al (2023) Development and characterization of a novel active and intelligent film based on pectin and betacyanins from peel waste of pitaya (Hylocereus undatus). Food Chem 404:134444. https://doi.org/10.1016/j.foodchem.2022.134444
Jorge AMS, Gaspar MC, Henriques MHF, Braga MEM (2023) Edible films produced from agrifood by-products and wastes. Innov Food Sci Emerg Technol 88:103442. https://doi.org/10.1016/j.ifset.2023.103442
Kandasamy S, Yoo J, Yun J et al (2021) Application of whey protein-based edible films and coatings in food industries: an updated overview. Coatings 11. https://doi.org/10.3390/coatings11091056
Kanmani P, Rhim J-W (2014) Antimicrobial and physical-mechanical properties of agar-based films incorporated with grapefruit seed extract. Carbohyd Polym 102:708–716. https://doi.org/10.1016/j.carbpol.2013.10.099
Karim A, Osse EF, Khalloufi S (2025) Innovative strategies for valorization of byproducts from soybean industry: a review on status, challenges, and sustainable approaches towards zero-waste processing systems. Heliyon 11:e42118. https://doi.org/10.1016/j.heliyon.2025.e42118
Karki DB, Kc Y, Khanal H et al (2020) Analysis of biodegradable films of starch from potato waste. AFSJ 28–40. https://doi.org/10.9734/afsj/2020/v14i330132
Karnwal A, Kumar G, Singh R et al (2025) Natural biopolymers in edible coatings: applications in food preservation. Food Chem: X 25:102171. https://doi.org/10.1016/j.fochx.2025.102171
Khin MN, Ahammed S, Kamal MdM et al (2024) Investigating next-generation edible packaging: protein-based films and coatings for delivering active compounds. Food Hydrocoll Health 6:100182. https://doi.org/10.1016/j.fhfh.2024.100182
Knorr D (2025) Food processing of chitin and chitosan: From waste to opportunities. J Food Eng 400:112634. https://doi.org/10.1016/j.jfoodeng.2025.112634
Kocira A, Kozłowicz K, Panasiewicz K et al (2021) Polysaccharides as edible films and coatings: characteristics and influence on fruit and vegetable quality—a review. Agronomy 11. https://doi.org/10.3390/agronomy11050813
Kokkuvayil Ramadas B, Rhim J-W, Roy S (2024) Recent progress of carrageenan-based composite films in active and intelligent food packaging applications. Polymers 16. https://doi.org/10.3390/polym16071001
Koshy RR, Reghunadhan A, Mary SK et al (2022) PH indicator films fabricated from soy protein isolate modified with chitin nanowhisker and Clitoria ternatea flower extract. Curr Res Food Sci 5:743–751. https://doi.org/10.1016/j.crfs.2022.03.015
Kozma M, Acharya B, Bissessur R (2022) Chitin, chitosan, and nanochitin: extraction, synthesis, and applications. Polymers 14. https://doi.org/10.3390/polym14193989
Kumar L, Deshmukh RK, Gaikwad KK (2022) Antimicrobial packaging film from cactus (Cylindropuntia fulgida) mucilage and gelatine. Int J Biol Macromol 215:596–605. https://doi.org/10.1016/j.ijbiomac.2022.06.162
Kumar N, Pratibha, Trajkovska Petkoska A et al (2021) Chitosan edible films enhanced with pomegranate peel extract: study on physical, biological, thermal, and barrier properties. Materials 14. https://doi.org/10.3390/ma14123305
Kumnerdsiri P, Sanprasert S, Seubsai A et al (2024) Properties of novel biodegradable film from gelatin capsule waste as influenced by various solvents and washing cycles. Future Foods 10:100485. https://doi.org/10.1016/j.fufo.2024.100485
Kurek M, Benbettaieb N, Ščetar M et al (2021) Novel functional chitosan and pectin bio-based packaging films with encapsulated Opuntia-ficus indica waste. Food Biosci 41:100980. https://doi.org/10.1016/j.fbio.2021.100980
Kuttiyatveetil JRA, Mitra P, Goldin D et al (2019) Recovery of residual nutrients from agri-food byproducts using a combination of solid-state fermentation and insect rearing. Int J Food Sci Technol 54:1130–1140. https://doi.org/10.1111/ijfs.14015
Lan X, Zhang X, Wang L et al (2023) A review of food preservation based on zein: the perspective from application types of coating and film. Food Chem 424:136403. https://doi.org/10.1016/j.foodchem.2023.136403

Lawton JW (2002) Zein: a history of processing and use. Cereal Chem 79:1–18. https://doi.org/10.1094/CCHEM.2002.79.1.1

Lee H-G, Lee SY, Yoo S (2024) Innovative food-upcycling solutions: comparative analysis of edible films from kimchi-extracted cellulose, sorbitol, and citric acid for food packaging applications. Food Chem 450:139267. https://doi.org/10.1016/j.foodchem.2024.139267

Licciardello F, Wittenauer J, Saengerlaub S et al (2015) Rapid assessment of the effectiveness of antioxidant active packaging—study with grape pomace and olive leaf extracts. Food Packag Shelf Life 6:1–6. https://doi.org/10.1016/j.fpsl.2015.08.001

Liu M, Zhang X, Wei A et al (2024) Protein-based active films: raw materials, functions, and food applications. Compr Rev Food Sci Food Saf 23:e13302. https://doi.org/10.1111/1541-4337.13302

Liu Y, Ahmed S, Sameen DE et al (2021) A review of cellulose and its derivatives in biopolymer-based for food packaging application. Trends Food Sci Technol 112:532–546. https://doi.org/10.1016/j.tifs.2021.04.016

Łopusiewicz Ł (2024) Scientific Interventions for Improving the Utilization of Palm Kernel Meal. In: Kumar M, Punia Bangar S, Panesar PS (eds) Oilseed meal as a sustainable contributor to plant-based protein: paving the way towards circular economy and nutritional security. Springer International Publishing, Cham, pp 187–204

Łopusiewicz Ł, Dmytrów I, Mituniewicz-Małek A et al (2022a) Natural gum from flaxseed by-product as a potential stabilizing and thickening agent for acid whey fermented beverages. Appl Sci 12. https://doi.org/10.3390/app122010281

Łopusiewicz Ł, Macieja S, Śliwiński M et al (2022b) Alginate biofunctional films modified with melanin from watermelon seeds and zinc oxide/silver nanoparticles. Materials 15. https://doi.org/10.3390/ma15072381

Łopusiewicz Ł, Drozłowska E, Trocer P et al (2020) Whey protein concentrate/isolate biofunctional films modified with melanin from watermelon (Citrullus lanatus) seeds. Materials 13. https://doi.org/10.3390/ma13173876

Łopusiewicz Ł, Jędra F, Bartkowiak A (2018a) New active packaging films made from gelatin modified with fungal melanin

Łopusiewicz Ł, Jędra F, Bartkowiak A (2018b) The application of melanin modified gelatin coatings for packaging and the oxidative stability of pork lard

Łopusiewicz Ł, Kowalczewski PŁ, Baranowska HM et al (2023) Effect of flaxseed oil cake extract on the microbial quality, texture and shelf life of gluten-free bread. Foods 12. https://doi.org/10.3390/foods12030595

Luangtana-anan M, Soradech S, Saengsod S et al (2017) Enhancement of moisture protective properties and stability of pectin through formation of a composite film: effects of shellac and plasticizer. J Food Sci 82:2915–2925. https://doi.org/10.1111/1750-3841.13956

Luís Â, Ramos A, Domingues F (2020) Pullulan films containing rockrose essential oil for potential food packaging applications. Antibiotics 9. https://doi.org/10.3390/antibiotics9100681

Macieja S, Lepczyński A, Bernaciak M et al (2023) Effect of melanin on the stability of casein films exposed to artificially accelerated UV aging. Coatings 13. https://doi.org/10.3390/coatings13071262

Martínez-Sanz M, Martínez-Abad A, López-Rubio A (2019) Cost-efficient bio-based food packaging films from unpurified agar-based extracts. Food Packag Shelf Life 21:100367. https://doi.org/10.1016/j.fpsl.2019.100367

Maryam Adilah ZA, Jamilah B, Nur Hanani ZA (2023) Storage stability of mayonnaise packaged in soy protein isolate films incorporated with mango kernel extract. Food Packag Shelf Life 40:101216. https://doi.org/10.1016/j.fpsl.2023.101216

Matloob A, Ayub H, Mohsin M et al (2023) A review on edible coatings and films: advances, composition, production methods, and safety concerns. ACS Omega 8:28932–28944. https://doi.org/10.1021/acsomega.3c03459

Medina-Jaramillo C, Quintero-Pimiento C, Gómez-Hoyos C et al (2020) Alginate-edible coatings for application on wild Andean blueberries (Vaccinium meridionale Swartz): effect of the addition of nanofibrils isolated from cocoa by-products. Polymers 12. https://doi.org/10.3390/polym12040824

Melini V, Melini F, Luziatelli F, Ruzzi M (2020) Functional ingredients from agri-food waste: effect of inclusion thereof on phenolic compound content and bioaccessibility in bakery products. Antioxidants 9. https://doi.org/10.3390/antiox9121216

Mishra B, Mohanta YK, Varjani S et al (2023) A critical review on valorization of food processing wastes and by-products for pullulan production. J Food Sci Technol 60:2121–2131. https://doi.org/10.1007/s13197-022-05490-5

Morseletto P (2020) Targets for a circular economy. Resour Conserv Recycl 153:104553. https://doi.org/10.1016/j.resconrec.2019.104553

Mostafa H, Airouyuwaa JO, Hamed F et al (2023) Structural, mechanical, antioxidant and antibacterial properties of soy protein isolate (SPI)-based edible food packaging films as influenced by nanocellulose (NC) and green extracted phenolic compounds from date palm leaves. Food Packag Shelf Life 38:101124. https://doi.org/10.1016/j.fpsl.2023.101124

Moura-Alves M, Esteves A, Ciríaco M et al (2023) Antimicrobial and antioxidant edible films and coatings in the shelf-life improvement of chicken meat. Foods 12. https://doi.org/10.3390/foods12122308

Munasinghe KA, Schwarz JG, Whittiker M (2015) Utilization of chicken by-products to form collagen films. J Food Process 2015:247013. https://doi.org/10.1155/2015/247013

Nargotra P, Sharma V, Tsai M-L et al (2023) Recent advancements in the valorization of agro-industrial food waste for the production of nanocellulose. Appl Sci 13. https://doi.org/10.3390/app13106159

Nayak B, Jain P, Kumar L et al (2024) UV blocking edible films based on corn starch/moringa gum incorporated with pine cone extract for sustainable food packaging. Int J Biol Macromol 267:131545. https://doi.org/10.1016/j.ijbiomac.2024.131545

Nemazifard M, Kavoosi G, Marzban Z, Ezedi N (2017) Physical, mechanical, water binding, and antioxidant properties of cellulose dispersions and cellulose film incorporated with pomegranate seed extract. Int J Food Prop 20:1501–1514. https://doi.org/10.1080/10942912.2016.1219369

Nemes SA, Călinoiu LF, Dulf FV et al (2022) Integrated technology for cereal bran valorization: perspectives for a sustainable industrial approach. Antioxidants 11. https://doi.org/10.3390/antiox11112159

Oliveira Filho JG de, Bezerra CC de ON, Albiero BR, et al (2020) New approach in the development of edible films: The use of carnauba wax micro- or nanoemulsions in arrowroot starch-based films. Food Packag Shelf Life 26:100589. https://doi.org/10.1016/j.fpsl.2020.100589

Oliveira Filho JG de, de Sousa TL, Bertolo MRV et al (2023) Next-generation food packaging: Edible bioactive films with alginate, mangaba pulp (Hancornia speciosa), and Saccharomyces boulardii. Food Biosci 54:102799. https://doi.org/10.1016/j.fbio.2023.102799

Oliveira Filho JG de, Rodrigues JM, Valadares ACF et al (2019) Active food packaging: alginate films with cottonseed protein hydrolysates. Food Hydrocoll 92:267–275. https://doi.org/10.1016/j.foodhyd.2019.01.052

Ortiz-Sanchez M, Inocencio-García P-J, Alzate-Ramírez AF, Alzate CA (2023) Potential and restrictions of food-waste valorization through fermentation processes. Fermentation 9. https://doi.org/10.3390/fermentation9030274

Osorio LL, Flórez-López E, Grande-Tovar CD (2021) The potential of selected agri-food loss and waste to contribute to a circular economy: applications in the food cosmetic and pharmaceutical industries. Molecules 26. https://doi.org/10.3390/molecules26020515

Othman SH, Edwal SAM, Risyon NP et al (2017) Water sorption and water permeability properties of edible film made from potato peel waste. Food Sci Technol 37:63–70. https://doi.org/10.1590/1678-457x.30216

Pan J, Li C, Liu J et al (2024) Polysaccharide-based packaging coatings and films with phenolic compounds in preservation of fruits and vegetables—a review. Foods 13. https://doi.org/10.3390/foods13233896

Panzella L, Moccia F, Nasti R et al (2020) Bioactive phenolic compounds from agri-food wastes: an update on green and sustainable extraction methodologies. Front Nutr 7:60. https://doi.org/10.3389/fnut.2020.00060

Papadaki A, Kachrimanidou V, K. Lappa I et al (2022) Tuning the physical and functional properties of whey protein edible films: Effect of pH and inclusion of antioxidants from spent coffee grounds. Sustain Chem Pharm 27:100700. https://doi.org/10.1016/j.scp.2022.100700

Pashova S (2023) Application of plant waxes in edible coatings. Coatings 13. https://doi.org/10.3390/coatings13050911

Paudel S, Regmi S, Bhattarai S et al (2025) Valorization of grapevine agricultural waste into transparent and high-strength biodegradable films for sustainable packaging. Sustain Food Technol. https://doi.org/10.1039/D5FB00211G

Pawde SV, Kaewprachu P, Kingwascharapong P et al (2025) A comprehensive review on plant protein-based food packaging: Beyond petroleum-based polymers. Curr Res Food Sci 10:101104. https://doi.org/10.1016/j.crfs.2025.101104

Pedreiro S, Figueirinha A, Silva AS, Ramos F (2021) Bioactive edible films and coatings based in gums and starch: phenolic enrichment and foods application. Coatings 11. https://doi.org/10.3390/coatings11111393

Petraru A, Amariei S (2023) A novel approach about edible packaging materials based on oilcakes—a review. Polymers 15. https://doi.org/10.3390/polym15163431

Peydayesh M, Bagnani M, Soon WL, Mezzenga R (2023) Turning food protein waste into sustainable technologies. Chem Rev 123:2112–2154. https://doi.org/10.1021/acs.chemrev.2c00236

Phannarangsee Y, Jiawkhangphlu B, Thanonkeo S et al (2024) Sorbitol production from mixtures of molasses and sugarcane bagasse hydrolysate using the thermally adapted Zymomonas mobilis ZM AD41. Sci Rep 14:5563. https://doi.org/10.1038/s41598-024-56307-8

Pillai ARS, Eapen AS, Zhang W, Roy S (2024) Polysaccharide-based edible biopolymer-based coatings for fruit preservation: a review. Foods 13. https://doi.org/10.3390/foods13101529

Pires AF, Díaz O, Cobos A, Pereira CD (2024) A review of recent developments in edible films and coatings-focus on whey-based materials. Foods 13. https://doi.org/10.3390/foods13162638

Pirsa S, Mahmudi M, Ehsani A (2023) Biodegradable film based on cress seed mucilage, modified with lutein, maltodextrin and alumina nanoparticles: physicochemical properties and lutein controlled release. Int J Biol Macromol 224:1588–1599. https://doi.org/10.1016/j.ijbiomac.2022.10.244

Priyadarshi R, Riahi Z, Rhim J-W (2022) Antioxidant pectin/pullulan edible coating incorporated with Vitis vinifera grape seed extract for extending the shelf life of peanuts. Postharvest Biol Technol 183:111740. https://doi.org/10.1016/j.postharvbio.2021.111740

Procopio FR, Ferraz MC, Lourenço RV et al (2024) Soy protein-based films incorporated with co-encapsulated cinnamon and paprika oleoresins. J Food Meas Charact 18:5145–5156. https://doi.org/10.1007/s11694-024-02562-2

Purewal SS, Kaur A, Bangar SP et al (2024) Protein-based films and coatings: an innovative approach. Coatings 14. https://doi.org/10.3390/coatings14010032

Rakesh B, Mahendran R (2024) Upcycling of food waste and food loss—a sustainable approach in the food sector. Trends Food Sci Technol 143:104274. https://doi.org/10.1016/j.tifs.2023.104274

Ramos ÓL, Fernandes JC, Silva SI et al (2012) Edible films and coatings from whey proteins: a review on formulation, and on mechanical and bioactive properties. Crit Rev Food Sci Nutr 52:533–552. https://doi.org/10.1080/10408398.2010.500528

Rani R, Gosh T, Badwaik LS (2023) Optimization of mustard, soybean and flaxseed meal blend formulation for development of biopolymeric film and its characterization. Sustain Chem Pharm 33:101147. https://doi.org/10.1016/j.scp.2023.101147

Rashid A, Qayum A, Liang Q et al (2024) Exploring the potential of pullulan-based films and coatings for effective food preservation: a comprehensive analysis of properties, activation strategies and applications. Int J Biol Macromol 260:129479. https://doi.org/10.1016/j.ijbiomac.2024.129479

Rashwan AK, Bai H, Osman AI et al (2023) Recycling food and agriculture by-products to mitigate climate change: a review. Environ Chem Lett 21:3351–3375. https://doi.org/10.1007/s10311-023-01639-6

Rațu RN, Veleșcu ID, Stoica F et al (2023) Application of agri-food by-products in the food industry. Agriculture 13:1559. https://doi.org/10.3390/agriculture13081559

Rhim J-W, Reddy JP, Luo X (2015) Isolation of cellulose nanocrystals from onion skin and their utilization for the preparation of agar-based bio-nanocomposites films. Cellulose 22:407–420. https://doi.org/10.1007/s10570-014-0517-7

Ribeiro LS, Órfão JJ de M, Pereira MFR (2017) Direct catalytic production of sorbitol from waste cellulosic materials. Bioresour Technol 232:152–158. https://doi.org/10.1016/j.biortech.2017.02.008

Rochín-Medina JJ, Ramírez-Medina HK, Rangel-Peraza JG et al (2018) Use of whey as a culture medium for Bacillus clausii for the production of protein hydrolysates with antimicrobial and antioxidant activity. Food Sci Technol Int 24:35–42. https://doi.org/10.1177/1082013217724705

Rodrigues DC, Cunha AP, Brito ES et al (2016) Mesquite seed gum and palm fruit oil emulsion edible films: influence of oil content and sonication. Food Hydrocoll 56:227–235. https://doi.org/10.1016/j.foodhyd.2015.12.018

Rojas-García A, Fernández-Ochoa Á, Cádiz-Gurrea MD et al (2023) Neuroprotective effects of agri-food by-products rich in phenolic compounds. Nutrients 15. https://doi.org/10.3390/nu15020449

Romruen O, Karbowiak T, Tongdeesoontorn W et al (2022) Extraction and characterization of cellulose from agricultural by-products of Chiang Rai province, Thailand. Polymers 14. https://doi.org/10.3390/polym14091830

Rovera C, Carullo D, Bellesia T et al (2023) Extraction of high-quality grade cellulose and cellulose nanocrystals from different lignocellulosic agri-food wastes. Front Sustain Food Syst 6:1087867. https://doi.org/10.3389/fsufs.2022.1087867

Roy S, Chawla R, Santhosh R et al (2023a) Agar-based edible films and food packaging application: a comprehensive review. Trends Food Sci Technol 141:104198. https://doi.org/10.1016/j.tifs.2023.104198

Roy S, Deshmukh RK, Tripathi S et al (2023b) Recent advances in the carotenoids added to food packaging films: a review. Foods 12. https://doi.org/10.3390/foods12214011

Roy S, Priyadarshi R, Łopusiewicz Ł et al (2023c) Recent progress in pectin extraction, characterization, and pectin-based films for active food packaging applications: a review. Int J Biol Macromol 239:124248. https://doi.org/10.1016/j.ijbiomac.2023.124248

Ryan MP, Walsh G (2016) The biotechnological potential of whey. Rev Environ Sci Bio/technol 15:479–498. https://doi.org/10.1007/s11157-016-9402-1

Sabater C, Ruiz L, Delgado S et al (2020) Valorization of vegetable food waste and by-products through fermentation processes. Front Microbiol 11:581997. https://doi.org/10.3389/fmicb.2020.581997

Sadh PK, Kumar S, Chawla P, Duhan JS (2018) Fermentation: a boon for production of bioactive compounds by processing of food industries wastes (by-products). Molecules 23. https://doi.org/10.3390/molecules23102560

Salas-Millán JÁ, Aguayo E (2024) Fermentation for revalorisation of fruit and vegetable by-products: a sustainable approach towards minimising food loss and waste. Foods 13. https://doi.org/10.3390/foods13223680

Sales A, Paulino BN, Pastore GM, Bicas JL (2018) Biogeneration of aroma compounds. Curr Opin Food Sci 19:77–84. https://doi.org/10.1016/j.cofs.2018.03.005

Salim NV, Madhan B, Glattauer V, Ramshaw JAM (2024) Comprehensive review on collagen extraction from food by-products and waste as a value-added material. Int J Biol Macromol 278:134374. https://doi.org/10.1016/j.ijbiomac.2024.134374

Sarma PP, Barman K, Baruah PK (2024) Rice starch-agar composite film decorated with green synthesized silver nanoparticles with superior properties. Inorg Chem Commun 170:113387. https://doi.org/10.1016/j.inoche.2024.113387

Sayanjali S, Lu Y, Howell K (2024) Extraction and characterization of cellulose from broccoli stems as a new biopolymer source for producing carboxymethyl cellulose films. Int J Food Sci 2024:7661288. https://doi.org/10.1155/2024/7661288

Schmid M, Müller K (2019) Chapter 11—Whey protein-based packaging films and coatings. In: Deeth HC, Bansal N (eds) Whey Proteins. Academic Press, pp 407–437

Shanbhag C, Shenoy R, Shetty P et al (2023) Formulation and characterization of starch-based novel biodegradable edible films for food packaging. J Food Sci Technol 60:2858–2867. https://doi.org/10.1007/s13197-023-05803-2

Sharma SK, Bansal S, Mangal M et al (2016) Utilization of food processing by-products as dietary, functional, and novel fiber: a review. Crit Rev Food Sci Nutr 56:1647–1661. https://doi.org/10.1080/10408398.2013.794327

Silveira YDO, Franca AS, Oliveira LS (2025) Cassava waste starch as a source of bioplastics: development of a polymeric film with antimicrobial properties. Foods 14. https://doi.org/10.3390/foods14010113

Singh P, Krishnaswamy K (2022) Sustainable zero-waste processing system for soybeans and soy by-product valorization. Trends Food Sci Technol 128:331–344. https://doi.org/10.1016/j.tifs.2022.08.015

Singh RS, Kaur N, Kennedy JF (2019) Pullulan production from agro-industrial waste and its applications in food industry: a review. Carbohyd Polym 217:46–57. https://doi.org/10.1016/j.carbpol.2019.04.050

Socas-Rodríguez B, Álvarez-Rivera G, Valdés A et al (2021) Food by-products and food wastes: are they safe enough for their valorization? Trends Food Sci Technol 114:133–147. https://doi.org/10.1016/j.tifs.2021.05.002

Song D-H, Hoa VB, Kim HW et al (2021) Edible films on meat and meat products. Coatings 11. https://doi.org/10.3390/coatings11111344

Soradech S, Nunthanid J, Limmatvapirat S, Luangtana-anan M (2017) Utilization of shellac and gelatin composite film for coating to extend the shelf life of banana. Food Control 73:1310–1317. https://doi.org/10.1016/j.foodcont.2016.10.059

Sorrentino C, Di Gisi M, Gentile G et al (2022) Agri-food by-products in cancer: new targets and strategies. Cancers 14. https://doi.org/10.3390/cancers14225517

Sothornvit R, Krochta JM (2005) Chapter 23—Plasticizers in edible films and coatings. In: Han JH (ed) Innovations in food packaging. Academic Press, London, pp 403–433

Sousa AMM, Gonçalves MP (2015) Strategies to improve the mechanical strength and water resistance of agar films for food packaging applications. Carbohyd Polym 132:196–204. https://doi.org/10.1016/j.carbpol.2015.06.022

Sultan M, Hafez OM, Saleh MA, Youssef AM (2021) Smart edible coating films based on chitosan and beeswax–pollen grains for the postharvest preservation of Le Conte pear. RSC Adv 11:9572–9585. https://doi.org/10.1039/D0RA10671B

Susmita Devi L, Kalita S, Mukherjee A, Kumar S (2022) Carnauba wax-based composite films and coatings: recent advancement in prolonging postharvest shelf-life of fruits and vegetables. Trends Food Sci Technol 129:296–305. https://doi.org/10.1016/j.tifs.2022.09.019

Tafa TG, Engida AM (2022) Preparation of green film with improved physicochemical properties and enhanced antimicrobial activity using ingredients from cassava peel, bamboo leaf and rosemary leaf. Heliyon 8. https://doi.org/10.1016/j.heliyon.2022.e10130

Tallawi M, Amrein D, Gemmecker G et al (2023) A novel polysaccharide/zein conjugate as an alternative green plastic. Sci Rep 13:13161. https://doi.org/10.1038/s41598-023-40293-4

Tedeschi G, Benitez JJ, Ceseracciu L et al (2018) Sustainable fabrication of plant cuticle-like packaging films from tomato pomace agro-waste, beeswax, and alginate. ACS Sustain Chem Eng 6:14955–14966. https://doi.org/10.1021/acssuschemeng.8b03450

Thakur R, Singhi H, Suryavanshi VR et al (2024) PH-sensitive intelligent packaging films from Kodo millet (Paspalum scrobiculatum) starch and gum tragacanth incorporated with beetroot peel extract for monitoring shrimp freshness. Food Packag Shelf Life 46:101405. https://doi.org/10.1016/j.fpsl.2024.101405

Thiviya P, Gamage A, Liyanapathiranage A et al (2023) Algal polysaccharides: structure, preparation and applications in food packaging. Food Chem 405:134903. https://doi.org/10.1016/j.foodchem.2022.134903

Thivya P, Bhosale YK, Anandakumar S et al (2022) Study on the characteristics of gluten/alginate-cellulose/onion waste extracts composite film and its food packaging application. Food Chem 390:133221. https://doi.org/10.1016/j.foodchem.2022.133221

Thongsomboon W, Baimark Y, Srihanam P (2023) Valorization of cellulose-based materials from agricultural waste: comparison between sugarcane bagasse and rice straw. Polymers 15. https://doi.org/10.3390/polym15153190

Tivano F, Chiono V (2023) Zein as a renewable material for the preparation of green nanoparticles for drug delivery. Front Biomater Sci 2:1156403. https://doi.org/10.3389/fbiom.2023.1156403

Tonolo F, Folda A, Ferro S et al (2023) Fermentation of corn by-products: from agrifood waste to higher value antioxidant products. Fermentation 9. https://doi.org/10.3390/fermentation9040373

Uribarrena M, Cabezudo S, Núñez RN et al (2025) Development of smart films based on soy protein and cow horn dissolved in a deep eutectic solvent: physicochemical and environmental assessment. Int J Biol Macromol 291:139045. https://doi.org/10.1016/j.ijbiomac.2024.139045

Uribarrena M, Rovira-Cal E, Urbina L et al (2024) Valorization of cheese whey: closing the loop from protein extraction to whey protein film composting. Green Chem 26:4103–4111. https://doi.org/10.1039/d3gc04304e

Usman M, Sahar A, Inam-Ur-Raheem M et al (2022) Gelatin extraction from fish waste and potential applications in food sector. Int J Food Sci Technol 57:154–163. https://doi.org/10.1111/ijfs.15286

Vallejo M, Esteves B, Carvalho P et al (2025) Upcycling onion peels for enhanced gas barrier and antioxidant bioplastics based on starch from potato processing slurries. Waste Manage 205:115005. https://doi.org/10.1016/j.wasman.2025.115005

Vallejos-Jiménez A, Cadena-Chamorro EM, Santa JF et al (2025) Development of novel pectin-based films from coffee waste: mucilage and pulp. Waste Biomass Valor. https://doi.org/10.1007/s12649-025-02926-9

Vergel-Alfonso AA, Arias-Avelenda R, Casariego-Año A et al (2025) Development and characterization of pectin and beeswax-based coatings enhanced with anthocyanins and its antioxidant and antifungal properties. Processes 13. https://doi.org/10.3390/pr13020542

Wang Q, Li H, Feng K, Liu J (2020) Oriented fermentation of food waste towards high-value products: a review. Energies 13. https://doi.org/10.3390/en13215638

Wani SM, Mir SA, Khanday FA, Masoodi FA (2021) Advances in pullulan production from agro-based wastes by aureobasidium pullulans and its applications. Innov Food Sci Emerg Technol 74:102846. https://doi.org/10.1016/j.ifset.2021.102846

Yadav S, Malik K, Moore JM et al (2024) Valorisation of agri-food waste for bioactive compounds: recent trends and future sustainable challenges. Molecules 29. https://doi.org/10.3390/molecules29092055

Yang M, Chen L, Wang J, Msigwa G, Osman AI, Fawzy S, Rooney DW, Yap PS (2023) Circular economy strategies for combating climate change and other environmental issues. Abstr Environ Chem Lett 21(1):55–80. https://doi.org/10.1007/s10311-022-01499-6

Yang F, Hanna MA, Sun R (2012) Value-added uses for crude glycerol—a byproduct of biodiesel production. Biotechnol Biofuels 5:13. https://doi.org/10.1186/1754-6834-5-13

Zhang B, Guo B, Wang S et al (2024a) A soy protein-based film based on chemical treatment and microcrystalline cellulose reinforcement obtained from corn husk byproducts. ACS Omega 9:15845–15853. https://doi.org/10.1021/acsomega.3c07907

Zhang M, Xiao D, Zhu C et al (2024b) Shelf-life extension and quality changes of fresh-cut apple via sago and soy-oil-based edible coatings. Coatings 14. https://doi.org/10.3390/coatings1409 1202

Zhang W, Li X, Jiang W (2020) Development of antioxidant chitosan film with banana peels extract and its application as coating in maintaining the storage quality of apple. Int J Biol Macromol 154:1205–1214. https://doi.org/10.1016/j.ijbiomac.2019.10.275

Zhu W, Chen J, Dong Q et al (2023) Development of organic-inorganic hybrid antimicrobial materials by mechanical force and application for active packaging. Food Packag Shelf Life 37:101060. https://doi.org/10.1016/j.fpsl.2023.101060

Chapter 14
Prospects and Challenges of Nanotechnology in Functional Food Packaging

Raphael Kanyire Seidu, George Kwame Fobiri, Jyoti Rani, Benjamin Tawiah, and Shouxiang Jiang

Abstract The emergence of nanotechnology has been a transformative approach in the development of food packaging materials with functional abilities, offering innovative solutions to enhance food safety, preservation, and quality. The incorporation of nanocomposites and nanoparticles in food packaging has improved the mechanical strength, barrier properties, and antimicrobial activity of packaging materials, thereby maintaining shelf-life extension and quality of food during storage and transportation. However, the application of nanotechnology in food packaging is not without challenges. Concerns regarding the safety and potential migration of nanoparticles into food products pose significant regulatory and consumer acceptance issues. The need for comprehensive safety assessments and standardised regulations is critical to addressing these concerns and ensuring the safe use of nanotechnology in food applications. In this chapter, we discuss the properties and functionalities of packaging with nanoparticles for food. There is further discussion on the production of

R. K. Seidu (✉) · S. Jiang (✉)
School of Fashion and Textiles, The Hong Kong Polytechnic University, Kowloon, Hong Kong, China
e-mail: raphael-kanyire.seidu@connect.polyu.hk; seiduraphael@gmail.com

S. Jiang
e-mail: kinor.j@polyu.edu.hk

G. K. Fobiri
Department of Fashion Design and Textiles Studies, Kumasi Technical University, Kumasi, Ghana

J. Rani
Parul Institute of Design, Parul University, Vadodara, Gujarat, India

B. Tawiah
Department of Industrial Art, Kwame Nkrumah University of Science and Technology, Kumasi, Ghana

S. Jiang
Research Centre of Textiles for Future Fashion, The Hong Kong Polytechnic University, Kowloon, Hong Kong, China

S. Roy and M. Shahid (eds.), *Biomaterials for Sustainable Food Packaging*, Environmental Footprints and Eco-design of Products and Processes,
https://doi.org/10.1007/978-3-032-08027-1_14

intelligent and active packaging systems for food using nanoparticles and nanocomposites. Several challenges are also highlighted that limit the implementation of these packaging materials. Key regulatory and safety issues, such as economic feasibility for nanotechnology-based packaging materials, together with consumer acceptance, toxicity and life cycle assessments, have been discussed. This chapter concludes by discussing the innovative approaches to the applications of nanotechnology in creating sustainable and effective packaging of food, making a compelling case for continued research and development.

Keywords Nanotechnology · Toxicity · Shelf life · Food safety · Food packaging

14.1 Introduction

The quality and appearance of edible products are crucial in the food industry for marketing and proper preservation. During storage and distribution, effective preservation and safety of food products remain important parameters for product quality and long shelf life. To ensure consumer satisfaction and improve food quality and delivery, companies have invested in research and development geared towards quality packaging (Ansari 2023). Food packaging materials are engineered to ensure the proper preservation and safety of food products while preventing any possible health threats to consumers. Over the years, plastics from synthetic petrochemical-based materials have been widely applied for various food packaging because of their low cost, good mechanical properties, and availability. However, these petroleum-based plastics have been found to pose health and environmental threats (Shaikh et al. 2021). Although recent developments have led to significant advancements in food packaging materials, most of these materials lack critical functional properties such as barrier, moisture, UV barriers, gas, and antimicrobial, amongst others (Azeredo 2009; Nath et al. 2022), to ensure the prolonged shelf life and quality of food products for consumption on the market.

Nanotechnology has been leveraged in recent years to provide efficient active, intelligent, and smart food packaging solutions to ensure food safety and preservation for suppliers and consumers (He et al. 2019). Nanotechnology involves the manipulation of nanoparticles at the nanoscale from 1 to 100 nm, which exhibit unique properties (such as thermal, chemical stability, and improved strength) for varying application areas (Tawiah et al. 2024). The application of nanotechnology has advanced the fabrication of active and intelligent food packaging materials that offer several advantages, such as improved durability, mechanical strength, antimicrobial and barrier properties (del Rosario Herrera-Rivera et al. 2024; Honarvar et al. 2016), effective real-time monitoring solutions, safety, and prolonged shelf life (de Sousa et al. 2023; Newton 2023). These innovations not only enhance the preservation of food products but also contribute to sustainability and consumer safety. For instance, nanomaterials such as nano-silica, nano clays, and silver nanoparticles are increasingly utilised to improve barrier properties, mechanical strength, and antimicrobial activity (Mihindukulasuriya and Lim 2014). These enhancements play a crucial role in extending the shelf life of food products and maintaining their quality. Also, the incorporation of nano-silica in polymer matrices can significantly improve oxygen and moisture barrier properties, reducing the permeability of packaging materials (Thanakkasaranee et al. 2012). This is particularly significant for perishable items, as it minimises spoilage and waste. Additionally, silver nanoparticles exhibit potent antimicrobial properties, which can inhibit the growth of pathogens and spoilage organisms on packaging surfaces, thereby ensuring food safety (Simbine et al. 2019). The combination of nanotechnology and digital technologies has made it possible to improve the functional properties of packing materials as well as effectively trace, track, and monitor preservation quality remotely to improve food delivery and ensure consumer safety (Rossi et al. 2017; Polyakov and Gordeeva 2020). Digital technologies in this regard encompass a range of innovations, including the Internet of Things (IoT), smart sensors, and blockchain. These tools facilitate real-time monitoring and tracking of food products throughout the supply chain. Smart packaging, equipped with sensors, can provide information on temperature, humidity, and gas composition, allowing for better management of food quality and safety (Lydekaityte and Tambo 2018). For instance, temperature-sensitive labels can alert consumers if a product has been exposed to conditions that compromise its safety. Also, blockchain technology enhances traceability, enabling consumers to verify the origin and handling of food products (Yang et al. 2024). This transparency fosters trust and accountability in the food supply chain. In light of this exciting development, numerous research on the application of nano and digital technologies in food packaging materials for improved food delivery has been on the rise in recent years as shown in Fig. 14.1. Data extracted from the Scopus database shows an increase in research in the field to advance food packaging materials using nanotechnology, with the top contributing countries being India, the United States, and China amongst others.

Aside from these advantages of food packaging with nanomaterials, there are several key challenges and safety concerns surrounding the incorporation of nanoparticles in food packaging materials (Enescu et al. 2019). These are important since the issue of toxicity and migration of nanoparticles into foods, as well as their possible

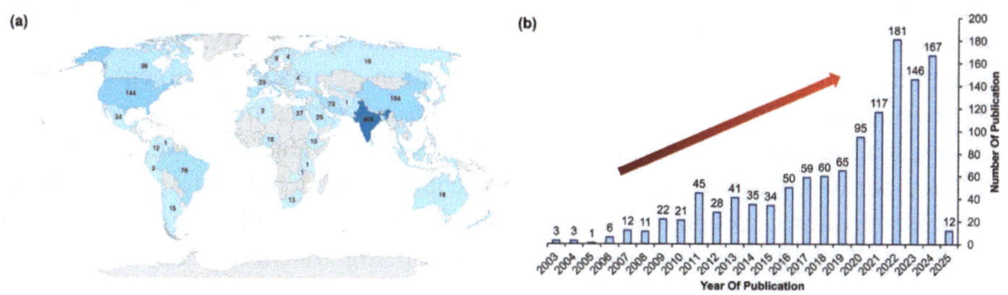

Fig. 14.1 **a** Top contributing countries in the field, **b** publication trends in food packaging materials using nanotechnology

environmental impact and regulatory requirements, remain unexplored in most cases (Alfei et al. 2020). There are concerns about the possible short-term to long-term effects of nanotechnology-based food packaging materials, although the data available does not explicitly support such fears (Ahmad et al. 2024). Therefore, it is imperative to further investigate any potential long-term harmful effects of nanotechnology applications in food packaging to effectively unlock key sustainable approaches to adopt to help limit any potential negative threats posed by this novel technology to advance food safety and preservation in the markets. This chapter thus highlights the prospects and applications of nanoparticles and nanocomposites in food packaging. Further discussion on the safety and regulatory issues of applying nanotechnology is provided. Lastly, key research opportunities and directions are provided, which form part of efforts to overcome challenges and foster advanced frontiers to sustainably apply nanoparticles in packaging for food.

14.2 Prospects of Nanotechnology in Food Packaging

Nanotechnology holds significant promise for revolutionising food packaging by enhancing quality, shelf life and safety. The incorporation of nanoparticles into packaging materials offers unique functional properties that address various safety and preservation challenges in the food industry. These advancements improve the shelf life extension and safety of food products by preventing microbial contamination and spoilage (Pattnaik et al. 2024). Applications of nanotechnology in food packaging include active packaging systems that release preservatives or absorb unwanted substances like oxygen or moisture to maintain food quality. Intelligent packaging incorporates nano-based sensors that monitor food conditions, providing real-time information about freshness and spoilage, helping consumers make informed decisions, and reducing food waste (Talambedu et al. 2025). Additionally, nanotechnology enhances biodegradable packaging, making it a viable alternative to traditional plastics by engineering materials that decompose safely and reduce environmental impact (Shrivastav et al. 2025).

There are significant benefits to applying nanotechnology in packaging. Nanoparticles with antimicrobial properties can inhibit bacteria and pathogens' growth, enhancing food safety (Astha et al. 2024). The barrier property and mechanical strength are improved, ensuring effective protection against environmental factors and physical damage. Furthermore, nanotechnology creates moisture and oxygen barriers, helping maintain the nutritional and freshness value of food products for extended periods. Despite these advantages, challenges such as safety and toxicity concerns, consumer acceptance, and regulatory issues still need to be resolved. The possibility of engineered nanoparticles (ENPs) being transferred into food from packaging raises safety concerns, necessitating thorough toxicological assessments. Public perception and acceptance of food packaging with nanoparticles also play a crucial role in its successful commercialisation. Regulatory and ethical issues, including safety, labelling, and transparency, require robust frameworks to build consumer trust.

14.2.1 Enhanced Barrier Properties

Nanotechnology has significantly enhanced the barrier properties of food packaging materials, offering improved protection against environmental factors such as oxygen, moisture, and microbial contamination. This advancement is crucial to prolong and safeguard food's shelf life and quality. The integration of nanomaterials into biopolymer matrices has led to innovative packaging solutions that are not only effective but also environmentally friendly. Below, the key aspects of these enhancements are discussed.

- *Oxygen and Moisture Barrier*: The incorporation of nanomaterials, such as silver-embedded silica particles, into polylactic acid (PLA) films has been shown to significantly reduce oxygen and water vapour permeability, thereby enhancing the barrier properties of the packaging. This improvement is crucial for preserving the freshness of fruits like *Muntingia calabura* over extended periods (Pradhan et al. 2024).
- *Gas Barrier Improvements*: Nanocomposite films, such as those made from modified cellulose nanocrystals and polycaprolactone, have demonstrated a dramatic reduction in the permeability of oxygen, carbon dioxide, and water vapour, with reductions of up to 97.6%, 96.7%, and 49.8%, respectively. This is achieved through the dense packing and structural integrity provided by the nanomaterials (Alkassfarity et al. 2024).
- *UV and Thermal Barrier*: The addition of chitosan nanoparticles to chitosan-based films enhances their UV-blocking and thermal properties, which are essential for protecting food from light-induced degradation and maintaining stability under varying temperatures (Wardana et al. 2024).

14.2.2 Antimicrobial and Mechanical Strength Enhancement

The integration of nanoparticles into its polymer matrix leads to improved mechanical strength and antimicrobial properties, which increases the application of fabricated materials for food packaging. This leads to the extended life, protection, and safety of food items within the food industry.

Antimicrobial Properties: The use of nanotechnology for food packaging replaces the traditional approach of salting, heating, fermentation, drying and freezing (Gustafson and Wilkinson 2005). Nanocomposites incorporating antimicrobial agents such as bacteriocins and silver nanoparticles have shown increased effectiveness against foodborne pathogens. For instance, silver nanoparticles combined with bacteriocins have demonstrated a 99.6% increase in antibacterial activity against

Staphylococcus aureus (Correia et al. 2024). The use of these nanoparticles is either integrated into the packaging products or directly on food to limit the growth of microbial pathogens (Malhotra et al. 2015). Integrating the nanoparticles into packaging ensures their periodic release for shelf-life extension and food quality. This draws several concerns from consumers on the safety of the food items; hence, advancements are made for antimicrobial agents from natural sources (Jayakumar et al. 2024). A variety of natural sources serving antimicrobial agents, such as modified clay minerals, isothiocyanates, lactic and sorbic acids, antimicrobial peptides and essential oils, are utilised to fabricate active food packaging materials to limit and inhibit the growth of bacteria and pathogens (Duda-Chodak et al. 2023; Popa et al. 2022).

Mechanical Strength: The integration of nanoparticles into the polymer matrix enhances the mechanical strength of the food packaging materials. This improves the structural integrity, making it resistant and durable to any damage. Integrating nanomaterials like hydroxyapatite nanoparticles derived from eggshells into chitosan films has resulted in a significant increase in tensile strength, enhancing the mechanical robustness of the packaging. This is vital for maintaining the integrity of the packaging during handling and transportation (Upadhyay and Ullah 2024). Other studies have reported that the incorporation of chitosan nanocomposite increased the tensile strength of fabricated films for food packaging (Xu et al. 2018; Carvalho et al. 2018).

14.2.3 Environmental and Sustainability Considerations

Biodegradability and Sustainability: The use of biopolymers such as PLA and chitosan, combined with nanotechnology, addresses environmental concerns associated with traditional plastic packaging. These materials are biodegradable and derived from renewable resources, offering a sustainable alternative to petrochemical-based plastics (Mishra et al. 2024).

Cost-Effectiveness: The incorporation of cost-effective materials like rice husk silica in nanocomposites not only enhances barrier properties but also reduces the overall cost of production, making these advanced packaging solutions more accessible (Pradhan et al. 2024). While the advancements in nanotechnology for food packaging are promising, some challenges and considerations need to be addressed. The possible risks of the migration of nanoparticles into foods and their impact on human health require thorough investigation. Additionally, the cost of nanotechnology and the scalability of production processes remain significant hurdles for widespread adoption. Further research advancement and development are essential to overcome these challenges and fully realise the potential of nanotechnology in creating safe, effective, and sustainable food packaging solutions (Correia et al. 2024).

14.3 Applications of Nanotechnology in Food Packaging

The application of nanomaterials offers unique solutions that enhance packaging properties, which significantly influence innovation in the food industry. These properties are imperative since a packaging material Couch et al. (2016) must have efficient strength, moisture, and gas permeability. The application of nanoparticles enhanced the properties of food packaging materials with improved durability, mechanical strength, antimicrobial, and barrier properties (del Rosario Herrera-Rivera et al. 2024; Honarvar et al. 2016). Aside from these, the effective use of nanoparticles enhances the development of critical active, intelligent and smart systems or solutions that interact with food products for functional performance and improved food quality. The latter ensures real-time monitoring solutions, safety, and extends their shelf life (de Sousa et al. 2023; Newton 2023). The release of preservatives extends food quality and prevents contamination, which poses harmful threats to consumers. These packaging materials further provide information about their functionality to limit food spoilage. Such benefits are critical to revolutionising packaging materials for food products in the industry. Herein, key highlights are provided for the use of nanocomposites and nanoparticles, active and smart solutions, and some implementation challenges in food packaging.

14.3.1 Nanocomposites and Nanoparticles in Food Packaging

The shortcomings in traditional packaging materials for food products, such as reduced shelf life and lack of functional properties like barrier, moisture, antimicrobial, gas and UV barrier, amongst others (Azeredo 2009; Nath et al. 2022). To overcome these, nanoparticles and nanocomposites are fabricated for improved properties of food packaging to enhance food quality and preservation. The use of composite materials for nanocomposites has advanced the technology in the application of nanoparticles, where at least 100 nm is found in the filler (Arora and Padua 2010) or about 10^{-9} m nanoscopic size for one of the components (Shalan et al. 2022). Two or more composite materials are combined or blended to form nanocomposites with improved properties and characteristics. Alternatively, nanoparticles are incorporated with a polymer matrix to enhance performance. These nanocomposites when applied to food packaging materials, enhanced their functionality or performance, making them a good alternative for food preservation. Furthermore, their shelf life is improved (Nath et al. 2022), barrier, thermal and mechanical properties are also improved with the application of nanocomposites (Avella et al. 2005). The adoption of appropriate fabrication methods for film-forming and coatings of nanocomposites is imperative to produce packaging materials for food with unique mechanical and chemical properties. Fabrication methods such as in-situ polymerisation, extrusion, layer-by-layer assembly and direct casting (Table 14.1) are among the most promising approaches to combine additives and polymers for nanocomposite coatings or films (Wang et al. 2018).

Table 14.1 Some fabrication methods for nanocomposite films

Fabrication methods	Description	Refs.
In-situ polymerisation	Herein, nanoparticles are mixed with monomer in a solvent, where there is a unique dispersion of the monomer in the nanoparticles Furthermore, the application of temperature and time causes the start of the polymerisation process, which enhances a stronger interaction	Shen et al. (2002), Nath et al. (2022)
Layer-by-layer assembly	This assembly technique requires non-specialised equipment for multicomponent film synthesisation, where the material and functional properties of the nanocomposite films are controlled In the production process, a critical parameter, pH, that "governs the electrical charge present in each layer", influences the roughness, modulus and thickness of the nanocomposites	Acevedo-Fani et al. (2015), Priyadarshi and Rhim (2020)
Extrusion	This is a process that ensures highly transparent nanocomposites, requires low energy and is a low-cost technique. The fabrication of the nanocomposite films undergoes key stages as; • Film composition preparation • Thorough mixing of the solution • Extruding the mixture solution guided by moisture and temperature conditions • Forming of pallets with a pelletizer • Effective drying of the pallets under a fixed temperature • Loading the dried pallets into the extruder to form a flat sheet • Forming of the blown film from the blown film extruder	Wang et al. (2018), Nath et al. (2022)
Direct casting	This is a widely common technique used for the fabrication of stable nanocomposite films with limited cracks or defects. The fabrication of the nanocomposite films undergoes key stages as; • The dissolution of bio-polymers in solvents with active bio-compounds like fillers or plasticizers • Thorough mixing of the solution to achieve homogeneity • The removal of insoluble particles and air bubbles from the solution through filtration • Transfer the solution into a mould • The solution is dried using the required humidity and temperature • The cast films are removed from the surface and stored effectively	Siemann (2005), Shankar and Rhim (2018)

These fabrication techniques are used with several nanoparticles such as silver nanoparticles (Ag), titanium dioxide (TiO_2), zinc oxide (ZnO), and silica (SiO_2), amongst others, to produce nanocomposites with unique characteristics and properties for food packaging materials. The adoption of greener methods for AGNPs effectively increases their safe application in food packaging materials (Shankar and Rhim 2017). Aside from this, Ag nanoparticles incorporated in the matrix of PLA nanocomposites limited micro-organism growth, delayed ripening, and enhanced the quality and shelf life of mangoes for about 15 days after post-harvest (Chi et al. 2019). There is also a prolonged shelf life and inhibition of bacterial growth in chicken sausage when nanocomposite food packaging materials are fabricated from polyvinyl alcohol/clay/silver composition (Mathew et al. 2019). Aside from the microbial properties, water vapour barrier and UV-blocking properties are further confirmed on food packaging materials fabricated from pectin/AgNPs composite films (Shankar et al. 2016), which promotes food safety and quality. Further investigations suggested that the incorporation of PLA polymer with AgNPs produced composite films with enhanced UV-screening abilities and water vapour barrier for food safety (Shankar et al. 2018). UV-blocking properties were further reported for food packaging from guar gum/Ag-copper alloy nanocomposites (Arfat et al. 2017). It is thus imperative that, due to the inherent properties of antibacterial or antimicrobial and UV blocking of Ag nanoparticles, their incorporation in nanocomposites enhances the performance abilities of packaging material for food products. Alternatively, zinc oxide (ZnO) has effective barrier properties, thermal stability, mechanical strength, high transparency, and antimicrobial properties for use in food packaging (Espitia et al. 2012; Bradley et al. 2011). For example, there are enhanced thermal, mechanical and barrier properties with ZnO incorporated in nanocomposites (Kumar et al. 2019). Aside from this, there is bacterial growth retardation and shelf life extension for food products such as orange juice using AG/ZnO nanocomposites (Emamifar et al. 2011), and cut apples using PVC-ZnO nanocomposites (Li et al. 2011). Table 14.2 further provides investigations on the incorporation of nanoparticles for enhanced performance of food packaging.

Subsequently, nanoparticles are composed of a nano-scale size range of 1–100 nm (Kim et al. 2010; Seidu et al. 2024) with unique characteristics. Owing to the distinctive characteristics of nanoparticles, such as oxidation prevention, antibacterial properties and UV-blocking abilities, their application for food packaging has been advanced in recent times to enhance the quality, safety, and long shelf life of food products (Handford et al. 2014; Silvestre et al. 2011; Barage et al. 2022). With such unique properties, nanoparticles are good for producing UV protection, antibacterial and oxidation protection food packaging materials. Nanomaterials such as starch nanocrystals, cellulose nano whiskers, carbon nanotubes, nano zinc oxide, nano titanium dioxide, nano clay and silver nanoparticles are commonly applied as functional additives for packaging materials in the food industry (Barage et al. 2022). ZnO is less harmful to people and animals, less expensive, has bacterial destruction properties, electrical conductivity and a high surface-to-volume ratio (Duncan 2011; Emamifar et al. 2010). Thus, microbial growth in food is reduced when ZnO nanoparticles are introduced into packaging materials for food products (Kazemi et al. 2020). For

Table 14.2 Nanocomposites for improved performance of food packaging

Composition	Materials and fabrication process	Properties	Refs.
LDPE-SiO_2 nanocomposite	Materials used are Low-density polyethylene (LDPE) and Nano-silicon dioxide (nano-SiO_2) Herein, modified nano-SiO_2 particles treated with NDZ-201 titanate cross-link reagent were mixed with LDPE for 10 min using a high-speed mixer. Upon cooling, the blend was extruded using a twin-screw extruder and further granulated	Freshness and Shelf life of pacific white shrimp	Luo et al. (2015)
Nanocomposite incorporated with Ag/TiO_2 and SiO_2	Materials used are Low-density polyethylene (LDPE) and nano-powders such as nano-SiO_2, attapulgite, nano-TiO_2 and nano-Ag Herein, there is a uniform mixture of the nanoparticles in the polymer matrix (PE) using melt blending, which is subsequently heated into a molten state. Fine packaging materials are formed by extruding, cooling and solidifying to form thin films	Preservation quality and Shelf life of mushrooms	Donglu et al. (2016)
Soy protein isolate (SPI)/ nano-silica (nano-SiO_2)	Materials used are nano-SiO_2 and soy protein Isolate (SPI) Herein, a film solution is developed by mixing water (600 ml) with glycerol and SPI under a 10.0 pH using 1.0 N sodium hydroxide. The dispersion of nano-SiO_2 was done by adding nano-SiO_2 into the mixture, which was heated for 2 h at 60 °C using a 200 rpm stirrer in a water bath. Subsequently, casting films are formed by adding the mixed solution into the Silastic plates, which are then dried for 12 h at 60 °C. The dried films were removed and conditioned with 50% RH at 25 °C for 24 h	10 days shelf-life extension of *Flammulina velutipes*	Xu et al. (2019)

(continued)

Table 14.2 (continued)

Composition	Materials and fabrication process	Properties	Refs.
Black plum peel extract, TiO_2 incorporated in chitosan	Materials used are chitosan (CS), black plum peel extract (BPPE), nano-sized TiO_2, and 2,2-Diphenyl-1-picrylhydrazyl (DPPH) Herein, acetic acid (1%) of about 160 mL was dissolved with chitosan of 3.2 g to form a CS solution. The latter is then added with BPPE (of 8 g), nano-sized TiO_2 (of 8 g) and a combination of both at equal quantities separately. These solutions are further added with glycerol and stirred continuously. This is then dried for 2 days with RH of 50% and 35 °C in a chamber to form the films, which are then stored at RH of 50% and 20 °C until equilibrium	Improved water barrier and mechanical properties	Zhang et al. (2019)
TiO_2/low-density polyethylene nanocomposites	The materials used are TiO_2 and low-density polyethylene (LDPE) Herein, a melt blending technique was used to mix LDPE with TiO_2 with a twin screw extruder. The molten mixture is then extruded via a die, forming films which are then solidified	Improved barrier properties	Nasiri et al. (2012)
Nanocomposite films from PLA/ZnO/essential oil	The materials used are PLA, ZnO nanoparticles and essential oils such as *Menthe piperita* L. essential oil (MEO)and *Zataria multiflora* Boiss. essential oil (ZEO) Herein, the appropriate concentration of ZnO nanoparticles is mixed with PLA to form a uniform solution after dissolving in a solvent. Subsequently, the PLA/ZnO solution is blended with the essential oils. The thin films are formed by casting the mixture solution on a flat surface and dried in controlled conditions for solvent evaporation	Improved shelf life and quality if fish fillers	Heydari-Majd et al. (2019)

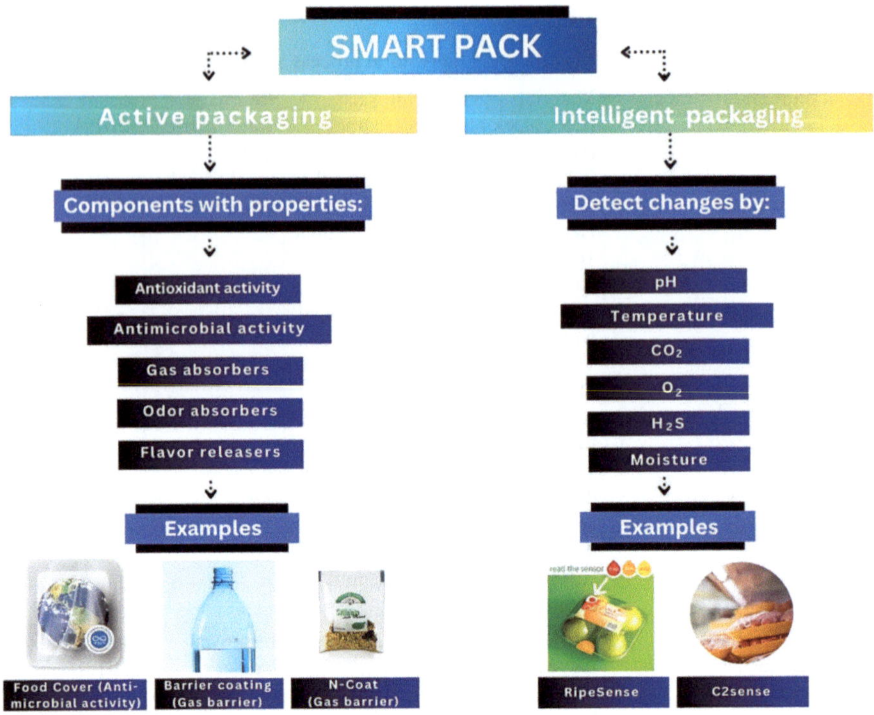

Fig. 14.2 Features of active and intelligent food packaging materials (de Sousa et al. 2023). This article is an open-access article distributed under the terms and conditions of the Creative Commons Attribution (CC BY) license

example, ZnO nanoparticles exhibited good antibacterial properties to limit bacterial growth in poultry meat (Akbar and Anal 2014).

The incorporation of nanoparticles and nanocomposites into packaging materials for the food industry has developed food packaging systems that are intelligent and active (Fig. 14.2) to monitor, improve quality, limit contamination and extend the shelf life of food products.

14.3.2 Active Food Packaging Systems

The use of packaging materials for food products plays a critical role in providing information for consumers and improving food quality and shelf life for consumers on the market. Due to the benefits of nanotechnology, active food packaging systems have been developed to effectively interact with the food, sense it, enhance its safety and promote its shelf life (Ahmed et al. 2022; Kadirvel et al. 2025). The application of nanoparticles helps in facilitating the provision of unique characteristics such as antimicrobial abilities, reduction of oxidative reactions and moisture absorption ability to prolong shelf life and limit spoilage. As such, these active food packaging systems are designed with key functionality for improved food quality and safety (Kraśniewska et al. 2020). These packaging systems are incorporated with active ingredients (Fig. 14.3) which boost the shelf life and quality of food products (Kadirvel et al. 2025). Furthermore, key strategies such as pads, sachets, coating and film are used to deliver active ingredients to the packaging materials. These active ingredients are released from the matrix using a common mechanism called diffusion. This release mechanism is further influenced by environmental, intermolecular interactions and microstructure factors.

14.3.3 Intelligent Food Packaging

Aside from the related benefits of food packaging systems, intelligent food packaging systems are developed to effectively provide data on food safety and quality through real-time monitoring, hence limiting food wastage (Kahyaoglu 2024). By adopting an active tracking system, indicators and sensors (Bhatlawande et al. 2024), intelligent food packaging is used for "recognition, monitoring, logging, detection and communication" (Yam et al. 2005), of key information about the food products for the consumers on the market.

These packaging solutions utilise interactive packaging, quality and gas sensors, tracking systems, time–temperature indicators and intelligent indicators for effective monitoring. For example, using key intelligent indicators like colour-changing labels, changes in the pH levels, composition of gas, humidity and temperature in a package can be monitored in real-time for effective detection and information provision. These alerts for consumers provide key information that may influence food quality. Alternatively, using time–temperature indicators provides a critical temperature history of perishable food products transported. This provides information for informed decisions by the consumers and suppliers on the safety and freshness of these products to prevent quality loss and microbial growth (Altaf et al. 2018). Furthermore, recorded history provides visual indicators that highlight any exposure of the food products to unfavourable conditions that could lead to food spoilage and the risk of consumption or sale to consumers. Similarly, the use of tracking systems like near-field communication and radio frequency identification technologies are integrated into food packaging to monitor, track and provide information on how these products have been distributed through the supply chain. This tracking information facilitates food safety, counterfeit reduction and traceability. Subsequently, the use of quality and gas sensors can easily measure and detect the presence of gases like carbon dioxide and oxygen, and other parameters like pH and humidity. This information influences the quality, ripeness and freshness of food products. Consumers can make informed choices on the freshness and quality of these products before purchase and consumption.

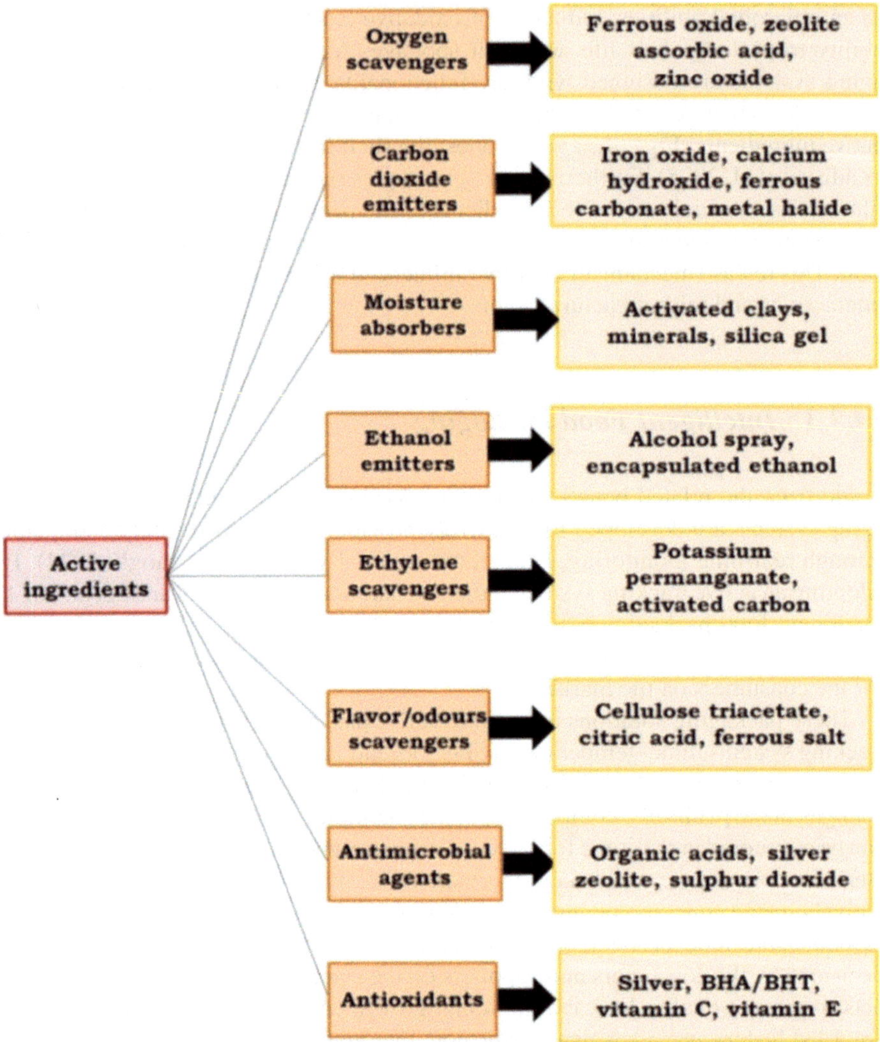

Fig. 14.3 Classification and examples of active ingredients for packaging systems. Reprinted/Adapted from ref (Kadirvel et al. 2025), Copyright 2024 John Wiley and Sons

Advancements have been made to develop intelligent food packaging systems to track, trace and provide critical information on food products for informed decisions by suppliers and consumers on the markets. For example, a colorimetric sensor fabricated with polydiacetylene/zinc oxide (PDA/ZnO) was used to measure real-time data on the freshness of milk (Weston et al. 2020). Herein, effective tracing of changes in pH and other parameters are monitored that could influence its fresh or spoiled state (Fig. 14.4). Other key studies detected milk spoilage using sensors from functionalised silver nanoparticles (Lakade et al. 2017), *salmonella* in milk using

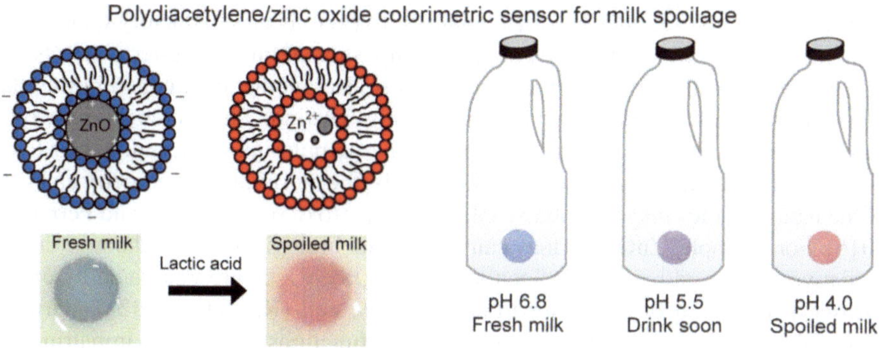

Fig. 14.4 Polydiacetylene/zinc oxide (PDA/ZnO) sensor to measure the freshness of milk. Reprinted/Adapted from ref (Weston et al. 2020) Copyright 2020 Elsevier

biosensor from Fe_3O_4 magnetic nanoparticles (Zou et al. 2019) and shelf life of milk using a fabricated colorimetric sensor from Schiff's reagent and SiO_2 nanoparticles (Ziyaina et al. 2019).

14.4 Challenges in Implementing Nanotechnology in Food Packaging

The key advantages of the advancement of nanocomposites and nanoparticles for packaging cannot be overemphasised. However, several challenges still limit the full utilisation of nanotechnology in food packaging for the industry. Key challenges such as safety concerns (migration of nanoparticles into foods), regulatory issues, toxicity and effects on human health influence the full implementation of nanotechnology.

Food quality is imperative to help limit any threats to human health. The issue of migration of nanoparticles into foods for consumers has remained a potential health safety concern (Bradley et al. 2011). There is possible oxidative damage, and inflammatory reactions due to the ability of nanoparticles to cross cellular barriers, when these NPs are ingested, penetrate through the skin and are inhaled by any organism (Kim et al. 2022). As such, the use of NPs for packaging materials exposes the possibility of these NPs encountering food products for migration, which are subsequently consumed by people. Microwave cooking and ingestion are the common routes for nanoparticle migration into foods, which are then consumed into the human body. The use of microwave heating for food packaging causes the rapid release of nanoparticles into foods (Echegoyen and Nerín 2013). This implies that, with an increase in temperature, there is a higher migration of nanoparticles due to nanoparticles solubility in aqueous solution (Song et al. 2011). This thus raised critical questions on the handling of these food packaging materials from nanoparticles to limit the migration of nanoparticles into foods for possible contamination that leads to potential health threats to humans. Another key area would be to establish a satisfactory quantity of nanoparticle migration into foods to prevent any short-term to long-term effects.

Subsequently, the toxicity of nanoparticles is also another challenge limiting the application of nanotechnology in food packaging. The individual's sensitivity, exposure duration, concentration and type of the nanoparticles may determine their toxicity levels (Dimitrijevic et al. 2015). Additionally, the morphology, geometry and size of the nanoparticles may potentially influence their toxicology (Souza and Fernando 2016). For example, ZnO NPs may cause potential biological system damage due to oxidative stress and reactive oxygen species (ROS), which are generated by endocytosis or diffusion of the nanoparticles into the cell (Kim et al. 2022). Furthermore, DNA damage, inflammation, organelle dysfunctions, and protein impairments, amongst others, are the biological impacts of ZnO nanoparticles. These nanoparticles further have environmental impacts on plants, microbial activity of soil, animals and the soil (Smaoui et al. 2023). Figure 14.5a, b illustrate the impact on soil and the human body from nanoparticles.

14.5 Safety Concerns and Regulatory Issues

The application of nanotechnology in the food industry has seen some critiques regarding safety as far as environmental and human health are concerned. A recent study posits that the use of nanomaterials such as silver nanoparticles (AgNPs), zinc oxide nanoparticles (ZnO NPs) and titanium dioxide nanoparticles (TiO_2 NPs), in food packaging causes health disorders, including inflammation, oxidative stress, genotoxicity, etc. (Gupta et al. 2024a). Also, graphene oxide is known for its ability to serve as reinforcement materials in food packages and preserve food from spoilage through its antioxidant and antimicrobial properties (Han Lyn and Nur Hanani 2022). Nonetheless, graphene-based nanomaterials cause health disorders regarding pulmonary toxicology upon excessive exposure and inhalation (Roberts et al. 2023). With several cutting-edge innovations observed in nanotechnology and food packaging, ethical and regulatory concerns remain, as far as oral consumption of these materials comes with health implications (Singh and Kumar 2023). This also has some implications for consumers' willingness to accept the technology in advanced food packaging. It is established that the uncertainties about the health impact of nanomaterials hinder their acceptance in food packaging by many consumers (Dimitrijevic et al. 2015). Allan et al. (2021) posit that there is a lack of harmonised regulatory definitions globally concerning the safe adoption of nanotechnology in the food industry, leading to difficulty in accepting the concept as far as safety is concerned. Rathore and Mahesh (2021) however, emphasise that steps are being taken globally to establish regulations to ensure the safe use of nanotechnology, with the United States of America leading in this initiative. Apart from safety concerns, social norms, and psychological factors come into play in consumer acceptance of nanotechnology

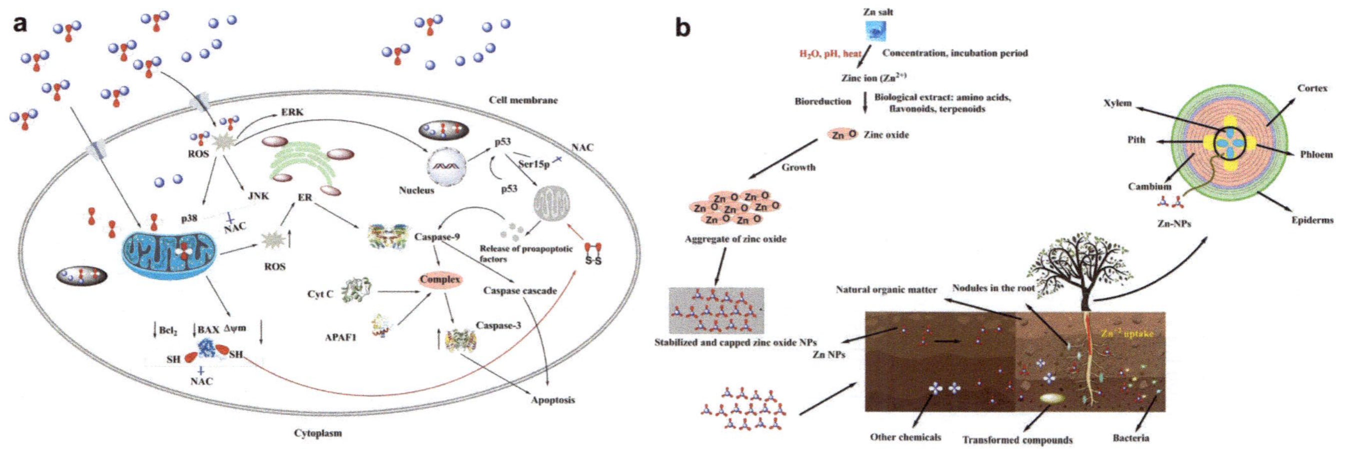

Fig. 14.5 **a** Toxicity mechanism effects of ZnO nanoparticles in the cells of the human body, **b** Effects of ZnO nanoparticles in the soil and plant stem. Reprinted/Adapted from Ref. Smaoui et al. (2023). Copyright 2023 Elsevier

in food packaging (Siddiqui et al. 2022). It is also established that the lack of knowledge regarding nanotechnology and the habit of opposing new technologies influence food packaging prospects in the industry (dos Santos Gonçalves et al. 2025). Addressing the regulatory and safety challenges surrounding nanomaterials incorporation for food packaging is crucial to unearthing their future applications. Focusing on consumer acceptance, economic feasibility, the life cycle, and toxicity assessment of packaging would help position the concept of nanotechnology in a favourable and attractive state for increased adoption.

14.5.1 Consumer Acceptance

Nanotechnology applications in food packaging hope to see the light of day upon full acceptance by consumers with a conviction that their health is not threatened. Safety assessment is a crucial phase of product development, especially for human consumption (Allan et al. 2021). This is evident in developed and developing countries where trust in technology positively influences the acceptance of a given technology (Rathore and Mahesh 2021). In situations where society remains sceptical about good health, environmental safety, and the authenticity of regulatory frameworks established to guide and approve innovative concepts in the food industry, alternative products other than nanotechnology-enhanced packages would be consumed. It is important to remain as open as possible concerning the safety of consumers regarding nanotechnology packaging for a sustainable food industry. This goes to say that food choices could be influenced by the technology adopted for packaging (Henchion et al. 2019). It is established that food packages designed with nanotechnology have gained massive acceptance over the years by farmers, food package manufacturers, and consumers, resulting in a larger portion of the food market witnessing nanotechnology-based packages with a focus on enhancing the protection/safety, ease of use, containment, and the informative abilities of food packages (Pathiraja and Munaweera 2024; Sharma et al. 2017; Unalan et al. 2014; Vanderroost et al. 2014). A study investigating consumer acceptance of nanotechnology in food packaging revealed that the perceived benefits of nanotechnology foods strongly influence consumers' willingness to accept such foods, whereas perceived risks have a weak influence (Stampfli et al. 2010). A similar study conducted by (Gupta et al. 2012) gave rise to the same results, where the societal acceptance of nanomaterial-based products is influenced by perceived usefulness. This presupposes that, although consumers have concerns about the risks involved in consuming food and products packaged with nanomaterials, their focus mainly lies in the benefits offered by such packages, including shelf life extension, food quality preservation, serving as barriers to permeability of O_2, CO_2, and moisture, aesthetic appeal, etc. (Biswas et al. 2022; Sridhar et al. 2021). It should be noted that the wide acceptance of nanotechnology-packaged foods does not override the need to establish regulatory issues to ensure the safe use of nanomaterials. The migration of particles into foods, a major drawback, must be addressed to curtail health disorders caused by this phenomenon.

14.5.2 Economic Feasibility

Nanomaterials are expensive as they need special treatment to be transported and stored for manufacturing larger-scale products (Yadav et al. 2023). Apart from the cost involved in transporting and storing nanomaterials, maintaining safe manufacturing grounds and production processes for nanomaterial-based products demands special attention, which does not come without financial implications. Also, the cost of investing in safe approaches to nanotechnology adoption in food packaging significantly affects regulatory compliance and consumer acceptance. Pre-consumption research and testing are important in combating the release of fake products into the market. The requisite limit of nanomaterials and toxicity rates is essential when adopting nanotechnology for food packaging (Ashfaq et al. 2022). However, the economic implications of conducting extensive research on toxicity and other significant tests, such as migration tests on packaged food, are very high. This goes a long way to influencing the price of products packaged through nanotechnology concepts. Products that have undergone a series of these pre-consumption tests are likely to be expensive, which subsequently would cause a decline in acceptance in a cost-sensitive market. It was predicted in 2011 that nanotechnology would be worth about $3 trillion, while manufacturers would require more than 6 million labourers by 2020 (Ill-Min et al. 2017; Roco et al. 2011). This shows how cost-demanding the technology is. To integrate nanomaterials into the mainstream of food packaging, the economic implications raise concerns but are significantly compensated with desirable features of the smart package including, barrier for gas, moisture UV rays, and ease of use of products (Kumar et al. 2021; Pathiraja and Munaweera 2024). Irrespective of the high cost of developing and manufacturing nanotechnology-based packaging materials, strategic approaches to ensure the mass production of such packages would cut down the cost involved. Generally, it is posited that using nanomaterials in product design is expensive, but strategic cost management significantly reduces the cost of manufacturing (Ali et al. 2022). Relocating indirect production costs is a strategic management step to make nanotechnology adoption economically feasible.

14.5.3 Life Cycle and Toxicity Assessments for Food Packaging

The life cycle assessment of food packages manufactured through nanotechnology has been studied recently as part of ethical and safety approaches to food processing and commercialisation. Life cycle assessment (LCA) is a tool designed to evaluate the challenges posed to the environment when nanomaterials are applied in product manufacturing, focusing on the harm caused by production through the various life stages till it is disposed of Oniha and Oyejide (2024). LCA provides several answers to questions regarding the impact of nanotechnology on human health and the environment, considering different products serving a common purpose (Klöpffer et al. 2007; Nizam et al. 2021). As a broad framework, it measures the environmental and human impacts of a product, considering its complete life cycle (Grant et al. 2015; Hemachandra et al. 2024; Hischier and Walser 2012). LCA is grounded on the International Organisation for Standardisation (ISO) standards 14,044, which projects four important components, including the definition of goal and scope, life cycle inventory analysis, life cycle impact assessment, and interpretation (Nizam et al. 2021; Pallas et al. 2020). Figure 14.6 shows a generic LCA framework. From the figure, the assessment begins by stating the goal for conducting the analysis and its scope. This helps to know the essence of the results, such as for product designing, policy-making, etc. The scope also defines the functional unit and determines the system boundaries, such as the processes, inputs, and outputs to be included. The second component (Life cycle inventory) involves a quantitative approach to assessing the impact of the products by compiling the inputs and outputs throughout their life cycle (Islam and Kumar 2019; Ramanathan et al. 2022; Suh and Huppes 2009). In the third phase, life cycle impact assessment, data collected under the inventory is translated into a set of possible impacts (Kikuchi and Kanematsu 2020; Singal et al. 2023; Torkayesh et al. 2022). The last phase of the framework (interpretation) involves results analysis and conclusions to help improve production, design strategic policies, and ensure sustainability (Mao et al. 2024; Omolayo et al. 2021). It could be said that the LCA framework is worth following to assess the environmental and human health impact of food packages designed through nanotechnology.

Several scholars have added their voices to the LCA in diverse fields as steps to evaluating the ecological effect and human health impact of products. Among these is a study evaluating the environmental impact of biofuels and their blends, focusing on biofuels obtained from waste vegetable oil (Dey and Ray 2022), LCA of bioenergy products made of microalgal and lignocellulosic biomass produced using various thermochemical processes (Ubando et al. 2019), the role of LCA and circularity indicators toward improved circular decision-making (Rigamonti and Mancini 2021), LCA in renewable energy, focusing on solar and wind, and stressing the need to consider social, economic, and environmental impacts in renewable energy assessment (Portillo et al. 2024), integrating qualitative scenarios into LCA to assess future technologies in a broader context, focusing on the agri-food sector (Voglhuber-Slavinsky et al. 2022).

There was further focus on the LCA of products manufactured specifically with nanotechnology. In relation to nanomaterial-based products, Pallas et al. (2020) performed LCA on nanowire-based solar cells and suggested using fluorine-free etching materials among other strategies to ensure sustainable production. Zhang et al. (2017) evaluated three packaging films with agents including silver (Ag) nanoparticles, titanium dioxide (TiO_2) nanoparticles, and Ag/TiO_2 nanoparticles composite using LCA. The results indicated a low environmental impact of the film comprising Ag/TiO_2 nanoparticle composites compared to the separate films (Ag nanoparticles and TiO_2 as distinct films). In the food packaging industry, Ros-Lis and Benitez (2024) analyse the advancements made in LCA regarding food packages

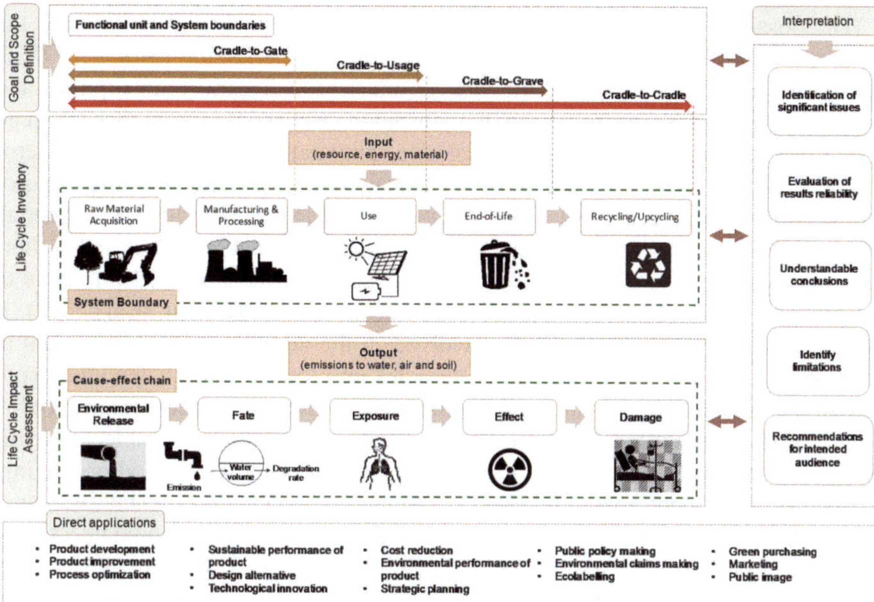

Fig. 14.6 The life cycle assessment framework (Nizam et al. 2021). This article is an open-access article distributed under the terms and conditions of the Creative Commons Attribution (CC BY) license

manufactured with nanomaterials. It is revealed from the study that nanoparticles such as titanium dioxide, clay, and silver can result in weight reduction of nanocomposite materials, shelf-life extension of fresh foods, and reduced food loss. With the cradle-to-grave approach, Camps-Posino et al. (2021) conducted LCA to understand the impact of food delivery packages on climate change in China. The findings indicate about 13 million tons of CO_2eq emissions. Further assessment indicates a possible reduction of 16% emission of single-use packages when the recycling rate is increased to 35%. As such, assessing the recyclability of nano-based food packages is relevant for increased adoption of and safe application of nanotechnology in food packaging. It is worth noting that other LCA techniques such as the social life cycle assessment (S-LCA) have been adopted to assess smart food packages for sustainable development (Kalita et al. 2024). The S-LCA measures the social and economic impact of products in their whole life cycle from material selection to final disposal (Bouillass et al. 2021; Yang et al. 2020). The S-LCA could be adopted to assess the economic, social acceptance, and impact of nanotechnology on sustainable food packaging as far as consumers are concerned. S-LCA broadly considers several impacts directly on people and not through the environment, as seen in LCA (Dyson 2024). The LCA framework can equally be used to devise several circular economy strategies toward sustainable production and consumption of nano-based products, of which food packages are not an exception, for wider acceptance (Peña et al. 2021; Rigamonti and Mancini 2021).

The possibility of particles migrating into food from packages necessitates the toxicity assessment aspect of nanotechnology in food packaging (Simonetti et al. 2024). A recent study reports the risk of nutrient indigestion, absorption, and metabolism regarding the migration of nano and microplastics in food (Kaseke et al. 2023). Several factors, including material type, temperature, contact time, the micromorphology of composite film, and food consumption, influence the migration of particles from food packages to food (Bastardo-Fernández et al. 2024; Gupta et al. 2024b; Hannon et al. 2016). These factors demand special attention and a thorough assessment to avoid health disorders caused by nanostructured materials to the human body and the environment upon accumulation (Bajpai et al. 2018). It is noted that the toxicological risks involved in the adoption of nanomaterials in food packaging lie in the migration of particles into food substances and the exposure of the human body to nanoparticles in the manufacturing process (Agriopoulou et al. 2021; Rovera et al. 2020). Toxicity assessment becomes valuable in identifying safe measures for applying and adopting nanotechnology in food packaging. In toxicity assessment, methodologies including investigating accidental disclosure to a compound, analysing with in vitro assays, and in vivo analysis could be adapted to test food packages from nanomaterials (Bhuia et al. 2023; Dorato and Buckley 2007). Following this, Carrillo-Romero et al. (2024) analysed the toxicity of metal and carbon-based nanomaterials and the risks posed to human health upon increased exposure, using the in vitro assay. The study successfully used the in vitro method to identify high toxicity and reactive oxygen species production of 3T3 and A549 cell lines when exposed to Zinc oxide. A comprehensive analysis of in vivo and in vitro toxicity assessment methods reveals their effective use in identifying the impact of nanoparticles on bacteria, humans, microalgae, zebrafish, fish, rats, mice, pigs, etc., which are worth adopting in the toxicity assessment of food packages (Kumar et al. 2017).

14.6 Conclusions and Research Opportunities

Safety concerns and regulatory issues surface when nanotechnology is mentioned. Health disorders, including inflammation, oxidative stress, and genotoxicity, have been linked to nanotechnology applications in food packaging (Gupta et al. 2024a). This is a result of the phenomenon of particle migration in nanomaterial-based food packages (Agriopoulou et al. 2021; Kaseke et al. 2023; Rovera et al. 2020). It has also been revealed that nanotechnology adoption in food packaging is expensive, but it could be reduced through strategic measures such as relocating indirect production costs (Ali et al. 2022). Notwithstanding these challenges, findings show the wide acceptance of nanotechnology in food packaging due to several benefits, including protection/safety, ease of use, containment, informative abilities, etc. The wide acceptance of nanotechnology calls for life cycle and toxicity assessment to ensure a safe and sustainable approach to nanomaterial-based food package design. Following this, the LCA framework is found to be an appropriate tool to ensure safe food packages are produced for consumers. Key phases of the framework such as goal and scope, life cycle inventory analysis, life cycle impact assessment, and interpretation help to enhance the environmental and human health impact of products, including nanomaterial-based food packages. Again, findings revealed that methodologies such as investigating accidental disclosure to a compound, analysing with in vitro assays, and in vivo are appropriate means of assessing the toxicity of food packages. These provide key opportunities for innovative solutions and sustainable approaches to advance the use of nanotechnology in food packaging. These key innovations in nanomaterials, such as enhanced barrier and mechanical properties, active and intelligent packaging solutions, and enhanced functional performance, have advanced their application for food packaging.

As a result, it is recommended for future research trends to comprehensively focus on five (5) key areas bio-degradable materials, safety considerations, reduction in material usage, functional performance and consumer awareness, as shown in Fig. 14.7. Environmental impacts from conventional food packaging materials remain a critical problem for the ecosystem. The use of biodegradable materials as one of the components for the fabrication of nanocomposites would drive a reduction in the threats to the environment at disposal. Future studies could advance the use of biodegradable materials such as starch, chitosan and cellulose in place of plastics to produce sustainable food packaging with functional performance for improved food quality and shelf life. Other areas as discovering large-scale production procedures at reduced cost, as proposed by Li et al. (2024) could also be advanced for the food industry. These studies should be guided by critical regulatory policies in the field to ensure that consumers are less likely to encounter health threats when they consume such food products wrapped in food packaging. Safety considerations for the application of nanotechnology in food packaging are important to limit the migration of nanoparticles into food products. Further research is needed to propose comprehensive safety frameworks that inform and guide manufacturers and consumers on the exposure levels of food products to nano-based food packaging. Alternatively, studies should also focus on the toxicological effects of nano-based food packaging on the markets will provide informed data on the appropriate temperature, pH and other environmental factors that could promote the migration of nanoparticles into the food products. This confirms future research studies by Primožič et al. (2021) and Bahadar et al. (2016). This way, consumers and suppliers are well informed on the ways to store these products to limit any health threats. Still, on the issue of limiting threats to human health and the environment, studies should be advanced to the fabrication of lighter but durable food packaging using nanotechnology without any compromise to functional performance. Herein, materials used for fabrication should be reduced but still achieve the functional performance of antibacterial ability, prolonged shelf life, and effective monitoring and tracking, amongst others. This leads to future studies to advance the frontiers of intelligent and active food packaging that effectively monitors conditions in real-time for improved food safety and preservation for suppliers and consumers. Herein, studies should develop integrated nanosensors for food packaging to ensure the full benefits are realised for the industry.

Fig. 14.7 Key research areas for the application of nanotechnology in food packaging

Lastly, future studies on consumer awareness of the effects of food packaging would foster effective understanding and utilisation for food safety.

Declaration Authors used generative AI tools to improve readability and language. The final content has been reviewed and verified by the authors, who take full responsibility for its accuracy and integrity.

References

Acevedo-Fani A, Salvia-Trujillo L, Soliva-Fortuny R, Martín-Belloso O (2015) Modulating biopolymer electrical charge to optimize the assembly of edible multilayer nanofilms by the layer-by-layer technique. Biomacromol 16(9):2895–2903. https://doi.org/10.1021/acs.biomac.5b00821

Agriopoulou S, Stamatelopoulou E, Skiada V, Tsarouhas P, Varzakas T (2021) Emerging nanomaterial applications for food packaging and preservation: safety issues and risk assessment. Proceedings 70 (1):7

Ahmad K, Li Y, Tu C, Guo Y, Yang X, Xia C, Hou H (2024) Nanotechnology in food packaging with implications for sustainable outlook and safety concerns. Food Biosci 103625

Ahmed MW, Haque MA, Mohibbullah M, Khan MSI, Islam MA, Mondal MHT, Ahmmed R (2022) A review on active packaging for quality and safety of foods: current trends, applications, prospects and challenges. Food Packag Shelf Life 33:100913. https://doi.org/10.1016/j.fpsl.2022.100913

Akbar A, Anal AK (2014) Zinc oxide nanoparticles loaded active packaging, a challenge study against Salmonella typhimurium and Staphylococcus aureus in ready-to-eat poultry meat. Food Control 38:88–95. https://doi.org/10.1016/j.foodcont.2013.09.065

Alfei S, Marengo B, Zuccari G (2020) Nanotechnology application in food packaging: a plethora of opportunities versus pending risks assessment and public concerns. Food Res Int 137:109664

Ali AMM, Jabir SM, Kadhim AA, Almagtome A (2022) Nanotechnology practices and cost restructure for effective cost management under industry 4.0 based manufacturing systems. TEM J 11(3)

Alkassfarity AN, Yassin MA, Abdel Rehim MH, Liu L, Jiao Z, Wang B, Wei Z (2024) Modified cellulose nanocrystals enhanced polycaprolactone multifunctional films with barrier, UV-blocking and antimicrobial properties for food packaging. Int J Biol Macromol 261:129871. https://doi.org/10.1016/j.ijbiomac.2024.129871

Allan J, Belz S, Hoeveler A, Hugas M, Okuda H, Patri A, Rauscher H, Silva P, Slikker W, Sokull-Kluettgen B, Tong W, Anklam E (2021) Regulatory landscape of nanotechnology and nanoplastics from a global perspective. Regul Toxicol Pharmacol 122:104885. https://doi.org/10.1016/j.yrtph.2021.104885

Altaf U, Kanojia V, Rouf A (2018) Novel packaging technology for food industry. Journal of Pharmacognosy and Phytochemistry 7(1):1618–1625

Ansari MA (2023) Nanotechnology in food and plant science: challenges and future prospects. Plants 12(13):2565

Arfat YA, Ejaz M, Jacob H, Ahmed J (2017) Deciphering the potential of guar gum/Ag-Cu nanocomposite films as an active food packaging material. Carbohydr Polym 157:65–71. https://doi.org/10.1016/j.carbpol.2016.09.069

Arora A, Padua GW (2010) Review: nanocomposites in food packaging. J Food Sci 75(1):R43–R49. https://doi.org/10.1111/j.1750-3841.2009.01456.x

Ashfaq A, Khursheed N, Fatima S, Anjum Z, Younis K (2022) Application of nanotechnology in food packaging: pros and cons. J Agric Food Res 7:100270. https://doi.org/10.1016/j.jafr.2022.100270

Astha S, Aakriti P, Neha M (2024) Nano innovation: enhancing food packaging through nanotechnology. Curr Nanosci 20:1–15. https://doi.org/10.2174/0115734137314364240920052006

Avella M, De Vlieger JJ, Errico ME, Fischer S, Vacca P, Volpe MG (2005) Biodegradable starch/clay nanocomposite films for food packaging applications. Food Chem 93(3):467–474. https://doi.org/10.1016/j.foodchem.2004.10.024

de Azeredo HMC (2009) Nanocomposites for food packaging applications. Food Res Int 42(9):1240–1253. https://doi.org/10.1016/j.foodres.2009.03.019

Bahadar H, Maqbool F, Niaz K, Abdollahi M (2016) Toxicity of nanoparticles and an overview of current experimental models. Iran Biomed J 20(1):1–11. https://doi.org/10.7508/ibj.2016.01.001

Bajpai VK, Kamle M, Shukla S, Mahato DK, Chandra P, Hwang SK, Kumar P, Huh YS, Han Y-K (2018) Prospects of using nanotechnology for food preservation, safety, and security. J Food Drug Anal 26(4):1201–1214. https://doi.org/10.1016/j.jfda.2018.06.011

Barage S, Lakkakula J, Sharma A, Roy A, Alghamdi S, Almehmadi M, Hossain MJ, Allahyani M, Abdulaziz O (2022) Nanomaterial in food packaging: a comprehensive review. J Nanomat 2022(1):6053922. https://doi.org/10.1155/2022/6053922

Bastardo-Fernández I, Chekri R, Oster C, Thoury V, Fisicaro P, Jitaru P, Noireaux J (2024) Assessment of TiO_2 (nano)particles migration from food packaging materials to food simulants by single particle ICP-MS/MS using a high efficiency sample introduction system. NanoImpact 34:100503. https://doi.org/10.1016/j.impact.2024.100503

Bhatlawande AR, Ghatge PU, Shinde GU, Anushree RK, Patil SD (2024) Unlocking the future of smart food packaging: biosensors, IoT, and nano materials. Food Sci Biotechnol 33(5):1075–1091. https://doi.org/10.1007/s10068-023-01486-9

Bhuia MS, Siam MSH, Ahamed MR, Roy UK, Hossain MI, Rokonuzzman M, Islam T, Sharafat R, Bappi MH, Mia MN, Emamuzzaman M, de Almeida RS, Coutinho HDM, Raposo A, Alturki HA, Islam MT (2023) Toxicity analysis of some frequently used food processing chemicals using allium cepa biomonitoring system. Biology 12(5):637

Biswas R, Alam M, Sarkar A, Haque MI, Hasan MM, Hoque M (2022) Application of nanotechnology in food: processing, preservation, packaging and safety assessment. Heliyon 8 (11). https://doi.org/10.1016/j.heliyon.2022.e11795

Bouillass G, Blanc I, Perez-Lopez P (2021) Step-by-step social life cycle assessment framework: a participatory approach for the identification and prioritization of impact subcategories applied to mobility scenarios. Int J Life Cycle Assess 26(12):2408–2435. https://doi.org/10.1007/s11367-021-01988-w

Bradley EL, Castle L, Chaudhry Q (2011) Applications of nanomaterials in food packaging with a consideration of opportunities for developing countries. Trends Food Sci Technol 22(11):604–610. https://doi.org/10.1016/j.tifs.2011.01.002

Camps-Posino L, Batlle-Bayer L, Bala A, Song G, Qian H, Aldaco R, Xifré R, Fullana-i-Palmer P (2021) Potential climate benefits of reusable packaging in food delivery services. A Chinese case study. Sci Total Environ 794:148570. https://doi.org/10.1016/j.scitotenv.2021.148570

Carrillo-Romero J, Mentxaka G, García-Salvador A, Katsumiti A, Carregal-Romero S, Goñi-de-Cerio F (2024) Assessing the toxicity of metal- and carbon-based nanomaterials in vitro: impact on respiratory, intestinal, skin, and immune cell lines. Int J Mol Sci 25(20):10910

Carvalho RA, Santos TA, de Azevedo VM, Felix PHC, Dias MV, Borges SV (2018) Bio-nanocomposites for food packaging applications: effect of cellulose nanofibers on morphological, mechanical, optical and barrier properties. Polym Int 67(4):386–392. https://doi.org/10.1002/pi.5518

Chi H, Song S, Luo M, Zhang C, Li W, Li L, Qin Y (2019) Effect of PLA nanocomposite films containing bergamot essential oil, TiO_2 nanoparticles, and Ag nanoparticles on shelf life of mangoes. Sci Hortic 249:192–198. https://doi.org/10.1016/j.scienta.2019.01.059

Correia LF, da Silva PG, da Cruz Neves TJ, de Oliveira Vieira KC, Maddela NR, Prasad R, Winkelstroter LK (2024) Nanotechnology innovation combined with bacteriocins as emerging strategy for the development of active and intelligent food packaging. Sustain Chem Pharm 39:101551. https://doi.org/10.1016/j.scp.2024.101551

Couch LM, Wien M, Brown JL, Davidson P (2016) Food nanotechnology: proposed uses, safety concerns and regulations. Agro Food Ind Hitech 27:36–39

de Sousa MS, Schlogl AE, Estanislau FR, Souza VGL, dos Reis Coimbra JS, Santos IJB (2023) Nanotechnology in packaging for food industry: past, present, and future. Coatings 13(8):1411

del Rosario H-R, Torres-Arellanes SP, Cortés-Martínez CI, Navarro-Ibarra DC, Hernández-Sánchez L, Solis-Pomar F, Pérez-Tijerina E, Román-Doval R (2024) Nanotechnology in food packaging materials: Role and application of nanoparticles. RSC Adv 14(30):21832–21858

Dey P, Ray S (2022) Chapter 32—Environmental impacts of biofuels and their blends: a case study on waste vegetable oil-derived biofuel blends. In: Gurunathan B, Sahadevan R (eds) Biofuels and bioenergy. Elsevier, pp 755–770. https://doi.org/10.1016/B978-0-323-90040-9.00017-5

Dimitrijevic M, Karabasil N, Boskovic M, Teodorovic V, Vasilev D, Djordjevic V, Kilibarda N, Cobanovic N (2015) Safety aspects of nanotechnology applications in food packaging. Procedia Food Sci 5:57–60. https://doi.org/10.1016/j.profoo.2015.09.015

Donglu F, Wenjian Y, Kimatu BM, Mariga AM, Liyan Z, Xinxin A, Qiuhui H (2016) Effect of nanocomposite-based packaging on storage stability of mushrooms (Flammulina velutipes). Innov Food Sci Emerg Technol 33:489–497. https://doi.org/10.1016/j.ifset.2015.11.016

Dorato MA, Buckley LA (2007) toxicology testing in drug discovery and development. Curr Protocols Toxicol 31 (1):19.11.11–19.11.35. https://doi.org/10.1002/0471141755.tx1901s31

dos Santos Gonçalves RA, Araújo Martins IB, de Alcantara M, Lima MF, de Oliveira DCR, Santos GN, Deliza R (2025) Exploring the Brazilian consumer perceptions of nanotechnology in food. J Future Foods 5(5):496–502. https://doi.org/10.1016/j.jfutfo.2024.08.007

Duda-Chodak A, Tarko T, Petka-Poniatowska K (2023) Antimicrobial compounds in food packaging. Int J Mol Sci 24(3):2457

Duncan TV (2011) Applications of nanotechnology in food packaging and food safety: Barrier materials, antimicrobials and sensors. J Colloid Interface Sci 363(1):1–24. https://doi.org/10.1016/j.jcis.2011.07.017

Dyson B (2024) Integration of life cycle and life cycle sustainability assessments into decision analytic approaches for sustainable technologies. In: Abraham MA (ed) Encyclopedia of sustainable technologies, 2nd Ed. Elsevier, Oxford, pp 643–655. https://doi.org/10.1016/B978-0-323-90386-8.00057-7

Echegoyen Y, Nerín C (2013) Nanoparticle release from nano-silver antimicrobial food containers. Food Chem Toxicol 62:16–22. https://doi.org/10.1016/j.fct.2013.08.014

Emamifar A, Kadivar M, Shahedi M, Soleimanian-Zad S (2010) Evaluation of nanocomposite packaging containing Ag and ZnO on shelf life of fresh orange juice. Innov Food Sci Emerg Technol 11(4):742–748. https://doi.org/10.1016/j.ifset.2010.06.003

Emamifar A, Kadivar M, Shahedi M, Soleimanian-Zad S (2011) Effect of nanocomposite packaging containing Ag and ZnO on inactivation of Lactobacillus plantarum in orange juice. Food Control 22(3):408–413. https://doi.org/10.1016/j.foodcont.2010.09.011

Enescu D, Cerqueira MA, Fucinos P, Pastrana LM (2019) Recent advances and challenges on applications of nanotechnology in food packaging. A literature review. Food Chem Toxicol 134:110814

Espitia PJP, Soares NdFF, Coimbra JSdR, de Andrade NJ, Cruz RS, Medeiros EAA (2012) Zinc oxide nanoparticles: synthesis, antimicrobial activity and food packaging applications. Food Bioprocess Technol 5(5):1447–1464. https://doi.org/10.1007/s11947-012-0797-6

Grant T, Barichello V, Fitzpatrick L (2015) Accounting the impacts of waste product in package design. Procedia CIRP 29:568–572. https://doi.org/10.1016/j.procir.2015.02.062

Gupta N, Fischer ARH, van der Lans IA, Frewer LJ (2012) Factors influencing societal response of nanotechnology: an expert stakeholder analysis. J Nanopart Res 14(5):857. https://doi.org/10.1007/s11051-012-0857-x

Gupta RK, Guha P, Srivastav PP (2024a) Investigating the toxicological effects of nanomaterials in food packaging associated with human health and the environment. J Hazard Mater Lett 5:100125. https://doi.org/10.1016/j.hazl.2024.100125

Gupta RK, Pipliya S, Karunanithi S, Eswaran UGM, Kumar S, Mandliya S, Srivastav PP, Suthar T, Shaikh AM, Harsányi E, Kovács B (2024b) Migration of chemical compounds from packaging materials into packaged foods: interaction, mechanism, assessment, and regulations. Foods 13(19):3125

Gustafson J, Wilkinson B (2005) Staphylococcus aureus as a food pathogen: the staphylococcal enterotoxins and stress response systems. Understand Pathog Behav Virulence, Stress Response Resist 111:331

Han Lyn F, Nur Hanani ZA (2022) Graphene-based polymer nanocomposites in food packaging and factors affecting the behaviour of graphene-based materials: a review. J Nanopart Res 24(9):179. https://doi.org/10.1007/s11051-022-05558-5

Handford CE, Dean M, Henchion M, Spence M, Elliott CT, Campbell K (2014) Implications of nanotechnology for the agri-food industry: opportunities, benefits and risks. Trends Food Sci Technol 40(2):226–241. https://doi.org/10.1016/j.tifs.2014.09.007

Hannon JC, Kerry JP, Cruz-Romero M, Azlin-Hasim S, Morris M, Cummins E (2016) Assessment of the migration potential of nanosilver from nanoparticle-coated low-density polyethylene food packaging into food simulants. Food Addit Contam Part A 33(1):167–178. https://doi.org/10.1080/19440049.2015.1114184

He X, Deng H, Hwang H-m (2019) The current application of nanotechnology in food and agriculture. J Food Drug Anal 27(1):1–21. https://doi.org/10.1016/j.jfda.2018.12.002

Hemachandra S, Hadjikakou M, Pettigrew S (2024) A scoping review of food packaging life cycle assessments that account for packaging-related food waste. Int J Life Cycle Assess 29(10):1899–1915. https://doi.org/10.1007/s11367-024-02349-z

Henchion M, McCarthy M, Dillon EJ, Greehy G, McCarthy SN (2019) Big issues for a small technology: Consumer trade-offs in acceptance of nanotechnology in food. Innov Food Sci Emerg Technol 58:102210. https://doi.org/10.1016/j.ifset.2019.102210

Heydari-Majd M, Ghanbarzadeh B, Shahidi-Noghabi M, Najafi MA, Hosseini M (2019) A new active nanocomposite film based on PLA/ZnO nanoparticle/essential oils for the preservation of refrigerated Otolithes ruber fillets. Food Packag Shelf Life 19:94–103. https://doi.org/10.1016/j.fpsl.2018.12.002

Hischier R, Walser T (2012) Life cycle assessment of engineered nanomaterials: state of the art and strategies to overcome existing gaps. Sci Total Environ 425:271–282. https://doi.org/10.1016/j.scitotenv.2012.03.001

Honarvar Z, Hadian Z, Mashayekh M (2016) Nanocomposites in food packaging applications and their risk assessment for health. Electron Physician 8(6):2531–2538. https://doi.org/10.19082/2531

Ill-Min C, Rajakumar G, Gomathi T, Seung-Kyu P, Seung-Hyun K, Thiruvengadam M (2017) Nanotechnology for human food: advances and perspective. Front Life Sci 10(1). https://doi.org/10.1080/21553769.2017.1365775

Islam Su, Kumar A (2019) 7—Life cycle assessment of rubber dam. In: Thomas S, Rane AV, V.K A, Kanny K, Dutta A (eds) Hydraulic rubber dam. William Andrew Publishing, pp 99–122. https://doi.org/10.1016/B978-0-12-812210-5.00008-0

Jayakumar A, Radoor S, Kim JT, Parameswaranpillai J, Siengchin S (2024) Chapter 20—Antimicrobial properties of nanocomposite films for food packaging applications. In: Muthukumar C, Thiagamani SMK, Krishnasamy S, Parameswaranpillai J, Siengchin S (eds) Biocomposites for industrial applications. Woodhead Publishing, pp 405–423. https://doi.org/10.1016/B978-0-323-91866-4.00017-2

Kadirvel V, Palanisamy Y, Ganesan Nandhini D (2025) Active packaging system—an overview of recent advances for enhanced food quality and safety. Packag Technol Sci 38(2):145–162. https://doi.org/10.1002/pts.2863

Kahyaoglu LN (2024) Intelligent packaging system: a promising approach to reduce food waste. Curr Food Sci Technol Rep 2(4):411–419. https://doi.org/10.1007/s43555-024-00044-6

Kalita S, Nath D, Pesaranhajiabbas E, Misra M, Mohanty AK, Mukherjee A, Kumar S (2024) Life cycle assessment of smart food packaging systems. In: Smart food packaging systems, pp 415–429. https://doi.org/10.1002/9781394189595.ch16

Kaseke T, Lujic T, Cirkovic Velickovic T (2023) Nano- and microplastics migration from plastic food packaging into dairy products: impact on nutrient digestion, absorption, and metabolism. Foods 12(16):3043

Kazemi MM, Hashemi-Moghaddam H, Mohammadi Nafchi A, Ajodnifar H (2020) Application of modified packaging and nano ZnO for extending the shelf life of fresh pistachio. J Food Process Eng 43(12):e13548. https://doi.org/10.1111/jfpe.13548

Kikuchi Y, Kanematsu Y (2020) Chapter 27—Life cycle assessment. In: Kozai T, Niu G, Takagaki M (eds) Plant factory, 2nd Ed. Academic Press, pp 383–395. https://doi.org/10.1016/B978-0-12-816691-8.00027-3

Kim BYS, Rutka JT, Chan WCW (2010) Nanomedicine. N Engl J Med 363(25):2434–2443. https://doi.org/10.1056/NEJMra0912273

Kim I, Viswanathan K, Kasi G, Thanakkasaranee S, Sadeghi K, Seo J (2022) ZnO nanostructures in active antibacterial food packaging: preparation methods, antimicrobial mechanisms, safety issues, future prospects, and challenges. Food Rev Intl 38(4):537–565. https://doi.org/10.1080/87559129.2020.1737709

Klöpffer W, Curran MA, Frankl P, Heijungs R, Köhler A, Olsen SI (2007) Nanotechnology and life cycle assessment. A systems approach to Nanotechnology and the environment: synthesis of results obtained at a workshop Washington, DC 2–3 October 2006

Kraśniewska K, Galus S, Gniewosz M (2020) Biopolymers-based materials containing silver nanoparticles as active packaging for food applications—a review. Int J Mol Sci 21(3):698

Kumar S, Basumatary IB, Sudhani HPK, Bajpai VK, Chen L, Shukla S, Mukherjee A (2021) Plant extract mediated silver nanoparticles and their applications as antimicrobials and in sustainable food packaging: a state-of-the-art review. Trends Food Sci Technol 112:651–666. https://doi.org/10.1016/j.tifs.2021.04.031

Kumar S, Boro JC, Ray D, Mukherjee A, Dutta J (2019) Bionanocomposite films of agar incorporated with ZnO nanoparticles as an active packaging material for shelf life extension of green grape. Heliyon 5(6). https://doi.org/10.1016/j.heliyon.2019.e01867

Kumar V, Sharma N, Maitra SS (2017) In vitro and in vivo toxicity assessment of nanoparticles. Int Nano Lett 7(4):243–256. https://doi.org/10.1007/s40089-017-0221-3

Lakade AJ, Sundar K, Shetty PH (2017) Nanomaterial-based sensor for the detection of milk spoilage. LWT 75:702–709. https://doi.org/10.1016/j.lwt.2016.10.031

Li S, Ren Y, Hou Y, Zhan Q, Jin P, Zheng Y, Wu Z (2024) Polysaccharide-based composite films: promising biodegradable food packaging materials. Foods 13(22):3674

Li XH, Li WL, Xing YG, Jiang YH, Ding YL, Zhang PP (2011) Effects of nano-ZnO power-coated PVC film on the physiological properties and microbiological changes of fresh-cut "Fuji" apple. Adv Mater Res 152–153:450–453. https://doi.org/10.4028/www.scientific.net/AMR.152-153.450

Luo Z, Xu Y, Ye Q (2015) Effect of nano-SiO2-LDPE packaging on biochemical, sensory, and microbiological quality of Pacific white shrimp Penaeus vannamei during chilled storage. Fish Sci 81(5):983–993. https://doi.org/10.1007/s12562-015-0914-3

Lydekaityte J, Tambo T (2018) Business perspectives of smart interactive packaging: digital transformation of brand's consumer engagement. In: Proceedings of the 8th international conference on the internet of things, pp 1–4

Malhotra B, Keshwani A, Kharkwal H (2015) Antimicrobial food packaging: potential and pitfalls. Front Microbiol 6. https://doi.org/10.3389/fmicb.2015.00611

Mao D, Yang S, Ma L, Ma W, Yu Z, Xi F, Yu J (2024) Overview of life cycle assessment of recycling end-of-life photovoltaic panels: a case study of crystalline silicon photovoltaic panels. J Clean Prod 434:140320. https://doi.org/10.1016/j.jclepro.2023.140320

Mathew S, Snigdha S, Mathew J, Radhakrishnan EK (2019) Biodegradable and active nanocomposite pouches reinforced with silver nanoparticles for improved packaging of chicken sausages. Food Packag Shelf Life 19:155–166. https://doi.org/10.1016/j.fpsl.2018.12.009

Mihindukulasuriya S, Lim L-T (2014) Nanotechnology development in food packaging: a review. Trends Food Sci Technol 40(2):149–167

Mishra B, Panda J, Mishra AK, Nath PC, Nayak PK, Mahapatra U, Sharma M, Chopra H, Mohanta YK, Sridhar K (2024) Recent advances in sustainable biopolymer-based nanocomposites for smart food packaging: a review. Int J Biol Macromol 279:135583. https://doi.org/10.1016/j.ijbiomac.2024.135583

Nasiri A, Shariaty-Niasar MS-N, Akbari Z (2012) Synthesis of LDPE/Nano TiO_2 nanocomposite for packaging applications. Int J Nanosci Nanotechnol 8(3):165–170

Nath D, Chetri R, Santhosh R, Sarkar P (2022) Chapter 7—Nanocomposites in food packaging. In: Pal K, Sarkar A, Sarkar P, Bandara N, Jegatheesan V (eds) Food, medical, and environmental applications of nanomaterials. Elsevier, pp 167–203. https://doi.org/10.1016/B978-0-12-822858-6.00007-8

Newton E (2023) Applications of nanotechnology in food packaging: a bright future ahead. Food saf Tech. https://foodsafetytech.com/column/applications-of-nanotechnology-in-food-packaging-a-bright-future-ahead/

Nizam NUM, Hanafiah MM, Woon KS (2021) A content review of life cycle assessment of nanomaterials: current practices, challenges, and future prospects. Nanomaterials 11(12):3324

Omolayo Y, Feingold BJ, Neff RA, Romeiko XX (2021) Life cycle assessment of food loss and waste in the food supply chain. Resour Conserv Recycl 164:105119. https://doi.org/10.1016/j.resconrec.2020.105119

Oniha MI, Oyejide SO (2024) Life cycle assessment of nanoparticles. In: Isibor PO, Devi G, Enuneku AA (eds) environmental nanotoxicology: combatting the minute contaminants. Springer Nature Switzerland, Cham, pp 317–324. https://doi.org/10.1007/978-3-031-54154-4_15

Pallas G, Vijver MG, Peijnenburg WJGM, Guinée J (2020) Life cycle assessment of emerging technologies at the lab scale: the case of nanowire-based solar cells. J Ind Ecol 24(1):193–204. https://doi.org/10.1111/jiec.12855

Pathiraja AS, Munaweera I (2024) Innovative nanotechnology-based sustainable food packaging: a brief review. JSFA Rep 4(1):19–32. https://doi.org/10.1002/jsf2.179

Pattnaik R, Panda SK, Biswas S, De S, Satahrada S, Kumar S (2024) Prospects and challenges of nanomaterials in sustainable food preservation and packaging: a review. Discov Nano 19(1):178. https://doi.org/10.1186/s11671-024-04142-1

Peña C, Civit B, Gallego-Schmid A, Druckman A, Pires AC, Weidema B, Mieras E, Wang F, Fava J, Canals LMi, Cordella M, Arbuckle P, Valdivia S, Fallaha S, Motta W (2021) Using life cycle assessment to achieve a circular economy. Int J Life Cycle Assess 26 (2):215–220. https://doi.org/10.1007/s11367-020-01856-z

Polyakov R, Gordeeva E (2020) Industrial enterprises digital transformation in the context of "Industry 4.0" growth: integration features of the vision systems for diagnostics of the food packaging sealing under the conditions of a production line. In: Digital transformation of the economy: challenges, trends and new opportunities. Springer, pp 590–608

Popa EE, Miteluţ AC, Râpă M, Popescu PA, Drăghici MC, Geicu-Cristea M, Popa ME (2022) Antimicrobial active packaging containing Nisin for preservation of products of animal origin: an overview. Foods 11(23):3820

Portillo F, Alcayde A, Garcia RM, Fernandez-Ros M, Gazquez JA, Novas N (2024) Life cycle assessment in renewable energy: solar and wind perspectives. Environments 11(7):147

Pradhan S, Aswathy NR, Sudha GS, Samal A, Mohapatra AK, Praharaj P, Pandey S (2024) Augmented barrier properties of poly (lactic acid) composite packaging film reinforced by antibacterial hybrid bio-filler: extending the shelf life of fruits. Polym Compos n/a (n/a). https://doi.org/10.1002/pc.29071

Primožič M, Knez Ž, Leitgeb M (2021) (Bio)Nanotechnology in food science—food packaging. Nanomaterials 11(2):292

Priyadarshi R, Rhim J-W (2020) Chitosan-based biodegradable functional films for food packaging applications. Innov Food Sci Emerg Technol 62:102346. https://doi.org/10.1016/j.ifset.2020.102346

Ramanathan A, Begum KMMS, Pereira AO, Cohen C (2022) Chapter 2—Biomass pyrolysis system based on life cycle assessment and Aspen plus analysis and kinetic modeling. In: Ramanathan A, Begum KMMS, Pereira AO, Cohen C (eds) A thermo-economic approach to energy from waste. Elsevier, pp 35–71. https://doi.org/10.1016/B978-0-12-824357-2.00006-1

Rathore A, Mahesh G (2021) Public perception of nanotechnology: a contrast between developed and developing countries. Technol Soc 67:101751. https://doi.org/10.1016/j.techsoc.2021.101751

Rigamonti L, Mancini E (2021) Life cycle assessment and circularity indicators. Int J Life Cycle Assess 26(10):1937–1942. https://doi.org/10.1007/s11367-021-01966-2

Roberts JR, Kodali VK, Stefaniak AB, Boots TE (2023) 31 analysis of the material properties of importance in the classification of toxicity of graphene nanomaterials. Ann Work Exposures Health 67 (Supplement_1):i48–i48. https://doi.org/10.1093/annweh/wxac087.116

Roco MC, Harthorn B, Guston D, Shapira P (2011) Innovative and responsible governance of nanotechnology for societal development. In: Nanotechnology research directions for societal needs in 2020: retrospective and outlook. Springer Netherlands, Dordrecht, pp 561–617. https://doi.org/10.1007/978-94-007-1168-6_14

Ros-Lis JV, Benitez M (2024) Chapter 25—Recent advances in life cycle assessment of nanomaterials for packaging applications. In: Jacob J, Cacciotti I, Thomas S (eds) Nanostructured materials for food packaging applications. Elsevier, pp 629–649. https://doi.org/10.1016/B978-0-323-99525-2.00009-8

Rossi M, Passeri D, Sinibaldi A, Angjellari M, Tamburri E, Sorbo A, Carata E, Dini L (2017) Nanotechnology for food packaging and food quality assessment. Adv Food Nutr Res 82:149–204

Rovera C, Ghaani M, Farris S (2020) Nano-inspired oxygen barrier coatings for food packaging applications: an overview. Trends Food Sci Technol 97:210–220. https://doi.org/10.1016/j.tifs.2020.01.024

Seidu RK, Doersam A, Ofori EA, Jiang S (2024) The Promise of nanotechnology in healthcare. In: George SC, Tawiah B (eds) Nanotechnology in societal development. Springer Nature Singapore, Singapore, pp 95–147. https://doi.org/10.1007/978-981-97-6184-5_3

Shaikh S, Yaqoob M, Aggarwal P (2021) An overview of biodegradable packaging in food industry. Curr Res Food Sci 4:503–520. https://doi.org/10.1016/j.crfs.2021.07.005

Shalan AE, Makhlouf ASH, Lanceros-Méndez S (2022) Nanocomposites materials and their applications: current and future trends. In: Shalan AE, Hamdy Makhlouf AS, Lanceros-Méndez S (eds) Advances in nanocomposite materials for environmental and energy harvesting applications. Springer International Publishing, Cham, pp 3–14. https://doi.org/10.1007/978-3-030-94319-6_1

Shankar S, Rhim J-W (2017) Preparation and characterization of agar/lignin/silver nanoparticles composite films with ultraviolet light barrier and antibacterial properties. Food Hydrocoll 71:76–84. https://doi.org/10.1016/j.foodhyd.2017.05.002

Shankar S, Rhim J-W (2018) Preparation of sulfur nanoparticle-incorporated antimicrobial chitosan films. Food Hydrocoll 82:116–123. https://doi.org/10.1016/j.foodhyd.2018.03.054

Shankar S, Rhim J-W, Won K (2018) Preparation of poly(lactide)/lignin/silver nanoparticles composite films with UV light barrier and antibacterial properties. Int J Biol Macromol 107:1724–1731. https://doi.org/10.1016/j.ijbiomac.2017.10.038

Shankar S, Tanomrod N, Rawdkuen S, Rhim J-W (2016) Preparation of pectin/silver nanoparticles composite films with UV-light barrier and properties. Int J Biol Macromol 92:842–849. https://doi.org/10.1016/j.ijbiomac.2016.07.107

Sharma C, Dhiman R, Rokana N, Panwar H (2017) Nanotechnology: an untapped resource for food packaging. Front Microbiol 8. https://doi.org/10.3389/fmicb.2017.01735

Shen Z, Simon GP, Cheng Y-B (2002) Comparison of solution intercalation and melt intercalation of polymer–clay nanocomposites. Polymer 43(15):4251–4260. https://doi.org/10.1016/S0032-3861(02)00230-6

Shrivastav A, Khapekar Y, Soni D, Sankathala S, Das S (2025) Application of nanotechnology in sustainable food packaging. In: Gaur N, Sharma E, Nguyen TA, Bilal M, Melkania NP (eds) Societal and environmental ramifications of plastic pollution. IGI Global, Hershey, PA, USA, pp 163–178. https://doi.org/10.4018/979-8-3693-9163-1.ch008

Siddiqui SA, Zannou O, Bahmid NA, Fidan H, Alamou A-F, Nagdalian AA, Hassoun A, Fernando I, Ibrahim SA, Arsyad M (2022) Consumer behavior towards nanopackaging—a new trend in the food industry. Future Foods 6:100191. https://doi.org/10.1016/j.fufo.2022.100191

Siemann U (2005) Solvent cast technology—a versatile tool for thin film production. In: Stribeck N, Smarsly B (eds) Scattering methods and the properties of polymer materials. Springer Berlin Heidelberg, Berlin, Heidelberg, pp 1–14. https://doi.org/10.1007/b107336

Silvestre C, Duraccio D, Cimmino S (2011) Food packaging based on polymer nanomaterials. Prog Polym Sci 36(12):1766–1782. https://doi.org/10.1016/j.progpolymsci.2011.02.003

Simbine EO, Rodrigues LDC, Lapa-Guimaraes J, Kamimura ES, Corassin CH, Oliveira CAF (2019) Application of silver nanoparticles in food packages: a review. Food Sci Technol 39:793–802

Simonetti G, Riccardi C, Pomata D, Acquaviva L, Fricano A, Buiarelli F, Senofonte M, Di Filippo P (2024) Studies of potential migration of hazardous chemicals from sustainable food contact materials. Foods 13(5):645

Singal SK, Goel V, Nautiyal H, Papantonis DE (2023) Chapter 10–Environmental impacts of small hydropower system. In: Singal SK, Goel V, Nautiyal H, Papantonis DE (eds) Small hydropower. Elsevier, pp 259–269. https://doi.org/10.1016/B978-0-323-91757-5.00010-5

Singh R, Kumar S (2023) Regulatory and safety concerns regarding the use of active nanomaterials in food industry. In: Nanotechnology advancement in agro-food industry. Springer Nature Singapore, Singapore, pp 269–306. https://doi.org/10.1007/978-981-99-5045-4_8

Smaoui S, Chérif I, Ben Hlima H, Khan MU, Rebezov M, Thiruvengadam M, Sarkar T, Shariati MA, Lorenzo JM (2023) Zinc oxide nanoparticles in meat packaging: a systematic review of recent literature. Food Packag Shelf Life 36:101045. https://doi.org/10.1016/j.fpsl.2023.101045

Song H, Li B, Lin QB, Wu HJ, Chen Y (2011) Migration of silver from nanosilver–polyethylene composite packaging into food simulants. Food Addit Contam: Part A 28(12):1758–1762. https://doi.org/10.1080/19440049.2011.603705

Souza VGL, Fernando AL (2016) Nanoparticles in food packaging: biodegradability and potential migration to food—a review. Food Packag Shelf Life 8:63–70. https://doi.org/10.1016/j.fpsl.2016.04.001

Sridhar A, Ponnuchamy M, Kumar PS, Kapoor A (2021) Food preservation techniques and nanotechnology for increased shelf life of fruits, vegetables, beverages and spices: a review. Environ Chem Lett 19(2):1715–1735. https://doi.org/10.1007/s10311-020-01126-2

Stampfli N, Siegrist M, Kastenholz H (2010) Acceptance of nanotechnology in food and food packaging: a path model analysis. J Risk Res 13(3):353–365. https://doi.org/10.1080/13669870903233303

Suh S, Huppes G (2009) Methods in the life cycle inventory of a product. In: Suh S (ed) Handbook of input-output economics in industrial ecology. Springer Netherlands, Dordrecht, pp 263–282. https://doi.org/10.1007/978-1-4020-5737-3_14

Talambedu U, Sushil Kumar M, Kadabagere Narayanaswamy H, Prakash H, Ravi H, Zeenat A, Nidhi D, Korthusheril Ajayakumar A, Arvind Kumar G, Manjeshwar Shrinath B (2025) Innovations in nanotechnology for food production and packaging: a review. Curr Nutr Food Sci 21:1–14. https://doi.org/10.2174/0115734013297685240820043042

Tawiah B, Ofori EA, George SC (2024) Nanotechnology in societal development. In: Nanotechnology in societal development. Springer, pp 1–64

Thanakkasaranee S, Pradittham A, Atong D, Pechyen C (2012) Effect of nano-silica loading on barrier and mechanical properties of food packaging based LLDPE film. Adv Mater Res 488:919–922

Torkayesh AE, Rajaeifar MA, Rostom M, Malmir B, Yazdani M, Suh S, Heidrich O (2022) Integrating life cycle assessment and multi criteria decision making for sustainable waste management: key Issues and recommendations for future studies. Renew Sust Energy Rev 168:112819. https://doi.org/10.1016/j.rser.2022.112819

Ubando AT, Rivera DRT, Chen W-H, Culaba AB (2019) A comprehensive review of life cycle assessment (LCA) of microalgal and lignocellulosic bioenergy products from thermochemical processes. Biores Technol 291:121837. https://doi.org/10.1016/j.biortech.2019.121837

Unalan IU, Cerri G, Marcuzzo E, Cozzolino CA, Farris S (2014) Nanocomposite films and coatings using inorganic nanobuilding blocks (NBB): current applications and future opportunities in the food packaging sector. RSC Adv 4(56):29393–29428

Upadhyay P, Ullah A (2024) Enhancement of mechanical and barrier properties of chitosan-based bionanocomposites films reinforced with eggshell-derived hydroxyapatite nanoparticles. Int J Biol Macromol 261:129764. https://doi.org/10.1016/j.ijbiomac.2024.129764

Vanderroost M, Ragaert P, Devlieghere F, De Meulenaer B (2014) Intelligent food packaging: the next generation. Trends Food Sci Technol 39(1):47–62. https://doi.org/10.1016/j.tifs.2014.06.009

Voglhuber-Slavinsky A, Zicari A, Smetana S, Moller B, Dönitz E, Vranken L, Zdravkovic M, Aganovic K, Bahrs E (2022) Setting life cycle assessment (LCA) in a future-oriented context: the combination of qualitative scenarios and LCA in the agri-food sector. Eur J Futures Res 10(1):15. https://doi.org/10.1186/s40309-022-00203-9

Wang H, Qian J, Ding F (2018) Emerging chitosan-based films for food packaging applications. J Agric Food Chem 66(2):395–413. https://doi.org/10.1021/acs.jafc.7b04528

Wardana AA, Wigati LP, Marcellino V, Kusuma G, Yan XR, Nkede FN, Jothi JS, Hang NPT, Tanaka F, Tanaka F, Liza C, Rifathin A, Zainuddin Z, Wahyuni NS, Van TT, Meng F, Laksmono JA, Wulandari R, Andiwinarto D (2024) The incorporation of chitosan nanoparticles enhances the barrier properties and antifungal activity of chitosan-based nanocomposite coating films. Int J Biol Macromol 280:135840. https://doi.org/10.1016/j.ijbiomac.2024.135840

Weston M, Kuchel RP, Ciftci M, Boyer C, Chandrawati R (2020) A polydiacetylene-based colorimetric sensor as an active use-by date indicator for milk. J Colloid Interface Sci 572:31–38. https://doi.org/10.1016/j.jcis.2020.03.040

Xu L, Cao W, Li R, Zhang H, Xia N, Li T, Liu X, Zhao X (2019) Properties of soy protein isolate/nano-silica films and their applications in the preservation of Flammulina velutipes. J Food Process Preserv 43(11):e14177. https://doi.org/10.1111/jfpp.14177

Xu Y, Willis S, Jordan K, Sismour E (2018) Chitosan nanocomposite films incorporating cellulose nanocrystals and grape pomace extracts. Packag Technol Sci 31(9):631–638. https://doi.org/10.1002/pts.2389

Yadav A, Yadav K, Abd-Elsalam KA (2023) Exploring the potential of nanofertilizers for a sustainable agriculture. Plant Nano Biol 5:100044. https://doi.org/10.1016/j.plana.2023.100044

Yam KL, Takhistov PT, Miltz J (2005) Intelligent packaging: concepts and applications. J Food Sci 70(1):R1–R10. https://doi.org/10.1111/j.1365-2621.2005.tb09052.x

Yang S, Ma K, Liu Z, Ren J, Man Y (2020) Chapter 5—Development and applicability of life cycle impact assessment methodologies. In: Ren J, Toniolo S (eds) Life cycle sustainability assessment for decision-making. Elsevier, pp 95–124. https://doi.org/10.1016/B978-0-12-818355-7.00005-1

Yang Y, Du Y, Gupta VK, Ahmad F, Amiri H, Pan J, Aghbashlo M, Tabatabaei M, Rajaei A (2024) Exploring blockchain and artificial intelligence in intelligent packaging to combat food fraud: a comprehensive review. Food Packag Shelf Life 43:101287

Zhang H, Hortal M, Dobon A, Jorda-Beneyto M, Bermudez JM (2017) Selection of nanomaterial-based active agents for packaging application: using life cycle assessment (LCA) as a tool. Packag Technol Sci 30(9):575–586. https://doi.org/10.1002/pts.2238

Zhang X, Liu Y, Yong H, Qin Y, Liu J, Liu J (2019) Development of multifunctional food packaging films based on chitosan, TiO_2 nanoparticles and anthocyanin-rich black plum peel extract. Food Hydrocoll 94:80–92. https://doi.org/10.1016/j.foodhyd.2019.03.009

Ziyaina M, Rasco B, Coffey T, Ünlü G, Sablani SS (2019) Colorimetric detection of volatile organic compounds for shelf-life monitoring of milk. Food Control 100:220–226. https://doi.org/10.1016/j.foodcont.2019.01.018

Zou D, Jin L, Wu B, Hu L, Chen X, Huang G, Zhang J (2019) Rapid detection of Salmonella in milk by biofunctionalised magnetic nanoparticle cluster sensor based on nuclear magnetic resonance. Int Dairy J 91:82–88. https://doi.org/10.1016/j.idairyj.2018.11.011

Chapter 15
Advances in Food Packaging with Nanotechnology-Enhanced Biomaterials

Garima Chetry, Swathi Deepak Nayak, Bader Alsubaie, Jameel M. Al-Khayri, Vasantha Veerappa Lakshmaiah, and Praveen Nagella

Abstract The growing concern about food safety and the need for environmental sustainability has triggered the shift from traditional plastic packaging to eco-friendly packaging materials derived from biological resources, enhanced with nanotechnology. Different challenges, such as low mechanical strength, thermal instability, barrier properties and microbial susceptibility, are being observed in traditional packaging. In this chapter, we discuss nanomaterials that are integrated with biopolymers for food packaging to address these issues and increase their physicochemical properties, functionalities, as well as practical applications. Biopolymers such as polylactic acid (PLA), chitosan, gelatin, cellulose and starch-based films, are increasingly being used and further reinforced with organic nanoparticles (nanocellulose, nano starch, carbon dots, and protein nanoparticles) and inorganic nanoparticles (silver, zinc oxide, titanium dioxide etc.) are addressed in the chapter. This chapter classifies and analyses nanomaterials according to their origin and function, demonstrating the enhancement of barrier properties, antimicrobial activity, UV shielding, and thermal resistance in nano-biocomposites. Special attention is paid to their application in smart packaging systems, including active systems that release antimicrobials or antioxidants and intelligent systems containing nano sensors that check for freshness and contamination. Examples of enhanced shelf life and quality preservation are discussed in fruits, vegetables, dairy, meat and bakery goods. The chapter also examines important safety, nanoparticle migration, toxicity regulatory issues, and the environmental impact, highlighting the emerging need for globally unified rules. To sum up, the world's targets for sustainable food systems, reducing waste, consumer

---

G. Chetry · S. D. Nayak · V. V. Lakshmaiah (✉) · P. Nagella (✉)
Department of Life Sciences, School of Sciences, Christ University, Bengaluru, India
e-mail: vasantha.vl@christuniversity.in

P. Nagella
e-mail: praveen.n@christuniversity.in

B. Alsubaie · J. M. Al-Khayri
Department of Agricultural Biotechnology, College of Agriculture and Food Sciences, King Faisal University, Al-Ahsa, Saudi Arabia

S. Roy and M. Shahid (eds.), *Biomaterials for Sustainable Food Packaging*,
Environmental Footprints and Eco-design of Products and Processes,
https://doi.org/10.1007/978-3-032-08027-1_15

health protection and safeguarding public health are enabled by nano-biomaterials, making it the most suitable innovative solution to the challenges of packaging.

Keywords Biopolymers · Nanoparticles · Active packaging · Smart packaging · Nano sensors

15.1 Introduction

All living things on Earth depend on food as their main source of nutrition to survive. The food sector has encountered challenges with safety, preservation, and extended shelf life, which significantly impact the global economy. More than $100 billion is lost annually in low- and middle-income nations because of poor storage practices, loss of quality, and contaminated food products. It has been estimated that foodborne diseases are responsible for approximately 600 million illnesses and 0.42 million deaths globally every year (World Health Organization 2015). Adding to this concern is the extensive use of plastics in the food packaging industry, which is considered deleterious to human well-being as well as significantly polluting the environment. Industrial production and application of plastics have elevated the carbon emissions (De Kock et al. 2020). To address these challenges, there is an urgent need for an effective, sustainable, and novel storage system or packaging material that can protect the food products from microbiological, chemical and physical deterioration.

In recent years, sustainable approaches have promoted advancement in packaging methodologies. To reduce pollution of the environment, biodegradable materials such as polylactic acid (PLA), polybutylene succinate (PBS), cellulose-based thermoplastics, and starch-based films have replaced traditional plastics, metals and glass. Edible films and coatings made up of natural polymers such as pectin, gelatin, chitosan, alginate, essential oils, etc., can enhance food protection and serve as an alternative to synthetic packaging materials and an eco-friendly alternative (Primožič et al. 2021). Integration of nanotechnology into these biomaterial matrices has further improved their characteristic and has significantly helped in resolving the typical limitations faced by the conventional packaging material (Herrera-Rivera et al. 2024).

Nanomaterials have a size ranging from 1 to 100 nm. They are gaining significant attention in the food industry for their widespread applications, ranging from pre-harvesting, food processing, packaging, quality and food safety (Khan et al. 2019; Singh et al. 2023). The greater surface area relative to volume of nanoparticles results in improved mechanical strength, durability, and barrier properties. Nano-based packaging solutions are equipped with antibacterial and antioxidant qualities which help in extending the shelf life of food products by preventing spoilage and tracking with extrinsic factors of spoilage like temperature, gas composition, and humidity (Valgimigli et al. 2018; Díez-Pascual 2018). The food industry faces a myriad of problems that nanomaterials with their physico-chemical properties can resolve. Inorganic nanoparticles including metal and metal oxides such as silver, zinc, copper, titanium oxide, zinc oxide are major among nanomaterials that have an antibacterial effect with enhanced thermal and mechanical properties of packaging material. They also aid in UV protection and the prevention of oxidative damage to food. Organic nanomaterials, including graphene oxide and carbon nanotubes, improve barrier and mechanical stress resistance, while nanometal oxides also protect against UV rays and oxidative damage of food products (Garcia 2022; Herrera-Rivera et al. 2024). Similarly, nanocellulose, nanostarch, and protein nanoparticles obtained from natural sources have remarkable properties that can enhance the relative permeability of water and tensile strength when incorporated in biopolymers while ensuring sustainability (Primožič et al. 2021).

There are various applications of using nanomaterials in food packaging, including the ability to create smart packaging systems like active and intelligent packaging systems. They interact with food to increase the food's shelf-life. Nano sensors made of NPs like gold nanoparticles (AuNPs) any fluctuation that can indicate food's quality or safety been compromised, provide antibacterial components that prevent the growth of pathogens that would otherwise make the food inedible, and have stronger barrier qualities in maintaining food hygiene. Furthermore, integration of nanomaterials such as AgNPs into PLA matrix, ZnONPs into chitosan films, and chitosan-xanthan gum blend has also shown enhanced biodegradability as compared to traditional packaging systems (Ramos et al. 2014; Tabassum et al. 2023; Sarojini et al. 2019). However, there are toxicological and ethical concerns regarding the use of nanoparticles, which needs to be properly addressed. Toxic particles are released during the fabrication process as well as post-synthesis, which can react with food, environment and human body. These particles can accumulate in the environment and disrupt the food chain in the aquatic and soil ecosystems. They can adversely affect human health by ROS production, denaturation of protein, DNA damage, and apoptosis, in the cells, potentially leading to various diseases (Jeevanandam et al. 2016; Tervonen et al. 2009). Thus, stringent regulation policies are required for its sustainabile application. The present chapter provides insights into the integration of nanomaterials in biopolymer-based packaging materials to explore their wider application in food industries as an eco-friendly and sustainable approach, and the challenges associated.

15.2 Types of Nanomaterials Used in Food Packaging

15.2.1 Inorganic Nanoparticles

The silver nanoparticles have been mainly used for food packaging purposès, since they exhibit excellent antimicrobial activity. These nanoparticles have proven to exhibit greater potential against a plethora of microbes, including bacteria (Gram-positive and Gram-negative), viruses, protozoa and fungi. These silver nanoparticles adopt various kinds of killing mechanisms against microorganisms which includes easy penetration through the outer, thicker lipopolysaccharide layer of Gram-negative bacteria, compared to a Gram-positive bacterium (Kumar et al. 2021). The antibacterial activity of the AgNPs derived from antibacterial peptide, bacteriocin, helped in effective preservation and improved shelf life of the tomato, when packaged with the paper coated with these nanoparticles (Sharma et al. 2023). Also, the addition of silver nanoparticles to biopolymers such as gelatin and chitosan, the cross-linking of the silver nanoparticles to the biopolymer matrix, had led to a reduction in the degree of swelling, water-retention abilities as well as a decrease in water vapour transmission rate. Besides, the nanoparticle in the composite film also helped in better reduction of bacterial activity when compared to plastic polyethene based packaging (Ediyilyam et al. 2021). In a particular study, the silver nanoparticles synthesized from Iturin-A, a lipopeptide from bacillus sps. was applied for food packaging owing to fungicidal property of Iturin A and the bactericidal activity of silver nanoparticles could help in reducing the microbial food contamination during storage (Zhao et al. 2021).

Besides AgNPs, zinc oxide and titanium oxide nanoparticles are also used in sustainable food packaging. Green synthesized zinc oxide nanoparticles from fruit extract of *Cassia fistula* infused along with polymer matrix of chitosan–gelatin was developed for its scope in food packaging as it exhibited amelioration in certain physical and structural properties such as improved elasticity, structural integrity as well as elevated thermal stability. The nanocomposite film could also exhibit microbicidal activity against *E. coli* (Kumar et al. 2020). Zinc oxide nanoparticles are also effective in providing protection against the UV rays, as observed in those food packaging systems that employed polyvinyl alcohol and ZnO NPs. Here, the zinc nanoparticles incorporated could block UV as they could absorb UV light at 380 nm wavelength, and the absorbance value elevated with an increase in the concentration of the nanoparticles (Channa et al. 2022). Similarly, in another study where the incorporation of zinc oxide nanoparticles into chicken skin gelatin/tapioca starch composite films has affected its mechanical properties such as improvement in its tensile strength, reduction in break at elongation, improved thermal properties, resistance to moisture as well as enhancement of microbicidal properties (Lee et al. 2021).

Titanium oxide nanoparticles (TiO_2NPs) have been recommended by the Food and Drug Administration (FDA) for their application in food packaging as they possess antimicrobial properties that help in the storage of life-improving edible goods (Venkatesan et al. 2024). TiO_2NPs synthesized from the leaf extract of *Mentha arvensis* could exert bactericidal and fungicidal activity over certain virulent microorganisms (Ahmad et al. 2020). Titanium oxide was incorporated in chitosan/cassava starch-based biofilms and was used in sustainable food packaging, where TiO_2NPs helped in ameliorating the mechanical properties of the biofilms and in reducing the permeability of water across the films. It also led to depletion of the transmission of UV radiation (de Menezes et al. 2021). Also, in the case of starch-based biofilms containing cellulose were incorporated with TiO_2 NPs, has led to modifications of various properties of the biofilms such as an increase in the tensile strength, increased protection against UV, and an increase in hydrophobicity (Picolotto et al. 2024), which are the main attractive features in the food packaging industry.

15.2.2 Organic Nanomaterials

Nanocellulose is usually obtained from numerous materials such as agricultural wastes, plant extracts, and so on. It has many beneficial properties such as being environment friendly, non-toxic, recyclable, cost effective, having high structural durability and surface area per unit mass, being light weighted and highly transparent as well as being highly resistant to chemical substances along with low coefficient of thermal expansion (Valsalan and Sivaranjana 2022; Barbash et al. 2022). These properties have therefore been useful when used as composite in food packaging as it helps in enhancing certain properties of packaging such as mechanical properties, thermal properties, barrier properties and its recyclability (Wang et al. 2022).

Nanostarch is an organic nanomaterial which have two forms including starch nanocrystals which are crystalline structures which are obtained by acid hydrolysis which facilitates the elimination of the non-crystalline portion from the starch granules and the other form is the starch nanoparticles which is spherical in shape and of different sizes within nanoscale having variable crystal structure. Nano starch has better thermal properties, antioxidant capability, and improved rate of biodegradability when compared to their macroscopic counterparts (Bajer 2023). They are found to enhance the mechanical barrier properties of edible vegetable-derived polymeric films and help to reduce the water permeability, protecting food from moisture, thus improving their longevity (Campelo et al. 2020).

Carbon nanotubes are cylindrical structures with both ends either closed or open, having diameters in the nanoscale range and lengths in the micron range, which are crimped by the graphite plane. Different types of carbon nanotubes based on the number of graphene layers include single-walled carbon nanotubes, double-walled carbon nanotubes, and multi-walled carbon nanotubes (Gao et al. 2023). The multiwalled carbon nanotubes have been used as a composite along with zinc oxide and polyvinyl alcohol for food packaging, as they can ameliorate different traits of food packaging, such as functionality of packaging, barrier, and mechanical properties (Wen et al. 2022). The carbon nanotubes, especially single-walled carbon nanotubes, have microbicidal properties as they can damage microbial cell membrane, disrupt metabolic pathways as well as disrupt morphological features of the cell (Azizi-Lalabadi et al. 2020).

Carbon dots are carbon-based, dimensionless nanomaterials having sizes below 10 nm (Cui et al. 2021). Carbon dots improved the mechanical properties and provide antioxidant properties with low cytotoxicity when integrated into gelatin/carrageenan-based food packaging biopolymer film (Roy et al. 2021). When used as composite with polyvinyl alcohol film carbon dots enhanced properties such as antimicrobial, mechanical and free-radical scavenging properties, resistance to moisture, improved thermal stability and UV blocking abilities (Zhao et al. 2022).

Certain food packaging systems have also been developed using protein-based nanoparticles as they help in enhancing strength and water-resistant properties.

Nanoparticles derived from peanuts were found to produce improved strength, resistance to humidity and temperature and zein nanoparticles could elevate the structural and moisture resistant traits of the whey protein-based film (Ashfaq et al. 2022). Table 15.1 provides a summary of various types of nanomaterials and their uses in food packaging.

Table 15.1 Types of nanomaterials used in food packaging and their properties

Type of nanomaterials	Properties of nanomaterials	References
Inorganic nanomaterials		
Silver nanoparticles	Antimicrobial action against microorganisms that are both gram positive and gram negative	Kumar et al. (2021)
Zinc oxide nanoparticles	Antimicrobial, UV barrier and enhanced mechanical properties	Kumar et al. (2020), Channa et al. (2022) Lee et al. (2021)
Titanium oxide nanoparticles	Antimicrobial, UV barrier and improvement in mechanical properties	de Menezes et al. (2021), Venkatesan et al. (2024)
Organic nanomaterials		
Nanocellulose	Biodegradability and amelioration of structural and thermal characteristics of packaging as well as recyclability	Barbash et al. (2022), Wang et al. (2022)
Nano starch	Ameliorated structural and thermal characteristic of packaging with reduced permeability to water with high biodegradability	Bajer (2023), Campelo et al. (2020)
Carbon nanotubes	Antimicrobial properties and enhancement of other important structural propertiessuch as mechanical and barrier	Wen et al. (2022), Azizi-Lalabadi et al. (2020)
Carbon dots	Improved mechanical properties, UV blocking ability and have antioxidant properties	Roy et al. (2021), Zhao et al. (2022)
Protein nanoparticles	Improved durability, temperature and water-resistant properties	Ashfaq et al. (2022)

15.3 Functional Properties and Mode of Action of Nanomaterials Utilized in Food Packaging Biomaterials

The demand for safer, more sustainable, and better preservation techniques is causing a major shift in the food packaging sector. The use of nanomaterials greatly improves the functional characteristics of biopolymers in food packaging, which allows for new applications and solves key limitations. Nanomaterials aid in food safety and prolong food freshness as they boost the antimicrobial effect by preventing microbial development (Crispín-Isidro et al. 2019). They also improve the composite's tensile strength and elasticity so that the polymer can easily endure physical pressure and shape changes while being packed, transported or stored (Yadav et al. 2020). Furthermore, nanomaterials enhance how well biopolymers block off gases, moisture, and oil, so that packaged food remains fresh and of high quality (Rhim and Wang 2014). Biopolymers with nanomaterials can operate at higher temperatures than their normal temperatures, which is an extremely beneficial property (Kasirga et al. 2012). Nanomaterials also enable active and intelligent packaging, allowing the packaging to release antimicrobials, absorb excess oxygen, or track and register changes in the food's condition (Silvestre et al. 2011). The sections that follow highlight some of the primary functional properties of nano packaging in biomaterials and their mode of action (Fig 15.1).

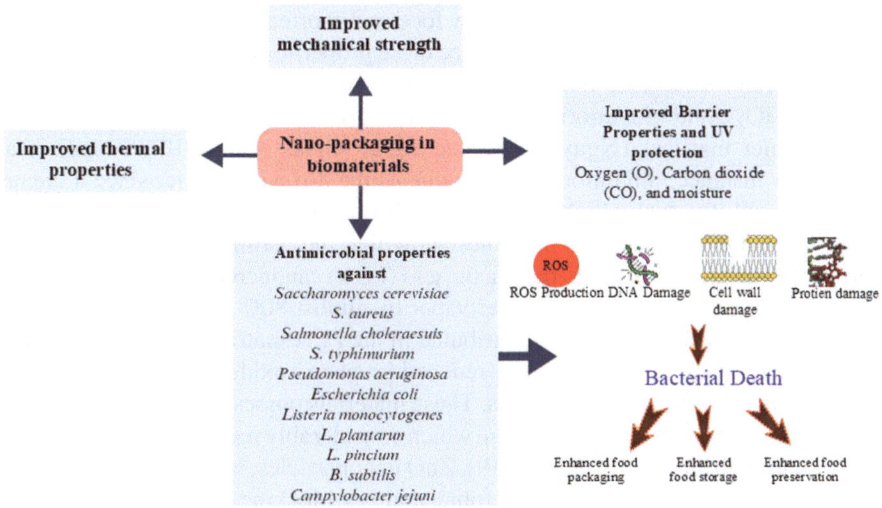

Fig. 15.1 Properties of nano packaging materials used in biomaterials for food packaging

15.3.1 Improved Mechanical and Barrier Properties

One of the most notable benefits of employing nanomaterials for food packaging is their ability to improve mechanical qualities. In particular, the elasticity and strength of conventional packaging materials are often insufficient to withstand various environmental conditions during storage or transport. Cellulose Nanocrystals (CNC) and Cellulose Nanofibers (CNF) are two types of nanocellulose that could significantly enhance the qualities of food packaging biopolymer films. Studies indicate that the mechanical, antibacterial, and barrier properties of edible packaging films are enhanced through the incorporation of CNC and CNF (Zhang et al. 2021). Incorporating around 5% CNC into polylactic acid (PLA) sheets has been found to increase their tensile strength by over 50% without compromising their flexibility (Yu et al. 2017). Nanofibers provide entangling effects that help boost mechanical and oxygen barrier properties in biopolymers such as polyhydroxyalkanoates (PHA) (Patel et al. 2022). The blend of PLA with poly(hydroxybutyrate) (PHB) with CNC and plasticizers produces optically transparent films with improved barrier to oxygen, UV-light-blocking, and heat-stability properties. This kind of transparency in packaging material is visually appealing (Arrieta et al. 2015). These improvements enable the production of thinner yet stronger packaging materials, reducing material usage, cost effectiveness and reducing wastage. The mechanical reinforcement provided by nanomaterials also improves the longevity of the packaging itself under diverse storage conditions. Food packaging must have barrier qualities to block ultraviolet (UV) light, gases (such as carbon dioxide and oxygen), and moisture from penetrating as these elements can oxidatively destroy food and shorten its shelf life (Mustafa and Andreescu 2020). Nanomaterials enhance these properties by altering the packaging material's microstructure. To improve barrier properties, zinc oxide (ZnO), titanium dioxide (TiO_2), and montmorillonite clay nanoparticles are frequently incorporated into polymer matrices (Nguyen and Lee 2022; Abbas et al. 2019; Şengül et al. 2017). For instance, montmorillonite augments the diffusion barrier since it creates a nanolayered structure which decreases the diffusion pathways for gases and water vapor (Tan and Thomas 2017). Studies show that the addition of montmorillonite clay in the range of 2–5% into polylactic acid (PLA) can increase its permeability of oxygen upto 49% and reducing water vapor by almost 80% depending on the type of clay that is used and how it is distributed in the PLA matrix (Şengül et al. 2017).

Graphene oxide (GO) along with reduced graphene oxide (rGO) are considered viable options for barrier applications. These materials possess an exceptionally high aspect ratio and impermeable structure which considerably restrict transmission rates for gas and moisture (Smith et al. 2019). ZnO nanoparticles, in addition to enhancing barrier properties, also serve antimicrobial and UV-blocking purposes which render them useful for active protective packaging (Silva-Leyton et al. 2019).

15.3.2 Antimicrobial Properties

In the food industry, microbial contamination-induced foodborne diseases continue to be a major concern (Long et al. 2023). When added to biomaterials for food packaging, nanoparticles provide a novel and efficient solution to increase food safety, prevent microbial growth, and extend the longevity of food goods. Numerous packaging-enhancing nanoparticles have been investigated for their antibacterial properties; zinc oxide (ZnO NPs), copper (CuNPs), titanium dioxide (TiO_2 NPs), gold (AuNPs), and silver (AgNPs) are among the frequently researched ones (Espitia et al. 2012; Gopinath et al. 2024; Xie and Hung 2018; Ahari et al. 2022; Sharma et al. 2023). These nanoparticles target bacteria using several mechanisms that include interactions with the microbial cell membrane level, generation of reactive oxygen species (ROS) and inhibition of some crucial biological activities (Feng et al. 2000; Vasiliev et al. 2023). Of all the different kinds of nanoparticles, silver nanoparticles (AgNPs) have the strongest antibacterial action against a variety of disease-causing bacteria, to name a few, *Listeria monocytogenes*, *Escherichia coli*, and *Staphylococcus aureus* (Patra and Baek 2017). The main mechanism by which AgNPs work is by releasing silver ions (Ag+), which causes microbial cell membranes to rupture, their permeability to rise, and the cell to die. Also, the Ag^+ ions enter the bacterial cells where they interfere with the synthesis of DNA as well as respiratory processes (Feng et al. 2000; Sondi and Salopek-Sondi 2004). The results have shown that the bacterial count on food surfaces can be reduced by up to 99.9% with the help of AgNPs, thus enhancing food safety (Fontecha-Umaña et al. 2020). Like AgNPs, CuNPs also have antibacterial qualities and can disrupt cell membranes, generate ROS and release copper ions ($Cu2^+$), which affect microbial metabolism and DNA replication. CuNPs cause oxidative damage to bacterial cells, resulting in protein and DNA damage that inhibits microbial growth. Previous studies show that CuNPs can reduce growth in *E. coli* by 99.7% after 24 h of exposure (Chatterjee et al. 2012). CuNPs can stop the growth of *E. coli* using a range of bactericidal pathways, and their effectiveness depends on both size and concentration (Lai et al. 2022). Zinc oxide nanoparticles have demonstrated considerable promise as antimicrobials, and so their application in food packaging has attracted significant interest. ZnO NPs exert their antimicrobial actions by releasing Zn^{2+} ions and by generating reactive oxygen species. The cell plasma membranes of the microorganisms accept the Zn^{2+} ions released from the nanoparticles on its surface and upon binding it causes alterations in the membrane structure resulting in the excretion of intracellular components. Reactive oxygen species, such as hydrogen peroxide and superoxide, cause stress in microorganisms. Oxidative damage to essential organs, primarily to lipids, proteins, and DNA, occurs in such stressed cells, leading to cell death (Jiang et al. 2016). In studies conducted, exposure to the minimal inhibitory concentrations (MIC) of ZnO NPs decreased the cell counts of *Staphylococcus aureus* and *Salmonella typhimurium* to zero within 4 and 8 h, respectively (Fontecha-Umaña et al. 2020; Tayel et al. 2011). A combination of ion release and ROS generation suppresses both cellular and bacterial growth due to the interference with the enzymatic function in cells and the inhibition of cellular respiration.

TiO_2 NPs are known for their photocatalytic activity, which provides them with an advantage of potent antibacterial properties when exposed to UV light (Jain and Vaya 2017). Exposure to UV light triggers TiO_2 NPs to create ROS, mainly superoxide anions and hydroxyl radicals, etc. The components of microbial cells are oxidized by this ROS, which causes membranes to rupture and cellular contents to leak out (Goñi-Ciaurriz et al. 2023; Phuinthiang et al. 2023). The photocatalytic effect exhibited by the TiO_2 NPs makes them sometimes more useful in the presence of UV light and thus provides more efficacy in antimicrobial protection in food packaging (Xie and Hung 2018). After 72 h of exposure to UV radiation, it was found that the TiO_2 NP coated films could significantly reduce the populations of *S. aureus* and *E. coli* by more than 85% (Jiang et al. 2016). TiO_2 NPs in cellulose acetate films have also shown 71% inhibition of *E. coli* and 88% of *S. aureus* (Goñi-Ciaurriz et al. 2023). Since TiO_2 NPs can generate more ROS when exposed to more UV radiation, they may be considered advantageous in food packaging applications, especially in locations that have easy access to UV light, adding an additional layer of bactericidal action (Xie and Hung 2018).

Gold nanoparticles (AuNPs) have been found to show a strong antibacterial effect using a variety of methods. AuNPs work mainly by collapsing the membrane potential, preventing ATPase activity, and interfering with ribosome function. It can also physically damage bacterial cell membranes (Cui et al. 2012; Hameed et al. 2020). It was previously believed that the release of Reactive oxygen species (ROS) is a stringent requirement for AuNPs to have bactericidal effects. which is not true as per recent studies (Cui et al. 2012). Rather, they cause mitochondrial malfunction and DNA damage, which sets off ROS-independent apoptotic pathways in bacteria and fungi (Lee et al. 2018). AuNPs' antibacterial efficiency was found to be influenced by their shape, with nano cubes exhibiting the maximum activity (Hameed et al. 2020).

15.3.3 Improved Thermal Properties

Due to reasons like low glass transition temperatures (T_g) and poor thermal stability of traditional biopolymers such as polyhydroxyalkanoates (PHAs), polylactic acid (PLA) and starch-based materials, improving its thermal properties has become a crucial area of research in food packaging (Van de Velde and Kiekens 2002). These disadvantages limit their use in high-temperature settings, including hot food storage or sterilization procedures (Van de Velde and Kiekens 2002; Rhim and Ng 2007). In order to address these problems, incorporating nanomaterials like metal-oxide nanoparticles (MONPs), nano clays, carbon-based nanoparticles, graphene-based fillers, has showed promise in improving the thermal performance of biopolymer composites (Gohary et al. 2023; Olusanya et al. 2023; Rocha et al. 2020). MONPs, for instance, TiO_2 and ZnO, improve the thermal conductivity of biopolymers, leading to greater heat dissipation. This uniform heat distribution reduces localized thermal stresses and delays thermal degradation (Li et al. 2009; Gohary et al. 2023). Addition of 1–5% TiO_2 NPs and ZnO NPs to PBAT/TPS bioplastic packaging films disrupted starch granules, increasing amorphous starch content. ZnO nanoparticles showed stronger hydrogen bonding with the TPS phase, modifying thermal degradation temperature. However, films with ZnO nanoparticles had increased migration levels, especially in acidic simulants (Phothisarattana et al. 2022). It has also been evidenced that adding TiO_2 NPs to starch films which had cellulose nanocrystals, significantly improved heat stability by almost 85 °C (Picolotto et al. 2024). The functional nanomaterials physically reinforce the polymer matrix of biomaterials to resist thermal stress-induced chain cleavage. Layered nano clays and derivatives of graphene reinforce the polymer by incorporating NPs of different species within the polymer, working as physical barriers that reduce exposure to oxygen and heat, two principal sources of thermal degradation. In PLA composites, whose heat degradation is prevented by the addition of nano clay up to 26 °C, this step has been particularly effective (Ojijo et al. 2014).

15.4 Smart Packaging

Smart packaging has been developed for increasing the longevity of the packaged food products. They are broadly categorised into two types: Active packaging and Intelligent packaging (Chen et al. 2020).

15.4.1 Active Packaging

Active packaging involves prolonging the storage life of food products either by the addition of any activating agent into the packaging system or by allowing the food item, packaging system, and environmental factors to interact (Almasi et al. 2021). Antibacterial active packaging involves the release of the antibacterial agent by the material that is being used for packaging, either on the food product or to the exterior surroundings, which would help in the inhibition of growth of those microorganisms in the surroundings of packaged edible goods. These microorganisms may be either considered disease-causing or those that contribute to the qualitative degradation of edible products. Certain inorganic nanoparticles such as copper, copper oxide, silver, zinc oxide, titanium oxide, and magnesium oxide nanoparticles, some organic nanoparticles such as carbon dots and graphene are used in active packaging due to their microbicidal characteristics developed by various mechanisms such as oxidative stress (ROS production and lipid peroxidation), cell membrane damage, enzyme inactivation, DNA damage and so on (Omerović et al. 2021). Most of the qualitative attributes of the food products like taste, aroma and appearance can be lost by oxidation of various components of the food such as oxidation of sulphur containing amino acids result in off-flavors in meat products and lipid oxidation can also lead to degradation of various aspects of food quality such as deterioration of nutritional value and rancidity of food products rich in lipid content (Lai 2022). To prevent this oxidation of food products, the addition of antioxidants has become necessary. Inorganic nanomaterials such as zinc oxide have been used as the antioxidant agent which is incorporated into the food packaging system facilitating active packaging (Lai 2022). Fe_2TiO_5 nanoparticle imparts antioxidant properties to the alginate films (Rizzotto et al. 2022). Selenium nanoparticles are ideal for usage in active food packaging as they have both antibacterial and antioxidant properties by their incorporation into the polymeric packaging films (Ndwandwe et al. 2021).

15.4.2 Intelligent Packaging

Intelligent packaging monitors both the inner and outer environmental conditions of food goods throughout storage and transit (Chen et al. 2020). Nano sensors are typically used for monitoring the presence of certain gases, small organic molecules and certain types of products. Nano sensors can detect and measure many external and internal characteristics such as flavor, pathogenic bacteria, allergens, colour, and so on that can contribute to food spoilage and so evaluates the condition of the food product without the requirement of determining its expiry date (Primožič et al. 2021). There are several types of nano sensors available, including electronic, microbiological indicators, and FFIs (food freshness indicators) which are internal sensors that respond to the release of chemical compounds by food products, and Time–Temperature Indicators (TTIs). Barcodes and radiofrequency-based identifying tags are examples of external sensors (Forghani et al. 2021). Gold nanoparticles (AuNPs) and carbon-based NPs have been commonly used in nano sensors for food packaging. AuNPs have been used to detect toxic substances produced by microorganisms in food products. AuNPs conjugated with anti-aflatoxins antibodies are utilized to detect aflatoxin in food products, a toxin generated by *Aspergillus* species. Besides toxins the gold nanoparticles are also used to detect adulterants such as melamine in food substances and for detection of carbamate and organophosphorus pesticides in packaged food stuff (Chelliah et al. 2021). Sensors made of carbon quantum dots have been used to find heavy metals in food, like copper in fruits and vegetables. They have also been used to monitor the level of arsenic in water when combined with gold nanoparticles as optical sensors. Additionally, gold nanoparticles have also been utilized to detect mercury in milk and water samples. Carbon nanotubes, including single and multiwall carbon nanotubes, are used to detect and test the quantities of folic acid and to track the nutritional value of food items. Immobilized gold-based particles have been used to identify antioxidant qualities in both red and white wine (Chelliah et al. 2021) (Fig. 15.2).

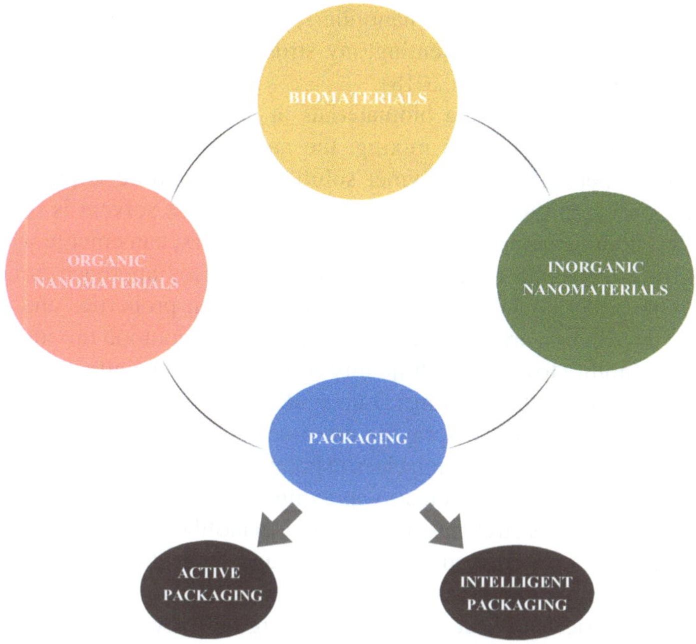

Fig. 15.2 Different food packaging strategies using organic and inorganic nanomaterials in biomaterials

15.5 Methods for Processing of Nano Packaging in Biomaterials

The construction of food packaging materials derived from nanomaterials requires the use of physical processing techniques. The popularity of this method stems from ease of use, effectiveness, functional properties enhancement, and the fact that they can be scaled for large volume production. Also, the construction method determines the distribution and concentration of nanosized fillers and nano reinforcements within the polymer (Lohat et al. 2020). Melt processing is one of the well-known industrial techniques for synthesizing polymer composites with nanoparticles. In this method, nanofillers are mixed with the polymers using single-screw or twin-screw extruders in a molten form. The still molten polymeric material is also subjected to shear and elongational stresses due to the rotation of screws which enhance the dispersion and alignment of the nanomaterials within the polymer matrix. In addition, the polymer is melted due to high temperature. Melt processing has good benefits because it is easy to scale up and compatible with many thermoplastic polymers. The final products, which include films, fibers, and molded parts can be directly manufactured from the blended materials using extrusion, injection molding, and blow molding. However, maintaining uniform distribution of nanofillers, preventing thermally sensitive materials from overheating, and preventing any structural or property changes are all important considerations (Vasile 2018).

Other methods of processing biomaterials at the nanoscale level often utilize solvent casting. During solution mixing, the nanoparticles are either agitated or ultrasonically dispersed into a polymer solution. The mixture is put into a mold to be casted, and then the solvent is removed. The excess solvent is removed and nanoparticles, homogeneously distributed within a matrix, can emerge whilst maintaining uniform distribution in the polymer matrix. This approach is convenient for developing coatings and thin films because uniform properties and controlled nanoparticle distribution are essential. At a smaller scale, solution mixing has gained notoriety for its ability to create high-quality composites, especially during prototype development or in laboratory settings (Luo et al. 2020).

Electrospinning and electro-spraying are processes akin to spraying or ejecting polymer solutions that require a high electric potential, done under standard temperature and pressure conditions. As electro-spraying creates tiny droplet like structures, electrospinning produces stretches of continuous nanofibers known as polymer jets. In both cases, precision is offered over the shape and size with the ability to measure from nanometers to micrometers. Nanofibrous structures are usually produced with the purpose of fabrics in tissue engineering, drug delivery, and filtration through electrospinning. On the other hand, electro-spraying is ideal for the fabrication of particle and coating nanostructures needed in medicine and electronics (Bavatharani et al. 2021). The process of making nano-based coatings is most often done using the sol–gel technique. This technique involves applying a thick colloidal sol at the surface using techniques like dipping and spraying and is mainly used to produce nano-based coatings. Through the sol's hydrolysis and polycondensation reactions, it can be transformed into gel state which can be dried or cured into a solid coating. It provides control over geometric dimensions and properties of the coating like thickness from a couple of nanometres up to several micrometres. Because of the flexibility and ease with which the sol–gel method can be combined with different substrates, it is frequently used to create thin uniform protective optical coatings and coatings for biomedical devices (Caruso et al. 2001) (Fig. 15.3).

15.6 Applications of Nanotechnology for Specific Food Categories

15.6.1 Packaging of Crops Produce

Certain qualitative traits of crop produce can get subjected to deterioration either due to microbial activity or due to oxidation caused because of their increased respiration rate. Usually, biopolymers derived from biologic sources are used as sustainable

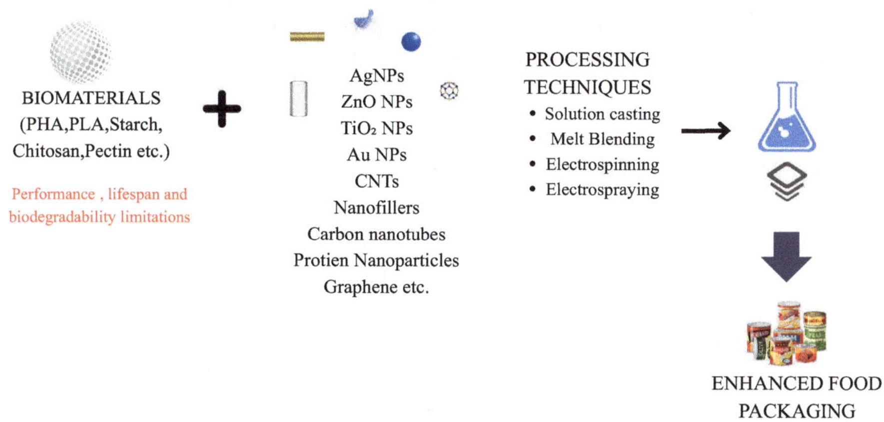

Fig. 15.3 Incorporation process of different nano sized materials into biomaterials

packaging material but to ameliorate their physical attributes, certain nanomaterials have also been used (Basumatary et al. 2022).

Silver nanoparticles synthesized from kaolin clay and *Ficus carica* have been incorporated into the chitosan film for apple slices which has ameliorated the moisture and light blocking properties of the chitosan film as well as its microbicidal and free-radical scavenging properties (Mouzahim et al. 2023). Oxidized starch derived metallic nano fiber has been used in packaging for preservation of bananas by adsorption of ethylene produced by banana thus delaying the ripening process (Du et al. 2024). A study was conducted where, the nanocellulose film derived from the banana pseudo stem has been used for storage of black grapes (Ghosh and Ghosh 2020). Nano composite packaging films developed by dispersing titanium oxide nanoparticles into the clay nanoparticles improved their mechanical and physical properties and therefore were used for preservation of tomato in cold storage (Bodaghi 2024). Gelatin based films containing Ti doped CuO nanoparticles were used in packaging of tomatoes which provided antimicrobial effect along with low cytotoxicity (Sooch and Mann 2021). Chitosan coating with nano films made of silicon and titanium oxide was formulated to prolong the storage life of the blue berry which not only helped in prevention of microbial growth but also helped in preservation of the nutrient content, ascorbic acid (Li et al. 2021). Similarly, in a study which involved the use of nano silicon with chitosan film containing nisin it was proved that the nanomaterial helped in preservation of blueberry fruit at ambient temperature for eight days by providing the antimicrobial effect and in texture maintenance (Eldib et al. 2020). For preservation of freshly cut carrots, nanocomposite packaging has been developed from whey protein isolates, jujube polysaccharide and spherical starch-based nanocrystals which has provided improved barrier properties, thereby by reducing permeability for both oxygen and water vapor which helped in improving physicochemical, microbiological and sensory properties (Gharibzahedi et al. 2023). Nano emulsions prepared from plant extracts have also been used for prolonging the storage life of certain vegetables such as zucchini by preserving the moisture content of the vegetable with enhanced antimicrobial and antioxidant effects (Amer 2023).

15.6.2 Packaging for Dairy Products

Dairy products have high nutritional value making them highly prone to contamination. The quality as well as the longevity of the dairy products depends on their packaging process after their manufacturing (Mirza Alizadeh et al. 2021). Milk packaging has been developed from incorporation of silica nanoparticles into the extraction of essential oil from *Pistacia atlantica* Gum and polypropylene. It showed antimicrobial properties against the pathogenic microbes *S. aureus*, *Listeria monocytogenes*, *Salmonella enterica* and *E. coli* (Ellahi et al. 2020). The biodegradable films made of alginate and lemongrass essential oil with silver nanoparticles incorporated into the film could preserve the texture, colour and softness of cheese as these incorporated silver particles and the essential oil extracted from lemongrass showed antimicrobial activity against the Gram-negative (*Escherichia coli* and *Salmonella* typhi) as well as the Gram-positive (*Bacillus cereus* and *Staphylococcus aureus*) (Motelica et al. 2021). The use of nano-Ag-graphene-TiO_2 Poly Lactic acid film for preservation of farmer cheese has improved mechanical properties and resistance to water hence extending its shelf life (Peter et al. 2021). Furcellaran nano composite containing silver nanoparticles have been used in packaging of two types of cheese, gouda and quark cheese, which helped in improving the overall microbiological properties of both cheeses (Pluta-Kubica et al. 2021).

15.6.3 Meat Packaging

The deterioration of physical and chemical attributes of the meat results in changes of texture and appearance of the meat such as off-odour, discoloration or slimy appearance of the meat product. Putrefaction of meat products has led to huge monetary losses, disease outbreaks, human sufferings and death (Shao et al. 2021). Therefore, during the food processing, to increase in certain qualitative traits of the food products substances such as antioxidants, binders and thickeners, humectants, sodium nitrate and nitrite as curing agent, flavor enhancers, and tenderizing enzymes are added. The use of these could be rendered harmful due to the presence of synthetic antioxidants in the conventional packaging system which could migrate to the packaged food items. The field of nanotechnology has thus been applied to increase the longevity of the meat products (Lamri et al. 2021). In a particular study a chitosan film was made into which was incorporated the cinnamaldehyde encapsulated in chitosan nanoparticle which was used in meat packaging which resulted in controlled release of cinnamaldehyde and hence providing antimicrobial effect (Mondéjar-López et al. 2024). The integration of zinc oxide nanoparticles in the packaging system could produce antimicrobial effect on packaged poultry meat thus preventing the growth of the pathogenic bacteria *Campylobacter jejuni* by increasing the amount of lactic acid been produced by *Lactobacillus* bacteria in meat (Hakeem et al. 2020). Similarly, the integration of the zinc oxide nanoparticles in the chitosan film could produce both antimicrobial and antioxidant effects on the poultry meat being packaged, thus ameliorating its shelf life (Souza et al. 2020).

15.6.4 Packaging of Bakery Products

Bakery produce are also susceptible to spoilage in the absence of the preservatives having a very low longevity of 3–5 days at room temperature and can undergo both physical and chemical deterioration such as loss of texture, taste and freshness and are also susceptible to contamination by microbiological agents, the most common being *Aspergillus, Penicillium, Neurospora, Mucor* and *Cladospora* species of fungi and bacterial species colonizing the bread products include *Bacillus subtilis* and *Bacillus licheniformis* (Qian et al. 2021). The storage life can thus be prolonged by the application of nanotechnology which can facilitate antimicrobial activity, resistance to moisture and prevent exchange of gaseous substances with the external environment (Onyeaka et al. 2022). Antimicrobial packaging system has been developed for preservation of toast bread where the silver nanoparticles have been incorporated into the gelatin-based film which could show microbicidal activity against bacteria and molds for 5 days of storage and also helped in controlling of the moisture content within the bread preventing the staleness or hardening of bread (Yousefi et al. 2024). The integration of silver nanoparticles synthesized from peels of citrus fruits and pomegranate into the cellulose wrapper used for packaging bread provided antimicrobial effect by delay in growth of microbes such as bacteria, molds and yeast for 7 days of storage, enhancing longevity (Gopalakrishnan et al. 2023). The polybutylene succinate film used for packaging of bread could produce antifungal effect against the two most common fungal species causing spoilage, *Aspergillus* and *Penicillium* species when incorporated with lignin nanoparticles (Moe et al. 2023) (Table 15.2).

15.7 Regulatory and Safety Considerations

Nanotechnology in food packaging offers new features including better barrier properties, shelf life, and even antimicrobial activity. These advantages, however, come with safety risks, particularly with regards to the environmental impacts brought on by the migration of nanoparticles to food products. Safety concerns have catalyzed the creation of regulatory frameworks globally, directed toward the evaluation and surveillance of nanomaterial-based food packaging. In more extreme conditions such as being acidic or high-temperature, studies have shown that AgNPs and CuNPs used

Table 15.2 Application of nanotechnology for packaging of different food products

Food products been packaged	Polymers used in packaging	Nanomaterials been applied to packaging	Beneficial effects of nanomaterials on food packaging or food product	References
Apple slices	Chitosan film	Silver nanoparticles from kaolin clay and Ficus carica	Improved antimicrobial, antioxicant properties as well as moisture and light barrier properties	Mouzahim et al. (2023)
Tomato	(i) Low density polyethylene (LDPE) (ii) Gelatin based	(i) Titanium oxide nanoparticles dispersed in clay nanoparticles (ii) Titanium doped CuO nanoparticles	(i) Improved physical and mechanical properties (ii) Enhanced antimicrobial effect	Bodaghi (2024), Sooch and Mann, (2021)
Blueberry	Chitosan	(i) Nanofilm made of titanium and silicon (ii) Nano silicon in chitosan film with nisin	(i) Antimicrobial effect and preservation of nutrient, ascorbicacid content (ii) Antimicrobial effect and texture maintenance	Li et al. (2021), Eldib et al. (2020)
Carrots	Jujube polysaccharide and whey protein isolates	Starch based nanocrystals	Improved resistance against oxygen andhumid conditions	Gharibzahedi et al. (2023)
Zucchini	Sodium alginate based edible coatings	Nano emulsions prepared from plant extract	Preservation of moisture along with antimicrobial and antioxidant properties	Amer (2023)
Milk	Polypropylene	Silica nanoparticles along with polypropylene and essential oil from Pistacia atlantica Gum	Antimicrobial property against the pathogenic microbes S. enterica, S. aureus, L. monocytogenes and E. coli	Ellahi et al. (2020)

(continued)

Table 15.2 (continued)

Food products been packaged	Polymers used in packaging	Nanomaterials been applied to packaging	Beneficial effects of nanomaterials on food packaging or food product	References
Cheese	(i) Alginate biofilm (ii) Poly Lactic acid film (iii) Furcellaran	(i) Silver nanoparticles and essential oils extracted from lemongrass (ii) Nano-Ag-graphene-TiO_2 Poly Lactic acid film (iii) Nano composite containing silver nanoparticles	(i) Antimicrobial activity on the Gram-positive (B. cereus and S. aureus) and Gram-negative (E. coli and S. typhi) (ii) Mechanical properties and water resistance enhancement for preservation of curd cheese (iii) Improvement of microbiological properties of gouda and quark cheese	Motelica et al. (2021), Peter et al. (2021), Pluta-Kubica et al. (2021)
Meat	(i) Chitosan (ii) Paper based products (3D paper tubes and 2D absorption pads)	(i) Cinnamaldehyde encapsulated in chitosan nanoparticles (ii) Zincoxide nanoparticles	(i) Controlled release of cinnamaldehyde producing antimicrobial effect (ii) Antibacterial effect against Campylobacter jejuni	Mondéjar-López et al. (2024), Hakeem et al. (2020)
Bread	(i) Gelatin based film (ii) Cellulose wrapper (iii) Polybutylene succinate film	(i) Silver nanoparticles (ii) Silver nanoparticles synthesized from citrusand pomegranate peel (iii) The incorporation of lignin nanoparticles to polybutylene succinate film	(i) Antimicrobial effect against bacteria and molds for 5 daysof storage (ii) Antimicrobial effect against bacteria, molds and yeast for 7 days of storage (iii) Fungicidal effect against Aspergillus and Penicillium species	Yousefi et al. (2024), Gopalakrishnan et al. (2023), Moe et al. (2023)

in nanocoated food packaging can migrate into foods (Hannon et al. 2016). Those active antimicrobial agents, ZnO nanoparticles, have been shown to partition and migrate vastly in acidic food simulants as well (Bumbudsanpharoke et al. 2024). These posited health concerns may include long-term strict safety boundaries of migration, toxicity, or gastrointestinal issues that could disrupt normal digestion.

One of the most advanced portions of legal legislation on the use of nanotechnology in food packaging is the European Union. Regulation (EU) No 10/2011 states that every single material including a nanomaterial that is meant to come in contact with food has to go through a safety assessment in totality. The regulation places an important emphasis on the assessment of migration potential, which is the study of whether nanoparticles can get transferred from a particular packaging material to food in certain conditions (de Sousa et al. 2023). The EFSA (European Food Safety Authority) gives detailed instructions and guidance on how to evaluate the such parameters of particle size, shape, chemical composition and stability that remain during the contact with food. According to European thresholds of safety nominated by the EU, EFSA has nanomaterials in polymers to show very low rates of migration roughly speaking less than 10 mg per kilogram of food. Also, the EU's labeling policies set forth by Regulation (EU) No 1169/2011 stipulate that any food incorporating nanomaterials must be explicitly labeled, thus increasing consumer awareness. A wide range of stakeholders both inside and beyond the EU are anticipated to be impacted by the labeling regulations in both good and negative ways (Grieger et al. 2016).

The United States Food and Drug Administration (USFDA) controls the application of nanomaterials in food packaging within the frame of the Federal Food, Drug, and Cosmetic Act (FFDCA). It is the manufacturer's responsibility to ensure that nanomaterials, especially those relevant to food contamination, do not pose any safety threats. The FDA does not create specific policies concerned with nanotechnology; it uses policies formulated for other purposes. The addition of nanomaterials used in food packaging requires approval of GRAS (Generally Recognized As Safe) status or prior permission from FDA (Bumbudsanpharoke et al. 2015).

During the release of these packaging materials into the environment, nanoparticles can cause threat to the environment by accumulating in aquatic and soil. Some regulatory bodies, like the European Food Safety Authority (EFSA), has called for the inclusion of systematic assessments of the potential unfavorable impact a technological innovation could have on the environment regarding the authorization of nanomaterials. For example, titanium dioxide nanoparticles that masquerade as UV filters used in food packages and pose ecological and toxicological threats along with the lack of adequate means for their disposal and the need for economically efficient solutions for their recycling (Rashid et al. 2021). Even though the European Union (EU) has established specific rules for marking and assessment of the safety of nanomaterials, other countries do not have similar policies. For instance, the differences in consumer transparency, safety regulation and obligatory labelling of the usage of nanoparticles in food packaging in the United States. Without compulsory labelling of the nanoparticles used in food packaging, consumers have no way of knowing what health risks they are exposed to. The absence of international regulations raises other concerns regarding the reliability of the safety measures and makes issues of international trade more complex (Grieger et al. 2016).

15.8 Challenges and Associated Risks

The distinct features of risk assessment concerning different nanomaterials (NMs) are related to their specific synthesis techniques and attributes. A significant determinant of nanoparticles' (NPs') is its rate of reactivity and agglomeration. One of the techniques used for maintaining stability is encapsulation, as smaller particles can agglomerate at much slower rates, but retaining size post-synthesis is sometimes difficult (Jeevanandam et al. 2016). NPs are different because they possess colloidal size-dependent chemistry; particles are often catalytically softened by cleavage bonds or functionalization (Tervonen et al. 2009). Agglomeration and aggregation are often prominent issues because they lead to unwanted phase changes, elevated solvate, and poor corrosion resistance which is detrimental to the structure itself (Jeevanandam et al. 2018). Moreover, due to their enhanced degree of reactivity, NPs undergo reactions with most contaminants with relative ease, hence, improved stability is necessary through encapsulation. Residual contaminants unavoidably pose risks concerning toxicity, while the unassured effects of encapsulation heighten the need for stringent regulations on recycling and disposal (Chaudhary et al. 2020; Jeevanandam et al. 2018).

Some of the negative impacts associated with NP exposure include phagocytosis interference, ROS production, denaturation of protein, mitochondrial dysfunction, and disruption. Classification of NPs considers the behaviour-related hazards such as dimensionality, arrangement, composition, clustering, symmetry, and homogeneity, which influence behavior. Nanoparticles (NPs) can significantly affect phagocytosis, an immune response that eliminates foreign particles from the body potentially leading to both augmentation and inhibition (Gustafson et al. 2015). They can interact with proteins from biological fluids, creating a "protein corona". Opsonins, coat this nanoparticle, making them more palatable for phagocytes, triggering phagocytosis (Saha et al. 2016). High NP doses can also cause chronic inflammation, inhibit macrophage activity, or increase immune response, potentially leading to autoimmunity. Antigen presentation disruption by immune cells raises concerns about NP safety in biomedical applications (Ray et al. 2021). Nanoparticles have also been found to influence changes in animals such as affecting cell growth and proteins in silkworm due to AgNPs, anxiogenic behaviour in mice due to accumulation of ZnO and AuNPs., disruption of dopaminergic neurochemical systems in zebrafish larvae etc. (de Souza et al. 2017; Meng et al. 2017; Hwang et al. 2022). Thus, to summarize the manufacturing, application, and disposal of products containing nanoparticles can altogether contribute significantly to releasing nanoparticulates into the environment, stressing the need for comprehensive risk management and reduction strategies (Chaudhary et al. 2020; Jeevanandam et al. 2016, 2018; Tervonen et al. 2009). The risk factors and mitigation strategies are summarized in Table 15.3.

Table 15.3 Properties, risk factors and mitigation strategy associated with nano packaging

Property	Risk factor	Mitigation approach	References
Particle size and chemical reactivity	Agglomeration due to size; instability post-synthesis	Using encapsulation techniques for stability	Jeevanandam et al. (2016), Tervonen et al. (2009)
Agglomeration	Structural breakdown, corrosion, and reduced solubility	Surface treatment, modification and functionalization improve stability	Jeevanandam et al. (2018)
Impurity contamination	Toxicity from lingering contaminants or by-products	Purification and encapsulation reduce toxic effects	Jeevanandam et al. (2018)
Environmental release	Pollution from nanoparticle production and disposal techniques	Focus on eco-friendly designs and recycling	Jeevanandam et al. (2016)
Toxicity risks	May induce oxidative stress, cellular dysfunction and protein damage	Evaluations of exposure reduction and toxicity	Chaudhary et al. (2020)

15.9 Conclusion

The food industry has employed nanoparticles packing of edible goods to prolong their longevity while maintaining their stability, flavor, and color. The use of nanoparticles has been proved to ameliorate the physical, mechanical and functional properties of packaging systems. It can also provide resistance to water and provide antimicrobial properties. Research towards the study of the toxicity and safety of these nanomaterials is required, and how their use in food packaging has grown rapidly, necessitating a characterization of their toxicity. Furthermore, their use in the food sector needs to be tightly regulated because they can leach out of the polymeric packaging matrix into the packaged food goods. In addition to toxicity, research must be done on the number of nanoparticles that must be incorporated to packaging systems to prevent harmful health effects from ingesting, skin absorption, and inhalation. The toxicity investigation involves the detection and quantification of the nanoparticles through various techniques such as mass spectroscopy, positron emission tomography, particle emission spectroscopy, and dynamic light scattering for studying several parameters including the size of the nanoparticles, size distribution of particles, size of nanoparticles, radio testing and for elemental analysis of trace elements. Overall, the application of nanotechnology in food packaging has facilitated a shift from conventional packaging involving additives and preservatives to a more sustainable packaging system that is considered natural and eco-friendly.

Acknowledgements The authors express their gratitude to the Head, Department of Life Sciences, and the management of Christ University for the technical support. The authors express appreciation for the support provided by the Deanship of Scientific Research, Vice Presidency for Graduate Studies and Scientific Research, King Faisal University, Saudi Arabia [Grant No. KFU253634].

Declaration AI tools were used to enhance the language and readability of this chapter. The authors have reviewed and verified the final content and take full responsibility for final content of the chapter.

References

Abbas M, Buntinx M, Deferme W, Peeters R (2019) (Bio)polymer/ZnO nanocomposites for packaging applications: a review of gas barrier and mechanical properties. Nanomaterials 9(10):1494. https://doi.org/10.3390/nano9101494

Ahari H, Fakhrabadipour M, Paidari S, Goksen G, Xu B (2022) Role of AuNPs in active food packaging improvement: a review. Molecules 27(22):8027. https://doi.org/10.3390/molecules27228027

Ahmad W, Jaiswal KK, Soni S (2020) Green synthesis of titanium dioxide (TiO_2) nanoparticles by using *Mentha arvensis* leaves extract and its antimicrobial properties. Inorg Nano-Metal Chem 50(10):1032–1038

Almasi H, Jahanbakhsh Oskouie M, Saleh A (2021) A review on techniques utilized for design of controlled release food active packaging. Crit Rev Food Sci Nutr 61(15):2601–2621

Amer MA (2023) Using effect of nano emulsion based on plant extracts as edible coating on prolonging the shelf life of fresh zucchini. Al-Azhar J Agric Res 48(1):12–30

Arrieta MP, Fortunati E, Dominici F, López J, Kenny JM (2015) Bionanocomposite films based on plasticized PLA–PHB/cellulose nanocrystal blends. Carbohydr Polym 121:265–275. https://doi.org/10.1016/j.carbpol.2014.12.05

Ashfaq A, Khursheed N, Fatima S, Anjum Z, Younis K (2022) Application of nanotechnology in food packaging: Pros and Cons. J Agric Food Res 7:100270

Azizi-Lalabadi M, Hashemi H, Feng J, Jafari SM (2020) Carbon nanomaterials against pathogens; the antimicrobial activity of carbon nanotubes, graphene/graphene oxide, fullerenes, and their nanocomposites. Adv Colloid Interface Sci 284:102250

Bajer D (2023) Nano-starch for food applications obtained by hydrolysis and ultrasonication methods. Food Chem 402:134489

Barbash VA, Yashchenko OV, Yakymenko OS, Zakharko RM, Myshak VD (2022) Preparation of hemp nanocellulose and its use to improve the properties of paper for food packaging. Cellulose 29(15):8305–8317

Basumatary IB, Mukherjee A, Katiyar V, Kumar S (2022) Biopolymer-based nanocomposite films and coatings: Recent advances in shelf-life improvement of fruits and vegetables. Crit Rev Food Sci Nutr 62(7):1912–1935

Bavatharani C, Muthusankar E, Wabaidur SM, Alothman ZA, Alsheetan KM, AL-Anazy M, Ragupathy D (2021) Electrospinning technique for production of polyaniline nanocomposites/nanofibres for multi-functional applications: a review. Synth Met 271:116609. https://doi.org/10.1016/j.synthmet.2020.116609

Bodaghi H (2024) Characterization and application of the nanocomposite packaging films containing clay and TiO_2 on preservation of tomato fruit under cold storage. BMC Plant Biol 24(1):521

Bumbudsanpharoke N, Ko S (2015) Nano-food packaging: an overview of market, migration research, and safety regulations. J Food Sci 80(5). https://doi.org/10.1111/1750-3841.12861

Bumbudsanpharoke N, Nurhadi RP, Chongcharoenyanon B, Kwon S, Harnkarnsujarit N, Ko S (2024) Effect of migration on the functionality of zinc oxide nanoparticle in polybutylene adipate terephthalate/thermoplastic starch films: a food simulant study. Int J Biol Macromol 263:130232. https://doi.org/10.1016/j.ijbiomac.2024.130232

Campelo PH, Sant'Ana AS, Clerici MTPS (2020) Starch nanoparticles: Production methods, structure, and properties for food applications. Curr Opin Food Sci 33:136–140

Caruso RA, Antonietti M (2001) Sol-Gel Nanocoating: an approach to the preparation of structured materials. Chem Mater 13(10):3272–3282. https://doi.org/10.1021/cm001257z

Channa IA, Ashfaq J, Gilani SJ, Shah AA, Chandio AD, Jumah MN (2022) UV blocking and oxygen barrier coatings based on polyvinyl alcohol and zinc oxide nanoparticles for packaging applications. Coatings 12(7):897. https://doi.org/10.3390/coatings12070897

Chatterjee AK, Sarkar RK, Chattopadhyay AP, Aich P, Chakraborty R, Basu T (2012) A simple robust method for synthesis of metallic copper nanoparticles of high antibacterial potency against E. coli. Nanotechnology 23(8):085103. https://doi.org/10.1088/0957-4484/23/8/085103

Chaudhary P, Fatima F, Kumar A (2020) Relevance of nanomaterials in food packaging and its advanced future prospects. J Inorg Organomet Polym Mater 30(12):5180–5192. https://doi.org/10.1007/s10904-020-01674-8

Chelliah R, Wei S, Daliri EBM, Rubab M, Elahi F, Yeon SJ, Oh DH (2021) Development of nanosensors based intelligent packaging systems: food quality and medicine. Nanomaterials 11(6):1515. https://doi.org/10.3390/nano11061515

Chen S, Brahma S, Mackay J, Cao C, Aliakbarian B (2020) The role of smart packaging system in food supply chain. J Food Sci 85(3):517–525

Crispín-Isidro G, Hernández-Rodríguez L, Ramírez-Santiago C, Sandoval-Castilla O, Lobato-Calleros C, Vernon-Carter EJ (2019) Influence of purification on physicochemical and emulsifying properties of tamarind (*Tamarindus indica* L.) seed gum. Food Hydrocoll 93:402–412. https://doi.org/10.1016/j.foodhyd.2019.02.046

Cui L, Ren X, Sun M, Liu H, Xia L (2021) Carbon dots: Synthesis, properties and applications. Nanomaterials 11(12):3419. https://doi.org/10.3390/nano11123419

Cui Y, Zhao Y, Tian Y, Zhang W, Lü X, Jiang X (2012) The molecular mechanism of action of bactericidal gold nanoparticles on *Escherichia coli*. Biomaterials 33(7):2327–2333. https://doi.org/10.1016/j.biomaterials.2011.11.057

De Kock L, Sadan Z, Arp R, Upadhyaya P (2020) A circular economy response to plastic pollution: Current policy landscape and consumer perception. S Afr J Sci 116(5/6). https://doi.org/10.17159/sajs.2020/8097

de Menezes FLG, de Lima Leite RH, dos Santos FKG, Aria AI, Aroucha EMM (2021) TiO_2-enhanced chitosan/cassava starch biofilms for sustainable food packaging. Colloids Surf A Physicochem Eng Asp 630:127661

de Souza JM, Mendes BO, Guimarães ATB, Rodrigues ASL, Chagas TQ, Rocha TL, Malafaia G (2017) Zinc oxide nanoparticles in predicted environmentally relevant concentrations leading to behavioral impairments in male Swiss mice. Sci Total Environ 605–606:863–874. https://doi.org/10.1016/j.scitotenv.2017.09.051

de Sousa MS, Schlogl AE, Estanislau FR, Souza VGL, dos Reis Coimbra JS, Santos IJB (2023) Nanotechnology in packaging for food industry: past, present, and future. Coatings 13(8):1411. https://doi.org/10.3390/coatings13081411

Díez-Pascual AM (2018) Antibacterial activity of nanomaterials. Nanomaterials 8(6):359. https://doi.org/10.3390/nano8060359

Du C, Jiang Y, Junejo SA, Jia X, Zhang B, Huang Q (2024) Metal-anchored oxidized starch-pullulan nanofiber films enhance ethylene adsorption and banana preservation. Int J Biol Macromol 282:137399

Eldib R, Khojah E, Elhakem A, Benajiba N, Helal M (2020) Chitosan, nisin, silicon dioxide nanoparticles coating films effects on blueberry (*Vaccinium myrtillus*) quality. Coatings 10(10):962. https://doi.org/10.3390/coatings10100962

Ellahi H, Khalili Sadrabad E, Hekmatimoghaddam S, Jebali A, Sarmast E, Akrami Mohajeri F (2020) Application of essential oil of *Pistacia atlantica* gum, polypropylene and silica nanoparticles as a new milk packaging. Food Sci Nutr 8(8):4037–4043

Ediyilyam S, George B, Shankar SS, Dennis TT, Wacławek S, Černík M, Padil VVT (2021) Chitosan/gelatin/silver nanoparticles composites films for biodegradable food packaging applications. Polymers 13(11):1680. https://doi.org/10.3390/polym13111680

Espitia PJP, Soares NFF, Coimbra JSR, Andrade NJ, Cruz RS, Medeiros EAA (2012) Zinc oxide nanoparticles: synthesis, antimicrobial activity and food packaging applications. Food Bioprocess Technol 5:1447–1464. https://doi.org/10.1007/s11947-012-0797-6

Feng QL, Wu J, Chen GQ, Cui FZ, Kim TN, Kim JO (2000) A mechanistic study of the antibacterial effect of silver ions on Escherichia coli and Staphylococcus aureus. J Biomed Mater Res 52(4):662–668. https://doi.org/10.1002/1097-4636(20001215)52:4%3c662::aid-jbm10%3e3.0.co;2-3

Fontecha-Umaña F, Ríos-Castillo AG, Ripolles-Avila C, Rodríguez-Jerez JJ (2020) Antimicrobial activity and prevention of bacterial biofilm formation of silver and zinc oxide nanoparticle-containing polyester surfaces at various concentrations for use. Foods 9(4):442. https://doi.org/10.3390/foods9040442

Forghani S, Almasi H, Moradi M (2021) Electrospun nanofibers as food freshness and time-temperature indicators: a new approach in food intelligent packaging. Innov Food Sci Emerg Technol 73:102804

Gao C, Guo M, Liu Y, Zhang D, Gao F, Sun L, Wang Y (2023) Surface modification methods and mechanisms in carbon nanotubes dispersion. Carbon 212:118133

Garcia CV (2022) Editorial: nanomaterials in food packaging. Front Nutr 9. https://doi.org/10.3389/fnut.2022.1083185

Gharibzahedi SMT, Ahmadigol A, Khubber S, Altintas Z (2023) Bionanocomposite films with plasticized WPI-jujube polysaccharide/starch nanocrystal blends for packaging fresh-cut carrots. Food Packag Shelf Life 36:101042

Goñi-Ciaurriz L, Rosas-Val P, Gamazo C, Vélaz I (2023) Photocatalytic and antibacterial performance of β-cyclodextrin-TiO_2 nanoparticles loading sorbic and benzoic acids. Colloid Interface Sci Commun. https://doi.org/10.1016/j.colcom.2023.100747

Gopalakrishnan K, Chandel M, Gupta V, Kaur K, Patel A, Kaur K, Sharma A (2023) Valorisation of fruit peel bioactive into green synthesized silver nanoparticles to modify cellulose wrapper for shelf-life extension of packaged bread. Food Res Int 164:112321

Gopinath K, Sathishkumar G, Xu L (2024) An overview of the copper oxide nanofillers integrated in food packaging systems. Coatings 14:81. https://doi.org/10.3390/coatings14010081

Ghosh M, Ghosh P (2020) Storage study of grapes (*Vitis vinifera*) using the nanocomposite biodegradable film from banana pseudostem. J Food Process Preserv 44(12):e14917

Grieger KD, Hansen SF, Mortensen NP, Cates S, Kowalcyk B (2016) International implications of labeling foods containing engineered nanomaterials. J Food Prot 79(5):830–842. https://doi.org/10.4315/0362-028X.JFP-15-335

Gustafson H, Holt-Casper D, Grainger D, Ghandehari H (2015) Nanoparticle uptake: the phagocyte problem. Nano Today 10:487–510. https://doi.org/10.1016/j.nantod.2015.06.006

Hakeem MJ, Feng J, Nilghaz A, Ma L, Seah HC, Konkel ME, Lu X (2020) Active packaging of immobilized zinc oxide nanoparticles controls *Campylobacter jejuni* in raw chicken meat. Appl Environ Microbiol 86(22):e01195-e1220

Hameed S, Wang Y, Zhao L, Xie L, Ying Y (2020) Shape-dependent significant physical mutilation and antibacterial mechanisms of gold nanoparticles against foodborne bacterial pathogens (*Escherichia coli*, *Pseudomonas aeruginosa* and *Staphylococcus aureus*) at lower concentrations. Mater Sci Eng C 108:110338. https://doi.org/10.1016/j.msec.2019.110338

Hannon JC, Kerry JP, Cruz-Romero M, Azlin-Hasim S, Morris M, Cummins E (2016) Human exposure assessment of silver and copper migrating from an antimicrobial nanocoated packaging material into an acidic food simulant. Food Chem Toxicol 95:128–136. https://doi.org/10.1016/j.fct.2016.07.004

Herrera-Rivera MR, Torres-Arellanes SP, Cortés-Martínez CI, Navarro-Ibarra DC, Hernández-Sánchez L, Solis-Pomar F, Pérez-Tijerina E, Román-Doval R (2024) Nanotechnology in food

packaging materials: role and application of nanoparticles. RSC Adv 14(30):21832–21858. https://doi.org/10.1039/D4RA03711A
Hwang KS, Son Y, Kim SS, Shin DS, Lim SH, Yang JY, Jeong HN, Lee BH, Bae MA (2022) Size-dependent effects of polystyrene nanoparticles (PS-NPs) on behaviors and endogenous neurochemicals in Zebrafish Larvae. Int J Mol Sci 23(18):10682. https://doi.org/10.3390/ijms231810682
Jain A, Vaya D (2017) Photocatalytic activity of TiO_2 nanomaterial. J Chil Chem Soc 62(4):3683–3690. https://doi.org/10.4067/s0717-97072017000403683
Jeevanandam J, Barhoum A, Chan YS, Dufresne A, Danquah MK (2018) Review on nanoparticles and nanostructured materials: history, sources, toxicity and regulations. Beilstein J Nanotechnol 9:1050–1074. https://doi.org/10.3762/bjnano.9.98
Jeevanandam J, Chan YS, Danquah MK (2016) Nano-formulations of drugs: recent developments, impact and challenges. Biochimie 128–129:99–112. https://doi.org/10.1016/j.biochi.2016.07.008
Jiang Y, Zhang L, Wen D, Ding Y (2016) Role of physical and chemical interactions in the antibacterial behavior of ZnO nanoparticles against *E. coli*. Mater Sci Eng C 69:1361–1366. https://doi.org/10.1016/j.msec.2016.08.044
Kasirga Y, Oral A, Caner C (2012) Preparation and characterization of chitosan/montmorillonite-K10 nanocomposites films for food packaging applications. Polym Compos 33(11).https://doi.org/10.1002/pc.22310
Khan I, Saeed K, Khan I (2019) Nanoparticles: properties, applications and toxicities. Arab J Chem 12(7):908–931. https://doi.org/10.1016/j.arabjc.2017.05.011
Kumar S, Basumatary IB, Sudhani HP, Bajpai VK, Chen L, Shukla S, Mukherjee A (2021) Plant extract mediated silver nanoparticles and their applications as antimicrobials and in sustainable food packaging: a state-of-the-art review. Trends Food Sci Technol 112:651–666
Kumar S, Mudai A, Roy B, Basumatary IB, Mukherjee A, Dutta J (2020) Biodegradable hybrid nanocomposite of chitosan/gelatin and green synthesized zinc oxide nanoparticles for food packaging. Foods 9(9):1143. https://doi.org/10.3390/foods9091143
Lai WF (2022) Design of polymeric films for antioxidant active food packaging. Int J Mol Sci 23(1):12. https://doi.org/10.3390/ijms23010012
Lai MJ, Huang YW, Chen HC, Tsao LI, Chang Chien CF, Singh B, Liu BR (2022) Effect of size and concentration of copper nanoparticles on the antimicrobial activity in *Escherichia coli* through multiplemechanisms. Nanomaterials 12(21):3715. https://doi.org/10.3390/nano12213715
Lamri M, Bhattacharya T, Boukid F, Chentir I, Dib AL, Das D, Djenane D, Gagaoua M (2021) Nanotechnology as a processing and packaging tool to improve meat quality and safety. Foods 10(11):2633. https://doi.org/10.3390/foods10112633
Lee H, Lee DG (2018) Gold nanoparticles induce a reactive oxygen species-independent apoptotic pathway in *Escherichia coli*. Colloids Surf B Biointerfaces 167:1–7. https://doi.org/10.1016/j.colsurfb.2018.03.049
Lee SW, Said NS, Sarbon NM (2021) The effects of zinc oxide nanoparticles on the physical, mechanical and antimicrobial properties of chicken skin gelatin/tapioca starch composite films in food packaging. J Food Sci Technol 58(11):4294–4302. https://doi.org/10.1007/s13197-020-04904-6
Li Y, Rokayya S, Jia F, Nie X, Xu J, Elhakem A, Helal M (2021) Shelf-life, quality, safety evaluations of blueberry fruits coated with chitosan nano-material films. Sci Rep 11(1):55
Li Y, Yang M, Wang DN, Lu J, Sun T, Grattan KT (2009) Fiber Bragg gratings with enhanced thermal stability by residual stress relaxation. Opt Express 17(22):19785–19790. https://doi.org/10.1364/OE.17.019785
Luo Y, Wang Q, Zhang Y (2020) Biopolymer-based nanotechnology approaches to deliver bioactive compounds for food applications: a perspective on the past, present, and future. J Agric Food Chem 68(46):12993–3000. https://doi.org/10.1021/acs.jafc.0c00277
Lohat SK, Kumar S, Gaba P (2020) An overview: preparation, characterization and applications of nanoparticles. J Drug Deliv Ther 10(6-s):159–167. https://doi.org/10.22270/jddt.v10i6-s.4398

Long J, Du G, Chen J, Xie C, Xu J, Yuan J (2023) Bacteria and poisonous plants/fungi were the primary causative hazards of foodborne disease outbreaks: a five-year survey from Guangzhou, Guangdong. Int J Food Microbiol 400:110264. https://doi.org/10.1016/j.ijfoodmicro.2023.110264

Meng X, Abdlli N, Wang N, Lü P, Nie Z, Dong X, Lu S, Chen K (2017) Effects of Ag nanoparticles on growth and fat body proteins in silkworms (*Bombyx mori*). Biol Trace Elem Res 180:327–337. https://doi.org/10.1007/s12011-017-0999-y

Mirza Alizadeh A, Masoomian M, Shakooie M, Zabihzadeh Khajavi M, Farhoodi M (2021) Trends and applications of intelligent packaging in dairy products: a review. Crit Rev Food Sci Nutr 62(2):383–397

Moe NC, Winotapun C, Hararak B, Wanmolee W, Suwanamornlert P, Leelaphiwat P, Chonhenchob V (2023) Application of lignin nanoparticles in polybutylene succinate based antifungal packaging for extending the shelf life of bread. Food Packag Shelf Life 39:101127

Mondéjar-López M, Castillo R, Jiménez AJL, Gómez-Gómez L, Ahrazem O, Niza E (2024) Polysaccharide film containing cinnamaldehyde-chitosan nanoparticles, a new eco-packaging material effective in meat preservation. Food Chem 437:137710

Motelica L, Ficai D, Oprea OC, Ficai A, Ene VL, Vasile BS, Andronescu E, Holban AM (2021) Antibacterial biodegradable films based on alginate with silver nanoparticles and lemongrass essential oil–Innovative packaging for cheese. Nanomaterials 11(9):2377. https://doi.org/10.3390/nano11092377

Mouzahim ME, Eddarai EM, Eladaoui S, Guenbour A, Bellaouchou A, Zarrouk A, Boussen R (2023) Effect of Kaolin clay and *Ficus carica* mediated silver nanoparticles on chitosan food packaging film for fresh apple slice preservation. Food Chem 410:135470

Mustafa F, Andreescu S (2020) Nanotechnology-based approaches for food sensing and packaging applications. RSC Adv 10:19309–19336. https://doi.org/10.1039/d0ra01084g

Ndwandwe BK, Malinga SP, Kayitesi E, Dlamini BC (2021) Advances in green synthesis of selenium nanoparticles and their application in food packaging. Int J Food Sci Technol 56(6):2640–2650

Nguyen S, Lee B (2022) PVA/CNC/TiO_2 nanocomposite for food-packaging: improved mechanical, UV/water vapor barrier, and antimicrobial properties. Carbohydr Polym 298:120064. https://doi.org/10.1016/j.carbpol.2022.120064

Ojijo V, Sinha Ray S, Sadiku R (2014) Concurrent enhancement of multiple properties in reactively processed nanocomposites of polylactide/poly[(butylene succinate)-co-adipate blend and organoclay. Macromol Mater Eng 299(5):596–608. https://doi.org/10.1002/mame.201300306

Olusanya JO, Mohan TP, Kanny K (2023) Effect of nanoclay (NC) on mechanical and thermal properties of cellulose nanoparticles (CNPs) filled cornstarch bioplastic film. J Polym Res 30(7):262. https://doi.org/10.1007/s10965-023-03646-1

Omerović N, Djisalov M, Živojević K, Mladenović M, Vunduk J, Milenković I, Vidić J (2021) Antimicrobial nanoparticles and biodegradable polymer composites for active food packaging applications. Compr Rev Food Sci Food Saf 20(3):2428–2454

Onyeaka H, Passaretti P, Miri T, Al-Sharify ZT (2022) The safety of nanomaterials in food production and packaging. Curr Res Food Sci 5:763–774

Patel MK, Zaccone M, De Brauwer L, Nair R, Monti M, Martinez-Nogues V, Frache A, Oksman K (2022) Improvement of poly(lactic acid)-poly(hydroxy butyrate) blend properties for use in food packaging: processing, structure relationships. Polymers 14(23):5104. https://doi.org/10.3390/polym14235104

Patra JK, Baek KH (2017) Antibacterial activity and synergistic antibacterial potential of biosynthesized silver nanoparticles against foodborne pathogenic bacteria along with its anticandidal and antioxidant effects. Front Microbiol 8:167. https://doi.org/10.3389/fmicb.2017.00167

Peter A, Cozmuta LM, Nicula C, Cozmuta AM, Talasman CM, Drazic G, Silvi S (2021) Chemical and organoleptic changes of curd cheese stored in new and reused active packaging systems made of Ag-graphene-TiO2-PLA. Food Chem 363:130341

Picolotto A, Rossa Beltrami LV, Dallé D, Ornaghi Júnior HL, Zattera AJ, Catto AL, Borsoi C (2024) Development and evaluation of biodegradable starch-based films containing cellulose nanocrystals/titanium dioxide nanoparticles as an alternative for food packaging. J Thermoplast Compos Mater 37(6):1991–2012

Phothisarattana D, Harnkarnsujarit N (2022) Migration, aggregations and thermal degradation behaviors of TiO2 and ZnO incorporated PBAT/TPS nanocomposite blown films. Food Packag Shelf Life 33:100901. https://doi.org/10.1016/j.fpsl.2022.100901

Phuinthiang P, Channei D, Ratananikom K, Nakaruk A, Khanitchaidecha W (2023) Antibacterial properties of TiO_2 nano coating on food packaging surfaces against Escherichia coli and Salmonella typhimurium. Surf Eng 39:433–444. https://doi.org/10.1080/02670844.2023.2232969

Pluta-Kubica A, Jamróz E, Khachatryan G, Florkiewicz A, Kopel P (2021) Application of furcellaran nanocomposite film as packaging of cheese. Polymers 13(9):1428

Primožič M, Knez Ž, Leitgeb M (2021) (Bio)Nanotechnology in food science—food packaging. Nanomaterials 11(2):292. https://doi.org/10.3390/nano11020292

Qian M, Liu D, Zhang X, Yin Z, Ismail BB, Ye X, Guo M (2021) A review of active packaging in bakery products: applications and future trends. Trends Food Sci Technol 114:459–471

Ramos M, Fortunati E, Peltzer M, Dominici F, Jiménez A, del Carmen GM, Kenny JM (2014) Influence of thymol and silver nanoparticles on the degradation of poly(lactic acid) based nanocomposites: thermal and morphological properties. Polym Degrad Stab 108:158–165. https://doi.org/10.1016/j.polymdegradstab.2014.06.006

Rashid MM, Forte Tavčer P, Tomšič B (2021) Influence of titanium dioxide nanoparticles on human health and the environment. Nanomaterials 11(9):2354. https://doi.org/10.3390/nano11092354

Ray P, Haideri N, Haque I, Mohammed O, Chakraborty S, Banerjee S, Quadir M, Brinker AE, Banerjee SK (2021) The impact of nanoparticles on the immune system: a gray zone of nanomedicine. J Immunol Sci 5(1):19–33

Rhim JW, Ng PK (2007) Natural biopolymer-based nanocomposite films for packaging applications. Crit Rev Food Sci Nutr 47(4):411–433. https://doi.org/10.1080/10408390600846366

Rhim JW, Wang LF (2014) Preparation and characterization of carrageenan-based nanocomposite films reinforced with clay mineral and silver nanoparticles. Appl Clay Sci 97–98:174–181. https://doi.org/10.1016/j.clay.2014.05.025

Rizzotto F, Vasiljevic ZZ, Stanojevic G, Dojcinovic MP, Jankovic-Castvan I, Vujancevic JD, Nikolic MV (2022) Antioxidant and cell-friendly Fe2TiO5 nanoparticles for food packaging application. Food Chem 390:133198

Rocha DB, de Souza AG, Szostak M, Rosa DD (2020) Polylactic acid/Lignocellulosic residue composites compatibilized through a starch coating. Polym Compos 41(8):3250–3259. https://doi.org/10.1002/pc.25616

Roy S, Ezati P, Rhim JW (2021) Gelatin/carrageenan-based functional films with carbon dots from enoki mushroom for active food packaging applications. ACS Appl Polym Mater 3(12):6437–6445

Saha K, Rahimi M, Yazdani M, Kim ST, Moyano DF, Hou S, Das R, Mout R, Rezaee F, Mahmoudi M, Rotello V (2016) Regulation of macrophage recognition through the interplay of nanoparticle surface functionality and protein corona. ACS Nano 10:4421–4430. https://doi.org/10.1021/acsnano.6b00053

SarojiniS,IndumathiMP,RajarajeswariGR(2019)Mahuaoil-based polyurethane/chitosan/ZnO composite for food nano films biodegradable packaging applications. Int J Biol Macromol 124:163–174. https://doi.org/10.1016/j.ijbiomac.2018.11.195

Şengül B, El-Abassy R, Materny A, Dilsiz N (2017) Poly(lactic acid)/organo-montmorillonite nanocomposites: synthesis, structures, permeation properties and applications. Polym Sci SerA 59:891–901. https://doi.org/10.1134/S0965545X17060098

Shao L, Chen S, Wang H, Zhang J, Xu X, Wang H (2021) Advances in understanding the predominance, phenotypes, and mechanisms of bacteria related to meat spoilage. Trends Food Sci Technol 118:822–832

Sharma S, Sharma N, Kaushal N (2023) Utilization of novel bacteriocin synthesized silver nanoparticles (AgNPs) for their application in antimicrobial packaging for preservation of tomato fruit. Front Sustain Food Syst 7:1072738

Silva-Leyton R, Quijada R, Bastías R, Zamora N, Olate-Moya F, Palza H (2019) Polyethylene/graphene oxide composites toward multifunctional active packaging films. Compos Sci Technol 184:107888. https://doi.org/10.1016/j.compscitech.2019.107888

Silvestre C, Duraccio D, Cimmino S (2011) Food packaging based on polymer nanomaterials. Prog Polym Sci 36(12):1766–1782. https://doi.org/10.1016/j.progpolymsci.2011.02.003

Singh A, Imtiyaz A (2023) Applications of nanotechnology in agriculture and food science: a review. Asian J Chem 35(5):1049–1062. https://doi.org/10.14233/ajchem.2023.27735

Smith A, LaChance A, Zeng S, Liu B, Sun L (2019) Synthesis, properties, and applications of graphene oxide/reduced graphene oxide and their nanocomposites. Nano Mater Sci. https://doi.org/10.1016/J.NANOMS.2019.02.004

Sondi I, Salopek-Sondi B (2004) Silver nanoparticles as antimicrobial agent: a case study on *Escherichia coli* as a model for Gram-negative bacteria. J Colloid Interface Sci 275(1):177–182. https://doi.org/10.1016/j.jcis.2004.02.012

Sooch BS, Mann MK (2021) Nano reinforced biodegradable gelatin based active food packaging film for the enhancement of shelf life of tomatoes (*Solanum lycopersicum* L.). Food Control 130:108322

Souza VGL, Rodrigues C, Valente S, Pimenta C, Pires JRA, Alves MM, Santos CF, Coelhoso IM, Fernando AL (2020) Eco-friendly ZnO/chitosan bionanocomposites films for packaging of fresh poultry meat. Coatings 10(2):110. https://doi.org/10.3390/coatings10020110

Tabassum Z, Girdhar M, Kumar A, Malik T, Mohan A (2023) ZnO nanoparticles-reinforced chitosan–xanthan gum blend novel film with enhanced properties and degradability for application in food packaging. ACS Omega 8(34):30987–30997. https://doi.org/10.1021/acsomega.3c03763

Tan B, Thomas N (2017) Tortuosity model to predict the combined effects of crystallinity and nano-sized clay mineral on the water vapour barrier properties of polylactic acid. Appl Clay Sci 141:46–54. https://doi.org/10.1016/J.CLAY.2017.02.014

Tayel AA, El-Tras WF, Moussa S, El-Baz AF, Mahrous H, Salem MF, Brimer L (2011) Antibacterial action of zinc oxide nanoparticles against foodborne pathogens. J Food Saf 31(2):211–218. https://doi.org/10.1111/j.1745-4565.2010.00287.x

Tervonen T, Linkov I, Figueira JR, Steevens J, Chappell M, Merad M (2009) Risk-based classification system of nanomaterials. J Nanopart Res 11(4):757–766. https://doi.org/10.1007/s11051-008-9546-1

Valgimigli L, Baschieri A, Amorati R (2018) Antioxidant activity of nanomaterials. J Mater Chem B 6(14):2036–2051. https://doi.org/10.1039/C8TB00107C

Van de Velde K, Kiekens P (2002) Biopolymers: overview of several properties and consequences on their applications. Polym Test 21:433–442. https://doi.org/10.1016/S0142-9418(01)00107-6

Valsalan A, Sivaranjana P (2022) Development of biodegradable films using nanocellulose for food packaging application. Anal Methods Environ Chem J 5(4):20–42

Vasiliev G, Kubo A-L, Vija H, Kahru A, Bondar D, Karpichev Y, Bondarenko O (2023) Synergistic antibacterial effect of copper and silver nanoparticles and their mechanism of action. Sci Rep 13:9202. https://doi.org/10.1038/s41598-023-36304-y

Vasile C (2018) Polymeric nanocomposites and nanocoatings for food packaging: a review. Materials 11(10):1834. https://doi.org/10.3390/ma11101834

Venkatesan R, Dhilipkumar T, Kiruthika A, Ali N, Kim SC (2024) Green composites for sustainable food packaging: exploring the influence of lignin-TiO_2 nanoparticles on poly (butylene adipate-co-terephthalate). Int J Biol Macromol 277:134511

Wang J, Han X, Zhang C, Liu K, Duan G (2022) Source of nanocellulose and its application in nanocomposite packaging material: a review. Nanomaterials 12(18):3158. https://doi.org/10.3390/nano12183158

Wen YH, Tsou CH, de Guzman MR, Huang D, Yu YQ, Gao C, Wang ZH (2022) Antibacterial nanocomposite films of poly(vinyl alcohol) modified with zinc oxide-doped multiwalled carbon nanotubes as food packaging. Polym Bull 1–20

World Health Organization (2015) Estimates of the global burden of foodborne diseases. WHO Geneva. https://www.who.int/news/item/03-12-2015-who-s-first-ever-global-estimates-of-foodborne-diseases-find-children-under-5-account-for-almost-one-third-of-deaths. Accessed 28 May 2025

Xie J, Hung YC (2018) UV-A activated TiO_2 embedded biodegradable polymer film for antimicrobial food packaging application. LWT 96:376–382. https://doi.org/10.1016/j.lwt.2018.05.050

Yadav M, Behera K, Chang YH, Chiu FC (2020) Cellulose nanocrystal reinforced chitosan based UV barrier composite films for sustainable packaging. Polymers 12(1):202. https://doi.org/10.3390/polym12010202

Yousefi A, Hosseini E, Sharifan A, Pourahmad R (2024) Application of gelatin-silver nanocomposite film as an active packaging material in toast bread. J Food Meas Charact 18(3):2397–2408

Yu H-Y, Zhang H, Song M-L, Zhou Y, Yao J, Ni Q-Q (2017) From cellulose nanospheres, nanorods to nanofibers: various aspect ratio induced nucleation/reinforcing effects on polylactic acid for robust-barrier food packaging. ACS Appl Mater Interfaces 9(50):43920–43938

Zhang W, Zhang Y, Cao J, Jiang W (2021) Improving the performance of edible food packaging films by using nanocellulose as an additive. Int J Biol Macromol 166:288–296. https://doi.org/10.1016/j.ijbiomac.2020.10.185

Zhao L, Zhang M, Mujumdar AS, Adhikari B, Wang H (2022) Preparation of a novel carbon dot/polyvinyl alcohol composite film and its application in food preservation. ACS Appl Mater Interfaces 14(33):37528–37539

Zhao X, Wang K, Ai C, Yan L, Jiang C, Shi J (2021) Improvement of antifungal and antibacterial activities of food packages using silver nanoparticles synthesized by Iturin A. Food Packag Shelf Life 28:100669. https://doi.org/10.1016/j.fpsl.2021.100669

Chapter 16
Safety Prospects and Environmental Impact of Biodegradable Packaging Materials

Varee Tyagi, Sushil K. Verma, Manoj K. Dhal, and Md. Adil

Abstract Plastic materials dominate the packaging sector in all domains due to their versatile properties. However, with a surge in environmental issues caused by the extensive usage of plastics, a shift towards adopting eco-friendly alternatives has been ongoing. One such sustainable alternative taking over the global market is Biodegradable packaging materials (BPMs). This chapter explores the processing parameters of commonly used BPMs while clearly differentiating between biobased and biodegradable plastics. Besides, a summary of its presence in the global market has been expanded with regard to the world's major economies, while also listing out the ongoing trends and factors affecting consumer acceptability in this market. Further, the environmental impact of these materials, along with an overview of the general mechanism of the biodegradation process, has been detailed. Last but not least, the life cycle assessment of BPMs, as well as their safety aspects, outlining the standards and guidelines being formulated across the globe, have been briefly discussed while also raising the potential issues that should be given due consideration to avoid adverse consequences.

Keywords Biodegradable plastics · Processing · Global market · Waste management · Life cycle assessment · Regulations

---

V. Tyagi (✉) · S. K. Verma
Centre for Sustainable Polymers, Indian Institute of Technology, Guwahati, Assam, India
e-mail: vareetyagi@rnd.iitg.ac.in

S. K. Verma
e-mail: sushilnano@gmail.com

M. K. Dhal · Md. Adil
Department of Chemical Engineering, Indian Institute of Technology Guwahati, Guwahati, Assam, India
e-mail: manojkumar@iitg.ac.in

Md. Adil
e-mail: m.adil@iitg.ac.in

S. Roy and M. Shahid (eds.), *Biomaterials for Sustainable Food Packaging*, Environmental Footprints and Eco-design of Products and Processes,
https://doi.org/10.1007/978-3-032-08027-1_16

16.1 Introduction

Diversity in packaging applications has led to a surge in packaging materials as well, from the utilization of glass, metals, and paper to a variety of synthetic/plastic packaging materials. Ever since the discovery of petroleum-based plastic materials, there has been a revolution in the packaging industry, as these materials offer robust packaging protection and are the go-to packaging material for all kinds of products. The cause of their dominance across all sectors (Plastics Europe and European Association of Plastics Recycling and Recovery Organizations 2021) (Fig. 16.1) is owing to their significant as well as beneficial properties, such as pliable mechanical properties, low cost, abundance, lightness, transparency, excellent barrier characteristics against oxygen, aromatic compounds, and resistance to chemicals (Stoica et al. 2020). A major application of such plastics is in the food and beverage sector, which constitutes the majority of the packaging sector in Fig. 16.1, including their use as primary/secondary packaging, in transportation, and labelling/branding.

However, environmental concerns caused by the production and disposal of these materials are shifting the momentum towards finding better options. Their reliance on non-renewable resources, with the use of toxic additives, chronic persistence in

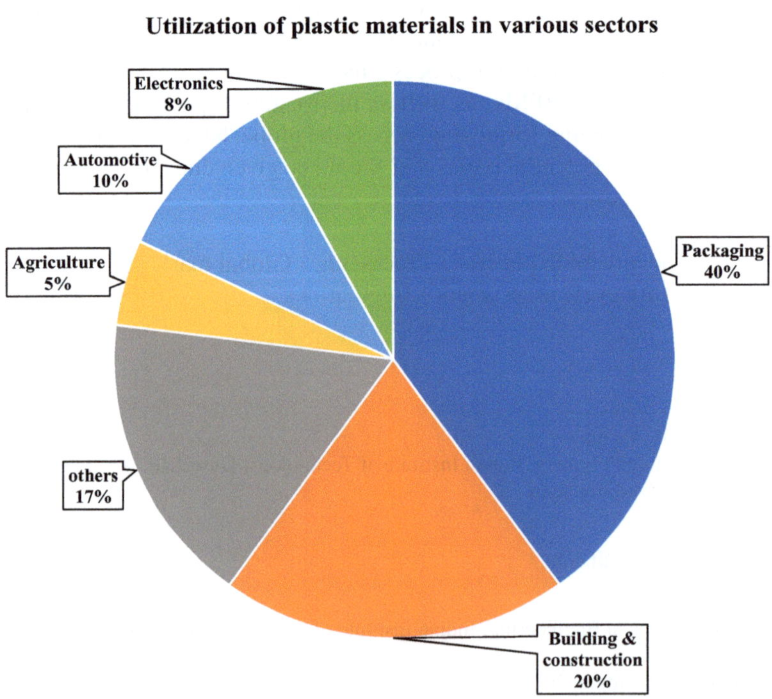

Fig. 16.1 Utilization of plastic materials in various sectors (Plastics Europe and European Association of Plastics Recycling and Recovery Organizations 2021)

the environment even after disposal, has contributed to microplastic pollution as well as migration of toxic substances from the packaging material to the product/ environment, putting human health at risk as well. Recently, reports of microplastics and various chemicals contaminating marine life have surfaced, indicating their entry into the food chain (Baak et al. 2020; Sequeira et al. 2020). Moreover, the presence of microplastics and toxic plasticizers has now been observed in the human system as well, such as in lungs and placenta, signifying grave adverse consequences (Ragusa et al. 2021; Jenner et al. 2022).

With such growing concerns over conventional plastic utilization, efforts are being concentrated on the development and adoption of sustainable packaging solutions. Among all, biodegradable packaging materials have garnered significant popularity, opening a new avenue for tackling environmental concerns as well as promoting a circular economy for packaging, waste reduction, and mitigating the adverse effects of conventional plastics. However, for a genuine appreciation of this subject, it is vital to understand the differences between various seemingly similar terminologies. The terms biobased plastics and biodegradable plastic are used synonymously, which is an incorrect practice. The difference between the two terms is based on the use of the raw material in each. Biobased plastics are materials that are derived from biomass feedstock, whereas biodegradable plastics are derived from polymers that are degraded via biological processes (microbes, enzymes). Moreover, the biodegradation of materials is equally dependent on the raw material as well as the structure and chemical composition of the final product, along with the environment under which they are expected to degrade. Therefore, it is quite likely that a biobased plastic product may not be biodegradable or a biodegradable plastic product is not based on biobased material. The European Bioplastics—a European umbrella organization for bioplastics, has demonstrated this differentiation in an axis model for easy comprehension while also devising the term bioplastics, which are biodegradable, biobased, or both (Rujnić-Sokele and Pilipović 2017; Goel et al. 2021) (Fig. 16.2).

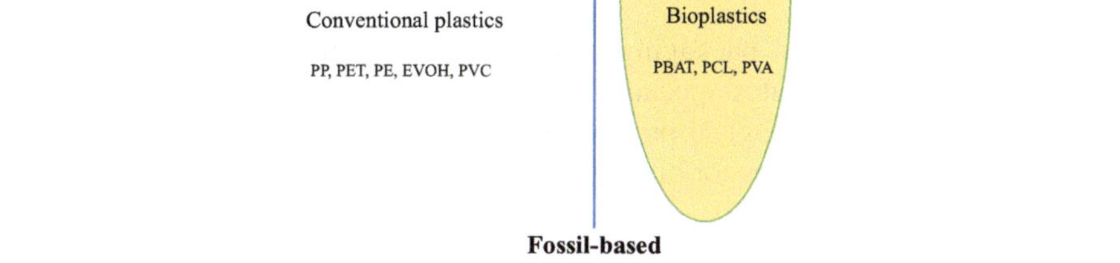

Fig. 16.2 European bioplastic-material coordinate system

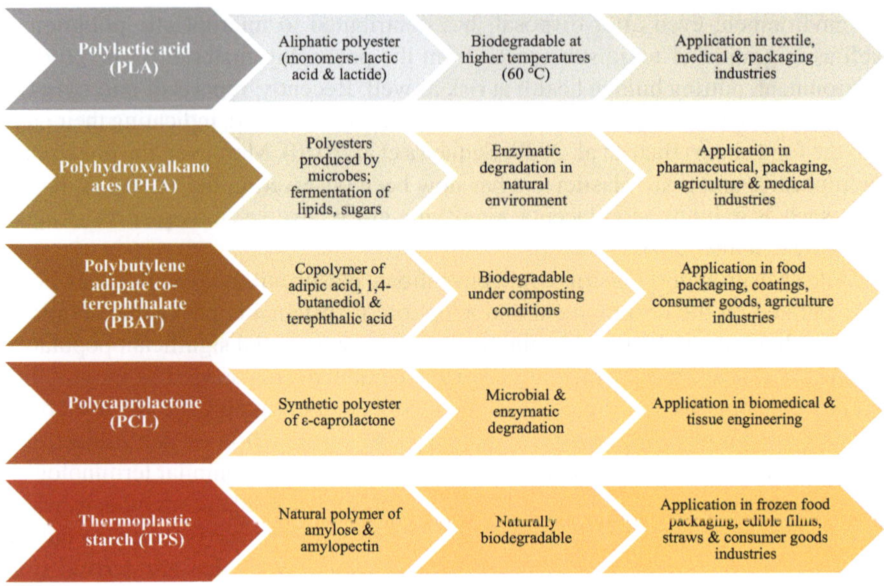

Fig. 16.3 Biodegradable plastics and their applications

The demand for biodegradable packaging material is growing quickly worldwide and is projected to grow more than threefold by 2032, due to the rising awareness among the public and stringent regulations against conventional plastics owing to their adverse impact both on the environment and human health (Statista Research Department 2024). This chapter focuses on the commonly used biodegradable plastics based on biomass or fossil-based raw materials (Fig. 16.3) along with their processing techniques and impact on their properties due to various modifications. Further, a global overview of the biodegradable packaging market has been covered to understand its growth trends and acceptability perceptions. The current impact of BPMs on the environment has also been showcased, along with life cycle assessment, giving an insight into their contribution towards reduced carbon footprint levels.

16.2 Processing of BPMs

Transforming BPMs into packaging materials such as films, trays, bottles, and other containers necessitates specialized processing techniques that cater to their unique physical and rheological properties. Below are some key considerations in the manufacturing processes for bioplastic packaging.

a. **Injection moulding**

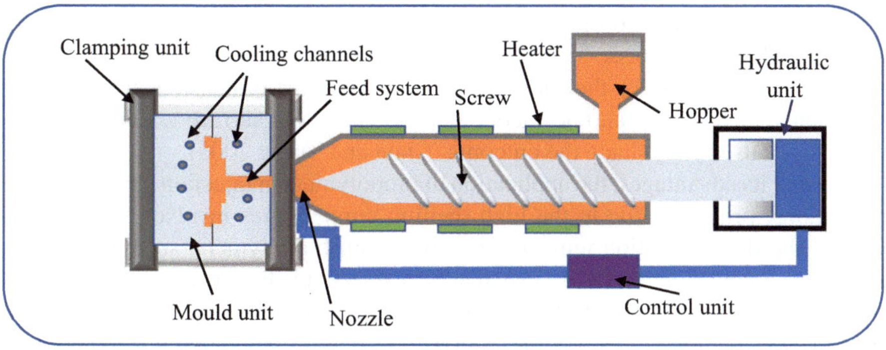

Fig. 16.4 Schematic diagram of injection moulding process

Injection moulding is a widely employed technique for manufacturing products using biopolymers. The process, as illustrated in Fig. 16.4, is typically divided into three distinct stages (Pantani et al. 2005):

Filling: During this stage, the polymer, melted at an elevated temperature, is rapidly injected into a cold mould, forming a cavity that matches the desired shape of the product.

Packing/Holding: This phase involves increasing the pressure to inject additional material into the mould. This compensates for the effects of cooling, such as changes in density caused by temperature reduction and crystallization during solidification.

Cooling: The cooling stage begins when the material at the cavity's entrance solidifies, preventing further movement of material into or out of the mould. Once a sufficiently thick solid layer forms on the product's surface, providing structural integrity, the retaining pressure is released, and the product can be ejected.

This systematic process ensures the production of biopolymer-based products with precise shapes and properties.

Similar to other manufacturing techniques, injection moulding induces significant changes in the rheological and thermomechanical properties of biopolymers. These changes result from variations in stress during the process, elevated temperatures, and rate of cooling of the final product (Kashyap and Datta 2015). Therefore, a comprehensive analysis of the factors influencing injection moulding is essential before proceeding with the production of a biopolymer-based product. These factors are broadly categorized as follows (Chen and Turng 2005):

Machine Parameters: These include the temperatures of the nozzle, barrel and coolant. Pressure parameters such as packaging pressure, back pressure, and injection pressure also play a crucial role. Additional considerations involve the movement and switching point of the machine, injection speed, and shot volume.

Process Parameters: These factors encompass the mould temperature, as well as the melting and cooling temperatures. Key elements such as the injection, holding, and cooling times, mould opening speed, and the dissipation of heat or cooling are also critical to the process.

Quality indicators for the final product, such as part dimensions, resistance to weld lines, sink marks, and aesthetic characteristics like surface texture or visible defects, are used to evaluate the outcome (Chen and Turng 2005).

Despite its advantages, the application of biopolymers in injection moulding faces limitations, primarily due to the high melting temperatures involved, which often exceed the decomposition temperatures of certain biopolymers, such as chitosan. However, advancements have been made in this field. For instance, a polymeric matrix based on chitosan and thermoplastic starch was developed and successfully processed using injection moulding, followed by compression moulding (Balla et al. 2021).

b. **Compression moulding**

The compression moulding technique, often referred to as press moulding, involves placing the desired biopolymer, typically in the form of granules or sheets, between two moulds that are preheated to elevated temperatures. Under applied pressure, the material conforms to the mould's shape and is then cooled and removed after curing to obtain the final matrix (Park and Lee 2012; Udayakumar et al. 2021). Figure 16.5 illustrates the compression moulding process.

This method is considered a secondary process as the biopolymer materials require pre-processing before undergoing compression moulding (Du et al. 2010). The process generally comprises 3 key steps (Park and Lee 2012):

Pre-Charge and Placement Preparation: A specific amount of biopolymer, referred to as the pre-load or load, is prepared and weighed before being placed into the preheated mould. The charge dimensions are designed to cover approximately 50% of the mould surface, and its positioning within the mould is critical. Improper placement can negatively affect the quality of the final product.

Mould Enclosure and Curing: Once the pre-load is placed in the mould, the upper mould descends at a controlled speed typically 5–10 mm/s to compress the material. Factors such as mould temperature and the speed of mould closure are pivotal to

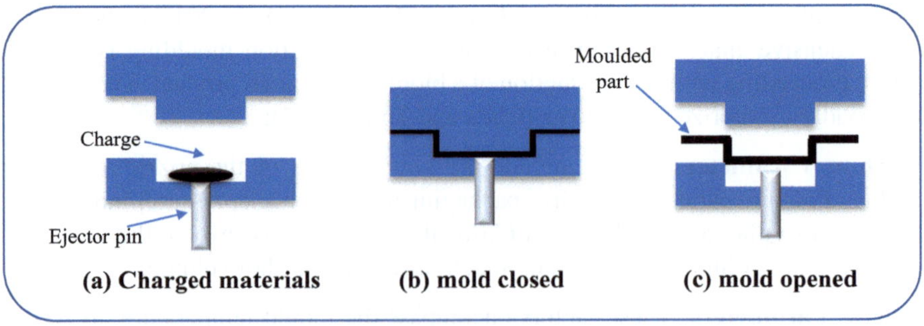

Fig. 16.5 Schematic diagram of compression moulding process

the process's efficiency and the quality of the resulting product. After the material is fully inserted into the mould cavity, the mould remains closed, and a consistent moulding pressure is maintained for a predetermined duration. This step allows the biopolymer to cure, resulting in the formation of the final product.

Cooling and Removal: Once the biopolymer has cured, the mould is cooled, and the product is carefully removed, completing the process.

Compression moulding offers several advantages, including high reproducibility, the ability to create complex shapes, minimal material wastage, and a relatively high rate of production, as each mould cycle is completed within a few minutes. This makes the technique viable for industrial-scale applications. However, one of the notable drawbacks is the relatively high cost of compression moulds compared to equipment used in alternative manufacturing methods (Park and Lee 2012; Jaafar et al. 2019).

c. **Extrusion process**

The extrusion process involves introducing polymers into a barrel through a hopper, where they are melted and forced through a die. Figure 16.6 shows the extrusion process. During this process, the material is subjected to thermal and shear energies, leading to changes in its chemical, structural, and nutritional properties. These alterations include starch gelatinization and degradation, protein denaturation, and lipid oxidation (Alam et al. 2016).

Extrusion is extensively used in the polymer industry and has notable applications in the food and pharmaceutical sectors. In these fields, it serves as a technique to alter the microstructure or chemical composition of polymers. For instance, starch, one of the most commonly extruded food materials, undergoes granule breakdown during extrusion, enhancing its digestibility. Additionally, the method is frequently employed to encapsulate flavors, nutrients, and pharmaceuticals (Wolf 2010).

The extrusion process involves several critical parameters, which can be categorized into three main groups (Muang et al. 2021):

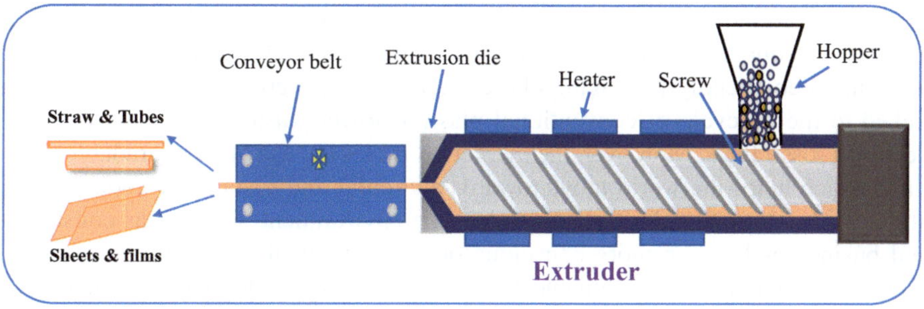

Fig. 16.6 Schematic diagram of extrusion process

- **Process Parameters**: These include factors related to the raw material, such as its moisture content, barrel temperature, screw speed, screw configuration, and the dimensions of the die.
- **System Parameters**: These encompass mechanical energy input, temperature, viscosity, pressure, and the distribution of residence time within the extrusion system.
- **Product Parameters**: Characteristics like color, nutritional content, texture, flavor, and fiber formation fall under this category.

The properties of the final product are significantly governed by these extrusion parameters. A variation of this technique, branded as cast film extrusion, is particularly useful for producing polymeric films, commonly used in packaging and coating applications. This method involves extruding a film through a hanger die and stretching it in air between the die and a cooling roller or a water bath. Typically, the stretching distance is a few centimeters (Agassant et al. 2005).

Starch is a commonly utilized material in extrusion. For example, starch/gelatin films have been developed with the addition of glycerol and sorbitol as plasticizers. These plasticizers interact at the molecular level with starch, forming hydrogen bonds that enhance the film's structure (Fakhouri et al. 2013). In a specific application, thermoplastic corn starch was blended to produce films. These films were analyzed and found to be homogeneous, with no visible granules both from starch or chitosan. Furthermore, the films exhibited improved extensibility and thermal stability, although there was a reduction in their mechanical properties (Mendes et al. 2016). In addition to starch, biopolyesters like polylactic acid (PLA) are widely used in extrusion. PLA extrusion is particularly favored for high-throughput production in packaging applications due to its efficiency and compatibility (Rhim et al. 2006).

16.3 Market Overview

Materials used in biodegradable packaging naturally decompose through biological processes, reducing waste and its negative environmental effects. They focus on providing eco-friendly packaging choices that decrease greenhouse gas emissions linked to the lifecycle of conventional plastics, lower waste, and decrease dependence on fossil fuels. Moreover, the market promotes the use of biodegradable materials, which helps the world move towards a circular and sustainable economy and aligns with international efforts to protect the environment. Further, as consumers and businesses become more conscious of environmental issues, this market has grown significantly as a sustainable alternative to conventional plastic packaging.

The worldwide biodegradable packaging market is being propelled by the growing need for environment-friendly and sustainable solutions. Businesses are adopting environmentally conscious methods to strengthen their brand reputation and meet customer expectations due to consumers' increased desire for eco-friendly packaging options. Additionally, as authorities work to reduce plastic waste and promote sustainable alternatives, government laws and efforts that support the use of biodegradable materials accelerate market growth. However, even with rising demand, the market still confronts several obstacles that could impede its rate of expansion (Research 2024):

- The price difference between biodegradable and conventional plastics prevents firms, especially small and medium-sized ones, from affording them.
- The broad use of biodegradable packaging materials in a variety of industries could be constrained by their limited availability and scalability.
- Technological challenges in producing biodegradable materials with the required qualities for a range of packaging applications hampers growth.
- Advising consumers about the benefits of biodegradable packaging over traditional plastics is a major concern. Public education and awareness about their advantages is still a challenge.
- Furthermore, in some areas, the lack of uniform definitions and laws pertaining to biodegradable packaging causes disputes and can result in the abuse or exploitation of biodegradable claims, which can erode consumer confidence and market integrity.

Despite these obstacles, the market for biodegradable packaging has a wealth of chances for innovation, investigation, and advancement, opening the door to more environmentally friendly possibilities. Developments in biodegradable materials, such as fibers produced by plants and bio-based polymers, have the potential to produce packaging solutions that are more effective and sustainable. Collaborations between manufacturers of packaging, academic institutions, and governmental agencies can spur innovation and get over technical obstacles, which will help the biodegradable packaging sector expand. The growing awareness of environmental issues, more stringent regulations, and an evolution toward sustainable consumption are all contributing to the biodegradable packaging sector's significant expansion globally.

United States: The market for biodegradable packaging is growing in the U.S., as a result of severe environmental legislation and customer desire for sustainable products. The region's innovative packaging technologies contribute to the growth of this industry (Yin and Woo 2024). The US Biodegradable Packaging Market is poised to grow at a sustainable compound annual growth rate (CAGR) of 6% for the next few years (Insights 2025).

European Union: Europe has been a pioneer in sustainability activities leading to significant growth in the biodegradable packaging sector. European nations have adopted rigorous regulations that limit the use of single-use plastics and encourage the utilization of biodegradable alternatives, causing particularly high demand for biodegradable packaging in the retail and food service industries. Germany and France are at the forefront in carrying out the EU's robust regulations that support biodegradable materials. With a CAGR of 6.4% from 2022 to 2032, the European

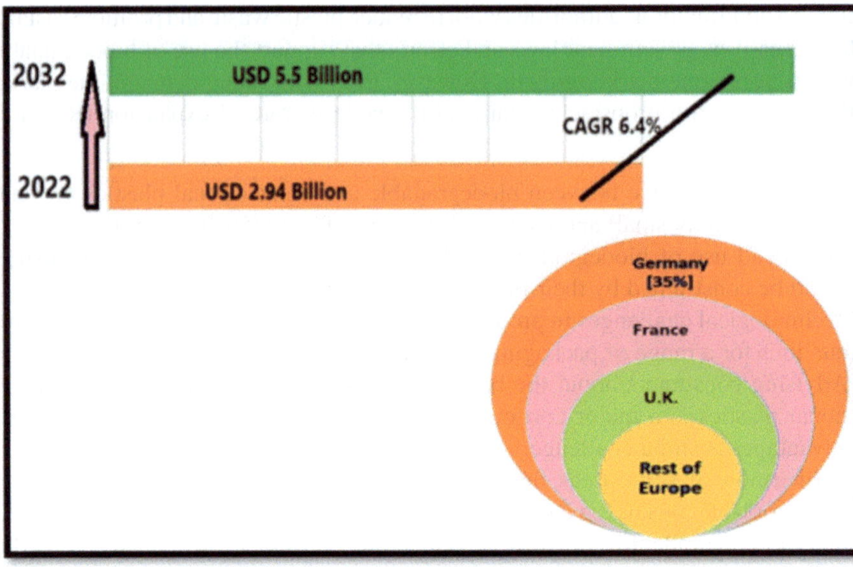

Fig. 16.7 European biodegradable packaging market (2022–2032) (Fact.Mr 2022)

biodegradable packaging market is likely to reach USD 5.5 billion by 2032 from an estimated USD 2.94 billion in 2022 (Fact.Mr 2022) (Fig. 16.7).

The biodegradable packaging material market is expanding significantly in Europe, with demand rising by a factor of 1.3. A higher priority on minimizing the use of problematic packaging materials, such plastics, and a growing need for environmentally friendly packaging across sectors like food and beverage and pharmaceuticals are the primary causes of this expansion (Fact.Mr 2022).

Asia–Pacific: Here, the biodegradable packaging market is growing steadily. In 2023, the eco-friendly packaging market in Asia Pacific was estimated to be worth USD 67.69 billion. It is projected to increase at a CAGR of 12.1% from 2024 to 2030, reaching USD 116.36 billion (https://www.blueweaveconsulting.com/report/asia-pacific-eco-friendly-packaging-market) (Fig. 16.8). Growing consumer awareness of environmentally friendly packaging and the implementation of strict government laws, such as prohibitions on single-use plastic packaging, are major factors powering this industry. Reducing product packaging, encouraging the use of reusable and renewable materials, cutting packaging costs, getting rid of hazardous elements from manufacturing, and making recycling more effective are all objectives of eco-friendly packaging solutions. Several countries in the Asia–Pacific region have banned single-use plastics and are switching to environmentally friendly packaging. It is anticipated that this strategic change would have a favorable impact on market dynamics and illustrate the region's rising appreciation for the significance of ecologically sustainable practices.

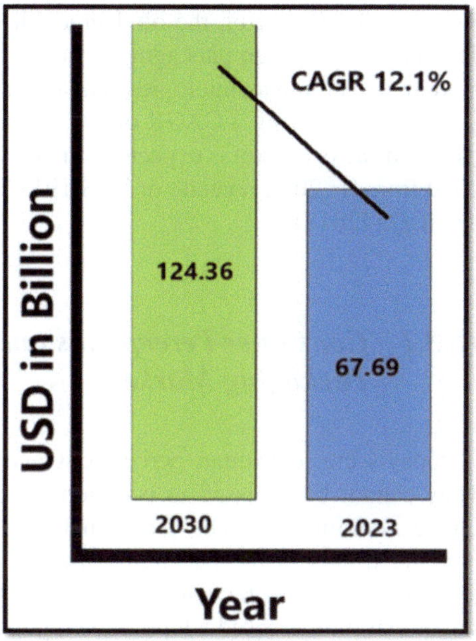

Fig. 16.8 Asia Pacific eco-friendly packaging market size (2023–2030)

China: China's biodegradable packaging industry is expected to expand by more than USD 4 billion between 2023 and 2028, according to the Bonafide Research report (p. https://www.bonafideresearch.com/product/6310697904/china-biodegradable-packaging-market). In this sector, the use of smart packaging technologies is growing in importance. Transparency, traceability, and safety are improved by features like RFID tags, sensors, and QR codes that give consumers up-to-date information on the origin, freshness, and authenticity of products. China, which emphasizes waste reduction and ethical material usage, is also a fervent supporter of the circular economy. By encouraging methods like recycling, composting, and material repurposing, biodegradable packaging fits in perfectly with this strategy and promotes more environmentally friendly waste management and consumption patterns.

India: India is becoming a major player as a result of growing market demand for sustainable packaging solutions and government initiatives. In order to decrease the usage of single-use plastics, the Indian government is putting in place a number of programs that encourage companies to utilize eco-friendly and sustainable packaging options. This contributes to the industry's expansion. One of the biggest end users in India is e-commerce, which is fueled by a youthful population, internet access, and better economic conditions. India's eco-friendly packaging market is also expanding due to growing consumer awareness of environmental preservation. The Government of India (GoI) has initiated a number of efforts to promote the use of environmentally friendly packaging materials, which further supports this. With the adoption of these packaging solutions in various enterprises such as the food and beverage (F&B) sector, the biodegradable packaging sector in India is on the rise. The market for green packaging is being positively impacted by the integration of modern packaging technologies enabled by India's rapid urbanization and economic growth. Considering a CAGR of 4.0% until 2029, the Indian biodegradable plastic bags and sacks market is expected to reach a readjusted size of US$ 2690 million by 2029 from its 2022 revenue of US$ 1914.3 million (p. https://qyresearch.in/report-details/7542618).

16.3.1 Consumer Perceptions and Trends in Biodegradable Packaging Market

In today's era, consumer perceptions and trends in the market go hand in hand. They cannot be assessed in isolation since their interdependence stems from and impacts common sources. Consumer acceptability of BPMs relies heavily on the ongoing trends in the packaging material sector as well as consumer awareness towards environmental concerns choices (Yin and Woo 2024). Current trends in this sector exhibit increased adoption owing to sustainable and eco-friendly movements across the globe, technological advancements in materials, composites, processes etc. and the streamlining of regulatory affairs for better governance and monitoring of overall developments in this sector (Vaishnav and Choudhary 2021) (Fig. 16.9).

These trends, in turn, are powerful indicators of consumer acceptability shaping preferences, perceptions and eventually influencing purchase decisions. In this domain of biodegradable packaging materials, the biggest consumer apprehension is with regards to the packaging material performance compared to the conventional plastics (Vaishnav and Choudhary 2021); however, that is being overcome by raising awareness and highlighting the advantages of biodegradable materials (Sabet 2024). Moreover, consumer perception towards BPMs have been turning favorably more so after being aware of the environmental havoc the conventional plastics has caused globally. This has not only changed the convenience mindset

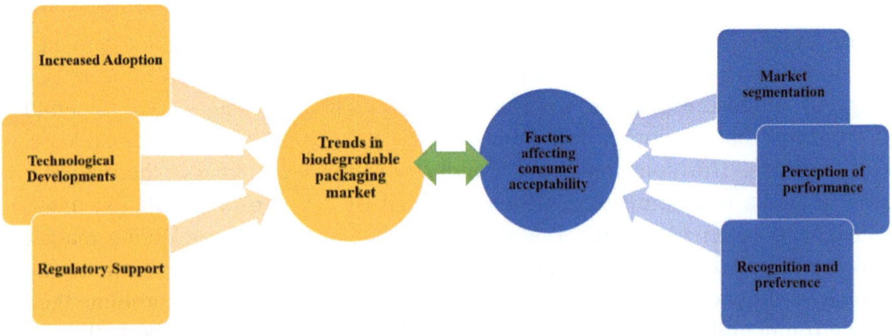

Fig. 16.9 Trends and factors affecting consumer acceptability in biodegradable packaging market

but is also nudging them towards paying extra for a sustainable alternative- for a better tomorrow. Further, gradual adoption of these materials in various industries have been shown to have varying level of consumer acceptance based on their direct bearing on the environment (Yin and Woo 2024).

16.4 Environmental Impact of BPMs: Life Cycle Assessment (LCA) and Safety Prospects

The waste management system for all sorts of plastic is very crude at the moment. Plastics are usually recycled, landfilled, or incinerated, thereby ending up in the environment one way or the other. The rising environmental awareness mandates both packaging materials and processes to have user- and eco-friendly traits, making biodegradability both a functional requirement as well as an important environmental attribute. Petrochemical-based polymers are produced from monomers derived from petrochemical refining, which possess a limited degree of inherent biodegradability (Song et al. 2009). Moreover, their synthesis as well as reutilization process is tedious, costly, and not all types of plastics can be reused. Besides, various additives used for their processing have been shown to be toxic (Tyagi and Bhattacharya 2019; Stoica et al. 2020). A broad classification of the environmental bearing of plastics can be summarized as follows:

- Ecosystem pollution: Accumulation, leaching, microplastic formation and persistent nature of conventional plastics has resulted in adverse health effects across all spectrum of living forms thereby also disrupting the food chain
- Biodiversity impact: Large-scale production and utilization of conventional plastics have been adversely affecting the overall biodiversity of the planet, species, genetic, and the ecosystem
- Climate change: Utilization and improper waste management of these plastics have immensely contributed to climate change by causing deforestation for providing raw material and producing greenhouse gases, respectively
- Economic impact: Plastic pollution has increased waste management and cleanup costs while destroying tourist locations, hitting national economies

Quite recently, microplastic formation in the environment has garnered a lot of attention. When exposed, all plastics undergo degradation, either physicochemical and/or biological. Physicochemical degradation includes weathering- that is degradation due to winds, water or sunlight and hydrolysis/oxidation. Such conditions are the prime cause of microplastic formation being reported these days. Moreover, huge amounts of mismanaged plastic waste further aides the microplastic contamination buildup in various ecosystems while also gradually entering human/animal food chain and accumulating within living beings (Rujnić-Sokele and Pilipović 2017; Goel et al. 2021; Moshood et al. 2022) (Fig. 16.10).

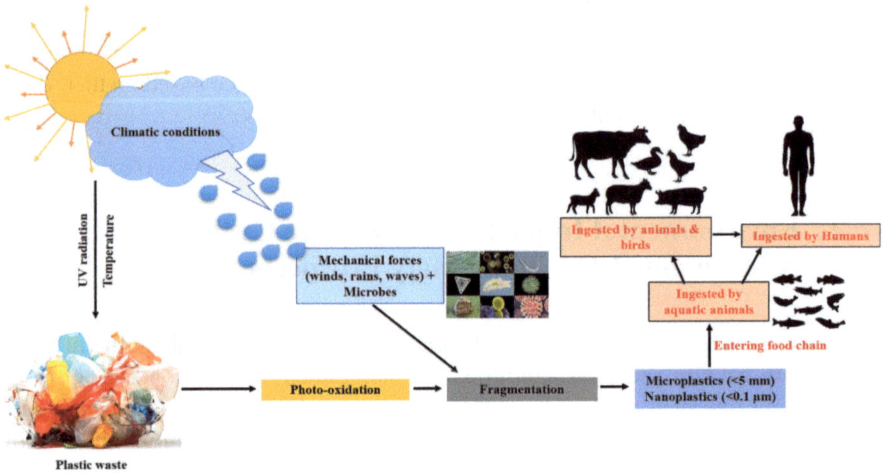

Fig. 16.10 Fragmentation process of plastics and their entry into the environment

To mitigate such problems of conventional plastics, BPMs have been under scrutiny as a potential alternative with regards to their robustness and biodegradability. At present, biodegradable plastic waste is usually tackled through industrial composting process. Here, the waste undergoes degradation under controlled conditions such as high temperatures, controlled oxygen and moisture levels as well as via specific microorganisms (bacteria, fungi). Biodegradation rate may be modified as per requirement according to the waste being processed via changing the environmental conditions such as temperature, light, pH, humidity, oxygen and enzyme levels etc. This biodegradation process can be divided into five different stages: biodeterioration, bio-fragmentation, depolymerization, assimilation and mineralization (Haider et al. 2019) as shown in Fig. 16.11.

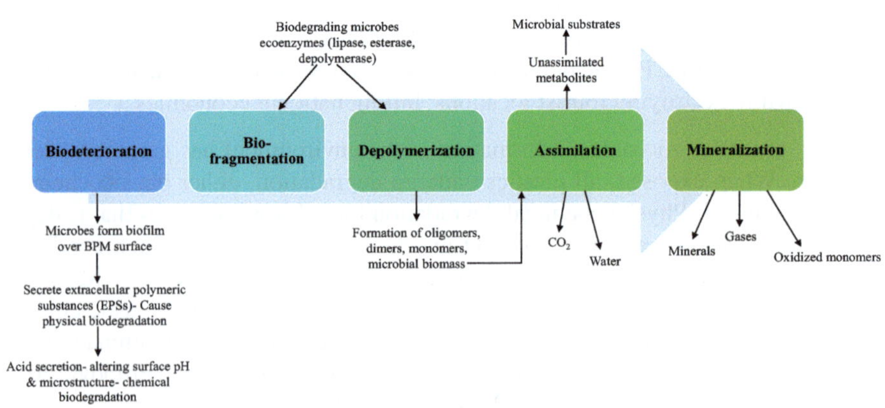

Fig. 16.11 Stages of biodegradation process

As already discussed, a large part of the global market is adopting BPMs. This due to their biodegradability feature, which is quite appealing when compared to conventional plastics owing to the reduction in waste management problems adversely impacting the environment. Further, if the feedstock of such plastics is made biomass alone, a huge carbon footprint reduction can be achieved. However, a thorough life cycle assessment (LCA) of these materials is necessary while putting them through different conditions as well as environments (Filiciotto and Rothenberg 2021). So far, it has only been possible to test these materials under simulated conditions which do provide a basis for comprehensive understanding of this process (Table 16.1) but still lack the evidence of a complete polymer biodegradation profile intended for a specific application, within an environment.

16.4.1 Life Cycle Assessment (LCA) of Biopolymers

LCA serves as a critical tool for evaluating the sustainability and environmental performance of materials. The life cycle of biopolymers closely resembles that of biomass (Seydim and Sarikus 2006; Mangaraj et al. 2019) and undergoes degradation through microbial and enzymatic activity, eventually converting into carbon dioxide and water. The LCA process for biopolymers is depicted in Fig. 16.12.

The biodegradation rate is determined by calculating the quantity of carbon dioxide released during the analysis and dividing it by the theoretical carbon dioxide content in the sample. A parameter known as ^{Ed}K is used to quantitatively assess the biodegradation potential of biodegradable polymers in natural environments. To measure ^{Ed}K, natural soil samples are introduced into bioreactors, and the biodegradation rates of reference materials are assessed over a two-week period. Starch and polyethylene serve as the reference materials, representing ^{Ed}K values of 100 and 0, respectively. These values are established using the ISO 14852 method, which quantifies the carbon dioxide released as the primary analytical parameter (Guo et al. 2010). Table 16.2 presents various biopolymers, detailing their conversion from monomers to polymers, environmental impact, and end-of-life (EoL) considerations based on LCA.

16.4.2 Safety and Regulatory Aspects of BPMs

Safety standards of BPMs are regulated by various regional regulatory bodies and their frameworks (Fig. 16.13). These bodies approve materials usually based on their migration and toxicological evaluation, which determines their safety while being used. However, threshold limits for each parameter, such as mechanical, thermal, barrier, microbial, chemical safety, migration, biodegradability, and sustainability of the material, vary with application, material being used, and local/domestic regulatory standards. Safety standards and regulations of BPMs are being derived and

Table 16.1 Biodegradation properties of BPMs

BPM matrix	Analysis method	Observations	References
PLA with jute, flax	Soil burial	Before burial, the samples weighed 1 g (PLA), 1.5 g (Jute), and 1.35 g (flax). After 30 days, the weights decreased to 0.925 g (7.5%), 1.462 g (2.6%), and 1.331 g (1.43%), respectively. The weight loss indicated biodegradability	Ejaz et al. (2020)
PHA with Pinus radiata wood flour	Natural weathering	Neat PHA lost 2.6% weight after 12 months of soil burial. Higher wood content increased degradation, while PHA with 20 wt% wood and PHA with 50 wt% wood losing ~2.5× and 5× more than neat PHA, respectively	Chan et al. (2019)
PBAT with anionic surfactant	Simulated soil biodegradation	Water absorption influenced biodegradability by attracting fungi and microorganisms that produce enzymes, accelerating degradation. This was evident from the higher biodegradation rate of modified PBAT samples	Camani et al. (2021)
PBAT with lignin	Soil biodegradation	Grafted lignin (VL) enhanced PBAT biodegradation, with degradation efficiencies after 90 days as follows: PBAT (44.9%), PBAT/VL-10% (77.3%), PBAT/VL-20% (89.2%), and PBAT/VL-30% (94.8%)	Liu et al. (2021)
PCL with inorganic clay, Organic additives (GMO, Oleic Acid)	Simulated soil biodegradation	Small (0.1–0.4 wt%) inorganic additives delayed biodegradation by enhancing crystal perfection. ≥1 wt% clay formed imperfect crystals, accelerating degradation. 3 wt% organoclay and glycerol monooleate (GMO) reduced PCL biodegradation from 2 years to 6 months. These additives could accelerate LDPE, HDPE, and PP degradation	Cesur et al. (2018)
Corn Starch, Sodium Alginate	Soil Biodegradation	Corn-starch–sodium-alginate liquid mulch degraded entirely in 25 days. Microbial analysis showed decreased bacterial diversity, increased fungal abundance, and favored dominant microorganisms	Gao et al. (2022)

amended from the existing regulations of conventional plastics, especially in the case of food packaging applications. Further, to demonstrate compliance with safety and environmental standards, these materials have to be certified via testing agencies (such as SGS (Switzerland), TUV (Germany), Intertek (Great Britain), Bureau veritas (France) and many more) that help manufacturers and consumers to trust the sustainability claims (Buchholz 2020).

Besides, to ensure BPMs meet specific criteria for compostability, biodegradation and environmental safety, certain certifications and standards have been formulated such as ASTM D6400 (Certification for compostability in industrial settings), EU standard EN 13,432 (Industrial composting certification), ISO 17088 (International

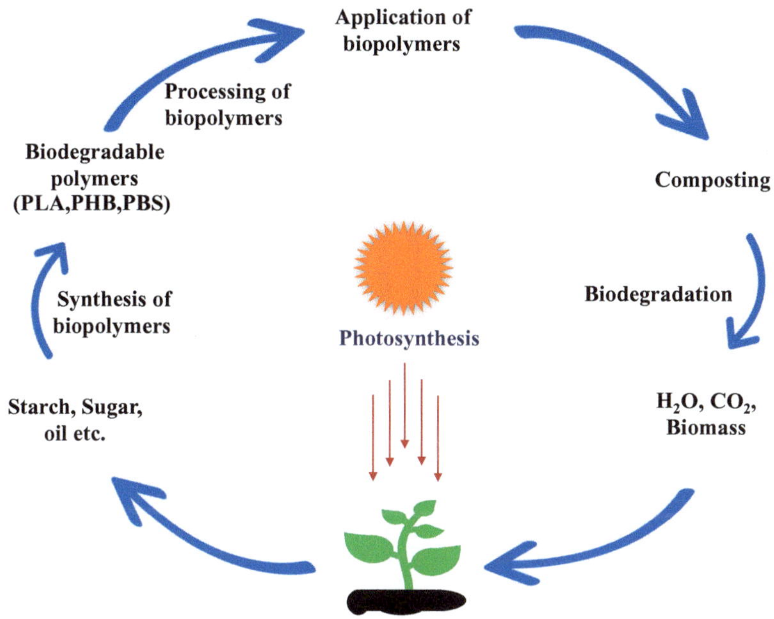

Fig. 16.12 Life cycle assessment of biopolymers

standard for biodegradable plastics), BPI (Biodegradable products institute) certification (Compositability certification in North America) (Filiciotto and Rothenberg 2021; Sabet 2024). These regulations provide assurance to consumers and industries regarding the eco-friendly contribution and effectiveness of BPMs in lowering the impact on the surroundings.

As mentioned in this chapter, BPMs do have an edge over conventional plastics via offering reduced carbon footprint, however, all plastics appear the same to the eye. This not only clouds consumer perception but also gives room to misrepresentation in the market, eventually misleading the customers. This misrepresentation further leads to misuse and misinterpretation of the limited standards in place for BPMs (Goel et al. 2021). Manufacturers label their products as biodegradable/compostable without accurately conforming to the standards due to lack of specific regulations for BPMs, while some simply claim their product to be biodegradable plastic when they are actually not due to lack of proper monitoring and regulatory governance (Filiciotto and Rothenberg 2021). This demands the need of immediate attention and rectification.

Bearing in mind the serious concerns, various safety regulations and guidelines with regard to the differentiation of BPMs from conventional plastics, their usage, and disposal are being gradually understood, drafted, and framed at domestic levels (USFDA, European Commission, etc.) to cater to and contain the nation-specific impact of these materials. However, these varied standards and regulatory frameworks across nations create inconsistencies in the global application and positioning

Table 16.2 Key aspects of different BPMs and end-of-life (EoL) in terms of LCA

BPMs	Characteristics	Conversion pathways	Applications	Environmental impact	End of life options	References
Polylactic acid (PLA)	Tensile strength: 50 MPa Impact strength: 96.1 J m^{-1} Density: 1.24 g cm^{-3}	PLA production: fermentation of plant-derived carbohydrates. Glucose is converted to lactic acid (LA) that polymerizes to low molecular weight PLA. Further, polymerized LA depolymerized to yield lactide that is utilized to produce high molecular weight PLA	• Agriculture • Tissue engineering • Biomedicine • 3D printing • Biomedicine	Carbon uptake of PLA only for biopolymers Advantageous for environment, compared to fossil-based plastics Throughout its lifecycle PLA emits 2.8 kg CO_2 kg^{-1} by natural conversion	• Biodegradable • Compostable • Recycling • Landfill • Incineration	Ebrahimi and Ramezani Dana (2022), Sudhakar and Dharani (2022), Dhal et al. (2023)
Polybutylene succinate (PBS)	Melting point: 90–120 °C Glass transition temp. (T_g): −45 to −10 °C	Lignocellulosic biomass (non-edible) hydrolysis produces PBS; polymerization initiated when 1,4-butanediol reacts with succinic acid that are produced by microbial strains generating PBS oligomers	• Biomedicine	Lignocellulosic biomass and food wastes from production decreases PBS environmental impact	• Biodegradable • Compostable • Recycling • Landfill • Incineration	Patel et al. (2018), Dickson et al. (2021), Elias Guckert et al. (2022), Kang et al. (2022)

(continued)

Table 16.2 (continued)

BPMs	Characteristics	Conversion pathways	Applications	Environmental impact	End of life options	References
Cellulose acetate (CA)	Melting temp.: 170 −240 °C Tensile strength: 30 MPa Water absorption: 2.2% Density: 1.28 g cm^{-3}	CA production: cellulose acetylation of hydroxyl groups	• Textile industries • Plastic films • Packaging • Cigarette filter	CA biodegraded/hydrolyzed into cellulose and acetic acid in natural environment without projecting any adverse effects to environment	• Biodegradable • Compostable • Recycling • Landfill	Araújo et al. (2020), Foroughi et al. (2021), Vatanpour et al. (2022), Yadav and Hakkarainen (2022)
Starch-based polymers (SBPs)	Different type of starch and blends used in variations were studied	SBPs production: done by extrusion of granular native starch and plasticizer agents	• Textile • Packaging • Biomedicines	Starch utilized for bioplastics production reduces GHG emissions (>80%) and fossil fuel consumption (>60%) SBPs shows better environmental profiles than that of conventional plastics	• Biodegradable • Compostable • Recycling • Landfill	Du et al. (2008), Calambás Pulgarin et al. (2022)

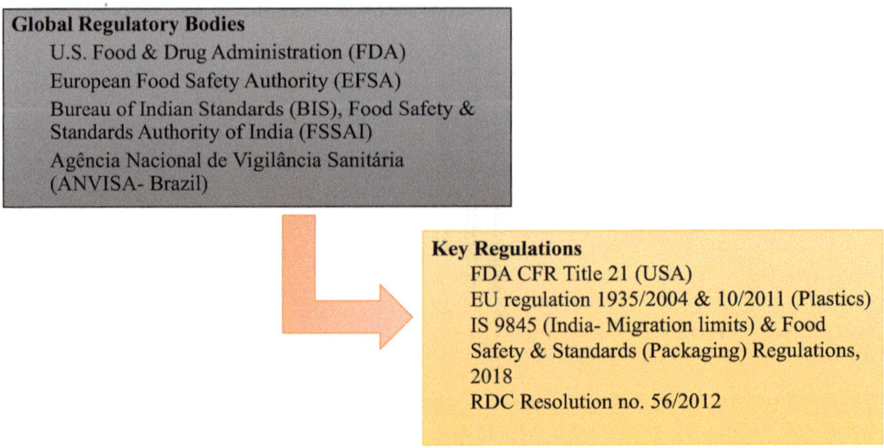

Fig. 16.13 Regulatory framework

of BPMs. Additionally, limited infrastructure for testing along with their high cost and complexity in securing certifications cause a major hindrance in promoting BPMs in the global market (Haider et al. 2019; Goel et al. 2021). Therefore, international organizations like the FAO/WHO Codex Alimentarius, Organization for Economic Co-operation and Development (OECD), and International Organization for Standardization (ISO) are working towards bringing these national regulatory bodies together under a single umbrella to formulate synchronized testing methods, threshold definitions, and values. Such a globally coherent regulatory framework would serve better as the guiding light for ensuring effectiveness, safety, and modulating the environmental impact of BPMs as a sustainable alternative for the future.

16.5 Conclusion and Prospects

In recent times, BPMs have held great potential as an alternative to conventional plastics. They are being widely used as a substitute in various fields, such as consumer goods, including the food and beverage sector. Moreover, BPMs have been projected to have reduced carbon emissions compared to conventional plastics due to their biodegradation characteristic. Based on this eco-friendly forecast of BPMs, they have the potential to become a disruptive innovation, taking over the packaging market worldwide; though a key focal point for consideration is the clear distinction between BPMs and conventional plastics that would help serve as a solid foundation for laying down specific regulatory standards and practices, monitoring misuse and misinformation. The principle of 'one size fits all' cannot be imposed simply because BPMs and conventional plastics appear to be similar when they are fundamentally different. Therefore, various aspects of these materials need undivided attention in order to showcase their true potential:

- One of the primary aspects is their high cost of production compared to conventional plastics, which includes technological problems with respect to their large-scale production.
- Another major area of focus is the biodegradation process of BPMs, which needs further exploration and understanding with regard to different environmental conditions and appropriate monitoring methods. Moreover, this process will be greatly affected by waste management systems that will need to be tuned for optimum degradation results.
- Regulatory standards and protocols explicitly governing BPMs have to be formulated and put in place to minimize misrepresentation and ambiguity.
- Last but not least, risk assessment of BPMs will have to be performed according to their applications and disposal in the ecosystem to formulate specific guidelines and practices.

Funding Sources No funds or grants were used in the preparation of this manuscript.

Declaration of Interest The authors declare no conflict of interest.

Declaration AI tools were used only to improve language and readability. The authors have reviewed the final manuscript and take full responsibility for the final content of the chapter.

References

Agassant JF, Demay Y, Sollogoub C, Silagy D (2005) Cast film extrusion: an overview of experimental and theoretical approaches. Int Polym Proc 20:136148. https://doi.org/10.3139/217.1874

Alam MS, Kaur J, Khaira H, Gupta K (2016) Extrusion and extruded products: changes in quality attributes as affected by extrusion process parameters: a review. Crit Rev Food Sci Nutr 56:445–473. https://doi.org/10.1080/10408398.2013.779568

Araújo D, Castro MCR, Figueiredo A et al (2020) Green synthesis of cellulose acetate from corncob: physicochemical properties and assessment of environmental impacts. J Clean Prod 260. https://doi.org/10.1016/j.jclepro.2020.120865

Asia Pacific Eco-Friendly Packaging Market Size, Share & Trend 2030_BlueWeave. https://www.blueweaveconsulting.com/report/asia-pacific-eco-friendly-packaging-market. https://www.blueweaveconsulting.com/report/asia-pacific-eco-friendly-packaging-market. Accessed 17 Feb 2025

Baak JE, Provencher JF, Mallory ML (2020) Plastic ingestion by four seabird species in the Canadian arctic: comparisons across species and time. Mar Pollut Bull 158:111386. https://doi.org/10.1016/j.marpolbul.2020.111386

Balla B, Bartos A, Kun D et al (2021) Improving mechanical and water sorption properties of thermoplastic starch by incorporating chitosan filler. Polym Test 101. https://doi.org/10.1016/j.polymertesting.2021.107278

Buchholz O (2020) Bioplastics successfully meet all EU safety standards. Eur Bioplast

Calambás Pulgarin HL, Caicedo C, López EF (2022) Effect of surfactant content on rheological, thermal, morphological and surface properties of thermoplastic starch (TPS) and polylactic acid (PLA) blends. Heliyon 8. https://doi.org/10.1016/j.heliyon.2022.e10833

Camani PH, Souza AG, Barbosa RFS et al (2021) Comprehensive insight into surfactant modified-PBAT physico-chemical and biodegradability properties. Chemosphere 269. https://doi.org/10.1016/j.chemosphere.2020.128708

Cesur S, Köroğlu C, Yalçın HT (2018) Antimicrobial and biodegradable food packaging applications of polycaprolactone/organo nanoclay/chitosan polymeric composite films. J Vinyl Add Tech 24:376–387. https://doi.org/10.1002/vnl.21607

Chan CM, Vandi LJ, Pratt S et al (2019) Insights into the biodegradation of PHA/wood composites: micro- and macroscopic changes. Sustain Mater Technol 21. https://doi.org/10.1016/j.susmat.2019.e00099

Chen Z, Turng LS (2005) A review of current developments in process and quality control for injection molding. Adv Polym Technol 24:165–182

China Biodegradable Packaging Market Size, Share, Analysis, Trends. https://www.bonafideresearch.com. https://www.bonafideresearch.com/product/6310697904/china-biodegradable-packaging-market. Accessed 17 Feb 2025

Dhal MK, Madhu K, Banerjee A et al (2023) Polylactic acid/polycaprolactone/sawdust based biocomposites trays with enhanced compostability. Int J Biol Macromol 253. https://doi.org/10.1016/j.ijbiomac.2023.126977

Dickson R, Mancini E, Garg N et al (2021) Sustainable bio-succinic acid production: superstructure optimization, techno-economic, and lifecycle assessment. Energy Environ Sci 14:3542–3558. https://doi.org/10.1039/d0ee03545a

Du Y, Zhang J, Wang C et al (2010) Kenaf bast fiber bundle-reinforced unsaturated polyester composites. II: water resistance and composite mechanical properties improvement. For Prod J 60

Du YL, Cao Y, Lu F et al (2008) Biodegradation behaviors of thermoplastic starch (TPS) and thermoplastic dialdehyde starch (TPDAS) under controlled composting conditions. Polym Test 27:924–930. https://doi.org/10.1016/j.polymertesting.2008.08.002

Ebrahimi F, Ramezani Dana H (2022) Poly lactic acid (PLA) polymers: from properties to biomedical applications. Int J Polym Mater Polym Biomater 71:1117–1130

Ejaz M, Azad MM, Shah AUR et al (2020) Mechanical and biodegradable properties of jute/flax reinforced PLA composites. Fibers Polym 21:2635–2641. https://doi.org/10.1007/s12221-020-1370-y

Elias Guckert F, Sayer C, de Oliveira D et al (2022) Synthesis of polybutylene succinate via Lipase-Catalyzed transesterification: enzyme stability, reuse and PBS properties in bulk polycondensations. Eur Polym J 179. https://doi.org/10.1016/j.eurpolymj.2022.111573

Fact.Mr (2022) Europe biodegradable packaging market statistics report to 2032. https://www.factmr.com/. https://www.factmr.com/report/europe-biodegradable-packaging-market. Accessed 20 Feb 2025

Fakhouri FM, Costa D, Yamashita F et al (2013) Comparative study of processing methods for starch/gelatin films. Carbohydr Polym 95:681–689. https://doi.org/10.1016/j.carbpol.2013.03.027

Filiciotto L, Rothenberg G (2021) Biodegradable plastics: standards, policies, and impacts. Chemsuschem 14:56–72. https://doi.org/10.1002/cssc.202002044

Foroughi F, Ghomi ER, Dehaghi FM et al (2021) A review on the life cycle assessment of cellulose: from properties to the potential of making it a low carbon material. Materials 14:1–23

Gao X, Fu C, Li M, Qi X, Jia X (2022) Effects of biodegradation of corn-starch–sodium-alginate-based liquid mulch film on soil microbial functions. Int J Environ Res Public Health 19(14):8631. https://doi.org/10.3390/ijerph19148631

Global and India Eco-friendly Plastic Bags Industry Research Report and Market Forecast 2024–2030. https://qyresearch.in/report-details/7542618. https://qyresearch.in/report-details/7542618. Accessed 17 Feb 2025

Goel V, Luthra P, Kapur GS, Ramakumar SSV (2021) Biodegradable/bio-plastics: myths and realities. J Polym Environ 29:3079–3104. https://doi.org/10.1007/s10924-021-02099-1

Guo W, Tao J, Yang C et al (2010) The rapid evaluation of material biodegradability using an improved ISO 14852 method with a microbial community. Polym Test 29:832–839. https://doi.org/10.1016/j.polymertesting.2010.07.004

Haider TP, Völker C, Kramm J et al (2019) Plastics of the future? the impact of biodegradable polymers on the environment and on society. Angewandte Chemie Int Ed 58:50–62. https://doi.org/10.1002/anie.201805766

Insights FB (2025) Sustainable packaging market. In: https://www.towardspackaging.com/insights/sustainable-packaging-market-sizing. https://www.towardspackaging.com/insights/sustainable-packaging-market-sizing. Accessed 20 Feb 2025

Jaafar J, Siregar JP, Tezara C et al (2019) A review of important considerations in the compression molding process of short natural fiber composites. Int J Adv Manuf Technol 105:3437–3450. https://doi.org/10.1007/s00170-019-04466-8

Jenner LC, Rotchell JM, Bennett RT et al (2022) Detection of microplastics in human lung tissue using μFTIR spectroscopy. Sci Total Environ 831:154907. https://doi.org/10.1016/j.scitotenv.2022.154907

Kang JH, Kang SW, Kim WJ et al (2022) Anaerobic co-digestion of bioplastics and food waste under mesophilic and thermophilic conditions: synergistic effect and biodegradation. Fermentation 8. https://doi.org/10.3390/fermentation8110638

Kashyap S, Datta D (2015) Process parameter optimization of plastic injection moulding: a review. Int J Plast Technol 19:1–18. https://doi.org/10.1007/S12588-015-9115-2

Liu Y, Liu S, Liu Z et al (2021) Enhanced mechanical and biodegradable properties of PBAT/lignin composites via silane grafting and reactive extrusion. Compos B Eng 220. https://doi.org/10.1016/j.compositesb.2021.108980

Mangaraj S, Yadav A, Bal LM et al (2019) Application of biodegradable polymers in food packaging industry: a comprehensive review. J Packag Technol Res 3:77–96. https://doi.org/10.1007/s41783-018-0049-y

Mendes JF, Paschoalin RT, Carmona VB et al (2016) Biodegradable polymer blends based on corn starch and thermoplastic chitosan processed by extrusion. Carbohydr Polym 137:452–458. https://doi.org/10.1016/j.carbpol.2015.10.093

Moshood TD, Nawanir G, Mahmud F (2022) Sustainability of biodegradable plastics: a review on social, economic, and environmental factors. Crit Rev Biotechnol 42:892–912. https://doi.org/10.1080/07388551.2021.1973954

Muang TT, Gu BY, Ryu GH (2021) Influence of extrusion process parameters on specific mechanical energy and physical properties of high-moisture meat analog. Int J Food Eng 17:149–157. https://doi.org/10.1515/IJFE-2020-0042

Pantani R, Coccorullo I, Speranza V, Titomanlio G (2005) Modeling of morphology evolution in the injection molding process of thermoplastic polymers. Prog Polym Sci (Oxford) 30:1185–1222

Park CH, Lee W I (2012) Compression moulding in polymer matrix composites. In: Advani Suresh G, Hsiao Kuang-Ting (eds) Manufacturing techniques for polymer matrix composites (PMCs): a volume in woodhead publishing series in composites science and engineering. Woodhead Publishing, pp 47–94

Patel MK, Bechu A, Villegas JD et al (2018) Second-generation bio-based plastics are becoming a reality—non-renewable energy and greenhouse gas (GHG) balance of succinic acid-based plastic end products made from lignocellulosic biomass. Biofuels, Bioprod Biorefin 12:426–441. https://doi.org/10.1002/bbb.1849

Plastics Europe, European Association of Plastics Recycling and Recovery Organizations (2021) Plastics-the facts 2021 an analysis of European plastics production, demand and waste data

Ragusa A, Svelato A, Santacroce C et al (2021) Plasticenta: first evidence of microplastics in human placenta. Environ Int 146:106274. https://doi.org/10.1016/j.envint.2020.106274

Research GV (2024) Biodegradable plastic market size & trends report coverage & deliverables, 2024–2030. In: https://www.grandviewresearch.com/horizon/outlook/biodegradable-plastic-market-size/global. https://www.grandviewresearch.com/horizon/outlook/biodegradable-plastic-market-size/global. Accessed 17 Feb 2025

Rhim JW, Mohanty AK, Singh SP, Ng PKW (2006) Effect of the processing methods on the performance of polylactide films: thermocompression versus solvent casting. J Appl Polym Sci 101:3736–3742. https://doi.org/10.1002/app.23403

Rujnić-Sokele M, Pilipović A (2017) Challenges and opportunities of biodegradable plastics: a mini review. Waste Manage Res 35:132–140. https://doi.org/10.1177/0734242X16683272

Sabet M (2024) Exploring biodegradable polymer composites for sustainable packaging: a review on properties, manufacturing techniques, and environmental impacts. Iran Polym J (English Edition). https://doi.org/10.1007/s13726-024-01365-y

Sequeira IF, Prata JC, da Costa JP et al (2020) Worldwide contamination of fish with microplastics: a brief global overview. Mar Pollut Bull 160:111681. https://doi.org/10.1016/j.marpolbul.2020.111681

Seydim AC, Sarikus G (2006) Antimicrobial activity of whey protein based edible films incorporated with oregano, rosemary and garlic essential oils. Food Res Int 39:639–644. https://doi.org/10.1016/j.foodres.2006.01.013

Song JH, Murphy RJ, Narayan R, Davies GBH (2009) Biodegradable and compostable alternatives to conventional plastics. Philos Trans Roy Soc B: Biol Sci 364:2127–2139. https://doi.org/10.1098/rstb.2008.0289

Statista Research Department (2024) Global biodegradable plastics market value 2022–2032. https://www.reportlinker.com/p06284289/Biodegradable-Plastics-Global-Market-Report.html?utm_source=GNW. https://www.reportlinker.com/p06284289/Biodegradable-Plastics-Global-Market-Report.html?utm_source=GNW. Accessed 20 Feb 2025

Stoica M, Marian Antohi V, Laura Zlati M, Stoica D (2020) The financial impact of replacing plastic packaging by biodegradable biopolymers—a smart solution for the food industry. J Clean Prod 277:124013. https://doi.org/10.1016/j.jclepro.2020.124013

Sudhakar MP, Dharani G (2022) Evaluation of seaweed for the production of lactic acid by fermentation using Lactobacillus plantarum. Bioresour Technol Rep 17. https://doi.org/10.1016/j.biteb.2021.100890

Tyagi V, Bhattacharya B (2019) Role of plasticizers in bioplastics. MOJ Food Process Technol 7:128–130. https://doi.org/10.15406/mojfpt.2019.07.00231

Udayakumar GP, Muthusamy S, Selvaganesh B et al (2021) Ecofriendly biopolymers and composites: Preparation and their applications in water-treatment. Biotechnol Adv 52. https://doi.org/10.1016/j.biotechadv.2021.107815

Vaishnav A, Choudhary DK (2021) Microbial polymers: applications and ecological perspectives. Springer Nature Singapore

Vatanpour V, Pasaoglu ME, Barzegar H et al (2022) Cellulose acetate in fabrication of polymeric membranes: a review. Chemosphere 295. https://doi.org/10.1016/j.chemosphere.2022.133914

Wolf B (2010) Polysaccharide functionality through extrusion processing. Curr Opin Colloid Interface Sci 15:50–54. https://doi.org/10.1016/j.cocis.2009.11.011

Yadav N, Hakkarainen M (2022) Degradation of cellulose acetate in simulated aqueous environments: one-year study. Macromol Mater Eng 307. https://doi.org/10.1002/mame.202100951

Yin Y, Woo MW (2024) Transitioning of petroleum-based plastic food packaging to sustainable bio-based alternatives. Sustain Food Technol 2:548–566. https://doi.org/10.1039/d4fb00028e

MIX
Papier aus verantwortungsvollen Quellen
Paper from responsible sources
FSC® C105338

If you have any concerns about our products,
you can contact us on
ProductSafety@springernature.com

In case Publisher is established outside the EU,
the EU authorized representative is:
Springer Nature Customer Service Center GmbH
Europaplatz 3, 69115 Heidelberg, Germany

Printed by Libri Plureos GmbH
in Hamburg, Germany